Pearl Millet in the 21st Century

Vilas A Tonapi •
Nepolean Thirunavukkarasu • SK Gupta •
Prakash I Gangashetty • OP Yadav
Editors

Pearl Millet in the 21st Century

Food-Nutrition-Climate resilience-Improved livelihoods

Springer

Editors
Vilas A Tonapi
Indian Institute of Millets Research
Hyderabad, India

Nepolean Thirunavukkarasu
Indian Institute of Millets Research
Hyderabad, India

SK Gupta
ICRISAT
Telangana, India

Prakash I Gangashetty
ICRISAT
Telangana, India

OP Yadav
ICAR
Central Arid Zone Research Institute
Jodhpur, India

ISBN 978-981-99-5889-4 ISBN 978-981-99-5890-0 (eBook)
https://doi.org/10.1007/978-981-99-5890-0

This Springer imprint is published by the registered company Springer Nature Singapore Pte Ltd.
The registered company address is: 152 Beach Road, #21-01/04 Gateway East, Singapore 189721, Singapore

This book is dedicated to the memory of the late
Dr. M Elangovan
The Scientist who dedicated his life to the Conservation of Millet Genetic Resources For the Benefit of Mankind

Foreword I

Global food and nutrition security in the twenty-first century is a challenging task due to climate change and economic vulnerabilities. It is important to explore more native crops like pearl millet, sorghum, and small millets to address food and nutrition security. The International Year of Millets 2023 calls for scaling up the production of these alternative nutritious food crops, which also offer nutritious fodder to ensure livestock productivity. Pearl millet is called as "Powerhouse of Nutrition" due to its richness in essential nutrients that are vital for leading a healthy life. It has various macronutrients as well as micronutrients like iron, zinc, magnesium, calcium, phosphorous, copper, manganese, riboflavin, and folic acid. Owing to such excellent nutritional values, it is gaining popularity and is preferred by people all over the world including developed countries. Indeed, multidisciplinary approaches, including breeding, genomics, bioinformatics, biotechnology, nutrition, and genetics encompassing trait-specific breeding, are required to exploit and harness the beneficial attributes of this nutricereal to make it more climate resilient and to combat hidden hunger and malnutrition.

Pearl millet is the sixth major cereal crop in the world followed by maize, rice, wheat, barley, and sorghum. In Africa, the west/central Africa (WCA) region (Nigeria, Niger, Chad, Mali, and Senegal), and east/southern Africa (ESA), which includes Sudan, Ethiopia, and Tanzania, are the two main areas of pearl millet cultivation. In India, it is the fourth most extensively grown cereal crop after rice, wheat, and maize. Rajasthan, Maharashtra, Uttar Pradesh, Gujarat, and Haryana are the major pearl millet-growing states, contributing 90% of the total production in India. Out of this, Rajasthan contributes a maximum of around 5–6 million tonnes. Pearl millet being a climate-resilient crop along with high nutritional value, it can be exploited for improving nutritional quality and combating malnutrition. Genetic improvement program was carried out for the selection of traditional germplasm to the development of high-yielding hybrids possessing tolerance to climatic stresses such as drought and heat along with different diseases. A total of 21,392 germplasm accessions, including 750 accessions of wild species of genera Pennisetum and Cenchrus, collected from 50 countries are conserved at the ICRISAT Genebank, while 8284 accessions are conserved at the National Bureau of Plant Genetic Resources (NBPGR), New Delhi, India. These lines have been widely used in breeding programs in both the public and private sectors for the development and commercialization of a large number of hybrids by both public institutions and private seed companies.

The book on *Pearl Millet in the 21st Century: Food-Nutrition-Climate Resilience-Improved Livelihoods* is being published during the International Year of Millets 2023. I am sure this book will enable new dimensions for pearl millet and encourage students, scientists, administrators, and policy makers to address climate change and nutrition security.

Department of Agriculture Research and Education (DARE), Himanshu Pathak
Indian Council of Agricultural Research (ICAR),
Ministry of Agriculture and Farmers Welfare
New Delhi, India 6th October 2023

Foreword II

In the twenty-first century, as our world confronts complex challenges related to food security, climate change, and sustainability, the humble pearl millet emerges as a beacon of resilience, adaptability, and nutritional significance. This book, dedicated to exploring pearl millet's role in our contemporary world, comes at a crucial juncture in our journey towards a more sustainable and nourished future.

Pearl millet, often underappreciated, holds the potential to address some of the most pressing issues of our time. As we face the daunting task of feeding a growing global population while mitigating the effects of climate change, this grain offers hope. Its remarkable ability to thrive in arid conditions, resist pests and diseases, and require minimal resources makes it a key player in sustainable agriculture.

The following pages provide an in-depth look at pearl millet in the twenty-first century, with insights from experts in understanding this remarkable crop. From cutting-edge research on its genetics and breeding to innovative agronomic management practices and its vital role in crop diversification, this book offers a comprehensive view of pearl millet's relevance today. This book provides a rich tapestry of knowledge, unveiling the secrets of pearl millet's forage and biofuel opportunities, impacting biotic and abiotic stresses, and innovative simulation modeling techniques. Additionally, the gene bank diversity of pearl millet is a treasure trove of genetic resources, preserving the legacy of this crop for future generations. The research leads and scientific thoughts shared in this book shed light on the importance of conserving these genetic treasures and their potential for breeding resilient and nutritious pearl millet varieties.

One of the most exciting aspects of pearl millet is its potential to combat malnutrition and improve human health. As the world grapples with the dual burden of undernutrition and diet-related diseases, the nutritional benefits of pearl millet are gaining recognition. This book explores the grain's rich nutritional profile and its contribution to enhancing the well-being of communities worldwide. Moreover, in an age when culinary diversity is celebrated and global cuisines intermingle, the versatility of pearl millet shines through. From traditional recipes passed down through generations to innovative culinary creations, this book showcases how pearl millet can be incorporated into our diets, enriching our culinary experiences.

I applaud the authors for their dedication to bringing pearl millet into the spotlight of the twenty-first century. Their collective expertise and passion have produced a comprehensive resource that will undoubtedly serve as a catalyst for positive change. As we navigate the challenges of the modern world, it is essential that we recognize and harness the potential of resilient and nutritious crops like pearl millet.

The story of pearl millet in the twenty-first century is a story of innovation, adaptability, and hope. Its capacity to address our most pressing global challenges cannot be overstated. May this book, jointly edited by researchers from ICAR-IIMR, ICRISAT, and ICAR-CAZRI, inspire a renewed appreciation for pearl millet and encourage us all to explore its possibilities in shaping a sustainable, climate-resilient and nourished twenty-first century. I firmly believe that this book will be of great practical use to scientists, administrators, and policymakers in developing strategies and research programs to improve production, productivity, and utilization of this climate-smart crop.

International Crops Research Institute for the Semi-Arid Tropics (ICRISAT)
Patancheru, Telangana, India
Jacqueline Hughes

Foreword III

For India, South Asia (SA), and Sub-Saharan Africa (SSA), pearl millet is the most important staple food crop that ensures nutrition, sustainability, and climate resilience in the semiarid and arid ecologies. The imminent issues of climate change, depleting water resources, droughts, and widespread malnutrition call for accelerated efforts to enhance the production of this superfood by enhancing the genetic gains in productivity. The stable cytoplasmic male sterility system with potent restorers from India and SSA laid the foundation of hybrid breeding in pearl millet. The decades of crop improvement efforts have resulted in a yield increase of 4% annually, but there is a huge scope to further increase efforts on hybrid breeding for drought-prone areas in SA and SSA. Heterotic grouping of hybrid parental lines is essential to sustain long-term genetic gains.

At Advanta, we are thrilled to join the celebration of the International Year of Millet 2023, a first step towards the journey dedicated to celebrating the importance of this versatile and nutritious grain in our diets and communities. Out of all the millets, sorghum and pearl millet are the most widely grown and consumed major millets in the world. They are among the few crops that cope with climate warming and attendant water problems, are adaptable in all conditions, and can thrive well even with low inputs and are highly nutritious grains that offer many health benefits. As a leading global supplier of grain sorghum and pearl millet genetics, we at Advanta provide high-quality seed solutions through our unmatched commitment to elite germplasm development. Our world-renowned breeding programs ensure reliable yields and stress tolerance. Our efforts align seamlessly with the global vision for a greener, self-reliant world. Join us on this incredible journey as we honor the importance of millets, foster sustainable farming practices, and continue to a brighter, healthier future for our planet. Together, we can make a difference. Time is now ripe for mainstreaming of the nutritional trait's improvement in pearl millet breeding programs. New opportunities are emerging to improve the efficiency and precision of breeding. Development and application of high-throughput genomic tools, speed breeding, and precision phenotyping protocols need to be intensified to exploit a huge wealth of native genetic variation available in pearl millet to accelerate the genetic gains.

The major challenges encountered in the twenty-first century are climate change, sustainability, malnutrition, and economic vulnerabilities. The climate-resilient crops like pearl millet, sorghum, and small millets are the answers to address global hunger and support most of the United Nation's sustainable development goals. I am really pleased to write this foreword for the book on *Pearl Millet in the 21st Century: Food-Nutrition-Climate Resilience-Improved Livelihoods* edited jointly by Dr Vilas Tonapi and the team of researchers from ICAR and ICRISAT. I hope that this publication will be a resource based on pearl millet for students, scientists, administrators, and policymakers to draw future action plans to enable food and nutrition security in the world.

Global CEO, Advanta Seeds
Dubai, United Arab Emirates

Bhupen Dubey

Contents

1 **Current Trends and Future Prospects in Global Production, Utilization, and Trade of Pearl Millet** 1
Kumara Charyulu Deevi, Nedumaran Swamikannu, Parthasarathy Rao Pingali, and Murali Krishna Gumma

2 **Status and Utility of Pearl Millet Germplasm for Crop Improvement** 35
M. Vetriventhan, Vinod Kumar, Naresh Reddy, Raja Srinivas, Kommineni Jagadeesh, Anil Kumar, Ovais Peerzada, M. Elangovan, K. Venkatesh, and Kuldeep Singh

3 **Milestones in Biology, Genetics, and Breeding of Pearl Millet** 61
O. P. Yadav, S. K. Gupta, and K. N. Rai

4 **Advances in Pearl Millet Hybrid Breeding and Development of Parental Lines** 87
S. K. Gupta, O. P. Yadav, and P. Sanjana Reddy

5 **Trait Mapping, Marker-Assisted Selection, and Introgression Breeding in Pearl Millet** 111
Divya Ramesh, Swati Singh, Priya Shah, Animikha Chakraborty, Tara C. Satyavathi, and Nepolean Thirunavukkarasu

6 **Genomic Selection and Its Application in Pearl Millet Improvement** 141
Bharathi Raja Ramadoss, Ameena Premnath, Thiruvengadam Venkatesan, and Nepolean Thirunavukkarasu

7 **Genome Editing and Opportunities for Trait Improvement in Pearl Millet** 163
Debasmita Panda, Mirza J. Baig, and Kutubuddin A. Molla

8 **Omics-Based Approaches in Improving Drought Stress Tolerance in Pearl Millet** 179
Swati Singh, Animikha Chakraborty, Aswini Viswanath, Renuka Malipatil, and Nepolean Thirunavukkarasu

9 Genetic Biofortification of Pearl Millet: Trait Priority, Breeding and Genomic Progress 221
Mahalingam Govindaraj, Mahesh Pujar, Rakesh Srivastava, S. K. Gupta, and Wolfgang H. Pfeiffer

10 Physiological and Molecular Bases of Drought and Heat Tolerance in Pearl Millet 247
B. Sine, Aïda Ndiaye, G. Kanfany, A. Grondin, Adama Ndiaye, A. Faye, O. Sy, O. Diack, N. A. Kane, and L. Laplaze

11 Forage Pearl Millet: Issues and Strategies for Genetic Improvement 279
P. Kaushal, A. K. Roy, D. R. Malaviya, N. R. Bhardwaj, and R. K. Agrawal

12 The Major Diseases of Pearl Millet in the Indian Sub-continent: Current Scenarios in Resistance and Management Strategies 305
Shilpi Bansal, Alexander Balamurugan, Mallana Gowdra Mallikarjuna, Sumer Pal Singh, S. Chandra Nayaka, and Ganesan Prakash

13 Pearl Millet Breeding for Enhancing Yield and Stability: Strategies, Achievements, and Perspectives 331
O. P. Yadav, S. K. Gupta, and P. Sanjana Reddy

14 Salinity Stress in Pearl Millet: From Physiological to Molecular Responses 361
Srividhya S, R. Swarna, J. Jinu, C. Dheeraj, and H. S. Talwar

15 Weed and *Striga* Management in Pearl Millet Production Systems in Sub-Saharan Africa 395
Djibril Yonli, Hamidou Traore, and Boubacar A. Kountche

16 Crop Simulation Models for Climate Change Adaptation in Pearl Millet 415
R. Swarna, S. Srividhya, C. Dheeraj, B. Bhargavi, C. Deepika, and B. Bhaskar

17 Modern Crop Management Practices for Pearl Millet Cultivation in Semi-Arid Africa 445
Dougbedji Fatondji, Hakeem A. Ajeigbe, and Ali Ibrahim

18 Modern Crop Management Practices for Pearl Millet Cultivation in Asia 479
B. Gangaiah and A. K. Yadav

19 Hybrid Seed Generation System Management to Ensure the Seed Quality in Pearl Millet 513
N. Kannababu, B. V. Venkatesh Bhat, Sooganna, K. Raghunath, R. Venkateswarlu, Jinu Jacob, V. M. Malathi, A. Srikanth, T. Nepolean, and Vilas A. Tonapi

20 Traditional Varieties of Pearl Millet and Food Diversity 531
M. Elangovan, Karnam Venkatesh, and B. Sangappa

21 Enhancing Shelf Life of Pearl Millet Flour 549
P. G. Padmaja, R. Venkateswarlu, S. P. Singh, and Vilas A. Tonapi

22 Biofuel Opportunities in Pearl Millet 567
A. V. Umakanth, P. Sanjana Reddy, Subburamu Karthikeyan, Desikan Ramesh, Geetha Thanuja Kalyanasundaram, and Sivakumar Subbarayan

23 An Ecosystem Approach to Promoting Pearl Millet: Balancing Demand and Supply 589
Ashok Dalwai

About the Editor

Vilas A. Tonapi is Former Director of ICAR-Indian Institute of Millets Research (IIMR), Hyderabad, with 35 years of Research, Administrative and Academic experience. He also served as Head, Division of Seed Science and Technology, Indian Agricultural Research Institute, New Delhi. His special attainments include planning and managing crop improvement programs for food and nutrition security, participatory seed system development, protection of plant varieties, and IPR management.

His team's dedicated efforts lead to the celebration of "National Year of Millets" in 2018–2019, notification of "Millets" as "Nutri-cereals," and international year of Millets 2023. He established "Technology Business Incubator" called "Nutrihub," which has created 500+ Millet start-ups. In his tenure IIMR bagged prestigious "Sardar Patel Outstanding ICAR Institution Award 2018," "Chaudhary Devi Lal Outstanding All India Coordinated Research Project Award 2019," and "Kirloskar Vasundhara Mitra Award 2022."

He received RV Swaminathan Gold medal for his Ph.D. and distinguished service award by Seedsmen Association. He has 60 research papers in peer-reviewed journals, 30 books, and 75 papers in national and international seminars. He is now working with Advanta Seeds, a UPL group company.

Nepolean Thirunavukkarasu has played a significant role in national and international crop improvement programs. He has contributed to both basic and strategic research through molecular breeding, quantitative genetics, and genomics approaches.

His research has led to the identification of important QTLs and functional genes responsible for biotic and abiotic stress tolerances, nutritional qualities, and yield traits. His work has also facilitated technology development in staple crops through the creation of genetic stocks, mapping populations, RILs, genetic linkage maps, and genome-wide SNPs.

Furthermore, he has demonstrated the proof of concept of linkage disequilibrium-based association mapping and genomic selection models for the first time in national research programs. This research has helped to better understand the genetics of drought tolerance in cereals.

He has also contributed to planning and implementing marker-assisted breeding strategies. His work has led to the development of maize hybrid products that sustain yield in rain-fed ecologies while eliminating the scourge of micronutrient malnutrition.

S. K. Gupta, Principal Scientist, heads pearl millet breeding research at ICRISAT-India. Since 2008, his focus has been on developing hybrid parents (both seed and restorer parents) of pearl millet with high productivity, rich in nutrition, and along with abiotic and biotic stress tolerances. He coordinated breeding activities of the South-Asian and African pearl millet breeding programs in different projects. He collaborated with several Central Asian countries to identify pearl millet cultivars suitable for salinity-affected areas. While partnering with multiple ARIs, he coordinated the standardization of genomic selection model for pearl millet, identified mechanisms of flowering period heat tolerance in pearl millet, development of heterotic pools, and the development of high biomass pearl millet for forage and 2G biofuels. He coordinates with about 30 international and regional seed companies to enhance productivity of pearl millet in different ecologies. He has won several research awards, mentored several Ph.D. scholars, and published about 120 research papers.

Prakash I. Gangashetty PhD, PDF (UK) is recipient of Monsanto's Beachell-Borlaug international fellowship. He led Africa's pearl millet improvement program from January 2015 to 2021. His major achievements have been the identification and utilization of the striga-resistant line in pearl millet; the development and successful release of Africa's first Fe-biofortified millet variety "CHAKTI" and hybrid "Nafagnon" in West Africa. He has been the leading principal investigator for BMGF-funded Striga control in pearl millet and sorghum, HarvestPlus challenge biofortification program for West Africa. UK innovates funded a project on Delivering low GI pearl millet to control type 2 diabetics in the African population & SMIL lab of the Feed the Future funded a project. He has received several awards such as the Bayer Ted Crosbie-Beachell-Borlaug impact award in 2020, the promising young scientist award in 2018 for outstanding pearl millet breeding work in West Africa, and the Resource mobilizer award in 2019. He has guided 1 PDF, 4 PhD scholars, and 5 master's students. He has published more than 65 journal articles with citations exceeding 350.

O. P. Yadav has more than 34 years of experience in research and management in India and abroad. Presently, he is the Director of the ICAR-Central Arid Zone Research Institute (CAZRI), Jodhpur, since 2015. Under his leadership, CAZRI has won the Best Institute Award of the Indian Council of Agricultural Research, Department of Agricultural Research & Education, Government of India. Earlier, he worked as the Director of the ICAR-Indian Institute of Maize Research and Coordinator of Pearl Millet Improvement Programme of India. He is an accomplished plant breeder, and his team's research efforts have led to the release of 11 commercial cultivars, including bio-fortified hybrids and varieties that are largely grown in India.

He has extensive experience in seed production of crops and management of natural resources, especially in drylands.

He has been awarded the Rafi Ahmed Kidwai Award for Outstanding Research in Agricultural Sciences, the highest award in India for conducting exceptional research work with high impact at farmers' field. He is a Fellow of the Indian National Science Academy (INSA), National Academy of Sciences, India (NASI), National Academy of Agricultural Sciences (NAAS), and several other academies. He has more than 300 publications in high-impact international journals.

Current Trends and Future Prospects in Global Production, Utilization, and Trade of Pearl Millet

1

Kumara Charyulu Deevi, Nedumaran Swamikannu, Parthasarathy Rao Pingali, and Murali Krishna Gumma

Abstract

Millets are a group of small-seeded annual grasses that occupy the sixth position after rice, wheat, maize, barley, and sorghum globally. Among the millets, pearl millet (*Pennisetum glaucum R. Br.*) represents approximately 75% of the global millet area. Around 70% of total millet is produced under warm tropics dryland climates with Africa and Asia accounting for 98% of the global area and 97% of global production. Among the countries, India is the largest producer followed by China, Niger, and Nigeria. The area under millets has come down globally from 35 to 32 million ha between 2007 and 2019 mainly due to the decline in Asia. Yields, however, increased significantly in Asia from 1100 kg/ha to 1292 kg/ha while it declined marginally in Africa. Globally, nearly 3/4 of the domestic supply of millet is used for food, and the rest is used for feed and other uses. Developed countries mainly use millet for animal feed/bird feed. The nutritive and gluten-free nature of millet has provided ample scope across the globe for developing several nutrition-based products, beverages, and baby foods. The enhanced demand for processed foods augers well for millet though still in a nascent stage. Millets are thinly traded with less than 2% of total millet production

K. C. Deevi (✉) · N. Swamikannu
Research Program on Enabling System Transformation (EST), ICRISAT, Patancheru, Hyderabad, India
e-mail: d.kumaracharyulu@cgiar.org; s.nedumaran@cgiar.org

P. R. Pingali
Former Assistant Program Director, Markets Institutions and Policies, ICRISAT, Patancheru, Hyderabad, India

M. K. Gumma
Geospatial Sciences and Big Data, Research Program on Resilient Farm and Food Systems (RFFS), ICRISAT, Patancheru, Hyderabad, India
e-mail: m.gumma@cgiar.org

V. A. Tonapi et al. (eds.), *Pearl Millet in the 21st Century*,
https://doi.org/10.1007/978-981-99-5890-0_1

being exported. Despite this based on International Food Policy Research Institute's (IFPRI's) IMPACT (International Model for Policy analysis of Agricultural Commodities and Trade) model projections, global millet demand is set to increase from 48.5 to 66.5 million tons between 2030 and 2050, and in Asia projected supply would be lower than the demand indicating a widening gap with scope for enhanced trade to meet the shortfall. Since millets are grown under poor soil and marginal environmental conditions, their yields are unstable due to biotic and abiotic stresses besides socio-economic constraints like linkage to markets, access to credit, etc. Addressing these constraints would enhance not only the competitiveness of millets but also the scope for expanding to more favorable areas.

Keywords

Production domains · Production trends · Utilization · Trade · Production constraints

Abbreviations

APEDA	Agricultural and Processed Food Products Export Development Authority
CIS	Commonwealth of Independent States
ESA	Eastern Southern Africa
FAO	Food and Agriculture Organization
GI	Glycemic Index
IFPRI	International Food Policy Research Institute
IMPACT	International Model for Policy analysis of Agricultural Commodities and Trade
LGP	Length of Growing Period
PD	Production Domain
SA	South Asia
SSA	Sub-Saharan Africa
WCA	Western Central Africa

1.1 Introduction

Millets belong to a group of annual grasses mainly cultivated in Asia and Africa. They are grown in marginal soils with low fertility, which support any other crop. They play a pivotal role in promoting the food and nutritional security of the people living in Africa and Asia. Their presence is quite predominant in developing countries (McDonough et al. 2000) located in Africa (Niger, Nigeria, Sudan, Mali, Burkina Faso) and Asia (India, China, Pakistan, Myanmar, Nepal) for diverse (food, feed, and fodder) uses. Millets contribute to <1% of global cereal production and approximately 3% of coarse cereal production. Roughly <2% of total millet production is traded globally. This clearly indicates that millets are grown mainly for self-sufficiency. The most important millets are pearl millet, finger millet, proso millet, and foxtail millet (see Box 1.1). The global area covered by these millets is 76% under pearl millet, along with finger millet (19%), teff (9%), and fonio (4%) (Obilana 2003). Pearl millet has the largest area among all the millets and occupies sixth position in total crops, respectively, after rice, wheat, maize, barley, and sorghum. With this background, the present chapter highlights the importance of millet in Asia and Africa, especially pearl millet. It also describes the global millet domains, major producing regions, and trends in the area, production, and productivity across major countries worldwide. The pattern of millet utilization and its trade among regions is also furnished. The projected millet demand and supply between 2030 and 2050 are summarized, and the widening gap among regions is identified. Major production constraints and scope for future expansion in demand among major regions are discussed at the end.

1.1.1 Importance of Pearl Millet

Pearl millet (*Pennisetum glaucum R. Br.*) is an important crop in the semi-arid and arid environments of South Asia (SA) and sub-Saharan Africa (SSA) that are typically challenged by erratic rainfall, high mean temperature, low soil organic carbon content, and poor water-holding capacity (Serba et al. 2020). Pearl millet is highly valued for its nutrient-rich grain for human consumption (Parthasarathy Rao et al. 2006), and its green fodder and dry stover for livestock (Andrews and Kumar 1992; Parthasarathy Rao and Hall 2003). The crop forms the source of livelihood and nutritional security for >90 million people in SSA and SA regions (Serba et al. 2020). The demand for pearl millet is anticipated to increase in the future due to increasing human and livestock populations in these regions, expanded industrial uses, and growing healthy food consciousness (Rai et al. 2008). The cultivation of pearl millet may extend to new areas where maize and sorghum are cultivated because of depleting water resources and anticipated climate change impacts. Pearl millet production is expected to become more challenging due to predicted intensive droughts, rise in day temperature, and more outbreaks of diseases in sub-Saharan Africa (SSA) (Sultan et al. 2013) and SA (Rama Rao et al. 2019). Hence, its production must be increased at a much quicker rate and more so in challenging agro-ecologies.

Currently, millets are grown in about 20 million ha in Africa and around 11 million ha in Asia. India has the largest area under this crop (~10 million ha) among Asian countries. Global area, production, and productivity data for millets are not available separately for all kinds of millets in Food and Agriculture Organization Corporate Statistical Database (FAOSTAT). Hence, the present study used millets data that combined pearl millet and other small millets. Approximately, pearl millet represents about 3/4 of the global millet area. Africa accounts for about 64% of the global millet area, while Asia represents nearly 34%. The rest only occupies 2% of the global millet area. In most parts of the world, pearl millet is cultivated as a subsistence crop for local consumption. Commercial millet production is risky, especially in Africa, because the absence of large market outlets means that fluctuations in output cause significant price fluctuations, particularly in areas where millet is the main food crop. Besides grain production, millet is cultivated for grazing, green fodder, or silage. Livestock is an important component of most millet production systems, and millet crop residues contribute significantly to fodder supplies (Parthasarathy Rao and Hall 2003).

1.2 Global Millet Production Domains

Concept of research domain or production zone has proven to be useful for targeting crop improvement research as well as priority setting. Identification of target countries in the similar production domain (PD) helps in defining appropriate research strategies, better resource allocation, and establishing potential research collaborations worldwide (Maredia et al. 1996; Hartkamp et al. 2000; Lubulwa et al. 2000). Homogenous research domains[1] for millets were initially defined during the early 1990s (ICRISAT 1992) to help breeders to manage genotype–environment interactions and to facilitate the transfer of technology from the region of origin to places where it might be beneficial or of potential use. These domains are identified to reflect the crop growth characteristics (such as length of growing period [LGP], major production constraints and cropping systems) in different countries of Africa and Asia (the main target regions of International Crops Research Institute for the Semi-Arid Tropics (ICRISAT)). Even though they are useful today, their accuracy has been limited (Mausch and Bantilan 2012) because they did not consider important factors such as temperature (both minimum and maximum), latitude, crop suitability, and distribution. This has led to the exclusion of certain important millet production regions around the world as well as the underestimation of its area. Further, the quantification of corresponding global welfare benefits, including spillover effects, was undervalued.

Mausch and Bantilan (2012) delineated 17 millet homogenous production domains using available spatial information such as millet production (You et al. 2011), agro-ecological zones defined by Food and Agriculture Organization (FAO), land cover images attributed to cropland, and population density as a proxy to market access. The characteristics of these production domains of millet have been described in Table 1.1. About 70% of the world's millets are produced in the warm tropics dryland climate (Fig. 1.1). Within warm tropics dryland climate, nearly 26.1% of that is produced within LGP (length of growing period) between 120 and

[1]The homogenous research domains for millets were drawn based on scientists and experts' judgments on climate, length of growing periods, and biotic and abiotic stress in the particular domains.

Table 1.1 Characteristics of millet production domains

S. no.	Production domains (PD)	PD characteristics, climate, and length of growing period (LGP)	Production ('000 tons)[a]	Production share (%)	Major countries	Major constraints
1	PD_1	Deserts	654.08	2.6	Pakistan, Sudan, Mai, Niger, Burkina Faso, Saudi Arabia, Iran, Iraq, Yemen, Morocco, Libya, Australia	Heat and drought, head caterpillars, *Striga*
2	PD_2	Warm tropics drylands, <60 days	730.23	2.9	Chad, Niger, Nigeria, Mali, Sudan, Zimbabwe, Australia	Downy mildew, drought
3	PD_3	Warm tropics drylands, 60–89 days	2842.61	11.2	Chad, Mali, Niger, Kenya, Namibia	Downy mildew, drought, photoperiod sensitivity
4	PD_4	Warm tropics drylands, 90–119 days	3687.02	14.6	India, Cameroon, Chad, Mali, Niger, Nigeria, Senegal, Burkina Faso, Namibia, Zimbabwe	Downy mildew, smut, need for reduced photoperiod sensitivity
5	PD_5	Warm tropics drylands, 120–149 days	6600.22	26.1	India, Cameroon, Chad, Benin, Gambia, Mali, Nigeria, Senegal, Burkina Faso, Angola, Mozambique, Zimbabwe, Zambia	Downy mildew, smut, drought
6	PD_6	Warm tropics drylands, >150 days	3098.35	12.2	India, Cameroon, Chad, Gambia, Angola, Malawi, Tanzania, Zambia	Drought, stem borer, and *Striga*
7	PD_7	Warm tropics sub-humid, >150 days	2716.09	10.7	Myanmar, Cameroon, Central African Republic, Benin, Ghana, Guinea, Togo, Malawi, Mozambique, Uganda, Zambia	Drought, stem borer
8	PD_8	Subtropical drylands, >150 days	355.70	1.4	Nepal, Pakistan, Argentina, Mexico	Stem borer, ergot
9	PD_9	Subtropical humid, <60 days	165.01	0.7	Pakistan, Zaire, Ivory Coast, Mexico	
10	PD_{10}	Subtropical humid, 60–89 days	334.63	1.3	Pakistan, Ethiopia, South Africa	Drought and heat

(continued)

Table 1.1 (continued)

S. no.	Production domains (PD)	PD characteristics, climate, and length of growing period (LGP)	Production ('000 tons)[a]	Production share (%)	Major countries	Major constraints
11	PD_{11}	Subtropical humid, 90–119 days	583.25	2.3	India, Pakistan, Bangladesh, Bhutan, South Korea, Australia	Downy mildew, drought, and heat
12	PD_{12}	Subtropical humid, 120–149 days	695.97	2.7	India, Pakistan, Bangladesh, Bhutan, South Korea, South Africa, Australia	Downy mildew, drought, and heat
13	PD_{13}	Subtropical humid, >150 days	342.55	1.4	Nepal, Australia	Downy mildew, drought, and heat
14	PD_{14}	Temperate drylands, <60 days	281.89	1.1	China, Hungary, Japan, Spain, Russia	Stem borer, ergot
15	PD_{15}	Temperate drylands, 60–89 days	537.17	2.1	China, Romania, USA, Russia	Stem borer, ergot
16	PD_{16}	Temperate drylands, 90–119 days	1293.96	5.1	China, Russia, Spain	Stem borer, ergot
17	PD_{17}	Temperate humid, >150 days	411.76	1.6	China, North Korea, Australia	Stem borer, ergot

Source: Nedumaran et al. (2015)
[a] The SPAM (2010) spatially distributed production map of millets is used to estimate the production level in each PD

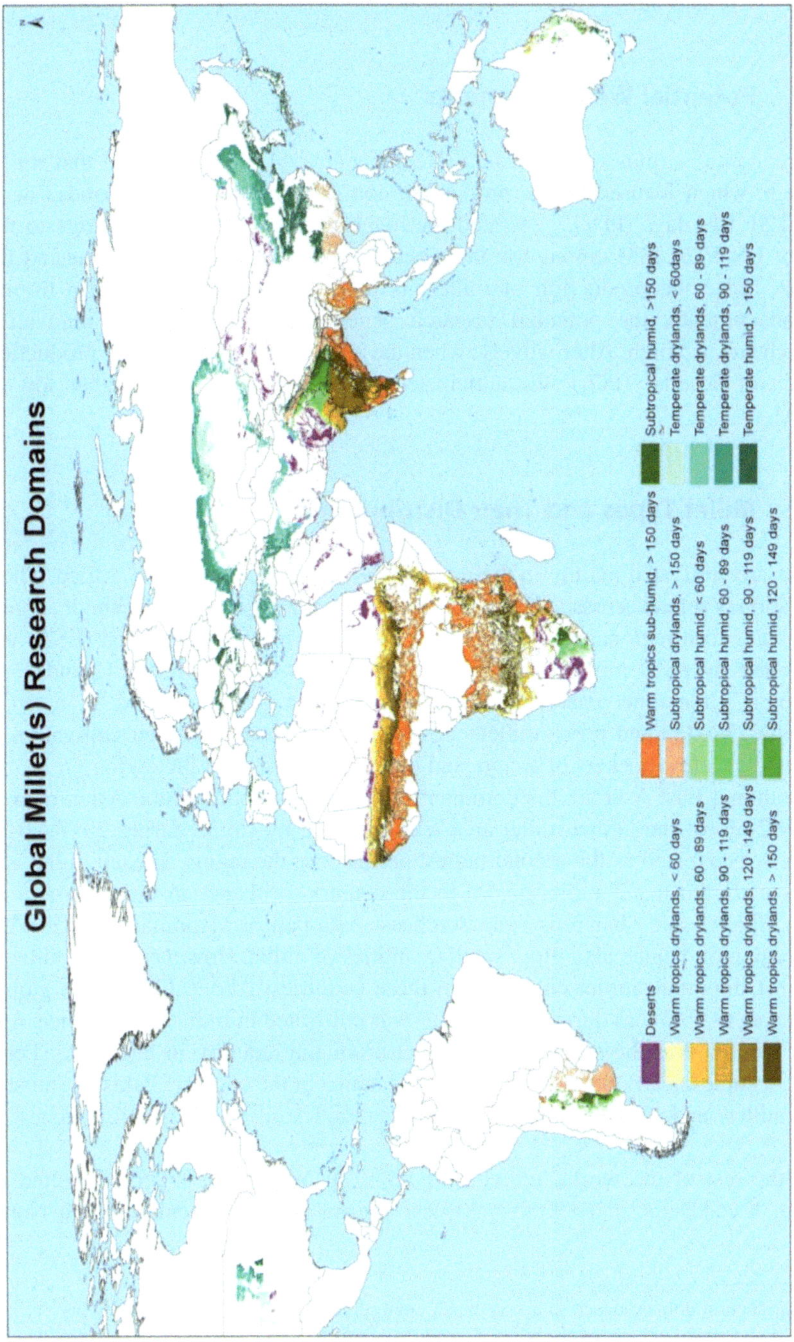

Fig. 1.1 Global millet research domains. (Source: Nedumaran et al. (2015))

149 days, and 14.6% and 1.2% are produced respectively under LGP between 90–119 days and 60–89 days.

1.2.1 Potential Welfare Benefits

The economic-surplus model results (Nedumaran et al. 2015) showed that millet research, which focused on the production domain "warm tropics drylands," with LGP 120–149 days (PD_5), generated the highest expected welfare benefits to the tune of US $720.48M among the identified 17 production domains delineated for millets. Since the production of millets is highly concentrated in warm tropics drylands climate, the potential research benefits accrued from the innovative research are also high. Alternatively, when the research is focused on the production domain of deserts[2] (PD_1), the anticipated research benefits were rather low at $326M.

1.2.2 Millet Types and Their Distribution

Millets are cultivated mainly in the developing countries of Asia and Africa. Their cultivation is spread across most African regions, with Niger being the single largest contributor to a regional area. The Sahel and Eastern Southern Africa (ESA) are dominated by pearl millet cultivation. Finger millet is preferred in Uganda and Tanzania, while other eastern and southern African countries grow it in small quantities. Foxtail and proso millets are not widely grown, and their cultivation is restricted to a few pockets of Kenya and other upland areas in the region.

South and East Asia are the dominant millet-growing parts of the Asian region. Nearly 2.5% of the global millet area was occupied by China during 2017–2019, which is recognized as the second largest producer in the region. Foxtail millet is a highly preferred major millet grown in the country, primarily in the provinces of Hebei, Shanxi, and Shandong. The Southeast Asia part of Myanmar and the West Asia region of Yemen also grow small quantities of millet. However, pearl millet is the most important millet cultivated in these countries. About 30% of the global millet area (8.59 m ha) during 2017–2019 was cultivated in India alone. Hence, it is being identified as the single largest millet-producing country in the world. Pearl millet occupies nearly 88% (7.52 m ha) of that area in the country. Pakistan (mostly pearl millet) and Nepal (dominated by finger millet) also grow small patches of millet.

In the rest of the world, millet cultivation, mainly proso millet, is limited to Russia, Kazakhstan, and Ukraine. Millets are not commonly cultivated in North

[2]The production domain desert is a very harsh environment with zero length of growing periods (LGP), high temperature, scanty rainfall, and poor soil fertility, which does not suit crop production. But about 2.6% of millet is produced in this production domain. This is mainly because millet is the only crop that grows with very little water and withstands high temperature. So it is also grown wherever little irrigation is available, for example in the deserts of North Africa, Middle East countries, Pakistan, and western part of India.

Table 1.2 Area planted to millets by region, 2017–2019

Region	Area ('000 ha)	Production ('000 tons)	Productivity (kg/ha)
Global	32,147 (100)	29,649 (100)	922
Africa	20,645 (64.2)	14,184 (47.8)	685
Eastern Africa	1394 (4.3)	1891 (6.4)	1358
Western Africa	14,310 (44.5)	9727 (32.8)	679
Americas	172 (0.5)	330 (1.1)	1911
Asia	10,943 (34.0)	14,612 (49.3)	1334
Europe	351 (1.1)	487 (1.6)	1375
Oceania	36 (0.1)	36 (0.1)	1016

Note: Figures in the parenthesis indicate % to column total

America, Latin America, Australia, and Europe. Millets are grown for bird feed in Europe and North America, but their cultivation is highly limited.

Box 1.1 Types of Millets and Their Distribution

Pearl millet accounts for over half of global millet population. It is grown across most of Africa, mainly in the Sahel; and in Asia, mainly in South and East Asia. It can be grown on poor soils in dry areas that are unsuitable for the cultivation of the other coarse cereals.

Finger millet is an important staple food in India, Nepal, and eastern Africa. It has a slightly higher water requirement than pearl millet and is typically grown on higher latitudes.

Proso or common millet cultivation is concentrated in the temperate parts of Russia, Ukraine, Kazakhstan, USA, Argentina, and Australia.

Foxtail millet is grown mainly in China where it is used both as food and feed. It is also grown in India, Indonesia, Korea, and some parts of southern Europe.

Teff is grown in Ethiopia where it is used primarily for food.

White fonio, black fonio, and Guinea millet are all minor millets that are important regionally in Africa.

Barnyard millet, little millet, Job's tears, and Kodo millet are minor millets that are important regionally in southern and south eastern Asia. Kodo millet is also grown in western Africa.

Source: FAO and ICRISAT (1996); Bhagavatula et al. (2013)

1.3 Global Millet Production Regions

Globally, developing countries of Africa and Asia are dominated by millet cultivation. African countries account for 64% of the global millet area and about 48% of global millet production (Table 1.2). Much of the crop is grown on marginal lands

Table 1.3 Changes in global millet area and production over last decade

Region	Area ('000 ha)			Production ('000 ha)		
	2017–2019	2007–2009	% Change	2017–2019	2007–2009	% Change
Global	32,147	35,064	−8.3	29,649	31,326	−5.4
Africa	20,645	20,985	−1.6	14,184	16,914	−16.1
Eastern Africa	1394	1542	−9.6	1891	1398	35.2
Western Africa	14,310	15,616	−8.4	9727	14,122	−31.1
Americas	172	177	−2.8	330	320	3.2
Asia	10,943	13,321	−17.9	14,612	13,347	9.5
Europe	351	543	−35.4	487	708	−31.2
Oceania	36	37	−2.7	36	37	−2.0

with low inputs and erratic weather conditions. Relatively, the yields, as expected in the African region, are low. The second most important block of millet producers is in the Asia region. They collectively contribute 34% of the global millet area and nearly 49% of the global millet production. Yields are nearly double compared to Africa, as improved/hybrid seeds are widely used (Kumara Charyulu et al. 2014). Recently, the total millet area in these countries has been falling as farmers shift to other remunerative crops.

Globally, the area under millets decreased from 35 to 32 million ha during the last decade (Table 1.3). The millet-cropped area is much more stable in Africa, while it has declined significantly in Asia. Except in Europe, millet-cropped areas under the Americas and Oceania regions exhibited stability between 2007–2009 and 2017–2019. Global millet production has declined marginally due to a reduction in area. African region (specifically West Central Africa) indicated a remarkable decline in crop production during the study period. The reasons might be a significant millet area reduction in Nigeria and a decline in mean productivity levels in the region. Contrary to the area trend in Asia, millet production has increased marginally due to an enhancement in productivity over the last decade (Kumara Charyulu et al. 2017). The production has significantly gone down in Europe when compared with the Americas and Oceania regions. Overall, millet area and production have come down globally and across major regions during the last decade.

1.3.1 Region-Wise Production and Productivity Trends

Region-wise long-term trends of millet production and productivity levels across major regions are summarized in Figs. 1.2 and 1.3, respectively. The global millet production has been almost stable (hovering around 28 m tons) during the last 20 years. The production in Africa went up and peaked during 2008 (19 m tons), and afterward, it came down to 13 m tons. The production in the Asia region has not changed much (13 m tons) over the study period (2000–2019). In Western Central

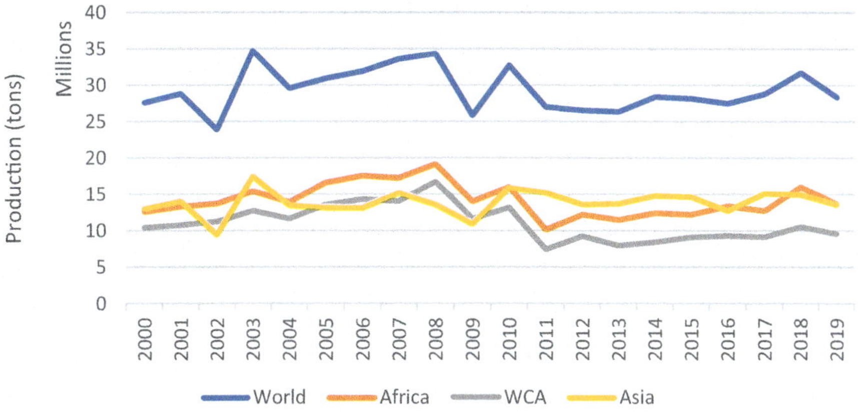

Fig. 1.2 Region-wise production trends, 2000–2019

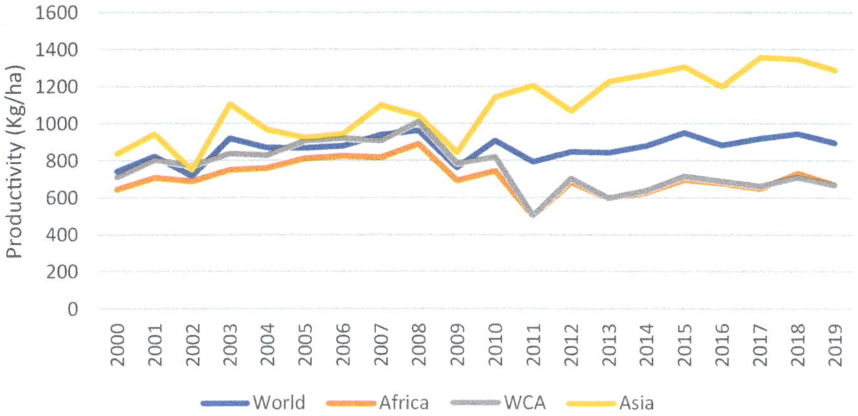

Fig. 1.3 Region-wise productivity trends, 2000–2019

Africa (WCA) region, there was a significant decline in millet production (from 10.4 to 9.5 m tons) during the two-decade period (Fig. 1.2).

Historical region-wise productivity trends during 2000–2019 across major millet-growing regions are furnished in Fig. 1.3. Among all, the Asia region exhibited consistent productivity growth and has increased from 800 to nearly 1400 kg per ha. This may be due to access and adoption of improved cultivars combined with improved management practices in Asia (Yadav et al. 2021; Venkata Rao et al. 2018). Millet productivity level in world is stable and hovering between 800 and 900 kg per ha. The mean productivity scenario in Africa and WCA has consistently declined over time (from 800 to 600 kg per ha). The decline in the cropped area during the study period, coupled with a decrease in productivity levels in Africa, led to a significant decline in crop production in the region (Jukanti et al. 2016; Pucher 2018).

1.3.2 Major Millet-Producing Countries

The performance of millet over the last decade across major producing countries in the world is furnished in Table 1.4. With respect to acreage, the top ten countries include seven in Africa (Niger, Nigeria, Sudan, Mali, Burkina Faso, Chad, and Senegal) and three in Asia (India, China, and Pakistan). Among all major countries, India stood at the top in the cropped area, followed by Niger and Nigeria. There was a significant decline (about 25%) in the millet-cropped area in India during the study period. The production has marginally reduced due to an enhancement in productivity by 20%. The millet-cropped area is stable in Niger (7 m ha) with marginal improvement in productivity. Nigeria has lost almost 50% of the cropped area and production due to a decline in productivity (−20%) during the study period. Millet-cropped area under Mali, Sudan, Angola, Burkina Faso, Chad, China, and Ethiopia has expanded compared to the base period (2009–2011). On the contrary, the area under millets has declined marginally in Nepal, Pakistan, and Senegal during the study period. The Russian Federation and Tanzania nearly lost one-quarter of the millet-cropped area over the decade.

The highest productivity is recorded in China (2841 kg/ha) followed by Ethiopia (2349 kg/ha), India (1248 kg/ha), Russian Federation (1247 kg/ha), and Tanzania (1206 kg/ha) during the 2017–2019 period. Between 2009–2011 and 2017–2019, there was a significant jump in productivity growth in Ethiopia, China, and Chad. The mean productivity growth (from 1037 to 1248 kg/ha) in India during the study period was around 20%.

1.4 Global Utilization of Millets

Globally, the availability of millets has gradually increased due to the expansion of the area and improved productivity over time. Being a self-sufficient crop, millets are consumed primarily for food purposes in most developing countries. They are the rich sources of high-energy and nutritious foods on the planet. The form in which millets are consumed varies across regions—as a thick porridge or as flatbreads (rotis) (see Box 1.2). Millet fodder is an important feed resource in the dryland systems of Africa and Asia, particularly in the post-monsoon seasons when other feed resources are not available or scarce. In a few developed countries, millet usage is largely restricted to bird feed. However, in recent times, their utilization pattern (alcohol manufacturing, livestock feed, and poultry feed) is significantly changing even in developing countries. In recent years, their importance has been recognized in processed baby foods. The diverse uses of millet amplify its demand across study countries and regions.

Table 1.4 Millet performance across major countries, 2009–2019

Country	2009–2011			2017–2019			% Change		
	Area	Production	Yield	Area	Production	Yield	Area	Production	Yield
India	11,158	11,576	1037	8921	11,142	1248	−25.1	−3.8	20.3
Mali	1660	1428	913	2101	1737	830	21.0	21.6	−9.1
Niger	6939	3092	444	6954	3639	523	0.2	17.7	17.8
Nigeria	3659	3790	979	2421	1873	781	−51.1	−50.6	−20.2
Sudan	NA	NA	NA	3093	1553	477	NA	NA	NA
Angola	197	43	217	274	63	232	39.1	46.5	6.9
Burkina Faso	1253	982	780	1264	995	785	0.9	1.3	0.6
Chad	1067	469	421	1189	711	598	11.4	51.6	42.0
China	780	1455	1867	846	2396	2841	8.5	64.7	52.2
Ethiopia	403	603	1494	452	1064	2349	12.2	76.5	57.2
Nepal	268	298	1112	263	311	1183	−1.9	4.4	6.4
Pakistan	494	314	636	488	357	733	−1.2	13.7	15.3
Russian Federation	355	425	1058	258	324	1247	−27.3	−23.8	17.9
Senegal	954	701	724	917	860	937	−3.9	22.7	29.4
Tanzania	365	353	975	282	337	1206	−22.7	−4.5	23.7

Note: Area in '000 ha, production in '000 tons, and yield in kg/ha

Table 1.5 Utilization pattern of pearl millet in different regions, 2016–2018 ('000 tons)

		% Share to domestic supply			
Region	Domestic supply	Food	Feed	Other uses[a]	Losses
Global	27,936	76.9	12.0	4.6	6.9
Africa	13,252	73.0	9.9	7.1	10.7
Eastern Africa	1717	70.5	2.3	22.0	5.6
Western Africa	9543	71.9	11.8	5.3	11.8
Americas	254	0.0	93.3	0.0	6.7
Asia	13,823	84.1	10.2	2.4	3.3
Europe	571	30.8	61.5	3.9	5.4
Oceania	36	0.0	94.4	0.0	5.6

[a] Includes non-food, seed, and processing

Box 1.2 Utilization of Different Types of Millets

In West Africa, a local food called "*degue*" is produced from pearl millet. Pearl millet is also used to produce a beer similar to the beer produced from maize. In India pearl millet grain is consumed in a variety of ways ranging from a thin gruel to a thick porridge. Pearl millet flour is also used to make *rotis*.

Finger millet in India and Africa is ground into flour and eaten as porridge or as an accompaniment to *dal* and curries.

Common or Proso millet can be eaten as whole grains like rice, or boiled and eaten as gruel. Ground into flour, it can be made into porridge or *chapati*. Mixed with wheat flour and yeast, it can be made into leavened bread. The green plant provides good green fodder, but the mature straw is of poor quality and cannot be used for thatching or dry fodder.

Little millet has a grain that has to be dehusked as well as threshed, and is consumed in the same way as other millets. The straw has advantages in that, when ripe and dry, it is palatable to cattle and horses. The green plant is a superior fodder, which may be grazed more than once.

Kodo millet is safe as a green plant and a useful fodder, but the grain of some varieties is toxic to both humans and livestock.

Foxtail millet is important in Asia, and is used as a rice substitute in both China and Japan. It can also be eaten as a gruel, porridge, or pudding. In Russia, it used to be consumed as beer. In the UK, it is used as bird seed while in the USA it is used for hay and silage in dryland areas.

Source: Bhagavatula et al. (2013)

The recent millet utilization pattern across major regions of the world is furnished in Table 1.5. FAOSTAT data available for the triennium period 2016–2018 are used for a better understanding of the utilization scenario of millets across regions. Globally, nearly 3/4 of the domestic supply is used for food, and the rest is used for feed and other uses. The utilization pattern is also quite similar in African regions

Table 1.6 Utilization pattern of millets across major countries, 2016–2018 ('000 tons)

		% Share to domestic supply			
Country	Domestic supply	Food	Feed	Other uses[a]	Losses
Angola	55	74.5	5.5	14.5	3.6
Burkina Faso	971	92.1	0.0	3.4	4.6
Chad	706	84.0	0.0	1.7	14.3
China	1698	46.6	46.1	2.4	4.9
Ethiopia	981	79.3	1.8	15.6	3.3
India	10,914	93.4	1.5	2.3	2.8
Mali	1711	62.8	13.3	12.2	11.7
Nepal	312	88.8	0.0	1.9	9.3
Niger	3846	71.7	13.8	3.4	13.1
Nigeria	1724	81.7	5.0	3.2	10.1
Pakistan	344	43.9	48.5	1.7	5.8
Russian Federation	327	40.1	48.3	4.9	7.0
Senegal	643	61.0	19.6	3.6	15.7
Sudan	1031	79.2	11.4	1.6	7.8
Tanzania	321	42.1	1.9	43.6	12.1

[a] Includes non-food, seed, and processing

as well. Relatively, the proportion of utilization toward food in domestic supply is slightly lower in the ESA region (70%) compared to the WCA region (72%). Americas and Oceania regions allocate almost their domestic supply for feed purposes. Among all regions, the Asia region allocates a significant chunk (84%) of their domestic supply for food, followed by feed.

The details of the country-wise millet utilization pattern have been summarized in Table 1.6. In general, the major share of domestic supply among study countries has been diverted toward food purpose. It was followed up by feed and other uses. The extent of utilization toward food is more than 90% in India and Burkina Faso countries. Nepal, Chad, Nigeria, Sudan, Angola, and Ethiopia have allocated nearly 80% of their millet production to food. Niger, Senegal, and Mali have diverted about 70–60% of their production for food, followed by feed purposes. China, Pakistan, Russia, and Tanzania had a diverse pattern of utilization of millet production beyond food uses.

1.5 Global Millet Trade

Globally, trade in millets is very thin with less than 2% of total production being traded. This clearly reveals their importance as a self-sufficiency crop in respective regions/countries. Traded volumes among countries have increased marginally from 0.2 million ton during 1980–1982 to 0.4 million ton by 2017–2019. Historically, the leading exporters in the world are India and the USA. India's exports largely comprise pearl millet while US exports primarily focus on proso millet. Latin

America, particularly Argentina, was a significant exporter in the early 1980s but in more recent times has all but ceased to export millet. Much of the exported millets are traded within the subregions—for example, much of India's exports go toward West Asia. Europe was a significant importer in the early 1980s, but in recent years, the net trade balance has evened out with higher exports from the CIS (Commonwealth of Independent States) countries.

In India, inter-state trade movement of pearl millet from major pearl millet growing states to primarily other urban centers have been observed. Pearl millet grain is exported from Gujarat and Uttar Pradesh to Rajasthan in years when there is drought in Rajasthan. Trade also happens within Rajasthan and with the eastern regions supplying grain to the western regions of the state. Alcohol industries in Haryana and Punjab import grain from Gujarat and Rajasthan when there is surplus production there.

1.5.1 The Regional Trend in Millet Imports

Globally, millet imports are consistently increasing (from 0.2 to 0.4 m ton) since 1995 (see Fig. 1.4). The peaks in global imports were observed between 1973 and 1993 to the tune of about 600,000 tons. Asia region closely followed the global trend and contributed nearly half of the total share in imports, particularly in the last two decades, while in Africa, the WCA region dominated the ESA region in terms of imports until 2002. However, the share of the ESA region in total African region imports has increased during the last two decades (2002–2019). Overall, Asia plays an important role in millet imports trade when compared to Africa region.

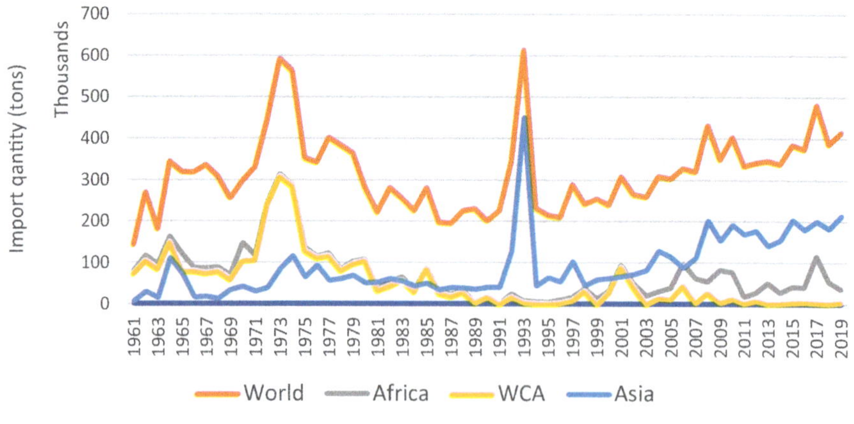

Fig. 1.4 Millet imports trend among major regions

1.5.2 Regional Millet Export Trend

The global millet export trade has been expanding (from 0.20 to 0.45 m ton) since the 1980s, with a peak observed in 1993 (see Fig. 1.5). Among the two regions, Asia has been more proactive during the last two decades (1999–2019) and has occupied the lion's share in global exports. The contribution of the African region has significantly declined from 1993 onward. WCA region played a dominant role in total Africa's export trade until 2006 and declined afterward. The USA and Europe might have accounted more for global exports when compared with Asia and African regions during the recent period.

1.5.3 Major Millet Export and Import Countries in the World

Based on FAOSTAT (2022) data, the top five exporting countries of millet (*mean of 2016–2020*) in the world are the USA, followed by India, Ukraine, Russia, and France (see Fig. 1.6). The mean export quantities of these countries are hovering close to 0.1 m ton only. This clearly emphasizes diversified exports with no single country dominating the export market.

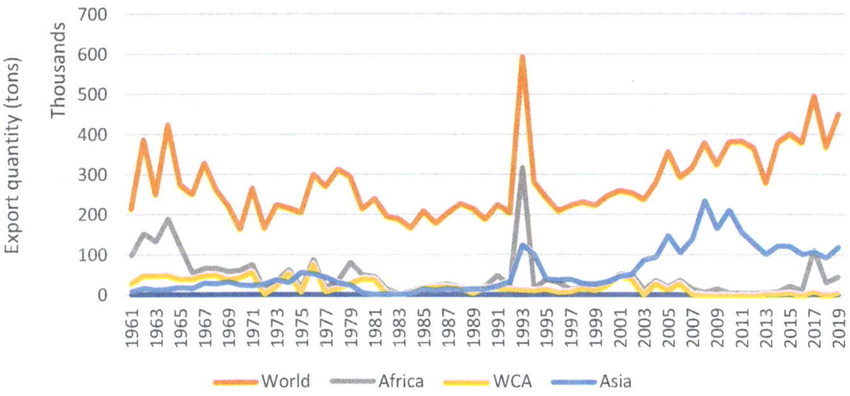

Fig. 1.5 Millet exports trend among major regions

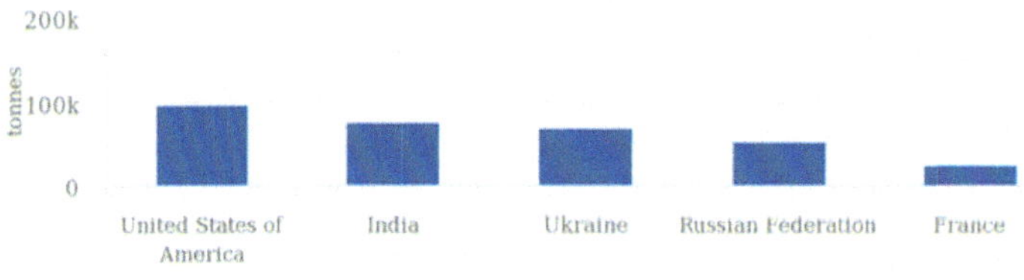

Fig. 1.6 Top five millet exporting countries in the world, 2016–2020. (Source: FAOSTAT (Feb 25, 2022))

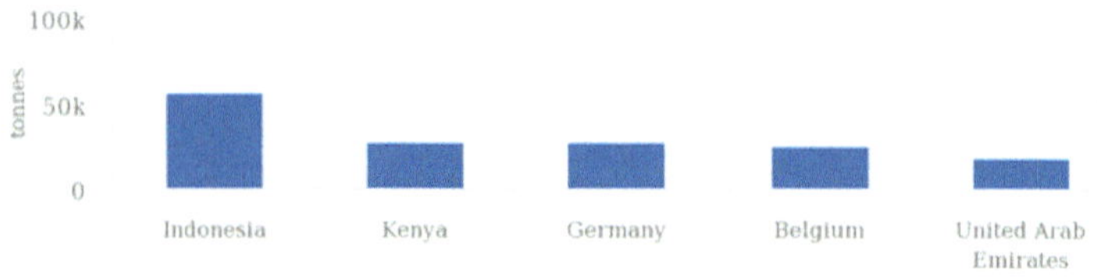

Fig. 1.7 Top five millet importing countries in the world, 2016–2020. (Source: FAOSTAT (Feb 25, 2022))

FAOSTAT (2022) showed that the top five millet importing countries (*based on the mean of 2016–2020*) in the world are Indonesia, Kenya, Germany, Belgium, and the United Arab Emirates (see Fig. 1.7). The quantum of imports is much lower (equal to less than 50,000 tons) per annum.

1.6 Future Prospects in Global Millet Supply and Demand

The International Model for Policy Analysis of Agricultural Commodities and Trade (IMPACT)[3] was developed by International Food Policy Research Institute (IFPRI) to make long-term global food projections among regions. The IFPRI's IMPACT (Robinson et al. 2015) model has been used for making the baseline projections (2005 base), which includes several commodities and not just the dryland cereals (millets). This gives a realistic picture because of the interactions between the demand and supply of all the crops. The effect of the fall in the prices of competing crops and the shift in demand for livestock products will have substantial implications for the demand and supply of dryland cereals (millets). The effect of trade policies has also been included in the model to reflect the price differential that exists between the country-level prices and the single-world price that the IMPACT model calculates. The export and import quantities of the countries are endogenous values just as the world price is calculated based on global demand and supply. A stock change variable is introduced to capture any mismatch between net trade and surplus/deficit production. The water scarcity on rainfed and irrigated lands is calculated through separate equations. Livestock production helps explicitly calculate the feed demand for feed crops. Demand for livestock products and demand for all food crops are calculated by one equation.

The global millet demand is increasing from 48.5 million tons in 2030 to 66.5 million tons by 2050 (Table 1.7). Correspondingly, the global supply projections are anticipated to increase from 49.1 million tons in 2030 to 67.1 million tons by 2050 with the available technologies. Africa and Asia are the major regions contributing to significant millet demand and supply in the world. Africa alone contributes nearly two-thirds of the share in global millet demand and supply. Western and Central Africa (WCA) region have a lion's (75%) share in the total

[3]Further details on IFPRI's IMPACT model are furnished in the Annexure.

Table 1.7 Demand and supply projections of millets among regions in the world

	2030	2040	2050	2030	2040	2050
Region	*Demand (000 tons)*			*Supply (000 tons)*		
Global	48,505	57,091	66,524	49,139	57,725	67,158
Africa	30,153	37,584	46,032	31,784	39,882	48,907
Eastern Africa	3068	3834	4677	3726	5042	6630
Western Africa	23,701	29,772	36,651	24,610	30,347	36,604
Americas	313	328	347	466	593	737
Asia	17,008	18,180	19,171	15,154	15,240	15,220

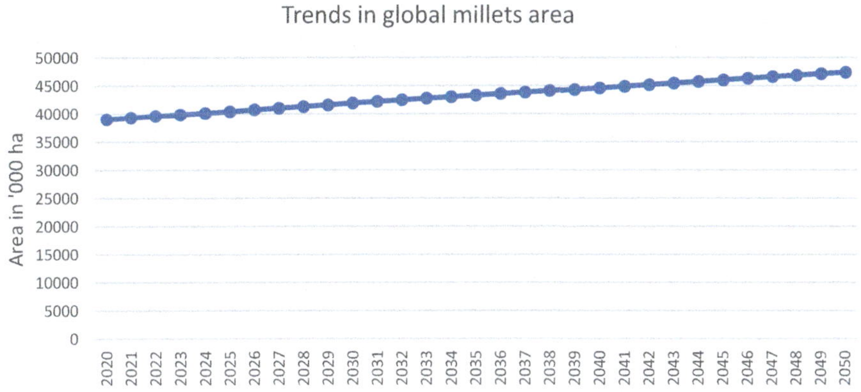

Fig. 1.8 Global millet area projections for 2020–2050

African region millet demand and supply. Asia region has only one-third of the share in the global millet demand and supply.

Africa region is projected to supply the burgeoning millet demand in the region during the study period. Except for the Asia region, the rest of the millet-growing regions in the globe are expected to supply the projected demand for millets in their respective regions. In Asia, the projected supply is lower than the regional demand in all the study periods. This clearly indicates the widening gap between projected supply and demand in the study period. There is a need to enhance millet production in the region by introducing improved cultivars and management practices.

The global millet area projections for the world, Asia, and Africa regions have been summarized in Figs. 1.8, 1.9, and 1.10 for the period 2020–2050. The millet area is globally expected to increase (21%) from 39 to 47 m ha (Fig. 1.8) over the next 30 years. The millet area projections for the Asia region indicate a decline of 14.5% from 13.1 to 11.2 m ha during the study period (Fig. 1.9). This might be due to severe competition from other economically high-value crops. However, the productivity in the region is expected to increase (Fig. 1.11) with improvements in genetic gains and enhanced access to improved seeds. The millet production is expected to stabilize (Table 1.7) over time (2030–2050) without any big quantum

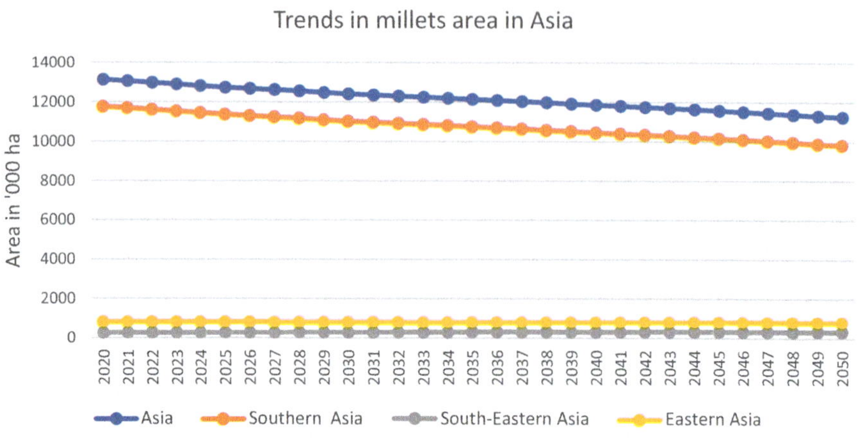

Fig. 1.9 Millet area projections for Asia from 2020 to 2050

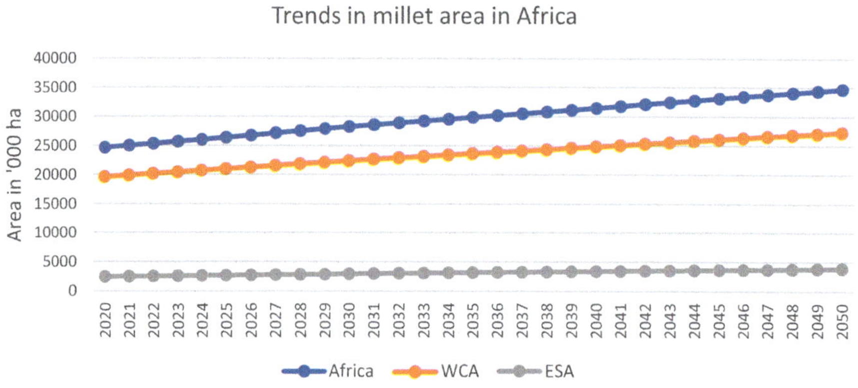

Fig. 1.10 Millet area projections for Africa from 2020 to 2050

jumps. Southern Asia has the lion's share (nearly 88%) in the total millet area in the region when compared to Eastern Asia and South-Eastern Asia regions.

Contrarily, the millet area in Africa is projected to expand from 24.7 to 34.8 m ha during the same period (Fig. 1.10). Western–Central Africa region played a crucial role in millet area expansion in the region compared to Eastern–Southern Africa. Relatively, the projected productivity gains in the African region are marginally lower when compared to the Asia region (Fig. 1.11). However, remarkable production gains in millets were observed in the WCA region, followed by the ESA region (Table 1.7). In the long term (2050), the millet productivity per ha is projected to plateau in Asia compared to Africa.

The country-wise projected demand and supply of millets for respective time periods are summarized in Table 1.8. India and Nigeria stood on the top for projected demand and supply of millets globally. Among WCA countries, Nigeria and Burkina

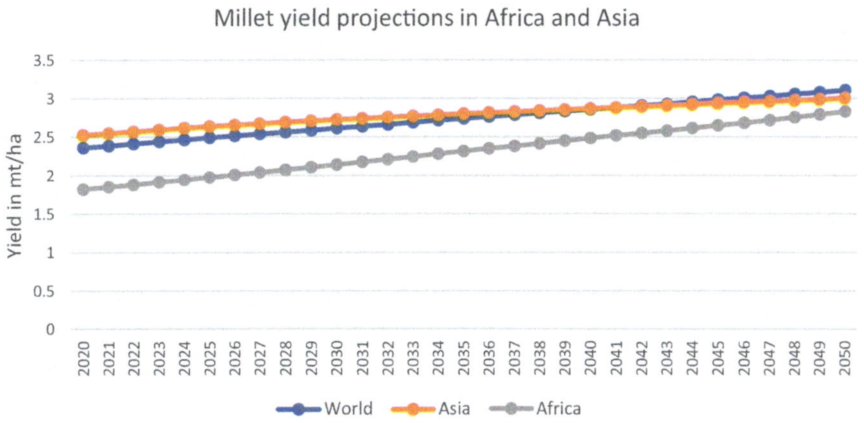

Fig. 1.11 Millet yield projections for Africa and Asia, 2020–2050

Table 1.8 Projected demand and supply of millets in major countries of the world

	2030	2040	2050	2030	2040	2050
Country	*Demand (000 tons)*			*Supply (000 tons)*		
India	13,430	14,452	15,341	**11,946**	**11,895**	**11,747**
Nigeria	12,200	15,178	18,603	**12,178**	**14,146**	**15,973**
Niger	5113	6784	8811	5364	6939	8769
Burkina Faso	2227	2734	3233	**2068**	**2519**	**2991**
China	2023	1994	1923	**1990**	**1991**	**1978**
Mali	1925	2354	2788	1975	2531	3259
Sudan	1600	1831	2024	**1374**	**1661**	**1905**
Uganda	1370	1762	2209	1436	1840	2288
Chad	857	1079	1324	1184	1577	2041
Ethiopia	767	880	977	1066	1434	1864
Senegal	721	842	959	1093	1519	2043
Russia	692	662	633	1182	1348	1511
Guinea	585	741	903	678	968	1279
Angola	532	597	812	**340**	**485**	**684**
Tanzania	439	542	649	615	906	1288
Nepal	430	476	515	**335**	**353**	**374**
Ghana	358	444	532	436	596	788
Pakistan	339	399	465	**275**	**304**	**326**
USA	274	284	299	434	552	683
Myanmar	234	232	227	299	346	401
Gambia	220	258	290	318	430	562
Ukraine	165	160	157	381	453	530
Yemen	147	178	208	**93**	**107**	**119**
Kenya	138	177	216	164	232	321

Their projected supply is lower than the projected demand respectively in 2030, 2040 and 2050

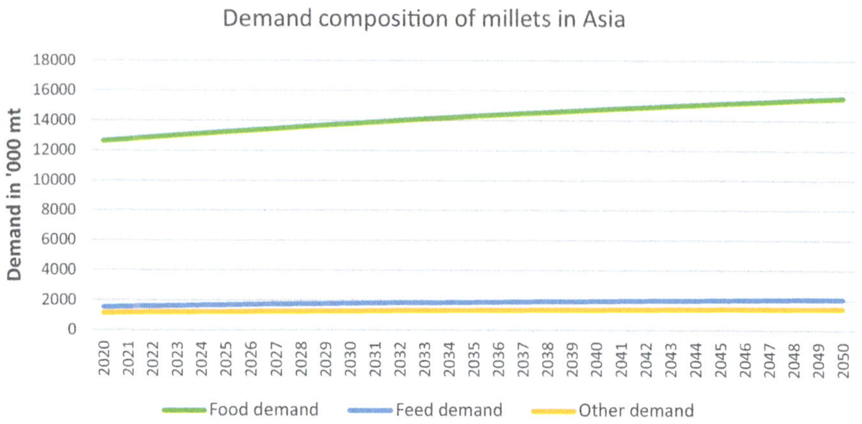

Fig. 1.12 Millet demand projections for Asia, 2020–2050

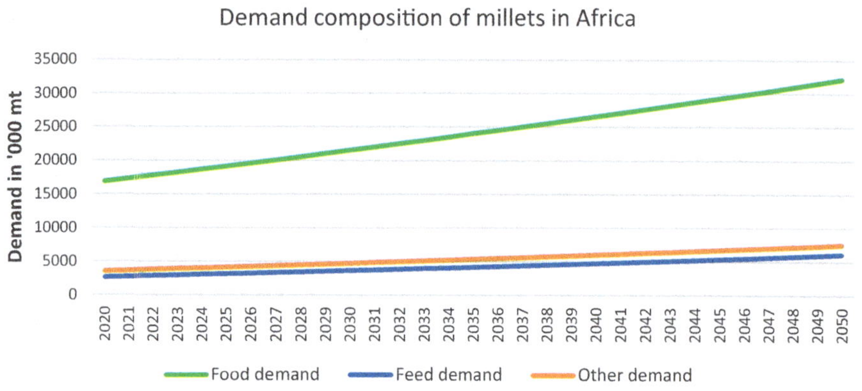

Fig. 1.13 Millet demand projections for Africa, 2020–2050

Faso showed the marginal gap between projected demand and supply over the study period. Sudan and Angola countries in the ESA region exhibited anticipated supply is lower than the projected demand of millets during the study period. Among all millet-growing countries, the conspicuous difference between millet demand and supply was noticed in India. Other Asian countries such as China, Nepal, and Pakistan showed a widening demand and supply gap of millets over the study period. All other millet-growing countries in the globe are expected to meet their millet demand through their supplies over time. There is a strong need for enhancing millet production in a few targeted countries in Africa and Asia regions for achieving global millet self-sufficiency.

The millet demand projections for Asia and Africa regions, along with their break-ups (food, feed, and other uses), are furnished in Figs. 1.12 and 1.13 for the period 2020–2050. Relatively, the food, feed, and other demand for millets are significantly expanding in the African region when compared to Asia during the study period. Millets showed ample scope for expansion, especially in the WCA region, where the demand for food and other uses (beverages, etc.) is very high. The feed demand for millets is third in priority in Africa, while it is second in priority in Asia.

1.7 Constraints in Millet Production

Agriculture, being a biological production system, faces a host of constraints that have the potential to affect the production and productivity of the system. Weather or climate plays a pivotal role in determining the extent of crop production. Land topography, soil fertility, water availability, incidence of pests and diseases, etc. are influenced by the climate of that region. Additionally, the socio-economic conditions of a farming community, like access to inputs, credit, and output markets, impact production decisions. All these factors are intricately related and impact the production and productivity of an agricultural production system. This section discusses the major constraints related to millet-based agricultural production systems across different ecologies in the world.

1.7.1 Biotic Constraints

Living organisms, such as fungi, oomycetes, bacteria, mycoplasma, nematodes, insects, birds, weeds, and parasitic plants, cause the biotic stress in millet production. The intensity of biotic stress varies depending on the weather, cropping system, cultivation practices, crop varieties, and their resistance levels. In general, high-input intensive cultivation, poor crop management, and hot and humid weather conditions aggravate its impact. Since millets are grown under dry climates, the extent of biotic stress on millets is less when compared with other crops.

Millets suffer significantly from fungal disease than bacterial, viral, and nematode diseases. The important diseases in millets are downy mildew, blast, smut, rust, and ergot diseases (Strange and Scott 2005; Das 2013). The impact of viral diseases is very low on millets. *Striga*, a parasitic weed, is a serious problem to cereal production in Africa (Ethiopia, Zimbabwe, Uganda, Rwanda, and Kenya), causing extensive yield losses in millets. The crop yields poor grain due to attacks on the roots of young plants and starves them nutritionally. Its impact is more severe in pearl millet when compared with other millets. An estimated 100 million ha of the African savannah zones are infested with *Striga* (Ejeta 2007) every year.

Bird damage to millet panicles is considered a potential threat to millet growers. The small-grain crops (millets) are highly prone to bird damage than large-grain crops. The extent of damage varies depending on growing conditions as well as the stage of the crop. There will be 100% loss in isolated fields and unprotected conditions. This is the most severe problem in African countries, specifically Ethiopia, Kenya, and Rwanda. It is estimated that Quelea and other birds can cause yield loss of about 1.6 million tons/year in eastern and southern Africa (Wortmann et al. 2009).

1.7.2 Abiotic Constraints

Soil fertility and environmental issues, in general, contribute to abiotic constraints in the production and productivity of millets. Abiotic constraints include the prevalence of recurrent droughts, insufficient rainfall, soil moisture, terminal heat stress (high temperatures), poor soil fertility status, and soil salinity issues. Among all, drought is the most critical abiotic constraint for millet production worldwide. Most of the millet-growing areas often receive scarce rainfall along with prolonged dry spells and highly erratic distribution. The standing crop in different agro-ecologies frequently receives insufficient soil moisture to grow optimally. Moisture stress may occur at different stages (seedling stage, vegetative stage, panicle stage, grain-filling stage, etc.) in different agro-ecologies. Drought and water stress are among the most important stresses of millet production in many African countries, such as Eritrea, Ethiopia, Ghana, Kenya, Namibia, Sudan, Tanzania, Uganda, and Zimbabwe (Matanyaire 1996; Gebretsadik et al. 2014). Soil-water deficits during millet-crop establishment and grain-fill stages are common in Ethiopia. But they are quite common during the mid-season in Uganda and Kenya. Hence, location-specific abiotic stress mitigation strategies are the need of the hour for sustainable millet production.

Poor soil fertility is another major limiting factor for enhancing the productivity of millets. In general, millets are grown on marginalized soils with low water-holding capacity and soil organic carbon. With the anticipated risks involved in millets' cultivation, farmers hardly apply fertilizers or follow crop rotations. These stresses, in combination with drought, severely limit the production potential of the millets in the semi-arid tropics. Soil salinity is another dimension that also severely constrains millet production. This is a major issue in problematic soils and not as widespread as drought and soil fertility. Poor drainage facility coupled with providing poor-quality irrigation water to millets exacerbates the soil salinity issue. The higher levels of soluble salts interfere with the seed germination process and cause a reduction in germination. Further, crop growth will be retarded and achieve poor yields.

1.7.3 Socio-Economic Constraints

Poor socio-economic development, limited infrastructure, more pressure on natural resources, lack of institutional capacity, poor marketing and credit facilities, etc. are the key factors that adversely impact agricultural development in any region. All these socio-economic drivers influence the extent of productivity of different agricultural systems in that region. Lack of sufficient demand (both quantity and quality) and other value-added opportunities for the product adversely impact its production.

All these factors vary from location to location and cropping system to cropping system. The socio-economic constraints play a greater role in the millet production of Africa than in other regions of the world. Relatively, all these factors have a more pronounced effect in Africa when compared with Asia.

Specifically, millet farmers in Africa use traditional varieties and landraces for cultivation, which results in low productivity levels in Africa. The farmers often undertake mixed cultivation of diverse crops and millet that meet their domestic needs. Women are primarily responsible for agricultural production, livestock rearing, post-harvest handling, and marketing. Lack of access to improved millet production technologies limits their production and productivity levels in the region. On the other hand, prevailing poor prices for millets and inadequate demand for them compel small and marginal farmers to sell their produce at lower prices. Sometimes, small-scale agri-business entrepreneurs sell their products on loan, a debt that later turns out to be a bad debt (Ja'afar-furo et al. 2011). However, the socio-economic constraints in India are different: shifting of millet-based croplands to more remunerative/commercial crops, which ultimately impacted millet production (Nagaraj et al. 2012; Seetharam 2015). A review was carried out to systematically synthesize the key constraints across major pearl millet growing ecologies worldwide, which has been furnished in Annexure Table 1.9.

1.8 Opportunities for Demand Expansion

The pearl millet market is set to grow from its current market value of more than $9 billion to over $12 billion by 2025. Favorable government initiatives aim to flourish the global millets market size over 2019–2025. Year 2023 has been identified as an "International Year of Millets" by the FAO, which is creating more awareness and knowledge about millets across different regions. Many developed and developing countries are giving enough impetus for promoting the millets in their respective countries. The pearl millet market is projected to grow with a compound annual growth rate of >3% over the forecasted period (2019–2025). Asia Pacific market, driven by India, procured more than 45% of the global share in the year 2018 (APEDA 2021). Revenues from sales of millets across the globe are forecasted to increase from over US$ 10,000 Mn in 2017 to exceed US$ 13,500 Mn by 2022-end. The projected diversity in its utilization is summarized in Fig. 1.14.

Pearl millet has more protein than rice and is rich in vitamins A and B, iron, phosphorus, magnesium, and manganese. The usage of millets in infant food and nutrition products is increasing, and many manufacturers are expanding their business operations by acquiring smaller firms. It has also found usage in beverages like beer. Breweries have started to add gluten-free beer options to increase their consumer base. Numerous developing nations (such as India) and federal governments are framing long-term policies to roll out nutrition programs to tackle malnutrition.

Millets have a limited shelf life compared with other coarse cereals because of rancidity issues in flour. This factor is estimated to hamper the market growth.

Table 1.9 Key constraints in major pearl millet growing countries in the world

Country	Constraint	Reference study
Global/India	Higher air temperatures (often >42 °C) coincide with flowering season. The crop suffers from reproductive sterility, leading to drastic reduction in seed set and finally lesser grain yield per ha	Gupta et al. (2015), Djanaguiraman et al. (2018)
Global	Climate change models projected that the pearl millet yield in SSA and SA will decrease by 6–17% by 2050	Knox et al. (2011)
India	Pearl millet-cropped area in India has contracted due to market constraints. Non-availability of improved and quality seeds at the right time, lack of proper seed marketing channels, and low returns relative to other crops under irrigated conditions are reasons commonly cited for the low adoption of high-yielding hybrids/varieties. On the demand side, changing food preferences owing to rising income and increased urbanization is leading to a substitution of coarse grains with fine cereals and livestock products. Downy mildew, blast, and leaf blight infestation are other major production constraints noticed in India	Bhagavatula et al. (2013)
Global	One of the major concerns is the diversion of millet-growing lands to the more remunerative crops, resulting in the shrinking of global areas under millet production. Among several, drought is the most important constraint in the millet-producing areas of the world. Drought and water stress are one of the most important stresses of millet production in many African countries, such as Eritrea, Ethiopia, Ghana, Kenya, Namibia, Sudan, Tanzania, Uganda, and Zimbabwe. The bird damage of millet grains is now considered a potential threat to millet growers, especially in Africa	Das and Rakshit (2016)
Ghana	Low yields and lack of improved seeds were ranked highest (ranked 1) in all regions, an indication that these two constraints were the driving force for improved and increased production and productivity of pearl millet in Ghana. Downy mildew and bird destruction were also ranked as important constraints in the Upper East region. The least important constraints in the Northern region were *Striga* infestation and ease of threshing, while in the Upper East, land availability and ease of threshing were mentioned. Low soil fertility, bird destruction, and land	Asungre et al. (2021)

(continued)

Table 1.9 (continued)

Country	Constraint	Reference study
	availability were the least important constraints in Upper West region. *Striga* infestation and low soil fertility were ranked low by farmers in the regions. In fact, in Upper West region, *Striga* did not feature at all as a constraint, while in the Northern and Upper East regions it was ranked 5 and 3, respectively	
Burkina Faso including Sahel and Sudan regions	Studies have indicated that drought is the most important constraint followed by the parasitic weed *Striga hermonthica*, head miner, and bird damage. Four diseases were listed: downy mildew, ergot, rust, and smut, among which downy mildew was ranked as the disease causing important yield losses in pearl millet. Other constraints such as cantharid attack, rust, and ergot were considered as minor constraints. Very long panicle and early maturity crop cycle were more preferred in the Sahel zone, whereas in the North-Sudan zone medium panicle length and medium maturity cycle were more preferred by farmers	Drabo et al. (2019), Rouamba et al. (2021)
Nepal	Lack of domain-specific high-yielding varieties and poor marketing infrastructure particularly in marginal areas are the major constraints. Uncontrolled weed infestation significantly reduces the crop yield between 16% and 94% in pearl millet depending on crop cultivars and management practices. The high hills of Nepal are facilitated with fragmented lands creating difficulty for mechanization. Increasing labor charges and scarcity of labor at transplanting, and intercultural operations have increased the workload to farmers. Ergot, smut, and blast are major diseases occurring on the millets in the country	Pramod (2021)
Senegal	Study revealed that parasitic *Striga* weed was the most constraining factor to pearl millet production across the rural communes. This was followed by low soil fertility and insect pests in that order. Other constraints included lack of machinery for sowing, plant diseases, drought, seed-eating birds, limited access to land for pearl millet cultivation, and limited seed availability.	Kanfany et al. (2020)
Nigeria	Major production constraint of the crop in the sub-Saharan Africa, particularly in Nigeria, is yield reduction and/or total crop failure	Azare et al. (2020), Maryam et al. (2017)

(continued)

Table 1.9 (continued)

Country	Constraint	Reference study
	caused by erratic seasonal rainfalls, floods, failing soil fertility, and poor crop management practices. Biotic stress due to weeds such as *Striga* spp. as well as insect pests and disease pestilences is a constraint. Downy mildew and labor availability also are other production constraints in the country. Therefore, appropriate climate change mitigation efforts to improve pearl millet production and enhance food security should be adopted	
West Africa	The lack of financial resources, high prices, risk aversion, and insufficient infrastructure inhibit many West Africa (WA) smallholder farmers from using fertilizers in millet cultivation. Low P input especially will become an increasing constraint, since resources of P fertilizer are scarce and non-renewable. Thus, increasing prices will make it unfeasible for poor pearl millet farmers to use such fertilizer. Pearl millet breeding in WA has therefore the task of developing varieties that are highly adapted to low input conditions	Pucher (2018)

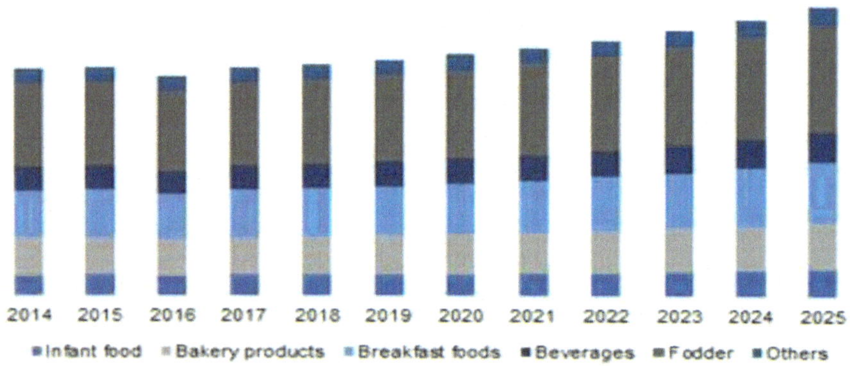

Fig. 1.14 Projected trends in global pearl millet utilization. (Source: Market Intelligence Report for millets; APEDA (2021))

However, the gluten-free property is expected to provide lucrative opportunities to produce gluten-free and low Glycemic Index (GI) food products. With steadily increasing demand from the urban population in Asia Pacific, mainly India, and China, farmers have begun drifting toward millet cultivation over rice and wheat.

High product prices in comparison with the commonly consumed grains are acting as a hindrance to penetrating the urban food market.

1.9 Conclusions

Millets are a group of small-seeded annual grasses grown mainly in Africa and Asia. Millets are recognized as photo-insensitive, resilient to climate change hardy crops that have a low carbon and water footprint. They can withstand high temperatures and grow on poor soils with little or no external inputs. Around 70% of total millet is produced under warm tropics dryland climate under poor soil and fragile environmental conditions with an LGP window ranging from 60 to 150 days. Globally, areas under millets occupy the sixth position after rice, wheat, maize, barley, and sorghum crops. Approximately, pearl millet (*Pennisetum glaucum R. Br.*) represents 75% of the global millet area. India is the world's largest producer of millets, followed by China, Niger, and Nigeria. During the last decade, the global area under millets has come down by three million ha, mainly due to a decline in Asia led by India. However, owing to rising productivity in Asia, production levels continued to be robust while they declined in Africa. Is the decline in millet area an indication of slowing demand for the crop? Projections based on IFPRI's IMPACT model indicate a growing demand for millets between 2030 and 2050, from 48.5 to 66.5 million tons. Africa accounts for 75% of the demand share and can meet its demand with domestic supplies over demand. For Asia, however, there will be a shortfall in supply as production will not keep pace with demand, thus opening trade opportunities for millets that are presently thinly traded, with less than 2% of its production entering global trade. While food demand remains the main source of millet demand, its demand for animal feed and industrial uses would increase, although from a low base.

Globally, nearly 3/4 of the domestic supply of millet is used for food and the rest for feed and other uses. Developed countries mainly use millet for animal/bird feed. The nutritive and gluten-free nature of millet has provided ample scope across the globe for developing several nutrition-based products, beverages, and baby foods. The enhanced demand from this sector is projected to expand the millet crop area and production across the globe.

Since millets are grown in marginal environments, they face several biotic and abiotic constraints. Fungal diseases include downy mildew, blast, smut, rust, and ergot. *Striga*, a parasitic weed, is another serious constraint to African millet production, causing extensive yield losses. Bird damage of millet grains is now considered a potential threat to growers. Among abiotic constraints, drought is the most important constraint, followed by poor soil fertility and low water-holding capacity, which are the very characteristics of the environments under which millets are grown. Poor market access for inputs and output, and lack of credit facilities are among the several socio-economic constraints that millet farmers face. It is thus pertinent that researchers and policy makers address the constraints facing millet production so the crop can be competitive with stable yields and acceptable quality standards for grain and fodder for consumers and processors.

Annexure: IMPACT Model

The IMPACT model is designed to examine alternative futures for global food supply, demand, trade, prices, and food security. It provides both fundamental, global baseline projections of agricultural commodity supply, demand, trade, prices, and malnutrition outcomes, and cutting-edge research results on quickly evolving topics such as bioenergy, climate change, changing diet/food preferences, and many other themes. IMPACT model covers 30 commodities, which account for virtually all of world food production and consumption, including all cereals, soybeans, roots and tubers, meats, milk, eggs, oils, meals, vegetables, fruits, and sugar and sweeteners in a partial equilibrium framework. It is specified as a set of 115 country-level supply and demand equations where each country model is linked to the rest of the world through trade. The model is written in the General Algebraic Modeling System (GAMS) programming language. The solution of the system of equations is achieved using the Gauss–Seidel algorithm. This procedure minimizes the sum of net trade at the international level at a world market price for a commodity that clears the world commodity market.

The basic IMPACT model is combined with the Water Simulation Model (WSM) to estimate the interactions between water supply and demand and food supply, demand, and trade. The scenarios for water are downscaled from and calibrated to Global Circulation Models (GCM) that represents future climates in the different IPCC SRES (Intergovernmental Panel on Climate Change Special Report on Emissions Scenarios). This allows for separate area and yield functions for rainfed and irrigated crops, water allocation among crops, and yield and area reductions from lack of water. Water demand is accounted for from irrigation for agriculture, livestock, domestic/municipal, and industrial sectors. This provides a further supply-side spatial disaggregation of production for irrigated and rainfed crop across 126 water basins. Combining these 126 basins with the 115 geopolitical regions gives 281 "food-producing units."

References

Andrews DJ, Kumar KA (1992) Pearl millet for food, feed and forage. Adv Agron 48:89–139. https://doi.org/10.1016/S0065-2113(08)60936-0

APEDA (2021). https://agriexchange.apeda.gov.in/Weekly_eReport/Millets_Report.pdf. Accessed 29 Apr 2022

Asungre PA, Akromah R, Kena AW, Gangashetty P (2021) Assessing production constraints, management and use of pearl millet in the Guinea Savanna Agro-ecology of Ghana. African J Plant Sci 15(11):288–298. https://doi.org/10.5897/AJPS2021.2183

Azare IM, Dantata IJ, Abdullahi MS, Adebayo AA, Aliyu M (2020) Effects of climate change on pearl millet (*Pennisetum glaucum* [L. R. Br.]) production in Nigeria. J Appl Sci Manag 24(1): 157–162

Bhagavatula S, Parthasarathy Rao P, Basavaraj G, Nagaraj N (2013) Sorghum and millet economies in Asia—facts, trends and outlook. International Crops Research Institute for the Semi-Arid Tropics, Patancheru. 80 pp. ISBN: 978-92-9066-557-1. Order code: BOE 060

Das IK (2013) Disease management in grain, forage and sweet sorghum. In: Chapke RR, Bhagwat VR, Patil JV (eds) Sorghum cultivation for value-added diversified products and sweet sorghum perspectives. Directorate of Sorghum Research, Hyderabad, pp 99–104

Das IK, Rakshit S (2016) Millets, their importance, and production constraints. In: Das IK, Padmaja PG (eds) Biotic stress resistance in millets. Elsevier, pp 1–22

Djanaguiraman M, Perumal R, Ciampitti IA, Gupta SK, Prasad PVV (2018) Quantifying pearl millet response to high temperature stress: thresholds, sensitive stages, genetic variability and relative sensitivity of pollen and pistil. Plant Cell Environ 41:993–1007. https://doi.org/10.1111/pce.12931

Drabo I, Zangre RG, Danquah EY, Ofori K, Witcombe JR, Tom Hash C (2019) Identifying farmers' preferences and constraints to pearl millet production in the Sahel and North-Sudan Zones of Burkina Faso. Exp Agric 55(5):765–775

Ejeta G (2007) The Striga scourge in Africa—a growing pandemic. In: Ejeta G, Gressel J (eds) Integrating new technologies for Striga control towards ending the witch-hunt. World Scientific Publishing Co., Singapore, pp 3–16

FAO and ICRISAT. 1996. The world sorghum and millet economies: facts, trends and outlook. ISBN 92-5-103861-9

FAOSTAT (2022) Various years. www.faostat.fao.org. Food and Agricultural Organization, Rome

Gebretsadik R, Shimelis H, Laing MD, Tongoona P, Mandefro N (2014) A diagnostic appraisal of the sorghum farming system and breeding priorities in Striga infested agro-ecologies of Ethiopia. Agric Syst 123:54_61

Gupta SK, Rai KN, Singh P, Ameta VL, Gupta SK, Jayalekha AK et al (2015) Seed set variability under high temperatures during flowering period in pearl millet (Pennisetum glaucum L. (R.), Br.). Field Crops Res 171:41–53. https://doi.org/10.1016/j.fcr.2014.11.005

Hartkamp AD, White JW, Rodríguez Aguilar A, Bänziger M, Srinivasan G, Granados G, Crossa J (2000) Maize production environments revisited: a GIS-based approach G. CIMMYT, Mexico, D.F.

ICRISAT (1992) ICRISAT medium term plan, 1994–1998. ICRISAT, Patancheru

Ja'afar-Furo MR, Bello K, Sulaiman A (2011) Assessment of the prospects of value addition among small scale rural enterprises in Nigeria: evidence from North-Eastern Adamawa State. J Dev Agric Econ 3(3):144_149

Jukanti AK, Laxmipathi Gowda CL, Rai KN, Manga VK, Bhatt RK (2016) Crops that feed the world 11. 2016. Pearl millet (Pennisetum glaucum L.): an important source of food security, nutrition and health in the arid and semi-arid tropics. Food Secur 8(2):307–329. https://doi.org/10.1007/s12571-016-0557-y

Kanfany G, Diack O, Kane NA, Gangashetty P, Sy O, Fofana A, Cisse N (2020) Implications of farmer perceived production constraints and varietal preferences to pearl millet breeding in Senegal. Afr Crop Sci J 28(3):2020. https://doi.org/10.4314/acsj.v28i3.6

Knox JW, Hess TM, Daccache A, Perez Ortola M (2011) What are the projected impacts of climate change on food crop productivity in Africa and South Asia? DFID Systematic Review Final Report, 77. Cranfield University, Cranfield

Kumara Charyulu D, Bantilan C, Rajalaxmi A, Rai KN, Yadav OP, Gupta SK, Singh NP, Moses Shyam D (2014) Development and diffusion of pearl millet improved cultivars in India: impact on growth and yield stability. Working Paper Series No. 52. International Crops Research Institute for the Semi-Arid Tropics, Patancheru. 76 pp

Kumara Charyulu D, Shyam DM, Bantilan C, Borikar ST, Gupta SK, Rai KN (2017) Pearl millet technology adoption and impact study in Maharashtra. Research Report 71. http://oar.icrisat.

org/9993/1/Research%20Report%20No%2071%20-%20Pearl%20Millet%20Technology%20Adoption.pdf
Lubulwa ASG, Menz K, White DH, Chudleigh P, McMeniman S, Hill D (2000) Determining international agricultural research priorities. ACIAR IAP Working Paper Series No. 37
Maredia MK, Ward R, Byerlee R (1996) Econometric estimation of a global spillover matrix for wheat varietal technology. Agric Econ 14:159–173
Maryam A, Dawud I, Angarawai I, Pangarayi B, Tongoona KO, Eleblu JSY, Beatrice EI (2017) Farmers' production constraints, knowledge of Striga and preferred traits of pearl millet in Jigawa State, Nigeria. Global J Sci Front Res (D) 17(3):23–28
Matanyaire CM (1996) Pearl millet production system(s) in the communal areas of northern Namibia: priority research foci arising from a diagnostic study. In: Leuschner K, Manthe CS (eds) Drought-tolerant crops for southern Africa: proceedings of the SADC/ICRISAT regional sorghum and pearl millet workshop, 25_29 July 1994, Gaborone, Botswana. ICRISAT, Patancheru, pp 43–58
Mausch K, Bantilan C (2012) Global homogenous groundnut zones—a tool to estimate the regional and international impact of agricultural research. Working Paper Series no. 31. International Crops Research Institute for the Semi-Arid Tropics, Patancheru. 28 pp
McDonough CM, Rooney LW, Serna-Saldivar SO (2000) The millets. Food science and technology: handbook of cereal science and technology, 2nd edn. CRC Press, Boca Raton, FL, pp 177–210
Nagaraj N, Basavaraj G, Parthasarathy Rao P, Bantilan MCS (2012) Future outlook and options for target crops: the sorghum and pearl millet economy of India. Policy Brief No. 15. RP Markets, Institutions and Policy (MIP), ICRISAT, Patancheru
Nedumaran S, Bantilan C, Gupta SK, Ahmed MI, Davis J (2015) Potential welfare benefit of millets improvement research to inform decision making: multi country—economic surplus model approach. Conference paper, presented at International Association of Agricultural Economists (IAAE) held at Milan, Italy
Obilana AB (2003) Overview: importance of millets in Africa. http://www.afripro.org.uk/papers/Paper02Obilana.pdf. Accessed 10 Jul 2015.
Parthasarathy Rao P, Hall AJ (2003) Importance of crop residues in crop-livestock systems in India and farmers perceptions of fodder quality in coarse cereals. Field Crop Res 84:189–198. Order code 787
Parthasarathy Rao P, Birthal PS, Reddy BVS, Rai KN, Ramesh S (2006) Diagnostics of sorghum and pearl millet grain-based nutrition. Int Sorghum Pearl Millet News Lett (ISMN):45–47
Pramod G (2021) Production trend, constraints, and strategies for millet cultivation in Nepal: a study from review perspective. Int J Agric Appl Sci 2(1):30–40. https://doi.org/10.52804/ijaas2021.213
Pucher AI (2018) Pearl millet breeding in West Africa—steps towards higher productivity and nutritional value. Doctoral Thesis (unpublished) submitted to Stuttgart Hohenheim, Germany
Rai KN, Gowda CLL, Reddy BVS, Sehgal S (2008) The potential of sorghum and pearl millet in alternative and health food uses. Comp Rev Food Sci Food Saf 7:340–352. https://doi.org/10.1111/j.1541-4337.2008.00049
Rama Rao CA, Raju BMK, Rao AVMS, Reddy DY, Meghana YL, Swapna N et al (2019) Yield variability of sorghum and pearl millet to climate change in India. Indian J Agril Econ 74:350–362
Robinson S, d'Croz DM, Islam S, Sulser TB, Robertson RD, Zhu T, Gueneau A, Pitois G, Rosegrant M (2015) The International Model for Policy Analysis of Agricultural Commodities and Trade (IMPACT): model description for version 3. IFPRI Discussion Paper 1483. International Food Policy Research Institute (IFPRI), Washington, DC. http://ebrary.ifpri.org/cdm/ref/collection/p15738coll2/id/129825
Rouamba A, Shimelis H, Drabo I, Laing M, Gangashetty P, Mathew I, Mrema E, Shayanowako AIT (2021) Constraints to pearl millet (Pennisetum glaucum) production and farmers'

approaches to Striga hermonthica management in Burkina Faso. Sustainability 2021(13):8460. https://doi.org/10.3390/su13158460

Seetharam A (2015) Genetic improvement in small millets. In: Tonapi VA, Patil JV (eds) Millets: ensuring climate resilience and nutritional security. Daya Publishing House, New Delhi, p 649

Serba DD, Yadav RS, Varshney RK, Gupta SK, Mahalingam G, Srivastava RK et al (2020) Genomic designing of pearl millet: a resilient crop for arid and semi-arid environments. In: Kole C (ed) Genomic designing of climate-smart cereal crops. Springer, Cham, pp 221–286. https://doi.org/10.1007/978-3-319-93381-8_6

SPAM (2010) Global spatially-disaggregated crop production statistics data for 2010 version 2.0 - IFPRI HarvestChoice dataverse (harvard.edu)

Strange RN, Scott PR (2005) Plant disease: a threat to global food security. Annu Rev Phytopathol 43:83_116

Sultan B, Roudier P, Quirion P, Alhassane A, Muller B, Dingkuhn M et al (2013) Assessing climate change impacts on sorghum and millet yields in the Sudanian and Sahelian savannas of West Africa. Environ Res Lett 8:014040. https://doi.org/10.1088/1748-9326/8/1/014040

Venkata Rao N, Rao KPC, Gupta SK, Mazvimavi K, Kumara Charyulu D, Nagaraj N, Singh RN, Surinder Singh S, Singh SP (2018) Impact of ICRISAT Pearl Millet Hybrid Parents Research Consortium (PMHPRC) on the livelihoods of farmers in India. Research Report 75. International Crops Research Institute for the Semi-Arid Tropics, Patancheru. 100 pp. ISBN 978-92-9066-595-3

Wortmann CS, Mamo M, Mburu C, Letayo E, Abebe G, Kayuki KC et al (2009) Atlas of sorghum production in eastern and southern Africa. University of Nebraska, Lincoln, NE

Yadav OP, Gupta SK, Govindaraj M, Sharma R, Varshney Rajeev K, Srivastava Rakesh K, Rathore A, Mahala RS (2021) Genetic gains in pearl millet in India: insights into historic breeding strategies and future perspective. Front Plant Sci 12:645038. https://www.frontiersin.org/article/10.3389/fpls.2021.645038

You L, Crespo S, Guo Z, Koo J, Ojo W, Sebastian K, Tenorio MT, Wood S, Wood-Sichra U (2011) Spatial production allocation model (SPAM) 2000 Version 3 Release 2. http://MapSPAM.info. Last Accessed 15 Mar 2011

Status and Utility of Pearl Millet Germplasm for Crop Improvement

2

M. Vetriventhan, Vinod Kumar, Naresh Reddy, Raja Srinivas, Kommineni Jagadeesh, Anil Kumar, Ovais Peerzada, M. Elangovan, K. Venkatesh, and Kuldeep Singh

Abstract

Millets assume greater importance in the context of changing climate and increasing demand for highly nutritious food and animal feed. The plant genetic resources are of greater value for crop improvement, which is essential in meeting future demands of food security and nutrition. Pearl millet is a climate-resilient and nutrient-dense crop enriched with a huge germplasm diversity that holds massive potential for crop improvement in the form of several unexplored genes for various traits. Globally, several genebanks together conserve over 73,000 pearl millet germplasms including wild and weedy relatives. The ICRISAT Genebank conserves the largest collection of *Pennisetum* spp. germplasm (24,663 accessions, originating from 51 countries). The characterisation and evaluation of the existing germplasm can unveil the trait diversity and can result in the identification of trait-specific sources. Several trait-specific sources identified for various biotic and abiotic stress resistance and quality traits can improve the utility of pearl millet germplasm for crop improvement. This chapter describes the status of pearl millet germplasm conservation ex situ and in situ, promising trait-specific sources for various biotic and abiotic traits and their utilisation in crop improvement.

M. Vetriventhan (✉) · V. Kumar · N. Reddy · R. Srinivas · A. Kumar · O. Peerzada · K. Singh
Genebank, International Crops Research Institute for the Semi-Arid Tropics (ICRISAT), Hyderabad, Telangana, India
e-mail: Vetriventhan.Mani@icrisat.org

K. Jagadeesh
Prof. Jayashankar Telangana State Agricultural University (PJTSAU), Rajendranagar, Hyderabad, Telangana, India

M. Elangovan · K. Venkatesh
ICAR-Indian Institute of Millets Research, Rajendranagar, Hyderabad, Telangana, India

V. A. Tonapi et al. (eds.), *Pearl Millet in the 21st Century*,
https://doi.org/10.1007/978-981-99-5890-0_2

2.1 Introduction

Pearl millet (*Pennisetum glaucum* (L.) R. Br., *Syn. Cenchrus americanus (L.)* Morrone) is an important climate-resilient nutrient-dense crop. It has wider adaptation to extreme climatic events such as drought and heat stresses and can play a significant role in ensuring global food, fodder and nutritional security in changing climate scenarios. Globally, pearl millet is the sixth most important cereal after rice (*Oryza sativa*), wheat (*Triticum aestivum*), maize (*Zea mays*), barley (*Hordeum vulgare*) and sorghum (*Sorghum bicolor*). Pearl millet is cultivated mostly on a subsistence basis in the marginal lands in semi-arid parts of Africa and Asia (Srivastava et al. 2020; Satyavathi et al. 2021). It is a tall robust quick-growing annual cereal crop with high photosynthetic efficiency and dry matter production. Being a hardy crop, it can be cultivated in low-rainfall areas (300–500 mm) as well as in high-temperature (up to 42 °C) regions (Vadez et al. 2012; Gupta et al. 2015). It is a staple food for around 90 million people in Africa and Asia, while its stalks are used as green and dry fodder for livestock. Worldwide, millets (including other types of millets) are grown in about 32 m ha and with a production of 30.46 million tons in 2020, and the major portion of it is occupied by pearl millet (https://www.fao.org/faostat/en/#data accessed on 13 October 2022). The major challenge of the twenty-first century would be global food and nutritional security, the most pressing issue for humanity (Godfray et al. 2010). Food and nutritional security is a multi-dimensional problem that faces several challenges arising from climate change, water scarcity, rapid population growth, rising food prices and persistent economic inequality severely affecting the smallholder farmers of their income and food supply. Millets assume greater importance in the context of changing climate and increasing demand for highly nutritious food and feed. Plant breeding primarily relies on the genetic variation that exists in the germplasm of crop species. Globally, over 66,000 germplasm accessions of pearl millet and its wild and weedy relatives are conserved in the ex situ genebanks. This chapter describes the status of pearl millet germplasm conservation ex situ and in situ, promising trait-specific sources for various biotic and abiotic traits and their utilisation in crop improvement.

2.2 Gene Pool and Races of Pearl Millet

Pearl millet genus is divided into five sections, viz. (1) *Penicillaria* (Tropical Africa and India), (2) *Brevivalvula* Döll (Pan-tropical), (3) *Gymnothrix* (Pan-tropical), (4) *Heterostachya* (Northeast Africa), and (5) *Eu-pennisetum* (Tropical and Sub-tropical Africa and India), each section having a variable number of species with different chromosome numbers, ploidy level (diploid to octoploid), reproductive behaviour (sexual or apomictic) and life cycle (annual, biennial or perennial) (Sharma et al. 2021). The annual diploid cultivated pearl millet *P. glaucum*, its wild species *P. glaucum* ssp. *monodii*, weedy species *P. glaucum* ssp. *stenostachyum* and the perennial tetraploid *Pennisetum purpureum* belong to the section *Penicillaria* (Pattanashetti et al. 2016; Sharma et al. 2021). Based on the cross-compatibility relationship between cultivated pearl millet and its crop wild relatives (CWRs), these species were grouped into primary, secondary and tertiary gene pools (Harlan and de Wet 1971). The Primary gene pool (GP1) includes domesticated diploid species *Pennisetum glaucum* ssp. *glaucum* ($2n = 2x = 14$ with AA genome), its wild progenitor *P. glaucum* ssp. *monodii* with its two ecotypes *Pennisetum violaceum* and *Pennisetum mollissimum* ($2n = 2x = 14$, AA), and the weedy forms, *shibras* (= *P. glaucum* ssp. *stenostachyum* Kloyzesh ex. Müll. Berol. Brunken; $2n = 2x = 14$ with AA genome). The absence of callus formation at maturity is the single and distinct character that separated species *P. glaucum* from *monodii* and *stenostachyum*. The GP1 species are easily crossed under sympatric conditions and form fertile hybrids with normal chromosome pairing (Harlan and de Wet 1971) and thus have a high possibility of successful introgression of genes among species in the GP1 into cultivated pearl millet. The Secondary gene pool (GP2) includes an allotetraploid rhizomatous perennial species *P. purpureum*, also known as Napier grass or elephant grass ($2n = 4x = 28$ with A′A′BB genome), and the apomictic and octoploid species *Pennisetum squamulatum* Fresen ($2n = 8x = 56$) (Kaushal et al. 2008; Sharma et al. 2021). *Pennisetum purpureum* and *P. squamulatum* can be easily crossed with cultivated pearl millet but their hybrids are highly sterile (Kaushal et al. 2008; Sharma et al. 2021). The Tertiary gene pool (GP3) includes the remaining species that are cross-incompatible with cultivated pearl millet. *Pennisetum schweinfurthii* (=*Pennisetum tetrastachyum*) Pilg. is the only Pennisetum species in the GP3 reported to have $2n = 2x = 14$ chromosomes and an annual growth habit, but its chromosomes are non-homologous to cultivated pearl millet (Hanna and Dujardin 1986; Martel et al. 2004) with different genomic localisations of rDNA probes (Martel et al. 1996). There are strong reproduction barriers between the members of GP3 and GP1 or GP2, and gene transfer is only possible by radical manipulations involving in vitro techniques or by using complex hybrid bridges (Martel et al. 1996; Sharma et al. 2020). Based on seed shapes, the cultivated pearl millet can be classified into four races (Brunken et al. 1977): *typhoides* (obovate in both frontal and profile views), *nigritarum* (obovate but angular in cross-section with three and six facets per grain), *globosum* (globular/ spherical, with each of its dimensions being approximately equal), and *leonis* (an acute, oblanceolate, terete caryopsis). It was suggested that migration events, in the very early period of domestication, followed by a combination of geographic and ethnographic isolation, are responsible for the variation in seed morphology of pearl millet (Brunken et al. 1977).

2.3 Germplasm Resources Conservation Status

2.3.1 Conservation Strategy

Biodiversity of the plant genetic resources for food and agriculture (PGRFA) is conserved following in situ and ex situ conservation strategies. In situ conservation includes the conservation of crop wild relatives in their natural habitat/genetic reserves and on-farm conservation of traditional landraces or farmer varieties. In situ conservation aims to protect, manage and monitor the biological diversity in their natural habitats, which supports the natural evolutionary processes to continue allowing new variations to be generated in the gene pool. Ex situ conservation aims to conserve biological diversity outside their natural habitats under the appropriate storage conditions such as seed storage, in vitro storage, DNA bank, and field and botanical gardens.

2.3.2 In Situ Conservation

In situ conservation of biological diversity particularly in the centre of origin and diversity is important and can complement ex situ conservation. In situ conservation aims to maintain target species and the collective genotypes they represent under evolution. A major rationale is based on the likelihood that continued exposure to changing selective forces will generate and favour new genetic variation and an increased likelihood that rare alleles of value to future agriculture are maintained. In situ conservation helps to conserve the process of evolution and adaptation of crops to their environments. It conserves the diversity at ecosystem, species and within species level and also conserves the ecosystem services critical to the functioning of the earth's life support system. In situ conservation integrates the farmers into a national plant genetic resources system, which helps to improve the livelihood of farmers through economic and social development.

Many crop wild relatives (CWRs) and landraces are under increasing threat of extinction due to several factors including urbanisation, habitat fragmentation and intensification of farming practices and most importantly climate change. The CWRs of pearl millet are conserved in situ by establishing biosphere reserves. There are currently 738 biosphere reserves in 134 countries that belong to the world network of biosphere reserves, promoting the conservation of biodiversity with its sustainable use (https://en.unesco.org/biosphere). For example, the Arly biosphere in Burkina Faso, which is situated in the West African Savannah, presents a wide variety of natural landscapes, where the cereal crops (millet and sorghum), peanut and cotton are the main crops (https://en.unesco.org/biosphere). On-farm conservation of landraces by farmers is largely practiced in pearl millet and other locally important crops. The establishment of community seed banks (CSB) supports the on-farm conservation of pearl millet landraces. There are many examples of the conservation of pearl millet and other locally important crops by farmers by establishing CSB (Vernooy et al. 2015). Many non-governmental organisations (NGOs) promote CSB by maintaining local germplasm collection and on-farm conservation of farmer varieties. Examples of such NGOs include the M.S. Swaminathan Foundation and Navdanya International, and Local Initiatives for Biodiversity, Research and Development (LI-BIRD).

2.3.3 Ex Situ Conservation

2.3.3.1 Global Status

A total of 66,682 accessions of *Pennisetum* spp. germplasm including 4581 wild and weedy relatives are conserved in 97 genebanks globally spread across 65 countries (https://www.fao.org/wiews-archive/wiews.jsp accessed on 24 October 2021). The major portion of cultivated, wild and weedy species accessions held by various genebanks in the world is presented in Table 2.1.

The ICRISAT (International Crops Research Institute for the Semi-Arid Tropics) Genebank conserves 24,663 accessions originating (collection site) from 52 countries and is the largest collection of *Pennisetum* spp. germplasm in the world. Landraces constitute the major position of this collection (87%) followed by advanced cultivars and/or breeding lines (10%) and wild and weedy relatives (3%). The major portion of the pearl millet collection at ICRISAT is from India (33%), followed by Nigeria (10.1%), Niger (9%) and Zimbabwe (7%), while the remaining countries represent <5%. The collection is conserved as an active collection (cold room at 4 °C, 20% relative humidity, stored in aluminium cans with screw caps having rubber gaskets) and base collection (cold room at −20 °C, seed moisture content at 3–7%, seeds vacuum sealed in an aluminium foil pouches). Several accessions of wild species, which either do not produce seed or produce very less seed, are being maintained in the field genebank as live plants. Untill 2023, the ICRISAT Genebank supplied over 168,000 seed samples to 86 countries globally

Table 2.1 The major collections of pearl millet germplasm conserved ex situ in the genebanks globally

Institute/Genebank	Country	Cultivated	Wild	Total
International Crops Research Institute for the Semi-arid Tropics, India	Global	23,847	816	24,663
National Bureau of Plant Genetic Resources	India	8196	185	8381
Embrapa Milho e Sorgo	Brazil	7225		7225
Laboratoire des Ressources Génétiques et Amélioration des Plantes Tropicales, ORSTOM	France	3620	789	4405
Plant Gene Resources of Canada, Saskatoon Research Centre, Agriculture and Agri-Food Canada	Canada	3543	273	3816
Serere Agriculture and Animal Production Research Institute	Uganda	2142		2142
Plant Genetic Resources Conservation Unit, Southern Regional Plant Introduction Station, University of Georgia, USDA-ARS	United States of America	1090	973	2063
Institut national de la recherche agronomique du Niger	Niger	2052		2052
National Plant Genetic Resources Centre, National Botanical Research Institute	Namibia	1419	2	1421
Plant Genetic Resources Program	Pakistan	1272	105	1377

Source: https://www.fao.org/wiews-archive/wiews.jsp, http://genebank.icrisat.org/

following the standard material transfer agreement (SMTA) of the International Treaty on Plant Genetic Resources for Food and Agriculture (ITPGRFA) of the Food and Agriculture Organisation (FAO). Out of the total accessions conserved at ICRISAT Genebank, 85% of them were distributed at least once, indicating the range of diversity distributed for pearl millet improvement. Besides germplasm distributions, the global collections held at the ICRISAT Genebank serve the purpose of restoration of germplasm to the source countries when national collections are lost due to natural calamities, civil strife, etc. The ICRISAT Genebank has repatriated 8773 pearl millet accessions to India (7189), Cameroon (922), Sudan (594), Chad (42) and Niger (26); thus the national programmes have regained their precious plant germplasm heritage that could have been lost if this was not conserved in the ICRISAT Genebank. As a safety backup, ICRISAT has submitted over 92% of its total pearl millet collection to Svalbard Global Seed Vault (SGSV), Norway. The ICAR-National Bureau of Plant Genetic Resources (ICAR-NBPGR), New Delhi, is conserving the second largest collections of *Pennisetum* spp. (8381 accessions) in the world (http://genebank.nbpgr.ernet.in/).

Globally over 4500 accessions of wild and weedy relatives of pearl millet are conserved ex situ in genebanks. Seven species represent over 100 accessions while others represent below 100 accessions (Table 2.2). The highest number of wild *Pennisetum* species of 973 accessions was held by Plant Genetic Resources Conservation Unit, Southern Regional Plant Introduction Station, University of Georgia, USDA-ARS, S9 (USA0160) with the highest being *Pennisetum ciliare* (862 accessions); followed by ICRISAT Genebank (831 accessions), which includes *Pennisetum glaucum* spp. *violaceum* (335 accessions, representing 12 countries [Fig. 2.1]), *P. mollissimum* (48 accsessions), *Pennisetum pedicellatum* (134 accessions), *Pennisetum polystachion* (97 accessions), *P. purpureum* 39 accs, *Pennisetum* orientale 34 accs, *Pennisetum ramosum* 25 accs, *P. ciliare* 13 accs, *Pennisetum mezianum* and *Pennisetum setaceum* 11 accs, *Pennisetum divisum* 7 accs, *Pennisetum hohenackerii* 7 accs, *Pennisetum flaccidum* 6 accs, *P. schweinfurthii* and *Pennisetum cenchroides* 5 accs, *Pennisetum thunbergii* 3 accs, *Pennisetum sphacelatum* 2 accs, *P. squamulatum* 2 accs, *Pennisetum villosum* 2 accs and remaining species *Pennisetum trachyphyllum*, *Pennisetum alopecuroides*, *Pennisetum hordeoides*, *Pennisetum lanatum*, *Pennisetum macrostachyum* and *Pennisetum macrourum* representing one accession each; and Laboratoire des Resources Génétiques et Amélioration des Plantes Tropicales, ORSTOM, ORSTOM-MONTP (FRA202) holds 798 accessions, which include *Pennisetum glaucum* ssp. *monodii* (346 accessions) and *P. glaucum ssp. stenostachyum* (452 accessions).

2.3.3.2 Gap Analysis in the Ex Situ Collection

Germplasm gap identification in the ex situ germplasm collection helps to identify and map the regions that are poorly represented by the particular crop germplasm collection held in the genebank. Gap analysis helps to collect and conserve the rarely distributed landraces and other germplasm accessions that are under the threat of extinction due to socio-anthropological reasons. An ideal collection includes

Table 2.2 Species-wise number of *Pennisetum* wild and weedy relatives conserved in the genebanks globally

Row labels	Number of accessions
Pennisetum ciliare	902
Pennisetum violaceum	765
Pennisetum purpureum	730
Pennisetum glaucum ssp. stenostachyum	452
Pennisetum pedicellatum	397
Pennisetum monodii	346
Pennisetum polystachion	298
Un classified species	107
Pennisetum orientale	56
Pennisetum unisetum	50
Pennisetum clandestinum	48
Pennisetum mollissimum	48
Pennisetum setigerum	44
Pennisetum sphacelatum	40
Pennisetum setaceum	38
Pennisetum thunbergii	34
Pennisetum mezianum	25
Pennisetum stramineum	20
Pennisetum ramosum	16
Pennisetum hohenackeri	13
Pennisetum trachyphyllum	13
Pennisetum procerum	12
Pennisetum squamulatum	12
Pennisetum divisum	11
Pennisetum flaccidum	11
Other species <10 accessions each: *Pennisetum alopecuroides, P. macrourum, P. purpureum* × *glaucum, P. villosum, Pennisetum massaicum, P. schweinfurthii, Pennisetum sieberianum, Pennisetum americanum, P. cenchroides, Pennisetum trisetum, Pennisetum basedowii, P. hordeoides, P. lanatum, Pennisetum prieurii, Pennisetum bambusiforme, Pennisetum dichotomum, Pennisetum dowsonii, Pennisetum foermeranum, P. glaucum* spp. *stenostachyon, Pennisetum latifolium, Pennisetum longissimum, Pennisetum macrostachys, P. macrostachyum, Pennisetum petiolare, Pennisetum riparium*	93
Total wild species accessions	**4581**

Source: https://www.fao.org/wiews-archive/wiews.jsp

accessions collected from the centre of diversity of species as well as representative areas where the crop is extensively cultivated. Explorer should cover the widest possible agroecological regions and the number of samples should be large enough to allow less frequent alleles to be represented in the collection. The ICRISAT Genebank conserves the largest collection of pearl millet but considerable gaps

Fig. 2.1 Plant with many tillers and panicles of the wild species *Pennisetum violaceum*. (Photo by Mr. D. Naresh, ICRISAT, India)

exist to enrich the collection. The gap analysis of the ICRISAT collection from the West and Central Africa (WCA) region identified 62 districts in 13 provinces of Nigeria, 50 districts in 16 provinces of Burkina Faso, 9 districts in 6 provinces each of Mali and Mauritania, 8 districts in 8 provinces of Chad and 7 districts in 3 provinces of Ghana as the major geographical gaps in the pearl millet germplasm conserved at ICRISAT Genebank (Upadhyaya et al. 2009b). Similarly, 34 districts located in 18 provinces of four East African countries and 76 districts located in 34 provinces of seven Southern African countries were identified as geographical gaps in the pearl millet collection at the ICRISAT Genebank (Upadhyaya et al. 2012). The gap analysis of germplasm from Asia revealed 134 districts in 14 provinces in India, 12 districts of Punjab province in Pakistan, Southern parts of North Yemen and Lahiz provinces in Yemen as gaps in the collection (Upadhyaya et al. 2010). Crop wild and weedy relatives contribute significantly to trait improvement in various crops including pearl millet. The gap analysis of the CWR collection indicated 194 provinces in 21 countries of Asia and Africa as geographical gaps in the *P. pedicellatum*, while in *P. glaucum* subsp. m*onodii*, 354 districts located in 86 provinces of 8 countries (158 districts in 35 provinces in Mali, 67 districts in 14 provinces in Burkina Faso, 59 districts in 7 provinces in Nigeria, 24 districts in 8 provinces in Mauritania, 15 districts in 4 provinces in Senegal, 14 districts in 7 provinces in Sudan, 12 districts in 6 provinces in Niger, and 5 districts in 5 provinces in Chad) were indicated in the ICRISAT Genebank pearl millet CWRs collection (Upadhyaya et al. 2014b). Enriching the world collection through collection missions and assembly of missing diversity is being done at ICRISAT considering the gaps identified through gap analysis.

2.3.4 Germplasm Subsets

Though large and diverse collections are available in many crops, their use in crop improvement will be limited mainly when there exists non-availability of information on traits of economic interest, linkage load with undesirable genes and assumed risks, and restricted access to the germplasm collection due to regulations governing international exchange, etc. (Upadhyaya et al. 2009c). To enhance the utilisation of diversity in crop improvement, the concepts of core (~10% of the entire collection; Frankel 1984) and mini-core (~10% of the core collection; Upadhyaya and Ortiz 2001) collections were advocated. In pearl millet, a core collection (2084 accessions; Bhattacharjee et al. 2007; Upadhyaya et al. 2009a) and a mini-core (238 accessions; Upadhyaya et al. 2011) collection have been developed. List of accessions in the subset are available in the Genesys PGR database (https://www.genesys-pgr.org/subsets/v2757WpAPbW). Evaluation of core/mini-core sets has led to the identification of new sources of variation for grain and fodder traits (Khairwal et al. 2007), resistance to blast and downy mildew (Sharma et al. 2013, 2015) and grain nutritional traits (Govindaraj et al. 2020a). Since pearl millet is completely outcrossing crop, landraces of pearl millet show high level of within accession diversity. Only limited attempts were made towards genomic investigation of germplasm for trait discovery using genome-wide association study and genomic selection. Therefore, ICRISAT in association with Aberystwyth University developed a world association mapping panel called the Pearl Millet inbred Germplasm Association Panel (PMiGAP). This PMiGAP comprises 346 lines generated by repeated rounds of selfing (S0 through S11) from 1000 accessions originating from 27 countries, representing diverse cultivars, landraces and mapping population parents. Thus, PMiGAP could be an excellent genetic resource for genomic investigation in pearl millet for traits discovery. This PMiGAP was re-sequenced and it has a repository of circa 29 million genome-wide Single nucleotide polymorphisms (SNPs), and is being used to map traits related to drought tolerance, grain nutrients, nitrogen use efficiency, etc. (Srivastava et al. 2020).

2.3.5 Characterisation and Evaluation of Pearl Millet

The characterisation and evaluation of germplasm is a significant step towards effective and enhanced utilisation in crop improvement. Germplasm without information on the characterisation and evaluation data results in poor utilisation of diversity in crop improvement. With the emerging high-throughput phenotypic characterisation and evaluation, a large number of germplasms could be phenotyped for various traits to develop trait-specific subsets for use in crop improvement. Literature suggests tremendous phenotypic and genetic diversity in pearl millet for various traits that help in the mining of genes and alleles for important traits of interest. The efforts made by researchers in pearl millet germplasm have created a pool of trait-specific germplasm that are promising sources for grain yield, stover yield, grain nutrition, photoperiod sensitivity, heat tolerance, phenological traits, shelf life, resistance to biotic and abiotic stresses and specific grain characteristics as preferred by consumers in accordance with the market demand (Dwivedi et al. 2012; Pattanashetti et al. 2016).

2.4 Trait Variability and Promising Sources

2.4.1 Morpho-Agronomic Traits

Characterisation of germplasm conserved at the ICRISAT Genebank following the descriptors of pearl millet is available to researchers globally to access desirable germplasm based on various traits (https://genebank.icrisat.org/). Variability in the qualitative traits indicates nine panicle shapes (cylindrical, conical, spindle, club, candle, dumb-bell, lanceolate, oblanceolate and globose), five seed shapes (obovate, oblanceolate, elliptical, hexagonal and globular), ten seed colours (ivory, cream, yellow, grey, deep grey, grey-brown, brown, purple, purplish-black and mixture of white and grey) and three classes of each of bristle length (short, medium and long), spike density (loose, medium and compact) and endosperm texture (mostly corneous, partly corneous and mostly starchy). Accessions with candle-shaped panicles, short-bristled panicles, globular seed shape, grey seed colour and seeds with corneous endosperm texture are more predominant in the ICRISAT collection (Upadhyaya et al. 2007).

The geographical distribution of pearl millet germplasm conserved at the ICRISAT Genebank revealed the adaptation of pearl millet to latitudes ranging between 33.00°S and 36.91°N (Upadhyaya et al. 2017a). Accessions with early flowering (33–40 days) were predominant in Pakistan, Ghana, Togo and India; very late flowering (121–159 days) in Sierra Leone and the Central African Republic; short plant types (80–100 cm) in India, Zambia and Sudan; tall types (401–490 cm) in Chad, Burkina Faso, Nigeria and the Central African Republic; high tillering types (11–35) in India and Yemen; with high panicle exertion (11–29 cm) in Ghana, Chad, India and Yemen; with long panicles (75–135 cm) in Nigeria and Niger; with thick panicles (41–58 mm) in Namibia, Togo and Zimbabwe; and those with large seeds (16–19 g 1000 seeds weight) were predominant in Togo, Benin, Ghana and Burkina Faso. Collections from Ghana for flowering (36–150 days); Burkina Faso for plant height (80–490); India and Yemen for total (1–35) and productive (1–19) tillers per plant; Niger for panicle exertion (−45 to 21.0), panicle length (9–135 cm) and thickness (12–55 mm); and Zimbabwe for 1000 seed weight (3.5–19.3 g) were found as important sources for trait diversity (Upadhyaya et al. 2017a). Promising trait-specific sources reported for various quantitative traits are listed in Table 2.3. Evaluation of a large set of germplasms (3059 accessions including 504 accessions in the core collection) for grain yield and its contributing traits indicated a large

Table 2.3 Promising trait-specific sources reported in pearl millet for phenological, yield, quality and stress tolerance traits in the global collection conserved at ICRISAT, India

Trait	Accessions (IP numbers)	Reference
Early flowering	Early flowering in Kharif season: Bchwaddawana (IP# 1417, 1418, 1425, 1427), Chadi local (IP# 3155, 3172, 3175, 3220, 3226, 3231, 3235, 3239, 3252, 3253, 3435, 13586), Desert type (IP# 3230, 3251), Pitta ganti (IP# 1249, 1251, 1252, 11782–11784, 11786, 11787, 11789–11792, 11794, 11867, 11868, 11873, 11877, 11878, 13460, 13465), Jakhrana (IP# 1478–1480, 3123 and 11885), Kala cumbu (IP 1315), Manavari (IP 1304), Peria cumbu (IP 1307), Adchi (IP# 9101, 9105, 17867), Arisi (IP# 15329 and 15412), Babapuri (IP# 1179, 3840, 12556), Gaorani (IP# 9107, 9113, 9115, 9121, 9122, 9146, 13554), Kattu cumbu (IP# 15335, 15373), Kullan (IP 1306), Pedda sajja (IP# 11810, 11811), Sadguru bajra (IP# 1266, 1267, 7813) and Vellai (IP 13600); Early flowering in Rabi season: Gaorani (IP# 4144, 4148 and 4149)	Upadhyaya et al. (2016b)
	OPY 197, SR 15, SR 17, SR 54, IP 4006, IP 550, IP 3549, IP 4066, IP 9496 and IP 9426 were extra-early flowering types; OPY 197, IP 3549 and SNV26 for early maturity	Khairwal et al. (2007)
High grain yield potential	Chadi local (IP# 3233, 3234), desert type (IP# 3244, 3245), Peria cumbu (IP# 3477) and Gaorani (IP# 9119) for grain yield	Upadhyaya et al. (2016b)
	IP 1763 and IP 1745 promising for long panicles; NSS 7776, NSS 7799 and IP 4394 for thick panicles; IP 399 for bold grains; IP 128, IP 3416, IP 33, NSS 7911, IP 17350, IP 3150, IP 17945 and IP 7095 for high grain yield; RMFB 35, RMFB 34, SR 75, SR 74, Desi Chothina, IP 13645, IP 15273 and IP 17493 for higher productive tillers	Khairwal et al. (2007)
	Gero (IP 12188), Guerguera (IP 5311), Haini (IP 10621, IP 10630 and IP 10678), Maewa (IP 13129), Mela (IP 11614), Nyali (IP 17681), Sanio (IP 10177 and IP 10576) and Sounari (IP 10623)	Upadhyaya et al. (2016a)
Fodder yield	Tall plants: IP# 11839, 11840, 17287, 20439, 20509, 20538, 20539, 20540, 20544, 20550, 20563, 20571, 20574, 20584, 20585; Higher number of tillers: IP# 3080, 3476, 3604, 3613, 3625, 3627, 3628, 3636, 3645, 3663, 3665, 6857, 6892, 8190, 8327, 11838, 13599, 13613, 15257, 15285, 15287, 15288, 15289, 15290, 15301, 15302, 15306, 15307, 15320, 15321, 15322, 15341, 15342, 15343, 15344, 15348, 15351, 15369, 15438, 15556, 20273,	Khairwal et al. (2007), Upadhyaya et al. (2016b, 2017c)

(continued)

Table 2.3 (continued)

Trait	Accessions (IP numbers)	Reference
	20339, 20344, 20346, 20347, 20348, 20349, 20350, 20379; IP 3416, IP 17880 and IP 104 promising for high fodder yield per plant. Green fodder yield: Chadi local (IP 3188 and IP 3246), Desert type (IP 3173 and IP 3190), Eravai (IP 3517), Kala cumbu (IP 3586), Peria cumbu (IP# 1305, 3610), Amreli (IP 1174), Gaorani (IP 9074 and IP 9076), Kattu cumbu (IP 3479 and IP 3596) and Sadguru bajra (IP 16283); Aria (IP 6062), Gaouri (IP 10574), Maewa (IP 17436), Souma (IP 10022 and IP 10030), Sounari (IP 10014 and IP 10034), Tiotioni (IP 9997 and IP 10085), Yadiri (IP 14258) and Zanfaroua (IP 9208). IP 11839 and IP 11840 for tall plant heights and more number of leaves per plant; IP 15710, IP 15735 and IP 15752 for stem thickness and leaf width; and IP 3628, IP 15285, IP 15288, IP 15302, IP 15342, IP 15351, IP 15290, IP 20347 and IP 20350 for total tillers per plant. Sweet stalk: IP 3471, IP 3481, IP 3509 and IP 3593	
High grain nutrients	High Fe: IP# 3329, 7536, 8972, 9301, 10394, 12682, 15817, 17620, 17690, 17707; High Zn: IP# 3329, 3749, 4454, 10471, 11584, 11784, 12181, 13900, 15614, 17217; High Mn: IP# 7536, 8972, 9572, 10394, 13384, 15402, 15614, 17217, 17620, 17707; High Cu: IP# 3329, 3626, 7208, 9351, 10394, 12507, 12682, 13384, 13900, 15614; High Mo: IP# 3329, 13900, 15614, 3749, 10471, 11316, 11353, 11584, 11784, 17217; High Ca: IP# 15402, 9416, 17707, 9351, 15614, 17620, 11584, 14148, 7536, 5316; High Mg: IP# 15614, 9572, 7208, 17217, 7536, 3626, 3329, 4454, 12507, 10471; High K: IP# 9572, 10471, 12682, 11584, 3329, 12507, 14148, 15614, 11316, 11353; High P: IP# 9572, 15614, 3329, 7536, 12507, 14148, 7208, 10471, 12682, 17217; High S: IP# 9572, 17217, 12507, 7536, 7838, 3329, 4454, 15614, 3626, 11784; Low Na: IP# 5316, 13384, 3626, 12181, 17690, 7838, 4454, 11353, 10471, 12682; Low Ni: IP# 11316, 12862, 13384, 7536, 11320, 9407, 11784, 14148, 9416, 10471. Multiple nutrients: IP# 3329, 3749, 4454, 5316, 7208, 7536, 9351, 9407, 9496, 9572, 11316, 11784, 12507, 12939, 14148	Govindaraj et al. (2020a)

(continued)

Table 2.3 (continued)

Trait	Accessions (IP numbers)	Reference
Downy mildew resistance	IP# 9, 29, 55, 87, 104, 172, 253, 262, 283, 336, 346, 352, 364, 365, 396, 486, 498, 505, 517, 545, 556, 558, 575, 682, 718; IP# 14537 resistant to 7 pathotypes; IP# 9645, 11943, 14542, 14599, 21438, 11930, 12374, 14522, 20715, 21187, 21201, 21244 resistant to 5-6 pathotypes	Sharma et al. (2007, 2015)
Blast resistance	IP# 21525, 21531, 21536, 21540, 21594, 21610, 21640, 21706, 21711, 21716, 21719, 21720, 21721, 21724, 21987, 21988 and 22160 (all *P. violaceum* wild species accessions); IP# 7846, 11036 and 21187 (cultivated pearl millet)	Sharma et al. (2013, 2020)
Rust resistance	Single plant selection from IP# 21629, 21645, 21658, 21660, 21662, 21711, 21974, 21975 and 22038 (all *P. violaceum* wild species accessions); ICML 11, ICML 17, ICML 18, ICML 19, ICML 20, ICML 21	Singh et al. (1990), Singh et al. (1987), Sharma et al. (2020)
Smut	ICML 5, ICML 6, ICML 7, ICML 8, ICML 9, ICML 10	Thakur et al. (1992)
Drought tolerance	IP# 3243, 3228, 3424, 3296, 3362, 3180, 3272, 3303, 3252, 3258, 11141, 3318, 3123, 3363 and 3244; IP# 133, 177, 164, 142, 120, 160, 136, 166, 192, 195, 106, 126, 121, 110, and 117	Yadav et al. (2003), Choudhary et al. (2021)
Salinity tolerance	IP# 22269, 6098, 6105, 3616, 6112, 6104, 3757	Kulkarni et al. (2006), Krishnamurthy et al. (2007)
Heat stress tolerance	IP# 19799, 19877, 19743, 21517, 3175	Yadav et al. (2012), Sehgal et al. (2015), Upadhyaya et al. (2016b)

variability, and promising sources for early maturity, tall plants, total tillers, productive tillers, grain yield, fodder yield, etc. (Khairwal et al. 2007) (Table 2.3). Evaluation of named landraces originating from India (692) and Africa (2929) identified promising landraces for yield and yield-contributing traits (Upadhyaya et al. 2016a, b) (Table 2.3).

2.4.2 Grain Nutrients and Quality Traits

Pearl millet grain is rich in nutrients and health-beneficial phenolic compounds, making it suitable for food and feed. Several studies have been conducted around the world to assess the genetic diversity of grain nutrients in pearl millet using different sets of germplasm accessions and breeding lines (Bashir et al. 2014; Kanatti et al. 2014; Rai et al. 2015; Anuradha et al. 2017; Govindaraj et al. 2019, 2020a, b) Grain nutrients assessment using a large number of pearl millet germplasms (428 accessions, part of core collection) conserved at the ICRISAT Genebank indicated large variability: Fe (34–90 mg/kg), Zn (30–74 mg/kg), Ca (85–249 mg/kg), Mn (10–17 mg/kg), Cu (4–7 mg/kg), Mo (0.6–2.2 mg/kg), Ni (0.7–1.9 mg/kg), Ca (85–249 mg/kg), Mg (1127–1837 mg/kg), Na (8–22 mg/kg), K (3167–5133 mg/kg), P (2900–4733 mg/kg) and S (1123–1703 mg/kg); these were tested using Inductively Coupled Plasma Optical Emission Spectrometer (ICP-OES) method (Govindaraj et al. 2020a) and identified top 10 accessions for each nutrient and 15 accessions for multiple nutrients, and 17 of them exceeded the biofortification breeding target for Fe 72 mg/kg (Table 2.3). A large variability for grain nutrients was also reported in 225 Sudanese germplasms of pearl millet: Fe (19.7–86.4 mg/kg), Zn (13.5–82.4 mg/kg), Ca (73.5–603.5 mg/kg), Na (3.6–44.5 mg/kg), Cu (1.6–12.4 mg/kg), Mg (118.6–1933.1 mg/kg), Mn (6.6–39.1 mg/kg), K (1444.1–8583.5 mg/kg), P (626.7–5705.5 mg/kg), S (867.5–2237.7 mg/kg) and b-carotene content (1.0–12.6 μg /100 g) (Bashir et al. 2014). These studies indicate significant variability in grain nutrient improvement. However, grain nutrient traits were reported to be significantly influenced by genotype, environment and their interactions; therefore, multi-environment testing considering soil types, season and locations are important to identify stable nutrient-dense accessions. In pearl millet, despite its nutritional superiority, the utilisation of pearl millet flour in the commercial market is limited because of the poor keeping quality of the flour and the development of off-odour during storage. In a recent study, 93 diverse germplasms of pearl millet (freshly milled and 10 days after milling) were evaluated for their hydrolytic and oxidative degradation of lipids using different biochemical parameters such as comprehensive acid value, comprehensive peroxide value, activities of lipase and lipoxygenase, and classified accessions into low, medium and high rancid groups, and rancidity matrix was developed (Goswami et al. 2020). This will support in identification of low-rancid pearl millet germplasm for varietal development.

2.4.3 Variation for Fodder Yield and Quality Traits

Forage pearl millet includes the annual pearl millet, the perennial Napier grass (*P. purpureum*) and Napier-bajra (*P. glaucum* × *P. purpureum*) hybrids. Other wild species such as *P. pedicellatum* (Dinanath grass) is also used for forage. The variability in germplasm for fodder traits plays a significant role in developing forage varieties. Several landraces were found to be promising for fodder traits indicating intensive selection by farmers for high fodder-yielding pearl millet. Based on the characterisation database of the ICRISAT Genebank and also visual selection, a total of 326 pearl millet accessions originating from 23 countries were evaluated for fodder yield and its contributing traits, and promising sources were identified for tall plants, more leaves, total tillers per plant, etc. (Upadhyaya et al. 2017c) (Table 2.3). Similarly, promising sources for fodder yield potential were identified in the pearl millet named landraces originating from India and Africa, conserved at

ICRISAT (Upadhyaya et al. 2016a, b). Some pearl millet germplasms identified as having sweet-stalk types are preferred as a fodder crop because fodder yield is mainly dependent on tiller number, leaf number and size, plant height, maturity and stem thickness (Rao et al. 1982). The sweet-stalk type germplasms (SAR 696, SAR 713, SAR 763 and SAR 856) are characterised by long narrow leaf blades, profuse nodal tillering with asynchronous maturity, short thin spikes, and very small grains as compared with the normal types. The sweet-stalk types could be easily identified by chewing them at the dough stage. At maturity, they contain more than twice the amount of soluble sugars than the normal types (Rao et al. 1982). IP 3471, IP 3481, IP 3509 and IP 3593 are also identified as promising sources for sweet-stalks with high stalk-soluble sugar content (17–19%) (Upadhyaya et al. 2016b).

2.4.4 Abiotic Stress Tolerance

Crop productivity is adversely affected primarily by abiotic stress conditions such as drought, salinity and extremely high or low temperatures. Severe moisture stress during emergence and early seedling phase cause seedling death, which results in poor crop establishment and a significant decline in crop yield. Moisture stress during the vegetative stage affects pearl millet growth, tillering ability and flowering while maximum grain yield penalty occurs when drought strikes during the flowering and post-flowering stages. Early flowering, high tillering and asynchrony of tillering contribute to adaptation to drought stress during the vegetative growth phase. The landraces representing diverse regions of West Africa, Ghana and western part of India were identified as a rich source of diversity for abiotic stress tolerance in pearl millet. The several cultivars developed from landraces, which are cultivated in arid zones of India (Rajasthan, Gujarat and Haryana) and named landrace *Iniadi* from West Africa, are endowed with inherent potential to escape terminal drought stress owing to early flowering and high-yielding ability under harsh environmental conditions (Shivhare and Lata 2017). The most widely used sources of earliness are the *Iniadi*-type landraces from western Africa, and cultivars bred by using *Iniadi* landraces have been widely adopted by farmers in India and Africa. Landraces from the drier regions are a good source for drought adaptation. In a study, 105 landraces from three Indian states namely Rajasthan, Gujarat and Punjab, were evaluated for drought tolerance response under stress and non-stress conditions and 15 landraces were identified having a high degree of drought tolerance for use in crop improvement (Yadav et al. 2003). Based on biochemical parameters, the promising source for drought tolerance germplasms is identified in pearl millet (IP 133, IP 177, IP 164, IP 142, IP 120, IP 160, IP 136, IP 166, IP 192, IP 195, IP 106, IP 126, IP 121, IP 110 and IP 117) (Choudhary et al. 2021).

Heat stress affects plant growth throughout its ontogeny, though heat-threshold level varies considerably at different developmental stages. The optimum temperature required for the growth and development of pearl millet is 30–35 °C (Garcia-Huidobro et al. 1985; Bhuva and Detroja 2018). However, pearl millet could thrive in hot environmental conditions and put forth optimal growth without affecting yield. Pearl millet seedlings are most vulnerable to high temperatures during the first 10 days of sowing. The germination of the seed occurs at 35–45 °C and starts decreasing at 47 °C and almost stops at 50 °C (Garcia-Huidobro et al. 1982). Thus, high-temperature stress at the seedling and reproductive stages has an impact on crop establishment and yield of pearl millet. Genetic variation has been observed for heat tolerance at the seedling and reproductive stages among the germplasm. The study conducted for reproductive stage heat tolerance over 3–4 years could identify the tolerant breeding and germplasm line (IP 19877) having an equivalent seed set as that of tolerant check 9444 (Gupta et al. 2015). The accessions, viz. IP 21517 and IP 3175, were reported as heat tolerant (Sehgal et al. 2015). Three lines, IP 19799, IP 19877, and IP 19743, were also identified as heat tolerant (seed set of >50%), and can also be further utilised for diversifying the genetic base of heat-tolerant materials in pearl millet. However, the mechanism of heat tolerance is yet to be investigated in these materials (Yadav et al. 2012).

Pearl millet, generally considered comparatively tolerant to salinity, could be an alternative crop option for salt-affected areas (Krishnamurthy et al. 2007). It has considerable effects to restrict the growth and development of pearl millet at different stages, i.e. germination, seedling growth, vegetative and reproductive stages, and yield. Kulkarni et al. (2006) screened pearl millet germplasm under field conditions and found that IP 22269, IP 6098, IP 6105, IP 3616, IP 6112 and IP 6104 yielded high dry fodder under saline conditions indicating salinity-tolerant ability of these accessions. Indian named landraces *Chadi* local and *Peria cumbu* (IP 3616) were reported to be salinity tolerant (Upadhyaya et al. 2013). IP 19584, IP 19585, IP 10586 and IP 19612 of *Enele* (named landrace) mostly grown in low-rainfall (126 mm) areas were also found to be tolerant to soil salinity (Upadhyaya et al. 2017b). IP 6112, IP 6104 and IP 3616 recorded low stress susceptible index values as compared to other genotypes at 9 dS m^{-1} indicating higher ability of tolerance in particular salinity level in respect of dry matter (Nadaf et al. 2010). Substantial variation was reported for salinity tolerance among 100 ICRISAT breeding lines tested with 250 mM NaCl. On the basis of shoot biomass ratio (shoot biomass under salinity/non-salinity control), 22 lines with a wide range of tolerance responses varying from highly tolerant to sensitive lines were identified (Krishnamurthy et al. 2007) (Table 2.3).

2.4.5 Biotic Stress Resistance

The pearl millet is prone to various biotic stresses such as fungi and bacterial infections causing downy mildew, rust, smut, blast and ergot diseases that impact crop productivity. The tolerance to most of the diseases has been reported in pearl millet genetic resources. The disease screening trials conducted at ICRISAT helped in the identification of the large number of accessions as either tolerant or resistant sources for pests and diseases in pearl millet. A brief list of germplasm accessions identified as promising sources for various traits is presented in Table 2.3, and these details can be accessed through the ICRISAT Genebank database http://genebank.icrisat.org/.

The blast disease caused by *Pyricularia grisea* has emerged as a serious disease in major pearl-millet-growing areas. In the process of identification and further utilisation of genetic resources for blast resistance, 238 mini-core accessions were evaluated against five different pathotypes of *Magnaporthe grisea* isolates (Pg118, Pg119, Pg56, Pg53 and Pg45), and it was identified that three accessions (IP 7846, IP 11036 and IP 21187) exhibited resistance to four of the five pathotypes (Sharma et al. 2013). Similarly, 305 accessions of *Pennisetum violaceum* were screened for resistance against five pathotype isolates of *M. grisea* and a local isolate of *Pennisetum substriata* var. *indica*. Among those, 17 accessions were found resistant (score $\leq$3.0) to all five pathotypes, and 24 accessions were resistant to four pathotypes of *M. grisea* (Sharma et al. 2020) (Table 2.3).

The downy mildew caused by the pathogen *Sclerospora graminicola* is a highly destructive and widespread disease in most pearl-millet-growing areas in Asia and Africa. The screening of a mini-core collection of 238 accessions for 8 pathotypes (Sg 384, Sg 409, 445, 457, 510, 519, 526, 542) of *S. graminicola* collected from different geographical locations in India, resulted in the identification of 62 accessions with resistance for two or more pathotypes (Sharma et al. 2015). This includes IP 14537 resistant against seven pathotypes, five accessions (IP 9645, IP 11943, IP 14542, IP 14599 and IP 21438) being resistant to any six pathotypes, and seven accessions (IP 11930, IP 12374, IP 14522, IP 20715, IP 21187, IP 21201 and IP 21244) resistant to any five pathotypes. Nineteen accessions were resistant to three to four pathotypes, while 30 accessions were resistant to any two pathotypes. IP 11930 and IP 11943 (Sierra Leone) were found resistant against Sg 457, Sg 510, Sg 519, Sg 526 and Sg 542 pathotypes, and IP 11943 was found to be resistant to pathotype Sg 445 (Sharma et al. 2015). In another study, 147 germplasm accessions were screened during two rainy seasons in the Mandor location at the Rajasthan state of India, in a downy mildew sick plot using infector row. Out of 147 germplasms, 25 were highly resistant, 32 were resistant, 52 were susceptible and 38 were highly susceptible to downy mildew. Among the downy mildew resistance accessions, ten were completely free from downy mildew infection at both seasons (Table 2.3) (Sharma et al. 2007). Promising genetic stocks registered as resistant to downy mildew include ICML 12, ICML 13, ICML 14, ICML 15 and ICML 16 (Thakur et al. 2006). Inbred progenies with high levels of resistance to downy mildew were reported that were developed from a highly downy mildew susceptible landrace (IP 2696) by pedigree selection for five generations (Singh et al. 1988). Besides this, large number of germplasm accessions were screened at ICRISAT and found resistant sources for downy mildew.

Rust is also having a significant impact on the pearl millet cropping system, which is caused by the organism *Puccinia sps.* Based on the mean rust severity of 305 accessions of *P. violaceum* accessions, six accessions were found to be resistant (<10% rust severity), 52 were moderately resistant (10–20% rust severity) and the remaining were susceptible (>20% rust severity) (Sharma et al. 2020). Significant variation in rust severity was reported in the pearl millet wild species *P. violaceum* accessions; none of them were found to be completely free from rust. However, single plant selections from nine accessions (IP 21629, IP 21645, IP 21658, IP 21660, IP 21662, IP 21711, IP 21974, IP 21975 and IP 22038), after four generations of pedigree selection and subsequent screening for rust resistance, were found highly resistant (0% rust severity) (Sharma et al. 2020). The ICML 11, ICML 17, ICML 18, ICML 19, ICML 20, and ICML 21 derived from single plant selection from IP 2696, IP 5030, IP 537 B, IP 11776, IP 2084, and IP 6132, respectively, were registered as promising stable genetic stocks for rust resistance (Singh et al. 1987, 1990).

Smut in pearl millet causes direct loss of grain by replacing them with smut sori, and yield losses of up to 30% have been reported (Thakur and King 1988a). Out of 1747 accessions screened at ICRISAT, 12 accessions from India, 10 from Nigeria, 8 from Mali, 4 each from Senegal and Cameroon, 2 each from Uganda and Niger and 1 each from Lebanon and Togo were smut resistant (Thakur et al. 1992). Selections from six germplasm lines were found to be resistant to smut and registered as genetic stock (ICML 5, ICML 6, ICML 7, ICML 8, ICML 9, ICML 10) (Thakur and King 1988b, c). *Ergot*, also called sugary disease, is an important disease most common in pearl-millet-growing regions in India and many African countries. Screening 2752 germplasm accessions conserved at the ICRISAT Genebank for ergot resistance resulted in the identification of 27 accessions from different countries including India (10), Nigeria (11), Togo (3) and Uganda (3) having varying frequencies of plants with 0–10% ergot severity and >75% selfed seed set (Thakur et al. 1992).

Striga is a serious constraint for pearl millet production in West Africa. Few landraces with less susceptibility and some *monodii* accessions with resistance to *Striga* have been identified (Pattanashetti et al. 2016). Six landraces (M141, M239, M029, M197, M017 and KBH) that show less sensitivity to the *Striga* and are relatively high-yielding compared to susceptible controls have been reported in pearl millet (Kountche et al. 2013). Using these resistant landraces, *Striga*-resistant gene pool was developed, through both full-sib progenies (intra- and inter-population) and S1 (self-pollinated) progenies being subjected to screening for *Striga* resistance, which showed the presence of genetic variation for *Striga* resistance in pearl millet and significant response to recurrent selection in a *Striga*-resistant gene pool (Kountche et al. 2013).

2.4.6 Crop Wild Relatives as a Source of Important Traits

The *Pennisetum* wild species offers diverse genetic resources that could be the source of genes of interest for tolerance to abiotic and biotic stresses, forage yield and forage quality, and male sterility and fertility restoration (Sharma et al. 2021). Pearl millet wild species accessions (529) belonging to *P. violaceum, P. mollissimum, P. purpreum, P. pedicellatum, P. polystachion* and *P. schweinfurthii* were screened for downy mildew resistance in the greenhouse and field-disease nurseries (Singh and Navi 2000). Among these wild species, *P. violaceum* was found to be most susceptible while the accessions belonging to

P. schweinfurthii were reported to have a high level of downy mildew resistance that can be exploited in disease resistance breeding as it has cross-compatibility with cultivated pearl millet (Singh and Navi 2000). Screening of 305 accessions of the wild species *P. violaceum* for resistance to blast under glasshouse conditions against five pathotype isolates (Pg 45, Pg 53, Pg 56, Pg 118 and Pg 119) resulted in the identification of 17 accessions (IP 21525, IP 21531, IP 21536, IP 21540, IP 21594, IP 21610, IP 21640, IP 21706, IP 21711, IP 21716, IP 21719, IP 21720, IP 21721, IP 21724, IP 21987, IP 21988 and IP 22160) that were resistant to all the five pathotypes of the blast (score #3.0) while 24 accessions were resistant to four pathotypes (Sharma et al. 2020). Rust is another important leaf disease in pearl millet that adversely affects biomass and the quality of forage. The 305 accessions of *P. violaceum* that were screened for blast resistance were also screened for rust resistance (Sharma et al. 2020). Single plant selections from nine accessions (IP 21629, IP 21645, IP 21658, IP 21660, IP 21662, IP 21711, IP 21974, IP 21975 and IP 22038) were found highly resistant to rust (0% rust severity) after four generations of pedigree selection and subsequent screening (Sharma et al. 2020). Wild species such as *P. hordeoides*, *P. pedicellatum* and *P. polystachion* were reported as promising sources for the blast, downy mildew and rust resistance (Sharma et al. 2020). The blast-resistant accessions and rust-resistant genetic stocks are being utilised in a pre-breeding programme at ICRISAT for introgressing resistance genes from the wild into the parental lines of cultivated and potential pearl millet hybrids and varieties. Wild pearl millet accessions belonging to *P. violaceum* were reported to have resistance to the parasitic weed *Striga* and identified to have accessions with consistently low *Striga* emergence (Wilson et al. 2000, 2004). The wild *Pennisetum* species – *P. ciliare, P. orientale, P. setaceum* and *P. squamulatum* for apomixes; *P. cenchroides*, *P. hordeoides*, *P. pedicellatum*, *P. purpureum* and *Pennisetum setosum* for fodder yield and quality traits; *P. alopecuroides*, *Pennisetum flassidum*, *P. orientale* and *P. setaceum* for ornamentals; and *P. sheweinfurthii* for large seed size – are some of the sources for important traits for pearl millet improvement (Sharma et al. 2020). *Pennisetum* wild species contributed significantly to improving forage yield and quality. Inter-specific hybridisation between *P. glaucum* and *P. purpureum* has led to the development of forage hybrids with high biomass and better quality.

Pennisetum pedicellatum withstands multi-cuts for several years for green fodder and is generally used as a cut and carry green forage at panicle emergence but it can also be made into silage and hay. When compared with other *Pennisetum* species, P. *pedicellatum* produces more tillers than that of ssp. *monodii*, and grows tall facilitating more green fodder yield in a short period (Upadhyaya et al. 2014a). Most of the *P. pedicellatum* accessions conserved at the ICRISAT Genebank produced green stems and the remaining produced purple stems. The wild species accessions belonging to *P. pedicellatum* with stem of >7.5 mm thickness (IP 21850, IP 21861 and IP 21846), longer leaves (>50 cm; IP 21849, IP 22072 and IP 21836), promising for fodder (IP 21850, IP 21853, IP 21853, IP 21861, IP 21850, IP 21849, IP 21862, IP 21845 and IP 21864) and maximum tillers (IP 21821, IP 21866, IP 22220, IP 22071 and IP 22091) have been reported (Upadhyaya et al. 2014a). In comparison to *Pennisetum glaucum* ssp. *monodii*, which has the weak stem and is susceptible to lodging, *P. pedicellatum* was found to be having early flowering, high tillering habit, tallness and a higher level of resistance to downy mildew, making it important in future pearl millet research in the development of high fodder and grain yield cultivars (Upadhyaya et al. 2014a). There are continuous efforts to introgression of disease-resistance genes from crop wild relatives in the development of high-yielding and disease-resistant cultivars (Sharma et al. 2020). The deep understanding of the wild species genetic resources and prioritisation of traits for their utilisation, trait discovery using high-throughput phenotyping and molecular tools, the introgression of traits with minimal linkage drag, and the continuous supply of the new and diverse genetic variability derived from CWRs in the breeding pipeline are important for further deployment in breeding programmes.

2.5 Germplasm Utilisation in Crop Improvement

Pearl millet germplasm resources have been extensively used to develop improved composites, open-pollinated varieties and breeding lines with high grain yield; disease and pest resistance; heat, drought and salinity tolerance; and high iron and high zinc content, and distributed globally for their utilisation in pearl millet improvement (Singh et al. 1987; Rai et al. 2014; Yadav et al. 2021). The best example of the direct use of landraces is the development of ICTP 8203 by selection within *Iniadi* landrace from northern Togo (Rai et al. 1990). The ICTP 8203 is a large-seeded and high-yielding open-pollinated variety bred at ICRISAT, Patancheru, and this variety was released as MP 124 in Maharashtra and Andhra Pradesh and as PCB 138 in Punjab states of India for commercial cultivation in 1988, and also as Okashana 1 in Namibia during 1990 and as Nyankhombo (ICMV 88908) in Malawi in 1996. Direct selection within the same landrace from Togo led to the development of a large-seeded and downy mildew-resistant male sterile line ICMA 88004, a seed parent of an early maturing hybrid (ICMH 356) released in India in 1993 (Rai et al. 1995). Further, utilising the intra-population variability within ICTP 8203, the high Fe and Zn biofortified varieties of pearl millet 'Dhanshakti' and 'Chakti' were released in India and Africa (Rai et al. 2014; Govindaraj et al. 2019). Other examples of direct selection from the landraces that were released as varieties include IKMP 3 (IP 11381) and IKMP 5 (IP 11317) in Burkina Faso; open-pollinated variety ICMV-IS 88102 (IP 6426) in Burkina Faso, while as Benkadi Nio in Mali; and IP 6104 and IP 19586 released in Mexico as high forage-yielding varieties (Upadhyaya et al. 2007; Pattanashetti et al. 2016). Earlier in the 1970s and 1980s, the focus was developing a wide and diverse range of improved trait-specific breeding lines and hybrid parents with having higher grain yield and downy mildew resistance as the main selection criteria, and with a limited focus towards lines with resistance to smut and ergot. Later, the pearl millet improvement programme adopted a country/regional strategy and made a strategic shift towards developing and distributing a diverse range of high-yielding and downy mildew-resistant trait-based breeding lines and hybrid parents (seed parents and restorers) to National

Agricultural Research System (NARS) and private seed industries for their utilisation in hybrid development and commercialisation. In the last two to three decades, about 60–70% of the pearl millet hybrids cultivated on farms in India are directly or indirectly based on ICRISAT-bred hybrid parental lines. The use of germplasm diversity representative subsets such as core and mini-core collection resulted in the identification of new sources of variation for use in pearl millet improvement (Sharma et al. 2015; Govindaraj et al. 2020a). Wild species accessions have been used in developing pre-breeding lines to expand the cultigen gene pool in pearl millet. Inter-subspecific crosses involving cultivated pearl millet with its wild species *P. glaucum ssp. monodii* and *Pennisetum stenostachyum* (weedy relative) have been successful in transferring desirable traits like rust resistance, male sterility and alleles for enhancing yield components to cultivated pearl millet (Pattanashetti et al. 2016; Sharma et al. 2021). High biomass photoperiod sensitive germplasms are largely utilised in developing high biomass multi-cut forage pearl millet cultivars and have been released in India and Central Asian Countries, and interspecific hybridisation between *P. glaucum* and *P. purpureum* has led to the development of forage hybrids with high biomass and better quality.

2.6 Summary

The genetic enhancement of pearl millet depends on several factors such as available genetic resources with the breeder, and inheritance and stability of the traits that need to be improved. Globally, a considerable number of germplasm resources are conserved ex situ in genebanks, and the ICRISAT Genebank conserves the largest collection of pearl millet in the world. Characterisation and evaluation of germplasm for various morpho-agronomic, productivity, quality and stress tolerance traits revealed the presence of significant variability for trait improvement. Being a highly cross-pollinated crop, the landraces of pearl millet are highly heterozygous and heterogeneous; therefore an appropriate strategy is required to use the existing variability in the crop improvement. Large-scale screening of germplasm for important traits following high-throughput phenotypic and genomic tools and use of crop wild relatives for trait introgression are required for accelerating the genetic improvement of pearl millet. The advanced high-throughput next-generation sequencing approach could be used to sequence the large number of pearl millet germplasms and can support the identification and tracking of natural genetic variations present within the germplasm collection, as well as discovery of genes and molecular markers associated with important agronomic traits.

References

Anuradha N, Satyavathi CT, Bharadwaj C, Nepolean T, Sankar SM et al (2017) Deciphering genomic regions for high grain iron and zinc content using association mapping in pearl millet. Front Plant Sci 8:1–17. https://doi.org/10.3389/fpls.2017.00412

Bashir EMA, Ali AM, Ali AM, Melchinger E, Parzies HK et al (2014) Characterization of Sudanese pearl millet germplasm for agro-morphological traits and grain nutritional values. Plant Genet Resour Charact Util 12:35–47. https://doi.org/10.1017/S1479262113000233

Bhattacharjee R, Khairwal IS, Bramel PJ, Reddy KN (2007) Establishment of a pearl millet [Pennisetum glaucum (L.) R. Br.] core collection based on geographical distribution and quantitative traits. Euphytica 155(1–2):35–45. https://doi.org/10.1007/s10681-006-9298-x

Bhuva HM, Detroja AC (2018) Thermal requirement of pearl millet varieties in Saurashtra region. J Agrometeorol 20(4):329–331

Brunken J, De Wet JMJ, Harlan JR (1977) The morphology and domestication of pearl millet. Econ Bot 31:163–174

Choudhary ML, Tripathi MK, Gupta N, Tiwari S, Tripathi N et al (2021) Screening of pearl millet [Pennisetum glaucum [L.] R. Br.] germplasm lines against drought tolerance based on biochemical traits. Curr J Appl Sci Technol 40(23):CJAST.73335. https://doi.org/10.9734/CJAST/2021/v40i2331483

Dwivedi S, Upadhyaya HD, Senthilvel S, Hash CT, Fukunaga K et al (2012) Millets: genetic and genomic resources. Plant Breed Rev 35:247–375

Frankel OH (1984) Genetic perspective of germplasm conservation. In: Arber W, Illmensee K, Peacock WJ, Starlinger P (eds) Genetic manipulations: impact on man and society. Cambridge University Press, Cambridge, pp 161–170

Garcia-Huidobro J, Monteith JL, Squire GR (1982) Time, temperature and germination of pearl millet (Pennisetum typhoides S. & H.). I. Constant temperature. J Exp Bot 33(133):288–296

Garcia-Huidobro J, Monteith JL, Squire GR (1985) Time, temperature and germination of pearl millet (Pennisetum typhoides S. & H.): III. Inhibition of germination by short exposure to high temperature. J Exp Bot 36(163):338–343. https://doi.org/10.1093/jxb/33.2.288

Godfray HCJ, Beddington JR, Crute IR, Haddad L, Lawrence D et al (2010) Food security: the challenge of feeding 9 billion people. Science 327(5967):812–818. https://doi.org/10.1126/science.1185383

Goswami S, Asrani P, Ansheef Ali TP, Kumar RD, Vinutha T et al (2020) Rancidity matrix: development of biochemical indicators for analysing the keeping quality of pearl millet flour. Food Anal Methods 13(11):2147–2164. https://doi.org/10.1007/s12161-020-01831-2

Govindaraj M, Rai KN, Cherian B, Pfeiffer WH, Kanatti A et al (2019) Breeding biofortified pearl millet varieties and hybrids to enhance millet markets for human nutrition. Agriculture 9(5):106. https://doi.org/10.3390/agriculture9050106

Govindaraj M, Rai KN, Kanatti A, Upadhyaya HD, Shivade H et al (2020a) Exploring the genetic variability and diversity of pearl millet core collection germplasm for grain nutritional traits improvement. Sci Rep 10(1):21177. https://doi.org/10.1038/s41598-020-77818-0

Govindaraj M, Yadav OP, Rajpurohit BS, Kanatti A, Rai KN et al (2020b) Genetic variability, diversity and interrelationship for twelve grain minerals in 122 commercial pearl millet cultivars in India. Agric Res 9(4):516–525. https://doi.org/10.1007/s40003-020-00470-7

Gupta SK, Rai KN, Singh P, Ameta VL, Gupta SK et al (2015) Seed set variability under high temperatures during flowering period in pearl millet (Pennisetum glaucum L. (R.) Br.). Field Crop Res 171:41–53. https://doi.org/10.1016/j.fcr.2014.11.005

Hanna WW, Dujardin M (1986) Cytogenetics of Pennisetum schweinfurthii Pilger and its hybrids with pearl millet 1. Crop Sci 26(3):449–453. https://doi.org/10.2135/cropsci1986.0011183x002600030001x

Harlan JR, de Wet JMJ (1971) Toward a rational classification of cultivated plants. Taxon 20(4): 509–517. https://doi.org/10.2307/1218252

Kanatti A, Rai KN, Radhika K, Govindaraj M, Sahrawat KL et al (2014) Grain iron and zinc density in pearl millet: combining ability, heterosis and association with grain yield and grain size. Springerplus 3:763

Kaushal P, Khare A, Zadoo SN, Roy AK, Malaviya DR et al (2008) Sequential reduction of Pennisetum squamulatum genome complement in P. glaucum (2n=28) x P. squamulatum

(2n=56) hybrids and their progenies revealed its octoploid status. Cytologia (Tokyo) 73(2): 151–158. https://doi.org/10.1508/cytologia.73.151

Khairwal I, Yadav SK, Rai KN, Upadhyaya HD, Kachhawa D et al (2007) Evaluation and identification of promising pearl millet germplasm for grain and fodder traits. SAT eJournal 5(1):1–6

Kountche BA, Hash CT, Dodo H, Laoualy O, Sanogo MD et al (2013) Development of a pearl millet Striga-resistant genepool: response to five cycles of recurrent selection under Striga-infested field conditions in West Africa. Field Crop Res 154:82–90. https://doi.org/10.1016/j.fcr.2013.07.008

Krishnamurthy L, Serraj R, Rai KN, Hash CT, Dakheel AJ (2007) Identification of pearl millet [Pennisetum glaucum (L.) R. Br.] lines tolerant to soil salinity. Euphytica 158(1–2):179–188. https://doi.org/10.1007/s10681-007-9441-3

Kulkarni V, Rai K, Dakheel A, Ibrahim M, Hebbara M et al (2006) Pearl millet germplasm adapted to saline conditions millets research genetic resources. ISMN 47:103–106

Martel E, Ricroch A, Sarr A (1996) Assessment of genome organization among diploid species (2n = 2x = 14) belonging to primary and tertiary gene pools of pearl millet using fluorescent in situ hybridization with rDNA probes. Genome 39(4):680–687. https://doi.org/10.1139/g96-086

Martel E, Poncet V, Lamy F, Siljak-Yakovlev S, Lejeune B et al (2004) Chromosome evolution of Pennisetum species (Poaceae): implications of ITS phylogeny. Plant Syst Evol 249(3–4): 139–149. https://doi.org/10.1007/s00606-004-0191-6

Nadaf SK, Al-hinai SA, Al-farsi SM, Al-lawati AH, Al-bakri AN (2010) Differential response of salt tolerant pearl millet genotypes to irrigation water salinity. In: A monograph on management of salt-affected soils and water for sustainable agriculture, pp 47–60

Pattanashetti SK, Upadhyaya HD, Dwivedi SL, Vetriventhan M, Reddy KN (2016) Pearl millet. Academic Press, Elsevier

Rai KN, Kumar KA, Andrews DJ, Rao AS, Raj AGB et al (1990) Registration of 'ICTP 8203' pearl millet. Crop Sci 30(4):959–959. https://doi.org/10.2135/cropsci1990.0011183x003000040048x

Rai KN, Gupta SK, Sharma R, Govindaraj M, Rao AS et al (2014) Pearl millet breeding lines developed at ICRISAT: a reservoir of variability and useful source of non-target traits. SAT eJournal 12:1–13

Rai KN, Velu G, Govindaraj M, Upadhyaya HD, Rao AS et al (2015) Iniadi pearl millet germplasm as a valuable genetic resource for high grain iron and zinc densities. Plant Genet Resour Character Util 13(1):75–82. https://doi.org/10.1017/S1479262114000665

Rai KN, Rao AS, Hash CT (1995) Registration of pearl millet parental lines ICMA 8804 and ICMB 8804. Crop Sci 35:1242. https://doi.org/10.2135/cropsci1995.0011183X003500040108x

Rao SA, Mengesha MH, Subramanian V (1982) Collection and preliminary evaluation of sweet-stalk pearl millet (Pennisetum). Econ Bot 36(177):286–290

Satyavathi CT, Ambawat S, Khandelwal V, Srivastava RK (2021) Pearl millet: a climate-resilient nutricereal for mitigating hidden hunger and provide nutritional security. Front Plant Sci 12: 659938. https://doi.org/10.3389/fpls.2021.659938

Sehgal D, Skot L, Singh R, Srivastava RK (2015) Exploring potential of pearl millet germplasm association panel for association mapping of drought tolerance traits. PLoS One 10(5): e0122165. https://doi.org/10.1371/journal.pone.0122165

Sharma YK, Yadav SK, Khairwal IS, Bajri-chomu D, Bajri-osian D (2007) Evaluation of pearl millet germplasm lines against downy mildew incited by Sclerospora graminicola in western Rajasthan. SAT eJournal 3(1):1–2

Sharma R, Upadhyaya HD, Manjunatha SV, Rai KN, Gupta SK et al (2013) Pathogenic variation in the pearl millet blast pathogen, Magnaporthe grisea and identification of resistance to diverse pathotypes. Plant Dis 97(2):189–195. https://doi.org/10.1094/PDIS-05-12-0481-RE

Sharma R, Upadhyaya HD, Sharma S, Gate VL, Raj C (2015) New sources of resistance to multiple pathotypes of Sclerospora graminicola in the pearl millet mini core germplasm collection. Crop Sci 55(4):1619–1628. https://doi.org/10.2135/cropsci2014.12.0822

Sharma R, Sharma S, Gate VL (2020) Tapping pennisetum violaceum, a wild relative of pearl millet (Pennisetum glaucum), for resistance to blast (caused by Magnaporthe grisea) and Rust (caused by Puccinia substriata var. Indica). Plant Dis 104(5):1487–1491. https://doi.org/10.1094/PDIS-08-19-1602-RE

Sharma S, Sharma R, Govindaraj M, Mahala RS, Satyavathi CT et al (2021) Harnessing wild relatives of pearl millet for germplasm enhancement: challenges and opportunities. Crop Sci 61(1):177–200. https://doi.org/10.1002/csc2.20343

Shivhare R, Lata C (2017) Exploration of genetic and genomic resources for abiotic and biotic stress tolerance in pearl millet. Front Plant Sci 7:1–17. https://doi.org/10.3389/fpls.2016.02069

Singh S, Navi SS (2000) Genetic resistance to pearl millet downy mildew ll. Resistance in wild relatives. J Mycol Plant Pathol 30(2):167–171. https://www.mendeley.com/catalog/genetic-resistance-pearl-millet-downy-mildew-ll-resistance-wild-relatives/

Singh SD, Andrews DJ, Rai KN (1987) Registration of ICML 11 rust-resistant pearl millet germplasm. Crop Sci 27:367–368

Singh SD, Williams RJ, Reddy PM (1988) Isolation of downy mildew resistant lines from a highly susceptible cultivar of pearl millet. Indian Pytopathol 41(3):450–456

Singh SD, King SB, Reddy PM (1990) Registration of five pearl millet germplasm sources with stable resistance to rust. Crop Sci 30:1165. https://doi.org/10.1515/9783112517581

Singh SD, Alagrswamy G, Talukdar BS, Hash CT (1994) Registration of ICML 22 photoperiod insensitive, downy mildew resistant pearl millet germplasm. Crop Sci 34:1421

Srivastava RK, Singh RB, Pujarula VL, Bollam S, Pusuluri M et al (2020) Genome-wide association studies and genomic selection in pearl millet: advances and prospects. Front Genet 10:1389. https://doi.org/10.3389/fgene.2019.01389

Thakur RP, King SB (1988a) Smut disease of pearl millet. Information Bulletin no. 25. Patancheru, A.P. 502 324. International Crops Research Institute for the Semi-Arid Tropics, India

Thakur RP, King SB (1988b) Registration of four ergot resistant germplasms of pearl millet. Crop Sci 28(2):382. https://doi.org/10.2135/cropsci1988.0011183x002800020058x

Thakur SR, King SB (1988c) Registration of six smut resistant germplasms of pearl millet. Crop Sci 2:382–383

Thakur RP, King SB, Rai KN, Rao VP (1992) Identification and utilization of smut resistance in pearl millet. International Crops Research Center for the Semi-Arid Tropics, Patancheru

Thakur RP, Shetty H, Khairwal I (2006) Pearl millet downy mildew research in India: progress and perspectives. SAT eJournal 2(1):1–6

Upadhyaya D, Ortiz R (2001) A mini core subset for capturing diversity and promoting utilization of chickpea genetic resources in crop improvement. Theor Appl Genet 102:1292–1298

Upadhyaya HD, Reddy KN, Gowda CLL (2007) Pearl millet germplasm at ICRISAT Genebank—status and impact. SAT eJournal 3(1):1–5

Upadhyaya HD, Gowda CLL, Reddy KN, Singh S (2009a) Augmenting the pearl millet core collection for enhancing germplasm utilization in crop improvement. Crop Sci 49(2):573–580. https://doi.org/10.2135/cropsci2008.06.0378

Upadhyaya HD, Reddy KN, Irshad Ahmed M, Gowda CLL, Haussmann BIG (2009b) Identification of geographical gaps in the pearl millet germplasm conserved at ICRISAT Genebank from West and Central Africa. Plant Genet Resour Charact Util 8(01):45–51. https://doi.org/10.1017/S147926210999013X

Upadhyaya HD, Pundir RPS, Dwivedi SL, Gowda CLL (2009c) Mini core collections for efficient utilization of plant genetic resources in crop improvement programs. Information. International Crops Research Institute for the Semi-Arid Tropics, Patancheru. 52 pp. ISBN 978-92-9066-519-9

Upadhyaya HD, Reddy KN, Irshad Ahmed M, Gowda CLL (2010) Identification of gaps in pearl millet germplasm from Asia conserved at the ICRISAT Genebank. Plant Genet Resour Charact Util 8(3):267–276. https://doi.org/10.1017/S1479262112000275

Upadhyaya HD, Yadav D, Reddy KN, Gowda CLL, Singh S (2011) Development of pearl millet minicore collection for enhanced utilization of germplasm. Crop Sci 51:217–223. https://doi.org/10.2135/cropsci2010.06.0336

Upadhyaya HD, Reddy KN, Irshad Ahmed M, Gowda CLL (2012) Identification of gaps in pearl millet germplasm from East and Southern Africa conserved at the ICRISAT Genebank. Plant Genet Resour Charact Util 10(3):202–213. https://doi.org/10.1017/S1479262112000275

Upadhyaya HD, Reddy KN, Singh S, Gowda CLL, Irshad Ahmed M et al (2013) Latitudinal patterns of diversity in the world collection of pearl millet landraces at the ICRISAT Genebank. Plant Genet Resour 12(01):91–102. https://doi.org/10.1017/S1479262113000348

Upadhyaya H, Reddy K, Singh S, Ahmed MI, Kumar V et al (2014a) Geographical gaps and diversity in Deenanath grass (Pennisetum pedicellatum Trin.) germplasm conserved at the ICRISAT Genebank. Indian J Plant Genet Resour 27(2):93–101. https://doi.org/10.5958/0976-1926.2014.00001.1

Upadhyaya HD, Reddy KN, Singh S, Gowda CLL, Ahmed MI et al (2014b) Diversity and gaps in Pennisetum glaucum subsp. monodii (Maire) Br. Germplasm conserved at the ICRISAT Genebank. Plant Genet Resour Charact Util 12(2):226–235. https://doi.org/10.1017/S1479262113000555

Upadhyaya HD, Reddy KN, Ahmed MI, Ramachandran S, Kumar V et al (2016a) Characterization and genetic potential of African pearl millet named landraces conserved at the ICRISAT Genebank. Plant Genet Resour Charact Util:1–15. https://doi.org/10.1017/S1479262116000113

Upadhyaya H, Reddy K, Ramachandran S, Kumar V, Ahmed MI (2016b) Adaptation pattern and genetic potential of Indian pearl millet named landraces conserved at the ICRISAT Genebank. Indian J Plant Genet Resour 29(2):97–113. https://doi.org/10.5958/0976-1926.2016.00015.2

Upadhyaya HD, Reddy KN, Ahmed MI, Kumar V, Gumma MK et al (2017a) Geographical distribution of traits and diversity in the world collection of pearl millet [Pennisetum glaucum (L.) R. Br., synonym: Cenchrus americanus (L.) Morrone] landraces conserved at the ICRISAT Genebank. Genet Resour Crop Evol 64(6):1365–1381. https://doi.org/10.1007/s10722-016-0442-8

Upadhyaya HD, Reddy KN, Ahmed MI, Ramachandran S, Kumar V et al (2017b) Characterization and genetic potential of African pearl millet named landraces conserved at the ICRISAT Genebank. Plant Genet Resour Charact Util 15(5):438–452. https://doi.org/10.1017/S1479262116000113

Upadhyaya HD, Reddy KN, Pattanashetti SK, Kumar V, Ramachandran S (2017c) Identification of promising sources for fodder traits in the world collection of pearl millet at the ICRISAT Genebank. Plant Genet Resour Charact Util 16(2):127–136. https://doi.org/10.1017/S147926211700003X

Vadez V, Hash T, Bidinger FR, Kholova J (2012) II.1.5 Phenotyping pearl millet for adaptation to drought. Front Physiol 3:386. https://doi.org/10.3389/fphys.2012.00386

Vernooy R, Shrestha P, Sthapit B (2015) Community seed banks: origins, evolution and prospects, 1st edn. Routledge

Wilson JP, Hess DE, Hanna WW (2000) Resistance to Striga hermonthica in wild accessions of the primary gene pool of Pennisetum glaucum. Phytopathology 90(10):1169–1172. https://doi.org/10.1094/PHYTO.2000.90.10.1169

Wilson JP, Hess DE, Hanna WW, Kumar KA, Gupta SC (2004) Pennisetum glaucum subsp. monodii accessions with Striga resistance in West Africa. Crop Prot 23(9):865–870. https://doi.org/10.1016/j.cropro.2004.01.006

Yadav OP, Weltzien-Rattunde E, Bidinger FR (2003) Genetic variation in drought response of landraces of pearl millet (Pennisetum glaucum (L.) R. Br.). Indian J Genet 63:37–40

Yadav OP, Rai KN, Gupta SK (2012) Pearl millet: genetic improvement in tolerance to abiotic stresses. In: Improving crop productivity in sustainable agriculture, pp 261–288

Yadav OP, Gupta SK, Govindaraj M, Sharma R, Varshney RK et al (2021) Genetic gains in pearl millet in India: insights into historic breeding strategies and future perspective. Front Plant Sci 12:645038. https://doi.org/10.3389/fpls.2021.645038

Milestones in Biology, Genetics, and Breeding of Pearl Millet

3

O. P. Yadav, S. K. Gupta, and K. N. Rai

Abstract

Pearl millet is a fascinating species for conducting basic research in biology and genetics; and for applied research in breeding. With a small number of large somatic chromosomes, pearl millet lends itself to investigation in classical and molecular cytogenetics. Its short life cycle, protogynous flowers and ability to set a large number of seeds per panicle make pearl millet highly suitable for studying flow of genes between cultivated annual species and related wild species. Centre of origin, domestication, primary and secondary gene pools of pearl millet helped in selection of suitable geographical area for collecting unique and diverse germplasm resources. The outcrossing nature of pearl millet provided the basis of exploitation of heterosis at commercial scale. Another important discovery related to pollination of pearl millet was role of pollen in reducing the infection of ovary by pathogens of ergot and smut. Knowledge of photoperiod response helped in extending the crop cultivation in new seasons and geographical regions; in controlling flowering in order to facilitate hybridization; and in selecting suitable sites for offseason nurseries. Outcrossing rate of above 85%, ease of inbred development, discovery of cytoplasmic male sterility and fertility restorer genes, lack of any negative association of cytoplasmic male sterility with growth and development, diseases and insect-pests, expression of positive and high magnitude of heterosis in productivity of hybrids and economic seed production provided a perfect platform for commercial exploitation of heterosis in pearl

O. P. Yadav (✉)
ICAR-Central Arid Zone Research Institute (CAZRI), Jodhpur, India
e-mail: director.cazri@icar.gov.in

S. K. Gupta · K. N. Rai
International Crops Research Institute for the Semi-Arid Tropics (ICRISAT), Patancheru, India
e-mail: s.gupta@cgiar.org

V. A. Tonapi et al. (eds.), *Pearl Millet in the 21st Century*,
https://doi.org/10.1007/978-981-99-5890-0_3

millet for the benefit of farming community. The genome of a reference genotype Tift 23D2B1-P1-P5 has been reported to contain an estimated 38,579 genes. Thus, a good understanding of biology and genetics of pearl millet has helped tremendously in breeding for higher productivity and stability.

3.1 Introduction

Pearl millet is of great importance in the farming systems of arid and semi-arid regions and is valued for its nutritious grains and ability to produce high amounts of grain and stover under some of the most challenging agro-ecosystems in South Asia and Sub-Sahara Africa. Its cultivation is also expanding in South America, Australia, and South Eastern America as a grain, forage, or alley crop due to its low water requirement and adaptation to high-temperature conditions. Being a C4 plant, pearl millet has very high photosynthetic conversion efficiency and high biomass production capacity even when grown under challenging agro-climatic conditions.

Pearl millet is a fascinating species for conducting basic research in biology and genetics, and applied research in breeding. With a small number of large somatic chromosomes ($2n = 14$), it lends itself to investigation in classical and molecular cytogenetics. Its short life cycle, protogynous flowers, and ability to set many seeds per panicle make pearl millet highly suitable for studying the flow of genes between cultivated annual species and related wild species. The outbreeding nature of pearl millet makes it an ideal species for studying and exploiting heterosis. Consequently, pearl millet has been researched to a great deal, and its genetic improvement in India is one of the most outstanding and successful stories of Indian agriculture (Yadav et al. 2019). There are several firsts in pearl millet among achievements in various crop breeding programs at the national and global levels. This could be possible due to deep learning and a greater understanding of pearl millet biology, genetics, physiology, taxonomy, genome sequencing, etc. The purpose of this chapter is to document such most important milestones and to understand how new knowledge has impacted the development of new breeding methods, improving the efficiency of breeding and creating new and better products.

3.2 Pearl Millet Biology

3.2.1 Center of Origin and Taxonomy

Pearl millet is believed to have been domesticated in Sub-Sahara Africa some 4000–5000 years ago (Munson 1975). However, there have been different opinions regarding its center of origin. Harlan (1971) suggested the center of origin in a belt stretching from western Sudan to Senegal. Based on pearl millet distribution, Brunken et al. (1977) considered the Sahel zone of West Africa to be pearl millet's original home, the view favored by Clegg et al. (1984) based on chloroplast DNA studies. Based on the available evidence, Appa Rao and de Wet (1999) concluded that pearl millet originated in western Africa some 4000 years ago.

Over the decades, pearl millet has received several different taxonomic treatments hence different Latin names. Thus, it was treated as a constituent of at least six different genera, viz., *Panicum*, *Holcus*, *Alopecuros*, *Cenchrus*, *Penicillaria*, and *Pennisetum* (Jauhar 1981). Three germplasm pools are recognized in cultivated pearl millet (Harlan and de Wet 1971). The primary pool contains all cultivated, weedy, and wild diploid ($2n = 14$) pearl millets that are freely cross-fertile. The secondary pool is solely *Pennisetum purpureum* (Shum.) ($2n = 28$). The tertiary pool contains numerous more distantly related *Pennisetum* species of various ploidy levels, which do not naturally interbreed with the primary pool but can potentially be accessed through various wide crosses (Dujardin and Hanna 1990).

Center of origin, domestication, and primary and secondary gene pools of pearl millet helped in the selection of suitable geographical areas for collecting unique and diverse germplasm resources; in understanding the gene synteny of pearl millet with other cereals; and in searching the genes conditioning adaptation to abiotic stresses, resistance to diseases, and sources of different economic traits.

3.2.2 Establishing Outcrossing Nature of Pearl Millet

Pearl millet floral biology was studied in detail in the 1920s, and 1930s (Godbole 1928), which established that individual spikelets in panicle bear two types of florets: bisexual and staminate. Ayyangar et al. (1933) concluded that the stigma emerged after the maturity of the spikelets, irrespective of whether the panicle has fully emerged from the boot or is still inside the boot. The stigma maturity is attained earlier than anther dehiscence, the phenomenon termed protogynous flowering. The time difference between stigma emergence and anther dehiscence varies from 24 to 48 h and is responsible for cross-pollination (>85%) in pearl millet. Cross-pollination is largely affected by wind (Leuck and Burton 1966) and, to some extent, by insects. Some selfing cannot be ruled out because of pollination of stigma from the pollen of panicles produced on the tillers of the same plant or from pollen produced by the same panicle on which stigma is born. Due to these flowering features, controlled pollination is very convenient in pearl millet. It is very easy to produce an adequate quantity of selfed, crossed, or sibbed seed in pearl millet by controlled pollination, even from a small number of plants, which makes breeding operations easy and quick. The outcrossing nature of pearl millet provided the basis for exploiting heterosis commercially.

Another important discovery related to the pollination of pearl millet was the role of pollen in reducing the infection of the ovary by pathogens of ergot and smut (Thakur and Williams 1980; Thakur et al. 1983). When both fungal spores and pollen grains land simultaneously on the stigma, the pollen germinates earlier and fertilizes the ovary leading to the drying of the stigma. Once the stigma is withered, the fungal spore cannot germinate on the stigma and protects against fungal infection. This knowledge has an important bearing on developing screening techniques against smut and ergot and also developing cultural measures to control the diseases. This has also explained the increased ergot and smut infection of pearl millet hybrids with uniform flowering than open-pollinated variety (OPV) with highly variable flowering.

3.2.3 Understanding Growth and Development Stages

Pearl millet is a fast-growing cereal even under the most adverse climatic conditions and is reported to have three major growth phases (Begg 1965; Maiti and Bidinger 1981). The first is the vegetative phase that includes growth from seedling emergence to panicle initiation of the main stem. The second phase is panicle development that starts from the panicle initiation and continues up to flowering in the main stem. The third and final stage is the grain-filling phase ranging from flowering to physiological maturity, which is judged by the appearance of a black layer on the seed.

Each of the three phases has been further subdivided to make a total of nine morphologically distinct and recognizable growth stages (Maiti and Bidinger 1981). The average duration of growth stage 1 (GS1) is 22–28 days, GS2 is 18–25 days, and GS3 is 22–25 days though there are huge differences in the duration of each phase among germplasms from different regions.

Such knowledge has been central in terms of understanding the changes in the pattern of growth and development under various abiotic and biotic stresses in different pearl millet germplasms and also for comparing pearl millet with other species. These facts have helped to develop improved pearl millet. The differential response of pearl millet to stress environments at three growth stages led to the identification of agronomic interventions that minimize the losses due to stress and maximize the production in optimum environments and helped in understanding the yield formation process under different stress environments.

3.2.4 Photoperiod Response and Day-Length Sensitivity

Photoperiod regulates the time of flowering, which determines adaptation, yield, and quality. Pearl millet is a photoperiod sensitive and largely a short-day plant in which flowering occurs in shorter-day lengths (Burton 1951, 1965b). Photoperiod sensitivity permits flowering and grain maturation at the end of the growing season, irrespective of the sowing date. Crossing a late-maturing genotype with an early-maturing genotype revealed that photoperiodism in pearl millet was controlled by several genes with additive effects and minimal dominance effects (Burton 1965b).

Late-maturing genotypes are reported to be short-day types, whereas the early-maturing, inbred-parental lines from the United States are mostly photoperiod insensitive and can flower in longer-day lengths (Burton 1951). On the other hand, almost all the germplasms from Africa and India are photoperiod sensitive. This feature has been exploited by the breeders to their advantage by growing pearl millet in the offseason or under greenhouse conditions to obtain flowering and grain maturity in their breeding programs as per the requirement of their program.

Bidinger et al. (1999) demonstrated that a recessive gene conditioning day-length-insensitive earliness can be effectively used to synchronize differential flowering of pearl millet hybrid parents. However, there are no reported cases of its use so far.

Knowledge of photoperiod response helped in extending crop cultivation in new seasons and geographical regions, in controlling flowering to facilitate hybridization, and in selecting suitable sites for offseason nurseries.

3.2.5 Environmental Adaptation

Pearl millet is cultivated in a challenging agro-environment, and its importance as one of the most drought-tolerant crops was recognized very early (Lahiri and Kharbanda 1965; Lahiri and Kumar 1966; Raymond 1968). Several growth processes, like the rate of germination, rate of coleoptile elongation, or the rate of photosynthesis, require rather high optimum temperatures in pearl millet (Garcia-Huidobro et al. 1982; Ong 1983; Soman and Peacock 1985), which is indicative of its good adaptation to the hot growing conditions in the Sahel and in many parts of India (Gupta et al. 2015b). Pearl millet also has a fairly good tolerance to soil salinity (Dua 1989), in addition to its good and noteworthy adaptation to nutrient-depleted soils (Gemenet et al. 2015).

New knowledge has been created regarding the mechanism of adaptation of pearl millet based on such initial understanding, which has eventually helped to develop more stress-tolerant pearl millet.

3.3 Pearl Millet Genetics

3.3.1 Heterosis: High Magnitude in a Positive Direction

The phenomenon of heterosis, which is expressed as the phenotypic and functional superiority exhibited in the first filial generation over the parents, has fascinated plant breeders ever since it was first described by Charles Darwin in 1876 in the vegetable kingdom and later elaborated by G.H. Shull and E.M. East in maize in 1908. Heterosis remains probably one singularity that is most exploited in crops despite being partially understood even after more than 100 years of its first echo.

Heterosis is more prevalent in outcrossing species like pearl millet than in the inbreeding species. Outcrossing rate of above 85%; ease of inbred development; the discovery of cytoplasmic male sterility (CMS) and fertility restorer genes; lack of any negative association of cytoplasmic male sterility (CMS) with growth and development, diseases, and insect-pests; and expression of positive and high magnitude of heterosis in productivity of hybrids and economic seed production provided a perfect platform for commercial exploitation of heterosis in pearl millet for the benefit of the farming community. The classical and strategic work undertaken during 1940–1960 at Tifton, Georgia, USA; Punjab Agricultural University (PAU) Ludhiana, India; and Indian Agricultural Research Institute (IARI), New Delhi, established a three-line breeding system requiring male-sterile (A-), maintainer (B-), and restorer (R-) lines that transformed pearl millet crop improvement program globally.

3.3.2 Discovery of Cytoplasmic-Nuclear Male Sterility

Male-sterile (MS) pearl millet was reported by Kajjari and Patil (1956). They suggested, but did not establish, that the observed male sterility could be of a cytoplasmic type and realized that it could be used to produce 100% hybrid seed. The male sterility discovered in the "inbred 556 × inbred 23" cross at the Tifton, Georgia, USA, by G.W. Burton was found to be cytoplasmic male sterility (Burton 1958). With the identification of restorers, it was confirmed to be cytoplasmic-nuclear in nature.

Both sterility maintenance of CMS lines and fertility restoration of hybrids are essentially required to make commercial use of CMS systems. Fortunately, there has been adequate availability of maintainers of male sterility and restorers of fertility in pearl millet germplasm and breeding materials.

3.3.3 Fertility Restorers in Hybrids

When CMS was discovered in pearl millet, Burton (1958) concluded that nearly two-thirds of available lines are complete fertility restorers for the cytoplasmic male-sterile plants. Adequate restorers were also found in Indian breeding material (Athwal 1961, 1966) with respect to this CMS source, later franchised as A1 CMS. The same was true for A2 and A3 CMS (Burton and Athwal 1967). Fertility restoration was dominant, and sterility was recessive.

While there was no dearth of restorers for A1 CMS, the frequency of restorers for later-discovered A4 CMS was low, and that for A5 CMS was found to be extremely low (Rai et al. 2006, 2009). The sources of fertility restoration were identified, and genetic stocks of A4 and A5 restorers were developed. The methodology to convert breeding lines into their A4 and A5 restorer versions was devised and fine-tuned, and its application led to the development of a series of A4 and A5 restorer versions of elite breeding lines from International Crops Research Institute for the Semi-Arid Tropics (ICRISAT) and Indian National Agricultural Research System (NARS) (Rai et al. 2006). The understanding of the genetics of A4 (Gupta et al. 2012) and A5 CMS (Gupta et al. 2018) helped in the development of restorers of A4 and A5 CMS systems. Recently, fertility restoration/maintainer frequency of 45 Asian and African origin/bred populations on three diverse CMS systems revealed that overall fertility restoration frequency was highest for A1 (86%), followed by A4 (37%) and for A5

(7%) CMS system, and populations were identified as potential sources for developing restorer lines for A1, A4, and A5 CMS systems (Patil and Gupta 2020).

3.3.4 Genetic Linkage Groups

The first report of the genetic linkage group in pearl millet (Gill 1969, 1971) was based on the relationship between various qualitative traits in which 44 pairs of characters were reported in seven linkage groups. Other studies followed in which linkage relationship between genes governing simply inherited traits was reported (Krishna Rao and Koduru 1979). In the era of molecular biology, Liu et al. (1994) constructed the first DNA marker-based genetic map in pearl millet, and it consisted of 181 Restriction Fragment Length Polymorphism (RFLP) markers and covered a total distance of 303 cM of seven linkage groups. Since then, several integrated genetic maps have been built using simple sequence repeat (SSR), Diversity Arrays Technology (DArT), and Single Nucleotide Polymorphism (SNP) markers, and the total map length has been extended up to 1749 cM with an inter-marker distance of 5.73 cM.

With the help of detailed marker-based genetic maps, several traits of agronomic importance, adaptation, and disease resistance could be linked with genomic regions paving the way for marker-assisted (MAS) selection of high-yielding cultivars.

3.3.5 Wide Hybridization

It is generally easier to produce interspecific hybrids in pearl millet using the protogynous nature of flowering using pearl millet as the female parent. The first report of interspecific hybrids of pearl millet was with Napier grass in the United States (Burton 1944) followed by that in India (Krishnaswamy and Raman 1949). Crossing of *Pennisetum glaucum* ($2n = 2x = 14$) and *P. purpureum* ($2n = 4x = 28$) produces a vigorous and sterile triploid ($2n = 3x = 21$) hybrid (Jauhar and Hanna 1998) that is used in improving pearl millet forage yield and quality. It thrives well on uncultivated lands with low water and nutrient requirements.

The wild, weedy pearl millet subspecies *monodii* and *stenostachyum* form the primary genepool for pearl millet and have been used as sources of disease and insect resistance, fertility restoration, cytoplasmic diversity, and several phenotypic traits.

Pennisetum glaucum subsp. *monodii* was found to be a source for new CMS, striga (*Striga* spp.) resistance, and rust and leaf spot resistance. *Pennisetum purpureum* proved to be a donor of forage yield and quality, stalk strength, and restorer genes of the A1 CMS system (Jauhar and Hanna 1998). Hybridization with *Pennisetum orientale* (Willd.) Rich. can help transfer genes for apomixis, drought tolerance, perennial growth habit, and pest resistance (Hanna and Dujardin 1982; Dujardin and Hanna 1987). *Pennisetum squamulatum* has the potential for apomictic gene transfer and *Pennisetum schweinfurthii* Pilg. for improved seed size.

Recently, new sources of blast resistance and flowering-stage heat tolerance have been developed in cultivated pearl millet backgrounds using wild *P. glaucum* subsp.

violaceum (Sharma et al. 2020). To harness the untapped diversity, traits-specific germplasm needs to be targeted, which requires detailed characterization for new and novel traits proposed by breeders and the crop product profiles in the near future.

3.3.6 d_2 Dwarfing Genes

Dwarfing genes were first reported by Burton and Fortson (1966) followed by Appa Rao et al. (1986). From among several sources of dwarfing genes discovered so far, the *d*2 dwarfing gene has emerged as the most preferred one as it reduces the plant height by 30–40%, and increases panicle length and tillering substantially with only small negative effect on grain yield (Rai and Rao 1991). The height is reduced by decrease in internode length rather than a reduction in the number of nodes resulting in higher leafiness. It has no negative effects on other development traits and the combining ability of lines carrying this gene. The *d*2 gene has been used extensively in converting tall fodder-type lines to dwarf grain types in the United States (Burton et al. 1969) and in breeding dwarf seed parents of hybrids in India as the gene reduces the risk of lodging in high-management conditions; helps in easy detection and rouging of off-type and pollen shedders; and facilitates easy seed harvest in hybrid seed production plots. The recessive inheritance of the d_2 gene allows breeding hybrids of variable heights depending upon the height of their male parent, making it a preferred choice of breeders.

3.3.7 Apomixis

Apomixis is an alternative path of reproduction that allows a plant to produce carbon copies of itself through progenies derived from either selfed or open-pollinated seed. Apomixis is a desirable trait in pearl millet because it can be used to produce true-breeding hybrids, or preserve hybrid vigor regardless of heterozygosity by self-seeding, or enable commercial production of hybrids across generations without CMS (Hanna and Bashaw 1987). Uniform progenies from heterozygous F1 or open-pollinated parents, maternal-type F1s in crosses, high seed set in unstable genotypes, and multiple ovules and seedings per ovule are indicators of the presence of apomixis (Hanna and Bashaw 1987). The three main mechanisms of apomixes—apospory, diplospory, and adventitious embryony—have been comprehensively described in the literature but only apospory is relevant to *Pennisetum* (Johar et al. 2006).

Apomixis has been explored in pearl millet largely for fixing hybrid vigor. The feasibility of the interspecific transfer of genes for apomixis from *P. orientale* and *P. squamulatum* into pearl millet was investigated by Hanna and Dujardin (1982). Some understanding of the genetics of apomixis has been developed, and it is generally believed to be simple genetic control (Hanna 1995) that could improve its chances of being transferred to a crop species.

Although the complete introgression of apomixis into sexual pearl millet continues to be plagued with problems, the genetic materials produced have contributed enormously to our understanding of apomictic reproduction at the developmental, cellular, and molecular levels (Johar et al. 2006). Molecular cytogenetic characterization of these lines has allowed the identification of a segment of a single chromosome that is most likely sufficient for the transmission of this reproductive trait. Whether near-obligate apomixis can yet be achieved in an agronomically acceptable genetic background remains to be determined.

The potential value of apomixis in producing new cultivars in pearl millet is high. Apomixis has many benefits, such as the ability to maintain the heterozygosity through seed, thereby eliminating the need to have two parental lines and distance isolation to produce commercial hybrids. Apomixis would also be beneficial in maintaining heterosis in forage hybrids through seed.

3.3.8 Doubled Haploids

Haploids are an important tool in cytogenetic research and have opened up the possibility of developing homozygous inbred lines in the shortest possible time in maize and rice. In pearl millet, haploid plants have been reported to occur simultaneously in natural populations, mutagen-treated inbred lines, progenies of trisomics, and near-tetraploid interspecific hybrids (Pantulu and Manga 1969; Jauhar 1970; Manga and Pantulu 1971; Gill et al. 1973). Unlike maize, where doubled haploid (DH) increased genetic gain drastically, there are no known reports of the development of inbred lines in pearl millet through chromosome doubling of haploids yet, though efforts are on in some advanced research institutes.

Since the frequency of spontaneous occurrence of haploids in pearl millet is reportedly very low, in vitro haploid production through anther culture technique was suggested as a potential technology for large-scale production of doubled haploids to accelerate the development of inbreds. Ha and Pernes (1982) were the first to report regeneration through in vitro anther cultures for doubled haploid production in pearl millet. The same year, Nitsch et al. (1982) obtained haploid plants through in vitro culture of anthers. A few attempts were made in this direction, and androgenic embryos were induced up to a frequency of 13.7% in pearl millet line 841-P3 (Sastry and Mallikarjuna 2014). The results so far have appeared promising, though it would require greater efforts further to develop a doubled haploid protocol for pearl millet breeding.

3.3.9 Genome Sequencing

Varshney et al. (2017) sequenced and analyzed the genome of a reference genotype Tift 23D2B1-P1-P5, which contains an estimated 38,579 genes, and highlighted the substantial enrichment for wax biosynthesis genes, which may contribute to heat and drought tolerance in this crop. When 994 pearl millet lines were resequenced and analyzed, it enabled insights into population structure, genetic diversity, and domestication; and established marker-trait associations that will be critically important in genomic selection, defining heterotic pools, and predicting hybrid performance. This "potential game changer event" would further pave the way for a new era of genomic-based breeding in pearl millet, as has already been seen in other crops like rice, wheat, maize, and many others, and contribute to a better understanding of the genetic basis of the exceptional drought and heat tolerance of pearl millet. A detailed understanding of pearl millet performance in hot, arid, and semi-arid regions might enable engineering of not only pearl millet but also other cereal crops like rice, maize, and wheat, which are currently able to provide only limited produce in arid or semi-arid regions.

3.4 Pearl Millet Breeding

3.4.1 Mass and Recurrent Selection

The mass selection was started as early as the 1930s to capitalize on naturally existing genetic variation within traditional landraces (Athwal 1961) and also in the introduced exotic material (Joshi et al. 1961; Krishnaswamy 1962). However, such efforts were sporadic, available material had a narrow genetic base, and there was very limited knowledge on the nature of inheritance of traits. In the course of time, new knowledge was created in different crops, especially maize, which had a great bearing on breeding methods (Athwal 1966).

The greater urgency for systematic population improvement programs started with the acquisition of a diverse range of germplasms from across the world in the 1970s with the establishment of the International Crops Research Institute for the Semi-Arid Tropics (ICRISAT) (Rachie and Majmudar 1980). Various recurrent selection methods (mass selection, gridded-mass selection, recurrent restricted phenotypic selection, S_1 and S_2 progeny selection, half-sib selection, and full-sib selection) were established and a diverse range of elite breeding materials was developed (Rai and Anand Kumar 1994; Witcombe 1999) with varying success depending upon the target trait under selection. A large number of composites of broad genetic base were developed (Singh et al. 1988).

3.4.2 Chance Hybrids Using Protogynous Flowering

The hermaphrodite nature of flowers of small size in pearl millet had put serious restrictions to exploit it at the commercial level. There were some innovative attempts in the late 1940s to exploit heterosis through developing "chance hybrids" (Burton 1948). The method included growing four inbreds of similar maturity in the mixture and allowing them to cross-pollinate to produce seeds that contained approximately 40% hybrid seed. The chance hybrids out-yielded local varieties marginally but could not become popular due to a lack of efficient seed production programs and their limited genetic superiority.

3.4.3 Introduction of Cytoplasmic Male-Sterile (MS) Lines in India

After the discovery of cytoplasmic-nuclear male sterility (Burton 1958), several male-sterile lines were developed and released. Two of them, viz., MS Tift 23A and Ms. Tift 18A, were introduced in India in the early 1960s by D.S. Athwal at the Punjab Agricultural University (PAU), Ludhiana. It was a landmark decision to use these male-sterile lines in the Indian pearl millet breeding programs. These lines were extensively utilized, both at the PAU and the Indian Agricultural Research Institute (IARI), New Delhi. Tift 23A was found to be more promising because of its short stature, profuse tillering, uniform flowering, and good combining ability. This laid the foundation of pearl millet hybrid breeding in India.

3.4.4 First CMS-Based Commercial Hybrid

With the availability of the male-sterile line Tift 23A, efforts were made to cross a large number of existing inbred lines available at the Punjab Agricultural University, Ludhiana. One such cross of Tift 23A and Bajra Inbred Line (BIL) 3B provided almost double yield than the best check, and this was released as "HB-1" (Hybrid Bajra-1) in 1965 (Athwal 1965). The release of HB-1 dramatically improved the grain yield productivity of the crop in dry areas of India. This initiated a new era of improvement of pearl millet. HB-1 has the distinction of being the first CMS-based hybrid of pearl millet in the world and of any crop in India. Since then, the current hybrid breeding program has been continuously building upon this strong foundation created almost six decades back (Yadav et al. 2021).

3.4.5 Alternative CMS Sources

Though the A_1 system of cytoplasmic-genic male sterility (CMS) continues to be the most utilized CMS source to date, the potential vulnerability of hybrids based on one source to unforeseen diseases and insect-pests has attracted attention in breeding programs since the beginning. As a result, continuing efforts were made to search for alternative CMS sources. This led to the identification of A_2 and A_3 CMS sources from genetic stocks and their derivatives (Burton and Athwal 1967); A_v and A_4 CMS sources from *P. glaucum* subspecies *monodii* (=*violaceum*) accessions (Marchais and Pernes 1985; Hanna 1989); and A_{egp} and A_5 CMS sources from gene pools (Sujata et al. 1994; Rai 1995). Based on the differential male-fertility restoration patterns of hybrids using common restorers, it has been established that the A_1, A_2, A_3, A_v, A_4, and A_5 are distinctly different CMS systems.

3.4.6 Cultivar Development

Research on genetic improvement and the development of cultivars of pearl millet started in the 1930s in India and the United States. Research in the United States was focused on studying the flowering habit, mode of pollination, and cytogenetics of the crop (Burton 1951). The main objective of the US pearl millet breeding program was to develop cultivars for forage purposes. Improvements were made in biomass productivity and dry matter digestibility by largely using dwarfing genes to increase the leaf-to-stem ratio. These efforts led to the development of inbred lines and improved cultivars at the Georgia Coastal Plain Experiment Station, Tifton, GA. In India, programs largely concentrated on improving the grain yield by mass selection in locally adapted material, which led to the development of some open-pollinated varieties (OPVs) (Table 3.1). The cultivar development program was strengthened with the establishment of the All India Coordinated Pearl Millet Improvement Program (AICPMIP) in 1965 at New Delhi.

The great push for developing OPVs as a part of the population improvement program started in the 1970s. A large number of composites of the broad genetic base were developed, which provided the base material to develop varieties that were widely grown in their niche areas.

Attempts were made to exploit the reported heterosis in the 1940s through "chance hybrids" (see Sect. 3.4.2), which could not become commercial due to a lack of methods for economic seed production. This limitation was circumvented with the discovery of cytoplasmic-nuclear male sterility and the release of male-sterile lines in the early 1960s. This initiated pearl millet hybrid breeding in India. As a result, five hybrids (HB 1, HB 2, HB 3, HB 4, and HB 5) were released between 1965 and 1969. Out of these, HB 3 became highly popular and was extensively cultivated because of its early maturity and adaptation to drought stress. There also existed limited variation in pollinator parents of hybrids. Cultivation of a few hybrids with a narrow genetic base on a large scale led to downy mildew (DM) epidemics offsetting impressive achievements in hybrid development in the mid-1970s. Hence, use of Tift 23A was discontinued, and new male-sterile lines were developed. Two new hybrids, BJ 104 and BK 560 produced on 5141A, became popular and were widely cultivated from 1977 to 1984. But a high incidence of downy mildew on 5141A and the resultant susceptibility of both hybrids caused 5141A to be phased out as a seed parent in 1985.

A critical appraisal of the situation reveals that downy mildew epidemics in the mid-1970s and mid-1980s were mainly due to a lack of diversity in the parental lines of hybrids. This led to the strengthening of research on the diversification of the genetic base of male-sterile lines and restorers. As a result, a large number of genetically diverse parental lines are being developed and utilized in hybrid breeding. Consequently, downy mildew (DM) was largely contained. Hybrids like MH 179, Pusa 23, GHB 30, HHB 60, HHB 67, MLBH 104, MBH 110, Eknath 301, ICMH 356, Shradha, Saburi, JKBH 26, 7686, 7688, HHB 67 Improved, RHB 121, GHB 538, GHB 558, GK 1004, Proagro 9444, 86M64, 86M82, Nandi

Table 3.1 Milestones in pearl millet biology, genetics, and breeding

Subject/area	Outcome	Reference
Biology		
Center of origin	• West Africa as primary center • Domesticated in Sub-Sahara Africa	Harlan (1971) Munson (1975)
Growth stages	• Seedling stage • Vegetative stage • Reproductive stage	Begg (1965) Maiti and Bidinger (1981)
Outcrossing nature	• Protogyny flowering • Cross-pollination	Godbole (1928) Ayyangar et al. (1933)
Photoperiod response/day-length sensitivity	• Extending the crop cultivation in new seasons and geographical regions • Selecting suitable sites for offseason nurseries	Burton (1965b)
Environmental adaptation	• Drought • High temperature • Salinity • Low nutrients in soils	Lahiri and Kharbanda (1965) Raymond (1968) Garcia-Huidobro et al. (1982) Ong (1983) Krishnamurthy et al. (2007) Gemenet et al. (2015)
Genetics		
Heterosis: High magnitude in positive direction	• Hybrid development	Burton (1948) Athwal (1961) Krishnaswamy (1962) Athwal (1966)
Cytoplasmic-nuclear male sterility	• Commercial exploitation of heterosis	Burton (1958) Burton (1965a, b)
Fertility restoration in hybrids	• Commercial exploitation of heterosis	Athwal (1961) Athwal (1966) Burton (1965a, b)
d2 Dwarfing genes	• Convenience in seed production • Reduced lodging in female parents of hybrids in seed production plots	Burton and Fortson (1966) Burton et al. (1969)
Apomixis	• Fixing heterosis	Hanna and Dujardin (1982) Hanna and Bashaw (1987)
Doubled haploids	• Faster inbred development	Pantulu and Manga (1969) Jauhar (1970)

(continued)

Table 3.1 (continued)

Subject/area	Outcome	Reference
		Manga and Pantulu (1971) Gill et al. (1973)
Anther culture	• Role in inbred development	Nitsch et al. (1982)
Genetic linkage groups	• Gene-trait association	Gill (1969) Gill (1971) Liu et al. (1994)
Wide hybridization	• Additional sources of specific traits • Genes conditioning adaptation	Burton (1944) Krishnaswamy and Raman (1949)
Genome sequencing	• Insights into population structure, genetic diversity, and domestication • Marker-trait associations for genomic selection and to predict hybrid performance	Varshney et al. (2017)
Breeding		
Mass and recurrent selection	• Superior local cultivars and composites	Joshi et al. (1961) Krishnaswamy (1962) Athwal (1966) Rachie and Majmudar (1980) Singh et al. (1988)
Chance hybrids using protogynous flowering	• Partial exploitation of heterosis	Burton (1948)
Introduction of cytoplasmic male sterile (MS) lines in India	• Tift 23A, Tift 18A • Utilization in hybrid development	Athwal (1961) Athwal (1966)
Restorer inbred	• BIL 3B	Burton and Athwal (1967)
First CMS-based commercial hybrids	• HB 1	Athwal (1965)
Alternative CMS sources	• A2, A3 • Av • A4 • A_{egp} • A5	Burton and Athwal (1967) Marchais and Pernes (1985) Hanna (1989) Sujata et al. (1994) Rai (1995)
Cultivar development	Most popular cultivars • HB 1 • HB 3 • BJ 104	Athwal (1965) Dave (1987) Gill (1991) Bhatnagar et al.

(continued)

Table 3.1 (continued)

Subject/area	Outcome	Reference
	• MH 179 • MBH 110 • HHB 67 • MLBH 104 • JKBH 26 • GK 1004 • Pusa 23/MH 169 • Pioneer 7686 • Proagro 9444 • RHB 173 • PB 106 • WC C75 • ICMS 7703 • ICTP 8203 • CZP 9802 • 86M86 • HHB 197 • GHB 538 • GHB 558 • HHB 146 • HHB 117 • Bio 70 • Nandi 70 • 86M01 • 86M82 • Nandi 75 • MH 1928 • Kaveri Super Boss • Proagro 9001 • 86M13 • MPMH 17	(1998) Khairwal et al. (2009) Yadav and Rai (2013) Satyavathi et al. (2018)
First marker-assisted selection (MAS) product in India	• HHB 67 improved	Hash et al. (2006) Khairwal et al. (2009)
First heat-tolerant hybrid for summer cultivation in India	• Proagro 9444	Yadav et al. (2015)
First salinity-tolerant OPV for Central Asian countries	• Hashaki 1	Yadav et al. (2012b)
First forage hybrid	• ProAgro 1	Satyavathi et al. (2018)
First biofortified variety	• Dhanshakti in India • Chakti in Africa	Rai et al. (2014a, b) Gangashetty et al. (2021)
Delineation of production area into mega-environments	• Three mega-environments	Gupta ct al. (2013)
Creation of genebank having world pearl millet collection	• Access to global germplasm (1972)	www.icrisat.org

(continued)

Table 3.1 (continued)

Subject/area	Outcome	Reference
Establishing core and mini-core collections	• Increased use of genetic resources	Bhattacharjee et al. (2007) Upadhyaya et al. (2009) Upadhyaya et al. (2011)
Heterotic pools	• Long-term enhancement in hybrid productivity	Rai et al. (2006) Gupta et al. (2020)
Hybrid breeding in Africa	• Extending hybrid technology to Africa • Better technologies for challenging environments	Sattler and Haussmann (2020) ICRISAT (2021)
Seed production innovations	• High female:male ratio in hybrid seed production plots • Staggered sowing	Khairwal et al. (1990) Yadav et al. (2015)

75, MH 1928, Kaveri Super Boss, Proagro 9001, and 86M13 became very popular in their cultivation domain (Table 3.1).

Cultivars for niche environments have also been developed and delivered for the target production ecologies. For instance, Proagro 9444, a highly heat-tolerant hybrid, was released for summer cultivation in India in 2005, and Hashaki 1 was released as the first salinity-tolerant OPV in salinity-affected central Asian countries in 2014. Nutrifeed is the first pearl millet forage hybrid released in India in 2014.

Hybrid breeding continues to put a very high emphasis on the strategic use of disease-resistant genetic resources and genetic and cytoplasmic diversification of hybrid parental lines, which ensured that no further major downy mildew epidemics could occur during the last 30 years, although individual hybrids have succumbed to this disease. Currently, the breeding program is laying great stress on adaptation to niche cultivation areas. With these strategies in place, ~60–70 hybrids have been cultivated on the farm at any point in time in India for the last two to three decades.

3.4.7 Delineation of Production Area into Mega-Environments

A single pearl millet cultivar cannot be expected to perform well under all the environmental conditions and a cultivar planted outside its adaptation zone would suffer yield reduction due to significant genotype × environment interactions. Therefore, breeding and evaluation require a subdivision of the testing environments into relatively more homogeneous groups of locations, called mega-environments, where specific genotypes can be targeted for individual mega-environment.

Using a multi-location and multi-year data, Gupta et al. (2013) found that all varietal test locations across all the pearl millet growing zones can be represented by two mega-environments, which is analogous to the All-India testing procedure (Yadav et al. 2012a) of testing of selected hybrids with specific adaptation in their respective adaptation zones. Different pearl millet breeding programs in India develop their product profiles depending on the need of their target mega-environment delineated by geographical locations and rainfall patterns.

3.4.8 Creation of Genebank Having World Pearl Millet Collection

Germplasm resources are a heritage requiring conservation to achieve sustainable development and food and nutritional security. Realizing this need, a global gene bank of six mandate crops, including pearl millet, was established in 1975 at ICRISAT, which serves as a world repository for the collection of germplasms. With ~23,000 germplasm accessions assembled from 52 countries through donations and collection missions, it is one of the largest international genebanks of pearl millet. The National Genebank at National Bureau of Plant Genetic Resources (NBPGR), New Delhi, established in 1986, has 7268 pearl millet accessions, including indigenous collection, exotic collection, wild species, genetic stocks, and released varieties.

The establishment of genebanks at global and regional levels during 1970–1980 ensured that breeders had easy access to global diversity, which could be useful in breeding programs. The collection serves as insurance against genetic erosion and a source of tolerance to diseases and pests, environmental stresses, higher nutritional quality, and traits related to yield for crop improvement. Accessions were also deposited at the Svalbard Global Seed Vault (SGSV), Norway, as a safety backup in 2008. The SGSV provides protection and safeguards against the loss of seeds in genebanks, as well as a refuge for seeds in the case of large-scale regional or global crises.

3.4.9 Establishing Core and Mini-Core Collections

Genetic variability is a bed-rock on which the success of crop improvement programs depends, which includes landraces, wild relatives, improved elite material, and genetic stocks. The germplasm resources remain underused. One of the main reasons for the low use of germplasm includes the larger size of collections. To enhance germplasm utilization, a "core collection" consisting of 10% of the entire collection (Upadhyaya et al. 2009) and a "mini-core collection" consisting of the 10% of core or 1% of the entire collection (Upadhyaya et al. 2011) have been developed. This is an efficient approach to enhancing the use of germplasm in breeding. The development of a mini-core collection representing the core and entire collection for diversity has resulted in the extensive evaluation of mini-core sets and the identification of several promising sources for agronomic and nutritional traits and biotic and abiotic stresses.

3.4.10 First Marker-Assisted Selection (MAS) Product in India

Pearl millet hybrid HHB 67, released in the 1990s, became highly popular because of its extra-early maturity (less than 65 days) and was grown over more than 500,000 ha in Haryana and Rajasthan. After its wide cultivation, it started having DM incidence of up to 30% in farmers' fields from 2000 onward. Hybrid HHB 67 was highly preferred and widely adopted by farmers, so its parental lines were chosen for DM resistance maintenance breeding. Marker-assisted backcrossing with elite donor parent ICMP 451 was used to add DM resistance to male parent H 77/833-2 (Hash et al. 2006). Additional genes for DM resistance were backcrossed into female parent 843A/B from donor ICML 22 using conventional backcross transfer of DM resistance. Greenhouse disease screening confirmed the improved DM resistance of the new versions of the parental lines and their hybrid, which was then tested in national trials. DM-resistant version of HHB 67 was released in 2005 as "HHB 67 Improved" (Khairwal et al. 2009). This is the first product of marker-assisted breeding in any crop to reach farmers in India. It is also among the first public-bred marker-assisted breeding products commercialized in developing countries globally. HHB 67 Improved is currently being grown over an area of 800,000 ha.

Efforts are underway to improve further the "HHB 67 Improved" with respect to its downy mildew and blast resistance and yielding ability for both grain and dry stover using genome-wide simple sequence repeat (SSR) DNA markers. Three DM resistance loci were moved to the male parent of HHB 67 Improved. The latest improvement has been christened "HHB 67 Improved 2-7" (meaning HHB 67 Improved second cycle improvement, 7th version). The second cycle version of HHB 67 Improved is being considered for release by the Indian Council of Agricultural Research (ICAR) for dry north and north-western India for boosting food, fodder, nutritional, and economic security.

3.4.11 First Biofortified Variety in India and Africa

The goal of core breeding in pearl millet has been to increase yield potential because pearl millet has been considered a highly nutritious cereal with higher levels of several minerals than other major cereals. Improving grain nutritional traits is a recent addition to the breeding objective, in view of global recognition of widespread deficiencies of iron (Fe) and zinc (Zn). A large variability for Fe and Zn content in germplasms and breeding lines is well documented. ICTP 8203, a released open-pollinated variety (OPV) under cultivation, having the highest level of Fe content, was subjected to three generations of progeny-based mass selection to improve Fe content in its grain. Eleven S3 progenies with high Fe content and 40–65% selfed seed set were selected, and randomly mated to constitute an improved higher-iron version of ICTP 8203, designated and tested as ICTP 8303 Fe10-2, and later named and released as Dhanashakti. These improved and original versions were evaluated in 10 field trials and 39 trials conducted by the All India Coordinated Pearl Millet Improvement Project (AICPMIP).

Dhanashakti was released and notified wide SO 1146(E) dated April 24, 2014, for cultivation and thus became the first biofortified crop cultivar in India (Rai et al. 2014a, b). Similarly, the variety Chakti was released for commercial cultivation in West Africa in 2018. This is the first biofortified variety of pearl millet released (Gangashetty et al. 2021).

3.4.12 Heterotic Pools

Historically, a diverse range of seed (A-lines) and restorer (R-lines) parental lines have been developed, utilizing a diverse range of genetic material, depending upon the specific phenotypic traits targeted in the development of A- and R-lines (Rai et al. 2006). Studies based on molecular diversity (Nepolean et al. 2012; Gupta et al. 2015a, b; Singh et al. 2018; Ramya et al. 2018) eventually confirmed the existence of two broad-based groups in hybrid parents—one each for seed parents and restorer parents. Heterotic grouping of hybrid parental lines has been recognized as an important strategy to increase the magnitude of heterosis on a long-term basis (Gupta et al. 2020) to enhance genetic gains in pearl millet (Yadav et al. 2021).

3.4.13 Hybrid Breeding in Africa

Unlike India and the United States, the development of open-pollinated varieties continued to be the primary pearl millet breeding objective in West Africa. Ease and economy of seed production, relatively less vulnerability to biotic and abiotic stresses, and the absence of a viable seed industry are contributing factors toward the preference of OPVs over hybrids in major pearl millet growing areas of Africa. Through interpopulation improvement, landraces with thicker stems and longer panicles are predominantly grown for dual purposes (grain and fodder) in West Africa.

Combining ability studies on population hybrids of African landraces reported the prevalence of a high level of heterosis that can potentially contribute to enhancing productivity in the Sahelian and Sudanian environments of West Africa (Ouendeba et al. 1993; Pucher et al. 2016). The hybrid breeding programs in West Africa focused on the development of population and top-cross hybrids to maintain the intervarietal heterogeneity for higher population buffering potential in stressed environments than genetically uniform single-cross hybrids (Haussmann et al. 2012). The success of hybrids in arid regions of India with small-scale farmers is prompting hybrid breeding to be upscaled in Africa in the recent past.

Hybrid Nafagnon (ICMH 147007) was released in 2021 in Burkina Faso, making it the country's first pearl millet hybrid approved in West and Central Africa (WCA) (ICRISAT 2021). Nafagnon matures early in 80–85 days and has a yield potential of about 3 tons per hectare; early maturity helps overcome terminal drought stress. It is a dual-purpose (grain and forage) hybrid resistant to downy mildew, the most harmful pearl millet disease in WCA. The seed size, yield potential, stay-green, and earliness of the hybrid are the traits highly preferred by farmers and end users in Burkina Faso, where the low yield of pearl millet varieties relative to other cereals is forcing farmers away from one of the best-suited crops for Sahel's harsh agro-ecologies.

The private sector is also showing interest in the production and marketing of hybrids, and it is, therefore, presumed that the development and adoption of hybrids would be potentially rewarding in the unpredictable production environments of Africa also (Sattler and Haussmann 2020).

3.4.14 Seed Production Innovations

The use of improved seed has made a tremendous impact on the productivity of pearl millet. The area under improved cultivars has increased considerably over the years, and currently >75% of pearl millet area in India is under improved cultivars, mostly hybrids.

The seed production program of improved pearl millet cultivars has evolved rapidly and extensively in India (Chopra 1982). The "seed village" concept involving the multiplication of specific cultivars in particular villages has been effectively used to undertake seed production programs in the non-traditional area for pearl millet, largely in southern India (Khairwal et al. 1990). Seed production by small farmers under a contractual arrangement with seed corporations and private seed companies is a unique example of the willingness of Indian farmers to embrace new and intricate technology to produce genetically pure seed with high seed vigor and quality.

During the last two decades, yield levels in seed production plots of hybrids have doubled, mainly due to the development of high-yielding parental lines and improved crop management skills of farmers, including staggered sowing (Yadav et al. 2015). The seed processing period has also been shortened considerably, reducing the time from harvesting to delivery of seed in target regions within a month. This circumvents the need to store seeds over seasons, resulting in reduced overhead costs. The male:female ratio of 1:6 is very common in certified seed production plots reducing the cost of seed production.

3.5 Institutionalization of Pearl Millet Research

In the beginning of the twentieth century, research in pearl millet improvement was conducted by departments of agriculture in different states of India. Such work mainly concentrated on understanding the pollination behavior of pearl millet and the improvement of local landraces. In the early 1950s, major work was initiated at the Indian Agricultural Research Institute. A further dimension in pearl millet improvement was added by the establishment of the All India Coordinated Millet

Improvement Project (AICMIP) in 1965, administered by the Indian Council of Agricultural Research (ICAR), for facilitating network interdisciplinary research across India with its headquarters at the Indian Agricultural Research Institute, New Delhi. Later, AICMIP was bifurcated in 1985 into two AICRPs (All India Coordinated Research Projects), one on pearl millet and the other on small millets. Currently, AICPMIP is operating from Jodhpur and is mandated to conduct and coordinate research activities in the improvement of pearl millet. In 2018, the Directorate of Sorghum research was upgraded to the Indian Institute of Millet Research at Hyderabad to undertake basic, strategic, and applied research on all aspects of pearl millet improvement, production, protection, value addition, and alternate uses.

The pioneering pearl millet breeding program was initiated at the College of Agriculture Experiment Stations, University of Georgia at Tifton, Georgia, USA, in the 1940s. The program was engaged in fundamental genetic studies, germplasm evaluation and enhancement, development of cytoplasmic male sterility (CMS), and identification and utilization of dwarfing genes.

The pearl millet improvement got global momentum with the establishment of the International Crops Research Institute for the Semi-Arid Tropics (ICRISAT) at Patancheru, India, in 1972, with the worldwide mandate of improving five crops, including pearl millet, with its regional hubs in several African countries.

3.6 Conclusions

A good understanding of the biology and genetics of pearl millet has helped tremendously in breeding for higher productivity and stability. The new tools and techniques are available now to generate large data from genome sequencing, transcriptome, proteome, metabolome, and epigenome in several crop species, though the progress in pearl millet is not as good as in other crops. The virtuous understanding and integration of data from omics and modeling in the near future would help to recognize the flow of biological information that underlies the complex nature of the adaptation of pearl millet to extreme stress environments (Pazhamala et al. 2021).

References

Appa Rao S, de Wet JMJ (1999) Taxonomy and evolution. In: Khairwal IS, Rai KN, Andrews DJ, Harinarayana G (eds) Pearl millet breeding. Science Publishers, Enfield, NH, pp 29–47

Appa Rao S, Mengesha MH, Reddy CR (1986) New sources of dwarfing genes in pearl millet (*Pennisetum americanum*). Theor Appl Genet 73:170–174

Athwal DS (1961) Recent developments in the breeding and improvement of *bajra* (pearl millet) in the Punjab. Madras Agric J 48:18–19

Athwal DS (1965) Hybrid bajra-1 marks a new era. Ind Farm 15:6–7

Athwal DS (1966) Current plant breeding research with special reference to Pennisetum. Indian J Genet Plant Breed 26A:73–85

Ayyangar RGN, Vijairaghvan C, Pillai GV (1933) Studies in *Pennisetum typhoides* (Rich.) The pearl millet. Part I. Anthesis. Indian J Agric Sci 3:688–694

Begg JE (1965) The growth and development of a crop of bulrush millet (*Pennisetum typhoides* S. and H.). J Agric Sci 65:341–349

Bhatnagar SK, Yadav OP, Gautam RC (1998) Research achievements in pearl millet (*Pennisetum glaucum*). Indian J Agric Sci 68:423–430

Bhattacharjee R, Khairwal IS, Bramel PJ, Reddy KN (2007) Establishment of a pearl millet [*Pennisetum glaucum* (L.) R. Br.] core collection based on geographical distribution and quantitative traits. Euphytica 155:35–45

Bidinger FR, Hash CT, Jayachandran R, Ratnaji RM, N. V. (1999) Recessive, day length-insensitive earliness to synchronize flowering of pearl millet hybrid parents. Crop Sci 39: 1049–1054

Brunken JN, de Wet JMJ, Harlan JR (1977) The morphology and domestication of pearl millet. Econ Bot 31:163–174

Burton GW (1944) Hybrids between Napier grass and cattail millet. J Hered 35:227–232

Burton GW (1948) The performance of various mixtures of hybrid and parent inbred pearl millet, *Pennisetum glaucum* (L.) R. Br J Am Soc Agron 40:908–915

Burton GW (1951) Quantitative inheritance in pearl millet (Pennisetum glaucum). Agron J 43:409–417

Burton GW (1958) Cytoplasmic male-sterility in pearl millet (*Pennisetum glaucum*) (L.) R. Br. Agron J 50:230

Burton GW (1965a) Pearl millet Tift 23 A released. Crops Soils 17:19

Burton GW (1965b) Photoperiodism in pearl millet, *Pennisetum typhoides*. Crop Sci 5:333–335

Burton GW, Athwal DS (1967) Two additional sources of cytoplasmic male sterility in pearl millet and their relationship to Tift 23 A. Crop Sci 7:209–211

Burton GW, Fortson JC (1966) Inheritance and utilization of five dwarfs in pearl millet. (*Pennisetum typhoides*) breeding. Crop Sci 6:69–72

Burton GW, Monson WG, Johnson JC, Lowrey RS, Chapman HD, Marchant WH (1969) Effect of the d2 dwarf gene on the forage yield and quality of pearl millet. Agron J 61:607–612

Chopra KR (1982) Technical guideline for sorghum and pearl millet seed production. Food and Agriculture Organization of the United Nations, Rome, p 110

Clegg MT, Rawson JRY, Thomas K (1984) Chloroplast DNA variation in pearl millet and related species. Genetics 106:449–461

Dave HR (1987) Pearl millet hybrids. In: Proceedings of the international pearl millet workshop, 7-11 April 1986. ICRISAT Center, Patancheru, pp 121–126

Dua RP (1989) Salinity tolerance in pearl millet. Indian J Agric Res 23:9–14

Dujardin M, Hanna WW (1987) Inducing male fertility in crosses between pearl millet and *Pennisetumorientale* Rich. Crop Sci 27:65–68

Dujardin M, Hanna WW (1990) Cytogenetics and reproductive behavior of 48 chromosome pearl millet x Pennisetum squamulatum derivatives. Crop Sci 30:1015–1016

Gangashetty PI, Riyazaddin M, Govindaraj DI, Issoufou K, Peter A, Ousmane S, Moussa DS, Ignatius A (2021) Identification of high-yielding iron biofortified open-pollinated varieties of pearl millet in West Africa. Front Plant Sci 12:688937

Garcia-Huidobro J, Monteith JL, Squire GR (1982) Time temperature and germination of pearl millet (*Pennisetum typhoides* S. & H.). constant temperature. J Exp Bot 33:288–296

Gemenet DC, Hash CT, Sanogo MD, Sy O, Zangre RG, Leiser WL et al (2015) Phosphorus uptake and utilization efficiency in West African pearl millet inbred lines. Field Crops Res. 171:54–66

Gill BS (1969) Genetic studies for establishment of linkage groups in *Pennisetum typhoides*. Final Project Report PL-480, PAU Ludhiana (cited by KS Gill, 1991)

Gill BS. 1971. Additional genetic stocks for linkage studies in *Pennisetum typhoides* (burm.) S.&H. J Res PAU 8: 291–299

Gill KS (1991) Pearl millet and its improvement. Indian Council of Agricultural Research, New Delhi. 305pp

Gill BS, Sharma HL, Dheshi JS (1973) Cytomorphological studies of haploid pearl millet. Cytologia 38:411–416

Godbole SV (1928) Studies on bajra (*Pennisetum typhoideum*). I. The morphology of *Pennisetum typhoideum*. Mem Dep Agric India, Bot Ser 14:247–268. (Cited by Gill, 1991)

Gupta SK, Rai KN, Govindaraj M, Rao AS (2012) Genetics of fertility restoration of the A4 cytoplasmic-nuclear male sterility system in pearl millet. Czech J Genet Plant Breed 48:87–92

Gupta SK, Rathore A, Yadav OP, Rai KN, Khairwal IS, Rajpurohit BS, Das RR (2013) Identifying mega-environments and essential test locations for pearl millet cultivar selection in India. Crop Sci 53:2444–2453

Gupta SK, Nepolean TV, Sankar SM, Rathore A, Das RR, Rai KN, Hash CT (2015a) Patterns of molecular diversity in current and previously developed hybrid parents of pearl millet [*Pennisetum glaucum* (L.) R. Br.]. Am J Plant Sci 06:1697–1712

Gupta SK, Rai KN, Singh P, Ameta VL, Gupta SK, Jayalekha AK, Mahala RS, Pareek S, Swami ML, Verma YS (2015b) Seed set variability under high temperatures during flowering period in pearl millet (*Pennisetum glaucum* L. (R.) Br.). Field Crops Res 171:41–53

Gupta SK, Yadav DV, Govindaraj M, Boratkar M, Kulkarni VN, Rai KN (2018) Inheritance of fertility restoration of A5 cytoplasmic-nuclear male sterility system in pearl millet [*Pennisetum glaucum* (L.) R. Br.]. Indian J Genet Plant Breed 78:228–232

Gupta SK, Patil KS, Rathore A, Yadav DV, Sharma LD, Mungra KD, Patil HT, Gupta SK, Kumar R, Chaudhary V, Das RR, Kumar A, Singh V, Srivastava RK, Gupta R, Moratkar M, Varshney RK, Rai KN, Yadav OP (2020) Identification of heterotic groups in South-Asian-bred hybrid parents of pearl millet. Theor Appl Genet 133:873–888

Ha BD, Pernes J (1982) Androgenesis in pearl millet—analysis of plants obtained from microspore culture. Z Pflanzenphysiol 108:317–327

Hanna WW (1989) Characteristics and stability of a new cytoplasmic-nuclear male sterile source in pearl millet. Crop Sci 29:1457–1459

Hanna WW (1995) Use of apomixis in cultivar development. Adv Agron 54:333–350

Hanna WW, Bashaw EC (1987) Apomixis: its identification and use in plant breeding. Crop Sci 27: 1136–1139

Hanna WW, Dujardin M (1982) Apomictic interspecific hybrids between pearl millet and *Pennisetumorientale* L.C. Rich. Crop Sci 22:857–859

Harlan JR (1971) Agricultural origins: centers and noncenters. Science 174:468–474

Harlan JR, de Wet JMJ (1971) Toward a rational classification of cultivated plants. Taxon 20:509–517

Hash CT, Sharma A, Kolesnikova-Allen MA, Singh SD, Thakur RP, Bhasker Raj AG, Ratnaji Rao MNV, Nijhawan DC, Beniwal CR, Sagar P, Yadav HP, Yadav YP, Srikant, Bhatnagar SK, Khairwal IS, Howarth CJ, Cavan GP, Gale MD, Liu C, Devos KM, Breese WA, Witcombe JR (2006) Teamwork delivers biotechnology products to Indian small-holder crop-livestock producers: pearl millet hybrid "HHB 67 Improved" enters seed delivery pipeline. SAT eJournal 2(1):1–3

Haussmann BIG, Rattunde HF, Weltzien-Rattunde E, Traoré PSC, vom Brocke K, Parzies HK (2012) Breeding strategies for adaptation of pearl millet and sorghum to climate variability and change in West Africa. J Agron Crop Sci 198:327–339

ICRISAT (2021). https://www.icrisat.org/burkina-faso-releases-its-first-pearl-millet-hybrid/. Accessed 22 Dec 2021

Jauhar PP (1970) Haploid meiosis and its bearing on eh phylogeny of pearl millet, *Pennisetum typhoides* Staft & Hubb. Genetica 41:407–427

Jauhar PP (1981) The eternal controversy on the Latin name of pearl millet. Indian J Bot 4:1–4

Jauhar PP, Hanna WW (1998) Cytogenetics and genetics of pearl millet. Adv Agron 64:1–26

Johar PP, Rai KN, Ozias-Akins P, Chen Z, Hanna WW (2006) Genetic improvement of pearl millet for grain and forage production: cytogenetic manipulation and heterosis breeding. In: Singh RJ, Jauhar PP (eds) Genetic resources, chromosome engineering, and crop improvement. CRC Taylor & Francis, pp 281–307

Joshi AB, Ahluwalia M, Shankar K (1961) 'Improved Ghana' is a better bajra. Indian Farm 11:12
Kajjari NB, Patil JA (1956) A male-sterile Bajri. Indian J Genet Plant Breed 16(2):146
Khairwal IS, Ram C, Chhabra AK (1990) Pearl millet seed production & technology. Manohar Publications, New Delhi, p 208
Khairwal IS, Rai KN, Yadav OP, Rajpurohit BS, Negi S (2009) Pearl millet cultivars: seeds of choice. All India Coordinated Pearl Millet Improvement Project, Mandor, Jodhpur. 36 pp
Krishna Rao M, Koduru PRK (1979) Genetics of five hairy phenotypes and a linkage group of *Pennisetum americanum*. Euphytica 28:445–452
Krishnamurthy L, Serraj R, Rai KN, Hash CT, Dakheel AJ (2007) Identification of pearl millet [*Pennisetum glaucum* (L.) R. Br.] lines tolerant to soil salinity. Euphytica 158:179–188
Krishnaswamy N (1962) Bajra: *Pennisetum typhoides* S & H. Indian Council of Agri-cultural Research, New Delhi, p 93
Krishnaswamy N, Raman VS (1949) Studies on the interspecific hybrid of *Pennisetum typhoides* Stapf and Hubb. × *P. purpureum* Schumach. III. The cytogenetics of the colchicine-induced amphidiploid. Genetica 27:253–272
Lahiri AN, Kharbanda BC (1965) Studies on plant water relationship: effects of moisture deficit at various development stages of bulrush millet. Proc Natl Acad Sci India B31:14–23
Lahiri AN, Kumar V (1966) Studies on plant water relationship III. Further studies on the drought mediated alterations in the performance of bulrush millet. Proc Natl Acad Sci India B 32:116–129
Leuck DB, Burton GW (1966) Pollination of pearl millet by insects. J Econ Bot 59:1308–1309
Liu CJ, Witcombe JR, Pittaway TS, Nash M, Hash CT, Busso CS, Gale MD (1994) An RFLP-based genetic map of pearl millet (*Pennisetum glaucum*). Theor Appl Genet 89:481–487
Maiti RK, Bidinger FR (1981) Growth and development of the pearl millet plant. Research Bulletin No. 6. International Crops Research Institute for the Semi-Arid Tropics, Patanchecu. 14 pp
Manga V, Pantulu JV (1971) The meiotic behaviour of a haploid pearl millet. Genetica 42:519–528
Marchais L, Pernes J (1985) Genetic divergence between wild and cultivated pearl millets (*Pennisetum typhoides*). I. Male sterility. Z Pflanzenzüchtg 95:103–112
Munson PJ (1975) Archaeological data on the origins of cultivation in the southwestern Sahara and its implications for West Africa. In: Harlan JR, DeWet JMJ, Stemler ABL (eds) The origins of African plant domestication. Mouton Press, The Hague, pp 187–210
Nepolean T, Gupta SK, Dwivedi SL, Bhattacharjee R, Rai KN, Hash CT (2012) Genetic diversity in maintainer and restorer lines of pearl millet. Crop Sci 52:2555–2563
Nitsch CS, Andersen M, Godard MG, Shedan WF (1982) Production of haploid plants of *Zea mays* and Pennisetum through androgenesis. In: Earle ED, Demanly Y (eds) Variability in plants regenerated from tissue culture. Praeger Publisher, New York, pp 69–91
Ong CK (1983) Response to temperature in a stand of pearl millet (*Pennisetum typhoides S. & H*) II. Reproductive development. J Exp Bot 34:337–348
Ouendeba B, Ejeta G, Nyquist WE, Hanna WW, Kumar A (1993) Heterosis and combining ability among African pearl millet landraces. Crop Sci 33:735–739
Pantulu JV, Manga V (1969) Meiosis in a haploid pearl millet. Curr Sci 38:143–144
Patil KS, Gupta SK (2020) Geographic patterns of genetic diversity and fertility restoration ability of Asian and African origin pearl millet populations. Crop J 10:468. https://doi.org/10.1016/j.cj.2021.04.013
Pazhamala LT, Kudapa H, Weckwerth W, Millar AH, Varshney RK (2021) Systems biology for crop improvement. Plant Genome 14(2):e20098. https://doi.org/10.1002/tpg2.20098
Pucher A, Sy O, Sanogo MD, Angarawai II, Zangre R, Ouedraogo M, Boureima S et al (2016) Combining ability patterns among West African pearl millet landraces and prospects for pearl millet hybrid breeding. Field Crops Res 195:9–20
Rachie KO, Majmudar JV (1980) Pearl millet. The Pennsylvania State University Press, University Park, London
Rai KN (1995) A new cytoplasmic-nuclear male sterility system in pearl millet. Plant Breed 114: 445–447

Rai KN, Anand Kumar K (1994) Pearl millet improvement at ICRISAT—an update. Intern Sorghum Millets News 35:1–29

Rai KN, Rao AS (1991) Effect of d2 dwarfing gene on grain yield and yield components in pearl millet near-isogenic lines. Euphytica 52:25–31

Rai KN, Kulkarni VN, Thakur RP, Haussmann BIG, Mgonja MA (2006) Pearl millet hybrid parents research: approaches and achievements. In: Gowda CLL, Rai KN, Reddy BVS, Saxena KB (eds) Hybrid parents research at ICRISAT. ICRISAT, Patancheru, pp 11–73

Rai KN, Khairwal IS, Dangaria CJ, Singh AK, Rao AS (2009) Seed parent breeding efficiency of three diverse cytoplasmic-nuclear male sterility systems in pearl millet. Euphytica 165:495–507

Rai KN, Patil HT, Yadav OP, Govindaraj M, Khairwal IS, Cherian B, Rajpurohit BS, Rao AS, Kulkarni MP (2014a) Dhanashakti: a high-iron pearl millet variety. Indian Farm 64:32–34

Rai KN, Patil HT, Yadav OP, Govindaraj M, Khairwal IS, Cherian B, Rajpurohit BS, Rao AS, Shivade H, Kulkarni MP (2014b) Pearl millet variety Dhanashakti. Indian J Genet Plant Breed 74:405–406

Ramya AR, Ahamed ML, Satyavathi CT, Rathore A, Katiyar P, Raj AG, Kumar S, Gupta R, Mahendrakar MD, Yadav RS, Srivastava RK (2018) Towards defining heterotic gene pools in pearl millet [*Pennisetum glaucum* (L.) R. Br.]. Front Plant Sci 8:1934

Raymond DC (1968) For a better understanding of growth and development of Pennisetum millet. L'Agronomie Tropicale 23:844–863

Sastry PS, Mallikarjuna N (2014) Induction of androgenesis in pearl millet. Univ J Agric Res 2(6): 216–223

Sattler FT, Haussmann BIG (2020) A unified strategy for West African pearl millet hybrid and heterotic group development. Crop Sci 60:1–13

Satyavathi CT, Khandelwal V, Rajpurohit BS, Supriya A, Beniwal BR, Kamlesh K, Sushila B, Shripal S, Mahesh CK, Yadav SL (2018) Pearl millet hybrids and varieties. ICAR-All India Coordinated Research Project on pearl millet, Mandor, Jodhpur, p 142

Sharma S, Sharma R, Pujar M, Yadav D, Yadav YP, Rathore A et al (2020) Utilization of wild Pennisetum species for improving biotic and abiotic stress tolerance in pearl millet (Pennisetum glaucum L.). Crop Sci 61:289–304

Singh P, Rai KN, Witcombe JR, Andrews DJ (1988) Population breeding methods in pearl millet improvement (*Pennisetum americanum*). L'Agron Trop 43:185–193

Singh S, Gupta SK, Thudi M, Das RR, Vemula A, Garg V, Varshney RK, Rathore A, Pahuja SK, Yadav DV (2018) Genetic diversity patterns and heterosis prediction based on SSRs and SNPs in hybrid parents of pearl millet. Crop Sci 58:2379–2390

Soman P, Peacock JM (1985) A laboratory technique to screen seedling emergence of sorghum and pearl millet at high soil temperature. Exp Agric 21:335–341

Sujata V, Sivaramakrishnan S, Rai KN, Seetha K (1994) A new source of cytoplasmic male sterility in pearl millet: RFLP analysis of mitochondrial DNA. Genome 37:482–486

Thakur RP, Williams RJ (1980) Pollination effects on pearl millet ergot. Phytopathology 70:80–84

Thakur RP, Subba Rao KV, Williams RJ (1983) Evaluation of a new field screening technique for smut resistance in pearl millet. Phytopathology 73:1255–1258

Upadhyaya HD, Gowda CLL, Reddy KN, Singh S (2009) Augmenting the pearl millet [*Pennisetum glaucum* (L.) R. Br.] core collection for enhancing germplasm utilization in crop improvement. Crop Sci 49:573–580

Upadhyaya HD, Yadav D, Reddy KN, Gowda CLL, Singh S (2011) Development of pearl millet mini core collection for enhanced utilization of germplasm. Crop Sci 51:217–223

Varshney RK, Shi C, Thudi M, Mariac C, Wallace J, Qi P, Zhang H, Zhao Y, Wang X, Rathore A, Srivastava RK, Chitikineni A, Fan G, Bajaj P, Punnuri S, Gupta SK, Wang H, Jiang Y, Couderc M, Katta MAVSK, Paudel DR, Mungra KD, Chen W, Harris-Shultz KR, Garg V, Desai N, Doddamani D, Kane N, Conner JA, Ghatak A, Chaturvedi P, Subramaniam S, Yadav OP, Berthouly-Salazar C, Hamiduo F, Wang J, Liang X, Clotault J, Upadhyaya HD, Cubry P, Rhoné B, Gueye MC, Sunkar R, Dupuy C, Sparvoli F, Cheng S, Mahala RS, Singh B, Yadav RS, Lyons E, Datta SK, Hash CT, Devos KM, Buckler E, Bennetzen JL, Paterson AH, Ozias-

Akins P, Grando S, Wang J, Mohapatra T, Weckwerth W, Reif JC, Xi L, Vigouroux Y, Xu X (2017) Pearl millet genome sequence provides a resource to improve agronomic traits in arid environments. Nat Biotechnol 35:969–974

Witcombe JR (1999) Population improvement. In: Khairwal IS, Rai KN, Andrews DJ, Harinarayana G (eds) Pearl millet breeding. Oxford and IBH, New Delhi, pp 213–256. 511 pp

Yadav OP, Rai KN (2013) Genetic improvement of pearl millet in India. Agric Res 2:275–292

Yadav OP, Rai KN, Rajpurohit BS, Hash CT, Mahala RS, Gupta SK, Shetty HS, Bishnoi HR, Rathore MS, Kumar A, Sehgal S, Raghvani KL (2012a) Twenty-five years of pearl millet improvement in India. All India Coordinated Pearl Millet Improvement Project, Jodhpur. 122 pp

Yadav OP, Rai KN, Gupta SK (2012b) Pearl millet: genetic improvement for tolerance to abiotic stresses. In: Tuteja N, GilL SS, Tuteja R (eds) Improving crop resistance to abiotic stress. WIley-VCH Verlag GmbH & Co. KGaA, pp 261–288

Yadav OP, Mahala RS, Rai KN, Gupta SK, Rajpurohit BS, Yadav HP (2015) Pearl millet seed production and processing. All India Coordinated Research Project on Pearl Millet, Indian Council of Agricultural Research, Mandor, Jodhpur. 60 pp

Yadav OP, Singh DV, Dhillon BS, Mohapatra T (2019) India's evergreen revolution in cereals. Curr Sci 116:1805–1808

Yadav OP, Gupta SK, Govindaraj M, Sharma R, Varshney RK, Srivastava RK, Rathore A, Mahala RS (2021) Genetic gains in pearl millet in India: insights into historic breeding strategies and future perspective. Front Plant Sci 12:645038. https://doi.org/10.3389/fpls.2021.645038

Advances in Pearl Millet Hybrid Breeding and Development of Parental Lines

4

S. K. Gupta, O. P. Yadav, and P. Sanjana Reddy

Abstract

Significant heterosis, stable male sterility systems, effective fertility restoration, and ease in selfing and crossing enabled breeders to enhance genetic gains in pearl millet through hybrid breeding. More than 50 years of breeding efforts in India and elsewhere utilizing germplasm from Asian and African regions have led to the development of a diverse array of trait- and adaptation-specific seed and restorer parental lines in productive backgrounds. Clarity on product profiles in breeding programs, maintaining significant genetic diversity among breeding lines, keeping seed and restorer parents in two separate heterotic pools, strong screening protocols for biotic and abiotic stresses, using molecular markers to select traits of interest, and utilization of tools to predict heterosis will be the key factors to move further in enhancing the hybrid potential of this crop in different regions of the world.

It is informed that consent has been taken for figures and graphs used in the current chapter.

S. K. Gupta (✉)
Pearl Millet Breeding, International Crops Research Institute for the Semi-Arid Tropics, Patancheru, Hyderabad, India
e-mail: s.gupta@cgiar.org

O. P. Yadav
ICAR-Central Arid Zone Research Institute, Jodhpur, India

P. S. Reddy
ICAR-Indian Institute of Millets Research, Hyderabad, India

V. A. Tonapi et al. (eds.), *Pearl Millet in the 21st Century*,
https://doi.org/10.1007/978-981-99-5890-0_4

4.1 Introduction: Hybrids, a Better Option in Pearl Millet

In pearl millet, two types of commercial cultivars are available due to the presence of a high degree of outcrossing; it can be open-pollinated varieties (OPVs) or hybrids to exploit the heterosis to enhance the productivity in this crop. Pearl millet hybrids display a high degree of heterosis for grain yield, though OPVs permit only partial exploitation of heterosis. Currently, hybrids have occupied the majority of the area in Indian ecologies having an annual rainfall of >400 mm where the pearl millet crop is supported with adequate doses of fertilizers. Also, decade-long trial data from the Indian national cultivar release system have indicated that hybrids have shown yield superiority of about 25% over landraces/OPVs in highly drought-prone ecologies (Yadav et al. 2011).

Hybrid breeding provides an excellent opportunity to use diverse parental lines to harness maximum heterosis, and yet produce a cultivar with a high degree of morphological uniformity. The open-pollinating breeding system of pearl millet, high degree of heterosis, and availability of stable cytoplasmic-nuclear male sterility (CMS) and its fertility restorers have made the hybrid option commercially viable in pearl millet. The first commercial pearl millet grain hybrid HB-1, developed at the Punjab Agricultural University (PAU), Ludhiana, and released in India in 1965, had twice as much grain yield as the then-improved OPVs (Athwal 1965). Hybrids also offer a better option to manage downy mildew (DM), the most destructive disease in pearl millet. In single-cross hybrids, DM resistance generally lasts for 4–8 years. In India, due to strong hybrid programs (in the public and private sectors), there is significant cultivar diversity with about 70–80 hybrids on the farmers' field at any point in time. A high hybrid replacement rate coupled with resistance breeding has been able to contain DM disease. Currently, India is the largest grower of pearl millet hybrids (dual-purpose) in the world cultivating them on about 4.5–5.0 m ha, followed by the USA (mostly forage), and recently some West African countries have also introduced dual-purpose hybrids in their region. Hence, this chapter mostly draws strategies and examples of hybrid breeding in relation to Indian pearl millet programs owing to their long experience of hybrid breeding in this crop.

4.2 Development of Hybrid Parents

4.2.1 Trait Prioritization Under Different Product Profiles/Market-Segments

Pearl millet is cultivated under diverse agro-ecologies ranging from near-optimum environments (with high use of irrigation and chemical fertilizers) to extremely challenging drought-prone environments (with little external inputs). A single cultivar cannot be expected to perform well under all the environmental conditions, and a cultivar planted outside its adaptation zone would suffer yield reduction due to significant genotype × environment interactions. Therefore, breeding and evaluation require a subdivision of the testing environments into relatively more homogeneous groups of locations, called "mega-environments" or termed alternatively "market-segments," where specific genotypes can be targeted for individual mega-environment. For instance, the pearl millet cultivation area in India has been divided into three mega-environments (designated as A_1, A, and B zones) considering the geographical location, rainfall pattern, local adaptation, and other environmental conditions (Gupta et al. 2013). Clarity about the priority traits needed in different segments is a must, to enhance the efficiency of the breeding program and to develop a relevant product for any ecology.

Product profiles are an important tool for defining market segments and identifying the traits to be prioritized in a particular specific segment. Most of the pearl millet breeding programs in the public and private sectors in India have developed their product profiles (which may not be formally documented sometimes), depending on the need of their target mega-environment. For instance, International Crops Research Institute for the Semi-Arid Tropics (ICRISAT), as a Consultative Group on International Agricultural Research (CGIAR) center with a global mandate on pearl millet improvement programs, based in India targeting all the mega-environments to support Indian national breeding programs, has four product profiles, and they are depicted as an example in Table 4.1. Each product profile should have a list of "must-have" traits; these traits should essentially be present in the breeding lines bred under this product profile. There can be "long-term priority traits," which need to be addressed in 5–10 years. High grain yield, disease resistance, and maturity duration of 75–85 days, as per the agro-ecological requirements, have been accorded the highest priority in pearl millet (Yadav and Rai 2013). Because of the growing importance of dry stover for fodder purposes, there has been a considerable emphasis, in recent times, on breeding for dual-purpose cultivars producing both high stover and grain yields (Yadav et al. 2012b).

4.2.1.1 Seed Parents

In the breeding of seed parents (A-lines), the high grain-yield potential of A-lines, both as lines per se and in hybrids (i.e., combining ability), is the most important consideration. Thus, high yield/agronomic potential is the first target trait for which selection can be made visually in un-replicated nurseries. High yield, coupled with other agronomic and farmer-preferred traits, can be targeted in A-line breeding.

Some traits like lodging resistance, compact panicles, good exertion, and seed set can be considered for all environments, while region-preferred traits, like maturity, plant height, tillering ability, seed color, and seed size, are specific to different agro-ecologies. Most of these agronomic traits have high heritability for which visual selection during advanced generations is fairly effective. The d_2 dwarf plant height is the most dominant plant type concept in seed parent breeding. This has several operational advantages: (1) it provides the option for breeding hybrids of varying heights, (2) it provides greater control on seed yield and quality by reducing the risk of lodging that can occur under high management seed production conditions, and (3) it allows a much rapid detection and efficient rouging of off-types and pollen shedders in A-line. In addition, the A-lines must have complete and stable male sterility, and the B-lines must have profuse pollen production ability across the seasons and sites.

Table 4.1 Product profiles of ICRISAT's pearl millet breeding program (Asia), based at Hyderabad, India

Product profile	Estimated area (ha)	% area/ effort	Target agro-ecologies	Product development goals
(1) Parent lines of medium to late maturing, dual-purpose hybrids with disease resistance, for adaptation to better-endowed environments	4.0 m ha (rainy season)	65%	**Target**: India: East Rajasthan, central and South Gujarat, Haryana, Uttar Pradesh (UP), Maharashtra, peninsular India (400–700 mm/ annum)	**Must-have traits:** 1. Parents with high productivity and good general combining ability (GCA) for grain yield to develop hybrids with at least a 10% increase in grain yield (3–4 tons/ha) over representative checks (GHB 905, RHB 173, 86M01, 86M86, PAC 909) with improved fodder yield 2. 76–90 days maturity 3. Downy mildew (≤10%) and blast resistance (<3.0 score) 4. High grain iron (≥60 ppm) and zinc (≥35 ppm) content **Long-term priority:** 1. Lodging tolerance 2. Rancidity 3. N-use efficiency 4. Stover digestibility
(2) Parent lines of early maturing, dual-purpose hybrids with disease resistance, for adaptation to drought-prone environments	The total area is 3.5 m ha, *but this PC targets about 1.5 m ha through hybrids*	20%	**Target:** India: Western Rajasthan and drier parts of Gujarat and Haryana (200–400 mm/ annum)	**Must have traits:** 1. Parents with high productivity and good GCA for grain yield to develop hybrids with at least a 10% increase in grain yield (2.0–2.5 tons/ ha) over

(continued)

Table 4.1 (continued)

Product profile	Estimated area (ha)	% area/ effort	Target agro-ecologies	Product development goals
				representative checks (RHB-177, MPMH 17, HHB-67-improved) with improved fodder yield (~2.0 t ha^{-1}) 2. 65–75 days maturity 3. Downy mildew (≤10%) and blast resistance (<3.0 score) 4. Improved terminal drought tolerance 5. High grain iron (≥50 ppm) and zinc (≥35 ppm) content and better fodder quality **Long-term priority:** 1. Stover digestibility 2. Seedling-stage heat tolerance
(3) Parent lines of medium to late maturing, dual-purpose hybrids with disease resistance, for adaptation to high heat stress ecology in the summer season	0.6–0.8 m ha (and increasing)	5%	**Target**: India: North Gujarat, South Rajasthan, Western UP (summer season)	**Must-have traits:** 1. Parents with high productivity and good GCA for grain yield to develop hybrids with at least a 10% increase in grain yield (5–6 tons/ha) over representative checks (86M11, 9444) with improved fodder yield 2. 85–90 days maturity 3. Downy mildew (≤10%) and blast resistance (<3.0 score) 4. Flowering-

(continued)

Table 4.1 (continued)

Product profile	Estimated area (ha)	% area/ effort	Target agro-ecologies	Product development goals
				stage heat tolerance (summer season; >42 °C) 5. High grain iron (≥60 ppm) and zinc (≥35 ppm) content
(4) Cultivars and hybrid parents exclusively for forage and high biomass	1.0 m ha	10%	**Target**: India: Gujarat, Punjab, Rajasthan, UP, Madhya Pradesh, peninsular India (both summer and rainy seasons)	**Must-have traits:** 1. 10% increase in biomass yield over the best check (PAC-981) with comparable or improved fodder quality traits 2. Green biomass of 40–55 tons/ha and dry biomass of 15–20 tons/ha 3. Non-hairy leaves with a leaf/stem ratio of 3:5 4. In vitro dry matter digestibility (IVDMD) of ≥50% with crude protein of ≥10% 5. Single cut (50–80 days); multi-cut (50–110 days) 6. Downy mildew (≤10%), rust (≤10%), and blast resistance (<3.0 score)

New plant types such as A-lines with long panicles (30–80 cm compared to standard normal of 10–20 cm), thick panicles (40–50 mm diameter compared to normal 20–30 mm), and large seed size (17–20 g of 1000-seed mass compared to standard normal of 9–12 g) are being developed with high productive ability in many pearl millet breeding programs in India. Progenies are evaluated for downy mildew and blast resistance during the generation advancement of breeding lines, which runs concurrent to agronomic evaluation to ensure that B- and R-lines finally produced are resistant to these diseases.

4.2.1.2 Restorer Parents

Restorer lines (R-lines) are used as pollinators to produce hybrids. These lines should (1) have good GCA, (2) restore complete male fertility in their grain hybrid, and (3) must produce profuse pollen that should remain viable at air temperatures as high as 42–44 °C. It is desirable to breed pollinators of 150–180 cm in height, but no shorter than the A-line with built-in attributes of panicle, maturity, and tillering that will be preferred by farmers in the hybrids. Besides being able to produce high-yielding hybrids, pollinators must have an acceptable level of lodging resistance and should also possess adequate levels of resistance to various diseases.

4.2.2 Genetic Diversification

A range of germplasms from India and Africa with diverse phenotypic characteristics, such as tillering, panicle size, earliness, grain size, grain color, and plant height, was strategically exploited to diversify the genetic base of both seed and restorer parents (Andrews and Kumar 1996; Rai et al. 2009a; Yadav et al. 2012c; Patil et al. 2020). In the last five decades, hybrid breeding in pearl millet has received a very high priority in India using genetically diverse parental lines targeting various production ecologies that have helped to intensify the genetic gains (Rai et al. 2009a; Yadav et al. 2012a, 2021).

Seed Parents Considering the agro-ecological conditions and farmers' preference for specific segments, a trait-based approach in breeding seed parents with high grain yield potential and high levels of DM resistance has been followed, resulting in the development of large diversity in seed parents. For instance, the utilization of diverse breeding materials in the development of 104 designated B-lines at ICRISAT from 2005 to 2018 has resulted in a wide range of variability. For instance, days to 50% flowering ranged from 41 to 64 days, plant height from 72 to 156 cm, number of productive tillers plant^{-1} from 1 to 9, panicle length from 14 to 37 cm, panicle diameter from 1.7 to 3.8 cm, and 1000-grain weight from 5.4 to 16.0 g in these 104 B-lines (Gupta et al. 2022a). Such a large morphological diversity originated from the focused selection for these traits. To enhance the genetic diversity, it is very important to continuously utilize diverse germplasms in the breeding pipeline; for instance, among the 203 A/B pairs developed at ICRISAT from 1981 to 2018, 17 B-lines were directly selected from the composites, 25 lines were derived from crosses that involved germplasms in their parentage, and 37 lines had composites in their parentage. Thus, 79 seed parental lines had some components of germplasm and/or composite in their parentage, indicating apparently substantial usage of germplasms/composites in the development of these seed parental lines. The remaining 124 seed parental lines were derived from crosses between elite inbred lines (Table 4.2).

New germplasms, like landraces and germplasms from gene banks or improved OPVs, can be strategically involved in the B-line breeding program based on the traits required in seed parents and also considering cytoplasmic maintenance

Table 4.2 Type of material used in the development of designated maintainer (B) lines (1981–2018) and restorer (R) lines (2006–2019) of pearl millet bred at ICRISAT-India

Type of genetic material	Number of lines derived		Remarks
	B-lines (203)	R-lines (122)	
Germplasm	3	2	Inbreeding and selection directly from germplasm
Composites	17	66	Includes composites and open-pollinated varieties
Germplasm × Elite line	22	6	Includes early-generation breeding lines derived from crosses between germplasm and elite lines
Composite × Elite line	37	6	Includes early-generation breeding lines derived from crosses of composites and elite lines
Elite line × Elite line	124	42	Includes cross between elite lines from advanced generations

reaction. Traits of importance, like desirable maturity period, panicle and grain traits contributing toward high yield, and photo-insensitiveness, help to identify germplasms for utilization in seed parent breeding programs. For example, germplasm from the Togo region of West Africa, known as "*iniari*" germplasm, having early maturity, considerable photo-insensitiveness (required for successful seed production and for stable hybrid performance in different agro-ecologies), bold grains, and dark gray color, was identified for its utilization in B-line breeding, and hence most of the programs developing B-lines have significantly used these genetic backgrounds (Andrews and Kumar 1996; Singh and Gupta 2019). At the beginning of a hybrid breeding era in pearl millet (the 1960s), trait-based composites and populations were observed for traits and fertility/sterility and were involved in seed parent or restorer parent development programs.

Restorer Parents Like seed parents, it is essential to enhance genetic diversity in the restorer lines. Breeding programs in the public and private sectors have introgressed significant genetic variability for different traits in the restorer lines. For instance, 122 R-lines bred at ICRISAT during 2006–2019 had plant height ranging from 100 to 179 cm and days to 50% flowering varied from 45 to 60 days, wherein 43% of lines were of medium duration, 35% were late, and 13% were very late, while only 9% of the lines were of early flowering. Panicle length varied from 12.7 to 31.1 cm wherein 53% of lines were small and 46% were medium. Panicle diameter varied from 1.4 to 3.8 cm wherein 74% were medium, 24% of lines were thick, and only 2% of lines showed thin panicles. Thousand-grain weights varied from 6.0 to 11.9 g wherein 72% of lines were medium, 20% were bold, and only 8% showed small grain size (Gupta et al. 2022b). Diverse germplasms should also be continuously utilized to enhance genetic diversity in the development of R-lines; for example, the parentage of mentioned 122 designated restorer lines at ICRISAT showed the utilization of a wide range of germplasms and improved lines in developing these R-lines. For instance, 67 R-lines were selected directly from trait or adoption-specific composites, and two R-lines were directly selected from the germplasm accessions (Table 4.2). In addition, six R-lines were derived from crosses that involved germplasms in their parentage, and seven lines had composites in their parentage. Thus, 82 restorer lines had some components of germplasm and/or composite in their parentage, indicating apparently substantial usage of germplasms and composites in the development of these R-lines. The remaining 44 R-lines were derived from crosses between elite inbred lines.

4.2.3 Cytoplasmic Diversification

4.2.3.1 CMS Search and Characterization

The entire hybrid development program has been (and still continues to be) based largely on the single CMS system (A1 system) in India and around the globe. This large-scale cytoplasmic uniformity exposes the hybrid pearl millet seed industry to the potential disease and insect-pest epidemics. Thus, efforts were made to assemble CMS sources developed elsewhere, initiate a search for new CMS sources, and characterize these for the stability of male sterility, their maintainer frequency in the germplasm, character association, and any possible impact on restorer line development. Initial evaluation of A-lines for pollen shedders (a measure of the stability of male sterility) had shown that the A2 and A3 sources were most unstable. Further evaluation of A-lines with other CMS sources during the rainy, and mild and hot summer environments showed that A-lines with Aegp cytoplasm (0.0–0.1% pollen shedders, depending on the nuclear genetic backgrounds and the environments in which A-lines were evaluated), and those with the A4 cytoplasm (0.0–0.3% pollen shedders) were more stable than those with the A1 cytoplasm (0.0–2.5% pollen shedders) (Rai et al. 2001). The A-lines with the A5 cytoplasm had no pollen shedders, implying that the A5 CMS system was the most stable. An additional advantage is that the genetic background of male-sterile lines in A_4 and A_5 cytoplasm does not affect the fertility restoration of hybrids, whereas the genetic background of A_1 cytoplasm has a significant effect on the fertility restoration (Rai et al. 2001, 2009b; Gupta et al. 2010). If restorer development for A4 and A5 CMS has to be enhanced, it would require a major shift in breeding efforts as most of the restorers in all the hybrid programs in India and elsewhere are of A1 CMS. But there is a serious lack of A_4 restorers and A_5 restorers in elite agronomic backgrounds (Gupta et al. 2012, 2018). However, excellent genetic stocks of A_4 and A_5 restorers have been developed at ICRISAT, which are used in backcross breeding to develop restorer lines of A_4 and A_5 CMS systems (Rai and Rao 2003; Rai et al. 2006).

New germplasms must be continuously involved in the breeding pipeline to enhance the genetic diversity of restorers. In one such effort, topcross hybrids developed between A-lines with the A1, A4, and A5 cytoplasm and several improved populations from Asia and Africa regions showed that the frequency of maintainers was highest for the A5 system (mean of 86%), followed by A4 system (37%), and it was the lowest for the A1 CMS system (7%), thus providing the prospects of greatest genetic diversification of A-lines with the A5 system, and then with the A4 and the A1 systems, in that order (Fig. 4.1) (Patil et al. 2020).

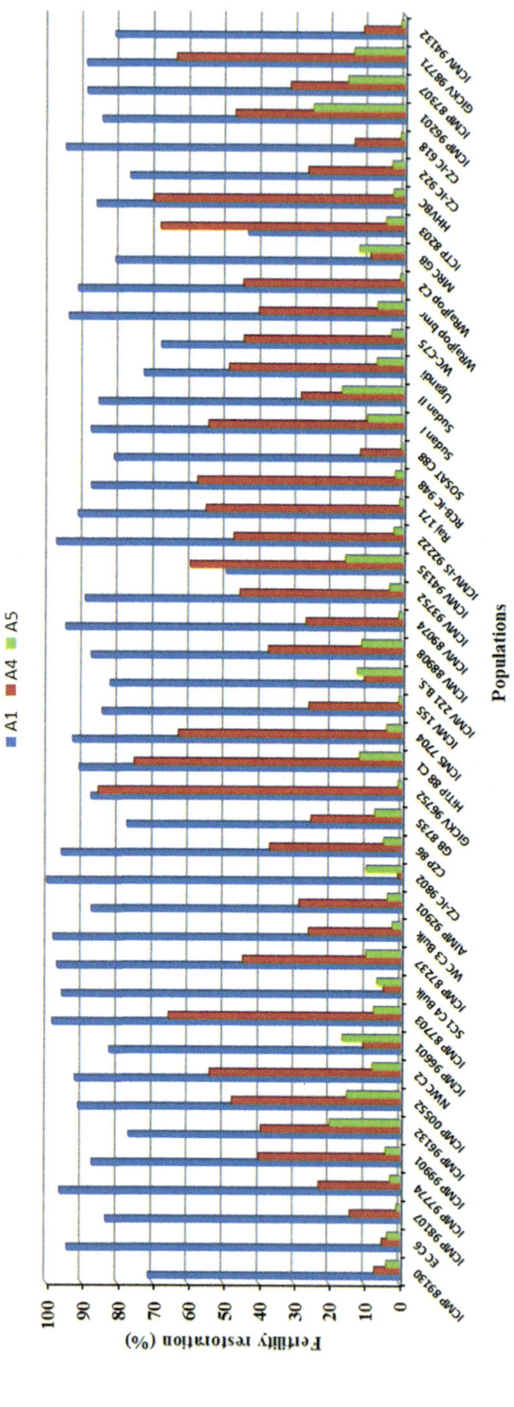

Fig. 4.1 Fertility restoration percentage in topcross hybrids of A_1, A_4, and A_5 made with 45 diverse African and Asian bred/origin pearl millet populations

4.2.3.2 Cytoplasmic Diversification of Seed and Restorer Parents

The public and private sector breeding programs continue using the A1 CMS system, due to the lesser availability of restorer parental lines of the A4 and A5 cytoplasm. There has been an increasing emphasis in different breeding programs on the utilization of the A4 and A5 CMS systems in seed parent's development, with a parallel effort on developing their restorers. For instance, of the 203 designated A-lines developed from 1981 to 2018 at ICRISAT-India, 108 are based on the A1 cytoplasm, 81 on the A4 cytoplasm, and 14 on the A5 cytoplasm. This provides a wide range of genetically diverse A-lines of these two new and most stable CMS systems. A/B pairs can be developed and designated on more than one CMS system, if possible. Out of the 104 A/B pairs developed from 2005 to 2018 at ICRISAT, 32, 41, and 10 were designated only on A_1, A_4, and A_5 CMS systems, respectively; while 7, 2, and 4 A/B pairs were developed simultaneously on 2 CMS systems: A_1 and A_4, A_1 and A_5, and A_4 and A_5, respectively. Four A/B pairs were developed simultaneously on three A_1, A_4, and A_5 CMS systems.

Considering the large availability of restorer lines of A1 CMS across the breeding programs in the public and private sectors and simultaneously to diversify the restorer base, the focus is on the order of A1 followed by A4 and A5 restoration. For instance, out of the 122 restorer parents bred at ICRISAT from 2006 to 2019, 102, 60, and 11 R-lines were restorers of A_1, A_4, and A_5 CMS (Cytoplasmic Male Sterility) systems, respectively (Gupta et al. 2022a, b). Some lines are restorers for more than one CMS system, like 41 out of mentioned 122 lines were dual restorers for A_1 and A_4, and 2 R-lines for both A_1 and A_5 CMS systems. Four R-lines were restorers of all three A_1, A_4, and A_5 systems. Following the same strategy, public and private sector breeding programs in India have diversified their cytoplasmic base and developed a large number of seed and restorer parental lines on different CMS systems.

4.2.4 Breeding Methods

Hybrid parent breeding program targets to develop productive B- and R-lines with combinations of numerous agronomic and adaptation traits. Once this is achieved in B-lines, conversion of B-lines into A-lines is followed through backcross conversion.

4.2.4.1 Line Breeding to Develop B- and R-Lines

For both B- and R-line breeding, the first planting of F2 populations and composites/ OPVs generally can have 500–1000 plants depending upon the diversity of parental lines involved in the cross and on the extent of variation in the population. For crosses or populations found outstanding, bigger populations can be planted in subsequent seasons to make more rigorous selections. Pedigree breeding should be followed generally up to F7/S5 stage after which bulk breeding might be carried out in lines that have attained high levels of uniformity for morphological traits and disease (DM and blast) resistance. At least one DM/blast screening of breeding lines should have been completed by the time the F7/S5 breeding stage is reached. The screening at this stage is done against the pathogen-isolate representing the ecology for which the cross is targeted to develop breeding lines. At each inbreeding stage, about 30–50% of breeding lines not found acceptable on the multiple-trait criteria can be discarded. However, lines found outstanding for even one trait are selected for further use in the crossing program. F5/S3 onward, selected lines can be evaluated in the plantings according to parentage, but highly selected lines from various populations are also classified into trait-specific groups. Inbred lines derived from different crosses can then be crossed to initiate a new cycle of inbreeding and selection for inbred line development.

While developing inbred lines for use as hybrid parents, there are three important considerations: (1) selection for performance per se with respect to yield potential, disease/insect pest resistance, and adaptation, (2) selection for fertility restoration ability in R-line breeding stream and sterility maintenance in B-line breeding stream, and (3) evaluation for combining ability. Selection for performance per se for priority traits (yielding ability, apparent grain quality traits, downy mildew, and blast resistance) in the breeding nurseries can be made at all inbreeding stages. In addition, selection for downy mildew or blast resistance using the greenhouse inoculation technique can be made at least at two inbreeding stages from F_5 onward. Test for fertility restoration/sterility maintenance reaction should be conducted at F_5/F_6 generation when the lines become reasonably uniform.

4.2.4.2 Selection for DM and Blast Resistance

Periodic evaluation of progenies for DM and blast resistance during the course of inbreeding and selection should run concurrent to agronomic evaluation to ensure that lines (B-lines and restorer lines [R-lines]) produced finally are resistant to this disease. Diverse pathotypes should be maintained in the greenhouse screening facilities. Results have shown that an un-replicated single-pot screening of breeding lines (approximately 30–40 seedlings per line) is highly effective for rejecting the susceptible lines, as the correlation between the single-pot screen and two-pot screen for DM incidence is highly significant ($r = 0.75^{**}$) (Rai et al. 2004).

Trait-specific breeding lines should be evaluated for DM and blast resistance against pathotypes from the region for which the lines are targeted. However, breeding lines in some trait-specific groups (e.g., early maturity, medium seed size, and average tillering with long panicles) should be evaluated in successive steps against more than one diverse pathotype because of the wider requirements of such materials.

The same trait-based selection approaches combined with high yield potential, DM, and blast resistance have also been followed for restorer line breeding as for the A-line breeding. One major difference is that in restorer breeding, there is a negligible effort on breeding dwarf R-lines. For DM and blast resistance, the same procedure is applied as in seed parent breeding.

4.2.4.3 Testing for Combining Ability

Early-generation testing has been recommended to select progenies for high general combining ability. B- and R-breeding lines at $F_{4/5}$ or $S_{2/3}$ stage are crossed to the tester(s) (from the opposite side of seed and restorer pool and identified for specific target ecology) in line × tester design, and test crosses should be evaluated in target ecology. Based on the multilocation evaluation of test crosses, lines with good GCA are identified. B-lines should be tested for GCA before they are used to develop A-lines to prevent unnecessary backcrossing efforts. The identified good combining B-lines are essentially converted to A-lines, while B-lines having high performance per se are also converted to A-lines. Selection is practiced for performance per se during inbreeding generations, and it has been observed in pearl millet that the correlation between performance per se of the lines and their general combining ability for grain yield is either non-significant or positive and significant (Rai and Virk 1999; Singh and Gupta 2019; Gupta et al. 2020). This would imply that high general combiners are more likely to be found in a high-yielding group of lines than in the low or average-yielding groups. Further, the high-yielding ability of parental lines, especially A-lines, is very important from the viewpoint of the seed production economy.

4.2.4.4 A-Line Development

Inheritance of male sterility maintenance, whether simple or complex, affects the efficiency and cost-effectiveness of the conversion process: the simpler the inheritance, the greater the selection response. A_1 CMS system is reported to be under single-gene control of male sterility maintenance and fertility restoration, but simultaneously the possibility of two- and three-gene models governing this trait is also not ruled out (Yadav et al. 2010). Also, dominant single-gene control of fertility restoration has been revealed for A_4 fertility restoration (Gupta et al. 2012), and trigenic inheritance of male fertility restoration was suggested for A5 CMS, where dominant alleles at any two of the three duplicate complimentary loci will lead to fertility restoration (Gupta et al. 2018).

Conversion Stage

Conversion of potential B-lines into A-lines can be initiated once it has been ascertained that they meet the multiple criteria of yield potential, agronomic traits, DM resistance, and high levels of morphological uniformity. Normally, at least three generations of progeny-based evaluations (i.e., at least up to F5/S3) must precede before undertaking them for conversion into A-lines, which translates to about 2 years × two seasons of prior field evaluation on a plot basis. In exceptional cases (either exceptional performance or any urgency to save critical time), a progeny with outstanding performance can be put into an A-line conversion scheme at an earlier inbreeding stage.

Conversion Method

Since highly uniform B-lines are used for conversion, crosses and backcrosses are made on a line basis rather than on an individual plant basis. Following this method, it has been observed that there is neither significant variability within B-lines for agronomic and adaptation traits nor any variability within the F1 and backcross progenies for male sterility, thus eliminating any need for within-plot selection. This method of conversion of B-lines into A-lines also reduces unnecessary work needed while dealing with sister A/B pairs in the plant × plant backcross conversion scheme and allows for the maintenance of greater diversity for the equivalent resources used in the plant × plant crossing scheme. Even with the use of advanced generation B-lines for conversion, lines found to have produced fully sterile F1s sometimes produce backcross progenies with low frequency of pollen shedders. First, such cases are very rare. Second, unless the B-line is truly outstanding, plant × plant crossing in backcross progenies is not done, and those backcross/B-line pairs are eliminated from the crossing program. Four to six backcrosses may be adequate if the new B-line and the cytoplasm donor CMS line are closely related, while up to eight backcrosses may be needed for more diverse lines.

Selection During the Conversion Process

During the conversion process, selection continues, but more among the B-lines than within the B-lines, especially during the advanced backcross stages. Generally, there can be about 20% rejection of the B-lines and their corresponding F_1 and backcross progenies at each backcross stage. Each plant in the F_1 and backcross progenies is also selfed (using the tiller panicle) to score for and confirm complete sterility. One or two generations of backcross time can be saved by selecting plants in the backcross progenies that have a greater resemblance to the basic plant morphology of the recurrent parent (B-line). This, however, would require planting larger plots of the backcross progeny (say 80–100 plants rather than the customary 20–25 plants) to make the selection more effective.

4.2.4.5 Improving Adaptation and Nutritional Traits

Drought and high heat stress challenge pearl millet in various production environments. While drought affects the crop across agro-ecologies, heat stress assumes importance in specific regions. Drought resulting from low rainfall or due to its erratic distribution is the primary abiotic production constraint. Drought affects the grain-filling stage the most, as both grain number and grain size are significantly reduced when the crop is exposed to drought stress at this stage (Fussell et al. 1991). Dissection of drought tolerance in terms of physiology, phenology, and morphology of the crop has led to the understanding of the yield formation process under drought (Van Oosterom et al. 2003; Yadav 2011), helping breeders to identify and target specific traits in different drought environments. It is very challenging to use physiological traits as selection criteria for drought tolerance, especially when dealing with many genotypes in breeding nurseries. Drought escape has been the most exploited mechanism by targeting early maturity for getting greater genetic gains in drought-prone regions (Yadav et al. 2011). Morphological traits such as high tillering, small grain size, shorter grain-filling periods, and high panicle harvest index are the most desirable and scorable traits that can be measured easily and have been manipulated successfully in breeding programs as there is an abundant variation available for these traits (Yadav et al. 2017).

Adapted landraces with elite genetic backgrounds should be involved in the hybridization to create new gene combinations that can lead to amalgamating of adaptation to stress environments and high productivity (Presterl and Weltzien 2003; Yadav and Rai 2011; Patil et al. 2020). Genome regions underlying drought tolerance-related traits have been identified and mapped (Yadav et al. 2002, 2004; Serba and Yadav 2016), and several such genomic regions are being manipulated to enhance drought tolerance (Bidinger et al. 2007; Sharma et al. 2014).

During the last decade, pearl millet has emerged as a highly productive and remunerative crop in the hot and dry summer season in the northern and western parts of India. With higher air temperatures (often >42 °C) coinciding with flowering in this season, the crop suffers from reproductive sterility, leading to drastic reductions in seed set and finally lesser grain yield (Gupta et al. 2015; Djanaguiraman et al. 2018). Heat tolerance at the reproductive stage has emerged as an important target trait to enhance genetic gains. Flowering-period heat stress screening protocols have been standardized for screening under controlled environment facilities (growth chambers) and field conditions in heat stress-prone target ecology (Gupta et al. 2015). Multilocational and multiyear field screening in the summer season involving a large number of hybrid parental lines, germplasm accessions, and improved populations established that: stigma is more heat-sensitive than pollen; large genetic variation exists between breeding lines and within open-pollinating populations; the boot-leaf stage is more heat-sensitive than a panicle-emergence stage; and heat tolerance behaves as a dominant trait. These screenings led to the identification of heat-tolerant breeding materials and can further be used to enhance heat tolerance (Gupta et al. 2016, 2019) and facilitate the pyramiding of heat tolerance in high-yielding hybrids.

Pearl millet has been considered a highly nutritious cereal with higher levels of proteins and several minerals than other major cereals; hence the goal of core breeding has been focused on increasing the yield potential. About 24.3% protein content has been reported in the germplasm and up to 19.8% in elite breeding lines (Singh et al. 1987). However, no serious efforts were made to improve protein content because of its negative correlations with grain yield (Singh and Nainawatee 1999). Recently, given the global recognition of widespread deficiencies of iron (Fe) and zinc (Zn), these nutritional traits have been added to the breeding objectives. A large variability has been indicated for Fe and Zn content in germplasms and breeding lines, suggesting the feasibility of genetic enhancement for these micronutrients in pearl millet. Screening of more than 120 Indian commercial hybrids has shown 46–56 ppm Fe and 37–44 ppm Zn (Rai et al. 2016). The daily recommended allowances for Indian adults are 17–21 and 10–12 mg/day for Fe and Zn, respectively. Fe and Zn contents in pearl millet are largely governed by additive genetic variance (Govindaraj et al. 2013; Kanatti et al. 2014), suggesting that both parental lines of hybrids would be required to improve for these micronutrients. Relatively lower G × E influences on the accumulation of Fe and Zn in pearl millet grains (Kanatti et al. 2014; Govindaraj et al. 2016) also indicated the effectiveness of progeny selection in the pedigree breeding to develop lines with increased levels of grain Fe and Zn densities. These two micronutrients are highly positively correlated to each other, and also have a positive and highly significant correlation with seed size (Gupta et al. 2009; Kanatti et al. 2014; Govindaraj et al. 2016). These associations would give breeders an opportunity to develop Fe- and Zn-rich cultivars with large grain size irrespective of their color and to allow enhancement of micronutrients in mainstream breeding.

4.3 Enhancing Magnitude of Heterosis

4.3.1 Heterotic Groups

A narrow genetic base is one of the major constraints to enhance the genetic gains for most of the hybrid breeding programs. To address this, a systematic assessment of genetic diversity (by phenotyping and/or genotyping) to develop heterotic groups is needed to develop hybrids with higher standard heterosis. The fundamental principle for the exploitation of heterosis is to characterize the germplasm into different heterotic groups, which helps breeders to develop inbred lines and use the available germplasm more efficiently for maximizing hybrid breeding outcomes (Melchinger and Gumber 1998). In pearl millet, a diverse range of breeding material has historically been used to develop either seed parents (B-lines) or restorer parents (R-lines), depending upon their specific phenotypic traits (Rai et al. 2006). Studies assessing molecular diversity classified such lines into genetically distinct groups and confirmed the existence of two broad-based pools in hybrid parents—one for seed parents and another for restorer parents (Nepolean et al. 2012; Gupta et al. 2015; Singh et al. 2018). The existence of B- and R-lines as separate groups has been found responsible for behaving as two separate broad heterotic pools, as B × R hybrids reported significantly higher levels of heterosis than B × B or R × R hybrids (Singh et al. 2018). Another study involving 320 R- and 260 B-lines derived from six major pearl millet breeding programs in India identified heterotic groups based on the heterotic performance and combining ability (Gupta et al. 2020) (Fig. 4.2). Hybrids

a

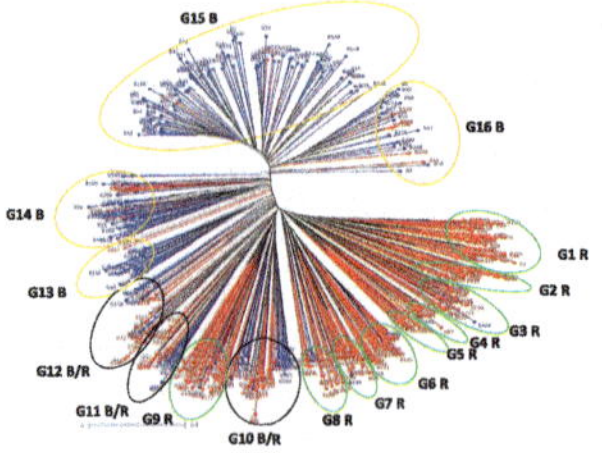

b

Sl.No.	Cross combination of marker groups	Genetic distance	Hybrid Yield (kg ha^{-1})	MPH (%)	BPH (%)	Yield advantage over ProAgro 9444 (%)	Yield advantage over Pioneer 86M86 (%)
1	G12B x G12R	0.25	4507.4	152.5	70.9	29.2	10.6
2	G10B x G12R	0.31	4456.1	96.0	64.1	27.7	9.3
3	G13B x G12R	0.31	4434.3	116.7	68.1	27.1	8.8

Fig. 4.2 (**a**) Clustering pattern based on restricted site-associated DNA (RAD) sequencing of 540 B- and R-lines; (**b**) identified heterotic groups. (Source: Gupta et al. 2020)

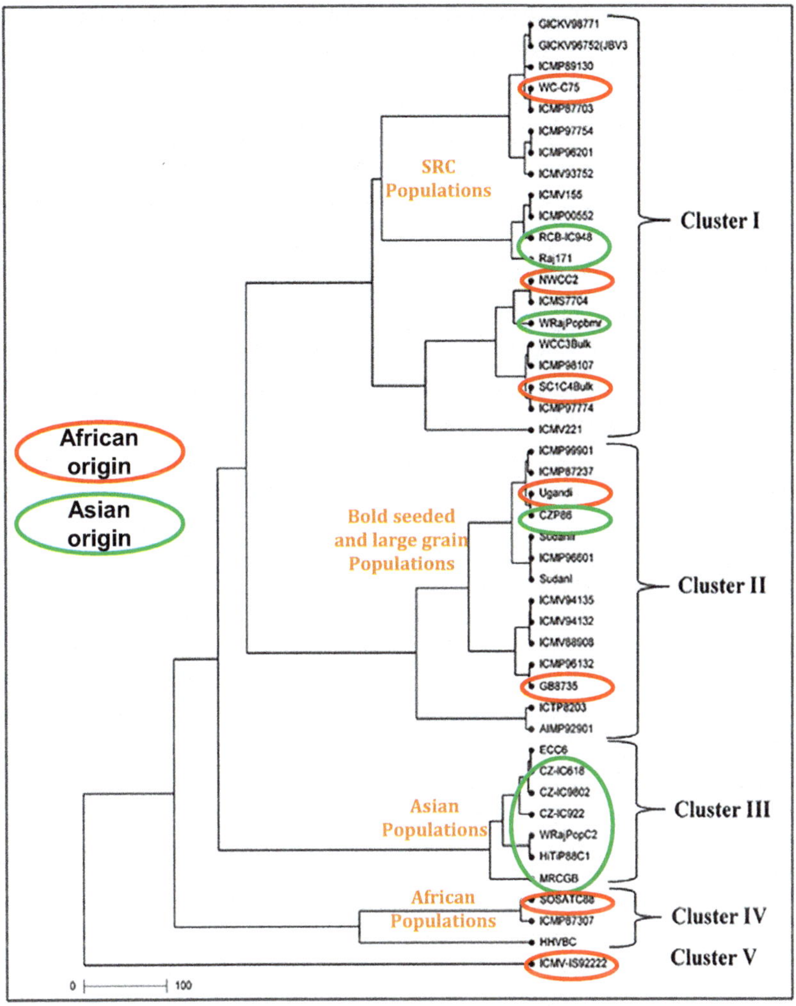

Fig. 4.3 Population heterotic pools formed among 45 African and Asian origin/bred pearl millet improved populations

from these identified B × R heterotic groups showed grain yield heterosis of more than 10% over the best available commercial hybrid checks. This study also indicated that distinct parental groups could be formed based on molecular markers, which can help in assigning the hybrid parental lines into heterotic groups to develop high-yielding hybrids. Now, appropriate testers to categorize the new hybrid parental lines or new germplasm into these heterotic groups have been identified to further enhance the heterosis (Unpublished PhD. Thesis 2022, P. Rakshith, ICRISAT). Moving beyond the existing variability in available inbreds, as shown in Fig. 4.3, heterotic pools have been formed among African and Asian origin/bred pearl millet improved populations to help breeders enhance genetic diversity in the future hybrid parental lines (Patil et al. 2020). New populations can be identified for developing heterotic parental lines using these results.

Most public and private sector pearl millet breeding programs have been advancing progenies based on performance per se of lines to develop high-yielding seed parental lines. Hence, there is not much information available about the extent of specific combining ability (SCA) and GCA variances in the existing B- and R-gene pools of pearl millet. A study has shown a high SCA:GCA ratio (about two times), indicating the predominance of SCA variance over GCA variance in pearl millet hybrid parents, which is quite different from many of the established maize hybrid breeding programs of the USA and Europe, where low SCA:GCA ratio was observed (Gupta et al. 2020). Higher values of GCA or high ratio of GCA:SCA enable breeders to predict the breeding value of their lines and hybrids with higher accuracy. Considering this scenario in other crops, there is a need to investigate SCA and GCA variances in the existing B- and R-line heterotic pools of pearl millet to better understand the contribution of GCA and SCA variances towards heterosis.

4.3.2 Predicting Heterosis

Identifying promising hybrids requires generating and testing large numbers of hybrids under different field conditions. At first, performing crosses to generate F1 hybrids is a labor-intensive process, and further evaluating hybrids across field trials under several environments requires significant time and resources. Hence, methods for selecting parental inbred lines and determining which crosses are likely to yield the best hybrids are a critical part of crop improvement. Traditionally, mid-parental values have been a common way to predict the performance of hybrids based on inbred values, combined with estimates of general combining ability (GCA) (Gowda et al. 2013; Xing et al. 2014; Mühleisen et al. 2015). However, for traits where significant heterosis exists, the phenotypes of hybrids can vary significantly from what would be predicted using mid-parent values and estimated GCA. In these cases, it can be necessary to estimate specific combining ability (SCA) values for each potential cross. The incorporation of genetic markers can improve the accuracy with which both GCA and SCA can be predicted by enabling the sharing of data across multiple tested lines carrying common haplotypes (Schrag et al. 2007). When applied to sets of genetic markers across the whole genome, this process is referred to as genomic prediction (GP), which can be used to implement breeding programs based on estimated breeding values from genome-wide sets of markers, a process known as genomic selection (GS). In a study on pearl millet, Liang et al. (2018), based on the data on inbreds and hybrids, reported prediction accuracy ranging from 0.73 to 0.74 for 1000-grain weight, 0.87 to 0.89 for days to flowering time, 0.48 to 0.51 for grain yield, and 0.72 to 0.73 for plant height. This study also indicated that even small numbers of selected single nucleotide polymorphisms (SNPs) can achieve relatively high prediction accuracy in the pearl millet population. The implementation of a hybrid GS/GP-guided pearl millet breeding program has the potential to improve the efficiency of breeding efforts significantly.

4.3.3 Molecular Breeding

Genomics and genomics-assisted breeding can help improve the precision and efficiency of the hybrid breeding program. Several efforts like sequencing of Pearl Millet Inbred Germplasm Association Panel, mapping population parents, and elite hybrid parental lines have led to the development of a huge (>32 million) repository of genome-wide SNPs, which can help in trait discovery, mapping, and deployment of quantitative trait loci (QTLs)/alleles/candidate genes linked to traits of economic interests. It also has helped with the development and implementation of prediction models for the pearl millet (Jarquin et al. 2020; Liang et al. 2018). Now, the need is to map and deploy genes of agronomic importance and re-sequence lines to mine and map genes of interest, and superior haplotypes for different traits can be identified to form the basis of haplotype-based breeding (Sinha et al. 2020). Many traits of agronomic importance have been mapped related to diseases, terminal drought, grain and fodder quality, combining ability loci, and heterotic gene pools (Kumar et al. 2018; Basava et al. 2019; Gupta et al. 2020; Srivastava et al. 2020). With simple sequence repeats (SSR) and SNP markers in pearl millet, these QTLs can be remapped using the currently available SNP-based high-throughput genotyping systems. This will allow integration into the modern breeding pipelines using high-throughput genotyping platforms available currently in pearl millet. Furthermore, it has become possible now to map breeding-related traits quickly with the availability of reference genome sequence, large-scale whole genome resequencing data, cost-effective genotyping platforms, and precise phenotyping platforms (Bohra et al. 2020). We believe the near future will witness the deployment of genomic breeding approaches such as haplotype-based breeding, forward breeding, genomic prediction, and gene editing for pearl millet improvement (Varshney et al. 2020).

4.4 Hybrid Breeding for Different Regions

Apart from India, many countries are now working toward the development of pearl millet hybrids, and hybrids have been released in a few countries. Grain and stover yield, and DM resistance are high-priority traits in both African regions (Eastern and Southern Africa [ESA], and West and Central Africa [WCA]). The parasitic weed *Striga hermonthica*, stem borer (*Coniesta ignefusalis*), and head miner (*Heliocheilus albipunctella*) are the additional biotic constraints in the WCA region, and thus need to be addressed.

Hybrids bred at ICRISAT-India have shown high yields and adaptation in the ESA (Eastern and Southern Africa) region; hence, India bred materials could be of direct and indirect use for hybrid breeding programs in the ESA region, and vice-versa to some extent. This is not true in the WCA region, so the breeding materials and parental lines developed in India will be of no direct use in the WCA region. African pearl millet programs will have to develop in-house potentially useful breeding lines and potential hybrid parents and test for their utility in the target regions. The initial thrust in African regions should be placed on addressing relatively better-endowed environments with fewer high-priority traits, considering all the factors that increase the probability of success in the medium term. Since hybrid cultivar programs of large size are not likely to be in place in the African regions soon, the frequent cultivar replacement strategy for DM management is not a likely option. The strategy for this region should be to strengthen the DM resistance breeding program (supported by marker-assisted selection [MAS]) and implement alternative hybrid options.

In the case of forage hybrids, which are quite popular in the USA, Brazil, and some central Asian countries, the use of A5 CMS should help to develop a range of seed parents due to its high maintenance potential in any genetic background, and also that crop homogeneity and fertility restoration is not required in forage hybrids.

4.5 Conclusions

Continuous enhancement of genetic diversity in hybrid parental lines has been the key driver of success behind pearl millet hybrid breeding in India. Strategic involvement of diverse germplasms to develop seed and restorer parental lines can help maintain sufficient genetic distance between seed and restorer parental lines to enhance heterosis. Cytoplasmic diversification is critical to broaden the genetic base of germplasms involved in hybrid breeding, while simultaneously providing protection to hybrids from the risk of cytoplasmic breakdown to the future virulent variants of pests and diseases. The breeding pipeline needs to be integrated with robust screening protocols for biotic and abiotic stresses to develop climatic resilient hybrids. Nutritional traits should be mainstreamed into the breeding pipeline to develop not only productive but also nutritionally rich hybrids. New tools should be tested rigorously and adopted in breeding programs to introgress new desirable alleles and also to predict heterosis to support the new-generation breeders. New hybrids would offer a robust solution to climate change by expanding pearl millet in new regions globally. To achieve this, the genetic gains in pearl millet can be further enhanced with the judicious use of conventional and modern tools.

References

Andrews DJ, Kumar KA (1996) Use of the West African pearl millet landrace Iniadi in cultivar development. Plant Genet Resourc Newsl 105:15–22

Athwal DS (1965) Hybrid bajra-1 marks a new era. Indian Farming 15(5):6–7

Basava RK, Hash CT, Mahendrakar MD, Kishor PBK, Satyavathi CT, Kumar S, Singh RB, Yadav RS, Gupta R, Srivastava RK (2019) Discerning combining ability loci for divergent environments using chromosome segment substitution lines (CSSLs) in pearl millet. PLoS One 14(8):e0218916

Bidinger FR, Nepolean T, Hash CT, Yadav RS, Howarth CJ (2007) Quantitative trait loci for grain yield in pearl millet under variable postflowering moisture conditions. Crop Sci 47(3):969–980

Bohra A, Chand Jha U, Godwin ID, Kumar Varshney R (2020) Genomic interventions for sustainable agriculture. Plant Biotechnol J 18(12):2388–2405

Djanaguiraman M, Perumal R, Ciampitti IA, Gupta SK, Prasad PVV (2018) Quantifying pearl millet response to high temperature stress: thresholds, sensitive stages, genetic variability and relative sensitivity of pollen and pistil. Plant Cell Environ 41(5):993–1007

Fussell LK, Bidinger FR, Bieler P (1991) Crop physiology and breeding for drought tolerance: research and development. Field Crop Res 27(3):183–199

Govindaraj M, Rai KN, Shanmugasundaram P, Dwivedi SL, Sahrawat KL, Muthaiah AR, Rao AS (2013) Combining ability and heterosis for grain iron and zinc densities in pearl millet. Crop Sci 53(2):507–517

Govindaraj M, Rai KN, Shanmugasundaram P (2016) Intra-population genetic variance for grain iron and zinc contents and agronomic traits in pearl millet. Crop J 4(1):48–54

Gowda M, Zhao Y, Maurer HP, Weissmann EA, Würschum T, Reif JC (2013) Best linear unbiased prediction of triticale hybrid performance. Euphytica 191(2):223–230

Gupta SK, Velu G, Rai KN, Sumalini K (2009) Association of grain iron and zinc content with grain yield and other traits in pearl millet (Pennisetum glaucum (L.) R. Br.). Crop Improv 6:4–7

Gupta SK, Rai KN, Kumar MS (2010) Effect of genetic background on fertility restoration of pearl millet hybrids based on three diverse cytoplasmic-nuclear male-sterility systems. J SAT Agric Res 8:1–4

Gupta SK, Rai KN, Govindaraj M, Rao AS (2012) Genetics of fertility restoration of the A_4 cytoplasmic-nuclear male sterility system in pearl millet. Czech J Genet Plant Breed 48(2): 87–92

Gupta SK, Rathore A, Yadav OP, Rai KN, Khairwal IS, Rajpurohit BS, Das RR (2013) Identifying mega-environments and essential test locations for pearl millet cultivar selection in India. Crop Sci 53(6):2444–2453

Gupta SK, Nepolean T, Sankar SM, Rathore A, Das RR, Rai KN, Hash CT (2015) Patterns of molecular diversity in current and previously developed hybrid parents of pearl millet [*Pennisetum glaucum* (L.) R. Br.]. Am J Plant Sci 6(11):1697–1712

Gupta SK, Ameta VL, Pareek S, Mahala RS, Jayalekha AK, Deora VS, Verma YS, Boratkar B, Atkari D, Rai KN (2016) Genetic enhancement for flowering period heat tolerance in peart millet (*Pennisetum glaucum* L. (R.) Br.). In: 7th International crop science congress (Beijing)

Gupta SK, Yadav DV, Govindaraj M, Boratkar M, Kulkarni VN, Rai KN (2018) Inheritance of fertility restoration of A_5 cytoplasmic-nuclear male sterility system in pearl millet [*Pennisetum glaucum* (L.) R. Br.]. Indian J Genet Plant Breed (TSI) 78(2):228–232

Gupta SK, Ameta VL, Pareek S, Singh I, Deora VS, Verma YS et al (2019) Enhancing flowering period heat tolerance in pearl millet through shuttle breeding. In: Presented at 3rd agriculture and climate change conference (Budapest)

Gupta SK, Patil KS, Rathore A, Yadav DV, Sharma LD, Mungra KD et al (2020) Identification of heterotic groups in South-Asian-bred hybrid parents of pearl millet. Theor Appl Genet 133(3): 873–888

Gupta SK, Patil S, Boratkar M, Pujar M (2022a) Morphological characteristics of ICRISAT-bred pearl millet hybrid seed parents (2005–2018). Technical Report. ICRISAT, Patancheru, Telangana, India

Gupta SK, Patil S, Kannan RL, Pujar M, Kumar P (2022b) Characterization of ICRISAT bred pearl millet restorer parents (2006–2019). Technical Report. ICRISAT, Patancheru, Telangana, India

Jarquin D, Howard R, Liang Z, Gupta SK, Schnable JC, Crossa J (2020) Enhancing hybrid prediction in pearl millet using genomic and/or multi-environment phenotypic information of inbreds. Front Genet 10:1294

Kanatti A, Rai KN, Radhika K, Govindaraj M, Sahrawat KL, Rao AS (2014) Grain iron and zinc density in pearl millet: combining ability, heterosis and association with grain yield and grain size. Springerplus 3(1):1–12

Kumar S, Hash CT, Nepolean T, Mahendrakar MD, Satyavathi CT, Singh G, Rathore A, Yadav RS, Gupta R, Srivastava RK (2018) Mapping grain iron and zinc content quantitative trait loci in an Iniadi-derived immortal population of pearl millet. Genes 9(5):248

Liang Z, Gupta SK, Yeh CT, Zhang Y, Ngu DW, Kumar R, Patil HT, Mungra KD, Yadav DV, Rathore A, Srivastava RK, Gupta R, Yang J, Varshney RK, Schnable PS, Schnable JC (2018) Phenotypic data from inbred parents can improve genomic prediction in pearl millet hybrids. G3 8(7):2513–2522

Melchinger AE, Gumber RK (1998) Overview of heterosis and heterotic groups in agronomic crops. In: Concepts and breeding of heterosis in crop plants, vol 25, pp 29–44

Mühleisen J, Piepho HP, Maurer HP, Reif JC (2015) Yield performance and stability of CMS-based triticale hybrids. Theor Appl Genet 128(2):291–301

Nepolean T, Gupta SK, Dwivedi SL, Bhattacharjee R, Rai KN, Hash CT (2012) Genetic diversity in maintainer and restorer lines of pearl millet. Crop Sci 52(6):2555–2563

Patil KS, Gupta SK, Marathi B, Danam S, Thatikunta R, Rathore A, Das RR, Kuldeep SD, Yadav OP (2020) African and Asian origin pearl millet populations: genetic diversity pattern and its association with yield heterosis. Crop Sci 60(6):3035–3048

Presterl T, Weltzien E (2003) Exploiting heterosis in pearl millet for population breeding in arid environments. Crop Sci 43(3):767–776

Rai KN, Rao AS (2003) ICMR 98001: a restorer stock of A5 cytoplasmic-nuclear male sterility in pearl millet. Int Sorghum Millets Newsl 44:136

Rai KN, Virk DS (1999) In: Khairwal IS, Rai KN, Andrews DJ, Harinarayana G (eds) Breeding methods in pearl millet breeding. Oxford & IBH Publishing Co., New Delhi, pp 185–212

Rai KN, Anand Kumar K, Andrews DJ, Rao AS (2001) Commercial viability of alternative cytoplasmic-nuclear male-sterility systems in pearl millet. Euphytica 121(1):107–114

Rai KN, Kulkarni VN, Thakur RP, Singh AK, Rao VP (2004) Effectiveness of within-progeny selection for downy mildew resistance in pearl millet. Int Sorghum Millets Newsl 45:45–47

Rai KN, Kulkarni VN, Thakur RP, Haussmann BIG, Mgonja MA (2006) Pearl millet hybrid parents research: approaches and achievements. In: Gowda CLL, Rai KN (eds) Potential welfare benefit of millets improvement research at ICRISAT: multi country-economic surplus model approach. International Crops Research Institute for the Semi-Arid Tropics, Patancheru. 212 pp. ISBN 92-9066-489-4; Order code BOE 039

Rai KN, Gupta SK, Bhattachajee R, Kulkarni VN, Singh AK, Rao AS (2009a) Morphological characteristics of ICRISAT-bred pearl millet hybrid parents. In: International crops research for the semi-arid tropics (Patancheru), 176

Rai KN, Khairwal IS, Dangaria CJ, Singh AK, Rao AS (2009b) Seed parent breeding efficiency of three diverse cytoplasmic-nuclear male-sterility systems in pearl millet. Euphytica 165:495–507

Rai KN, Yadav OP, Govindaraj M, Pfeiffer WH, Yadav HP, Rajpurohit BS, Patil HT, Kanatti A, Rathore A, Rao AS, Shivade H (2016) Grain iron and zinc densities in released and commercial cultivars of pearl millet (*Pennisetum glaucum*). Indian J Agric Sci 86(03):11–16

Schrag TA, Maurer HP, Melchinger AE, Piepho HP, Peleman J, Frisch M (2007) Prediction of single-cross hybrid performance in maize using haplotype blocks associated with QTL for grain yield. Theor Appl Genet 114(8):1345–1355

Serba DD, Yadav RS (2016) Genomic tools in pearl millet breeding for drought tolerance: status and prospects. Front Plant Sci 7:1724

Sharma PC, Singh D, Sehgal D, Singh G, Hash CT, Yadav RS (2014) Further evidence that a terminal drought tolerance QTL of pearl millet is associated with reduced salt uptake. Environ Exp Bot 102:48–57

Singh S, Gupta SK (2019) Formation of heterotic pools and understanding relationship between molecular divergence and heterosis in pearl millet [*Pennisetum glaucum* (L.) R. Br.]. PLoS One 14(5):e0207463

Singh F, Nainawatee HS (1999) Grain quality traits. In: Khairwal IS, Rai KN, Andrews DJ, Harinarayana G (eds) Pearl millet breeding. Oxford and IBH, New Delhi, pp 157–183

Singh P, Singh U, Eggum BO, Kumar KA, Andrews DJ (1987) Nutritional evaluation of high protein genotypes of pearl millet (*Pennisetum americanum* (L.) Leeke). J Sci Food Agric 38(1): 41–48

Singh S, Gupta SK, Thudi M, Das RR, Vemula A, Garg V, Varshney RK, Rathore A, Pahuja SK, Yadav DV (2018) Genetic diversity patterns and heterosis prediction based on SSRs and SNPs in hybrid parents of pearl millet. Crop Sci 58(6):2379–2390

Sinha P, Singh VK, Saxena RK, Khan AW, Abbai R, Chitikineni A, Desai A, Molla J, Upadhyaya HD, Kumar A, Varshney RK (2020) Superior haplotypes for haplotype-based breeding for drought tolerance in pigeonpea (Cajanus cajan L.). Plant Biotechnol J 18(12):2482–2490

Srivastava RK, Singh RB, Pujarula VL, Bollam S, Pusuluri M, Chellapilla TS, Yadav RS, Gupta R (2020) Genome-wide association studies and genomic selection in pearl millet: advances and prospects. Front Genet 10:1389

Van Oosterom EJ, Bidinger FR, Weltzien ER (2003) A yield architecture framework to explain adaptation of pearl millet to environmental stress. Field Crop Res 80(1):33–56

Varshney RK, Sinha P, Singh VK, Kumar A, Zhang Q, Bennetzen JL (2020) 5Gs for crop genetic improvement. Curr Opin Plant Biol 56:190–196

Xing N, Fan C, Zhou Y (2014) Parental selection of hybrid breeding based on maternal and paternal inheritance of traits in rapeseed (Brassica napus L.). PLoS One 9(7):e103165

Yadav OP (2011) Breeding dual-purpose pearl millet (Pennisetum glaucum) for north western India: understanding association of biomass and phenotypic traits. Indian J Agric Sci 81:816–820

Yadav OP, Rai KN (2011) Hybridization of Indian landraces and African elite composites of pearl millet results in biomass and stover yield improvement under arid zone conditions. Crop Sci 51(5):1980–1987

Yadav OP, Rai KN (2013) Genetic improvement of pearl millet in India. Agric Res 2(4):275–292

Yadav RS, Hash CT, Bidinger FR, Cavan GP, Howarth CJ (2002) Quantitative trait loci associated with traits determining grain and stover yield in pearl millet under terminal drought-stress conditions. Theor Appl Genet 104(1):67–83

Yadav RS, Hash CT, Bidinger FR, Devos KM, Howarth CJ (2004) Genomic regions associated with grain yield and aspects of post-flowering drought tolerance in pearl millet across stress environments and tester background. Euphytica 136(3):265–277

Yadav D, Gupta SK, Kulkarni VN, Rai KN, Behl RK (2010) Inheritance of A1 system of cytoplasmic-nuclear male sterility in pearl millet [*Pennisetum glaucum* (L). R. Br.]. Cereal Res Commun 38(2):285–293

Yadav OP, Rai KN, Khairwal IS, Rajpurohit BS, Mahala RS (2011) Breeding pearl millet for arid zone of North-Western India: constraints, opportunities and approaches. All India coordinated pearl millet improvement project, Jodhpur, p 28

Yadav OP, Rajpurohit BS, Kherwa GR, Kumar A (2012a) Prospects of enhancing pearl millet (Pennisetum glaucum) productivity under drought environments of North-Western India through hybrids. Indian J Genet Plant Breed 72(1):25–30

Yadav OP, Rai KN, Rajpurohit BS, Hash CT, Mahala RS, Gupta SK, Shetty HS, Bishnoi HR, Rathore MS, Kumar A, Sehgal S, Raghvani KL (2012b) Twenty-five years of pearl millet improvement in India. ICAR

Yadav OP, Rai KN, Bidinger FR, Gupta SK, Rajpurohit BS, Bhatnagar SK (2012c) Pearl millet (Pennisetum glaucum) restorer lines for breeding dual-purpose hybrids adapted to arid environments. Indian J Agric Sci 82:922–927

Yadav OP, Upadhyaya HD, Reddy KN, Jukanti AK, Pandey S, Tyagi RK (2017) Genetic resources of pearl millet: status and utilization. Indian J Plant Genet Resour 30(1):31–47

Yadav OP, Gupta SK, Govindaraj M, Sharma R, Varshney RK, Srivastava RK, Rathore A, Mahala RS (2021) Genetic gains in pearl millet in India: insights into historic breeding strategies and future perspective. Front Plant Sci 396

Trait Mapping, Marker-Assisted Selection, and Introgression Breeding in Pearl Millet

5

Divya Ramesh, Swati Singh, Priya Shah, Animikha Chakraborty, Tara C. Satyavathi, and Nepolean Thirunavukkarasu

Abstract

Pearl millet (*Pennisetum glaucum* (l.) R. Br.) is a staple food crop in the arid and semi-arid tropics that is gaining importance as a climate-resilient crop. Polygenes control important pearl millet traits, and the improvement of such traits requires the identification of the genomic regions underlying them. The genomic regions associated with the trait of interest are identified using molecular markers through quantitative trait locus (QTL) mapping and association studies. The results obtained from mapping studies are further improved by validation through fine mapping, gene expression studies, and using alternate mapping populations. The identified genomic regions are utilized in marker-assisted approaches such as marker-assisted backcross breeding, marker-assisted gene pyramiding, and marker-assisted recurrent selections. The marker-aided approaches can be integrated with other techniques, such as speed breeding and doubled haploid technology, to increase efficiency and reduce the breeding cycle time. Basic knowledge about the prerequisites for marker-assisted breeding, establishing marker-trait associations, QTL validation, and marker-assisted breeding approaches are reviewed.

Keywords

QTL · Mapping · Marker-assisted selection · Mapping population

D. Ramesh · S. Singh · P. Shah · A. Chakraborty · T. C. Satyavathi · N. Thirunavukkarasu (✉)
ICAR- Indian Institute of Millets Research, Hyderabad, Telangana, India
e-mail: nepolean@millets.res.in

V. A. Tonapi et al. (eds.), *Pearl Millet in the 21st Century*,
https://doi.org/10.1007/978-981-99-5890-0_5

Abbreviations

DH	Doubled haploid
GEI	Genotype-by-environment interaction
LD	Linkage disequilibrium
LG	Linkage group
MABC	Marker-assisted backcrossing
MARS	Marker-assisted Recurrent selection
MAS	Marker-Assisted selection
MTA	Marker-trait association
NIL	Near-isogenic lines
QTL	Quantitative trait loci
RIL	Recombinant Inbred Line

5.1 Introduction

Marker-assisted selection (MAS) has been developed as a solution to the problems associated with conventional breeding by making use of the genes instead of phenotype for the selection of the trait of interest. Marker-assisted breeding has been able to study both traits: one that is governed by a single gene and the other that are quantitative trait loci (QTLs). The MAS is based on the linkage between the molecular markers and desired trait. The desired quantitative trait locus (QTL) detection is made by the genetic linkage analysis, which is based on the principle of genetic recombination during meiosis. The molecular marker identified should be closely linked (beside or within the target gene) or co-segregate with the gene of interest (Tanksley 1983).

There have been continuous efforts made by plant breeders to develop superior crop varieties to the existing ones. In the conventional breeding methods, the selection of the superior genotypes was made by the phenotypic observation within the segregating progenies generated as the result of crosses. This phenotypic observation lacks accuracy as it does not consider the effect of genotype on environment interactions (G × E). Additionally, this is an expensive, time-consuming procedure.

Pearl millet (*Pennisetum glaucum* (l.) R. Br.) is a highly nutritious crop and ranks sixth among the important cereals. Pearl millet holds great nutritive importance for the population of drylands of sub-Saharan Africa and South Asia (Jukanti et al. 2016). It is a cross-pollinated species with a chromosome number of $2n = 22$ and a genome size of 2450 bp. The cultivation of this hardy crop is affected by various biotic and abiotic stresses. Globally, India is the largest producer of pearl millet in terms of both area and production, which accounts for 6.93 million ha and 8.61 million tons (Directorate of Millet Development 2020).

In India, pearl millet is cultivated in the three agro-ecological zones based on the annual rainfall, which is arid, semi-arid, and hyper-arid. The arid zone (A) is part of northern India and receives an annual rainfall of more than 400 mm. The semi-arid zone (B) consists of northwestern India, also receiving higher than 400 mm annual rainfall, while the hyper-arid (A1) zone includes the area in peninsular India with an annual rainfall of less than 400 mm. Arid and semi-arid zones account for about 75% of the pearl millet cultivation, while the hyper-arid zone contributes to 25%. As per these agro-ecologies, different breeding programs have been carried out in India where the targeted traits were high grain yield, disease resistance, and duration of maturity (75–85 days) (Yadav and Rai 2013).

Pearl millet holds better nutritive value than other cereals such as rice, wheat, maize, and sorghum. It possesses abundant levels of iron, zinc, calcium, lipids, proteins, vitamins, antioxidant, and more balanced essential amino acids and micronutrients. Hence, this makes pearl millet a highly demanding crop for the population belonging to the marginal area for alleviating the hidden hunger found to be commonly prevalent among poor village communities (Rai et al. 2008). Pearl millet holds numerous health benefits such as being gluten-free and having slow-digesting starch. Hence, the successful utility of the MAS for further improvement of this nutritionally relevant crop would enhance food and nutrition security for the population belonging to the agriculturally dissolute environments.

5.2 Marker-Assisted Breeding: A Tool for Crop Improvement

The success of any plant breeding program lies in the precision of selection. Selection based on phenotype per se is not precise because of several factors. The discovery of molecular markers eases the problem. Marker-assisted selection (MAS) is an indirect selection made using the markers (morphological, biochemical, or molecular markers) that are associated with the trait of interest. Plant breeders and breeding companies have widely used this tool in various forms for the generation of improved crop cultivars. The rates of genetic gain attained through the use of genomic tools have been two times higher than the genetic gain by phenotypic selection (Ragot et al. 2007). The success of marker-assisted breeding depends on the available germplasm collection, which represents the genetic diversity of the crop species and the identified important agronomic traits of the germplasm accessions. Availability of the whole genome sequence data has made a way nearer to precise selections. Marker-assisted breeding has several advantages over conventional breeding, like precision in the selection, including rapid, simple, early selection, and single plant selection (Morris et al. 2003). A major constraint in conventional plant breeding is time. It takes 8–10 years to come up with a new variety. Thus, moving from phenotype to genotype to speed up the process was of absolute importance (Lema 2018).

5.3 Understanding the Genetic Basis of Complex Traits

In the early 1900s, Mendel's work in pea explained the discrete phenotypic variations but failed to explain the continuous distribution of several traits. Later this issue was resolved when R. A. Fischer explained that the continuous variation of a few traits is due to the involvement of many small-effect genes (Fisher 1918). Fisher's work laid the foundation for understanding quantitative traits. The traits that are controlled by major genes/qualitative genes are easy to deal with since they are least affected by the factors like epistasis and genotype-by-environment interaction (GEI). These factors have confounded our understanding of complex quantitative traits since the number of genes controlling a trait and environmental complexity increases. These complex scenarios can be considered with the help of several computer simulation models while using marker-assisted selection for breeding such complex traits (Xu and Crouch 2008).

5.4 Prerequisites for a Successful Marker-Assisted Breeding Program

5.4.1 Navigating the Marker Systems

Before the advent of DNA marker technology, the idea of rapidly uncovering the loci controlling complex and multigenic traits seemed like a dream. The availability of molecular markers has played a prominent role in plant breeding in the last few decades. The concept of marker systems started with the utilization of morphological markers/visually distinguishable characters to identify the desirable phenotype (Morgan 1910; Sax 1923). Further biochemical constituents were used as the markers (Rick et al. 1974), whose availability to cover the entire genome was a constraint. The advent of DNA markers during the 1980s was a significant break-through in plant breeding. Several hybridization-based DNA markers were introduced, requiring hybridization with a probe-like restriction fragment length polymorphism (RFLP) (Botstein et al. 1980). With the advent of the polymerase chain reaction (PCR) technique by Kary B. Mullis in 1983 (Mullis 1990), next-generation PCR-based markers like random amplified polymorphic DNA (RAPD) (Williams et al. 1990; Welsh and McClelland 1990), amplified fragment length polymorphism (AFLP) (Vos et al. 1995), and single sequence repeat (SSR) (Litt and Luty 1989) were developed. SSR markers were a marker of choice for genotyping and MAS programs because of their codominant and multi-allelic nature, and ease of use (McCouch et al. 2002). Sequencing technology made it possible to develop specific sequence-based markers to identify polymorphism at the single nucleotide level (Sanger et al. 1977), making marker-assisted breeding cost-effective. Single nucleotide polymorphisms (SNPs) are the best marker system because of several reasons, like having high frequency throughout the genome, being ubiquitous, being amenable to high- and ultra-high-throughput automation (Mammadov et al. 2012), and ability to generate polymorphism due to variation at the single base level. These are the best choice for identifying marker-trait associations (MTAs) since they reveal functional variations as they are mostly developed from genes (Singh et al. 2013).

5.4.2 Marker-Trait Association

Another prerequisite for marker-assisted selection is a strong association between the marker and the gene/QTL controlling the trait of interest. Strong association means the marker and the QTL/gene should be very close to each other (<5 cM) so that no recombination occurs between them (Jaganathan et al. 2020). Marker-trait association (MTA) is significant in a way that a loose association might lead to recombination and ultimately end up with a product without the trait of interest (Prahalada et al. 2017).

Linkage mapping and association mapping are the two ways of establishing marker-trait associations. Both approaches share a common strategy in that, based on the recombination, the genome is broken into pieces that can be correlated to phenotypic variation, with the major difference being the control over recombination (Myles et al. 2009). Association mapping exploits the concept of linkage disequilibrium (LD). LD gets decayed after several rounds of recombination, leaving only markers closely associated with the trait of interest, thus revealing the marker-trait association (Vos et al. 2017). Six MTAs for grain iron, 10 for grain zinc content (Anuradha et al. 2017), 18 MTAs for iron, 43 for zinc, and 17 for protein content (Pujar et al. 2020) were identified in pearl millet through association mapping studies. A linkage map was constructed by genotyping the recombinant inbred lines (RILs) of pearl millet with diversity arrays technology (DArT) and SSR markers containing 321 loci and spanning 1148 cM (Supriya et al. 2011). Five marker-trait associations for three different isolates of pearl millet downy mildew resistance were identified using a linkage-based mapping population (RIL) (Chelpuri et al. 2019). Advantages of association mapping over linkage mapping in detecting MTA include analyzing the genetic variation in a broader genetic background, and that higher resolution, multi-trait phenotypic data can be included (Agrama and Scott 2006). QTLs associated with different pearl millet traits are represented in Figs. 5.1, 5.2, 5.3, 5.4, 5.5, 5.6, and 5.7.

While mapping the QTLs, certain QTLs are located very close to each other and are mapped together. Such co-localized QTLs are useful in simultaneously improving more than one QTL at a time. Co-mapping the QTLs of several traits is either due to pleiotropism or tight linkage between the genes/QTLs. Co-localization of several major genes/QTLs correlated to each other indicates that the QTL is robust and can be considered for marker-assisted breeding. The confidence in such QTLs is improved whenever it is co-mapped in multiple years and environments. Kumar et al. (2016) co-localized two QTLs for Zn and Fe content in pearl millet on Linkage group 3, and the QTLs showed a major effect and were stable across the two tested environments. The co-localization of both the QTLs could be because both nutrients might follow a common pathway right from the uptake to the deposition in the grain

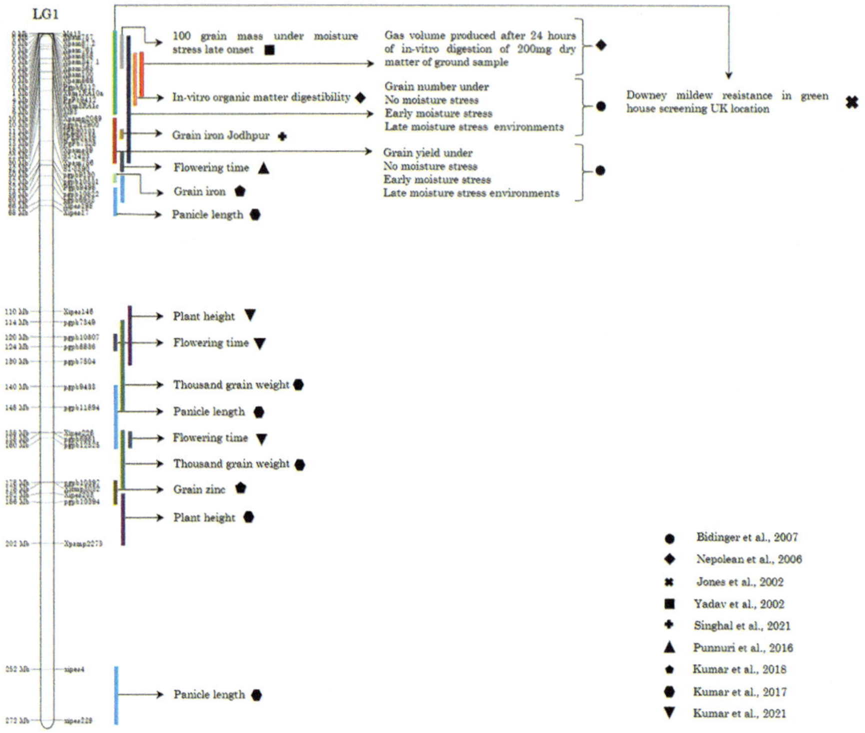

Fig. 5.1 QTL map of pearl millet linkage group 1. Colored bars represent the QTL region for the traits

(Grotz and Guerinot 2006). The association between the ability to maintain the grain and stover yield in terminal drought stress was revealed by analysis of co-mapping of the QTLs controlling the grain and the stover yield and the drought QTLs (Yadav et al. 2002). From the above studies, it can be inferred that such co-localized stable QTLs could be reliable targets for marker-assisted selection. The flow of marker-assisted selection is shown in Fig. 5.8.

5.5 Factors Influencing the Marker-Trait Association

5.5.1 Type of the Mapping Population

Two types of populations can exploit marker-trait associations: (1) linkage-based, and (2) linkage-disequilibrium-based (LD). Linkage-based mapping uses bi-parental mapping populations, which can detect few QTLs since genes from only two parents are available for mapping the QTL. In comparison, LD-based mapping approach has several advantages: detecting a large number of QTLs, analyzing the genetic

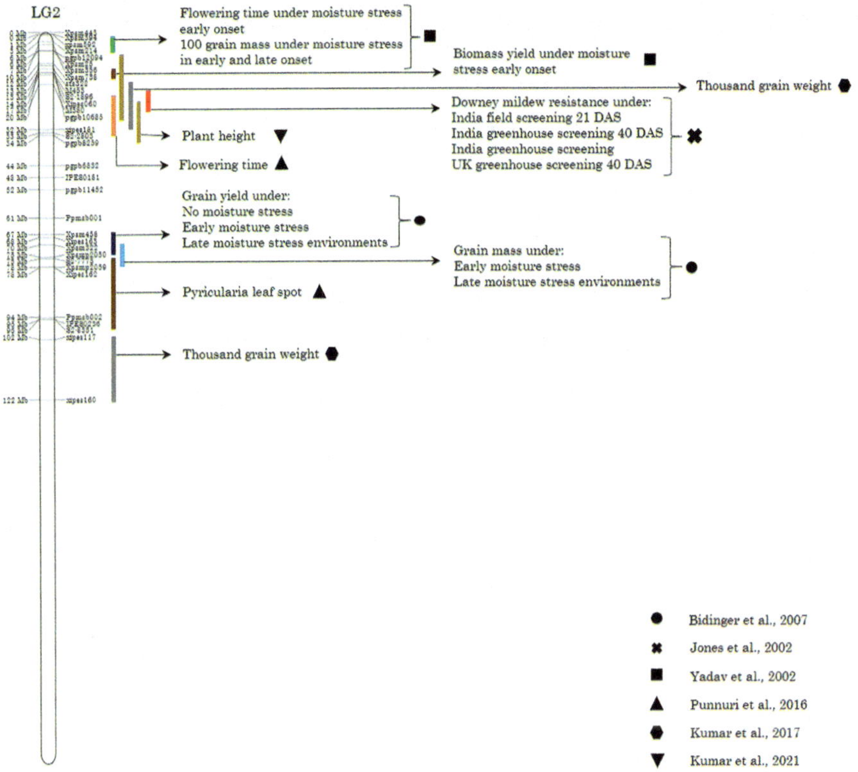

Fig. 5.2 QTL map of pearl millet linkage group 2. Colored bars represent the QTL region for the traits

variation in a broader genetic background, and that higher resolution, multi-trait phenotypic data can be included (Agrama and Scott 2006).

Among different types of linkage-based mapping populations for QTL mapping, recombinant inbred line (RIL) populations can be called the best for obtaining marker-trait associations because they are developed with more rounds of meiosis and recombination cycles (Singhal et al. 2021). Some residual heterozygosity seen in RILs could be overcome using the doubled haploid (DH) populations for mapping (Heun et al. 1991). RILs, near-isogenic lines (NILs), and DHs cannot detect the interaction effects except additive × additive interactions. LD-based mapping populations, nested association mapping populations, are more advantageous than linkage-based mapping populations in terms of QTL detection power and detection of interactions (Li et al. 2011).

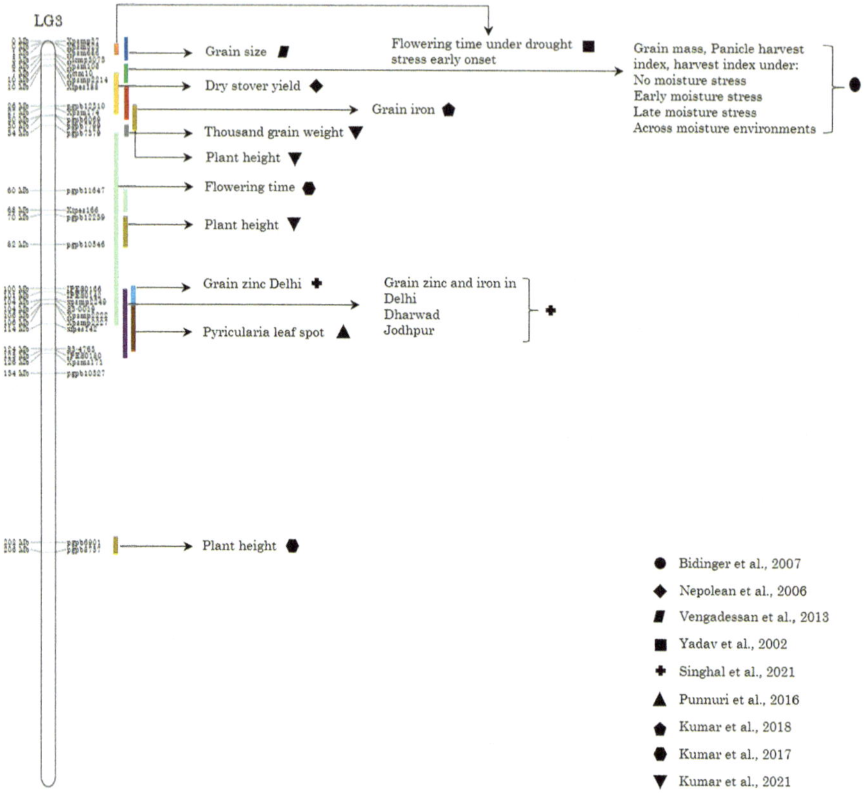

Fig. 5.3 QTL map of pearl millet linkage group 3. Colored bars represent the QTL region for the traits

5.5.2 Type of Marker System

Among all the available markers discussed in the previous discussion on marker systems, SNPs are the best marker system to obtain marker-trait associations because of their abundance, whole genome coverage, and high-throughput nature. Among the genic and flanking markers, marker-trait associations will be more accurate and reliable if genic markers are used because these markers are derived directly from the functional variants in the genic regions that govern the trait, and there is no question of recombination (Kage et al. 2015). The direct markers or genic markers reduce the dependency on the proximity between marker and gene, whereas this is not possible with the flanking markers. Using flanking markers in establishing marker-trait associations may sometimes fail due to the double cross-over and loss of the QTL/gene controlling the trait (Toojinda et al. 1998). Using flanking markers leads to the problem of linkage drag when there is no strong marker-trait association. Hence, genic markers help overcome such uncertainty and improve the precision of estimating marker-trait association (Molla et al. 2015).

Fig. 5.4 QTL map of pearl millet linkage group 4. Colored bars represent the QTL region for the traits

Few genic markers show multiple trait associations; hence these markers can be used to develop several characters simultaneously. For example, Nayak et al. (2022) identified three genic markers, M69, M55, and Sd1, showing multiple trait associations for grain size-related traits in rice. Hence these markers can be used to develop more than one associated character.

5.5.3 Marker Density

Densely genotyped markers are prerequisites for precise QTL mapping. In loosely genotyped populations, the ability to detect QTLs will be reduced. Marker density correlates to the number of QTLs detected, QTL size, logarithm of odds (LOD), and phenotypic variation produced by the QTL in the QTL mapping experiment. Sivasakthi et al. (2018) compared major QTL sizes for four plant-vigor-related traits

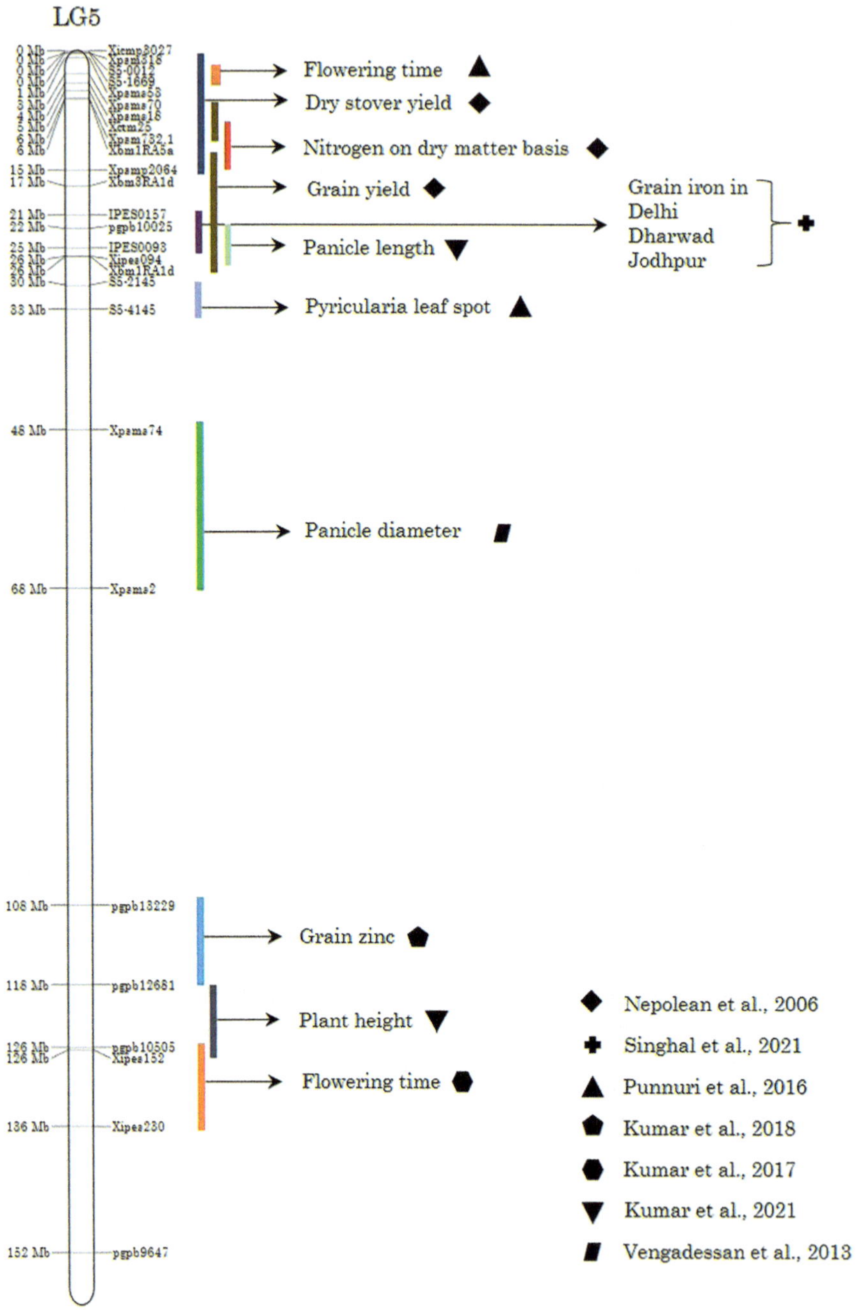

Fig. 5.5 QTL map of pearl millet linkage group 5. Colored bars represent the QTL region for the traits

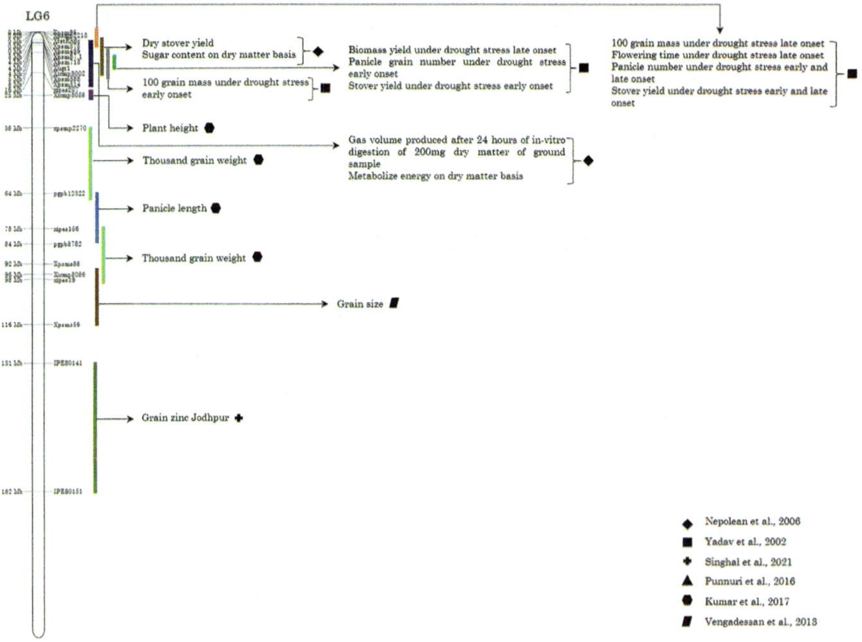

Fig. 5.6 QTL map of pearl millet linkage group 6. Colored bars represent the QTL region for the traits

of chickpeas using three different marker densities and types, viz., 241-SSR low-density markers, 1007-SSR + SNPs-high-density marker, and 1557-SNPs-ultra-high-density markers. The major-effect QTL sizes were very narrow, ranging from 0.14 cM to 0.15 cM for all the four traits, viz., plant vigor, leaf area, plant height, and shoot dry weight when ultra-high-density SNP markers were used for mapping. Also, the LOD and phenotypic variation explained (PVE%) increased with marker density. This is because the SNP markers have genome-wide coverage, and high-density markers narrow down the QTL size. Hence, using the densely genotyped markers gives better mapping results.

5.5.4 Population Size

The accuracy, the number, and the phenotypic variation of the QTL detected are influenced by the number of individuals in a mapping population used for QTL mapping. A population with a large number of individuals can detect most of the QTLs segregating in the population, which includes both major and minor QTLs. Smaller population sizes underestimate the number of QTLs detected, inflate the QTL effects, and cannot precisely detect interactions (Vales et al. 2005).

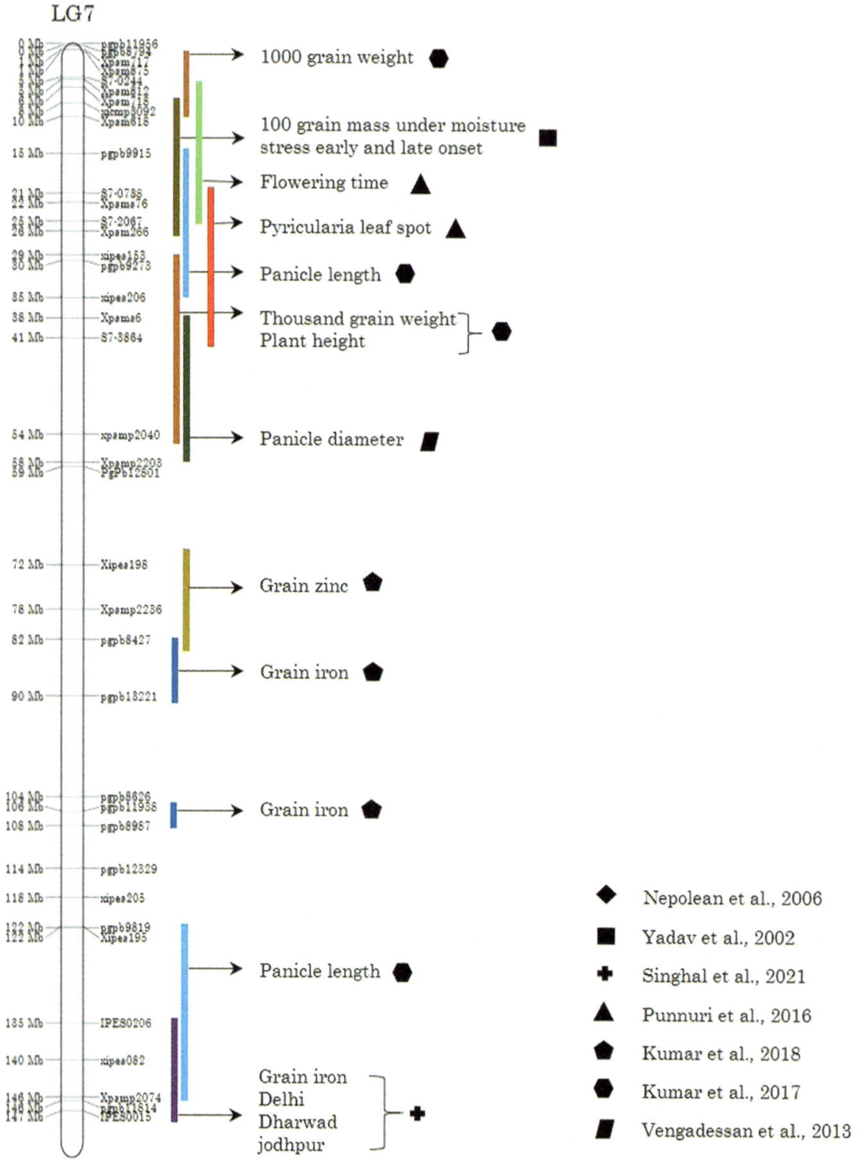

Fig. 5.7 QTL map of Pearl millet linkage group 7. Colored bars represent the QTL region for the traits

A simulation study was conducted by Beavis (1998) to determine how the size of the mapping population affects the detection of QTLs. When the population size was small (100), the statistical power of detecting a small-effect QTL was as low as 3%, and the estimated effects were inflated tenfolds. Small sample sizes may fail to detect

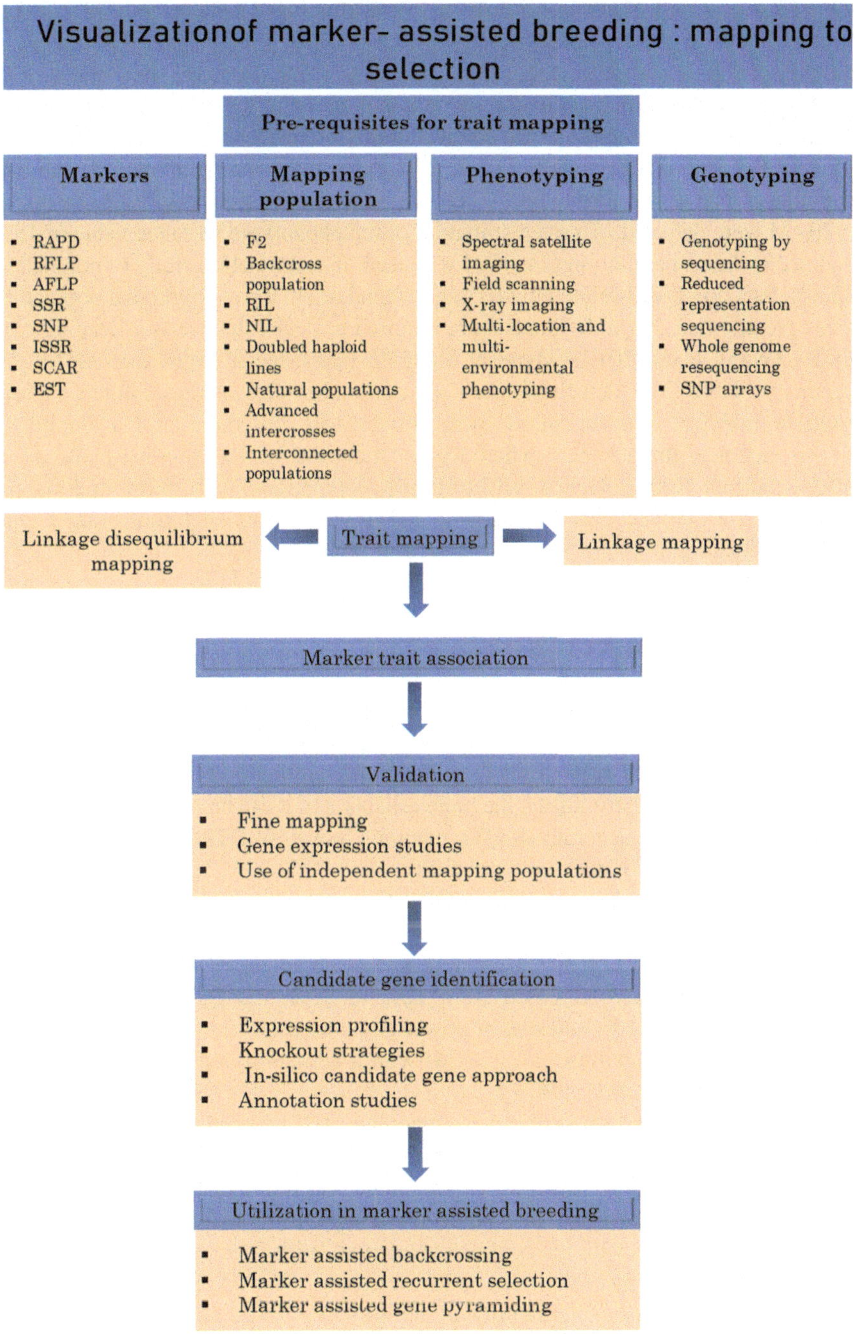

Fig. 5.8 Visualization of marker-assisted breeding: mapping to selection

the QTLs of small effect and result in overestimating the identified QTLs. This average estimate of the phenotypic variances associated with the QTLs in relation to the population size is called the Beavis effect.

Raghavan and Collard (2012) studied the effect of population sizes on the accuracy of QTL mapping, especially the QTL effects and position. Large ($N = 1000$) and small populations ($N = 94$, $N = 120$) were generated by random sampling, and it was found that when the small population was used, errors of >20 cM were detected. An overestimation of the phenotypic variance value of 15% was seen when a population size of 94 was used. In a simulation study of population size in detecting QTLs related to heading date in barley, it was found that population sizes greater than 384 individuals were required to detect three major heading date QTLs (Wang et al. 2012). Hyne et al. (1995) reported a slight increase in the accuracy of the QTL detection when the population size of the doubled haploid used in mapping was increased from 100 to 150, but there was a significant improvement in the accuracy when a population size of 400 was used instead of 100 (Charmet 2000). Hence selecting the populations with more individuals increases the power and accuracy of QTL detection, including major and minor QTLs.

5.5.5 Heritability of the Trait

The trait heritability shows how much of the phenotypic variance in the population is genetically controlled and how reliably a QTL can be mapped. Heritability varies from trait to trait. Qualitative traits have higher heritability compared to quantitative traits. The higher the heritability, the more reliably a trait can be mapped. Heritability of the trait affects the accuracy of QTL mapping, power of QTL detection, and QTL position. The effects of different heritability rates (10%, 30%, 50%, 70%, and 90%) on mapping accuracy in recombinant inbred lines were simulated by Su et al. (2015). The accuracy of QTL mapping was increased with increasing heritability levels. More than 90% accuracy of QTL mapping was obtained with heritability greater than 50%. Heritability increased from 10% to 90%, the QTL detection power increased, and up to 100% detection power could be achieved. The variation in the QTL position was maximum (47.15 cM) with a heritability of 10% and minimum (0 cM) with a heritability of 90%.

The effect of the trait heritability on mapping QTL was estimated by comparing the trait heritabilities of 0.02, 0.05, and 0.08 in the north carolina II (NCII) population of *Brassica napus* by Bu et al. (2015), and it was found that there was a general trend of increase in the power of QTL detection with the increase in trait heritability. Heritability correlates with the confidence interval. With an increase in the heritability, the confidence interval of the QTL decreases. The confidence interval was greater than 1 when the heritability of the trait was 0.05, and it reached nearly <0.5 with a heritability of 0.30 (Charmet 2000).

5.6 Interaction Effects of the QTLs

Apart from producing the major phenotypic effects, QTLs interact with the other QTLs and the environments and such interactions have a profound effect on phenotypic expression.

5.6.1 Epistasis

Epistasis refers to the interaction of two or more loci that results in a specific phenotypic outcome (Churchill 2013). This interaction is important to consider when mapping complex traits, as failure to do so can lead to biased estimations. Additive × additive, additive × dominance, and dominance × dominance interactions are all examples of epistasis. Mapping epistatic interactions is challenging due to the need for large populations and statistical procedures to detect significant interactions. Epistasis can either suppress or enhance the effects of other QTLs or change the direction of their effects.

In a study of barley, two models were compared, one excluding epistasis and the other including additive × additive interactions, to determine the amount of phenotypic variance produced by marker effects and interaction effects. The model that included epistasis produced a higher phenotypic variance than the model that excluded epistasis, demonstrating the importance of including epistasis in understanding the genetic control of phenotypic values (Bocianowski 2013).

In a mapping population of pearl millet, 21 digenic interactions were identified for panicle length and flowering time across two environments. Additive × additive interactions explained 18–40% of the phenotypic variance for flowering time and 13–19% for plant length in the recombinant inbred line population. The presence of significant epistasis indicated that it must be considered before utilizing the QTLs in further breeding programs (Kumar et al. 2017).

5.6.2 QTL-by-Environment Interactions (QEIs)

When developing varieties that can be grown in multiple locations, it is important to consider QTL by location interactions. This involves conducting QTL mapping experiments in each target location and selecting QTLs that are consistent across locations for further breeding. Additionally, different marker-assisted selection (MAS) schemes should be developed for different breeding objectives in the presence of genotype-by-environment interactions (Liu et al. 2006). Breeding for specific locations is typically given more emphasis than developing varieties suitable for multiple locations. However, QTL-by-environment interactions can also be estimated within a single location. When both QEI and epistasis are significant, QTL effects should be re-estimated frequently within the breeding program (Podlich et al. 2004). QTL-by-environment interactions within a location can be further classified into QTL-by-year, QTL-by-season, and QTL-by-population interactions.

If these interactions are substantial, specific QTLs must be selected for a particular season, location, year, or population.

5.6.2.1 QTL-by-Season Interactions

In order to select QTLs for use in breeding programs, it is important to evaluate their interaction effects across different seasons within a given location. QTL-by-environment interactions were estimated in a recombinant inbred line population of pearl millet for flowering time and several important agronomic traits in two seasons, Kharif and summer. The QTL-by-season interaction was found to be non-significant in both seasons, indicating that the selection of superior genotypes was less challenging, and these interactions could be ignored in marker-assisted breeding, as reported by Kumar et al. (2017). However, in a study by Yadav et al. (2003), the F3 population of pearl millet was mapped in two seasons (dry and rainy), and significant QTL-by-season interaction was observed for the flowering time QTLs mapped on LG4 and LG6. Specifically, the H77 allele delayed flowering time in the dry season, whereas the PRLT allele delayed flowering in the rainy season for the QTL on LG4. This suggests that different alleles have varying effects on the flowering trait QTLs in different seasons, and QTL-by-season interactions are pronounced, making certain QTLs suitable for selection in specific seasons.

5.6.2.2 QTL-by-Year Interactions

The detection of QTLs may vary from one crop year to another, and some QTLs identified in 1 year may not be detected in another. However, if the phenotypic variance of a trait remains significant across multiple years, then the QTL may be considered stable over time. For example, Bidinger et al. (2007) mapped QTLs that improved grain yield in moisture-stressed environments over 4 years and identified two promising QTLs that explained a substantial proportion (13–25%) of the phenotypic variance. Notably, there was no significant QTL-by-year interaction for grain yield, suggesting that the predicted effects were consistent across years. In contrast, Singhal et al. (2021) mapped QTLs related to zinc and iron content in pearl millet grains over 3 years and found a significant genotype-by-year interaction, indicating the need to repeat the mapping across multiple years to identify stable QTLs.

5.6.2.3 QTL-by-Population Interactions

When identifying QTLs, it is important to consider the interactions between the QTL and the population, as the genetic makeup of the mapping population can influence the detection of QTLs. For instance, in a study conducted by Kumar et al. (2018), a major QTL for grain iron content was mapped on LG1, whereas in a previous study by the same authors (Kumar et al. 2016), a major QTL for the same trait was mapped on LG3. Notably, these QTLs were detected in different RIL populations that were derived from different parental combinations, indicating that the genomic content of the parental combinations influenced the detection of the QTL. Therefore, it is essential to validate the presence of the QTL in specific populations before proceeding with further breeding efforts.

5.7 QTL Validation

Validating a QTL is crucial to determine its value in future breeding programs. It is important to ensure that the QTL effect is not lost due to factors such as false identification, recombination between involved genes, or loss of effect over time (Collard and Mackill 2008) before utilizing the identified QTL in marker-assisted selection. Below are some methods that can be used to validate a QTL.

5.7.1 Fine Mapping

Fine mapping increases the mapping resolution by narrowing down the QTL region to the shortest possible marker interval (Jaganathan et al. 2020). It is used to locate the specific QTL responsible for a particular trait of interest with high precision. Churchill introduced the concept of pooled mapping for high-resolution mapping, which involved screening a large mapping population with only two markers known to flank the target gene, selecting plants expressing the recessive phenotype of the target trait from a large segregating population of a suitable cross, and dividing them into several random pools, each analyzed with a large number of markers (Churchill et al. 1993). With the availability of the reference genome sequence data, it is now possible to fine-map the QTLs through bulk segregant analysis sequencing (BSA-seq), where the SNPs differing in the two parental lines are used to create the genetic map (Zhang et al. 2021). One of the key advantages of fine mapping is the identification of candidate genes or markers that can be used in marker-assisted selection (Ye et al. 2015). A modified double digest restriction site associated DNA (ddRAD) technique was used to fine-map the drought QTL contributing to the hybrid grain and stover yield potential under terminal drought stress on LG2 in pearl millet (Srivastava et al. 2017).

5.7.2 Gene Expression Studies

Once the QTL region is fine-mapped and the exact location of the QTL is validated, further identification of the candidate genes underlying the identified QTL increases the confidence in the position of the gene. This can be done by a combination of the mapping data and transcriptomic results (Marino et al. 2009). One approach to using gene expression studies in fine mapping QTLs is called expression quantitative trait locus (eQTL) mapping. This involves analyzing the relationship between genetic variation and gene expression levels in a population. By examining the association between genetic variation and gene expression levels, researchers can identify QTLs that affect gene expression and, thus, the trait of interest (Shi et al. 2007). Mahendrakar et al. (2020) validated the major QTL regions grain iron concentration (GFeC) and grain zinc concentration (GZnC) responsible for Zn and Fe metabolism in pearl millet through transcriptome analysis and revealed that the *PglZIP* and *PglNRAMP* gene families were the candidate gene families in linkage group

7. Zhao et al. (2021) fine mapped the QTL QTgw.caas-5B for thousand-grain weight in wheat using the recombinant heterozygous line. The QTL region was delimited to an approximately 2 Mb region (49.6–51.6 Mb) containing 17 high-confidence candidate genes. TraesCS5B02G044800 was identified to be a potential candidate gene for QTgw.caas-5B based on the expression studies.

5.7.3 Use of Alternate Mapping Population

QTL effect can be validated by using an alternate population; that is, if a recombinant inbred line (RIL) was used to map the QTL, validating the same QTL in another population like a near-isogenic line, BC_2 population would confirm the effect of the QTL (Kinkade and Foolad 2013; Farré et al. 2016). For example, among the two identified QTLs (lyc7.1 and lyc12.1) for the trait lycopene content in tomato using RILs (Ashrafi et al. 2012), QTL lyc7.1 did not show a significant effect on the fruit lycopene content in BC2 population derived from the same RIL, whereas the QTL lyc12.1 had a significant effect in RIL-derived BC2 and also near-isogenic line. Hence, the QTL lyc12.1 is considered to be a promising QTL for the high lycopene trait (Kinkade and Foolad 2013).

5.8 Molecular Marker-Based Breeding Strategies

5.8.1 Marker-Assisted Backcrossing (MABC)

Backcross breeding is a technique that facilitates the transfer of a desired trait from a donor parent to a desired genetic background. In cases where a particular line possesses all desirable traits except for one, a backcross can be performed to introduce that trait into the desired background. Marker-assisted backcross breeding offers the advantage of the precise transfer of complex traits, such as disease resistance, which are controlled by multiple genes. By tracking the inheritance of these genes, breeders can select plants that closely resemble the desired phenotype, reducing the time and effort required to develop improved varieties (Vogel 2009). Backcross breeding can be used to enhance pest or disease resistance, increase yield, improve abiotic stress tolerance, and enhance quality traits, as well as for the production of hybrids using existing material by incorporating cytoplasmic male sterility (CMS) or fertility restorers. However, traditional backcross breeding can be time consuming and unpredictable since breeders rely on visual evaluation and phenotypic analysis to determine desirable traits. Furthermore, the release of a new and improved line for a particular targeted trait might not necessarily be economically feasible. The use of marker technology accelerates the backcrossing process and increases its efficiency.

In 1989, Tanksley proposed the use of markers to maximize the recovery of the recurrent parent genome, which was later termed "background selection" by Hospital and Charcosset in 1997. By incorporating markers into the backcross breeding program, the number of backcrosses required could be reduced from the conventional five or more to as few as two or three, as demonstrated by Frisch et al. in 1999. Marker-assisted backcrossing serves to achieve effective foreground selection, competent background selection, and avoidance of linkage drag (Miah et al. 2015). Analysis of DNA markers covering the entire genome can be used to identify individuals with a high proportion of recurrent parent genome recovery, thereby reducing the number of backcross generations required. One notable success of marker-assisted breeding in pearl millet is the commercial release of a hybrid improved for downy mildew, "HHB 67 Improved" (Hash et al. 2006). Additionally, the drought tolerance of the HHB 226 pearl millet hybrid was improved using a marker-assisted backcross approach. This involved transferring drought tolerance quantitative trait loci (QTL) from the donor parent 863B to the recurrent parent HBL 226, resulting in the development of a drought-tolerant version of the parental line HBL 226 (Rani et al. 2022).

5.8.1.1 Foreground Selection

Foreground selection, also known as target locus selection, involves the use of linked markers to indirectly select for the target gene or quantitative trait locus (QTL) of interest. The primary aim of this approach is to maintain the heterozygous state until the final backcross, after which the selected plants are selfed to identify homozygous plants carrying the donor allele (Hasan et al. 2015). By reducing the length of the introgressed QTL, the number of genes involved can be minimized, and the degree of linkage drag can be reduced. Marker-assisted backcrossing is more efficient than conventional backcrossing, which typically involves transferring segments of 50 cM or more (Salina et al. 2003). In contrast, marker-assisted backcrossing aims to reduce the length of the fragments as much as possible (Neeraja et al. 2007).

A comparison study by Bai et al. (2006) suggested that direct selection of the QTL gene, when possible, is the most effective method for transferring the QTL or gene of interest. When direct selection is not feasible, indirect selection using flanking markers is more efficient than using a single marker. The frequency of favorable alleles was estimated to be 1.00, 0.97, and 0.75 after five intercross generations for direct selection and indirect selection using flanking markers and single markers, respectively.

The efficiency of marker-assisted backcrossing also depends on the genome size of the crop. Crops with smaller genomes have a much higher efficiency than crops with larger genome sizes. In a simulation study by Frisch and Melchinger (2005), it was observed that sugar beet required only 30% of the marker data points required in maize to achieve a 10% response to selection, indicating the significant role of genome size in determining the response to selection.

5.8.1.2 Recombinant Selection

Recombinant selection plays a critical role in marker-assisted backcrossing, as it involves the use of molecular markers to identify plants that have undergone recombination between target and donor genes. This results in the transfer of desirable traits from the donor to the target plant and can significantly increase the efficiency of MABC by reducing the time and resources needed to identify plants with the desired traits. By selecting recombinants, the size of the introgressed donor chromosome segment carrying the target gene can be minimized, which helps to mitigate linkage drag caused by undesirable genes linked to the target gene from the donor parent that could have negative effects on the crops. This selection requires a large population size that typically depends on the distance between the flanking marker and the target loci. Double recombinants are rare events, so this selection usually involves at least two backcross generations for accurate and precise selection (Hospital 2005; Frisch et al. 1999).

5.8.1.3 Background Selection

The most vital part of MABC is the recurrent parent genome recovery. The recurrent parent genome recovery ensures that undesirable genomic portions are eliminated while the gene of interest is preserved in the recombined progenies without further segregation (Chukwu et al. 2020). In conventional backcross breeding, several generations of backcrosses are needed to reduce the genome of the donor parent. Markers unlinked to the target traits, polymorphic between the parents, and covering the whole genome are ideal for background selection. With the availability of the marker technology, the number of backcrosses is reduced to as low as two, with as high as 95.2% of recurrent parents. The success depends on the availability of polymorphic markers covering the entire genome (Hasan et al. 2020).

The percentage recovery of the recurrent parent depends upon the number of markers per chromosome used. Benchimol et al. (2005) tested three criteria (two, four, and six markers per chromosome) for efficient recurrent parent recovery. Comparison among two, four, and six markers per chromosome revealed that the plants chosen based on the six markers per chromosome criterion was most similar to the recurrent parent. Bai et al. (2006) compared four different methods of background selection, viz., random selection, marker-assisted best linear unbiased prediction selection, genomic similarity selection, and index selection, and found that the highest recurrent parent recovery was attained with genomic similarity index (99.99%) method of background selection after four generations of backcrossing.

With the advent of genomics, background selection has been simplified through the use of breeding chips that rely on high-quality re-sequencing data (Chen et al. 2014). Broad-spectrum blast resistance genes Pi2 was successfully introgressed into the background of Feng39S through marker-assisted backcrossing, and the recovery of the recurrent genome was made using an SNP chip, RICE6K. This method even helped eliminate the linkage drag segments, and 99.67% of recurrent parent recovery could be achieved (Yang et al. 2019).

The percentage of recurrent parent genome recovery was compared between the conventional and marker-assisted backcross approaches in a simulation study conducted by Hospital (2003). It was observed that the percentage recovery of the recurrent parent genome at each generation of backcross was higher in the marker-assisted backcross approach. Backcross generation four had 99% recurrent parent genome recovery in the marker-assisted backcross approach, while it was only 96.9 in conventional breeding.

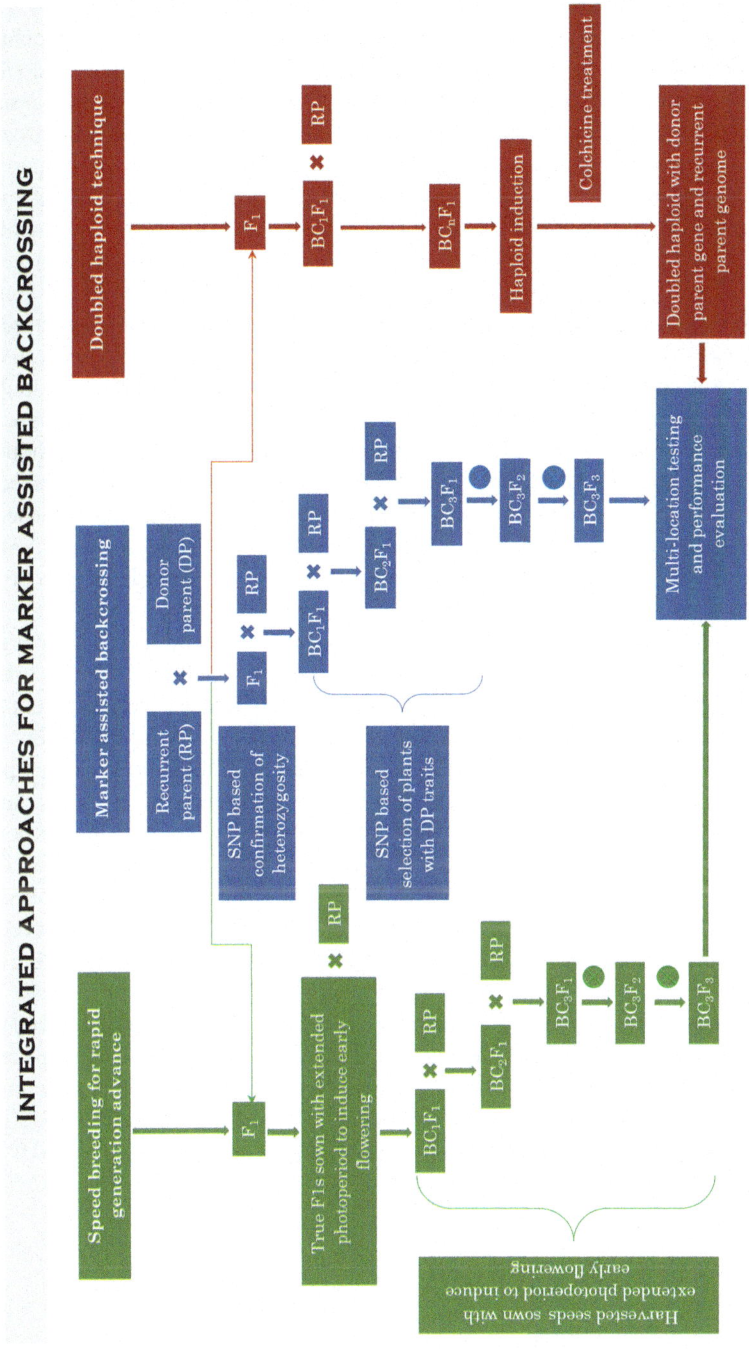

Fig. 5.9 Integrated approaches for marker-assisted backcrossing

The marker-assisted backcross approach can be integrated with other technologies like speed breeding and doubled haploid technology (Fig. 5.9). Doubled haploid technique could serve as an alternative to selfing the BC_nF_1 population in selecting the plants homozygous for the target allele. The usefulness of doubled haploid compared to BC_nF_2 population in the final step of marker-assisted backcrossing was compared in relation to the number of target genes for introgression, and it was found that the use of doubled haploid was advantageous compared to the selfed progeny when the number of independently segregating genes was more. The population size of BC_nF_2 required to find at least one favorable genotype homozygous for the target gene was found to be more, viz., 16 (one target gene), 71 (two target genes), 4713 (five target genes), and 4,847,545 (ten target genes), when compared to the doubled haploid population size, viz., 17 (one target gene), 16 (two target genes), 145 (five target genes), and 4713 (ten target genes), at 99% probability level (Lübberstedt and Frei 2012). This shows the usefulness of doubled haploid technology in reducing the population size by selecting plants with favorable traits. The backcross procedure was coupled with doubled haploid technique in wheat crops to incorporate the stripe rust resistance genes into new breeding materials. BC_nF_2 lines were used to develop the DH lines, which were used to evaluate the effective introgression of the stripe rust resistance (Bakhtiar et al. 2014). Such efforts can be made in pearl millet crops to harvest the advantages of DH techniques.

Speed breeding technology involves using specialized lighting and temperature regimes to accelerate the growth of crops and shorten the breeding cycle. Integrating this technique with marker-assisted backcrossing can lead to more efficient and precise breeding of crops. The day lengths are adjusted according to the crop photoperiodism, and flowering is induced earlier than usual. Altering the temperature, humidity, and light is another consideration. The generation cycle can be shortened two to three times the conventional approach, and a fast introgression of the trait can be achieved (Begna 2022). Eighty-seven $BC_1F_{3:4}$ introgression lines containing multiple disease resistance from four parents were developed in two-rowed barley in just 2 years by Alahmad et al. (2018). A biotron speed breeding system was used, to accelerate the breeding cycle in rice, for introgressing salt tolerance gene *hst1* to the background of a high-yielding line. A cross success rate of 54–69% at each generation of crossing could be seen. BC_3F_3 population carrying the desired allele in a heterozygous state could be developed in six generations and a period of 17 months. Further, at each step of selection, an SNP-based selection of plants containing the target gene further increases the efficiency (Rana et al. 2019). Standardizing speed breeding protocols in pearl millet will help to ensure develop new varieties more quickly and accurately.

5.9 Marker-Assisted Gene Pyramiding

Broad spectrum and durable resistance are major problems in breeding resistant varieties. Single major genes conferring resistance to disease might break due to changes in the race composition of the pathogens and insects. Hence, pyramiding several genes prevents such problems and could be done through marker-assisted backcrossing of several major broad-spectrum genes simultaneously (Jiang et al. 2012). While pyramiding multiple genes, the recovery of the recurrent parent genome in the backcrossing program and the number of marker data points required are important. Frisch and Melchinger compared the results of merging two genes in early generations (e.g., F2 or F3) versus merging them in later generations (e.g., F5 or F6). Their analysis considered both the level of recurrent parent genome recovery and the number of marker data points. They found that when the two genes were merged in the advanced generations, there was a higher rate of recurrent parent genome recovery and it required more marker data points (Frisch and Melchinger 2001).

Among all the three approaches for pyramiding genes/QTLs, convergent marker-assisted backcrossing is a relatively more acceptable strategy since it combines the advantages of both simultaneous and stepwise transfer (Jiang 2013). While transferring a large number of loci, the problem of linkage drag is commonly seen. This can be reduced by using a greater number of background markers (Suh et al. 2013; Pradhan et al. 2015). Also, when the donor is an improved line, the linkage drag will be very less compared to the wild donor lines. Flash-flood tolerance (*Sub1*) and bacterial blight resistance (*Xa4*, *xa5*, *xa13*, *Xa21*) were pyramided into mega variety Swarna background. Since both the donors used (IRBB60 and Swarna-Sub1) were improved already, the problem of linkage drag was not much evident (Pradhan et al. 2019).

5.10 Marker-Assisted Recurrent Selection (MARS)

For the improvement of quantitative traits, the recurrent selection is an efficient technique that includes continuous crossing, selection, and increasing the frequency of the favorable alleles through recombination. Phenotypic selection is influenced by the environment and takes a long time for each selection. Hence, to overcome this delimitation, the molecular markers associated with the targeted traits are utilized at each generation level. At every crossing and selection cycle, only selected individual plants, based on phenotypic data with marker scores, are made to cross. This ultimately leads to an increase in the efficiency and accuracy of the recurrent selection along with an accelerated selection cycle (Ribaut et al. 2010). In maize, four cycles of phenotypic selection would require 8 years, whereas marker-assisted recurrent selection requires only 3 years (Bernardo and Charcosset 2006). Marker-assisted recurrent selection overcomes the limitations of marker-assisted backcrossing in a way that it deals with even multiple QTLs controlling the expression of a trait.

In 2006, Bernardo and Charcosset demonstrated the advantages of two different marker-assisted recurrent selection models in maize: the QTL Per se model, which involves knowing the markers for the QTL themselves, and the Flanking marker model, which involves knowing the markers surrounding the QTL. Their study found that for a trait controlled by a moderately large number of QTLs (such as 40), the QTL Per se model resulted in a 50% increase in response to marker-assisted recurrent selection, whereas the Flanking marker model only resulted in a 17% increase. However, if the number of QTLs controlling the trait is smaller (for example, 10), the use of flanking markers alone is sufficient to accumulate a high frequency of favorable alleles.

Comparing the usefulness of F_2 and doubled haploid population in marker-assisted recurrent selection, it was concluded that when the number of QTLs is more, heritability is low, and population size is less, and doubled haploids are most useful in marker-assisted recurrent selection than F_2 (Mayor and Bernardo 2009). In 2006, Bernado and co-workers investigated whether the level of fitness of an individual, which refers to the number of progenies it contributes to the next selection cycle, impacts the response to recurrent selection. The study revealed that varying the fitness of selected individuals did not affect the response to recurrent selection in the short term (cycles 1–5). As marker-assisted recurrent selection is also a short-term procedure, the level of fitness of individuals did not have an impact on the process.

5.11 Conclusion

Breeding approaches that solely rely upon the phenotype is time consuming and inefficient. Moving from phenotype to genotype overcomes the drawbacks of conventional plant breeding approaches. The availability of high-density genetic maps and the development of efficient marker systems for pearl millet have facilitated the application of MAS in breeding programs, providing a powerful tool for selecting superior genotypes. QTL mapping results should be used to identify the candidate genes and narrow down the regions controlling the trait. With the advent of next-generation sequencing (NGS) technologies, it is now possible to genotype thousands of markers across the genome. The NGS technique has allowed researchers to map QTLs with higher resolution and greater accuracy, which has led to the identification of smaller QTLs and a more precise understanding of their effects (Singh et al. 2015). Developing multi-parent populations for mapping overcomes the limitations of bi-parental populations, increases the diversity of alleles, and improves the precision of QTL mapping (Scott et al. 2020). Using approaches complementary to QTL mapping, namely genome wide association studies (GWAS), renders advantages like detecting variants with smaller effects or associating genetic variants with phenotypic variation in a large, unrelated population, and it is not limited to detecting QTLs in crosses (Brachi et al. 2011). With advances in high-throughput techniques, it is now possible to collect data on multiple levels of biological organization (e.g., genomics, transcriptomics, proteomics, metabolomics). Integrating these different types of data can improve the precision and accuracy of QTL mapping and provide insights into the underlying biological mechanisms.

Other complementary approaches namely integrating MAS with speed breeding and doubled haploid technology can help improve the efficiency of the breeding programs and accelerate the development of new crop varieties that are better adapted to changing environmental conditions and meet the evolving needs of farmers and consumers. Despite the mapping of several traits in pearl millet, their practical application in breeding is not apparent. Therefore, there is a need to combine all available technologies to enhance essential pearl millet traits and create new varieties that can sustain the global food and nutritional security.

References

Agrama HA, Scott JW (2006) Quantitative trait loci for tomato yellow leaf curl virus and tomato mottle virus resistance in tomato. J Am Soc Hortic Sci 131(2):267–272

Alahmad S, Dinglasan E, Leung KM, Riaz A, Derbal N, Voss-Fels KP, Able JA, Bassi FM, Christopher J, Hickey LT (2018) Speed breeding for multiple quantitative traits in durum wheat. Plant Methods 14:1–15. https://doi.org/10.1186/s13007-018-0302-y

Anuradha N, Satyavathi CT, Bharadwaj C, Nepolean T, Sankar SM, Singh SP, Meena MC, Singhal T, Srivastava RK (2017) Deciphering genomic regions for high grain iron and zinc content using association mapping in pearl millet. Front Plant Sci 8:412

Ashrafi H, Kinkade MP, Merk HL, Foolad MR (2012) Identification of novel quantitative trait loci for increased lycopene content and other fruit quality traits in a tomato recombinant inbred line population. Mol Breed 30:549–567

Bai J-Y, Zhang Q, Jia X-P (2006) Comparison of different foreground and background selection methods in marker-assisted introgression. Acta Genet Sin 33(12):1073–1080

Bakhtiar F, Afshari F, Najafian G, Mohammadi M (2014) Backcross-breeding and doubled-haploid facilitated introgression of stripe rust resistance in bread wheat. Arch Phytopathol Plant Protect 47:1675. https://doi.org/10.1080/03235408.2013.854612

Beavis WD (1998) QTL analyses: power, precision and accuracy. In: Paterson AH (ed) Molecular analysis of complex tween-population information in addition to within-traits. CRC Press, Boca Raton, FL, pp 145–161

Begna T (2022) Speed breeding to accelerate crop improvement. Int J Agric Sc Food Technol 8 (2):178–186

Benchimol LL, Souza Jr CLD, Souza APD (2005) Microsatellite-assisted backcross selection in maize. Genet Mol Biol 28:789–797

Bernardo R, Charcosset A (2006) Usefulness of gene information in marker-assisted recurrent selection: a simulation appraisal. Crop Sci 46(2):614–621

Bernardo R, Moreau L, Charcosset A (2006) Number and fitness of selected individuals in marker-assisted and phenotypic recurrent selection. Crop Sci 46(5):1972–1980

Bidinger FR, Nepolean T, Hash CT, Yadav RS, Howarth CJ (2007) Identification of QTLs for grain yield of pearl millet [Pennisetum glaucum (L.) R. Br.] in environments with variable moisture during grain filling. Crop Sci 47(3):969–980

Bocianowski J (2013) Epistasis interaction of QTL effects as a genetic parameter influencing estimation of the genetic additive effect. Genet Mol Biol 36(1):93 100. https://doi.org/10.1590/S1415-47572013000100013. Epub 2013 Mar 4. PMID: 23569413; PMCID: PMC3615531

Botstein D, White RL, Skolnick M, Davis RW (1980) Construction of a genetic linkage map in man using restriction fragment length polymorphisms. Am J Hum Genet 32(3):314

Brachi B, Morris GP, Borevitz JO (2011) Genome-wide association studies in plants: the missing heritability is in the field. Genome Biol 12:232. https://doi.org/10.1186/gb-2011-12-10-232

Bu S, Xinwang Z, Yi C, Wen J, Tu J, Zhang Y-M (2015) Interacted QTL mapping in partial NCII design provides evidences for breeding by design. PLoS One 10:e0121034. https://doi.org/10.1371/journal.pone.0121034

Charmet G (2000) Power and accuracy of QTL detection: simulation studies of one-QTL models. Agronomie 20(3):309–323

Chelpuri D, Sharma R, Durga KK, Katiyar P, Mahendrakar MD, Singh RB, Yadav RS, Gupta R, Srivastava RK (2019) Mapping quantitative trait loci (QTLs) associated with resistance to major pathotype-isolates of pearl millet downy mildew pathogen. Eur J Plant Pathol 154:983–994

Chen H, Xie W, He H, Yu H, Chen W, Li J, Yu R, Yao Y, Zhang W, He Y, Tang X (2014) A high-density SNP genotyping array for rice biology and molecular breeding. Mol Plant 7(3):541–553

Chukwu SC, Rafii MY, Ramlee SI, Ismail SI, Oladosu Y, Muhammad II, Musa I, Ahmed M, Jatto MI, Yusuf BR (2020) Recovery of recurrent parent genome in a marker-assisted backcrossing against rice blast and blight infections using functional markers and SSRs. Plan Theory 9(11):1411

Churchill GA (2013) Epistasis 505–507

Churchill GA, Giovannoni JJ, Tanksley SD (1993) Pooled-sampling makes high-resolution mapping practical with DNA markers. Proc Natl Acad Sci 90(1):16–20

Collard BC, Mackill DJ (2008) Marker-assisted selection: an approach for precision plant breeding in the twenty-first century. Philos Trans R Soc B Biol Sci 363(1491):557–572

Directorate of Millet Development (2020) Crop profile. http://www.aicpmip.res.in/aboutus.html

Farré A, Sayers L, Leverington-Waite M, Goram R, Orford S, Wingen L, Mumford C, Griffiths S (2016) Application of a library of near isogenic lines to understand context dependent expression of QTL for grain yield and adaptive traits in bread wheat. BMC Plant Biol 16(1):1–13

Fisher RA (1918) The correlation between relatives on the supposition of Mendelian inheritance. Trans Roy Soc Edinb 52:399–433

Frisch M, Melchinger AE (2001) Marker-assisted backcrossing for simultaneous introgression of two genes. Crop Sci 41(6):1716–1725

Frisch M, Melchinger AE (2005) Selection theory for marker-assisted backcrossing. Genetics 170(2):909–917. https://doi.org/10.1534/genetics.104.035451. Epub 2005 Mar 31. PMID: 15802512; PMCID: PMC1450430

Frisch M, Bohn M, Melchinger AE (1999) Comparison of selection strategies for marker-assisted backcrossing of a gene. Crop Sci 39(5):1295–1301

Grotz N, Guerinot ML (2006) Molecular aspects of Cu, Fe and Zn homeostasis in plants. Biochim Biophys Acta (BBA) Mol Cell Res 1763(7):595–608

Hasan M, Hasibuzzaman ASM, Abdullah HM, Kallol MMH (2020) Genetic and genomic resources and their exploitation for unlocking genetic potential from the wild relatives. Rediscovery of Genetic and Genomic Resources for Future Food Security:193–210

Hasan MM, Rafii MY, Ismail MR, Mahmood M, Rahim HA, Alam MA, Ashkani S, Malek MA, Latif MA (2015) Marker-assisted backcrossing: a useful method for rice improvement. Biotechnol Biotechnol Equip 29(2):237–254

Hash CT, Sharma A, Kolesnikova-Allen MA, Singh SD, Thakur RP, Raj AGB, Rao MR, Nijhawan DC, Beniwal CR, Sagar P, Yadav HP (2006) Teamwork delivers biotechnology products to Indian small-holder crop-livestock producers: pearl millet hybrid "HHB 67 improved" enters seed delivery pipeline. J SAT Agric Res 2(1)

Heun M, Kennedy AE, Anderson JA, Lapitan NLV, Sorrells ME, Tanksley SD (1991) Construction of a restriction fragment length polymorphism map for barley (Hordeum vulgare). Genome 34(3):437–447

Hospital F (2003) Marker-assisted breeding. In: Newbury HJ (ed) Plant molecular breeding. Blackwell Publishing and CRC Press, Oxford and Boca Raton, pp 30–59

Hospital F (2005) Selection in backcross programmes. Philos Trans R Soc B Biol Sci 360(1459):1503–1511

Hospital F, Charcosset A (1997) Marker-assisted introgression of quantitative trait loci. Genetics 147(3):1469–1485

https://www.fao.org/3/w1808e/w1808e0c.htm

Hyne V, Kearsey MJ, Pike DJ et al (1995) QTL analysis: unreliability and bias in estimation procedures. Mol Breed 1:273–282. https://doi.org/10.1007/BF02277427

Jaganathan D, Bohra A, Thudi M, Varshney RK (2020) Fine mapping and gene cloning in the post-NGS era: advances and prospects. Theor Appl Genet 133(5):1791–1810

Jiang GL (2013) Plant marker-assisted breeding and conventional breeding: challenges and perspectives. Adv Crop Sci Technol 1:e106

Jiang H, Feng Y, Bao L, Li X, Gao G, Zhang Q, Xiao J, Xu C, He Y (2012) Improving blast resistance of Jin 23B and its hybrid rice by marker-assisted gene pyramiding. Mol Breed 30: 1679–1688

Jukanti AK, Gowda CLL, Rai KN, Manga VK, Bhatt RK (2016) Crops that feed the world 11

Kage U, Kumar A, Dhokane D, Karre S, Kushalappa A (2015) Functional molecular markers for crop improvement. Crit Rev Biotechnol 36:917. https://doi.org/10.3109/07388551.2015.1062743

Kinkade MP, Foolad MR (2013) Validation and fine mapping of lyc12. 1, a QTL for increased tomato fruit lycopene content. Theor Appl Genet 126(8):2163–2175

Kumar S, Hash CT, Thirunavukkarasu N, Singh G, Rajaram V, Rathore A, Senapathy S, Mahendrakar MD, Yadav RS, Srivastava RK (2016) Mapping quantitative trait loci controlling high iron and zinc content in self and open pollinated grains of pearl millet [Pennisetum glaucum (L.) R. Br.]. Front Plant Sci 7:1636

Kumar S, Hash CT, Nepolean T, Satyavathi CT, Singh G, Mahendrakar MD, Yadav RS, Srivastava RK (2017) Mapping QTLs controlling flowering time and important agronomic traits in pearl millet. Front Plant Sci 8:1731

Kumar S, Hash CT, Nepolean T, Mahendrakar MD, Satyavathi CT, Singh G, Rathore A, Yadav RS, Gupta R, Srivastava RK (2018) Mapping grain iron and zinc content quantitative trait loci in an Iniadi-derived immortal population of pearl millet. Genes 9(5):248

Lema M (2018) Marker assisted selection in comparison to conventional plant breeding. Agric Res Technol 14:555914

Li H, Bradbury P, Ersoz E, Buckler ES, Wang J (2011) Joint QTL linkage mapping for multiple-cross mating design sharing one common parent. PLoS One 6(3):e17573

Litt M, Luty JA (1989) A hypervariable microsatellite revealed by in vitro amplification of a dinucleotide repeat within the cardiac muscle actin gene. Am J Hum Genet 44:397–401

Liu P-Y, Zhu J, Yan L (2006) Impacts of QTL× environment interactions on genetic response to marker-assisted selection. Acta Genet Sin 33(1):63–71

Lübberstedt T, Frei UK (2012) Application of doubled haploids for target gene fixation in backcross programmes of maize. Plant Breed 131:449–452

Mahendrakar MD, Parveda M, Kishor PK, Srivastava RK (2020) Discovery and validation of candidate genes for grain iron and zinc metabolism in pearl millet [Pennisetum glaucum (L.) R. Br.]. Sci Rep 10(1):16562

Mammadov J, Aggarwal R, Buyyarapu R, Kumpatla S (2012) SNP markers and their impact on plant breeding. Int J Plant Genomics 2012:728398

Marino R, Ponnaiah M, Krajewski P, Frova C, Gianfranceschi L, Pè ME, Sari-Gorla M (2009) Addressing drought tolerance in maize by transcriptional profiling and mapping. Mol Gen Genomics 281:163–179

Mayor PJ, Bernardo R (2009) Doubled haploids in commercial maize breeding: one-stage and two-stage phenotypic selection versus marker-assisted recurrent selection. Maydica 54(4): 439–448

McCouch SR, Teytelman L, Xu Y, Lobos KB, Clare K, Walton M, Fu B, Maghirang R, Li Z, Xing Y, Zhang Q (2002) Development and mapping of 2240 new SSR markers for rice (Oryza sativa L.). DNA Res 9(6):199–207

Miah G, Rafii MY, Ismail MR, Puteh AB, Rahim HA, Latif MA (2015) Recurrent parent genome recovery analysis in a marker-assisted backcrossing program of rice (Oryza sativa L.). C R Biol 338(2):83–94

Molla KA, Debnath AB, Ganie SA, Mondal TK (2015) Identification and analysis of novel salt responsive candidate gene based SSRs (cgSSRs) from rice (Oryza sativa L.). BMC Plant Biol 15(1):1–11

Morgan TH (1910) Sex limited inheritance in Drosophila. Science 32(812):120–122

Morris M, Dreher K, Ribaut JM, Khairallah M (2003) Money matters (II): costs of maize inbred line conversion schemes at CIMMYT using conventional and marker-assisted selection. Mol Breed 11:235–247

Mullis KB (1990) The unusual origin of the polymerase chain reaction. Sci Am 262(4):56–65

Myles S, Peiffer J, Brown PJ, Ersoz ES, Zhang Z, Costich DE, Buckler ES (2009) Association mapping: critical considerations shift from genotyping to experimental design. Plant Cell 21(8): 2194–2202

Nayak AK, Anilkumar C, Behera S, Sah RP, Lavanya GR, Kumar A, Behera L, Tp MA (2022) Genetic dissection of grain size traits through genome-wide association study based on genic markers in rice. Rice Sci 29(5):462–472

Neeraja CN, Maghirang-Rodriguez R, Pamplona A, Heuer S, Collard BC, Septiningsih EM, Vergara G, Sanchez D, Xu K, Ismail AM, Mackill DJ (2007) A marker-assisted backcross approach for developing submergence-tolerant rice cultivars. Theor Appl Genet 115:767–776

Podlich DW, Winkler CR, Cooper M (2004) Mapping as you go: an effective approach for marker-assisted selection of complex traits. Crop Sci 44(5):1560–1571

Pradhan SK, Nayak DK, Mohanty S, Behera L, Barik SR, Pandit E, Lenka S, Anandan A (2015) Pyramiding of three bacterial blight resistance genes for broad-spectrum resistance in deepwater rice variety, Jalmagna. Rice 8:1–14

Pradhan SK, Pandit E, Pawar S, Baksh SY, Mukherjee AK, Mohanty SP (2019) Development of flash-flood tolerant and durable bacterial blight resistant versions of mega rice variety 'Swarna' through marker-assisted backcross breeding. Sci Rep 9(1):1–15

Prahalada GD, Shivakumar N, Lohithaswa HC, Sidde Gowda DK, Ramkumar G, Kim SR, Ramachandra C, Hittalmani S, Mohapatra T, Jena KK (2017) Identification and fine mapping of a new gene, BPH31 conferring resistance to brown planthopper biotype 4 of India to improve rice, Oryza sativa L. Rice 10(1):1–15

Pujar M, Gangaprasad S, Govindaraj M, Gangurde SS, Kanatti A, Kudapa H (2020) Genome-wide association study uncovers genomic regions associated with grain iron, zinc and protein content in pearl millet. Sci Rep 10(1):1–15

Raghavan C, Collard BCY (2012) Effect of small mapping population sizes on reliability of quantitative trait locus (QTL) mapping. Afr J Biotechnol 11(47):10661–10674

Ragot M, Lee M, Guimaraes E (2007) Marker-assisted selection in maize: current status, potential, limitations and perspectives from the private and public sectors. In: Marker-assisted selection, current status and future perspectives in crops, livestock, forestry and fish, pp 117–150

Rai KN, Hash CT, Singh AK, Velu G (2008) Adaptation and quality traits of a germplasm-derived commercial seed parent of pearl millet. Plant Genet Resour Newsl 154:20–24

Rana MM, Takamatsu T, Baslam M, Kaneko K, Itoh K, Harada N, Sugiyama T, Ohnishi T, Kinoshita T, Takagi H, Mitsui T (2019) Salt tolerance improvement in rice through efficient SNP marker-assisted selection coupled with speed-breeding. Int J Mol Sci 20(10):2585

Rani A, Taunk J, Jangra S, Yadav RC, Yadav NR, Yadav D, Yadav HP (2022) Development of advance pearl millet lines tolerant to terminal drought stress using marker-assisted selection. Vegetos 35(1):63–73

Ribaut JM, De Vicente MC, Delannay X (2010) Molecular breeding in developing countries: challenges and perspectives. Curr Opin Plant Biol 13(2):213–218

Rick CM, Zobel RW, Fobes JF (1974) Four peroxidase loci in red-fruited tomato species: genetics and geographic distribution. Proc Natl Acad Sci 71(3):835–839

Salina E, Dobrovolskaya O, Efremova T, Leonova I, Roder MS (2003) Microsatellite monitoring of recombination around the Vrn-B1 locus of wheat during early backcross breeding. Plant Breed 122:116–119

Sanger F, Nicklen S, Coulson AR (1977) DNA sequencing with chain-terminating inhibitors. Proc Natl Acad Sci U S A 74(12):5463–5467

Sax K (1923) The association of size differences with seed-coat pattern and pigmentation in Phaseolus vulgaris. Genetics 8(6):552

Scott MF, Ladejobi O, Amer S et al (2020) Multi-parent populations in crops: a toolbox integrating genomics and genetic mapping with breeding. Heredity 125:396–416. https://doi.org/10.1038/s41437-020-0336-6

Shi C, Uzarowska A, Ouzunova M, Landbeck M, Wenzel G, Lubberstedt T (2007) Identification of candidate genes associated with cell wall digestibility and eQTL (expression quantitative trait loci) analysis in a Flint x Flint maize recombinant inbred line population. BMC Genomics 8:22

Singh VK, Singh A, Singh SP, Ellur RK, Singh D, Gopala Krishnan S et al (2013) Marker-assisted simultaneous but stepwise backcross breeding for pyramiding blast resistance genes Piz5 and Pi54 into an elite Basmati rice restorer line 'PRR 78'. Plant Breeding 132(5):486–495

Singh VK, Singh AK, Singh S, Singh BD (2015) Next-generation sequencing (NGS) tools and impact in plant breeding. In: Al-Khayri J, Jain S, Johnson D (eds) Advances in plant breeding strategies: breeding, biotechnology and molecular tools. Springer, Cham. https://doi.org/10.1007/978-3-319-22521-0_20

Singhal T, Satyavathi CT, Singh SP, Kumar A, Sankar SM, Bhardwaj C et al (2021) Multi-environment quantitative trait loci mapping for grain iron and zinc content using bi-parental recombinant inbred line mapping population in pearl millet. Frontiers in Plant Science 12:659789

Sivasakthi K, Thudi M, Tharanya M, Kale SM, Kholová J, Halime MH, Jaganathan D, Baddam R, Thirunalasundari T, Gaur PM, Varshney RK (2018) Plant vigour QTLs co-map with an earlier reported QTL hotspot for drought tolerance while water saving QTLs map in other regions of the chickpea genome. BMC Plant Biol 18:1–18

Srivastava RK, Vadez V, Kholova J, Lachagari VBR, Mahendrakar MD, Katiyar P, Lekkala SP, Praveen M, Gupta R, Blumwald E (2017) Fine mapping of the linkage group 2 drought tolerance QTL in pearl millet [Pennisetum glaucum (L.) R. Br.]. In: InterDrought-V, February 21–25, 2017, Hyderabad, India

Su CF, Liu ZB, Li YG (2015) Influence of various quantitative trait loci (QTL) mapping methods on the mapping accuracy under varying heritability levels. Genet Mol Res 14(4):13003–13012. https://doi.org/10.4238/2015.October.21.21. PMID: 26505453

Suh JP, Jeung JU, Noh TH, Cho YC, Park SH, Park HS, Shin MS, Kim CK, Jena KK (2013) Development of breeding lines with three pyramided resistance genes that confer broad-spectrum bacterial blight resistance and their molecular analysis in rice. Rice 6:1–11

Supriya A, Senthilvel S, Nepolean T, Eshwar K, Rajaram V, Shaw R, Hash CT, Kilian A, Yadav RC, Narasu ML (2011) Development of a molecular linkage map of pearl millet integrating DArT and SSR markers. Theor Appl Genet 123:239–250

Tanksley SD (1983) Molecular markers in plant breeding. Plant Mol Biol Rep 1:3–8

Tanksley SD, Young ND, Paterson AH, Bonierbale MW (1989) RFLP mapping in plant breeding: new tools for an old science. Bio/Technology 7(3):257–264

Toojinda T, Baird E, Booth A, Broers L, Hayes P, Powell W, Thomas W, Vivar H, Young G (1998) Introgression of quantitative trait loci (QTLs) determining stripe rust resistance in barley: an example of marker-assisted line development. Theor Appl Genet 96(1):123

Vales MI, Schön CC, Capettini F, Chen XM, Corey AE, Mather DE, Mundt CC, Richardson KL, Sandoval-Islas JS, Utz HF, Hayes PM (2005) Effect of population size on the estimation of QTL: a test using resistance to barley stripe rust. Theor Appl Genet 111:1260–1270

Vogel KE (2009) Backcross breeding. Methods Mol Biol 526:161–169. https://doi.org/10.1007/978-1-59745-494-0_14. PMID: 19378006

Vos P, Hogers R, Bleeker R et al (1995) AFLP: a new technique for DNA fingerprinting. Nucleic Acids Res 23:4407–4414

Vos PG, Paulo MJ, Voorrips RE, Visser RG, van Eck HJ, van Eeuwijk FA (2017) Evaluation of LD decay and various LD-decay estimators in simulated and SNP-array data of tetraploid potato. Theor Appl Genet 130:123–135

Wang H, Smith KP, Combs E, Blake T, Horsley RD, Muehlbauer GJ (2012) Effect of population size and unbalanced data sets on QTL detection using genome-wide association mapping in barley breeding germplasm. Theor Appl Genet 124:111–124

Welsh J, McClelland M (1990) Fingerprinting genomes using PCR with arbitrary primers. Nucleic Acids Res 18:7213–7218

Williams JGK, Kubelik AR, Livak KJ et al (1990) DNA polymorphisms amplified by arbitrary primers are useful as genetic markers. Nucleic Acids Res 18:1631–1635

Xu Y, Crouch JH (2008) Marker-assisted selection in plant breeding: from publications to practice. Crop Sci 48(2):391–407

Yadav OP, Rai KN (2013) Genetic improvement of pearl millet in India. Agric Res 2:275–292. https://doi.org/10.1007/s40003-013-0089-z

Yadav RS, Hash CT, Bidinger FR, Cavan GP, Howarth CJ (2002) Quantitative trait loci associated with traits determining grain and stover yield in pearl millet under terminal drought-stress conditions. Theor Appl Genet 104:67–83

Yadav RS, Bidinger FR, Hash CT, Yadav YP, Yadav OP, Bhatnagar SK, Howarth CJ (2003) Mapping and characterisation of QTL× E interactions for traits determining grain and stover yield in pearl millet. Theor Appl Genet 106(3):512–520

Yang D, Tang J, Yang D, Chen Y, Ali J, Mou T (2019) Improving rice blast resistance of Feng39S through molecular marker-assisted backcrossing. Rice 12(1):1–16

Ye C, Tenorio FA, Redoña ED, Morales-Cortezano PS, Cabrega GA, Jagadish KS, Gregorio GB (2015) Fine-mapping and validating qHTSF4. 1 to increase spikelet fertility under heat stress at flowering in rice. Theor Appl Genet 128:1507–1517

Zhang B, Qi F, Hu G, Yang Y, Zhang L, Meng J, Han Z, Zhou X, Liu H, Ayaad M, Xing Y (2021) BSA-seq-based identification of a major additive plant height QTL with an effect equivalent to that of semi-dwarf 1 in a large rice F2 population. Crop J 9(6):1428–1437

Zhao D, Yang L, Liu D, Zeng J, Cao S, Xia X, Yan J, Song X, He Z, Zhang Y (2021) Fine mapping and validation of a major QTL for grain weight on chromosome 5B in bread wheat. Theor Appl Genet 134:3731–3741

Genomic Selection and Its Application in Pearl Millet Improvement

6

Bharathi Raja Ramadoss, Ameena Premnath, Thiruvengadam Venkatesan, and Nepolean Thirunavukkarasu

Abstract

Pearl millet [*Pennisetum glaucum* (L.) R. Br] is a staple grain for about 90 million people in India, sub-Saharan Africa, and South Asia. Genomic selection is a new tool that helps to identify better lines among experimental cultivars in plant breeding programs. Genomic selection examines the phenotypes and high-density marker scores of lines in a population to predict breeding values. The integration of all marker information in the prediction model contributes to the effectiveness of genomic selection by eliminating biased marker effect estimations and collecting more of the variance associated with small-effect quantitative trait loci (QTL). The whole genome sequence of pearl millet has recently been sequenced, allowing genomic selection models to be used to improve the selection process in the pearl millet breeding program. Genomic selection, which employs genomic-estimated breeding values of individuals obtained from genome-wide markers to identify candidates for the next breeding cycle, is a powerful tool for enhancing quantitative traits. Models used for genomic selection frequently encounter problems when the number of markers

B. R. Ramadoss
Saskatoon Research and Development Centre, Agriculture and Agri-Food Canada, Saskatoon, SK, Canada

A. Premnath
Department of Plant Biotechnology, Centre for Plant Molecular Biology and Biotechnology, Tamil Nadu Agricultural University, Coimbatore, Tamil Nadu, India

T. Venkatesan (✉)
Agricultual College and Research Institute, Tamil Nadu Agricultural University, Kudumiyanmalai, Tamil Nadu, India

N. Thirunavukkarasu
ICAR - Indian Institute of Millets Research, Hyderabad, Telangana, India

V. A. Tonapi et al. (eds.), *Pearl Millet in the 21st Century*,
https://doi.org/10.1007/978-981-99-5890-0_6

exceeds the number of phenotypic data. To address this issue and enhance prediction accuracy, genomic selection models and algorithms such as Bayesian, Gaussian, and machine learning have been used. This chapter focuses extensively on the transition from conventional selection techniques used in plant breeding to the genomic selection, the underlying statistical models and methods used for this purpose, the current state of genomic selection research in pearl millet, and the prospects for its successful application in the development of climate resilient pearl millet varieties suitable for different end users.

Keywords

Pearl millet · Genomic selection · Marker effect · GEBVs · Algorithm · Prediction models

6.1 Introduction

The primary objective of any crop improvement program is to develop varieties and cultivars with increased yield and biotic and abiotic stress tolerance. The process of developing varieties is continuous to meet the food demand of a growing global population. The current rate of genetic improvement in major agricultural crops is about 0.8–1.2%, which is insufficient to meet global food security in the future (Krishnappa et al. 2021). As the human population is expected to increase to 9.5 or 10 billion by 2050, the varietal improvement needs to be accelerated to 2.4% to feed the hungry stomach (Hickey et al. 2019; Ray et al. 2012, 2013).

In general, the development of high-yielding crop varieties has been achieved through forward genetics and/or conventional breeding approaches with difficulties. Because most of the yield and yield contributing traits are genetically complex and highly influenced by changing environmental conditions and climate change, the rate of genetic improvement is restricted (Bailey-Serres et al. 2019).

Over the decades, different molecular breeding approaches have been developed to speed up domestic and global breeding programs. After the discovery of molecular markers, marker-assisted selection/breeding (MAS/MAB) has been used to facilitate the development of crop varieties in different crop plants. MAS/MAB has been used successfully in crop improvement programs. This has led to the release of improved varieties of many crops, such as HHB67-Improved in pearl millet (Rai et al. 2008), C214 in chickpea (Varshney et al. 2014a), JL24 and TAG24 in groundnut (Varshney et al. 2014b), JTN5503 and DS880 in soybean (Arelli et al. 2006, 2009), HUW510 in wheat (Vasistha et al. 2017), Pusa Samba 1850 (Krishnan et al. 2019), Pusa Basmati 1728 (Singh et al. 2017a), Pusa Basmati 1637 (Singh et al. 2017b), Improved Pusa Basmati1 (Gopalakrishnan et al. 2008), Swarna-Sub1 (Neeraja et al. 2007), Improved Samba Mahsuri (Ratna Madhavi et al. 2016), and CR 1009 Sub 1(Robin et al. 2019) in rice. However, MAS/MAB is inefficient for improving traits like yield and biotic and abiotic stress tolerance because it is controlled by many genes or quantitative loci with small effects. The biggest problems with MAS/MAB are (1) that it uses a low-density marker system, (2) that it doesn't have a good statistical method for improving traits that are controlled by many loci with small effects, and (3) that it uses a certain type of population. Hence, we need an appropriate approach for improving polygenic traits such as yield and biotic and abiotic stress tolerance.

A genomic selection (GS) approach was proposed by Meuwissen et al. (2001) for a breeding population to facilitate the selection of polygenic traits. The GS approach is used to predict the individual marker effect along with the sum of all the marker effects, which is used to calculate the genomic-estimated breeding values (GEBV) of the individual genotype.

Pearl millet (*Pennisetum glaucum* (L) R. Br., syn. *Cenchrus americanus* (L.) Morrone) is a drought-tolerant, open-pollinated, climate-resilient, C4 plant that grows primarily in resource-limited or marginal soil and environmental conditions such as low soil fertility, high temperatures, and limited water availability (Srivastava et al. 2020). As a C4 plant, it has great photosynthetic and high biomass-producing potential, which makes pearl millet an important staple food for those who are living in poverty or developing countries. It was considered an orphan/neglected crop and limited efforts have been made to improve its yield and quality. Later, much emphasis was given to the development of genetic and genomic resources to breed high-yielding and climate-resilient pearl millet varieties/hybrids for marginal farmers. The discovery of the pearl millet reference sequence in 2017 (Varshney et al. 2017) aided in the advancement of the pearl millet crop.

In this chapter, we present the current status and promising prospects of genomic selection methods, prediction models, and trait improvement concerning their application in pearl millet improvement.

6.2 Prediction Methods and Models

Several different series of models and algorithms for genetic prediction have been proposed by several researchers in GS. Initially, GS prediction approaches were mainly proposed and performed for animal genetic prediction. Later, the development of GS models provided a platform to utilize the models in plant breeding programs.

Compared to conventional plant breeding, genomic prediction models can accelerate crop improvement per unit of time by reducing labor costs and shortening the generation interval in a breeding cycle. A fundamental prerequisite for genomic prediction is the distribution of markers across the genome, with at least one marker being in linkage disequilibrium (LD) with each QTL. When estimating effects with a "training" population, all markers are employed concurrently. Genomic selection (GS) uses a "training population" of individuals that have been both phenotyped and genotyped to train a prediction model for calculating genomic estimated breeding values (GEBVs). Genomic prediction can foretell the GEBVs of individuals for selection based on the information from the training population. There are two steps involved in estimating the GEBVs with high prediction ability

1. Determining the size of the training population and the number of markers utilized in the suitable prediction model;
2. Testing and model validation to foresee the phenotype of those lines that were not included in the training models.

The genetic diversity and size of the training population, marker density, trait heritability, marker or gene effects, and the extent and distribution of LD between markers and QTL are a few variables that impact the accuracy of genomic prediction.

The basic process of any genomic selection process starts with the creation of a training population, i.e., individuals having both genotypic and phenotypic information and this information is used to build a model, where the phenotype is used as a response and the genotype as a predictor. In "training" populations, the effects of each marker are used simultaneously to create prediction models. Fixed regression methods utilizing ordinary least squares cannot be utilized to create prediction models since the number of predictors (markers) is typically more than the sample size ($P \gg n$). Prediction models are created using statistical techniques that see marker effects as random, such as ridge regression best linear unbiased prediction (RR-BLUP) and various Bayesian models. One can evaluate the accuracy of genomic selection, by comparing GEBVs to the breeding values predicted using conventional techniques that depend on phenotypic data. For the purpose of predicting phenotypes utilizing a large number of markers, several GS models have been created. The key area where these models diverge is in the proportion of variance that is attributable to marker effects.

6.3 Methods Used in Genomic Selection

6.3.1 M1: General Combining Ability (GCA) Model ($E + G_{P1} + G_{P2}$)

For characterizing the ith hybrid, this model utilizes genomic information obtained from the inbreds via the GCA of the parents; hence, modeling of male and female effects can be performed. This model is composed of two genetic scores, which are derived from the main effects of the markers of those inbreds acting as parent 1 or B-lines (g_{P1i}) and parent 2 or R-lines (g_{P2i}), respectively (Technow et al. 2014; Kadam et al. 2016). Collecting the aforementioned results and assumptions, the linear predictor for modeling the hybrid performance via the GCA of inbreds is obtained as follows

$$y_{ij} = m + E_j + g_{P1i} + g_{P2i} + e_{ij}$$

where y_{ij} is the yield performance of the ith ($i = 1, 2, \ldots, I$) hybrid in the jth ($j = 1, 2, \ldots, J$) environment, m is the common mean, E_j is the main effects of the jth environments, $g_{P1} = \{g_{P1i}\} \sim N(0, G_{P1}\sigma^2_{P1g})$ and $g_{P2} = \{g_{P2i}\} \sim N(0, G_{P2}\sigma^2_{P2g})$ with $G_{P1} = X_{P1}X^1{}_{P1}/p$, $G_{P2} = X_{P2}X^1 0_{P2}/p$, $\sigma^2_{P1g} = p \times \sigma^2_{bP1}$ and $\sigma^2_{P2g} = p \times \sigma^2_{bP2}$ as the corresponding variance components of the parental effects, and $e_{ij} \sim N(0, \sigma^2_e)$ and

$\sigma 2_E$ and $\sigma 2_e$ represent the associated variance components of environments and residual terms. One of the disadvantages of this model is that it does not take into consideration the specific effect of crossing parent 1 with parent 2, but rather the average effects between both parents. Moreover, it returns a common genetic effect for the same hybrid in different environments (Jarquin et al. 2020).

6.3.2 M2: General Plus-Specific Combining Ability Model ($E + G_{P1} + G_{P2} + G_{P1 \times P2}$)

This model is an extension of model M1, and it not only accounts for the main effects of the genetic components of the inbreds but also includes the specific interaction effect of crossing inbred parent 1 and parent 2 (Acosta-Pech et al. 2017). The main effect is accounted for by the GCA component, and the interaction effect is accounted for by the SCA component. The SCA was modeled using the cell-by-cell product of the entries of the covariance structures from inbred parent 1 (G_{P1}) and inbred parent 2 (G_{P2}), such that $g_{P1xP2} = \{g_{P1ixP2i}\} \sim N(0, G_{P1xP2}\sigma^2_{P1gxP2g})$, where $G_{P1xP2} = (Z_{gP1}G_{P1}Z^1_{gP1})\cdot(Z_{gP2}G_{P2}Z^1_{gP2})$, $\sigma^2_{P1gxP2g}$ is the variance component associated with this interaction term, and Z_{gP1} and Z_{gP2} are the corresponding incidence matrices for parent 1 and parent 2 for the hybrids.

The model in which both the GCA and the SCA components are included can be written as.

$$y_{ij} = m + E_j + g_{P1i} + g_{P2i} + g_{P1ixP2i} + e_{ij}$$

Although this model considers the effects of crossing parent 1 with parent 2, it brings a common genetic effect across environments for the same hybrid in different environments similar to the previous model.

6.3.3 M3: General Plus-Specific Combining Ability in Interaction with Environments Model ($E + G_{P1} + G_{P2} + G_{P1 \times P2} + G_{P1} \times E + G_{P2} \times E + G_{P1 \times P2} \times E$)

This model is an extension of M2, in that it includes both the GCA and SCA components but also accounts for the interaction of the inbred markers with environments by including the interaction between GCA and SCA components and environments. The model can be written as

$$y_{ij} = m + E_j + g_{P1i} + g_{P2i} + g_{P1ixP2i} + gE_{P1ij} + gE_{P2ij} + gE_{P1ijxP2ij} + e_{ij}$$

where $gE_{P1} = \{gE_{P1ij}\} \sim N(0, (Zg_{P1}G_{P1}Z^1_{gP1})\cdot(Z_EZ^1_E)\sigma^2 g_{EP1})$, $gE_{P2} = \{gE_{P2ij}\} \sim N(0, (Z_{gP2}G_{P2}Z^1_{gP2})\cdot(Z_EZ^1_E)\sigma^2_{gEP2})$ and $gE_{P1xP2} = \{gE_{P1ijxP2ij}\} \sim N(0, (I_4 \otimes ((Z_{gP1}G_{P1}Z^1_{gP1})\cdot(Z_{gP2}\ G_{P2}Z^1_{gP2})))\cdot(Z_EZ^1_E)\sigma 2_{gEP1xP2})$; σ^2_{gEP1}, σ^2_{gEP2}, and $\sigma^2_{gEP1xP2}$ are the corresponding variance components for interaction terms between markers of inbred and environments for the GCA (parent 1 and parent 2) and SCA (P1 × P2) terms; Z_E is the corresponding incidence matrix for environments. The genetic effects of the genotypes derived from this model are particular to each environment.

6.4 Models Implied for Genomic Selection

The process of selecting suitable individuals in GS starts with a simple linear model also known as least-squares regression or ordinary least-squares regression (OLS):

$$Y = 1_n\mu + X\beta + \varepsilon$$

where $Y = n \times 1$ vectors of observations, μ is the mean, $\beta = p \times 1$ vectors of marker effects, $\varepsilon = n \times 1$ vectors of random residual effects, $X =$ design matrix of order $n \times p$ (where each row represents the genotype/individuals/lines (n), and each column corresponds to the marker (p)), and $\varepsilon \sim N(0, \sigma^2_e)$.

One major limitation in linear models using several thousands of genome-wide markers is that the number of markers (p) exceeds the number of observations (n), i.e., genotypes/individuals/lines and this creates the problem of over-parameterization (large "p" and small "n" problem ($p \gg n$)). Using a subset of significant markers can be an alternative for dealing with the large "p" and small "n" problem. Meuwissen et al. (2001) modified the least-squares regression for GS and performed least-squares regression analysis on each marker separately with the following model.

$$Y = X_j\beta_j + \varepsilon$$

where $X_j = j$th column of the design matrix of the markers and $\beta_j =$ genetic effect of the jth marker.

Markers with significant effects are selected using the log-likelihood of this model, and those are further used for estimating the breeding values. However, some key information may be lost by selection based on the subset of markers. Hence, an efficient solution for the over-parameterization problem in linear models is using ridge regression (RR), which is a penalized regression-based approach (Meuwissen et al. 2001). It also solves the problems of multicollinearity simultaneously (i.e., correlated predictors, e.g., SNP or markers). RR shrinks the coefficients of correlated predictors equally towards zero and solves the regression problem using $\ell 2$ penalized least squares. Here, the goal is to derive an estimator of parameter β with a smaller variance than the least-squares estimator. Similar to RR, the least absolute shrinkage and selection operator (LASSO) (Usai et al. 2009) is another variant of penalized regression, which uses the $\ell 1$ penalized least-squares criterion to obtain a sparse solution. However, sometimes LASSO may not work well with highly correlated predictors (e.g., SNPs in high linkage disequilibrium) (Ogutu et al. 2012). The elastic net (ENET) is an extension of the LASSO that is robust to extreme correlations among the predictors (Friedman et al., 2010), and it is a compromise between $\ell 1$ penalty (LASSO) and $\ell 2$ penalty (RR) (Zou and Hastie 2005).

The RR model considers that each marker contributes to equal variance, which is not true for all traits. Therefore, the variance of the markers based on the trait's genetic architecture has to be modeled. For this purpose, several Bayesian models have been proposed where it is assumed that there is some prior distribution of marker effects (Budhlakoti et al. 2022). Furthermore, inferences about model parameters are obtained based on posterior distributions of marker effects. There are several variants of Bayesian models for genomic prediction such as Bayes A, Bayes B, Bayes Cπ, and Bayes Dπ (Meuwissen et al. 2001; Habier et al. 2011) and other derivatives, e.g., Bayesian LASSO and Bayesian ridge regression (BRR). Besides the marker-based models, the best linear unbiased prediction (BLUP) (Henderson et al. 1959) is one of the most commonly used genomic prediction methods. There are many variants of BLUP available for this purpose, e.g., genomic BLUP (GBLUP), single-step GBLUP (ssGBLUP), ridge regression BLUP (RRBLUP), and GBLUP with linear ridge kernel regression (rrGBLUP), of which GBLUP is very frequently used. The GBLUP uses the genomic relationships calculated using markers instead of the conventional BLUP which uses the pedigree relationships to obtain the GEBV of the lines or individuals (Meuwissen et al. 2001).

The genomic prediction models perform well for traits with additive genetic architecture, but their performance becomes very poor in the case of epistatic genetic architecture. Hence, Gianola et al. (2006) first used nonparametric and semiparametric methods for modeling the complex genetic architecture. Subsequently, several statistical methods were implemented to model both additive and epistatic effects for genomic selection (Xu 2007; Cai et al. 2011; Legarra and Reverter 2018). Several nonparametric methods have been studied in relation to genomic selection, e.g., NW (Nadaraya–Watson) estimator (Gianola et al. 2006), RKHS (reproductive kernel Hilbert space) (Gianola et al. 2006), SVM (support vector machine) (Maenhout et al. 2007; Long et al. 2011), ANN (artificial neural network) (Gianola et al. 2011), and RF (random forest) (Holliday et al. 2012), among which SVM, NN, and RF are based on the machine learning approach.

Methods discussed earlier are based on genomic information where information is available for a single trait, i.e., single-trait genomic selection (STGS). As the performance of STGS-based methods may be affected significantly in the case of pleiotropy, i.e., one gene linked to multiple traits, a mutation in a pleiotropic gene may have an effect on several traits simultaneously. It was observed that low heritability traits could borrow information from correlated traits and consequently achieve higher prediction accuracy. However, STGS-based methods consider the information of each trait independently.

Hence, crucial information may be lost which may ultimately result in poor genomic prediction accuracy. Nowadays, as we are receiving data on multiple traits, multi-trait genomic selection (MTGS)-based methods may provide more accurate GEBV and subsequently a higher prediction accuracy. Several MTGS-based methods have been studied in relation to GS, e.g., the multivariate mixed model

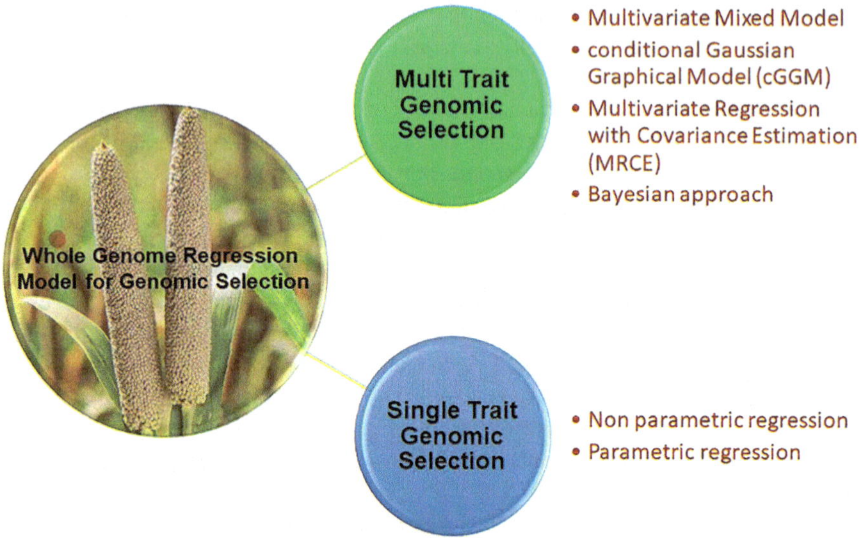

Fig. 6.1 Summary of models utilized in genomic selection

approach (Jia and Jannink 2012; Klápště et al. 2020), Bayesian multi-trait model (Jia and Jannink 2012; Cheng et al. 2018), MRCE (multivariate regression with covariance estimation) (Rothman et al. 2010), and cGGM (conditional Gaussian graphical model) (Chiquet et al. 2017). Jia and Jannink (2012) presented three multivariate linear models (i.e., GBLUP, Bayes A, and Bayes Cπ) and compared them to univariate models, and a detailed comparison of various STGS and MTGS-based methods has also been studied by Budhlakoti et al. (2019). A brief structure of different STGS- and MTGS-based methods used in GS studies is given in Fig. 6.1.

The architecture of the target traits may impact how well prediction models for complex traits function (Momen et al. 2018). At ICRISAT, India, whole genome resequencing (WGRS) along with phenotypic data for several important traits of PMiGAP lines was exploited for GWAS and GS. Different whole-genome prediction/genomic selection models were created and optimized based on the numerous target features identified by GWAS in pearl millet. Varshney et al. (2017) performed GS to predict grain yield for test crosses by ridge regression best linear unbiased prediction (RR-BLUP) at ICRISAT. GS strategy was also used to predict the performance of the hybrids derived from a CMS (Cytoplasmic male sterility) using both (RADseq and tGBS) techniques and four genomic prediction schemes (Liang et al. 2018). Pilot studies on the genomic selection with different prediction models in pearl millet are summarized in Table 6.1.

Table 6.1 Studies on genomic selection in pearl millet

Traits studied	Breeding material	Prediction model employed	References
Grain yield	170 hybrid combinations	RR-BLUP (ridge regression best linear unbiased prediction)	Varshney et al. (2017)
1000-grain weight, days to flowering, grain yield, and plant height	Two common control lines/hybrids and 13 experimental lines	RR-BLUP (ridge regression best linear unbiased prediction)	Liang et al. (2018)
Grain yield	320 pearl millet hybrids and 37 inbred parents	Bayesian generalized linear regression (BGLR)	Jarquin et al. (2020)

6.5 Trait Improvement

6.5.1 Yield

Modern breeding efforts have increased the productivity of almost all agricultural crops. Pearl millet productivity increased by almost four times from 1950 to 2019. This genetic improvement of pearl millet is divided into four phases (Yadav et al. 2019; Yadav and Rai 2013). Each phase of improvement and its emphasis are presented in Fig. 6.2. Pearl millet yielded 162% more after adopting high-yielding, pest and disease-resistant, and abiotic stress-tolerant standard agronomic management practices, while sorghum, wheat, rice, and maize yielded 26%, 59%, 69%, and 113%, respectively (Yadav et al. 2019; Yadav and Rai 2013).

The rate of genetic gain achieved in pearl millet and other cereals is the collective outcome of improved varieties and best crop management strategies. This quantum jump in the productivity of pearl millet was achieved in two different ways. First, almost 90% of pearl millet is grown under rainfed conditions; second, pearl millet has attracted fewer human resources and infrastructure than other crops (Yadav et al. 2019). In pearl millet association studies (Anuradha et al. 2017), Xibmsp11/AP6.1, a known SNP marker found on acetyl CoA carboxylase gene, was shown to be significantly linked with yield and yield components (grain harvest index and grain yield). Stay green and grain yield-related traits are closely linked to Xibmcp09/AP10.2 and Xibmcp09/AP10.1 InDels markers on chlorophyll a/b binding protein genes. (Anuradha et al. 2017).

6.5.2 Grain Quality

Since pearl millet is a particularly nutritious grain with higher quantities of protein and various minerals than other cereals, breeding has sought to focus on increasing the yield potential (Singh and Nainawatee 1999) along with other important traits (Fig. 6.3). Earlier studies revealed that the protein concentration of pearl millet was

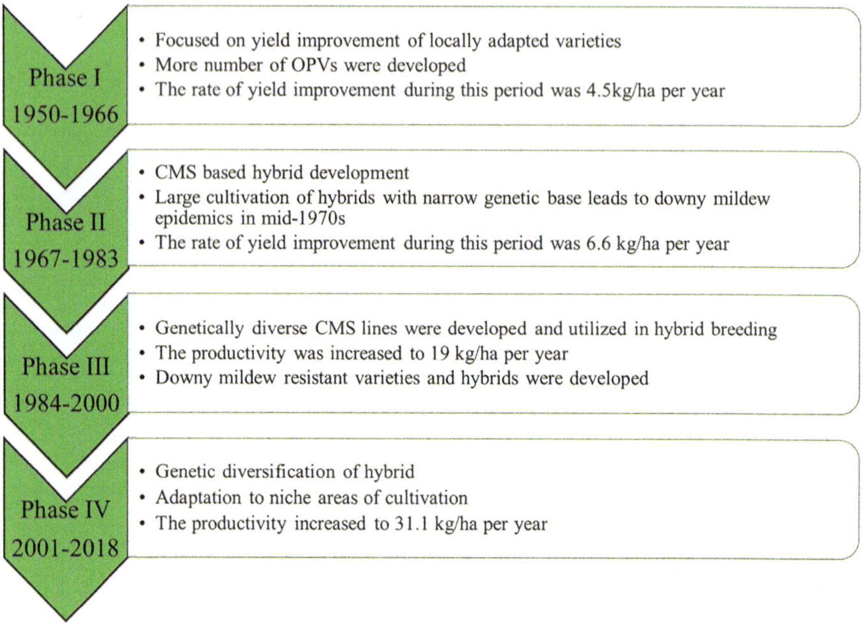

Fig. 6.2 Four phases of yield improvement in pearl millet

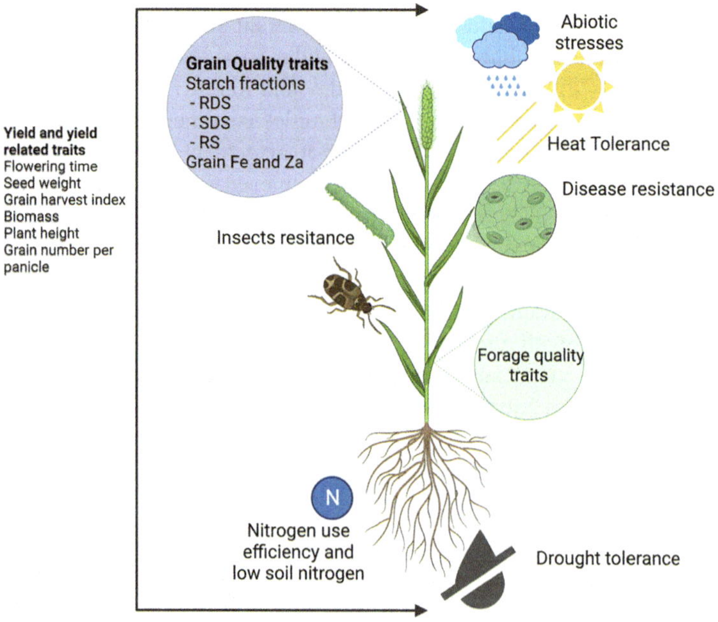

Fig. 6.3 Important traits improved in pearl millet

reported at up to 24.3% (Jambunathan and Subramanian 1988), with top breeding lines reaching 19.8% (Singh et al. 1987). However, due to the unfavorable associations between protein content and grain yield, no significant efforts to enhance it were made in most of the pear millet breeding programs (Singh and Nainawatee 1999). In light of the growing awareness of widespread iron (Fe) and zinc (Zn) deficiencies worldwide, improving grain nutritional characteristics has recently been included as a breeding target. The main areas focused on include the evaluation of genotype-environment interactions; correlations between grain minerals and agronomic features; genetic regulation of micronutrients; and assessment of the degree of genetic variation for grain Fe and Zn content.

There is a lot of variation in the amount of Fe and Zn in pearl millet germplasm and breeding lines. This suggests that these micronutrients could be improved through pre-breeding approaches. The mapping populations and germplasm accessions originating from the *Iniadi* landrace have the greatest levels of Fe and Zn, respectively (Velu et al. 2008; Govindaraj et al. 2016; Govindaraj et al. 2020a, b; Govindaraj et al. 2021). When more than 120 commercial Indian hybrids were tested, 46–56 ppm Fe and 37–44 ppm Zn were found (Rai et al. 2016). In 2018, the Indian national testing and cultivar release policy set 42 ppm Fe and 32 ppm Zn as a baseline for the mainstreaming of Fe and Zn in pearl millet (AICPMIP 2018). For Indian adults, the daily recommended amounts of Fe and Zn are 17–21 mg/d and 10–12 mg/day, respectively.

The presence of Fe and Zn in pearl millet is mostly determined by additive gene action, implying that both parental lines of hybrids would be required to enrich these minerals (Kanatti et al. 2014). Lower G × E influences interaction on Fe and Zn accumulation in pearl millet grains and also revealed the success of progeny selection in pedigree breeding to generate lines with higher grain Fe and Zn densities (Kanatti et al. 2014; Govindaraj et al. 2016). Between Fe and Zn, a significant and positive association has been found (Kanatti et al. 2014; Govindaraj et al. 2020a, b). Additionally, there was a significant positive correlation between these two micronutrients and grain size (Kanatti et al. 2014; Govindaraj et al. 2016). These relationships would help generate pearl millet cultivars rich in Fe and Zn without compromising their grain size, regardless of color, and also increase micronutrient content through crop improvement.

6.6 Biotic Stress Tolerance

6.6.1 Disease Resistance

The most important disease in pearl millet in India and Africa is Downy Mildew (DM), caused by *Sclerospora graminicola* (Sacc.) J. Schröt. DM in pearl millet causes significant yield loss in major growing countries, mainly in India and Africa (Yadav et al. 2021). Identifying various resistance sources through greenhouse and field screening of large panels of germplasm accessions and breeding lines has helped to advance the development of DM-resistant hybrids in pearl millet significantly. This hybrid population of diverse genetic backgrounds has been an important factor in successfully managing widespread DM outbreaks in pearl millet (Singh et al. 1987, 1990).

The severity of smut (*Moesziomyces penicillariae* Bref. Vanky) and ergot (*Claviceps fusiformis* Lov.) significantly reduces the grain yield in pearl millet. Both infections are soil-borne and infect the host through the stigma during flowering (Thakur and Williams 1980). Pollen wash is the main reason for the severity of smut and ergot diseases during the rainy season. The development of field and greenhouse screening methods was aided by knowledge of the biology and epidemiology of these diseases. Numerous lines have been examined to determine their resistance to these diseases. The lack of incredibly high ergot resistance in germplasm accessions has driven the development of ergot-resistant lines by interbreeding less susceptible germplasm lines and selecting and rescreening resistant progeny for multiple generations under intense disease pressure. Resistance to Smut is a dominant trait that is easy to pass on to the next generation. But there have also been reports of quantitative resistance incorporating both additive and nonadditive gene effects (Thakur et al. 2011). Smut and DM resistance has been discovered in several lines (Thakur et al. 1992, 2011).

It is common knowledge that rust (*Puccinia substriata* var. *indica* Ramachar & Cumm) has little impact on grain (Pearl millet) crops, but in the fodder industry, it has a significant impact on yield and quality. A combination of field and greenhouse screening resulted in the discovery of stable sources of resistance. DArT and SSR-based linkage maps were created using a mapping population of 168 F7 RILs, which was also screened for rust resistance (Ambawat et al. 2016). Three QTLs on linkage groups 1, 4, and 7 were identified for pearl millet rust resistance, accounting for 58% of the phenotypic variation in rust reactivity. A newly discovered QTL for rust resistance, Linkage Group 1 (LG1), confers a long-lasting slow rusting phenotype (Ambawat et al. 2016).

The leaf spot or blast disease, caused by *Pyricularia grisea* Sacc. (syn. *Magnaporthe grisea*), has emerged as a major pearl millet disease (Rai et al. 2012). The identification of blast-resistant lines from pearl millet breeding and germplasm accessions allowed for the development of hybrids that are resistant to the disease (Sharma et al. 2013; Goud et al. 2016). In pearl millet, a single dominant gene is responsible for resistance to Indian strains of *Magnoporthe grisea* (Gupta et al. 2012; Singh et al. 2018). A total of six blast-resistant pearl millet genotypes (ICMB 97222, ICMB93333, ICMR 11003, IP 21187-P1, and ICMR 06222) were crossed with two susceptible genotypes (ICMB 89111 and ICMB 95444) to study the inheritance pattern. Their generations and backcrosses were screened for resistance against Pg53 and Pg45 (*Magnoporthe grisea* isolates). Molecular markers are also being used to identify QTLs for blast disease pathotypes. Using SSR markers, two significant QTLs for blast resistance have been identified in linkage groups 1 (LG1) and 6 (LG6) (Maganlal et al. 2018). A large panel of germplasm accessions of pearl millet was collected from 13 different countries and tested against *Magnoporthe grisea* isolates Pg45, Pg53, Pg56, Pg118, and Pg119. The accessions were then classified according to the amount of resistance. It was found that

182 different accessions of pearl millet exhibited resistance against different pathotypes. (Sharma et al. 2021).

6.6.2 Insect Resistance

Although there have been reports of more than 100 insect pests linked with cropping systems based on pearl millet, only a small number of these insects are considered to have the ability to cause considerable damage to the crop. These include the stem borers (*Chilo partellus* in India and *Coniesta ignefusalis* in western Africa), shoot flies (*Atherigona approximata*), and white grubs (*Holotrichia consanguinea*) in India. Additionally, ear head worms (*Helicoverpa armigera*), gray weevils (*Myllocerus species*), and leaf rollers (*Marasmia trapezalis*) are found (Raghvani et al. 2008). Pest control techniques have been created based on research into the pests' life cycles and the types of harm they cause. Each region has its own unique pattern of insect-pest dispersion and devastation. Long-term research has shown that no single approach to pest management is successful against any kind of pest. Cultural and chemical management is needed as part of an integrated pest management strategy (Sharma and Youm 1999). Insect pest incidence on commercial cultivation and experimental test genotypes is regularly monitored, and no breeding initiatives are pursuing insect resistance as a goal trait in pear millet.

6.7 Abiotic Stress Tolerance

6.7.1 Drought Tolerance

Crop plants suffer from impaired growth and development when drought conditions persist due to insufficient rainfall and its unpredictable distribution pattern. It has been discovered that QTL for drought tolerance contributes to differences in photosynthetic pigments and ROS scavenging enzymes in different accessions of pearl millet. The ascorbate peroxidase (APX) activity was found to be elevated in tolerant genotypes, although the superoxide dismutase (SOD) and catalase (CAT) activities remained unaltered, according to the QTL that was investigated. It was shown that the presence or lack of drought-related QTL had no effect on the molecules that make up photosynthetic pigments (Kholová et al. 2011).

The pearl millet grain filling stage is the most susceptible stage to drought stress, which results in a decrease in grain size and grain test weight (Fussell et al. 1991). A pearl millet germplasm association panel was recently constructed and is being used for drought tolerance trait association mapping. A significant association was observed between an SNP in the acetyl-CoA carboxylase genes and panicle yield, grain harvest index, and grain yield, whereas an InDel was shown to be significantly connected with grain yield and stay-green phenotype traits under drought conditions (Sehgal et al. 2015). Debieu et al. (2018) used genotyping by sequencing (GBS) to identify QTLs linked with agronomic parameters in 188 inbred lines under drought conditions. Four marker-trait relationships for the stay-green trait were identified on chromosome 6, and two SNPs were shown to be significantly linked with biomass yield under early drought stress conditions. One of the two SNPs identified for biomass yield was mapped between two predicted genes, Pgl GLEAN 10037359 and Pgl GLEAN 10037360, while the other was mapped between two predicted genes, Pgl GLEAN 10036946 and Pgl GLEAN 10036945. Early drought stress in lines resulted in a decreased grain and biomass yield, although only minor changes were found in grain weight (Debieu et al. 2018). Shivhare et al. (2020) discovered 1129 DEGs on all seven pearl millet chromosomes except chromosome 4. Most genes were found on and mapped to chromosome 2 (196), followed by chromosome 3 (171), chromosome 5 (168), chromosome 6 (164), chromosome 7 (140), and chromosome 4 (108). Recent research on transcriptome analysis identified 2792 transcription factors, 1223 transcriptional regulators, 315 transcription factors, and 128 transcriptional regulators expressed under drought conditions. Using RNA-Pacibio sequencing, a total of 6484 genes for drought stress were identified (Sun et al. 2020). In recent research, Zhang et al. (2021) investigated the mechanism of drought resistance in pearl millet by comparing physiological and transcriptome data under drought and controlled conditions. It has been found that 12 genes were elevated under stress, some of which are connected with drought stress in other species, such as ADH1, FtsH, and CCCH. Also, the expression levels of genes like SnRK2 and PP2C involved in ABA signaling pathways were found to vary (Zhang et al. 2021).

6.7.2 Heat Tolerance

The ideal temperature for normal pearl millet development is 33–34 °C. The seedling and reproductive phases of pearl millet are equally susceptible to the effects of higher temperatures. By 2050, climate change models predict that sub-Saharan Africa (SSA) and South Asia (SA) pearl millet yields will fall by 6–17%. (Knox et al. 2011). High-temperature stress during the flowering stage of pearl millet produces sterility in the flowers, resulting in a drastic decrease in seed set and grain yield (Gupta et al. 2015; Djanaguiraman et al. 2018). Pearl millet has evolved as a very prolific and profitable crop throughout the hot and dry summer season in northern and western India during the last two decades. With increased air temperatures (typically >42 °C) coinciding with blooming during this season, the crop suffers from reproductive sterility, resulting in dramatic decreases in seed set and subsequently decreased grain production (Gupta et al. 2015; Djanaguiraman et al. 2018). During the reproductive stage, plants' ability to handle high temperatures has become an important trait in improving genetic gains.

Pearl millet roots have recently been studied to see how heat stress affects the plant's physiological and transcriptional systems. Trehalose concentration in the roots increased between 3 and 7 h of heat stress. Furthermore, peroxidase (POD) activity steadily increased from 3 to 7 h of heat stress. HSFs, bZIP, and bHLHs were the most frequently identified transcription factors expressed under heat stress. There was a total of 16 bZIPs, 7 HSFs, and 18 bHLH genes that showed different expressions under heat stress (Sun et al. 2021).

The ability to withstand high temperatures during the reproductive phase has emerged as an essential genetic characteristic of pearl millet. Guidelines for flowering-period heat stress screening in controlled environments (greenhouses and growth chambers) and the field have been established (Gupta et al. 2015). There is a substantial amount of genetic heterogeneity between breeding lines and among open-pollinating populations. Multi-location field experiments were carried out to understand the sensitive plant reproductive parts highly susceptible to heat stress and found that stigma is more heat-sensitive than the pollen in pearl millet, thus helping researchers to pyramid heat tolerance in high-yielding hybrids and open-pollinated varieties (OPVs) (Gupta et al. 2016, 2019).

6.8 Genomic Selection (GS) in Pearl Millet

Genomic selection (GS) is a promising method that has tremendous potential to investigate and enhance the genetic gain per selection (in a set of timeframes) in a breeding scheme and, as a result, speed and efficiency in breeding programs. Genomic selection can also be called genome-wide selection (Spindel et al. 2015). In cereals and numerous other crops, including pearl millet, GS has been shown to be a cost-effective and feasible alternative to marker-assisted selection (MAS) and phenotype selection (PS) for quantitative traits and rapid crop development initiatives (Zhong et al. 2009; Heffner et al. 2009; Crossa et al. 2010; Poland et al. 2012; Ornella et al. 2012; Spindel et al. 2015; Muleta et al. 2019; Liang et al. 2018; Jarquin et al. 2020).

Pearl millet's genome was recently assembled and is available in the public domain (Varshney et al. 2017). Because a reference genome is now widely available, genomic selection can be used to predict the general combining ability (GCA) of newly inbred parents and the specific performance of individual pearl millet hybrids (Liang et al. 2018). It has been demonstrated that adding information about the parents into genetic prediction can be advantageous (Massman et al. 2013). Genotyping based on sequencing makes it possible to produce thousands of SNPs that are used for differentiating breeding lines within the population. As stated in the introduction of this chapter, genomic selection-assisted breeding programs that include high-throughput genotyping and estimated breeding values have the ability to overcome all three of the most significant bottlenecks in pearl millet breeding efforts. GS uses genome-wide high-density DNA markers in linkage disequilibrium (LD) with QTL to predict the genomic estimated breeding values (GEBV) of the testing population, which only has genotypic data. The predicted GEBVs are used for selection (Meuwissen et al. 2001). Making selection decisions during the off-season allows for increases in genetic gain each year, which is one of the main benefits of GS (Heffner et al. 2009). The development and use of GS in pearl millet breeding programs make it possible to use resources well and predict how well hybrids will perform.

In ICRISAT, efforts are being undertaken to take advantage of the pearl millet inbred germplasm association (PMiGAP) panel, along with whole-genome resequencing (WGRS) and phenotyping data for several characteristics for GS. The development and optimization of several whole-genome prediction/genomic selection models in pearl millet are based on the various target features using genome-wide association studies (GWAS). To predict grain yield for test crosses in different environments, ICRISAT applied WGRS data for genomic selection by ridge regression best linear unbiased prediction (RR-BLUP). In this prediction, high prediction accuracies were found for the performance of genotypes across different environments. Additionally, it was reported that the GS strategy (additive and dominance effects) was used to analyze grain yield data with 302,110 SNPs to predict hybrid performance. One hundred and seventy promising hybrid combinations were discovered. Among them, more than ten hybrid combinations are already being utilized in heterosis breeding. Liang et al. (2018) analyzed the effectiveness of genomic selection and prediction using two potential genotyping techniques, RAD-seq and tGBS, to characterize a series of ICRISAT-developed inbred pearl millet lines. Twenty random rounds of fivefold cross-validation were performed for a tested SNP set, utilizing the projected hybrids from both (RADseq and tGBS) methodologies and four genomic prediction systems in pearl millet and assessing for each trait. The genomic prediction technique (RR-BLUP) was found to produce median prediction ranges (in parenthesis) for various traits, including 1000-grain weight (0.73–0.74), days to flowering (0.87–0.89), grain yield (0.48–0.51), and plant height (0.72–0.73), respectively, using hybrid data. It was also predicted that adding inbred phenotypic data sets and hybrid trait values relative to the mean trait values of that population made hybrid GEBVs slightly better.

Recently, Jarquin et al. (2020) implemented and compared three GS models utilizing grain yield and high-density molecular marker information from two distinct genotyping platforms for pearl millet (C [conventional GBS RAD-seq] and T [tunable GBS tGBS]). The three GS models were evaluated using different cross-validation (CV) schemes that mimic the scenarios of breeders encountered during the varietal developmental processes in breeding projects. While CV1 forecasts the performance of untested hybrids and CV0 predicts the performance of hybrids in unobserved environmental conditions, CV2 is similar to an incomplete field trial. Through this study, Jarquin et al. (2020) discovered that the addition of phenotypic information of the parental inbreds, consideration of genotype-by-environment interaction, and use of molecular markers derived from tGBS improved the predictive ability and performance of the GS models.

6.9 Implications and Future Prospects

Pearl millet, as a poor man's crop, has received less attention despite its enormous potential. Pearl millet breeding has lagged behind that of other crops. With a growing population, the worldwide demand and diversification of feed, energy, and food are on the rise, urging the development of efficient methods for the development of varieties for different end users. Pearl millet's genetic and genomic insights could be gained using genomics if the wide range of wild species and germplasm accessions were studied. On the other hand, heterozygosity, high outcrossing rates, inbreeding depression, and residual heterozygosity pose bottlenecks in inbred development programs, thus influencing the association mapping panels and parental line/cultivar development. Furthermore, populations for functional genomics, such as mapping populations, natural diversity panels, molecular modules, GWAS, and genetic engineering, could aid in dissecting the population's valuable variables. GS should be carried out based on different end users. Single-cross hybrids dominate the Indian market, whereas top-cross and three-way hybrids are prominent in Africa. Therefore, an appropriate GS model can be employed based on the type of populations and genotype x environment interactions.

Genome editing, a new method in the genomic era, is important for creating novel targeted mutations and identifying genes and their functions for crop improvement. However, limited efforts have been made in pearl millet breeding. The implementation of this approach is heavily reliant on transformation efficiency, which is still significantly lower in crops like pearl millet and sorghum than in other key crops. To enhance CRISPR/Cas9 use in pearl millet breeding, it is necessary to optimize the transformation system. Even though a lot of progress has been made in finding the genetic loci that control important agronomic and grain quality traits, epigenomes, pan-genomes, and other fields should be brought together for the dissection of genetic diversity and identification of superior alleles for the development of superior pearl millet varieties to achieve global food and nutritional security.

Acknowledgments We thank "Biorender.com" for providing tools to create a figure of important trait improvement in pearl millet.

References

Acosta-Pech R, Crossa J, de los Campos G, Teyssèdre S, Claustres B, Pérez-Elizalde S, Pérez-Rodríguez P (2017) Genomic models with genotype x environment interaction for predicting hybrid performance: an application in maize hybrids. Theor Appl Genet 130:1431–1440

AICPMIP (2018) Proceedings of the 53rd annual group meeting of ICAR - All India Coordinated Research Project on Pearl Millet (AICPMIP). Agriculture University, Jodhpur. http://www.aicpmip.res.in/pw2018.pdf. Accessed 30 Sept 2022

Ambawat S, Senthilvel S, Hash CT, Nepolean T, Rajaram V, Eshwar K et al (2016) QTL mapping of pearl millet rust resistance using an integrated DArT-and SSR-based linkage map. Euphytica 209(2):461–476

Anuradha N, Satyavathi CT, Bharadwaj C, Nepolean T, Sankar SM, Singh SP et al (2017) Deciphering genomic regions for high grain iron and zinc content using association mapping in pearl millet. Front Plant Sci 8:412

Arelli PR, Young LD, Mengistu A (2006) Registration of high yielding and multiple disease-resistant soybean germplasm JTN-5503. Crop Sci 46(6):2723

Arelli PR, Young LD, Concibido VC (2009) Inheritance of resistance in soybean PI 567516C to LY1 nematode population infecting cv. Hartwig. Euphytica 165(1):1–4

Bailey-Serres J, Parker JE, Ainsworth EA, Oldroyd GE, Schroeder JI (2019) Genetic strategies for improving crop yields. Nature 575(7781):109–118

Budhlakoti N, Mishra DC, Rai A, Lal SB, Chaturvedi KK, Kumar RR (2019) A comparative study of single-trait and multi-trait genomic selection. J Comput Biol 26:1100–1112

Budhlakoti N, Kushwaha AK, Rai A, Chaturvedi KK, Kumar A, Pradhan AK, Kumar U, Kumar RR, Juliana P, Mishra DC, Kumar S (2022) Genomic selection: a tool for accelerating the efficiency of molecular breeding for development of climate-resilient crops. Front Genet 13: 832153

Cai X, Huang A, Xu S (2011) Fast empirical Bayesian LASSO for multiple quantitative trait locus mapping. BMC Bioinformatics 12:211–213

Cheng H, Kizilkaya K, Zeng J, Garrick D, Fernando R (2018) Genomic prediction from multiple-trait Bayesian regression methods using mixture priors. Genetics 209:89–103

Chiquet J, Mary-Huard T, Robin S, Robin S (2017) Structured regularization for conditional Gaussian graphical models. Stat Comput 27:789–804

Crossa J, Campos GDL, Pérez P, Gianola D, Burgueno J, Araus JL et al (2010) Prediction of genetic values of quantitative traits in plant breeding using pedigree and molecular markers. Genetics 186(2):713–724

Debieu M, Sine B, Passot S, Grondin A, Akata E, Gangashetty P et al (2018) Response to early drought stress and identification of QTLs controlling biomass production under drought in pearl millet. PLoS One 13(10):e0201635

Djanaguiraman M, Perumal R, Ciampitti IA, Gupta SK, Prasad PVV (2018) Quantifying pearl millet response to high temperature stress: thresholds, sensitive stages, genetic variability and relative sensitivity of pollen and pistil. Plant Cell Environ 41(5):993–1007

Friedman J, Hastie T, Tibshirani R (2010) Regularization paths for generalized linear models via coordinate descent. J Stat Softw 33(1):1

Fussell LK, Bidinger FR, Bieler P (1991) Crop physiology and breeding for drought tolerance: research and development. Field Crop Res 27(3):183–199

Gianola D, Fernando RL, Stella A (2006) Genomic-assisted prediction of genetic value with semiparametric procedures. Genetics 173:1761–1776

Gianola D, Okut H, Weigel KA, Rosa GJ (2011) Predicting complex quantitative traits with bayesian neural networks: a case study with Jersey cows and wheat. BMC Genet 12:87

Gopalakrishnan S, Sharma RK, Anand Rajkumar K, Joseph M, Singh VP, Singh AK et al (2008) Integrating marker assisted background analysis with foreground selection for identification of superior bacterial blight resistant recombinants in basmati rice. Plant Breed 127(2):131–139

Goud TY, Sharma R, Gupta SK, Devi GU, Gate VL, Boratkar M (2016) Evaluation of designated hybrid seed parents of pearl millet for blast resistance. Indian J Plant Protect 44(1):83–87

Govindaraj M, Rai KN, Shanmugasundaram P (2016) Intra-population genetic variance for grain iron and zinc contents and agronomic traits in pearl millet. Crop J 4(1):48–54

Govindaraj M, Rai KN, Kanatti A, Upadhyaya HD, Shivade H, Rao AS (2020a) Exploring the genetic variability and diversity of pearl millet core collection germplasm for grain nutritional traits improvement. Sci Rep 10(1):1–13

Govindaraj M, Virk PS, Kanatti A, Cherian B, Rai KN, Anderson MS, Pfeiffer WH (2020b) Biofortified pearl millet cultivars offer potential solution to tackle malnutrition in India. In: Quantitative genetics, genomics and plant breeding. CABI, Wallingford, pp 385–396

Govindaraj M, Kanatti A, Rai KN, Pfeiffer WH, Shivade H (2021) Association of grain iron and zinc content with other nutrients in pearl millet germplasm, breeding lines, and hybrids. Front Nutr 8:746625

Gupta SK, Sharma R, Rai KN, Thakur RP (2012) Inheritance of foliar blast resistance in pearl millet (Pennisetum glaucum). Plant Breed 131(1):217–219

Gupta SK, Rai KN, Singh P, Ameta VL, Gupta SK, Jayalekha AK et al (2015) Seed set variability under high temperatures during flowering period in pearl millet (Pennisetum glaucum L. (R.) Br.). Field Crop Res 171:41–53

Gupta SK, Ameta VL, Pareek S, Mahala RS, Jayalekha AK, Deora VS et al (2016) Genetic enhancement for flowering period heat tolerance in peart millet (Pennisetum glaucum L.(R.)

Br.). In: 7th International Crop Science Congress, Beijing. http://oar.icrisat.org/9757/1/Page1.pdf. Accessed 30 Sept 2022

Gupta SK, Ameta VL, Pareek S, Singh I, Deora VS, Verma YS et al (2019) Enhancing flowering period heat tolerance in pearl millet through shuttle breeding. In: Presented at 3rd agriculture and climate change conference, Budapest. https://www.elsevier.com/__data/assets/pdf_file/0010/847882/AGRI-2019-Oral-Programme_v5_DE.pdf. Accessed 30 Sept 2022

Habier D, Fernando RL, Kizilkaya K, Garrick DJ (2011) Extension of the Bayesian alphabet for genomic selection. BMC Bioinformatics 12:186–197

Heffner EL, Sorrells ME, Jannink JL (2009) Genomic selection for crop improvement. Crop Sci 49(1):1–12

Henderson CR, Kempthorne O, Searle SR, von Krosigk CM (1959) The estimation of environmental and genetic trends from records subject to culling. Biometrics 15:192

Hickey LT, N Hafeez A, Robinson H, Jackson SA, Leal-Bertioli S, Tester M, et al (2019) Breeding crops to feed 10 billion. Nat Biotechnol 37(7):744–754

Holliday JA, Wang T, Aitken S (2012) Predicting adaptive phenotypes from multilocus genotypes in sitka spruce (*Picea sitchensis*) using random forest. G3 (Bethesda) 2(9):1085–1093

Jambunathan R, Subramanian V (1988) Grain quality and utilization of sorghum and pearl millet. In: Biotechnology in tropical crop improvement, pp 133–139. https://oar.icrisat.org/518/1/RA_00134.pdf

Jarquin D, Howard R, Liang Z, Gupta SK, Schnable JC, Crossa J (2020) Enhancing hybrid prediction in pearl millet using genomic and/or multi-environment phenotypic information of inbreds. Front Genet 10:1294

Jia Y, Jannink JL (2012) Multiple-trait genomic selection methods increase genetic value prediction accuracy. Genetics 192:1513–1522

Kadam DC, Potts SM, Bohn MO, Lipka A, Lorenz AJ (2016) Genomic prediction of single crosses in the early stages of maize hybrid breeding pipeline. Genes Genom Genet 6:3443–3453

Kanatti A, Rai KN, Radhika K, Govindaraj M, Sahrawat KL, Rao AS (2014) Grain iron and zinc density in pearl millet: combining ability, heterosis and association with grain yield and grain size. Springerplus 3(1):1–12

Kholová J, Hash CT, Kočová M, Vadez V (2011) Does a terminal drought tolerance QTL contribute to differences in ROS scavenging enzymes and photosynthetic pigments in pearl millet exposed to drought? Environ Exp Bot 71(1):99–106

Klápště J, Dungey HS, Telfer EJ, Suontama M, Graham NJ, Li Y, McKinley R (2020) Marker selection in multivariate genomic prediction improves accuracy of low heritability traits. Front Genet 11:499094

Knox JW, Hess TM, Daccache A, Ortola MP (2011) What are the projected impacts of climate change on food crop productivity in Africa and S Asia. DFID systematic review final report. Cranfield University, p 71

Krishnan SG, Singh AK, Rathour R, Nagarajan M, Bhowmick PK, Ellur RK et al (2019) Rice variety Pusa Samba 1850. Indian J Genet 79:109–110

Krishnappa G, Savadi S, Tyagi BS, Singh SK, Mamrutha HM, Kumar S et al (2021) Integrated genomic selection for rapid improvement of crops. Genomics 113(3):1070–1086

Legarra A, Reverter A (2018) Semi-parametric estimates of population accuracy and bias of predictions of breeding values and future phenotypes using the LR method. Genet Sel Evol 50:53–68

Liang Z, Gupta SK, Yeh CT, Zhang Y, Ngu DW, Kumar R et al (2018) Phenotypic data from inbred parents can improve genomic prediction in pearl millet hybrids. G3 (Bethesda) 8(7):2513–2522

Long N, Gianola D, Rosa GJM, Weigel KA (2011) Application of support vector regression to genome-assisted prediction of quantitative traits. Theor Appl Genet 123:1065–1074

Maenhout S, De Baets B, Haesaert G, Van Bockstaele E (2007) Support vector machine regression for the prediction of maize hybrid performance. Theor Appl Genet 115:1003–1013

Maganlal SJ, Sanghani AO, Kothari VV, Raval SS, Kahodariya JH, Ramani HR et al (2018) The SSR based linkage map construction and identification of QTLs for blast (Pyricularia grisea).

resistance in pearl millet (Pennisetum glaucum (L.) r. br.). J Pharmacogn Phytochem 7(2): 3057–3064

Massman JM, Gordillo A, Lorenzana RE, Bernardo R (2013) Genomewide predictions from maize singlecross data. Theor Appl Genet 126:13–22

Meuwissen THE, Hayes BJ, Goddard ME (2001) Prediction of total genetic value using genome-wide dense marker maps. Genetics 157:1819–1829

Momen M, Mehrgardi AA, Sheikhi A, Kranis A, Tusell L, Morota G, Rosa GJM, Gianola D (2018) Predictive ability of genome-assisted statistical models under various forms of gene action. Sci Rep 8:12309

Muleta KT, Pressoir G, Morris GP (2019) Optimizing genomic selection for a sorghum breeding program in Haiti: a simulation study. G3 (Bethesda) 9(2):391–401

Neeraja CN, Maghirang-Rodriguez R, Pamplona A, Heuer S, Collard BC, Septiningsih EM et al (2007) A marker-assisted backcross approach for developing submergence-tolerant rice cultivars. Theor Appl Genet 115(6):767–776

Ogutu JO, Schulz-Streeck T, Piepho HP (2012) Genomic selection using regularized linear regression models: ridge regression, lasso, elastic net and their extensions. BMC Proc 6:S10

Ornella L, Singh S, Perez P, Burgueño J, Singh R, Tapia E et al (2012) Genomic prediction of genetic values for resistance to wheat rusts. Plant Genome 5(3):136

Poland J, Endelman J, Dawson J, Rutkoski J, Wu S, Manes Y et al (2012) Genomic selection in wheat breeding using genotyping-by-sequencing. Plant Genome 5(3):103

Raghvani KL, Juneja RP, Ghelani YH, Parmar GM, Dangaria CJ (2008) Influence of abiotic factors on population fluctuations of major insect pest of pearl millet. Indian J Appl Entomol 22:48–50

Rai KN, Hash CT, Singh AK, Velu G (2008) Adaptation and quality traits of a germplasm-derived commercial seed parent of pearl millet. Plant Genet Resour Newsl 154:20–24

Rai KN, Yadav OP, Gupta SK, Mahala RS (2012) Emerging research priorities in pearl millet. J SAT Agric Res 10:1–5. http://ejournal.icrisat.org/index.htm

Rai KN, Yadav OP, Govindaraj M, Pfeiffer WH, Yadav HP, Rajpurohit BS et al (2016) Grain iron and zinc densities in released and commercial cultivars of pearl millet (Pennisetum glaucum). Indian J Agric Sci 86(3):11–16

Ratna Madhavi K, Rambabu R, Abhilash Kumar V, Vijay Kumar S, Aruna J, Ramesh S et al (2016) Marker assisted introgression of blast (Pi-2 and Pi-54) genes in to the genetic background of elite, bacterial blight resistant indica rice variety, Improved Samba Mahsuri. Euphytica 212(2): 331–342

Ray DK, Ramankutty N, Mueller ND, West PC, Foley JA (2012) Recent patterns of crop yield growth and stagnation. Nat Commun 3(1):1–7

Ray DK, Mueller ND, West PC, Foley JA (2013) Yield trends are insufficient to double global crop production by 2050. PLoS One 8(6):e66428

Robin S, Jeyaprakash P, Amudha K, Pushpam R, Rajeswari S, Manonmani S et al (2019) Rice CR1009 Sub 1 (IET 22187)-A new flood tolerant rice variety. Electron J Plant Breed 10(3): 995–1004

Rothman AJ, Levina E, Zhu J (2010) Sparse multivariate regression with covariance estimation. J Comput Graph Stat 19:947–962

Sehgal D, Skot L, Singh R, Srivastava RK, Das SP, Taunk J et al (2015) Exploring potential of pearl millet germplasm association panel for association mapping of drought tolerance traits. PLoS One 10(5):e0122165

Sharma HC, Youm O (1999) Host plant resistance in integrated pest management. In: Khairwal IS, Rai KN, Andrews DJ, Harinarayana G (eds) Pearl millet breeding. Oxford and IBH Co, Pvt. Limited, New Delhi, pp 381–418

Sharma R, Upadhyaya HD, Manjunatha SV, Rai KN, Gupta SK, Thakur RP (2013) Pathogenic variation in the pearl millet blast pathogen Magnaporthe grisea and identification of resistance to diverse pathotypes. Plant Dis 97(2):189–195

Sharma S, Sharma R, Pujar M, Yadav D, Yadav Y, Rathore A et al (2021) Use of wild Pennisetum species for improving biotic and abiotic stress tolerance in pearl millet. Crop Sci 61(1):289–304

Shivhare R, Asif MH, Lata C (2020) Comparative transcriptome analysis reveals the genes and pathways involved in terminal drought tolerance in pearl millet. Plant Mol Biol 103(6):639–652
Singh F, Nainawatee HS (1999) Grain quality traits. In: Khairwal IS, Rai KN, Andrews DJ, Harinarayana G (eds) Pearl millet breeding. Oxford & IBH, New Delhi, pp 157–183
Singh P, Singh U, Eggum BO, Kumar KA, Andrews DJ (1987) Nutritional evaluation of high protein genotypes of pearl millet (Pennisetum americanum (L.) Leeke). J Sci Food Agric 38(1): 41–48
Singh SD, King SB, Reddy PM (1990) Registration of five pearl millet germplasm sources with stable resistance to downy mildew. Crop Sci 30(5):1164
Singh AK, Gopala Krishnan S, Ellur RK, Bhowmick PK, Nagarajan M, Vinod KK et al (2017a) Notification of basmati rice variety, Pusa basmati 1728. Indian J Genet 77:584
Singh AK, Gopala Krishnan S, Nagarajan M, Bhowmick PK, Ellur RK, Haritha B et al (2017b) Notification of basmati rice variety Pusa basmati 1637. Indian J Genet 77:583–584
Singh S, Sharma R, Pushpavathi B, Gupta SK, Durgarani CV, Raj C (2018) Inheritance and allelic relationship among gene (s) for blast resistance in pearl millet [Pennisetum glaucum (L.) R. Br.]. Plant Breed 137(4):573–584
Spindel J, Begum H, Akdemir D, Virk P, Collard B, Redona E et al (2015) Genomic selection and association mapping in rice (Oryza sativa): effect of trait genetic architecture, training population composition, marker number and statistical model on accuracy of rice genomic selection in elite, tropical rice breeding lines. PLoS Genet 11(2):e1004982
Srivastava RK, Singh RB, Pujarula VL, Bollam S, Pusuluri M, Chellapilla TS et al (2020) Genome-wide association studies and genomic selection in pearl millet: advances and prospects. Front Genet 10:1389
Sun M, Huang D, Zhang A, Khan I, Yan H, Wang X et al (2020) Transcriptome analysis of heat stress and drought stress in pearl millet based on Pacbio full-length transcriptome sequencing. BMC Plant Biol 20(1):1–15
Sun M, Lin C, Zhang A, Wang X, Yan H, Khan I et al (2021) Transcriptome sequencing revealed the molecular mechanism of response of pearl millet root to heat stress. J Agron Crop Sci 207(4):768–773
Technow F, Schrag TA, Schipprack W, Bauer E, Simianer H, Melchinger AE (2014) Genome properties and prospects of genomic prediction of hybrid performance in a breeding program of maize. Genetics 197(4):1343–1355
Thakur RP, Williams RJ (1980) Pollination effects on pearl millet ergot. Phytopathology 70(2): 80–84
Thakur RP, King SB, Rai KN, Rao VP (1992) Identification and utilization of smut resistance in pearl millet. International Crops Research Institute for the Semi-Arid Tropics, Hyderabad
Thakur RP, Sharma R, Rao VP (2011) Screening techniques for pearl millet diseases. Information bulletin no 89
Usai MG, Goddard ME, Hayes BJ (2009) LASSO with cross-validation for genomic selection. Genet Res 91:427–436
Varshney RK, Mohan SM, Gaur PM, Chamarthi SK, Singh VK, Srinivasan S et al (2014a) Marker-assisted backcrossing to introgress resistance to Fusarium wilt race 1 and Ascochyta blight in C 214, an elite cultivar of chickpea. Plant Genome 7(1):plantgenome2013-10
Varshney RK, Pandey MK, Janila P, Nigam SN, Sudini H, Gowda MVC et al (2014b) Marker-assisted introgression of a QTL region to improve rust resistance in three elite and popular varieties of peanut (Arachis hypogaea L.). Theor Appl Genet 127(8):1771–1781
Varshney RK, Shi C, Thudi M, Mariac C, Wallace J, Qi P, Zhang H, Zhao Y, Wang X, Rathore A, Srivastava RK, Chitikineni A, Fan G, Bajaj P, Punnuri S, Gupta SK, Wang H, Jiang Y, Couderc M, Katta MAVSK, Paudel DR, Mungra KD, Chen W, Harris-Shultz KR, Garg V, Desai N, Doddamani D, Kane NA, Conner JA, Ghatak A, Chaturvedi P, Subramaniam S, Yadav OP, Berthouly-Salazar C, Hamidou F, Wang J, Liang X, Clotault J, Upadhyaya HD, Cubry P, Rhoné B, Gueye MC, Sunkar R, Dupuy C, Sparvoli F, Cheng S, Mahala RS, Singh B, Yadav RS, Lyons E, Datta SK, Hash CT, Devos KM, Buckler E, Bennetzen JL, Paterson AH, Ozias-

Akins P, Grando S, Wang J, Mohapatra T, Weckwerth W, Reif JC, Liu X, Vigouroux Y, Xu X (2017) Pearl millet genome sequence provides a resource to improve agronomic traits in arid environments. Nat Biotechnol 35:969–976

Vasistha NK, Balasubramaniam A, Mishra VK, Srinivasa J, Chand R, Joshi AK (2017) Molecular introgression of leaf rust resistance gene Lr34 validates enhanced effect on resistance to spot blotch in spring wheat. Euphytica 213(12):1–10

Velu G, Rai KN, Sahrawat KL, Sumalini K (2008) Variability for grain iron and zinc contents in pearl millet hybrids. J SAT Agric Res 6:4p

Xu S (2007) An empirical Bayes method for estimating epistatic effects of quantitative trait loci. Biometrics 63:513–521

Yadav OP, Rai KN (2013) Genetic improvement of pearl millet in India. Agric Res 2(4):275–292

Yadav OP, Singh DV, Dhillon BS, Mohapatra T (2019) India's evergreen revolution in cereals. Curr Sci 116(11):1805–1808

Yadav OP, Gupta SK, Govindaraj M, Sharma R, Varshney RK, Srivastava RK et al (2021) Genetic gains in pearl millet in India: insights into historic breeding strategies and future perspective. Front Plant Sci 12:645038

Zhang A, Ji Y, Sun M, Lin C, Zhou P, Ren J et al (2021) Research on the drought tolerance mechanism of Pennisetum glaucum (L.) in the root during the seedling stage. BMC Genomics 22(1):1–14

Zhong S, Dekkers JC, Fernando RL, Jannink JL (2009) Factors affecting accuracy from genomic selection in populations derived from multiple inbred lines: a barley case study. Genetics 182(1): 355–364

Zou H, Hastie T (2005) Regularization and variable selection *via* the elastic net. J R Stat Soc B 67: 301–320

Genome Editing and Opportunities for Trait Improvement in Pearl Millet

7

Debasmita Panda, Mirza J. Baig, and Kutubuddin A. Molla

Abstract

Pearl millet [*Cenchrus americanus* (L.) Morrone; also known as *Pennisetum glaucum*], originated 4900 years ago, is a C4 crop with high photosynthetic efficiency and fulfills the food and fodder needs of resource-poor farmers of sub-Saharan Africa, Southeast Asia, and the Indian subcontinent. Pearl millet is a climate-ready crop and grows well in poor and low-fertility soil. It is profoundly nutritious, fiber-rich, and non-glutinous. Based on transcriptome and bioinformatics studies, it is estimated that 1.79 Gb of the pearl millet genome consists of 38,579 genes. Unfortunately, functional genomics and genotype-phenotype association in pearl millet are poorly explored areas. Pearl millet suffers from low yield for many reasons. There is an urgent need to validate the functions of important genes to improve the crop and better utilize it for future agriculture in a changing climate scenario. In recent years, genome editing, especially CRISPR-Cas, has come under the spotlight for improving crop varieties. In this chapter, we discuss how the available genome editing tools can play a significant role in deciphering the functions of pearl millet genomic regions and crop improvement. We also highlight major bottlenecks to using genome editing in pearl millet and discuss possible ways to overcome those constraints.

Keywords

Climate-smart crop · C4 grass · CRISPR-Cas9 · Base editing · Prime editing · Abiotic stress · Biotic stress

D. Panda · M. J. Baig · K. A. Molla (✉)
National Rice Research Institute, Cuttack, Odisha, India
e-mail: kutubuddin.molla@icar.gov.in

V. A. Tonapi et al. (eds.), *Pearl Millet in the 21st Century*,
https://doi.org/10.1007/978-981-99-5890-0_7

Abbreviation

ABEs	Adenine base editors
CBEs	Cytosine base editors
CGBEs	C to G base editors
CRISPR-Cas	Clustered regularly interspaced short palindromic repeats/CRISPR-associated protein
DSBs	Double-strand breaks
HDR	Homology-directed repair
MMEJ	Microhomology mediated end joining
NHEJ	Non-homologous end joining
pegRNA	Prime editing gRNA
RFLP	Restriction fragment length polymorphism
SNPs	Single nucleotide polymorphisms
SSN	Sequence-specific nuclease
TALENs	Transcription activator like effector nucleases
ZFNs	Zinc finger nucleases

7.1 Introduction

Global shifts in climate and population increase are the two significant issues of concern in the present scenario. Estimates indicate that the human population will reach approximately 10 billion by 2050, which has a counter effect on food security (FAO 2017). Climate change could be a serious threat to global food security. About one-third of the total greenhouse gas emissions is contributed by our food system. We have been focusing heavily on a few cereal crops (wheat, rice, and maize) as sources of macro- and micronutrients, ignoring their potential for global warming. For example, wheat has the highest CO_2 emission capacity (4 tons CO_2eq/ha), followed by rice and maize (3.4 tons CO_2 eq/ha) (Jain et al. 2016). Compared to these cereals, there are orphan crops, like millets and sorghum, with lower carbon emission rates (Prasad and Staggenborg 2009). Therefore, focusing on improving orphan crop varieties could help ensure food security in the changing climate. Among these orphan crops, pearl millet is a diploid ($2n = 2x = 14$) C4 grass species having high photosynthetic activity and biomass production ability. It can grow on poor sandy soils. It is well suited for growth in unbearable conditions such as high soil pH, high temperature, low moisture, high salinity, and low rainfall compared with sorghum and maize (Vadez et al. 2012; Nambiar et al. 2011).

Millet grains are highly nutritious, low in starch, and rich in fiber (Nambiar et al. 2011; Kumar et al. 2016). Pearl millet has a broad range of phenotypic and genotypic variations because the cultivation varies in different agro-climatic conditions. Despite all the above advantages, only a limited proportion of its germplasm could be utilized for improving the pearl millet crop for different important traits. Pearl millet suffers from a low crop productivity rate with an average grain yield of just 900 kg/ha, the presence of anti-nutritional factors in the grains, and the vulnerability to several foliar diseases like downy mildew, blast, and smut. Although pearl millet grains are gluten free and could be a recommended diet for people with celiac disease, it is not popular due to their rapid rancidity upon storage.

Historically, crop plants have been improved by harnessing genetic diversity. Genetic diversity is an effect of mutations found in various regions of the genome, which can be natural or artificially induced. Natural mutations are extremely slow. To create rapid and useful genetic variations, researchers use four major techniques: conventional breeding through crossing, mutation breeding, transgenesis, and genome editing (Chen et al. 2019). Conventional breeding or cross-breeding is a targeted crossing between two plants via sexual recombination. Major restrictions of this technique are—it is time-consuming, only applicable for sexually compatible plants, the occurrence of linkage drag, and low genetic variability. Later, mutation breeding came to the forefront, relying on chemical and radiation treatment (Holme et al. 2019). This technique generates random mutations, necessitates extensive screening, and demands intensive labor. Subsequently, transgenic breeding technology empowered us to introduce beneficial genes from virtually any organism to crop plants for desired traits. While it has huge potential for crop improvement, developing crop varieties through transgenesis is expensive and faces stringent regulatory hurdles in many countries. For example, golden rice (vitamin A enriched transgenic rice), developed nearly two decades ago to combat malnutrition in the developing world, received approval for open cultivation in the Philippines in July 2021 (https://www.irri.org). Moreover, the integration site of the transgene in the genome is not controllable. Most recently, genome editing technology has emerged as the latest breeding technique, allowing for precise, targeted, and predictable mutations in crop varieties. Genome editing has been rapidly adopted across plant species to generate valuable genetic variations (Gao 2021; Molla et al. 2021). In this chapter, we will discuss the fundamentals of genome editing technology, advanced genome editing tools, a few noteworthy examples of crop improvement through genome editing, and the potential applications of genome editing for enhancing traits in orphan crops like pearl millet.

7.2 Basics of Genome Editing

Genome editing represents a significant advancement over conventional breeding, as it allows us to modify genetic regions of interest more precisely and rapidly in a targeted fashion. The fundamental of genome editing is to create a targeted double-strand break (DSB) in a desired genomic location. DSBs are created with different sequence-specific nucleases (SSN). For example, ZFNs (zinc finger nucleases) (Kim et al. 1996), TALENs (transcription activator-like effector nucleases) (Christian et al. 2010), and CRISPR-Cas (clustered regularly interspaced short palindromic repeats/CRISPR-associated protein) (Jinek et al. 2012) are well studied programmable SSN systems for recognizing and modifying specific DNA sequences. ZFNs and

TALENs systems rely on protein-based DNA recognition, whereas CRISPR-Cas depends on RNA-guided DNA recognition. Therefore, for targeting each new DNA sequence by ZFNs and TALENs, new proteins need to be engineered. However, new DNA sequences could be targeted by CRISPR-Cas by manipulating a small RNA sequence kmown as guide RNA (gRNA). This RNA-dependent DNA targeting platform gained rapid popularity for manipulating the genome of diverse organisms, including plants. The creation of a DSB by SSN triggers cellular pathways to repair the damaged DNA. There are two methods for repairing DSBs: one is more common but error-prone, known as non-homologous end joining (NHEJ), and the other is less frequent but errorless, called homology-directed repair (HDR) (Molla and Yang 2020). Besides canonical NHEJ and HDR, another type of error-prone repair system is called microhomology-mediated end joining (MMEJ). Out of these repair systems, NHEJ and MMEJ lead to the creation of indels (insertions and deletions) at the DSB sites of the target gene, potentially resulting in frameshift mutations and enabling effective gene knockout. In summary, genome editing through CRISPR-Cas methods generates indels in the genome, leading to genetic variation in organisms.

CRISPR-Cas9 genome editing tool is a two-component system: a customizable small guide RNA (gRNA) and a Cas9 nuclease. The 20-nucleotide sequence present at the 5′ end of the guide RNA is crucial for site-specific DNA recognition and subsequent cleavage by Cas9. Therefore, for each new target DNA, only the 20-nucleotide sequence of the guide RNA needs to be changed, while the other components remain the same. Although single guide RNA (gRNA)-based CRISPR is a common tool for targeting a single gene, multiple genes can be edited in a single experiment by including an array of guide RNAs with Cas9. Multiplex editing is a powerful technology for large-scale genome engineering, parallel targeting of gene family members, and metabolic engineering. A schematic diagram of the steps involved in plant genome editing through CRISPR-Cas9 is presented in Fig. 7.1. Available genome editing tools are listed in the Table 7.1.

7.3 Precise Editing with CRISPR-Cas9

CRISPR-Cas technique efficiently generates random insertions or deletions, which cannot be controlled. Although the mutation outcome can be predicted (Molla and Yang 2020), conventional CRISPR-Cas tools are unsuitable for generating precise sequence modifications. Base editing and prime editing, the two advanced CRISPR-Cas-based approaches, can create precise mutations in DNA without generating double-stranded breaks (DSBs) (Molla et al. 2021). Most of the agronomic characteristics are determined by single nucleotide polymorphisms (SNPs) or combinations of a few SNPs and indels. Introducing specific SNPs from wild to elite varieties through conventional breeding techniques is an extensive and time-consuming procedure; base editing and prime editing could accelerate this process.

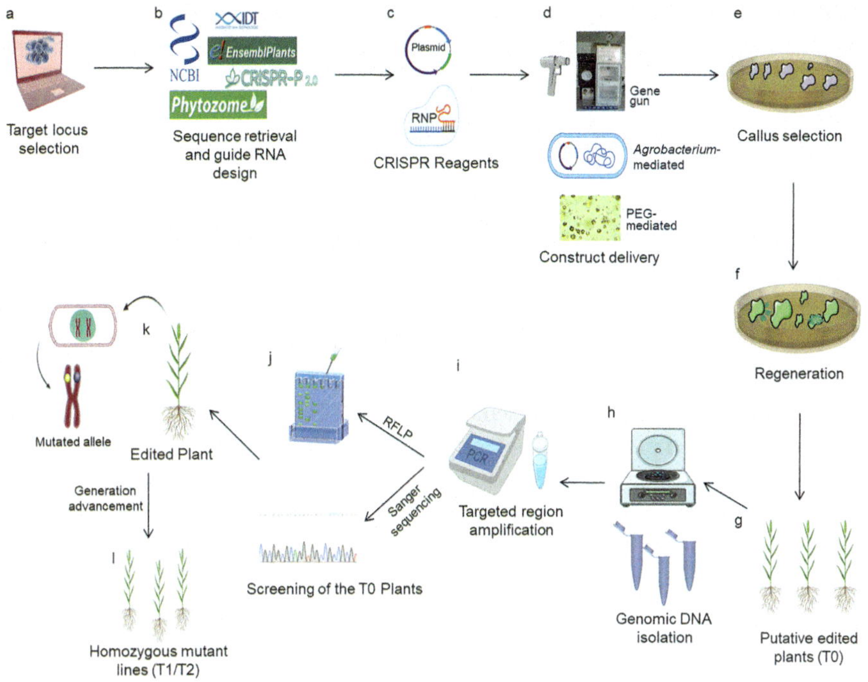

Fig. 7.1 Steps for performing CRISPR-Cas-based genome editing in plants. (**a**, **b**) Target selection and guide RNA design. After the target genes or genetic region is selected, the sequence is retrieved from the public database. The sequence is used in web tools for protospacer design (the 20 bp at the 5′ end of guide RNA). Web tools facilitate predicting suitable guides with high on-target efficiency and low off-target efficiency. (**c**, **d**) CRISPR-reagent (plasmid construct or RNPs) are delivered to explant by biolistic, *Agrobacterium*, or PEG-mediated transfection. For RNP, in-vitro transcribed guide RNA is complexed with Cas9 protein before transforming into plant cells. PEG-mediated transfection is done with isolated protoplast. (**e–g**) Selection and regeneration of the transformed calli. (**h**, **i**) Genomic DNA isolation and amplification of the targeted region from putative edited plants (T0). (**j**) Screening of the putative edited plants by restriction fragment length polymorphism (RFLP) and/or Sanger sequencing. (**k**, **l**) Generation advancement to obtain homozygous lines

7.3.1 Base Editing

Base editors were developed for installing single nucleotide changes in target loci. Recently, three types of DNA base editors have been developed, viz., cytosine base editors (CBEs), adenine base editors (ABEs), and C to G base editors (CGBEs). CBEs create C to T transitions, while ABEs create A to G transitions. CBE, the first among the base editors, contains four different components: a cytosine deaminase (which converts C to U), a modified Cas9 (nCas9 or Cas9 nickase D10A), UGI (uracil DNA glycosylase inhibitor), and a guide RNA (Komor et al. 2016; Nishida et al. 2016; Molla and Yang 2019). It induces the creation of C to T transition (G to A in the complementary strand) in the genome. ABEs are made up of nCas9 (D10A) fused with protein-engineered deoxyadenosine deaminase and can convert targeted

Table 7.1 Available tools for nuclear genome editing

Genome editing tool	Targeted modification possible
ZFN	Genetic variation, knock-out
TALEN	Genetic variation, knock-out
CRISPR-Cas9	Genetic variation, knock-out
Cytosine Base editors (CBEs)	C-to-T point mutation (G-to-A in the opposite strand)
Adenine base editors (ABEs)	A-to-G point mutation (T-to-C in the opposite strand)
C-to-G base editors (CGBEs)	C-to-G point mutation (G-to-C in the opposite strand)
Prime editors	All 12 types of point mutations, predefined small insertions and deletions (indels), and combinations of point mutations and indels

A to G (T to C in the complementary strand) (Gaudelli et al. 2017). Similarly, recently developed platforms enable us to execute C to G conversion (G to C in the opposite strand) in the genome (Molla et al. 2020a). Diverse plant-based editing platforms have evolved and been applied for performing precise base alteration in various plant species (Molla et al. 2021).

7.3.2 Prime Editing

Although base editors opened a new path to perform single nucleotide changes (C to T, A to G, and C to G) in the genome, they are restricted to transition mutations and two types of transversion mutations. The other six types of transversion mutations (C to A, G to T, A to T, T to A, A to C, and T to G) can not be installed using the available base editors at the time of writing this chapter. A recently developed genome editing technique called prime editing enables us to install all 12 types of base conversions, small insertions, deletions, and combinations of those mutations (Anzalone et al. 2019). Prime editors consist of a nickase Cas9 (nCas9-H840A) fused with reverse transcriptase (RT) and an extended guide RNA. The extended guide RNA is called prime editing gRNA (pegRNA), which contains information to rectify the target sequence (Anzalone et al. 2019). The 5′ end of the pegRNA carries the guide region, which binds to the target site of the genome, while the 3′ extension contains the primer binding site and RT template. nCas9-H840A creates a nick in the DNA. The primer binding site binds to the 3′ end of the nicked DNA strand, and the RT template encodes the desired edit that is directly copied to the genomic DNA by the RT enzyme. Prime editing is considered a highly promising tool in the field of genome editing because it can create transversion mutations and precise insertions and deletions. Although its efficiency is currently low in plant systems (Molla et al. 2021), with technical improvements, prime editing holds great promise for plant genome engineering in the near future.

7.4 Some Spotlight Case Studies on Crop Improvement Through Genome Editing

Genome editing technologies alternatively called targeted mutagenesis enable us to modify the gene of interest rapidly and precisely The repurposing of the bacterial CRISPR-Cas9 immune system in 2012 has brought about a revolution in the field of genome engineering. The rapid advancement of CRISPR-Cas technologies, from double strand break (DSB) to base and prime editing, has opened new paths of precise plant and animal genome editing. It would not be an exaggeration to say that we are now in an era where we can modify genes precisely predictably. In 2013, several groups began applying the CRISPR-Cas system to plants (Feng et al. 2013; Shan et al. 2013; Xie and Yang 2013). Since then, we have witnessed numerous succesful examples of utilizing this tool for crop improvement. For example, broad-spectrum bacterial blight-resistant rice (Oliva et al. 2019), low-gluten wheat (Sánchez-León et al. 2018), waxy corn (Gao et al. 2020), male sterile wheat (Li et al. 2020a), clonally propagated hybrid rice (Khanday et al. 2019; Wang et al. 2019), and high yielding tomato (Rodríguez-Leal et al. 2017) have been developed with conventional CRISPR-Cas9. Recently, high oleic acid soybean in the USA and high GABA tomato in Japan have been commercialized and are available for human consumption. This is remarkable that within a few years of its development, genome editing tools have provided us with commercialized improved crop varieties.

In the last few years, base editing evolved as a promising device for performing genome editing with single base resolution. CBEs have been demonstrated to alter C to T in several crops like rice (Shimatani et al. 2017), and maize (Zong et al. 2018). Similarly, A to G conversion was executed by ABE in rice (Li et al. 2018; Molla et al. 2020b) and wheat (Li et al. 2018). Recently, C to G base conversion in rice, tomato, and poplar was achieved (Sretenovic et al. 2021). Single base editing in acetolactate synthase (ALS), acetyl-CoA carboxylase (ACCase), and TubA2 genes have been found to confer herbicide resistance in rice, wheat, Arabidopsis, and maize (Shimatani et al. 2017; Zhang et al. 2019; Liu et al. 2021; Dong et al. 2020; Li et al. 2020b). Herbicide-tolerant germplasm generated through base editing will serve greatly in effective weed management. Base editors were also used to alter the nutritional composition of fruits. For example, the sugar content in strawberries and carotenoid content in tomatoes increased with base editors (Hunziker et al. 2020; Xing et al. 2020). In a recent review, Molla et al. (2021) comprehensively portrayed the current and potential crop improvement applications of base editors and prime editors. Therefore, along with canonical CRISPR-Cas tools, base, and prime editors hold great promise for pearl millet improvement.

7.5 Harnessing Climate-Resilience and Nutritional Value of Pearl Millet by Genome Editing

The C4 crop pearl millet was domesticated approximately 4900 years ago in West Africa, and it plays a central role in ensuring food security in arid and semi-arid regions of Asia and Africa (Burgarella et al. 2018). It is well adapted to abiotic and biotic stresses; for example, some wild pearl millet varieties can thrive in extreme environmental conditions (they can withstand temperatures exceeding 42 °C) (Gupta et al. 2015). Therefore, pearl millet could serve as a fascinating biological model for ensuring future food security in a changing climate. Varshney et al. (2017) reported that the 1.79 GB pearl millet genome has a high GC content (47.9%), which is a characteristic associated with desiccation tolerance in monocots. The pearl millet genome is also enriched with cutin, suberin, and wax biosynthetic genes, which might contribute to its drought and heat tolerance (Varshney et al. 2017). Furthermore, Sharma et al. (2020) reviewed that high genetic differentiation is found between cultivated and wild *Pennisetum* species, revealing that only 74% of the genetic diversity found in wild species is recognized in cultivated species. Thus, domestication has led to a reduction in genetic diversity. A group of researchers suggests that the wild *Pennisetum* species could serve as a repository for preserving new allelic variants associated with climate-resilient and disease-resistant traits (Oumar et al. 2008; Mariac et al. 2006). Although pearl millet is a good source of nutrition, including proteins, vitamins, and minerals, major concerns about it include low grain yield, the presence of anti-nutritional factors like polyphenols and phytic acid, and the rapid rancidity of pearl millet flour. Therefore, increasing genetic diversity will pave the way to develop pearl millet varieties that combine improved nutritional values with high yields and disease resistance.

Traditional breeding has the potential to harness allelic variants from wild germplasm to incorporate beneficial traits. However, challenges such as linkage drag and crossing barriers often impede this process. Therefore, to complement traditional breeding efforts, CRISPR-Cas-based genome editing tools can be used to remove undesired characterstics by knocking out genes and to mimic beneficial alleles through precise editing within the genetic background of cultivated species (Table 7.2).

7.5.1 Pearl Millet Traits Improvement for Nutrition

Tackling the hunger of the world population amid changing climate conditions is the biggest challenge facing humanity. Thus, improvements in cereal crops like pearl millet, which not only serves as a major source of carbohydrates, proteins, fats, minerals, and vitamins but also thrives in harsh climatic conditions, could play a crucial role in addressing these challenges. Although pearl millet is a nutritionally rich crop, it has remained underutilized and is not widely popular among the general people and food processing sectors. Here, present some futuristic ideas on how

Table 7.2 Potential target genes for genome editing for pearl millet improvement

Genes	Function	Traits	Types of genome editing	References
PglYSL2, PglZIP2, PglZIP8, PglNRAMP2, PglFER1	Zn and Fe metabolism	Functional genomics and nutritional improvements	Knock out	Mahendrakar et al. (2020)
Waxy	Amylose synthesizing gene	Starch quality	Knock out	Nakayama et al. (1998)
ACCase	Acetyl CoA carboxylase	High yield under drought	Base editing	Sehgal et al. (2015)
PHYC	Phytochrome C	Flowering time and spike length	Base editing/ prime editing	Saïdou et al. (2009)
PPR	Pentatricopeptide repeat	Flowering time	Base editing/ prime editing	Diack et al. (2020)
CAD	Cinnamyl alcohol dehydrogenase	Biofuel production	Knock out	Liu et al. (2019)
Dwarf 8	Dwarf protein	Reduce height	Knock out	Diack et al. (2020), Parvathaneni et al. (2019)
LOX	Lipoxygenase	Increased shelf life	Knock out	Sharma and Chugh (2017)

CRISPR-Cas-derived genome editing tools could accelerate the genetic improvement of pearl millet.

Pearl millet is inherently nutritious and contains an appropriate amounts of proteins, dietary fiber, macro-, and micronutrients. Alarmingly, in India, 54% of children and 74% of women suffer from iron (Fe) deficiency anemia, while 54% of children below 5 years are stunted due to zinc (Zn) deficiency (WHO 2020). Pearl millet is a rich source of the principal micronutrients iron and zinc compared to other cereals like rice and wheat (Tako et al. 2015). However, the availability of these micronutrients is limited due to the presence of antinutritional factors like phytic acid, which chelate them into multivalent cations (Fe^{2+}, Zn^{2+}, etc.). CRISPR-Cas9-mediated disruption of the *TaIPK1* gene has been shown to reduce phytic acid and enhance Iron and Zinc content in wheat grains (Ibrahim et al. 2021). Similarly, the knockout of the *BnITPK* gene has been found to reduce phytic acid in oilseed rape (Sashidhar et al. 2020). A similar strategy could be employed to reduce phytic acid content in pearl millet grain, thereby increasing the availability of those key nutrients.

Govindaraj et al. (2020) developed a core collection of 39 accessions of pearl millet with multiple nutrients. However, the genetic regions or genes associated with higher nutrients are largely unknown. Genetic studies aimed at identifying these genes and functional genomics studies would greatly facilitate the efficient use of this germplasm. Similarly, Mahendrakar et al. (2020) identified several putative candidate genes for grain iron and zinc content. In this regard, CRISPR-Cas9 would play a significant role in functional genomics to elucidate their exact function. Interestingly, a recent study identified single nucleotide polymorphisms (SNPs) that co-segregate with Fe and Zn (Pujar et al. 2020). Once the responsible and functional SNPs are pinpointed, these variations could be introduced into cultivars using precise editing tools like base and prime editors.

The starch composition of pearl millet can be altered using genome editing tools. For example, waxy grains of millet have nearly 0% amylose content, are recommended for infants because they are easily digestible. The waxy phenotype is produced by the recessive allele *wx*, which is a loss of function allele of the dominant *Wx* allele (Nakayama et al. 1998; Fukunaga et al. 2002). CRISPR-Cas-mediated knockout of this dominant *Wx* allele can create waxy grains in pearl millet.

One of the key reasons for pearl millet's limited popularity among consumers and the food processing industry is its rapid rancidity during storage. Pearl millet flour contains a high level of lipids, which are prone to oxidation by lipoxygenase (LOX) to produce lipid hydroperoxides (Vinutha et al. 2022). Subsequent nonenzymatic oxidation of lipid hydroperoxides leads to rancidity of flours during storage. Genome editing can be used to knock out suitable candidate *LOX* gene(s) to increase the shelf life of pearl millet flour. However, care should be taken to avoid generating any concomitant phenotypic anomalies.

A recent study on green millet (*Setaria viridis*) showed the use of CRISPR-Cas9 to knock out the *Less Shattering1* (*SvLes1*) gene, which controls seed shattering (Mamidi et al. 2020). The strategy and methodology with appropriate modifications could be beneficial for the successful application of CRISPR-Cas9 in pearl millet genotypes.

7.5.2 Pearl Millet Traits Improvement Related to Different Abiotic Stress

The orphan crop pearl millet thrives in adverse conditions such as water scarcity and high temperatures. Its high adaptability to various abiotic stresses makes it a useful crop for functional genomics studies to understand the molecular basis of stress adaptation and tolerance (Shivhare and Lata 2017). However, it is unfortunate that only a limited number of stress-related genes have been identified and validated in pearl millet to date. By combining transcriptome data and bioinformatics analysis, Varshney et al. 2017 predicted that a strong advancement of the cutin, suberin, and wax biosynthesis genes has occurred in pearl millet compared to other crop varieties (Debieu et al. 2017). The knockout approach using CRISPR-Cas9 will be useful in rapidly analyzing stress-responsive genes. Drought adversely affects crop yields by at least 25% (Choudhary et al. 2015). In such environmental conditions, improved drought-tolerant varieties of pearl millet could be a favorable choice for future generations (Habiyaremye et al. 2016). Sehgal et al. (2015) reported that an SNP in the *acetyl CoA carboxylase* gene is associated with grain yield and panicle number under drought stress. A single nucleotide change from C to T in acetyl CoA carboxylase alleles has shown a significant difference in grain yield; with the T-containing allele producing more grains than the C allele. In the future, using cytosine base editors (CBEs), researchers can target the C nucleotide in the acetyl CoA carboxylase gene to improve grain yield in pearl millet under drought.

Apart from drought stress, temperature stresses can affect the germination, photosynthesis, and groundwater uptake of pearl millet varieties (Dwivedi et al. 2012; Garcia-Huidobro et al. 1982). Although pearl millet has the inherent capacity to withstand abiotic stresses, some varieties are sensitive to these stresses. For example, H 77/833-2, a popular inbred line in northwestern India, is sensitive to drought stress but tolerates high-temperature stress (Yadav et al. 2014). Therefore, the C to T SNP mentioned earlier could be introduced by CBE into this genotype to transform it into a drought-tolerant variety.

Unlike other crops, pearl millet is also susceptible to biotic stresses such as fungal, bacterial, and viral infections. Fungal diseases like downy mildew, smut, rust, blast, and ergot are considered important pathogens of pearl millet. Downy mildew (DM), caused by *Sclerospora graminicola,* is a major concern for pearl millet farmers worldwide (Kumar and Manga 2011). It has been reported that the loss of the *Mildew Resistance Locus O (MLO)* results in resistance to powdery mildew in barley (Jacott et al. 2021). The *MLO* gene is a susceptibility gene for powdery mildew fungus and has been targeted by researchers using genome editing to develop resistance in wheat (Wang et al. 2014; Li et al. 2022) and grapevine (Wan et al. 2020). However, there have been no reports regarding the susceptibility genes of the downy mildew fungus in millets. *MLO* gene families could be identified through genome-wide analysis, and a similar strategy to knockout candidate *MLO* genes could be used for developing resistance. Numerous strategies are available to develop disease resistance in plants using the power of genome editing tools (Karmakar et al. 2022), which could also be harnessed for resistance breeding in pearl millet.

7.5.3 Pearl Millets for Biofuel

Like other cereals, pearl millet holds excellent potential for biofuel production using its lignocellulosic biomass. However, lignin is a major barrier to converting cellulose into biofuel. It would be interesting to mutate key lignin biosynthesis genes like *cinnamyl alcohol dehydrogenase (CAD)* and *caffeic acid O-methyltransferase (COMT)* to reduce lignin content and develop low-lignin pearl millet. A report by Liu et al. (2019) has already demonstrated CRISPR-Cas-mediated knock out of *CAD* gene in *Sorghum.*

7.6 Bottlenecks to Use Genome Editing in Pearl Millet Improvement

Successful genome editing in plants requires efficient genetic transformation and a regeneration system. To unlock the potential of genome editing in pearl millet, the availability of reliable methods for delivering genome editing reagents and regenerating transformed cells is crucial. Like other monocots, pearl millet faces challenges in tissue culture, transformation, and regeneration. Reports of successful transgenic pearl millet development are scarce in the literature (Latha et al. 2006; Ramineni et al. 2014). The recalcitrance bottleneck must be broken to apply genome editing tools for basic research and improving pearl millet.

In an earlier study, ectopic expression of developmental regulators like *Wuschel (Wus)* and *Baby boom (Bbm)* has been shown to enhance organogenesis efficiency and reduce the time required for regeneration in maize (Lowe et al. 2016). Similarly, a GRE-GRF chimeric protein enhanced the efficiency of regeneration in wheat (Debernardi et al. 2020). Co-delivery of these developmental regulators with gene editing reagents may be beneficial in reducing the recalcitrance of pearl millet to tissue culture and transformation. Along with the conventional biolistic and *Agrobacterium-mediated* gene transfer methods, nanoparticle-based and RNA virus-based delivery methods could also be explored.

7.7 Conclusion and Future Prospective

Pearl millet is a nutritious, climate-resilient crop that can thrive in harsh climatic conditions where other cereals may fail. In recent decades, we have seen tremendous growth in various genetic and genomic studies. Whole-genome sequencing and resequencing efforts of the pearl millet genome have provided insights into the genes and their associations with yield, abiotic and biotic stresses. Despite all these efforts, there is an urgent need to functionally validate genes important for their nutrient contents, stress resistance, and other beneficial traits. Rapid functional validation using genome editing tools would play a significant role in this direction. From knockout to base editing and prime editing, CRISPR-Cas-derived tools and techniques are valuable resources for the rapid characterization of genomic regions and their utilization in pearl millet improvement programs. However, to realize the full potential of genome editing, efficient genetic transformation and regeneration protocol needs to be established for different pearl millet genotypes. The small genetic changes made by editing are impossible to differentiate from those occuring naturally. Therefore, classifying genome-edited crops as GMOs seems irrational, and they should be regulated no differently from crop varieties derived from mutation or conventional breeding. A positive regulatory environment is crucial for the rapid adoption of these powerful new breeding tools. Recently, the Government of India exempted genome-edited crops (free from foreign-DNA) from stringent GMO regulations. There is no doubt that the availability of gene transformation protocols and rapid adoption of genome editing tools will open new avenues for pearl millet improvement, facilitating nutrient fortification and climate-smart agriculture.

References

Anzalone AV, Randolph PB, Davis JR, Sousa AA, Koblan LW, Levy JM et al (2019) Search-and-replace genome editing without double-strand breaks or donor DNA. Nature 576(7785): 149–157

Burgarella C et al (2018) A Western Sahara Centre of domestication inferred from pearl millet genomes. Nat Ecol Evol 2:1377–1380

Chen K, Wang Y, Zhang R, Zhang H, Gao C (2019) CRISPR/Cas genome editing and precision plant breeding in agriculture. Annu Rev Plant Biol 70:667–697

Choudhary M, Jayanand, Padaria JC (2015) Transcriptional profiling in pearl millet (*Pennisetum glaucum L.R. Br.*) for identification of differentially expressed drought responsive genes. Physiol Mol Biol Plants 21:187–196

Christian M, Cermak T, Doyle EL, Schmidt C, Zhang F et al (2010) Targeting DNA double-strand breaks with TAL effector nucleases. Genet 186:757–761

Debernardi JM, Tricoli DM, Ercoli MF, Hayta S, Ronald P, Palatnik JF et al (2020) A GRF–GIF chimeric protein improves the regeneration efficiency of transgenic plants. Nat Biotechnol 38: 1274–1279

Debieu M, Kanfany G, Laplaze L (2017) Pearl millet genome: lessons from a tough crop. Trends Plant Sci 22(11):911–913

Diack O, Kanfany G, Gueye MC, Sy O, Fofana A, Tall H et al (2020) GWAS unveils features between early-and late-flowering pearl millets. BMC Genomics 21(1):1–11

Dong H et al (2020) Generation of imidazolinone herbicide resistant trait in *Arabidopsis*. PLoS One 15:e0233503

Dwivedi S, Upadhyaya H, Senthilvel S, Hash C (2012) Millets: genetic, and genomic resources. In: Janick J (ed) Plant breeding reviews. Wiley, Hoboken, pp 247–375

FAO (2017) The future of food and agriculture: trends and challenges. FAO

Feng Z, Zhang B, Ding W, Liu X, Yang DL, Wei P et al (2013) Efficient genome editing in plants using a CRISPR/Cas system. Cell Res 23(10):1229–1232

Fukunaga K, Kawase M, Kato K (2002) Structural variation in the Waxy gene and differentiation in foxtail millet [*Setaria italica* (L.) P. Beauv.]: implications for multiple origins of the waxy phenotype. Mol Genet Genomics 268(2):214–222

Gao C (2021) Genome engineering for crop improvement and future agriculture. Cell 184(6): 1621–1635

Gao H, Gadlage MJ, Lafitte HR, Lenderts B, Yang M, Schroder M et al (2020) Superior field performance of waxy corn engineered using CRISPR–Cas9. Nat Biotechnol 38(5):579–581

Garcia-Huidobro T, Monteith T, Squire GR (1982) Time temperature and germination of pearl millet (*Pennisetum typhoides* S. *& H.*): I. Constant temperature. J Exp Bot 33:288–296

Gaudelli NM, Komor AC, Rees HA, Packer MS, Badran AH, Bryson DI, Liu DR (2017) Programmable base editing of A• T to G• C in genomic DNA without DNA cleavage. Nature 551(7681):464–471

Govindaraj M, Rai KN, Kanatti A, Upadhyaya HD, Shivade H, Rao AS (2020) Exploring the genetic variability and diversity of pearl millet core collection germplasm for grain nutritional traits improvement. Sci Rep 10(1):1–13

Gupta SK et al (2015) Seed set variability under high temperatures during flowering period in pearl millet (*Pennisetum glaucum* L. (R.) Br.). Field Crops Res 171:41–53

Habiyaremye C et al (2016) Proso millet (*Panicum miliaceum* L.) and its potential for cultivation in the Pacific northwest, US: a review. Front Plant Sci 7:1961

Holme IB, Gregersen PL, Brinch-Pedersen H (2019) Induced genetic variation in crop plants by random or targeted mutagenesis: convergence and differences. Front Plant Sci 10:1468

https://www.irri.org/news-and-events/news/philippines-becomes-first-country-approve-nutrient-enriched-golden-rice

Hunziker J et al (2020) Multiple gene substitution by target-AID base-editing technology in tomato. Sci Rep 10:20471

Ibrahim S, Saleem B, Rehman N, Zafar SA, Naeem MK, Khan MR (2021) CRISPR/Cas9 mediated disruption of inositol Pentakisphosphate 2-kinase 1 (TaIPK1) reduces phytic acid and improves iron and zinc accumulation in wheat grains, vol 37. J Adv Res, p 33

Jacott CN, Ridout CJ, Murray JD (2021) Unmasking mildew resistance locus O. Trends Plant Sci 26(10):1006–1013

Jain N, Arora P, Tomer R, Mishra SV, Bhatia A, Pathak H, Chakraborty D, Kumar V, Dubey D, Harit R et al (2016) Greenhouse gases emission from soils under major crops in Northwest India. Sci Total Environ 542:551–561

Jinek M, Chylinski K, Fonfara I, Hauer M, Doudna JA, Charpentier E (2012) A programmable dual-RNA–guided DNA endonuclease in adaptive bacterial immunity. Science 337(6096): 816–821

Karmakar S, Das P, Panda D, Xie K, Baig MJ, Molla KA (2022) A detailed landscape of CRISPR-Cas-mediated plant disease and pest management. Plant Sci 323:111376

Khanday I, Skinner D, Yang B, Mercier R, Sundaresan V (2019) A male-expressed rice embryogenic trigger redirected for asexual propagation through seeds. Nature 565(7737):91–95

Kim YG, Cha J, Chandrasegaran S (1996) Hybrid restriction enzymes: zinc finger fusions to Fok I cleavage domain. Proc Natl Acad Sci U S A 93:1156–1160

Komor AC, Kim YB, Packer MS, Zuris JA, Liu DR (2016) Programmable editing of a target base in genomic DNA without double-stranded DNA cleavage. Nature 533:420–424

Kumar A, Manga VK (2011) Downy mildew of pearl millet. Biores Bull 4:182–200

Kumar S, Hash CT, Thirunavukkarasu N, Singh G, Rajaram V, Rathore A et al (2016) Mapping quantitative trait loci controlling high iron and zinc in self and open pollinated grains of pearl millet [*Pennisetum glaucum* (L) R. Br.]. Front Plant Sci 7:16–36

Latha AM, Rao KV, Reddy TP, Reddy VD (2006) Development of transgenic pearl millet (*Pennisetum glaucum* (L.) R. Br.) plants resistant to downy mildew. Plant Cell Rep 25(9): 927–935

Li C, Zong Y, Wang Y, Jin S, Zhang D, Song Q et al (2018) Expanded base editing in rice and wheat using a Cas9-adenosine deaminase fusion. Genome Biol 19:1–9

Li J, Wang Z, He G, Ma L, Deng XW (2020a) CRISPR/Cas9-mediated disruption of TaNP1 genes results in complete male sterility in bread wheat. J Genet Genomics 47(5):263–272

Li Y et al (2020b) Precise base editing of non-allelic acetolactate synthase genes confers sulfonylurea herbicide resistance in maize. Crop J 8:449–456

Li S, Lin D, Zhang Y, Deng M, Chen Y, Lv B et al (2022) Genome-edited powdery mildew resistance in wheat without growth penalties. Nature 602(7897):455–460

Liu G, Li J, Godwin ID (2019) Genome editing by CRISPR/Cas9 in sorghum through biolistic bombardment. In: Sorghum. Humana Press, New York, pp 169–183

Liu L et al (2021) Developing a novel artificial rice germplasm for dinitroaniline herbicide resistance by base editing of Os*TubA2*. Plant Biotechnol J 19:5–7

Lowe K, Wu E, Wang N, Hoerster G, Hastings C, Cho MJ et al (2016) Morphogenic regulators Baby boom and Wuschel improve monocot transformation. Plant Cell 28(9):1998–2015

Mahendrakar MD, Parveda M, Kishor PK, Srivastava RK (2020) Discovery and validation of candidate genes for grain iron and zinc metabolism in pearl millet [*Pennisetum glaucum* (L.) R. Br.]. Sci Rep 10(1):1–16

Mamidi S, Healey A, Huang P, Grimwood J, Jenkins J, Barry K et al (2020) The Setaria viridis genome and diversity panel enables discovery of a novel domestication gene. bioRxiv 744557

Mariac C, Luong V, Kapran I, Mamadou A, Sagnard F, Deu M et al (2006) Diversity of wild and cultivated pearl millet accessions (*Pennisetum glaucum* [L.] R. Br.) in Niger assessed by microsatellite markers. Theor Appl Genet 11:49–58

Molla KA, Yang Y (2019) CRISPR/Cas-mediated base editing: technical considerations and practical applications. Trends Biotechnol 37:1121–1142

Molla KA, Yang Y (2020) Predicting CRISPR/Cas9-induced mutations for precise genome editing. Trends Biotechnol 38(2):136–141

Molla KA, Qi Y, Karmakar S, Baig MJ (2020a) Base editing landscape extends to perform Transversion mutation. Trends Genet 36:899–901

Molla KA, Shih J, Yang Y (2020b) Single-nucleotide editing for zebra3 and wsl5 phenotypes in rice using CRISPR/Cas9-mediated adenine base editors. aBIOTECH 1:106–118

Molla KA, Sretenovic S, Bansal KC, Qi Y (2021) Precise plant genome editing using base editors and prime editors. Nat Plants 7(9):1166–1187

Nakayama H, Afzal M, Okuno K (1998) Intraspecific differentiation and geographical distribution of Wx alleles for low amylose content in endosperm of foxtail millet, Setaria italica (L.) Beauv. Euphytica 102(3):289–293

Nambiar VS, Dhaduk JJ, Sareen N, Shahu T, Desai R (2011) Potential functional implications of pearl millet (*Pennisetum glaucum*) in health and disease. J Appl Pharm Sci 1:62–67

Nishida K, Arazoe T, Yachie N, Banno S, Kakimoto M, Tabata M et al (2016) Targeted nucleotide editing using hybrid prokaryotic and vertebrate adaptive immune systems. Science 353(6305): aaf8729

Oliva R, Ji C, Atienza-Grande G, Huguet-Tapia JC, Perez-Quintero A, Li T et al (2019) Broad-spectrum resistance to bacterial blight in rice using genome editing. Nat Biotechnol 37(11): 1344–1350

Oumar I et al (2008) Phylogeny and origin of pearl millet (Pennisetum glaucum [L.] R. Br) as revealed by microsat-ellite loci. Theor Appl Genet 117:489–497

Parvathaneni RK, Spiekerman JJ, Zhou H, Wu X, Devos KM (2019) Structural characterization of ABCB1, the gene underlying the d2 dwarf phenotype in pearl millet, Cenchrus americanus (L.) Morrone. G3: Genes, Genomes. Genetics 9(8):2497–2509

Prasad PV, Staggenborg SA (2009) Growth and production of sorghum and millets. In: Soils, plant growth and crop production, vol 2. EOLSS Publishers, Oxford

Pujar M, Gangaprasad S, Govindaraj M, Gangurde SS, Kanatti A, Kudapa H (2020) Genome-wide association study uncovers genomic regions associated with grain iron, zinc and protein content in pearl millet. Sci Rep 10(1):1–15

Ramineni R, Sadumpati V, Khareedu VR, Vudem DR (2014) Transgenic pearl millet male fertility restorer line (ICMP451) and hybrid (ICMH451) expressing *Brassica juncea* nonexpressor of pathogenesis related genes 1 (*BjNPR1*) exhibit resistance to downy mildew disease. PLoS One 9(3):e90839

Rodríguez-Leal D, Lemmon ZH, Man J, Bartlett ME, Lippman ZB (2017) Engineering quantitative trait variation for crop improvement by genome editing. Cell 171(2):470–480

Saïdou AA, Mariac C, Luong V, Pham JL, Bezancon G et al (2009) Association studies identify natural variation at *PHYC* linked to flowering time and morphological variation in pearl millet. Genetics 182:899–910

Sánchez-León S, Gil-Humanes J, Ozuna CV, Giménez MJ, Sousa C, Voytas DF, Barro F (2018) Low-gluten, nontransgenic wheat engineered with CRISPR/Cas9. Plant Biotechnol 16(4): 902–910

Sashidhar N, Harloff HJ, Potgieter L, Jung C (2020) Gene editing of three BnITPK genes in tetraploid oilseed rape leads to significant reduction of phytic acid in seeds. Plant Biotechnol 18(11):2241–2250

Sehgal D, Skot L, Singh R, Srivastava RK, Das SP, Taunk J et al (2015) Exploring potential of pearl millet germplasm association panel for association mapping of drought tolerance traits. PLoS One 10(5):e0122165

Shan Q, Wang Y, Li J, Zhang Y, Chen K, Liang Z et al (2013) Targeted genome modification of crop plants using a CRISPR-Cas system. Nat Biotechnol 31(8):686–688

Sharma B, Chugh LK (2017) Two isoforms of lipoxygenase from mature grains of pearl millet [Pennisetum glaucum (L.) R. Br.]: purification and physico-chemico-kinetic characterization. J Food Sci Technol 54(6):1577–1584

Sharma S, Sharma R, Govindaraj M, Mahala RS, Satyavathi CT, Srivastava RK et al (2020) Harnessing wild relatives of pearl millet for germplasm enhancement: challenges and opportunities. Crop Sci 61(1):177–200

Shimatani Z, Kashojiya S, Takayama M, Terada R, Arazoe T, Ishii H et al (2017) Targeted base editing in rice and tomato using a CRISPRCas9 cytidine deaminase fusion. Nat Biotechnol 35: 441–443

Shivhare R, Lata C (2017) Selection of suitable reference genes for assessing gene expression in pearl millet under different abiotic stresses and their combinations. Sci Rep 6(1):1–12

Sretenovic S, Liu S, Li G, Cheng Y, Fan T, Xu Y et al (2021) Exploring C-To-G base editing in rice, tomato, and poplar. Front Genome Ed 3:756766

Tako E et al (2015) Higher iron pearl millet (*Pennisetum glaucum* L.) provides more absorbable iron that is limited by increased polyphenolic content. Nutr J 14(1):1–9

Vadez V, Hash T, Bidinger FR, Kholova J (2012) II 1.5 phenotyping pearl millet for adaptation to drought. Front Physiol 3:386

Varshney RK, Shi C, Thudi M, Mariac C, Wallace J, Qi P et al (2017) Pearl millet genome sequence provides a resource to improve agronomic traits in arid environments. Nat Biotechnol 35:969

Vinutha T, Kumar D, Bansal N, Krishnan V, Goswami S, Kumar RR et al (2022) Thermal treatments reduce rancidity and modulate structural and digestive properties of starch in pearl millet flour. Int J Biol Macromol 195:207–216

Wan DY, Guo Y, Cheng Y, Hu Y, Xiao S, Wang Y, Wen YQ (2020) CRISPR/Cas9-mediated mutagenesis of VvMLO3 results in enhanced resistance to powdery mildew in grapevine (*Vitis vinifera*). Hortic Res 7(1):1–14

Wang Y, Cheng X, Shan Q, Zhang Y, Liu J, Gao C, Qiu JL (2014) Simultaneous editing of three homoeoalleles in hexaploid bread wheat confers heritable resistance to powdery mildew. Nat Biotechnol 32(9):947–951

Wang C, Liu Q, Shen Y, Hua Y, Wang J, Lin J et al (2019) Clonal seeds from hybrid rice by simultaneous genome engineering of meiosis and fertilization genes. Nat Biotechnol 37(3): 283–286

WHO (2020) Anaemia. https://www.who.int/health-topics/anaemia#tab=tab_1

Xie K, Yang Y (2013) RNA-guided genome editing in plants using a CRISPR–Cas system. Mol Plant 6(6):1975–1983

Xing S et al (2020) Fine-tuning sugar content in strawberry. Genome Biol 21:230

Yadav AK, Arya RK, Narwal MS (2014) Screening of pearl millet F1hybrids for heat tolerance at early seedling stage. Adv Agric, Article ID. 231301. https://doi.org/10.1155/2014/231301

Zhang R, Liu J, Chai Z, Chen S, Bai Y, Zong Y et al (2019) Generation of herbicide tolerance traits and a new selectable marker in wheat using base editing. Nat plants 5(5):480–485

Zong Y, Song Q, Li C, Jin S, Zhang D, Wang Y et al (2018) Efficient C-to-T base editing in plants using a fusion of nCas9 and human APOBEC3A. Nat Biotechnol 36:950–953

Omics-Based Approaches in Improving Drought Stress Tolerance in Pearl Millet

8

Swati Singh, Animikha Chakraborty, Aswini Viswanath, Renuka Malipatil, and Nepolean Thirunavukkarasu

Abstract

Pearl millet (*Pennisetum glaucum* (L.) R. Br), a climate-resilient Nutri-cereal grown primarily in Africa and South Asia's arid and semi-arid regions, is well adapted to adverse abiotic stresses such as drought and heat. Drought stress is an abiotic disaster that impacts yield and productivity globally. Plants have adapted pathways to survive under stress, but the advent of a new omics-based approach encompassing genomics, transcriptomics, proteomics, and metabolomics is critical for developing tolerant varieties. Acknowledging plant responses at different organizational levels, such as growth, physiology, metabolites (metabolomics), proteins (proteomics), and genes (genomics), will enable us to devise strategies for designing drought-tolerant crops. Understanding crop responses, particularly at the "omics" level, will improve the quality and significance of the biological information deduced to develop stress-resistant cultivars. This chapter entails the latest information about how drought stress affects various traits involving growth, physiology, genes, proteins, and metabolites in pearl millet and how we can develop drought-resistant varieties using these omics-based approaches.

Keywords

Abiotic stress · Omics-based · Genomics · Transcriptomics · Proteomics · Metabolomics · Drought-resistant

S. Singh · A. Chakraborty · A. Viswanath · R. Malipatil · N. Thirunavukkarasu (✉)
ICAR-Indian Institute of Millets Research, Hyderabad, India
e-mail: nepolean@millets.res.in

V. A. Tonapi et al. (eds.), *Pearl Millet in the 21st Century*,
https://doi.org/10.1007/978-981-99-5890-0_8

Abbreviations

2 DE	2 dimensional gel electrophoresis
2D-DIGE	2 dimensional difference gel electrophoresis
ABA	Abscisic acid
AFLP	Amplified fragment length polymorphism
DEG	Differentially expressed genes
DEP	Differentially expressed proteins
ELISA	Enzyme-linked immunosorbent assay
ESTs	Expressed sequence tags
FTIR	Fourier-transform infrared spectroscopy
GHI	Global Hunger Index
GWAS	Genome-wide association studies
HPLC	High-performance liquid chromatography
iTRAQ	Isobaric tags for relative and absolute quantitation
LCMS	Liquid chromatography-mass spectrometry
MALDI TOF	Matrix-assisted laser desorption ionization-time-of-flight
MAPK	Mitogen-activated protein kinase
MAS	Marker-assisted selection
MS	Mass spectrometry
NFHS-5	National Family Health Survey 5
NMR	Nuclear magnetic resonance
PAGE	Polyacrylamide gel electrophoresis
PMF	Peptide mass fingerprinting
QTL	Quantitative trait loci
SCRNA	Single-cell ribonucleic acid
SELDI	Surface-enhanced laser desorption ionization
SNP	Single nucleotide polymorphism
SSH	Suppression subtractive hybridization
SSR	Single sequence repeats
WGS	Whole genome sequence

8.1 Introduction

Plant breeding advances have resulted in the development of resistant and adaptable crops to adverse environmental conditions in recent decades, with pearl millet being one such crop. *Pennisetum glaucum (L.)* R. Br is a Poaceae family plant with a short growing season. It is a cross-pollinated warm-season C4 plant species with a ~1.79 Gb genome, high photosynthetic efficiency, and biomass production (Varshney et al. 2017). It is grown as a staple food, straw for fodder, and fuel source for more than 90 million poor people in Africa and South Asia's arid and semi-arid regions (Anuradha et al. 2017; Varshney et al. 2017). This crop has evolved through adaptive evolution, allowing it to thrive in these harsh conditions.

India is now the world's largest millet producer and fifth-largest exporter. According to 2020 statistics, India accounted for roughly 41% of total global millet production, rising from 14.52 million tons in 2015–2016 to 17.96 million tons in 2020–2021. In 2020, nearly one-third of the world's population (2.37 billion) lacked adequate food, a 320 million increase in just 1 year, a serious concern for humanity. The Global Hunger Index (GHI) 2021 ranks India 101st out of 116 countries, up from 94th in 2020, trailing its neighbors Pakistan, Bangladesh, and Nepal (indianexpress.com). According to the NFHS-5 (National Family Health Survey) report, 14% of India's population, or approximately 189.2 million people, are undernourished, 20% of children under five are underweight, and 34.7% are stunted (www.indiafoodbanking.org/hunger). Micronutrient deficiencies such as Fe and Zn lead to severe disorders like anemia; in India, 57% of women and 46% of girls under 15 are iron deficient (NFHS-5 India report).

Crops with higher nutritional value and dietary fibers are prioritized to meet rising food demand; pearl millet is one of the crops that will meet the demand. Pearl millet is also known as Nutri-cereal because it contains more minerals, micronutrients, and fiber than other cereals (Gazette of India, No 133, 13 April 2018). It is a high-energy cereal rich in amino acid content, vitamin A, fat content, high fiber, minerals like Fe, Mg, Mn, Zn, K, P, phytic acid, polyphenols, lipids, and protein 8–19%, which includes globulin, albumin, prolamin, and glutelin (Nambiar et al. 2011; Tako et al. 2015). Despite its valuable characteristics and high nutritional value, its productivity and yield are low, and it is not treated as crucially as other cereals such as wheat, rice, and maize (Choudhary et al. 2015).

The global population increase is intensifying food requirements, raising the demand for grain yield. Due to the dramatic rise in global population, natural sources are depleting, and the climate is adversely affected. Climate change significantly impacts crop yield and output, endangering dietary diversity now and in the future by exposing plants to various stresses (El Sabagh et al. 2019). Exposure to severe stresses leads to plant death by preventing flowering, seed formation, and induced senescence (Diallo et al. 2014; Wasternack 2007). Under biotic stress, living things deprive the plants/ host of nutrition, affecting production and yield, ultimately resulting in plant deaths. Pearl millet, like other cereals, is susceptible to a few biotic factors such as fungal (*Tolyposporium penicillariae, Puccinia penniseti Zimm*), bacterial, and viral diseases, as well as attack by major insects such as shoot fly, stem borer, cutworm, white grub resulting in conditions such as smut, rust, blast, and downy mildew, which affect crop yield and forages quality. Abiotic stress factors such as drought, heat, salinity, and cold affect plant growth rate and production.

8.2 Abiotic Stress

Among the several stress factors affecting the pearl millet mentioned in the above section, abiotic stress poses a major threat to this plant, and it disturbs its metabolism through metabolic enzyme inhibition and substrate shortage, resulting in a decline in grain and fodder production. Abiotic stress is defined as excess or deficit in the physical and chemical components of the environment, such as temperature, water, and salt concentration. This plant is usually grown in arid and semi-arid regions of Asia and Africa, so it is more exposed to abiotic stress such as drought, heat, and salinity. Various abiotic stresses, including drought, prevent plants from exploiting their full potential. A thorough understanding of signaling pathways and other plant responses to various abiotic stresses will aid in developing strategies for inculcating abiotic stress tolerance, particularly drought tolerance, in pearl millet, ensuring plant longevity and yield (Ahmad et al. 2016). Pearl millet has received comparatively more recognition among millets in gathering and analyzing data on drought, salinity, and high-temperature tolerance (Dwivedi et al. 2012). The following section will focus on how drought stress affects pearl millet globally.

8.3 Drought Stress

Drought is a primary abiotic constraint triggered by a confluence of hydrological, ecological, and physiological factors that result in inadequate water availability over an extended period of time (Bartels and Sunkar 2005). Drought is a complex trait involving numerous genes associated with various signaling and metabolic pathways, altering several physiological, anatomical, and molecular responses (Ahuja et al. 2010; Sultan and Gaetani 2016). Pearl millet is a novel source for investigating drought-related defense mechanisms and responses. However, due to adaptive evolution and natural selection, it has become one of the most drought and heat-tolerant cereals. Recent research has shown that plants can adjust to changing environmental factors *via* escape and avoid or tolerate mechanisms (Lata et al. 2015). These responses are genetically controlled and confer stress tolerance. Drought tolerance molecular mechanisms will aid in developing drought-tolerant varieties capable of maintaining grain yield under drought conditions.

8.3.1 Primary and Secondary Effects on Plants by Drought Stress

Reduced productivity results from the complex genetic trait known as drought stress, which negatively impacts plant growth, development, metabolism, photosynthetic activity, stomatal movement, and physiology (Lata and Prasad 2011). Drought causes plant growth to slow, shoot growth to cease, and metabolite demand to decrease. Drought triggers immediate harm and manifests in primary consequences such as wilting, scorching, and partial defoliation due to lost turgor in plant cells, irreversible cell membrane shrinkage, and increased ABA synthesis (Ahuja et al.

2010). Numerous effects of drought stress on plants include membrane damage that inhibits growth and ultimately results in cell death, as well as oxidative damage to DNA, proteins, lipids, and macromolecules. Pearl millet alters its morphological properties under drought stress, including increased root length and density, shorter shoots, early maturity, a high root-to-shoot ratio, and some leaf and root characteristics.

The root is the first organ to respond, which gets affected when there is a water deficit condition; cell water permeability gets altered, and root growth gets influenced to so much extent in search of absorbing water from the soil (Janiak et al. 2016). Increases in root length and root hair density allow for increased soil absorption of water and nutrients and drought resistance in some situations (Comas et al. 2013; Passot et al. 2016). Reduced leaf number, color change, leaf rolling to minimize water loss, leaf area reduction or reduced leaf size, increased leaf hair density, increased leaf waxiness, and stomatal closure are some of the leaf traits that are altered during a water deficit condition (Cal et al. 2019; Yoo et al. 2010). These changes are made to conserve water because it is crucial for many metabolic processes, including germination, cell division, and plant growth for survival.

Drought stress has varying effects and intensities at different stages of the life cycle. Early seedling-stage drought stress causes poor crop establishment and decreased production (Bidinger et al. 1987). Pearl millet grain yield is not negatively impacted by drought during the vegetative phase because plants produce more tillers to help them adapt to the stress of the drought. At this point, drought impacts pearl millet growth, flowering, and tillering. Drought stress is known as terminal drought stress during reproductive stages, including flowering and grain-filling (Yadav et al. 2004). During this time, it significantly impacts the loss of grain and stover yield due to a decrease in the number of seeds per panicle, a reduction in grain size, and a decrease in grain mass (Mahalakshmi and Bidinger 1986). Grain yield loss is maximum at this stage under drought stress. Plants adapt a primary adaptation mechanism, which strengthens the link between grain development and the transfer of assimilates from leaves to respond to terminal drought stress (Kholova et al. 2010). Compared to the reproductive stages, the post-flowering stage is less affected by drought and has fewer seeds per panicle (Table 8.1).

8.3.2 Response

Pearl millet is a crop on which extensive research has been conducted to better understand its response and adaptation to water scarcity at various stages of development. Plant organs show different reactions to drought and develop defense mechanisms caused by a group of genes by altering their expression pattern. Plants sense stress, get stimulated by drought, and generate appropriate responses evolved to impart drought tolerance. Plants have developed extensive coping mechanisms for drought, such as leaf rolling, wilting, and other behaviors (Kusaka et al. 2005) (Fig. 8.1). To avoid drought, pearl millet plants develop deep roots for effective water absorption, stomatal closure, and leaf rolling to prevent water loss, and staying

Table 8.1 Drought effects on various stages of pearl millet

S. No	Plants stage affected by drought	Physiological Impacts on pearl millet	Impacts on yield component	Responses
1	Early vegetative (up to flowering)	Reduced leaf area and root growth – Closed stomata and decreased transpiration rate – Lower photosynthetic rate and chlorophyll content – Lower carbohydrate storage and delayed germination (Anuradha et al. 2017)	Fewer tillers and plants per unit area – Fewer spikelets per panicle – Smaller seed size (Varshney et al. 2017)	Root plasticity (Oweis et al. 1999), Stomatal regulation (Wahid et al. 2007), Osmotic adjustments (Chaves et al. 2002)
2	Reproductive (flowering to maturity)	Lower pollen viability – Slower grain filling rate – Decreased grain water potential (Varshney et al. 2017)	Reduced number of filled grains per spikelet – Smaller 1000-grain weight (Varshney et al. 2017)	Anthesis-silking interval (ASI) adjustment (Tiwari et al. 2017), Reproductive plasticity (Lojkova et al. 2015), Antioxidant defense mechanisms Salazar-Cerezo et al. 2018)
3	Post-reproductive	Reduced grain moisture content, size, and weight (Varshney et al. 2017)	Shorter stature and Decreased grain yield (Tako et al. 2015)	Water-use-efficiency (Magwanga et al. 2018 and Xu et al. 2018), Osmotic adjustment (Chaves et al. 2002), reduced seed production (Lojkova et al. 2015)

green is another characteristic that helps (Djanaguiraman et al. 2018). With the help of mechanisms like early flowering and maturation before being subjected to stress, remobilization of assimilates, and a high photosynthetic capacity, plants have evolved ways to avoid the stress associated with water limitations (Vadez et al. 2013). Making osmotic adjustments, having high stomatal conductance, having high proline concentration, and maintaining photosynthesis all contribute to drought tolerance (Kusaka et al. 2005).

Plants have evolved a wide range of physiological and biochemical modifications, including ROS detoxification, membrane integrity, osmoregulation, macromolecules, and antioxidant defense, to adapt to drought (Caverzan et al. 2012). Plants have developed various defense mechanisms to survive the effects of drought stress. These include cuticle thickening, decreased leaf expansion, higher production of ROS-scavenging compounds, reduced water loss by increasing stomatal closure,

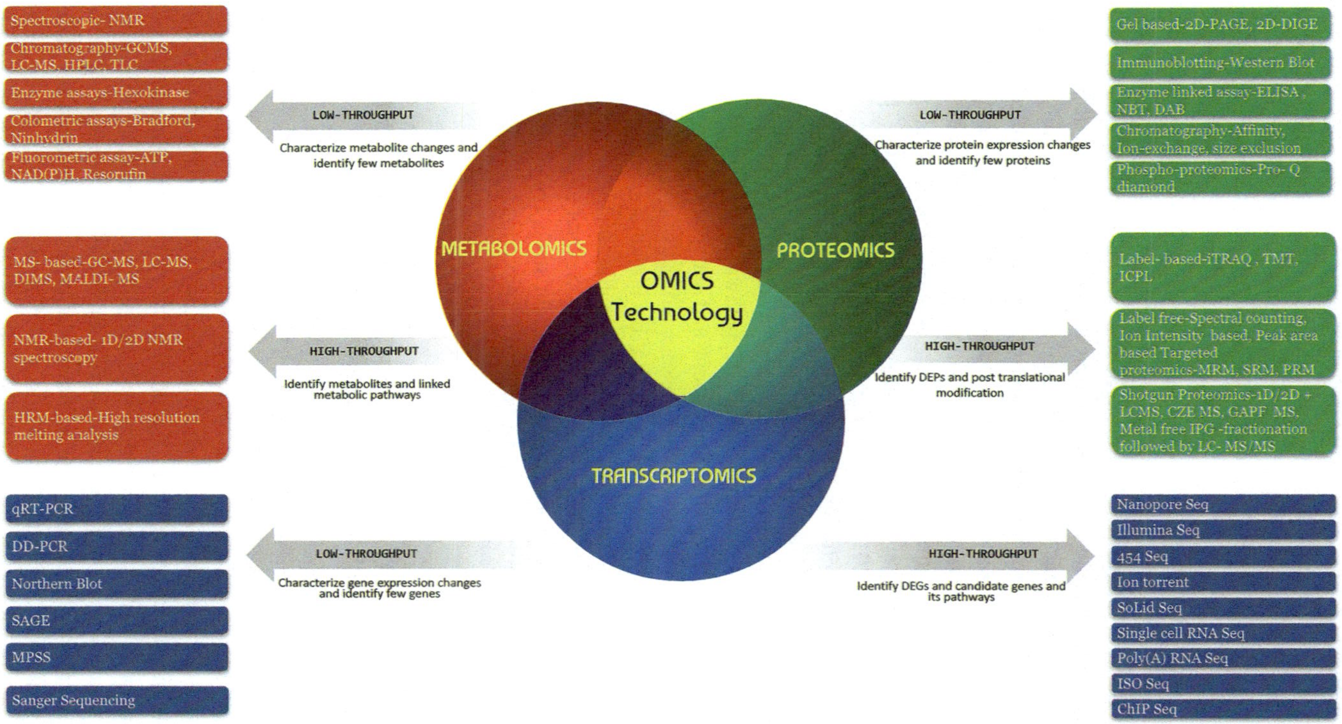

Fig. 8.1 Omics-based technologies to analyze different aspects of gene expression, protein expressions, and metabolite profile to improve drought tolerance in pearl millet

and increased production of phytohormones, such as ABA, amino acids, and alkaloids (Mahmood et al. 2019; Wilkinson et al. 2012). When exposed to drought, root length, and density increase, and xylem abundance rises to better absorb water from the surrounding and deeper soil. These processes promote drought stress tolerance to have a stronger effect on the yield and growth of pearl millet (Kusaka et al. 2005).

As discussed earlier, plant responses are caused by genes at the molecular level by causing modifications in gene expression, resulting in plants acclimating and facilitating drought tolerance. Plant drought tolerance is typically characterized by a combination of physiological and biochemical changes resulting from the coordinated expression of a gene hierarchy. The interaction of plant heredity and external environmental changes results in this complex mechanism. Pearl millet incorporates a power pack of genes responsible for conferring abiotic stress tolerance, and it can be employed as a genomic resource for extracting highly desired genes for producing advanced plants with superior quality, such as stress-tolerant varieties. Various stress-responsive genes at molecular and cellular levels regulate plant drought response (Jaiswal et al. 2018; Mukesh Sankar et al. 2021). Water scarcity produces ABA, a plant growth regulator essential in defense mechanisms. A high level of ABA is associated with stomata because it promotes stomatal closure and thus reduces the transcription rate, resulting in water conservation (Ahuja et al. 2010; Sirichandra et al. 2009). Various metabolic pathways, including phytohormone signaling pathways such as ABA-dependent and independent pathways, MAPK signaling pathways, and photosynthesis-associated pathways, function in drought stress response (Tuteja 2007; Liu et al. 2015a, b). Phytohormones are important in plant development, hormone regulation, and signaling pathways leading to biotic and abiotic stress tolerance (Wilkinson et al. 2012). Several efforts have been made in recent years to elucidate mechanisms of drought stress responses in non-model cereal crops using genomics, transcriptomics, proteomics, and metabolomics approaches.

8.4 Enhancement of Crop Productivity and Drought Tolerance Through Omics Technology Interventions

There is an immediate need to address the challenge and boost the productivity of crops in both quantitative and qualitative terms. With the advancement in science, the omics-based approach has allowed us to deal with a variety of serious challenges addressing the current agricultural system and has provided an opportunity to ensure food, nutrition, and energy security. Accomplishing rising food demand is a critical concern; on top of that, climate changes, such as rising temperatures, shifting rainfall patterns, and an increase in the frequency of extreme weather events, complicate the situation by causing biotic and abiotic threats, limiting crop production (Ma et al. 2013; Narsaiah et al. 2012; Jung and Main 2014; Singh et al. 2013; Hollister 2014). A comprehensive understanding of plant genetics and phenotypic diversity is required to develop crops with significantly improved agronomic traits.

The omics-based approach is a promising technology that first emerged in 1986 with the Human Genome Project (Wheelock and Miyagawa 2006). It is gaining popularity in agricultural sciences as it enables in-depth functional analyses of plants at the molecular and cellular levels and has shown to be a validating method for candidate gene expression. Omics-based approaches are extremely valuable in understanding the genetic and molecular basis of physiological and environmental stress responses by performing functional analyses of alterations at the molecular level, gene expression from transcriptional regulation to processing mRNA, translation, and post-translational modifications (Muthamilarasan et al. 2019). Omics technologies have influenced multiple domains of plant sciences and have become a significant component of the breeding program, assisting in decoding stress tolerance mechanisms in several crops so far, including millets acclimating to various physiological situations. It is opening up new avenues for increasing agricultural production in both qualitative and quantitative ways, as well as aiding in the study of the cause-and-effect link between genome and phenotype in an integrative and predictive way (Cowie et al. 2013). Several omics methods, including transcriptomics, genomics, metabolomics, proteomics, phenomics, and ionomics, can be used to uncover QTLs and candidate genes that regulate economic factors like grain yield. This collaboration offers the ability to characterize and assess gene functions and networks in response to varied environmental stresses (Salt et al. 2008; Houle et al. 2010; Wu et al. 2017).

An omics-based approach can address issues such as the rapid development of novel crop varieties resistant to various stressors and having high yield capacity, nutritional content, and disease resistance. Technological breakthroughs in the omics technology disciplines have assisted in gene identification and expression profiling of genes, proteins, and metabolites, offering complete knowledge about gene expression patterns, protein structure and interaction, and metabolite composition (Fig. 8.1) (Rhee et al. 2006). Omics methods can be beneficial in developing smart crops for sustainable agriculture to safeguard the environment from adverse weather conditions (Kumar et al. 2015). This technique has been included in many crop breeding programs and has shown to be an effective method for understanding growth, yield, senescence, and the biochemical response of plants to biotic and abiotic stress.

Implementing this technology in recent years has allowed researchers to perform a few comparison studies and reveals stress-related molecular responses, but the entire response pathway to many stresses remains unknown. A full grasp of complex characteristics' genetic and molecular basis may be researched using these methodologies, which will be useful in generating elite cultivars. Research on the genomic, metabolic, and proteomic responses of diverse crops, particularly tolerant genotypes, might considerably benefit continuing attempts to introduce drought/heat stress resistance. Using omics technology, we will acquire a better knowledge of the genes and metabolic processes that influence several essential agronomic properties, which will assist in producing superior resistant millets. In the following section, we will go through various omics-based methodologies for finding genes and pathways that can help with in-depth functional investigations of pearl millet resistance to drought stress.

8.5 Transcriptomics

Transcriptomics, a subset of functional genomics, is the study of a collection of transcripts, namely mRNA, and the analysis of gene expression patterns across a wide variety of cellular and metabolic responses and phenotypes at any given time. Charles Auffray coined the word in 1996, and it was first used in a scholarly publication in 1997. It adds to a thorough understanding of genes, their transcriptional roles, and the molecular mechanisms of cellular metabolism (Borovitz and Chory 2004). Unlike the genome, the transcriptome is dynamic, changing with different circumstances, and the same gene can create several transcripts owing to alternative splicing. Analyzing expressed genes helps identify key genes responsible for specific complex biological traits and primary enzymes involved in metabolic pathways (Mochida and Shinozaki 2010). This productive approach enables researchers to detect changes in gene expression in vitro for a broader understanding of a specific gene's first-layer activities.

8.5.1 Transcriptomics Techniques

The emergence of transcriptomics approaches and their implementation in genomic research has resulted in massive amounts of improved transcriptome data for various cereal crops. Traditionally, the dynamic transcriptome was studied using Sanger sequencing to sequence RNA transcripts known as ESTs, cDNA-AFLP, DD-PCR, SSH, Northern blotting method, and qRT-PCR, which only covered a small area of the transcriptome with low resolution (Lowe et al. 2017; Singh et al. 2020). The use of Sanger EST collections in conjunction with sequencing data obtained by NGS platforms of diverse crops led to the production of a wide-ranging transcriptome assembly of these crops. In 1995, the serial analysis gene expression (SAGE) approach was utilized to identify new genes expressed in the tissue under certain physiological circumstances by creating short sequence tags of 10–14 bp (Velculescu et al. 1995). SAGE has an advantage over EST approaches in that it does not require prior knowledge of the sequences to be studied. Molina et al. 2008 used super sage with fine resolution in conjunction with NGS technology in a study to delineate the full transcriptome of salt and drought-stressed plant roots and nodules. MPSS, a sequencing-based technique, assesses gene expression by measuring mRNA transcripts present in the sample.

Microarray is another burgeoning technology that has grown in popularity as it has been used to explore plant biological processes such as stress response and grasp DEG's significance at various plant stages (Aharoni and Vorst 2002). Its applicability is limited to characterizing the expression level of known genes since it requires antecedent knowledge of genome assembly and ESTs. The advancement of array technology has improved its accuracy and specificity. Array assays used in expression profiling investigations include Agilent, Exiqon, Miltenyi, nimble gene chip, advanced high-density array, and Affymetrix gene chip array. DEGs have been discovered in numerous crops during developmental and reproductive stages under drought stress using microarray analysis such as the Affymetrix gene chip array (Yang et al. 2021).

Following that, high throughput NGS technologies such as Roche 454, SOLid, Pacbio, Oxford Nanopore, Nanostring, and Illumina Seq emerged to give high-quality transcriptome data more accurately than existing approaches for sequencing mRNA transcripts (Darshan et al. 2020; Le Nguyen et al. 2019). The advent of NGS platforms has revolutionized transcriptomics in recent decades; it has great coverage, is cost-effective, and aids in discovering stress-responsive genes and gene families that are part of biological processes. NGS approaches revealed transcriptome assemblies that contain information on various markers, such as SSR and SNP, associated with stress-response mechanisms. RNA sequencing employing these technologies reveals underlying new regulatory RNAs, such as noncoding RNAs and long noncoding RNAs, which govern the drought tolerance characteristic across the plant genome (Huang et al. 2014; Kakumanu et al. 2012). Several RNA Seq investigations in various crops revealed several crosstalk pathways and tissue-specific expressions in response to drought stress, providing a deeper knowledge of the molecular mechanism of the drought stress response (Kogenaru et al. 2012). In-situ RNA sequencing and spatial transcriptomics, which include gene expression detection in living cells or tissues and gene expression analysis with spatial information within the cells and tissues, are becoming increasingly important in providing a better understanding of physiological processes in organisms at the molecular and cellular levels (Burgess 2015; Ke et al. 2013).

Alternate splicing transcriptomics, an approach developed to produce a substantial number of transcripts in response to water scarcity, has revealed the role of splicing variables in controlling multiple abiotic parameters (Laloum et al. 2018). Another NGS technology that aids in the characterization of individual cells and rare populations from early developmental stages, as well as their complex mechanisms in response to different abiotic conditions and uncovering the regulatory relationships between the genes, is single-cell RNA sequencing. However, scRNA Seq is low throughput and time-demanding approach, and the diversity in its data presents computational hurdles. The scRNA experiments have revealed cell-to-cell gene expression heterogeneity in several human tissues. Some scRNA seq methods utilized for single-cell transcriptome investigations include the Tang technique, Quartz Seq, SUPeR-seq, Smart-seq, MATQ-seq, Cell-seq, Seq-well, DroNC-seq, SPLiT-seq, and Quartz-Seq2 (Shaw et al. 2021). Using scRNA Seq can give insights into the heterogeneity and dynamics of gene expression in cells and boost single-cell transcriptomics investigations.

8.5.2 Transcriptomics Application in Imparting Drought Tolerance

Transcriptomics studies have proven to play an important role in understanding stress tolerance in plants. Several studies in cereals have been conducted to elucidate the drought stress tolerance mechanism and to identify the candidate genes responsible for resistance using transcriptomics approaches. These analyses also led to the discovery of a number of stress-responsive genes (Table 8.2), such as transcription factors and co-regulators. Transcription factors contribute significantly at the cellular and molecular levels in response to various abiotic stresses, including drought stress, by regulating gene expression patterns under the influence of different intracellular signals and governing target gene action via activation or repression (Fig. 8.2). ABF/AREB, bZIP, DREB/CBF, MYB, MYC, NAC, and WRKY have all been identified as TFs in many crops through expression profiling studies (Shikha et al. 2017; Nepolean et al. 2017).

Transcriptomic studies have indicated that NAC TFs such as OSNAC2, OSNAC05, TaNAC69, and SNAC1 aid in enhancing drought stress tolerance in several crops via ABA-dependent pathways, increased stress-related gene expression and by boosting root development and reducing water loss, respectively (Dudhate et al. 2021). PgNAC21 from pearl millet, when overexpressed in Arabidopsis, enhanced its salinity stress tolerance (Dudhate et al. 2021). Another research discovered that the WRKY gene is involved in plant response to environmental stress and in pearl millet; PgWRKY TF plays an important role in responding to both biotic and abiotic stress and imparting drought tolerance, as well as contributing to plant development (Chanwala et al. 2020). Ndiaye et al. (2022) demonstrated in their study that exogenous methyl jasmonate could improve crop drought tolerance by modulating signaling, photosynthesis, and metabolic pathways.

Whole genome transcriptome analysis in plants discovered genes associated with drought stress responses while undermining stress response molecules. The first whole genome transcriptome study in millets was done in foxtail millet, which revealed 81.7% expressed genes and 1367 pseudogenes using root and leaf RNA sequencing on the Illumina GAII platform (Muthamilarasan et al. 2016). Comparative transcriptomic analysis of resistant and susceptible pearl millet varieties identifies candidate genes and adaptation mechanisms under dehydration stress, providing a better understanding of differential gene expression patterns in response to these stress conditions. Mishra et al. 2007 used transcriptomic analysis to illustrate the underlying mechanism of pearl millet under drought stress. The incorporation of transcriptomics in pearl millet has facilitated the identification of PgDREB2A, PgNAC21, PgNHX1, PgDHN, PgLEA, PgRab7, PgHSP90, and PgGPX genes and pathways, as well as their roles in inducing drought tolerance (Zhou et al. 2010; Islam et al. 2015; Shinde et al. 2019; Reddy et al. 2012; Dudhate et al. 2018). A comprehensive understanding of drought stress and its impact on germination and vegetative stage transcriptomic studies was carried out, and it was discovered that at these stages, plants provide information about hormone signaling pathways, reproductive system interactions, and metabolic pathways.

Table 8.2 A comprehensive list of genes and their functions in response to drought in crops

S. NO	Genes	Biological function	Phenotypes	Tags	References
1	CrNCED1	ABA biosynthesis	Transgenic lines displayed enhanced tolerance to dehydration, drought, salt, and oxidative stresses	Ion and Osmotic Homeostasis; Stomatal Movement; Regulatory Proteins; Hormone Signaling; ABA Biosynthesis	Xian et al. (2014)
2	NCED	ABA biosynthesis key enzyme	Overexpressior: improved drought tolerance; antisense: drought-sensitive phenotype	Ion and Osmotic Homeostasis; Stomatal Movement; Regulatory Proteins; Hormone Signaling; ABA Biosynthesis	Iuchi et al. (2001)
3	CYP707A3	ABA catabolism; 8′-hydroxylation of ABA	cyp707a3 mutant exhibited high ABA levels even after transferring to high-humidity conditions	Ion and Osmotic Homeostasis; Stomatal Movement; Regulatory Proteins; Hormone Signaling	Okamoto et al. (2009)
4	CYP707A1	ABA catabolism; 8′-hydroxylation of ABA	Transcript increases in response to salt, osmotic, dehydration stress, and ABA; stomatal closure of the cyp707a1 mutant was ABA hypersensitive when epidermal peel was treated with exogenous ABA	Ion and Osmotic Homeostasis; Stomatal Movement; Regulatory Proteins; Hormone Signaling; ABA Degradation	Okamoto et al. (2009)
5	ABF3	ABA-responsive element (ABRE) binding bZIP factor	Overexpression: enhanced drought tolerance	Regulatory Proteins; Gene Expression; Transcription Factors; bZIP	Kang et al. (2002)
6	ABF4	ABA-responsive element (ABRE) binding bZIP factor	Overexpression: enhanced drought tolerance	Regulatory Proteins; Gene Expression; Transcription Factors; bZIP	Kang et al. (2002)
7	LLA23	ABA-, stress-, and ripening-induced protein	Expression in Ath: reduced sensitivity to ABA; stomata stay open longer under drought, but plants are more tolerant to drought!	Ion and Osmotic Homeostasis; Stomatal Movement	Yang et al. (2005)
8	ABCG22	ABC-transporter	Knockout: more susceptible to drought stress	Functional Proteins; Channels and Transporters; Transporters; Others	Kuromori et al. (2011)

(continued)

Table 8.2 (continued)

S. NO	Genes	Biological function	Phenotypes	Tags	References
9	ABCG25	ABC-transporter, ABA export from cells	Overexpressor: higher leaf temperature	Ion and Osmotic Homeostasis; Stomatal Movement; Functional Proteins; Channels and Transporters; ABA Exporter	Kuromori et al. (2010)
10	ABCG40	ABC-transporter, ABA import	Knockout: upregulation of ABA-responsive genes is delayed, stomata close more slowly	Ion and Osmotic Homeostasis; Stomatal Movement; Functional Proteins; Channels and Transporters; ABA Importer	Kang et al. (2010)
11	HvPIP2;1	aquaporin	Expression in rice: increased water loss and CO_2 assimilation	Ion and Osmotic Homeostasis; Functional Proteins; Channels and Transporters; Water Channels	Hanba et al. (2004)
12	OsPIP1-1	aquaporin	Expression in Arabidopsis: enhanced tolerance to salt and drought	Ion and Osmotic Homeostasis; Functional Proteins; Channels and Transporters; Water Channels	Guo et al. (2006)
13	OsPIP2-2	aquaporin	Expression in Arabidopsis: enhanced tolerance to salt and drought	Ion and Osmotic Homeostasis; Functional Proteins; Channels and Transporters; Water Channels	Guo et al. (2006)
14	RWC3	aquaporin	Overexpression: higher root osmotic hydraulic conductivity	Ion and Osmotic Homeostasis; Functional Proteins; Channels and Transporters; Water Channels	Lian et al. (2004)
15	ThbZIP1	bZIP TF	Expression of ThbZIP1 is induced by ABA, salt, drought, methyl viologen, and cold	Regulatory Proteins; Gene Expression; Transcription Factors; bZIP	Ji et al. (2013)
16	ABP9	bZIP TF, ABRE binding	Expression in Ath: enhanced tolerance to drought, high salt, freezing	Regulatory Proteins; Gene Expression; Transcription Factors; bZIP	Zhang et al. (2011)
17	AREB1	bZIP TF, ABRE binding	Needs ABA-dependent phosphorylation for activation; phosphor mimicking variant	Regulatory Proteins; Gene Expression; Transcription Factors; bZIP	Furihata et al. (2006)

			overexpressed: expression of ABA-inducible genes w/o ABA treatment		
18	OsbZIP23	bZIP transcription factor	Overexpressor: improved tolerance to drought and high-salinity stress	Regulatory Proteins; Gene Expression; Transcription Factors; bZIP	Xiang et al. (2008)
19	OsbZIP71	bZIP transcription factor	OsbZIP71 overexpressing (p35S:: OsbZIP71) rice significantly improved tolerance to drought, salt, and PEG osmotic stresses	Regulatory Proteins; Gene Expression; Transcription Factors; bZIP	Liu et al. (2014)
20	OsbZIP72	bZIP transcription factor	Overexpressor: ABA hypersensitive, enhanced drought tolerance	Regulatory Proteins; Gene Expression; Transcription Factors; bZIP	Lu et al. (2009)
21	ZFP1	C2H2-type zinc-finger protein	Overexpression in AT: greater tolerance to cold and drought	Regulatory Proteins; Gene Expression; Transcription Factors; Zinc Fingers	Luo et al. (2012)
22	AIRP1	C3H2C3-type RING E3 ubiquitin ligase	Overexpressor: tolerant to severe drought stress	Regulatory Proteins; Post-translational Modification; Ubiquitin Ligases	Ryu et al. (2010)
23	ZFP1	Cys2/His2 FZP protein	Heterologous expression in AT: improved tolerance to salinity and drought	Regulatory Proteins; Gene Expression; Transcription Factors; Zinc Fingers	Gao et al. (2012)
24	ZPT2-3/ EPF2-7	Cys2/His2-type zinc finger protein	Overexpression: increased tolerance to dehydration	Regulatory Proteins; Gene Expression; Transcription Factors; Zinc Fingers	Sugano et al. (2003)
25	DgZFP3	Cys2/His2-type zinc finger protein gene	The transgenic lines exhibited less accumulation of H_2O_2 under drought stress, more accumulation of proline, and greater activities of peroxidase (POD) and superoxide dismutase than the WT under both control conditions and drought stress	Regulatory Proteins; Gene Expression; Transcription Factors; Zinc Fingers	Liu et al. (2013a, b)
26	GsZFP1	Cys2/His2-type zinc-finger protein	Overexpression of GsZFP1 also enhanced the drought tolerance of alfalfa	Regulatory Proteins; Gene Expression; Transcription Factors; Zinc Fingers	Tang et al. (2013)

(continued)

Table 8.2 (continued)

S. NO	Genes	Biological function	Phenotypes	Tags	References
27	TAS14	Dehydrin, group2 LEA proteins	Overexpressors: improved long-term drought tolerance	Detoxification; Protection Factor; Functional Proteins; Protection Factors; LEA Proteins	Muñoz-Mayor et al. (2012)
28	IMT1	D-myo-inositol methyltransferase	Expression in tobacco: enhanced tolerance to drought and salinity	Detoxification; Osmolyte Production; Functional Proteins; Enzymes for Osmolyte Biosynthesis	Sheveleva et al. (1997)
29	CBF4	DREB family TF	Upregulated by drought stress, overexpressors are more tolerant to freezing and drought stress	Regulatory Proteins; Gene Expression; Transcription Factors; DREB	Haake et al. (2002)
30	DREB1A/ CBF3	DREB family TF	Overexpression in tobacco: improved drought stress tolerance	Regulatory Proteins; Gene Expression; Transcription Factors; DREB	Maruyama et al. (2004)
31	ZmDREB2.7	DREB TF	Natural variation in the promoter of ZmDREB2.7 contributes to maize drought tolerance, transgenic Arabidopsis overexpressing ZmDREB2.7 displayed enhanced tolerance to drought stress.	Regulatory Proteins; Gene Expression; Transcription Factors; DREB	Liu et al. (2013a, b)
32	OsDERF1	Drought-responsive ethylene response factor (ERF)	Knockdown: enhanced tolerance to drought; overexpressor: reduced tolerance	Regulatory Proteins; Gene Expression; Transcription Factors; ERF	Wan et al. (2011)
33	EDT1/ HDG11	Enhanced drought tolerance1, HD START TF	T-DNA activation tagged mutant: enhanced drought tolerance, more extensive root system, deeper roots, more lateral roots	Growth Control; Root/Leaf Development; Regulatory Proteins; Gene Expression; Transcription Factors; Others	Yu et al. (2008)
34	JERF1	Ethylene response factor (ERF)	Overexpression in tobacco: improved drought stress tolerance	Regulatory Proteins; Gene Expression; Transcription Factors; ERF	Zhang et al. (2010)

35	SlERF5 (ERF5)	Ethylene response factor (ERF) superfamily	Expression induced by salinity, drought; overexpression: higher tolerance to drought	Regulatory Proteins; Gene Expression; Transcription Factors; ERF	Pan et al. (2012)
36	HVA1	Group 3 late embryogenesis abundant protein	Transgenic wheat expressing HVA1 shows enhanced drought tolerance	Detoxification; Protection Factor; Functional Proteins; Protection Factors; LEA Proteins	Bahieldin et al. (2005)
37	HSFA1b	Heat shock transcription factor	Increased HSFA1b expression showed increased water productivity and harvest index under water-replete and water-limiting conditions	Regulatory Proteins; Gene Expression; Transcription Factors	Bechtold et al. (2013)
38	DREB2A	A homolog to arabidopsis DREB2A	Functional transcript induced by stress, overexpression leads to increased stress tolerance	Regulatory Proteins; Gene Expression; Transcription Factors; DREB	Qin et al. (2007)
39	OsDREB2A	A homolog to Arabidopsis DREB2A	Expression induced after dehydration; stress-inducible expression in rice variety: improved performance under severe drought	Regulatory Proteins; Gene Expression; Transcription Factors; DREB	Cui et al. (2011)
40	TaLEA	Late embryogenesis abundant	TaLEA gene may enhance salt and drought tolerance by protecting cell membranes from damage	Detoxification; Protection Factor; Functional Proteins; Protection Factors; LEA Proteins	Gao et al. (2013)
41	NPK1	MAPKKK from tobacco; stress tolerance in tobacco	Expression of tobacco NPK1 in maize: enhanced drought tolerance	Regulatory Proteins; Signal Transduction; Protein Kinases	Shou et al. (2004)
42	MIR168A	MicroRNA168	Overexpressor: ABA hypersensitive, enhanced drought tolerance	Regulatory Proteins; Gene Expression; miRNA	Li et al. (2012)
43	miR393	microRNA393	Overexpressor: reduced tolerance to drought	Regulatory Proteins; Gene Expression; miRNA	Xia et al. (2012)
44	GhMKK1	Mitogen-activated protein kinase kinases	Overexpression of GhMKK1 in *Nicotiana benthamiana* enhanced its tolerance to salt and drought stresses	Regulatory Proteins; Signal Transduction; Protein Kinases	Lu et al. (2013)

(continued)

Table 8.2 (continued)

S. NO	Genes	Biological function	Phenotypes	Tags	References
45	MYB4	MYB TF	Overexpression in AT: accumulation of compatible solutes; drought tolerance	Detoxification; Osmolyte Production; Regulatory Proteins; Gene Expression; Transcription Factors; MYB	Mattana et al. (2005)
46	CpMYB10	MYB TF	Overexpression in Ath: desiccation and salt tolerance	Regulatory Proteins; Gene Expression; Transcription Factors; MYB	Villalobos et al. (2004)
47	MdoMYB121	MYB TF	A transgenic analysis indicated that the overexpression of the MdoMYB121 gene remarkably enhanced the tolerance to high salinity, drought, and cold stresses in transgenic tomato and apple plants	Regulatory Proteins; Gene Expression; Transcription Factors; MYB	Cao et al. (2013)
48	OsMYB3R-2	MYB TF	Expression in Ath: increased tolerance to drought, cold and salt stress	Regulatory Proteins; Gene Expression; Transcription Factors; MYB	Dai et al. (2007)
49	MYB44	MYB type TF	Overexpression: more sensitive to ABA, more rapid ABA-induced stomatal closure	Regulatory Proteins; Gene Expression; Transcription Factors; MYB	Jung et al. (2008)
50	NAC2a	NAC TF	Tobacco overexpressing TaNAC2a: higher dry and fresh weight under drought conditions	Regulatory Proteins; Gene Expression; Transcription Factors; NAC	Tang et al. (2012)
51	PEPCK	PEP carboxykinase	Isoform that is specifically Expressed in GCs; knockout: reduced drought tolerance	Ion and Osmotic Homeostasis; Stomatal Movement	Penfield et al. (2012)
52	BnPIP1	PIP	Overexpression in tobacco: increased tolerance to water stress; antisense: decreased tolerance	Ion and Osmotic Homeostasis; Functional Proteins; Channels and Transporters; Water Channels	Yu et al. (2005)
53	PIP1;4	PIP	Overexpression: rapid water loss under dehydration stress	Ion and Osmotic Homeostasis; Functional Proteins; Channels and Transporters; Water Channels	Jang et al. (2007)

54	PIP2;5	PIP	Overexpression: rapid water loss under dehydration stress	Ion and Osmotic Homeostasis; Functional Proteins; Channels and Transporters; Water Channels	Jang et al. (2007)
55	NtAQP1	PIP1 plasma membrane aquaporin	Reduced expression: lower water stress resistance	Ion and Osmotic Homeostasis; Functional Proteins; Channels and Transporters; Water Channels	Siefritz et al. (2002)
56	OST2	Plasma membrane proton ATPase	Mutant: reduced ability to close stomata in response to drought; downregulated during drought	Ion and Osmotic Homeostasis; Stomatal Movement; Regulatory Proteins; Acid Anhydride Hydrolases	Merlot et al. (2007)
57	OsiSAP8	Stress-associated protein gene family, zinc finger domain	Overexpression in tobacco and rice: tolerance to cold, salt and drought	Regulatory Proteins; Gene Expression; Transcription Factors; Zinc Fingers	Kanneganti and Gupta (2008)
58	PDH45	Stress-responsive helicase	Overexpressor: Stress-induced root growth was also substantially higher; stay-green phenotype and increased chlorophyll stability under stress	Regulatory Proteins, Gene Expression; Chromatin	Manjulatha et al. (2014)
59	TERF1	Tomato ERF transcription activator	Expression in tobacco: ABA hypersensitivity and drought tolerance	Regulatory Proteins; Gene Expression; Transcription Factors; ERF	Zhang et al. (2005)
60	DREB2A	Transcription factor (dehydration-responsive element binding)	Expression induced by dehydration	Regulatory Proteins; Gene Expression; Transcription Factors; DREB	Liu et al. (1998)
61	OsDREB1A	Transcription factor (dehydration-responsive element binding)	Transgenic rice: improved drought, high-salt, and cold stress tolerance	Regulatory Proteins; Gene Expression; Transcription Factors; DREB	Ito et al. (2006)
62	MYB2	Transcriptional activator of ABA signaling	Overexpressor: improved stress tolerance	Regulatory Proteins; Gene Expression; Transcription Factors; MYB	Abe et al. (2003)
63	MYC2	Transcriptional activator of ABA signaling	Overexpressor: improved stress tolerance	Regulatory Proteins; Gene Expression; Transcription Factors; MYC	Abe et al. (2003)
64	WR1			Regulatory Proteins; Gene Expression; Transcription Factors; AP2-Domain	Chen et al. (2012)

(continued)

Table 8.2 (continued)

S. NO	Genes	Biological function	Phenotypes	Tags	References
		Wax synthesis regulatory gene 1, homology to AT SHN1/WIN1	Transcription induced by drought; overexpressor: improved drought tolerance		
65	OsWRKY45	WRKY type TF	Expression in Ath: enhanced tolerance to drought and biotic stress	Regulatory Proteins; Gene Expression; Transcription Factors; WRKY	Qiu and Yu (2009)
66	WRKY19	WRKY-type TF	Overexpression in AT: enhanced drought tolerance	Regulatory Proteins; Gene Expression; Transcription Factors; WRKY	Niu et al. (2012)
67	WRKY2	WRKY-type TF	Overexpression in AT: enhanced drought tolerance	Regulatory Proteins; Gene Expression; Transcription Factors; WRKY	Niu et al. (2012)

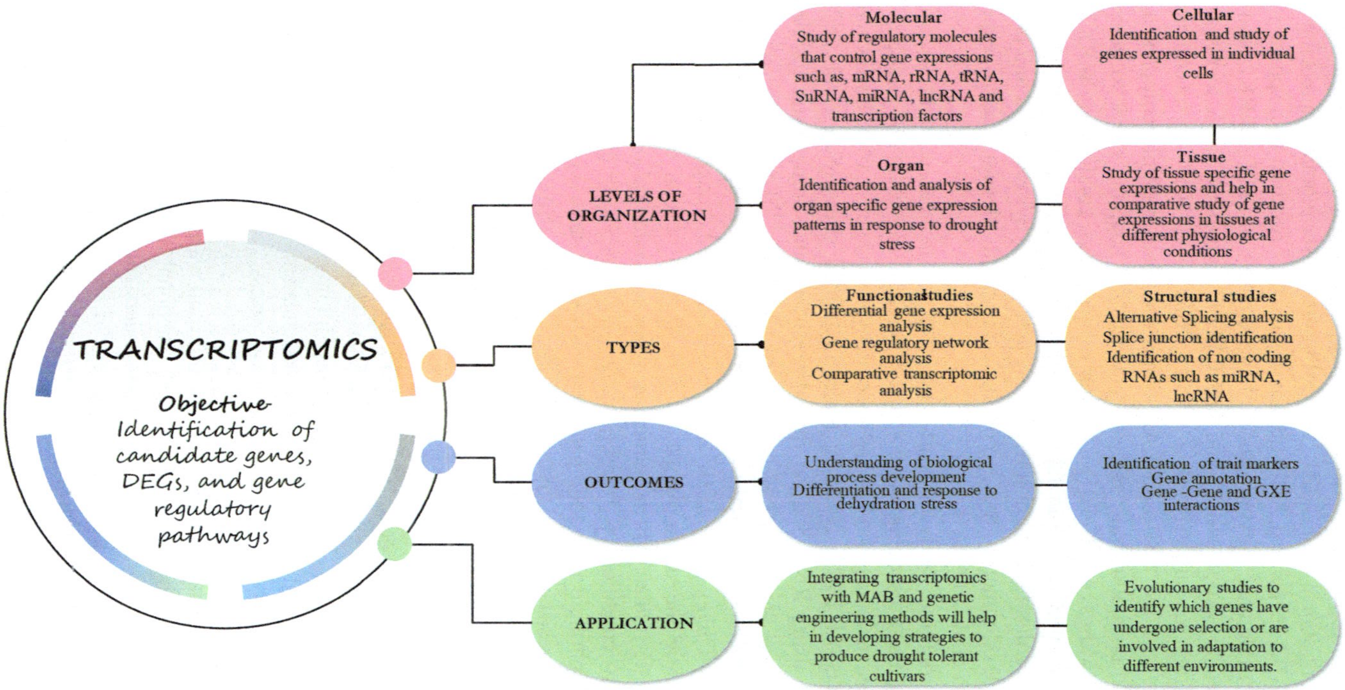

Fig. 8.2 From genes to function: a flowchart of transcriptomics at different levels of the organization

Transcriptomic strategies have asserted that miRNAs play an important role in drought stress response and that Auxin biosynthesis generates drought tolerance in cereals (Chakraborty et al. 2020). According to transcriptomic sequencing reports, increased ABA levels and ABA-mediated pathways function in response to stress and drought tolerance. Several studies have shown that PP2C, SnRK2, and ABRF are involved in aba signaling, and that OSGH32 regulates ABA levels and thus confers drought tolerance (Kulik et al. 2011; Xiang et al. 2017). In several crops, the GTL1 gene and the ASR and LAA genes modulate multiple complex traits by regulating transpiration rate and stomatal closure, adapting crops to dehydration stress (Magwanga et al. 2018). Under water-stress conditions, SnRK2.6/OST induces ABA-mediated stomatal closure, and multiple proteins have been studied that aid in regulating drought tolerance by acting as ABA receptors and promoting stomatal development, root growth, and plant development, respectively (Nakashima et al. 2009).

A transcriptomic study of pearl millet exposed to heat, drought, or combined stress using Pacbio sequencing revealed differential expression of multiple TF and coregulators, including 6484 DEGs under dehydration stress, 6920 DEGs under heat stress, and 1881 DEGs common under combined stress (Sun et al. 2020). An experiment using RNA-Seq on drought-resistant and drought-susceptible cereal crop cultivars at the seedling stage revealed 180 DEGs, 70 of which were unidentified novel genes or attributed to regulatory proteins and signal transduction under stress (Abdel-Ghany et al. 2020). Another Peg-induced drought experiment aided in comprehending the major regulatory pathways in several crop genotype-based transcriptomic studies, which delineated major physiological responses to dehydration stress conditions. Transcriptomic analysis has also disclosed that glycerophospholipid metabolic pathways and ATP Citrate synthase are implicated in conferring resistance to dehydration stress in plants (Sun et al. 2020).

Multiple traits in different plants have been enhanced using transcriptomic techniques, resulting in superior cultivars, which would include drought tolerance, fusarium wilt resistance, cyst nematode resistance, Rust resistance, blast resistance, downey mildew resistance, stemphylium blight resistance, striga resistance, aluminum toxicity (Kotresh et al. 2006; Cahill and Schmidt 2004; Saha et al. 2010; Omoigui et al. 2012; Varshney et al. 2014). Though sequencing reports in pearl millet are still limited, implementing transcriptomics strategies will provide novel insights into the plant's stress-responsive mechanisms and a better understanding of the correlation between terminal drought stress and yield, resulting in crop improvement.

Transcriptome data are highly dynamic and variable, so careful analysis is required. Another challenge is a limited functional analysis of target traits; more analysis is necessary for a complete understanding. The barriers to whole transcriptome analysis are a lack of access to high throughput and cost-effective phenotyping platforms and a lack of appropriate infrastructure. Although RNA-Seq is being prioritized over other technologies for improved genome coverage, the process is time consuming (Kaur et al. 2021). Further advancements in RNA-seq are revolutionizing plant transcriptomics studies, aiming to develop crop varieties that can withstand biotic and abiotic stress and produce a higher yield in the near future.

8.6 Proteomics

Proteomics is a sub-field of functional genomics that studies the proteome, or the entire set of proteins expressed by the genome of a cell, tissue, or organism at a given point in time (Renaut et al. 2006). In contrast to the genome, the proteome is dynamic in nature, undergoing translation and post-translational modifications (Rhee et al. 2006). It is a link between the transcriptome and the metabolome, and it is a powerful tool for analyzing the crop's complex defense response and biochemical pathways. Proteins are critical components of major biochemical and signaling pathways, and they help plants adapt to stress by protecting cellular proteins from denaturing, improving amino acid synthesis, and phospholipid metabolism (Rollins et al. 2013; Ashoub et al. 2015). This omic strategy intends to examine the global patterns of protein and its content, functions, and structure and provide a more detailed picture of protein networks and metabolic pathways typically involved in drought tolerance mechanisms (Fig. 8.3).

Protein expression profiling offers a thorough understanding of expressed proteins, beginning with identifying their primary amino acid sequence and progressing to measuring their relative content, post-translation state, structure, and interaction with other proteins and molecules (Barbrer-Brygoo and Joyard 2004). Expression analysis aids in evaluating plant stress responses by identifying

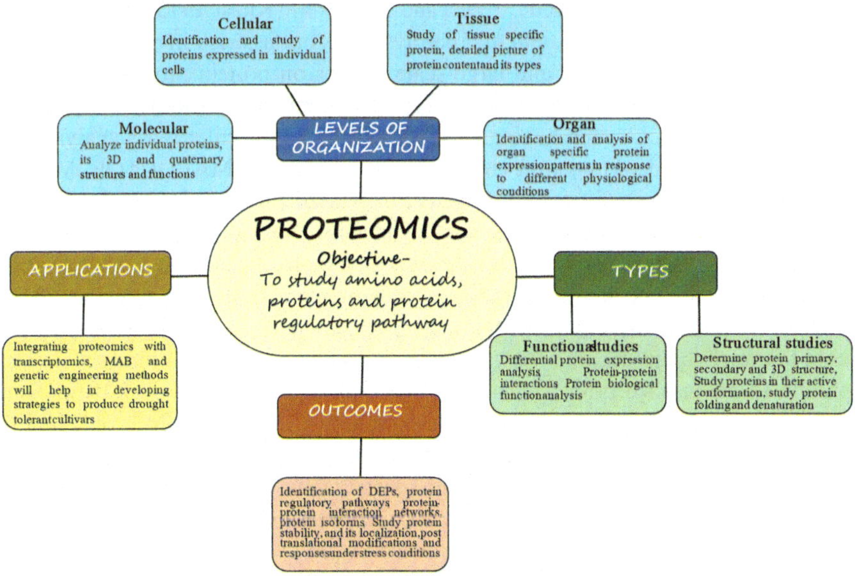

Fig. 8.3 Navigating proteomics: an overview of methods, applications, and analysis

relevant genes associated with nutrient uptake, transport, and accumulation, as well as a master regulator protein that plays a significant role in fundamental pathways, facilitating the development of genetically engineered crops (Aizat and Hassan 2018). This technology has proven advantageous for crop improvement because it allows a better understanding of complex biological mechanisms and cellular responses to environmental stress.

The application of proteomics in crop sciences has made it possible to comprehend the post-translational modification of proteins in drought-stressed plants. Dehydration stress induces functional proteins involved directly in protecting cells, such as ion channel proteins, stress-related proteins, heat shock proteins, late embryogenesis abundant proteins, and metabolic enzymes. Protein expression studies provide a thorough understanding of functional and regulatory proteins that are vital components of stress physiology and plant biochemistry fundamentals. Biotic and abiotic stress can bring out dramatic alterations in a plant's proteome.

8.6.1 Proteomics Techniques

Proteomics tools for performing large-scale protein studies have advanced over the last few decades. Earlier, conventional techniques such as Ion exchange chromatography, size exclusion chromatography, and affinity chromatography have been used to separate proteins based on protein charge, sample size, and sample binding affinity with the stationary phase (Yang et al. 2021) (Fig. 8.1). Later, western blotting and ELISA techniques were utilized for selective protein analysis, followed by advanced gel-based protein identification, separation, and visualization techniques, such as SDS PAGE, 2 DE, and 2D-DIGE (Klose 1975). Rapid protein expression analysis using protein microarrays/chips has assisted in detecting minute amounts of protein samples and large-scale studies on protein profiling and functional characterization (Yang et al. 2021; MacBeath 2002). Some of the major proteomics approaches that have gained attention for molecular mass analysis and molecular weight determination are MS, LCMS, LC, and MALDI TOF. MALDI TOF has the ability to identify and detect expressed proteins, which benefits the identification of relevant genes via reverse genetics (Tanaka et al. 1988; Baggerman et al. 2005; Washburn et al. 2001). LCMS development has led to a wide-scale analysis of protein metabolite and protein–protein interaction (Schubert et al. 2017).

MS can be used to analyze peptide sequences, and electrospray ionization can be used to convert peptides into ions. Another MS technique that can bind the desired protein to the surface prior to analysis is SELDI (Hashiguchi et al. 2010). Using multidimensional protein identification technology (Mudpit), LCMS, and HPLC, automated analysis of peptide mixtures, and some major high throughput approaches such as X-ray crystallography and NMR spectroscopy can be used for protein 3D structure determination (Washburn et al. 2001; Twyman et al. 2013; Woolfson 2018; Aslam et al. 2017). Flow cytometry is a technique that has gained popularity in proteomics due to technological advancements (Galbraith 2014). It facilitates protein screening, though its limited proteome coverage is a challenge that must be overcome. Functional proteomics includes yeast one hybrid (Y1H) and yeast two-hybrid (Y2H) techniques for studying extracellular signaling complexes and preparing genome-wide protein interaction mapping (Fields and Song 1989; Chen and Harmon 2006). These techniques have led to the discovery of several differentially expressed proteins that provide insights into the stress-responsive mechanisms of plants exposed to drought stress.

8.6.2 Application of Proteomics Technologies

Incorporating proteomics technologies in crop science studies will provide a better understanding of stress response mechanisms in plants under water deficit conditions as stress alters the expression of the proteins, which will help elucidate the function of the expressed protein. This powerful analytical tool can monitor developmental changes or influence environmental stimuli on protein patterns, providing insights into plants' functioning at the molecular and cellular levels. Proteomic analysis has unveiled known and novel differentially expressed proteins that contribute significantly to cell signaling, gene transcription, and regulation (Rhee et al. 2006). A protein expression profiling study discovered ROS scavengers, dehydrins, and molecular chaperons HSPs in different crops (Sun et al. 2020). This omics-based strategy has helped identify drought-sensitive and drought-tolerant varieties in some crops. Phosphoproteomics has revealed proteins associated with dehydration stress, and it aims to analyze protein phosphorylation and detect phosphorylated amino acid residues in both quantitative and qualitative aspects (Mosa et al. 2017; Baslam and Mitsui 2020).

Tissue-specific proteomic analysis of drought stress responses in pearl millet demonstrated that the leaf displayed consequential changes followed by root, and 2281 proteins were identified (Ghatak et al. 2016). Another tissue-specific comparative proteomic study of drought-tolerant and drought-sensitive cultivars of pearl millet and wheat led to the identification of 12,558 proteins (Ghatak et al. 2021). In research using LC-MS/MS, 75 differential expressed proteins involved in ROS catabolic enzyme and signal transduction pathways were identified in a dehydration stress cultivar. In some crops, comparative gene expression and proteome analysis using 2 DE+ LC-MS/MS have recognized DEPs associated with abiotic stress responses (Arefian et al. 2019). Using 2D Gel electrophoresis and ESI LCMS/MS, novel proteins and several DEPs in response to dehydration stress have been identified (Larrainzar et al. 2009; Ramalingam et al. 2015). Responsive pathways were studied in another experiment using MALDI TOF, SDS-PAGE, MS 2DE, and peptide mass fingerprinting (PMF) stress (Demirevska et al. 2008; Toorchi et al. 2009; Deeba et al. 2012). Implementing these strategies has resulted in improved separation resolution and detailed knowledge of target proteins' physical and chemical properties.

To examine the variation in the protein content of a cereal, protein profiling along with the SSR and RAPD PCR was carried out (Kumar et al. 2012). The organ-specific proteomic analysis aimed to study the proteins assembling specifically during abiotic stress (Hossain and Komatsu 2013). Using these techniques, several proteins have been identified that are responsible for male sterility traits, especially necessary for hybrid selections and more efficient light fixation. It elucidates the extremely complex genetic interactions involved in plant and microbe interactions for studying the symbioses in plants. Veeranagamallaiah et al. (2008) demonstrated that the seedling stage in a crop using 2DGE, 29 DEPs, contributed to photosynthesis pathways, signal transduction, and metabolic stress responses. Though proteomics-based research in pearl millet is still limited, applying these approaches will provide a better understanding of the physiological responses of this crop to water deficit conditions and the production of transgenic crops with desired characteristics, including drought tolerance.

Quantitative proteomics analysis using iTRAQ in some crops has illustrated differential expression of proteins under abiotic stress, identified a number of stress-responsive DEPs, and provided new insights into somatic embryogenesis (Liu et al. 2015b; Yang et al. 2020; Zhu et al. 2018). Another expression profiling study of protein unraveled a comprehensive understanding of cellular mechanisms underlying seed development during grain filling and seed maturation phases. Proteomics, compared to other biological systems, has not been studied globally in plants, limiting our knowledge of protein assemblies to certain families and pathways (Yang et al. 2020). This omics-based approach is facing some challenges that need to be overcome for appropriate stress-responsive studies in crop science, which include the dynamic nature of proteome results in biological variation in proteins that are responsible for less reproducibility of proteomic data and abundance of proteins extracted from samples could be a hindrance in the analysis of derived and targeted protein.

8.7 Metabolomics

Metabolomics is a robust technology gaining traction for its ability to provide a full suite of low molecular weight metabolites synthesized by organisms, cells, tissues, or organs at any point in the process under any physiological condition (Fig. 8.4). It is a complementary tool to proteomics and transcriptomics that will help researchers better understand the cellular responses of plants to different stresses. Dehydration stress has negative effects, and plants' primary response to stress is the production of ROS scavengers, which causes damage and a decrease in photosynthetic activity. This technology aids in the comprehensive understanding of plants' cellular responses to drought stress, such as changes in cell wall architecture, cell membrane system adjustments, and alterations in cell cycle and cell division. Plants respond adaptively and specifically to an external stimulus such as drought stress, which can be very spontaneous and bring out metabolic changes that are critical features of plants' stress-responsive mechanism. Plants' central and specialized metabolisms and metabolic networks in response to drought stress can be studied using metabolomics strategies, a new field in the post-genomic era.

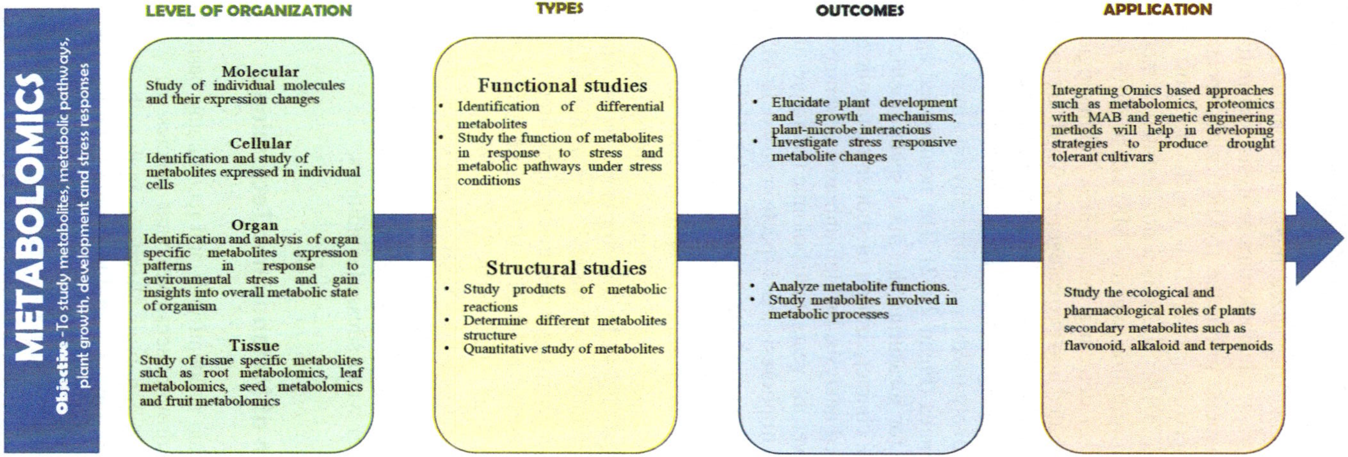

Fig. 8.4 Mapping the terrain of metabolomics: a comprehensive flow chart of objectives, studies, and applications

8.7.1 Metabolomics Techniques

Emerging scientific technologies in the field of metabolomics have offered remarkable opportunities to seize specialized metabolites and changes in metabolism in response to drought stress. Metabolomics analysis was initially compelled by the advent of nuclear magnetic resonance (NMR), mass spectrometry, and Fourier-transform infrared spectroscopy (FTIR), which are most widely used in crop sciences (Shalini et al. 2018). NMR is a nondestructive methodology used to get metabolite structural information, whereas MS, a destructive analysis, is popular due to its high sensitivity in detecting metabolites. Technologies used for analyzing metabolomes are segregated into two categories, which are targeted and nontargeted, on the basis of the purpose of the study. The targeted analysis is highly sensitive and involved in identifying and quantifying known metabolites, but it has low coverage. This analysis is extensively used to examine phytohormones and study the primary effects of genetic changes (Fiehn 2002). The nontargeted analysis is a discovery-based approach used for measuring the mass spectrometric features of unknown metabolites and for the characterization of sudden metabolic changes in response to drought; unlike targeted analysis, its high unbiased coverage is advantageous.

In crop sciences, over the years, metabolomic techniques that are gaining importance are capillary electrophoresis mass spectrometry (CE–MS) (Zhang et al. 2020), gas chromatography-mass spectrometry (GC–MS) (Casartelli et al. 2018), vibrational spectroscopy (VS) (Khakimov et al. 2014), and liquid chromatography-mass spectrometry LC-MS) (Cao et al. 2017). With evolving scientific technologies, other techniques such as ultraperformance liquid chromatography-mass spectrometry (UPLC–MS) (Zhao et al. 2014), high-performance liquid chromatography (HPLC) (Karre et al. 2017), capillary electrophoresis time-of-flight mass spectrometry (CE–TOF–MS) (Suharti et al. 2016), gas chromatography time-of-flight mass spectrometry (GC–TOF–MS) (Ramataur et al. 2019), and liquid chromatography high-resolution mass spectrometry (LC–HRMS) (Benincasa et al. 2020) that are being used in plant metabolomic studies. These technologies have enhanced the idea of complicated biosynthetic pathways and complex cellular and molecular responses and have led to identifying stress-responsive primary and secondary metabolites involved in plants under water-stressed conditions.

8.7.2 Application of Metabolomic Technology

Metabolomic studies aid in identifying and recognizing basic distinctions between drought-tolerant and drought-sensitive plant species, as well as the correlation between phenotypic and genotypic alterations in drought-stressed plants. Research on certain crops employing metabolite profiling found integrative biochemical networks of leaves responding to water deprivation circumstances (Michaletti et al. 2018). According to metabolic profiling and transcriptome analyses, the buildup of specialized root diterpenoids benefits drought tolerance in a crop (Tiedge et al. 2022). Gundaraniya et al. (2020) observed that pentitol, phytol, xylonic acid,

D-xylopyranose, stearic acid, and D-ribose were drought-sensitive metabolites among 46 metabolites in a drought-resistant and vulnerable crop. Another plant transcriptome and metabolomic investigation found that purine and phenylpropanoid metabolism play an important role in response to dehydration stress (Zhang et al. 2021). Metabolomics has gotten minimal attention compared to other biological systems, which remains a difficulty. Identifying important metabolic genes using a metabolomics approach will help improve drought tolerance and productivity in pearl millet.

8.8 Conclusion and Future Perspectives

Drought has a detrimental effect on pearl millet growth, development, yield, and metabolism, leading to the evolution of complicated mechanisms in some tolerant genotypes. Drought-responsive genes and pathways, such as genes involved in flavonoid and alkaloid biosynthesis and MAPK signaling pathways, plant hormone pathways, and ABA-dependent and independent pathways, have been identified using omics technologies such as genomics, transcriptomics, proteomics, and metabolomics. These methodologies indicated that proline accumulation and MOA content are some of the physiological factors contributing to oxidative stress tolerance in pearl millet seedlings. These methods have shown to be useful in finding and revealing relevant candidate genes and their pathways in pearl millet, despite the fact that most of this crop remains undiscovered.

An omics-based methodology is a viable tool for solving drought challenges that plague various different crops, including pearl millet. Most of these methodologies used high-throughput approaches to help find small RNAs, SNPs, and molecular markers to delineate functional analyses, genes, and metabolic networks in agricultural sciences. Omics will fill the void between gene and environment interactions, leading to the development of smart crops that will assure food and nutritional security. The combination of omics technology and genome-wide association studies (GWAS) has proven to be extremely effective in revealing metabolic processes and dehydration stress in pearl millet. This emerging genomics, transcriptomics, proteomics, metabolomics technologies, and computational tools, and the construction of models to anticipate agronomically important traits have revealed G-P-E interactions and may be employed in crop improvement breeding programs. Omics investigations in pearl millet can be exploited for MAS and mapping of drought stress QTLS in this crop, which is still in its early stages with limited information available.

Traditional QTL mapping, WGS, and GWAS techniques can assist in uncovering underpinning haplotypes and unexplored QTLs across the entire genome, which can be used to create superior cultivars through genome-assisted breeding (Sukumaran et al. 2018). Evolving omics methodologies with transgenic techniques will enhance India's existing agriculture system conditions. Omics methods like metabolomics and transcriptomics have been shown to be useful in building association maps in various crops. The increasing need and demand for improved crop variety and product characteristics are some of the critical elements driving the use of omics techniques for developing genetically modified crops with increased drought stress tolerance.

References

Abdel-Ghany SE, Ullah F, Ben-Hur A, Reddy AS (2020) Transcriptome analysis of drought-resistant and drought-sensitive sorghum (Sorghum bicolor) genotypes in response to peg-induced drought stress. Int J Mol Sci 21(3):772. https://doi.org/10.3390/ijms21030772

Abe H, Urao T, Ito T, Seki M, Shinozaki K, Yamaguchi-Shinozaki K (2003) Arabidopsis AtMYC2 (bHLH) and AtMYB2 (MYB) function as transcriptional activators in abscisic acid signaling. Plant Cell 15:63–78. https://doi.org/10.1105/tpc.006130

Aharoni A, Vorst O (2002) DNA microarrays for functional plant genomics. Plant Mol Biol 48:99–118. https://doi.org/10.1023/A:1013734019946

Ahmad P, Rasool S, Gul A, Sheikh SA, Akram NA, Ashraf M et al (2016) Jasmonates: multifunctional roles in stress tolerance. Front Plant Sci 7:813. https://doi.org/10.3389/fpls.2016.00813

Ahuja I, de Vos RCH, Bones AM, Hall RD (2010) Plant molecular stress responses face climate change. Trends Plant Sci 15:664–674. https://doi.org/10.1016/j.tplants.2010.08.002

Aizat WM, Hassan M (2018) Proteomics in systems biology. In: Aizat W, Goh HH, Baharum S (eds) Omics applications for systems biology. advances in experimental medicine and biology. Springer, Berlin, pp 31–49. https://doi.org/10.1007/978-3-319-98758-3_3

Anuradha N, Satyavathi CT, Chellapilla B, Thirunavukkarasu N, Sankar SM, Singh SP, Meena MC, Singhal T, Srivastava RK (2017) Deciphering genomic regions for high grain iron and zinc content using association mapping in pearl millet. Front Plant Sci 8:412. https://doi.org/10.3389/fpls.2017.00412

Arefian M, Vessal S, Malekzadeh-Shafaroudi S, Siddique KH, Bagheri A (2019) Comparative proteomics and gene expression analyses revealed responsive proteins and mechanisms for salt tolerance in chickpea genotypes. BMC Plant Biol 19(1):1–26. https://doi.org/10.1186/s12870-018-1592-2

Ashoub A, Baeumlisberger M, Neupaertl M, Karas M, Brüggemann W (2015) Characterization of common and distinctive adjustments of wild barley leaf proteome under drought acclimation, heat stress, and their combination. Plant Mol Biol 87:459–471. https://doi.org/10.1007/s11103-015-0327-3

Aslam B, Basit M, Atif NM, Khurshid M, Rasool MH (2017) Proteomics: technologies and their applications. J Chromatogr Sci 55(2):182–196. https://doi.org/10.1093/chromsci/bmw167

Baggerman G, Vierstraete E, De Loof A, Schoofs L (2005) Gel-based versus gel-free proteomics: a review. Comb Chem High Throughput Screen 8:669–677. https://doi.org/10.2174/138620705774962490

Bahieldin A, Mahfouz HT, Eissa HF, Saleh OM, Ramadan AM, Ahmed IA, Dyer WE, El-Itriby HA, Madkour MA (2005) Field evaluation of transgenic wheat plants stably expressing the HVA1 gene for drought tolerance. Physiol Plant 123(4):421–427. https://doi.org/10.1111/j.1399-3054.2005.00496.x

Barbrer-Brygoo H, Joyard J (2004) Introduction - focus on plant proteomics. Plant Physiol Biochem 42:913–917. https://doi.org/10.1016/j.plaphy.2004.11.003

Bartels D, Sunkar R (2005) Drought and salt tolerance in plants. CRC Crit Rev Plant Sci 24(1): 23–58. https://doi.org/10.1080/07352680590910410

Baslam M, Mitsui T (2020) Proteomic for quality: mining the proteome as a strategy to elucidate the protein complex applied for quality improvement. In: The future of rice demand: quality beyond productivity. Springer Nature, Singapore, pp 473–494

Bechtold U, Albihlal WS, Lawson T, Fryer MJ, Sparrow PA, Richard F, Persad R, Bowden L, Hickman R, Martin C, Beynon JL (2013) Arabidopsis heat shock transcription factora1b

overexpression enhances water productivity, resistance to drought, and infection. J Exp Bot 64(11):3467–3481. https://doi.org/10.1093/jxb/ert196
Benincasa P, D'Amato R, Falcinelli B, Troni E, Fontanella MC, Frusciante S et al (2020) Grain endogenous selenium and moderate salt stress work as synergic elicitors in the enrichment of bioactive compounds in maize sprouts. Agronomy 10(5):735. https://doi.org/10.3390/agronomy10050735
Bidinger F, Mahalakshmi V, Rao GDP (1987) Assessment of drought resistance in pearl millet (Pennisetum americanum (L.) Leeke). II. Estimation of genotype response to stress. Aust J Agr Res 38(1):49–59. https://doi.org/10.1071/AR9870049
Borovitz JO, Chory J (2004) Genomics tools for QTL analysis and gene discovery. Curr Opin Plant Biol 7:132–136
Burgess DJ (2015) Putting transcriptomics in its place. Nat Rev Genet 16:319. https://doi.org/10.1038/nrg3951
Cahill DJ, Schmidt DH (2004) Use of marker assisted selection in a product development breeding program. In: Proceedings of the 4th international crop science congress, Brisbane, Australia, 26 Sept–1 Oct
Cal AJ, Sanciangco M, Rebolledo MC, Luquet D, Torres RO, McNally KL, Henry A (2019) Leaf morphology, rather than plant water status, underlies genetic variation of rice leaf rolling under drought. Plant Cell Environ 42(5):1532–1544
Cao ZH, Zhang SZ, Wang RK, Zhang RF, Hao YJ (2013) Genome wide analysis of the apple MYB transcription factor family allows the identification of MdoMYB121 gene confering abiotic stress tolerance in plants. PloS One 8(7):e69955
Cao D, Lutz A, Hill CB, Callahan DL, Roessner U (2017) A quantitative profiling method of phytohormones and other metabolites applied to barley roots subjected to salinity stress. Front Plant Sci 7:2070. https://doi.org/10.3389/fpls.2016.02070
Casartelli A, Riewe D, Hubberten HM, Altmann T, Hoefgen R, Heuer S (2018) Exploring traditional aus-type rice for metabolites conferring drought tolerance. Rice 11(1):1–16. https://doi.org/10.1186/s12284-018-0209-9
Caverzan A, Passaia G, Rosa SB, Ribeiro CW, Lazzarotto F, Margis-Pinheiro M (2012) Plant responses to stresses: role of ascorbate peroxidase in the antioxidant protection. Genet Mol Biol 35:1011–1019. https://doi.org/10.1590/S1415-47572012000600020
Chakraborty A, Viswanath A, Malipatil R, Rathore A, Thirunavukkarasu N (2020) Structural and functional characteristics of miRNAs in five strategic millet species and their utility in drought tolerance. Front Genet 11:608421. https://doi.org/10.3389/fgene.2020.608421
Chanwala J, Satpati S, Dixit A et al (2020) Genome-wide identification and expression analysis of WRKY transcription factors in pearl millet (Pennisetum glaucum) under dehydration and salinity stress. BMC Genomics 21(1):231. https://doi.org/10.1186/s12864-020-6622-0
Chaves MM, Pereira JS, Maroco J, Rodrigues ML, Ricardo CPP, Osório ML et al (2002) How plants cope with water stress in the field? Photosynth Growth Ann Bot 89(7):907–916
Chen S, Harmon AC (2006) Advances in plant proteomics. Proteomics 6(20):5504–5516
Chen SC, Wu PH, Su YC, Wen TN, Wei YS, Wang NC, Hsu CA, Wang AHJ, Shyur LF (2012) Biochemical characterization of a novel laccase from the basidiomycete fungus Cerrena sp. WR1. Protein Eng Des Sel 25(11):761–769. https://doi.org/10.1093/protein/gzs064
Choudhary M, Jayanand, Padaria JC (2015) Transcriptional profiling in pearl millet (Pennisetum glaucum L.R. Br.) for identification of differentially expressed drought responsive genes. Physiol Mol Biol Plants 21(2):187–196. https://doi.org/10.1007/s12298-015-0299-8
Comas LH, Becker SR, Cruz VMV, Byrne PF, Dierig DA (2013) Root traits contributing to plant productivity under drought. Front Plant Sci 4:442. https://doi.org/10.3389/fpls.2013.00442
Cowie P, Ross R, MacKenzie A (2013) Understanding the dynamics of gene regulatory systems; characterisation and clinical relevance of cis-regulatory polymorphisms. Biology 2:64–84. https://doi.org/10.3390/biology2010064

Cui M, Zhang W, Zhang Q, Xu Z, Zhu Z, Duan F, Wu R (2011) Induced over-expression of the transcription factor OsDREB2A improves drought tolerance in rice. Plant Physiol Biochem 49(12):1384–1391. https://doi.org/10.1016/j.plaphy.2011.09.016

Dai X, Xu Y, Ma Q, Xu W, Wang T, Xue Y, Chong K (2007) Overexpression of an R1R2R3 MYB gene, OsMYB3R-2, increases tolerance to freezing, drought, and salt stress in transgenic Arabidopsis. Plant Physiol 143(4):1739–1751

Darshan K, Aggarwal R, Bashyal BM, Singh J, Shanmugam V, Gurjar MS, Solanke AU (2020) Transcriptome profiling provides insights into potential antagonistic mechanisms involved in Chaetomium globosum against Bipolaris sorokiniana. Front Microbiol 11:578115. https://doi.org/10.3389/fmicb.2020.578115

Deeba F, Pandey AK, Ranjan S, Mishra A, Singh R, Sharma YK et al (2012) Physiological and proteomic responses of cotton (Gossypium herbaceum L.) to drought stress. Plant Physiol Biochem 53:6–18. https://doi.org/10.1016/j.plaphy.2012.01.002

Demirevska K, Simova-Stoilova L, Vassileva V, Vaseva I, Grigorova B, Feller U (2008) Drought-induced leaf protein alterations in sensitive and tolerant wheat varieties. Gen Appl Plant Physiol 34:79–102. https://doi.org/10.7892/boris.110728

Diallo AO, Agharbaoui Z, Badawi MA, Ali-Benali MA, Moheb A, Houde M, Sarhan F (2014) Transcriptome analysis of an mvp mutant reveals important changes in global gene expression and a role for methyl jasmonate in vernalization and flowering in wheat. J Exp Bot 65(9): 2271–2286. https://doi.org/10.1093/jxb/eru102

Djanaguiraman M, Perumal R, Ciampitti IA, Gupta SK, Prasad PVV (2018) Quantifying pearl millet response to high temperature stress: thresholds, sensitive stages, genetic variability and relative sensitivity of pollen and pistil. Plant Cell Environ 41(5):993–1007. https://doi.org/10.1111/pce.13163

Dudhate A, Shinde H, Tsugama D, Liu S, Takano T (2018) Transcriptomic analysis reveals the differentially expressed genes and pathways involved in drought tolerance in pearl millet [Pennisetum glaucum (L.) R. Br]. PLoS One 13(4):e0195908. https://doi.org/10.1371/journal.pone.0195908

Dudhate A, Shinde H, Yu P et al (2021) Comprehensive analysis of NAC transcription factor family uncovers drought and salinity stress response in pearl millet (Pennisetum glaucum). BMC Genomics 22(1):70. https://doi.org/10.1186/s12864-021-07382-y

Dwivedi S, Upadhyaya H, Senthilvel S, Hash C, Fukunaga K, Diao X, Santra D, Baltensperger D, Prasad M (2012) Millets: genetic and genomic resources. In: Janick J (ed) Plant breeding reviews, vol 35. Wiley, Hoboken, pp 247–375

El Sabagh A, Hossain A, Islam MS, Barutçular C, Hussain S, Hasanuzzaman M, Akram T, Mubeen M, Jatoi W, Fahad S et al (2019) Drought and salinity stresses in barley: consequences and mitigation strategies. Aust J Crop Sci 13(6):810–820. https://doi.org/10.21475/ajcs.19.13.06.pne1367

Fiehn O (2002) Metabolomics—the link between genotypes and phenotypes. Funct Genomics 2(3): 155–171

Fields S, Song O (1989) A novel genetic system to detect protein-protein interactions. Nature 340(6230):245–246

Furihata T, Maruyama K, Fujita Y, Umezawa T, Yoshida R, Shinozaki K, Yamaguchi-Shinozaki K (2006) Abscisic acid-dependent multisite phosphorylation regulates the activity of a transcription activator AREB1. Proc Natl Acad Sci 103(6):1988–1993. https://doi.org/10.1073/pnas.0505667103

Galbraith DW (2014) Flow cytometry and sorting in Arabidopsis. In: Arabidopsis protocols. Springer, Berlin, pp 509–537

Gao H, Song A, Zhu X, Chen F, Jiang J, Chen Y, Sun Y, Shan H, Gu C, Li P, Chen S (2012) The heterologous expression in Arabidopsis of a chrysanthemum Cys2/His2 zinc finger protein gene confers salinity and drought tolerance. Planta 235:979–993. https://doi.org/10.1007/s00425-011-1566-1

Gao W, Bai S, Li Q, Gao C, Liu G, Li G, Tan F (2013) Overexpression of TaLEA gene from Tamarix androssowii improves salt and drought tolerance in transgenic poplar (Populus simonii× P. nigra). PloS One 8(6):e67462. https://doi.org/10.1371/journal.pone.0067462
Ghatak A, Chaturvedi P, Nagler M, Roustan V, Lyon D, Bachmann G, Postl W, Schröfl A, Desai N, Varshney RK, Weckwerth W (2016) Comprehensive tissue-specific proteome analysis of drought stress responses in Pennisetum glaucum (L.) R. Br. (Pearl millet). J Proteomics 143: 122–135. https://doi.org/10.1016/j.jprot.2016.02.032
Ghatak A et al (2021) Physiological and proteomic signatures reveal mechanisms of superior drought resilience in pearl millet compared to wheat. Front Plant Sci 11:600278. https://doi.org/10.3389/fpls.2020.600278
Gundaraniya SA, Ambalam PS, Tomar RS (2020) Metabolomic profiling of drought-tolerant and susceptible peanut (Arachis hypogaea L.) genotypes in response to drought stress. ACS Omega 5(48):31209–31219. https://doi.org/10.1021/acsomega.0c04409
Guo L, Wang ZY, Lin H, Cui WE, Chen J, Liu M et al (2006) Expression and functional analysis of the rice plasma-membrane intrinsic protein gene family. Cell Res 16(3):277–286
Haake V, Cook D, Riechmann J, Pineda O, Thomashow MF, Zhang JZ (2002) Transcription factor CBF4 is a regulator of drought adaptation in Arabidopsis. Plant Physiol 130(2):639–648. https://doi.org/10.1104/pp.006783
Hanba YT, Shibasaka M, Hayashi Y, Hayakawa T, Kasamo K, Terashima I, Katsuhara M (2004) Overexpression of the barley aquaporin HvPIP2;1 increases internal CO2 conductance and CO2 assimilation in the leaves of transgenic rice plants. Plant Cell Physiol 45(5):521–529
Hashiguchi A, Ahsan N, Komatsu S (2010) Proteomics application of crops in the context of climatic changes. Food Res Int 43(7):1803–1813
Hollister JD (2014) Genomic variation in Arabidopsis: tools and insights from next-generation sequencing. Chromosome Res 22:103–115. https://doi.org/10.1007/s10577-014-9406-5
Hossain Z, Komatsu S (2013) Contribution of proteomic studies towards understanding plant heavy metal stress response. Front Plant Sci 3:310. https://doi.org/10.3389/fpls.2012.00310
Houle D, Govindaraju DR, Omholt S (2010) Phenomics: the next challenge. Nat Rev Genet 11: 855–866. https://doi.org/10.1038/nrg2897
Huang L, Zhang F, Zhang F, Wang W, Zhou Y, Fu B, Li Z (2014) Comparative transcriptome sequencing of tolerant rice introgression line and its parents in response to drought stress. BMC Genomics 15:1026. https://doi.org/10.1186/1471-2164-15-1026
Islam T, Manna M, Reddy MK (2015) Glutathione peroxidase of Pennisetum glaucum (PgGPx) is a functional Cd2+ dependent peroxiredoxin that enhances tolerance against salinity and drought stress. PLoS One 10(12):e0143344. https://doi.org/10.1371/journal.pone.0143344
Ito Y, Katsura K, Maruyama K, Taji T, Kobayashi M, Seki M, Shinozaki K, Yamaguchi-Shinozaki K (2006) Functional analysis of rice DREB1/CBF-type transcription factors involved in cold-responsive gene expression in transgenic rice. Plant Cell Physiol 47(1):141–153. https://doi.org/10.1093/pcp/pci230
Iuchi S, Kobayashi M, Taji T, Naramoto M, Seki M, Kato T et al (2001) Regulation of drought tolerance by gene manipulation of 9-cis-epoxycarotenoid dioxygenase, a key enzyme in abscisic acid biosynthesis in Arabidopsis. Plant J 27(4):325–333
Jaiswal S, Antala TJ, Mandavia MK, Chopra M, Jasrotia RS, Tomar RS et al (2018) Transcriptomic signature of drought response in pearl millet (Pennisetum glaucum (L.) and development of web-genomic resources. Sci Rep 8(1):1–16. https://doi.org/10.1038/s41598-017-18612-5
Jang JY, Lee SH, Rhee JY, Chung GC, Ahn SJ, Kang H (2007) Transgenic Arabidopsis and tobacco plants overexpressing an aquaporin respond differently to various abiotic stresses. Plant Mol Biol 64:621–632
Janiak A, Kwaśniewski M, Szarejko I (2016) Gene expression regulation in roots under drought. J Exp Bot 67(4):1003–1014. https://doi.org/10.1093/jxb/erv466
Ji X, Liu G, Liu Y, Zheng L, Nie X, Wang Y (2013) The bZIP protein from Tamarix hispida, ThbZIP1, is ACGT elements binding factor that enhances abiotic stress signaling in transgenic Arabidopsis. BMC Plant Biol 13:1–13. https://doi.org/10.1186/1471-2229-13-151

Jung S, Main D (2014) Genomics and bioinformatics resources for translational science in Rosaceae. Plant Biotechnol Rep 8:49–64. https://doi.org/10.1007/s11816-013-0265-5

Jung C, Seo JS, Han SW, Koo YJ, Kim CH, Song SI, Nahm BH, Choi YD, Cheong JJ (2008) Overexpression of AtMYB44 enhances stomatal closure to confer abiotic stress tolerance in transgenic Arabidopsis. Plant Physiol 146:623–635

Kakumanu A, Ambavaram MR, Klumas C, Krishnan A, Batlang U, Myers E, Pereira A (2012) Effects of drought on gene expression in maize reproductive and leaf meristem tissue revealed by RNA-seq. Plant Physiol 160:846–867. https://doi.org/10.1104/pp.112.200444

Kang JY, Choi HI, Im MY, Kim SY (2002) Arabidopsis basic leucine zipper proteins that mediate stress-responsive abscisic acid signaling. Plant Cell 14(2):343–357

Kang J, Hwang JU, Lee M, Kim YY, Assmann SM, Martinoia E, Lee Y (2010) PDR-type ABC transporter mediates cellular uptake of the phytohormone abscisic acid. Proc Natl Acad Sci 107(5):2355–2360

Kanneganti V, Gupta AK (2008) Overexpression of OsiSAP8, a member of stress associated protein (SAP) gene family of rice confers tolerance to salt, drought and cold stress in transgenic tobacco and rice. Plant Mol Biol 66:445–462

Karre S, Kumar A, Dhokane D, Kushalappa AC (2017) Metabolo-transcriptome profiling of barley reveals induction of chitin elicitor receptor kinase gene (HvCERK1) conferring resistance against Fusarium graminearum. Plant Mol Biol 93(3):247–267. https://doi.org/10.1007/s11103-016-0552-2

Kaur B, Sandhu KS, Kamal R, Kaur K, Singh J, Röder MS, Muqaddasi QH (2021) Omics for the improvement of abiotic, biotic, and agronomic traits in major cereal crops: applications, challenges, and prospects. Plants 10(10):1989. https://doi.org/10.3390/plants10101989

Ke R, Marco M, Pacureanu A, Svedlund J, Botling J, Wählby C, Nilsson M (2013) In situ sequencing for RNA analysis in preserved tissue and cells. Nat Methods 10:857–860. https://doi.org/10.1038/nmeth.2563

Khakimov B, Bak S, Engelsen SB (2014) High-throughput cereal metabolomics: current analytical technologies, challenges and perspectives. J Cereal Sci 59(3):393–418. https://doi.org/10.1016/j.jcs.2014.03.008

Kholova J, Hash CT, Kakkera A, Kocova M, Vadez V (2010) Constitutive water-conserving mechanisms are correlated with the terminal drought tolerance of pearl millet Pennisetum glaucum (L.) R. Br J Exp Bot 61(2):369–377. https://doi.org/10.1093/jxb/erp304

Klose J (1975) Protein mapping by combined isoelectric focusing and electrophoresis of mouse tissues. A novel approach to testing for induced point mutations in mammals. Humangenetik 26: 231–243

Kogenaru S, Qing Y, Guo Y, Wang N (2012) RNA-seq and microarray complement each other in transcriptome profiling. BMC Genomics 13:629. https://doi.org/10.1186/1471-2164-13-629

Kotresh H, Fakrudin B, Punnuri S, Rajkumar B, Thudi M, Paramesh H, Lohithswa H, Kuruvinashetti MS (2006) Identification of two RAPD markers genetically linked to a recessive allele of a Fusarium wilt resistance gene in pigeonpea (Cajanus cajan L.) Millsp. Euphytica 149: 113–120. https://doi.org/10.1007/s10681-005-9054-4

Kulik A, Wawer I, Krzywińska E, Bucholc M, Dobrowolska G (2011) SnRK2 protein kinases—key regulators of plant response to abiotic stresses. OMICS 15(12):859–872. https://doi.org/10.1089/omi.2011.0062

Kumar A, Sharma N, Panwar P, Gupta AK (2012) Use of SSR, RAPD markers and protein profiles based analysis to differentiate Eleusine coracana genotypes differing in their protein content. Mol Biol Rep 39:4949–4960. https://doi.org/10.1007/s11033-011-1237-6

Kumar A, Pathak RK, Gupta SM, Gaur VS, Pandey D (2015) Systems biology for smart crops and agricultural innovation: filling the gaps between genotype and phenotype for complex traits linked with robust agricultural productivity and sustainability. OMICS 19:581–601

Kuromori T, Miyaji T, Yabuuchi H, Shimizu H, Sugimoto E, Kamiya A et al (2010) ABC transporter AtABCG25 is involved in abscisic acid transport and responses. Proc Natl Acad Sci 107(5):2361–2366

Kuromori T, Sugimoto E, Shinozaki K (2011) Arabidopsis mutants of AtABCG22, an ABC transporter gene, increase water transpiration and drought susceptibility. Plant J 67(5):885–894

Kusaka M, Ohta M, Fujimura T (2005) Contribution of inorganic components to osmotic adjustment and leaf folding for drought tolerance in pearl millet. Physiol Plant 125(4):474–489

Laloum T, Martín G, Duque P (2018) Alternative splicing control of abiotic stress responses. Trends Plant Sci 23:140–150. https://doi.org/10.1016/j.tplants.2017.09.019

Larrainzar E, Wienkoop S, Scherling C, Kempa S, Ladrera R, Arrese-Igor C et al (2009) Carbon metabolism and bacteroid functioning are involved in the regulation of nitrogen fixation in Medicago truncatula under drought and recovery. Mol Plant Microbe Interact 22:1565–1576. https://doi.org/10.1094/MPMI-22-12-1565

Lata C, Prasad M (2011) Role of DREBs in regulation of abiotic stress responses in plants. J Exp Bot 62:4731–4748. https://doi.org/10.1093/jxb/err210

Lata C, Muthamilarasan M, Prasad M (2015) Drought stress responses and signal transduction in plants. In: Pandey GK (ed) The elucidation of abiotic stress signaling in plants. Springer, New York, pp 195–225

Le Nguyen K, Grondin A, Courtois B, Gantet P (2019) Next-generation sequencing accelerates crop gene discovery. Trends Plant Sci 24(3):263–274. https://doi.org/10.1016/j.tplants.2018.12.004

Li W, Cui X, Meng Z, Huang X, Xie Q, Wu H, Jin H, Zhang D, Liang W (2012) Transcriptional regulation of Arabidopsis MIR168a and argonaute1 homeostasis in abscisic acid and abiotic stress responses. Plant Physiol 158(3):1279–1292

Lian H-L, Yu X, Ye Q, Ding X-S, Kitagawa Y, Kwak S-S et al (2004) The role of aquaporin RWC3 in drought avoidance in rice. Plant Cell Physiol 45(4):481–489

Liu Q, Kasuga M, Sakuma Y, Abe H, Miura S, Yamaguchi-Shinozaki K, Shinozaki K (1998) Two transcription factors, DREB1 and DREB2, with an EREBP/AP2 DNA binding domain separate two cellular signal transduction pathways in drought- and low-temperature-responsive gene expression, respectively, in Arabidopsis. Plant Cell 10(8):1391–1406

Liu QL, Xu KD, Zhong M, Pan YZ, Jiang BB, Liu GL, Jia Y, Zhang HQ (2013a) Overexpression of a novel chrysanthemum Cys2/His2-type zinc finger protein gene DgZFP3 confers drought tolerance in tobacco. Biotechnol Lett 35:1953–1959. https://doi.org/10.1007/s10529-013-1257-6

Liu S, Wang X, Wang H, Xin H, Yang X, Yan J, Li J, Tran LSP, Shinozaki K, Yamaguchi-Shinozaki K, Qin F (2013b) Genome-wide analysis of ZmDREB genes and their association with natural variation in drought tolerance at seedling stage of Zea mays L. PLoS Genet 9(9): e1003790. https://doi.org/10.1371/journal.pgen.1003790

Liu C, Mao B, Ou S, Wang W, Liu L, Wu Y et al (2014) OsbZIP71, a bZIP transcription factor, confers salinity and drought tolerance in rice. Plant Mol Biol 84:19–36. https://doi.org/10.1007/s11103-013-0124-y

Liu H, Sultan MARF, Liu XL, Zhang J, Yu F, Zhao HX (2015a) Physiological and comparative proteomic analysis reveals different drought responses in roots and leaves of drought-tolerant wild wheat (Triticum boeoticum). PloS One 10(4):e0121852. https://doi.org/10.1371/journal.pone.0121852

Liu B, Zhang N, Zhao S, Chang J, Wang Z, Zhang G et al (2015b) Proteomic changes during tuber dormancy release process revealed by iTRAQ quantitative proteomics in potato. Plant Physiol Biochem 86:181–190. https://doi.org/10.1016/j.plaphy.2014.12.003

Lojkova L, Datta R, Sajna M, Marfo TD, Janous D, Pavelka M, Formanek P (2015) Limitation of proteolysis in soils of forests and other types of ecosystems by diffusion of substrate. Soil Biol Biochem 47:1690–1691

Lowe R, Shirley N, Bleackley M, Dolan S, Shafee T (2017) Transcriptomics technologies. PLoS Comput Biol 13:e1005457. https://doi.org/10.1371/journal.pcbi.1005457

Lu G, Gao C, Zheng X, Han B (2009) Identification of OsbZIP72 as a positive regulator of ABA response and drought tolerance in rice. Planta 229:605–615

Lu W, Chu X, Li Y, Wang C, Guo X (2013) Cotton GhMKK1 induces the tolerance of salt and drought stress, and mediates defence responses to pathogen infection in transgenic Nicotiana benthamiana. PloS One 8(7):e68503

Luo X, Bai X, Zhu D, Li Y, Ji W, Cai H, Wu J, Liu B, Zhu Y (2012) GsZFP1, a new Cys2/His2-type zinc-finger protein, is a positive regulator of plant tolerance to cold and drought stress. Planta 235:1141–1155. https://doi.org/10.1007/s00425-011-1579-9

Ma NL, Rahmat Z, Lam SS (2013) A review of the "omics" approach to biomarkers of oxidative stress in Oryza sativa. Int J Mol Sci 14:7515–7541. https://doi.org/10.3390/ijms14047515

MacBeath G (2002) Protein microassays and proteomics. Nat Genet Suppl 32(3):526–532

Magwanga RO, Lu P, Kirungu JN, Lu H, Wang X, Cai X, Zhou Z, Zhang Z, Salih H, Wang K, Liu F (2018) Characterization of the late embryogenesis abundant (LEA) proteins family and their role in drought stress tolerance in upland cotton. BMC Genet 19:6

Mahalakshmi V, Bidinger FR (1986) Water deficit during panicle development in pearl millet: yield compensation by tillers. J Agric Sci 106(1):113–119. https://doi.org/10.1017/S0021859600068868

Mahmood T, Khalid S, Abdullah M, Ahmed Z, Shah MKN, Ghafoor A, Du X (2019) Insights into drought stress signaling in plants and the molecular genetic basis of cotton drought tolerance. Cells 9(1):105

Manjulatha M, Sreevathsa R, Kumar AM, Sudhakar C, Prasad TG, Tuteja N, Udayakumar M (2014) Overexpression of a pea DNA helicase (PDH45) in peanut (Arachis hypogaea L.) confers improvement of cellular level tolerance and productivity under drought stress. Mol Biotechnol 56:111–125

Maruyama K, Sakuma Y, Kasuga M, Ito Y, Seki M, Goda H, Shimada Y, Yoshida S, Shinozaki K, Yamaguchi-Shinozaki K (2004) Identification of cold-inducible downstream genes of the Arabidopsis DREB1A/CBF3 transcriptional factor using two microarray systems. Plant J 38(6):982–993. https://doi.org/10.1111/j.1365-313X.2004.02100.x

Mattana M, Biazzi E, Consonni R, Locatelli F, Vannini C, Provera S, Coraggio I (2005) Overexpression of Osmyb4 enhances compatible solute accumulation and increases stress tolerance of Arabidopsis thaliana. Physiol Plant 125(2):212–223

Merlot S, Leonhardt N, Fenzi F, Valon C, Costa M, Piette L, Vavasseur A, Genty B, Boivin K, Müller A, Giraudat J (2007) Constitutive activation of a plasma membrane H+-ATPase prevents abscisic acid-mediated stomatal closure. EMBO J 26(13):3216–3226

Michaletti A, Naghavi MR, Toorchi M, Zolla L, Rinalducci S (2018) Metabolomics and proteomics reveal drought-stress responses of leaf tissues from spring-wheat. Sci Rep 8(1):5710. https://doi.org/10.1038/s41598-018-23611-6

Mishra RN, Reddy PS, Nair S, Markandeya G, Reddy AR, Sopory SK, Reddy MK (2007) Isolation and characterization of expressed sequence tags (ESTs) from subtracted cDNA libraries of Pennisetum glaucum seedlings. Plant Mol Biol 64:713–732

Mochida K, Shinozaki K (2010) Genomics and bioinformatics resources for crop improvement. Plant Cell Physiol 51(4):497–523

Molina C, Rotter B, Horres R, Udupa SM, Besser B, Bellarmino L, Baum M, Matsumura H, Terauchi R, Kahl G, Winter P (2008) SuperSAGE: the drought stress-responsive transcriptome of chickpea roots. BMC Genomics 9:553. https://doi.org/10.1186/1471-2164-9-553

Mosa KA, Ismail A, Helmy M (2017) Omics and system biology approaches in plant stress research. In: Plant stress tolerance: an integrated omics approach. Springer, Berlin, pp 21–34

Mukesh Sankar S, Tara Satyavathi C, Barthakur S, Singh SP, Bharadwaj C, Soumya SL (2021) Differential modulation of heat-inducible genes across diverse genotypes and molecular cloning of a sHSP from pearl millet [Pennisetum glaucum (L.) R. Br.]. Front Plant Sci 12:659893. https://doi.org/10.3389/fpls.2021.659893

Muñoz-Mayor A, Pineda B, Garcia-Abellán JO, Antón T, Garcia-Sogo B, Sanchez-Bel P, Flores FB, Atarés A, Angosto T, Pintor-Toro JA, Moreno V (2012) Overexpression of dehydrin tas14 gene improves the osmotic stress imposed by drought and salinity in tomato. J Plant Physiol 169(5):459–468

Muthamilarasan M, Dhaka A, Yadav R, Prasad M (2016) Exploration of millet models for developing nutrient-rich graminaceous crops. Plant Sci 242:89–97

Muthamilarasan M, Singh NK, Prasad M (2019) Multi-omics approaches for strategic improvement of stress tolerance in underutilized crop species: a climate change perspective. Adv Genet 103: 1–38. https://doi.org/10.1016/bs.adgen.2018.12.001

Nakashima K, Fujita Y, Kanamori N, Katagiri T, Umezawa T, Kidokoro S, Maruyama K, Yoshida T, Ishiyama K, Kobayashi M et al (2009) Three Arabidopsis SnRK2 protein kinases, SRK2D/SnRK2.2, SRK2E/SnRK2.6/OST1 and SRK2I/SnRK2.3, involved in ABA signaling are essential for the control of seed development and dormancy. Plant Cell Physiol 50:1345–1363. https://doi.org/10.1093/pcp/pcp083

Nambiar VS, Dhaduk JJ, Sareen N, Shahu T, Desai RA (2011) Potential functional implications of Pearl millet (Pennisetum glaucum) in health and disease. J Appl Pharm Sci 1:62–67. https://doi.org/10.7324/JAPS.2011.11011

Narsaiah K, Jha SN, Bhardwaj R, Sharma R, Kumar R (2012) Optical biosensors for food quality and safety assurance—a review. J Food Sci Technol 49:383–406. https://doi.org/10.1007/s13197-011-0454-4

Ndiaye A, Diallo AO, Fall NC et al (2022) Transcriptomic analysis of methyl jasmonate treatment reveals gene networks involved in drought tolerance in pearl millet. Sci Rep 12(1):5158. https://doi.org/10.1038/s41598-022-09152-6

Nepolean T, Sharma R, Singh N, Shiriga K, Mohan S, Mittal S, Mittal S, Mallikarjuna MG, Rao AR, Dash PK et al (2017) Genomewide expression and functional interactions of genes under drought stress in maize. J Genom 2017(2017):2568706. https://doi.org/10.1155/2017/2568706. [PMC free article] [PubMed] [CrossRef] [Google Scholar]

Niu CF, Wei W, Zhou QY, Tian AG, Hao YJ, Zhang WK et al (2012) Wheat WRKY genes TaWRKY2 and TaWRKY19 regulate abiotic stress tolerance in transgenic Arabidopsis plants. Plant Cell Environ 35:1156–1170. https://doi.org/10.1111/j.1365-3040.2012.02480

Okamoto M, Tanaka Y, Abrams SR, Kamiya Y, Seki M, Nambara E (2009) High humidity induces abscisic acid 8′-hydroxylase in stomata and vasculature to regulate local and systemic abscisic acid responses in Arabidopsis. Plant Physiol 149(2):825–834

Omoigui LO, Kamara AY, Ishiyaku MF, Boukar O (2012) Comparative responses of cowpea breeding lines to Striga and Alectra in the dry savannah of northeast Nigeria. Afr J Agric Res 7: 747–754. https://doi.org/10.5897/AJAR11.1797

Oweis T, Hachum A, Kijne J (1999) Water harvesting and supplemental irrigation for improved water use efficiency in dry areas, vol 7. IWMI, New Delhi

Pan Y, Seymour GB, Lu C, Hu Z, Chen X, Chen G (2012) An ethylene response factor (ERF5) promoting adaptation to drought and salt tolerance in tomato. Plant Cell Rep 31:349–360. https://doi.org/10.1007/s00299-011-1182-9

Passot S, Gnacko F, Moukouanga D, Lucas M, Guyomarc'h S, Ortega BM et al (2016) Characterization of pearl millet root architecture and anatomy reveals three types of lateral roots. Front Plant Sci 7:829

Penfield S, Clements S, Bailey KJ, Gilday AD, Leegood RC, Gray JE, Graham IA (2012) Expression and manipulation of phosphoenolpyruvate carboxykinase 1 identifies a role for malate metabolism in stomatal closure. Plant J 69(4):679–688

Qin F, Kakimoto M, Sakuma Y, Maruyama K, Osakabe Y, Tran LSP, Shinozaki K, Yamaguchi-Shinozaki K (2007) Regulation and functional analysis of ZmDREB2A in response to drought and heat stresses in Zea mays L. Plant J 50(1):54–69. https://doi.org/10.1111/j.1365-313X.2007.03034.x

Qiu Y, Yu D (2009) Over-expression of the stress-induced OsWRKY45 enhances disease resistance and drought tolerance in Arabidopsis. Environ Exp Bot 65(1):35–47. https://doi.org/10.1016/j.envexpbot.2008.07.002

Ramalingam A, Kudapa H, Pazhamala LT, Weckwerth W, Varshney RK (2015) Proteomics and metabolomics: two emerging areas for legume improvement. Front Plant Sci 6:1116. https://doi.org/10.3389/fpls.2015.01116

Reddy PS, Reddy GM, Pandey P, Chandrasekhar K, Reddy MK (2012) Cloning and molecular characterization of a gene encoding late embryogenesis abundant protein from Pennisetum glaucum: protection against abiotic stresses. Mol Biol Rep 39:7163–7174. https://doi.org/10.1007/s11033-012-1558-y
Renaut J, Hausman JF, Wisniewski ME (2006) Proteomics and low temperature studies: bridging the gap between gene expression and metabolism. Physiol Plant 126:97–109. https://doi.org/10.1111/j.1399-3054.2006.00612.x
Rhee SY, Dickerson J, Xu D (2006) Bioinformatics and its applications in plant biology. Annu Rev Plant Biol 57:335–360
Rollins JA, Habte E, Templer SE, Colby T, Schmidt J, Von Korff M (2013) Leaf proteome alterations in the context of physiological and morphological responses to drought and heat stress in barley (Hordeum vulgare L.). J Exp Bot 64:3201–3212. https://doi.org/10.1093/jxb/ert164
Ryu MY, Cho SK, Kim WT (2010) The Arabidopsis C3H2C3-type RING E3 ubiquitin ligase AtAIRP1 is a positive regulator of an abscisic acid-dependent response to drought stress. Plant Physiol 154(4):1983–1997. https://doi.org/10.1104/pp.110.157794
Saha GC, Sarker A, Chen W, Vandemark GJ, Muehlbauer FJ (2010) Inheritance and linkage map positions of genes conferring resistance to stemphylium blight in lentil. Crop Sci 50:1831–1839. https://doi.org/10.2135/cropsci2009.09.0502
Salazar-Cerezo S, Martínez-Montiel N, García-Sánchez J, Pérez-y-Terrón R, Martínez-Contreras RD (2018) Gibberellin biosynthesis and metabolism: a convergent route for plants, fungi and bacteria. Microbiol Res 208:85–98
Salt DE, Baxter I, Lahner B (2008) Ionomics and the study of the plant ionome. Annu Rev Plant Biol 59:709–733. https://doi.org/10.1146/annurev.arplant.59.032607.092942
Schubert O, Röst H, Collins B et al (2017) Quantitative proteomics: challenges and opportunities in basic and applied research. Nat Protoc 12(7):1289–1294. https://doi.org/10.1038/nprot.2017.040
Shalini S, Singla A, Goyal M, Kaur V, Kumar P (2018) Omics in agriculture: applications, challenges and future perspectives. In: Kumar P, Kumar P (eds) Crop improvement for sustainability. Springer, Berlin, pp 343–360. https://doi.org/10.1007/978-981-13-1126-1_14
Shaw R, Tian X, Xu J (2021) Single-cell transcriptome analysis in plants: advances and challenges. Mol Plant 14(1):115–126. https://doi.org/10.1016/j.molp.2020.11.007
Sheveleva E, Chmara W, Bohnert HJ, Jensen RG (1997) Increased salt and drought tolerance by D-ononitol production in transgenic Nicotiana tabacum L. Plant Physiol 115(3):1211–1219. https://doi.org/10.1104/pp.115.3.1211
Shikha M, Kanika A, Rao AR, Mallikarjuna MG, Gupta HS, Nepolean T (2017) Genomic selection for drought tolerance using genome-wide SNPs in maize. Front Plant Sci 8:550
Shinde H, Dudhate A, Tsugama D, Gupta SK, Liu S, Takano T (2019) Pearl millet stress-responsive NAC transcription factor PgNAC21 enhances salinity stress tolerance in Arabidopsis. Plant Physiol Biochem 135:546–553
Shou H, Bordallo P, Wang K (2004) Expression of the Nicotiana protein kinase (NPK1) enhanced drought tolerance in transgenic maize. J Exp Bot 55(399):1013–1019. https://doi.org/10.1093/jxb/erh124
Siefritz F, Tyree MT, Lovisolo C, Schubert A, Kaldenhoff R (2002) PIP1 plasma membrane aquaporins in tobacco: from cellular effects to function in plants. Plant Cell 14(4):869–876
Singh HP, Kumar A, Parthasarathy VA, Singh B (2013) Advances in horticulture biotechnology nanotechnology in agriculture. Westville Publishing House, New Delhi
Singh J, Aggarwal R, Gurjar M, Sharma S, Jain S, Saharan M (2020) Identification and expression analysis of pathogenicity-related genes in Tilletia indica inciting Karnal bunt of wheat. Australas Plant Pathol 49:393–402. https://doi.org/10.1007/s13313-020-00709-1
Sirichandra C, Wasilewska A, Vlad F, Valon C, Leung J (2009) The guard cell as a single-cell model towards understanding drought tolerance and abscisic acid action. J Exp Bot 60(5):1439–1463. https://doi.org/10.1093/jxb/erp025

Sugano S, Kaminaka H, Rybka Z, Catala R, Salinas J, Matsui K, Ohme-Takagi M, Takatsuji H (2003) Stress-responsive zinc finger gene ZPT2-3 plays a role in drought tolerance in petunia. Plant J 36(6):830–841. https://doi.org/10.1046/j.1365-313X.2003.01930.x

Suharti WS, Nose A, Zheng S-H (2016) Metabolomic study of two rice lines infected by Rhizoctonia solani in negative ion mode by CE/TOF-MS. J Plant Physiol 206:13–24. https://doi.org/10.1016/j.jplph.2016.09.011

Sukumaran S, Reynolds MP, Sansaloni C (2018) Genome-wide association analyses identify QTL hotspots for yield and component traits in durum wheat grown under yield potential, drought, and heat stress environments. Front Plant Sci 9:81. https://doi.org/10.3389/fpls.2018.00081

Sultan B, Gaetani M (2016) Agriculture in West Africa in the twenty-first century: climate change and impacts scenarios, and potential for adaptation. Front Plant Sci 7:1262. https://doi.org/10.3389/fpls.2016.01262

Sun M, Huang D, Zhang A, Khan I, Yan H, Wang X, Zhang X, Zhang J, Huang L (2020) Transcriptome analysis of heat stress and drought stress in pearl millet based on Pacbio full-length transcriptome sequencing. BMC Plant Biol 20(1):1–15. https://doi.org/10.1186/s12870-019-2222-2

Tako E, Hoekenga OA, Kochian LV, Glahn RP (2015) Retraction note: high bioavailable iron maize (Zea mays L.) developed through molecular breeding provides more absorbable iron in vitro (Caco-2 model) and in vivo (Gallus gallus). Nutr J 14(1):1265. https://doi.org/10.1186/s12937-015-0109-x

Tanaka K, Waki H, Ido Y, Akita S, Yoshida Y, Yoshida T (1988) Protein and polymer analyses up to m/z 100000 by laser ionization time of-flight mass spectrometry. Rapid Commun Mass Spectrom 2:151–153. https://doi.org/10.1002/rcm.1290020802

Tang Y, Qin S, Guo Y, Chen Y, Wu P, Chen Y (2012) Molecular characterization of novel TaNAC genes in wheat and overexpression of TaNAC2a confers drought tolerance in tobacco. Physiol Plant 144:210–224

Tang L, Cai H, Ji W, Luo X, Wang Z, Wu J, Wang X, Cui L, Wang Y, Zhu Y, Bai X (2013) Overexpression of GsZFP1 enhances salt and drought tolerance in transgenic alfalfa (Medicago sativa L.). Plant Physiol Biochem 71:22–30. https://doi.org/10.1016/j.plaphy.2013.06.012

Tiedge K, Li X, Merrill AT, Davisson D, Chen Y, Yu P et al (2022) Comparative transcriptomics and metabolomics reveal specialized metabolite drought stress responses in switchgrass (Panicum virgatum). New Phytol 236(4):1393–1408. https://doi.org/10.1111/nph.17477

Tiwari B, Sellamuthu B, Ouarda Y, Drogui P, Tyagi RD, Buelna G (2017) Review on fate and mechanism of removal of pharmaceutical pollutants from wastewater using biological approach. Bioresour Technol 224:1–12

Toorchi M, Yukawa K, Nouri MZ, Komatsu S (2009) Proteomics approach for identifying osmotic-stress-related proteins in soybean roots. Peptides 30:2108–2117. https://doi.org/10.1016/j.peptides.2009.09.006

Tuteja N (2007) Abscisic acid and abiotic stress signaling. Plant Signal Behav 2(3):135–138. https://doi.org/10.4161/psb.2.3.4167

Twyman R, Cfe PD, George A (2013) Principles of proteomics. Garland Science, New York

Vadez V, Kholová J, Yadav RS, Hash CT (2013) Small temporal differences in water uptake among varieties of pearl millet (Pennisetum glaucum (L.) R. Br.) are critical for grain yield under terminal drought. Plant and Soil 371(1–2):447–462

Varshney RK, Pandey MK, Janila P, Nigam SN, Sudini H, Gowda MVC et al (2014) Marker-assisted introgression of a QTL region to improve rust resistance in three elite and popular varieties of peanut (Arachis hypogaea L.). Theor Appl Genet 127(8):1771–1781. https://doi.org/10.1007/s00122-014-2344-3

Varshney RK, Shi C, Thudi M, Mariac C, Wallace J, Qi P, Zhang H, Zhao Y, Wang X, Rathore A, Srivastava RK, Chitikineni A, Fan G, Bajaj P, Punnuri S, Gupta SK, Wang H, Jiang Y, Couderc M et al (2017) Pearl millet genome sequence provides a resource to improve agronomic traits in arid environments. Nat Biotechnol 35(10):969–976. https://doi.org/10.1038/nbt.3943

Veeranagamallaiah G, Jyothsnakumari G, Thippeswamy M, Reddy PCO, Surabhi G-K, Sriranganayakulu G, Mahesh Y, Rajasekhar B, Madhurarekha C, Sudhakar C (2008) Proteomic analysis of salt stress responses in foxtail millet (Setaria italica L. cv. Prasad) seedlings. Plant Sci 175:631–641. https://doi.org/10.1016/j.plantsci.2008.06.005

Velculescu VE, Zhang L, Vogelstein B, Kinzler KW (1995) Serial analysis of gene expression. Science 270(5235):484–487

Villalobos MA, Bartels D, Iturriaga G (2004) Stress tolerance and glucose insensitive phenotypes in Arabidopsis overexpressing the CpMYB10 transcription factor gene. Plant Physiol 135(1): 309–324

Wahid A, Gelani S, Ashraf M, Foolad MR (2007) Heat tolerance in plants: an overview. Environ Exp Bot 61(3):199–223. https://doi.org/10.1016/j.envexpbot.2007.05.011

Wan L, Zhang J, Zhang H, Zhang Z, Quan R, Zhou S, Huang R (2011) Transcriptional activation of OsDERF1 in OsERF3 and OsAP2-39 negatively modulates ethylene synthesis and drought tolerance in rice. PloS One 6(9):e25216. https://doi.org/10.1371/journal.pone.0025216

Washburn MP, Wolters D, Yates JR (2001) Large-scale analysis of the yeast proteome by multidimensional protein identification technology. Nat Biotechnol 19(3):242–247

Wasternack C (2007) Jasmonates: an update on biosynthesis, signal transduction and action in plant stress response, growth and development. Ann Bot 100(4):681–697. https://doi.org/10.1093/aob/mcm079

Wheelock CE, Miyagawa H (2006) The omicization of agrochemical research. J Pestic Sci 31(3): 240–244. https://doi.org/10.1584/jpestics.31.240

Wilkinson S, Kudoyarova GR, Veselov DS, Arkhipova TN, Davies WJ (2012) Plant hormone interactions: innovative targets for crop breeding and management. J Exp Bot 63(9):3499–3509. https://doi.org/10.1093/jxb/ers124

Woolfson M (2018) The development of structural X-ray crystallography. Phys Scr 93(1):014001. https://doi.org/10.1088/1402-4896/aa9c30

Wu S, Ning F, Zhang Q, Wu X, Wang W (2017) Enhancing omics research of crop responses to drought under field conditions. Front Plant Sci 8:174. https://doi.org/10.3389/fpls.2017.00174

Xia K, Wang R, Ou X, Fang Z, Tian C, Duan J, Wang Y, Zhang M (2012) OsTIR1 and OsAFB2 downregulation via OsmiR393 overexpression leads to more tillers, early flowering and less tolerance to salt and drought in rice. PloS One 7(1):e30039

Xian L, Sun P, Hu S, Wu J, Liu JH (2014) Molecular cloning and characterization of CrNCED1, a gene encoding 9-cis-epoxycarotenoid dioxygenase in Citrus reshni, with functions in tolerance to multiple abiotic stresses. Planta 239:61–77

Xiang Y, Tang N, Du H, Ye H, Xiong L (2008) Characterization of OsbZIP23 as a key player of the basic leucine zipper transcription factor family for conferring abscisic acid sensitivity and salinity and drought tolerance in rice. Plant Physiol 148(4):1938–1952. https://doi.org/10.1104/pp.108.129353

Xiang Y, Sun X, Gao S, Qin F, Dai M (2017) Deletion of an endoplasmic reticulum stress response element in a ZmPP2C-A gene facilitates drought tolerance of maize seedlings. Mol Plant 10(3): 456–469. https://doi.org/10.1016/j.molp.2016.10.003

Xu X, Zhang M, Li J, Liu Z, Zhao Z, Zhang Y et al (2018) Improving water use efficiency and grain yield of winter wheat by optimizing irrigations in the North China Plain. Field Crop Res 221: 219–227

Yadav RS, Hash CT, Bidinger FR, Devos KM, Howarth CJ (2004) Genomic regions associated with grain yield and aspects of post-flowering drought tolerance in pearl millet across stress environments and tester background. Euphytica 136(3):265–277. https://doi.org/10.1023/B:EUPH.0000032743.05662.f0

Yang CY, Chen YC, Jauh GY, Wang CS (2005) A lily ASR protein involves abscisic acid signaling and confers drought and salt resistance in Arabidopsis. Plant Physiol 139(2):836–846

Yang Y, Saand MA, Abdelaal WB, Zhang J, Wu Y, Li J et al (2020) TRAQ-based comparative proteomic analysis of two coconut varieties reveals aromatic coconut cold-sensitive in response to low temperature. J Proteomics 220:103766. https://doi.org/10.1016/j.jprot.2020.103766

Yang Y, Saand MA, Huang L, Abdelaal WB, Zhang J, Wu Y, Li J, Sirohi MH, Wang F (2021) Applications of multi-omics technologies for crop improvement. Front Plant Sci 12:563953. https://doi.org/10.3389/fpls.2021.563953

Yoo CY, Pence HE, Jin JB, Miura K, Gosney MJ, Hasegawa PM, Mickelbart MV (2010) The Arabidopsis GTL1 transcription factor regulates water use efficiency and drought tolerance by modulating stomatal density via transrepression of SDD1. Plant Cell 22(12):4128–4141. https://doi.org/10.1105/tpc.110.077353

Yu Q, Hu Y, Li J, Wu Q, Lin Z (2005) Sense and antisense expression of plasma membrane aquaporin BnPIP1 from Brassica napus in tobacco and its effects on plant drought resistance. Plant Sci 169(4):647–656

Yu H, Chen X, Hong YY, Wang Y, Xu P, Ke SD, Liu HY, Zhu JK, Oliver DJ, Xiang CB (2008) Activated expression of an Arabidopsis HD-START protein confers drought tolerance with improved root system and reduced stomatal density. Plant Cell 20:1134–1151. https://doi.org/10.1105/tpc.107.054975

Zhang X, Zhang Z, Chen J, Chen Q, Wang XC, Huang R (2005) Expressing TERF1 in tobacco enhances drought tolerance and abscisic acid sensitivity during seedling development. Planta 222:494–501

Zhang Z, Li F, Li D, Zhang H, Huang R (2010) Expression of ethylene response factor JERF1 in rice improves tolerance to drought. Planta 232:765–774. https://doi.org/10.1007/s00425-010-1209-x

Zhang X, Wang L, Meng H, Wen H, Fan Y, Zhao J (2011) Maize ABP9 enhances tolerance to multiple stresses in transgenic Arabidopsis by modulating ABA signaling and cellular levels of reactive oxygen species. Plant Mol Biol 75:365–378

Zhang W, Zhang Y, Qiu H, Guo Y, Wan H, Zhang X et al (2020) Genome assembly of wild tea tree DASZ reveals pedigree and selection history of tea varieties. Nat Commun 11(1):3719. https://doi.org/10.1038/s41467-020-17498-1

Zhang C, Chen J, Huang W, Song X, Niu J (2021) Transcriptomics and metabolomics reveal purine and phenylpropanoid metabolism response to drought stress in Dendrobium sinense, an endemic orchid species in Hainan Island. Front Genet 12:692702. https://doi.org/10.3389/fgene.2021.692702

Zhao Y-Y, Wu S-P, Liu S, Zhang Y, Lin R-C (2014) Ultra-performance liquid chromatography–mass spectrometry as a sensitive and powerful technology in lipidomic applications. Chem Biol Interact 220:181–192. https://doi.org/10.1016/j.cbi.2014.08.005

Zhou ML, Ma JT, Pang JF, Zhang ZL, Tang YX, Wu YM (2010) Regulation of plant stress response by dehydration responsive element binding (DREB) transcription factors. Afr J Biotechnol 9(54):9255–9269

Zhu HG, Cheng WH, Tian WG, Li YJ, Liu F, Xue F et al (2018) iTRAQ-based comparative proteomic analysis provides insights into somatic embryogenesis in Gossypium hirsutum L. Plant Mol Biol 96:89–102. https://doi.org/10.1007/s11103-017-0681-x

Genetic Biofortification of Pearl Millet: Trait Priority, Breeding and Genomic Progress

9

Mahalingam Govindaraj, Mahesh Pujar, Rakesh Srivastava, S. K. Gupta, and Wolfgang H. Pfeiffer

Abstract

Malnutrition is a result of one or more micronutrients and vitamins deficiency in the staple diets. Crop biofortification is increasingly recognized as a sustainable breeding strategy to address micronutrients and vitamin deficiency among rapidly growing populations. Pearl millet is a dryland cereal that possesses climate resilience and potentially contributes to the food and nutrition (largely iron and zinc) supply in semi-arid tropics. HarvestPlus supported the pearl millet biofortification programme at ICRISAT assessed the feasibility of iron biofortification in pearl millet and established baseline and breeding targets to achieve the health impact. These targeted breeding efforts accomplished a rapid micronutrient screening lab, pre-breeding, advanced high-iron genetic materials. To date, about 12 biofortified cultivars released for general cultivation in India and West Africa. These cultivars provide > 60% higher Fe than commercial cultivars potentially benefiting millions of people. Several high-Fe seed and restorer parents and more than 1000 high-Fe advanced breeding lines were developed through targeted biofortification breeding and disseminated to public

M. Govindaraj (✉)
Alliance of Bioversity International and the International Center for Tropical Agriculture (CIAT), Cali, Colombia

International Crops Research Institute for the Semi-Arid Tropics (ICRISAT), Patancheru, Telangana, India
e-mail: m.govindaraj@cgiar.org

M. Pujar · R. Srivastava · S. K. Gupta
International Crops Research Institute for the Semi-Arid Tropics (ICRISAT), Patancheru, Telangana, India

W. H. Pfeiffer
HarvestPlus, International Food Policy Research Institute (IFPRI), Washington, DC, USA

V. A. Tonapi et al. (eds.), *Pearl Millet in the 21st Century*,
https://doi.org/10.1007/978-981-99-5890-0_9

and private sectors through biofortification partnership projects and the HPRC model at ICRISAT. Breeding progress made in achieving the global breeding target of 77 mg/kg of Fe in pearl millet has set the stage for mainstreaming. Rapid screening tool and diagnostic markers encoding the high-Fe (4 SNPs) and drought tolerance (10 SNPs) will enhance the early generation selection and mainstreaming efficacy with stress tolerance in hybrid parents and cultivars breeding pipelines. Bioavailability studies have confirmed that biofortified pearl millet supplies 80% of daily Fe requirements, implying the significance of micronutrient traits in the pearl millet commercial product profiles in India and West Africa.

Keywords

Biofortification · Genetic gains · Hybrid breeding · Micronutrients · Iron · Zinc

Abbreviation

AAS	Atomic absorption spectrometry
CGIAR	Consultative Group for International Agricultural Research
FAO	Food and Agriculture Organization
Fe	Iron
g	Grams
ICAR	Indian Council of Agricultural Research
ICP-OES	Inductively Coupled Plasma- Optical Emission spectrometry (ICP)
kcal	Kilocalorie
kg/ha	Kilogram per hectare
kg/ha/year	Kilogram per hectare per year
mg/day	Milligrams per day
mg/day	Milligrams per day
NARS	National Agricultural Research System
RDA	Recommended dietary allowances
t/ha	Tonnes per hectare
Zn	Zinc
µg/g	Microgram per gram

9.1 Introduction

Pearl millet (*Pennisetum glaum* (L.) R. Br.) is a dryland cereal adapted to a variable climate and growing conditions in Africa, and India's arid and semi-arid tropical regions. It is a C4 plant that enables to faster carbon fixation, significantly improving the crop photosynthetic and water use efficiency under restricted photosynthesis and respiration process. This mechanism allows pearl millet to withstand a very high air

Table 9.1 Recommended daily allowance of iron and zinc in Indian context (ICMR 2010)

Category		Recommended daily allowance (RDA) (mg/day)	
		Iron	Zinc
Adult men	–	17	12
Adult women	Normal	21	10
	Pregnant	35	12
	Lactating	21	12
Children	1–9 years	9–16	5–8
Adolescents	Boys	21–28	9–12
	Girls	26–27	9–12

temperature (42 °C) along with maintaining high photosynthetic efficiency and dry matter production (Djanaguiraman et al. 2018). Pearl millet could grow well under low fertile soils and adverse agro-climatic conditions such as heat (42 °C), drought (annual rainfall ≤400) and salinity (~200 mM NaCl) have contributed to making pearl millet a climate-smart crop. About 90 million people across the world are dependent on pearl millet for their daily food and income, largely in north western India and sub-Saharan Africa. Globally, pearl millet is cultivated on more than 25 million hectare (FAOSTAT 2020). India has the largest area with seven million hectare and the highest average annual pearl millet grain production of 8.61 million tonnes (Directorate of Millets Development 2020; Project Coordinator Review 2020). The attractiveness of this crop in dryland is also due to dual purposes (grain and fodder). Pearl millet grains are a substantial energy source that provides 361 kcal/100 g of energy equivalent to other major cereals like wheat (346 kcal/100 g), rice (345 kcal/100 g), maize (125 kcal/100 g) and sorghum (349 kcal/100 g) as per the Nutritive value of Indian foods (NIN 2003). The lowest glycemic index (55) and highest fibre content (1.2 g/100 g) among cereals are found in pearl millet grains (Mani et al. 1993). In addition to these facts, pearl millet grains are also rich in amino acids like lysine and tryptophan; mineral elements such as iron (Fe), zinc (Zn), calcium, potassium, magnesium and manganese; and vitamins such as thiamine and niacin than other cereals. For example, pearl millet is a staple food crop in parts of India and the majority of sub-Saharan Africa, serving as a major source of Fe and Zn as well as higher energy. In states like Maharashtra, Gujarat and Rajasthan, pearl millet makes up 20–60% of all the cereal consumed in India. Therefore, it accounts for 19–63% of the Fe and 16–56% of the Zn intake from all food sources (Parthasarathy Rao et al. 2006). The Recommended Dietary Allowances (RDA) for Fe and Zn among different age groups in India are computed by the Indian Council of Medical Research (ICMR) (Table 9.1; ICMR 2010) that clearly emphasizes the fact that the pregnant and lactating women (Fe = 35 and 21 mg/day and Zn = 12 mg/day each) require relatively more amount of Fe and Zn on a daily basis than the other age group. The recommended daily allowance of Fe and Zn for adolescents among both boys and girls vary from 21 to 28 mg/day and 9 to 12 mg/day, respectively.

More than two billion people, or one in three people, globally suffer from hidden hunger (FAO 2019). Malnutrition is a result of one or more micronutrients and vitamins deficiency in the staple diets. Over the last two decades, total food grain production in India increased from 198 to 303 million tonnes in the 2020–2021 crop seasons (Ministry of Agriculture and Farmers Welfare 2021). Salutation to green revolution efforts and high-inputs responsive and high-yielding cultivars contributed by national (NARS) and international research and development organizations (CGIAR and partners). The critical observation is that the rate of malnutrition in India remains very high despite the historical food production record. The productivity of pearl millet has increased from 19.0 to 31.1 kg/ha/year (in 2018), representing a nearly 200% increase in productivity (in 2000) (Yadav et al. 2019). Similar production and productivity gains were not reflected till now in Africa where yield levels are significantly lower (<800 kg/ha) and consequently aggravated food insecurity and malnutrition prevalence. Therefore, crop biofortification is increasingly recognized as a sustainable breeding strategy to address micronutrients and vitamin deficiency in rapidly growing populations. Biofortification, the process which involves breeding nutrients into food crops, offers a comparatively affordable, long-term, sustainable method of delivering more micronutrients and has the potential to improve the nutritional status and health of undernourished populations in both rural and urban areas of developing countries (Bouis 2003). Consumers easily accept a biofortified cultivar with improved mineral content as it does not require a change in dietary habits. Biofortification programme in general targets to improve the micronutrient content of those cultivars which have with preferred agronomic and consumption traits like grain yield. HarvestPlus Challenge Programs of the CGIAR (a global research partnership for a secure food future) have started the development and promotion of numerous biofortified cultivars with increased micronutrient content in a variety of food crops along with pearl millet (Yadav et al. 2017). The present chapter deals with the importance of the biofortification programme of Fe and Zn micronutrients and their progress in genetic improvement through conventional and modern genomic breeding approaches in pearl millet.

9.2 Micronutrient Trait's Prioritization

Beginning of the pearl millet biofortification programme, so-called phase-I (2004–2008), the assessment and identification of key micronutrient variability in pearl millet are very critical; while in the process, it was realized that the variability for Fe content in pearl millet grain was found to be substantially larger and at higher values than for Zn content (Govindaraj et al. 2020). Furthermore, a strong and exceptionally high magnitude of positive correlation observed between the Fe and Zn contents in pearl millet (often $r > 0.70$ and higher) suggests that the selection for Fe can be acceptable and will be relatively successful for Zn as an associated characteristic (Rai et al. 2012; Govindaraj et al. 2019, 2020; Pujar et al. 2020a). On the other hand, compared to Zn deficiency, the anaemia caused by a deficit in Fe is far more common, its negative health effects are more severe and it is well understood. Based on these factors, pearl millet biofortification at ICRISAT began to focus primarily on germplasm screening and breeding for Fe content, with Zn serving as the secondary related trait. This priority was also deliberated in various national and international pearl millet technical forums and the priority of breeding for the Fe trait in pearl millet is communicated to the partner's breeding organizations. During the initial screening and trait identification process, many researchers indicated the importance of β-carotene for human health and its levels found in yellow colour seed pearl millet germplasm (referred to as golden pearl millet). Further research has shown that it was not adequate (up to 1.73 μg/g) to undertake genetic enhancement through conventional breeding approaches. Additionally, these golden pearl millet lines also have significant issues with seedling emergence, and accurate β-carotene assays and phenotyping are time-consuming and extremely expensive (Muthusamy et al. 2014). Nevertheless, attempts to find additional sources within the germplasm proved unsuccessful. These factors led to the abandonment of β-carotene research and the entire breeding programme were switched with the focus on improvement in pearl millet Fe content in grains as well as other agronomic qualities.

9.3 Variability for Fe and Zn Content

Compared to other cereal grains, pearl millet contains a considerably higher concentration of Fe and Zn. A staple crop's per capita consumption level will determine its potential contribution to individual's daily energy and nutrition supply. Pearl millet per capita in India is 4–5 kg while it was significantly higher in west African countries (Yadav et al. 2019). To increase any given attribute through plant breeding, it is important to evaluate genetic variances within the targeted gene pool. Initial studies indicated wider variability for Fe and Zn contents in pearl millet. Global germplasm collection of pearl millet has more than 26,000 accessions; a set of these accessions, so-called core collections (representing 10% of entire pearl millet collections) shown huge variability for Fe (30–90 mg/kg) and Zn (30–74 mg/kg) in pearl millet that forms the best source of selection for improvement (Govindaraj et al. 2020). Similar variability search for Fe and Zn in Sudanese germplasm collection demonstrated fourfold variations (19.7–86.4 mg/kg) and sixfold variations (13.5–82.4 mg/kg), respectively (Bashir et al. 2014) (Table 9.2). Twofold variability was observed among WCA germplasm collection for Fe (24.2–48.7 mg/kg) and Zn (19.8–43.4 mg/kg) (Pucher et al. 2014). The origin of iniadi germplasm is Togo and neighbouring areas of west Africa, which are a superb reservoir of Fe (51–121 mg/kg) and Zn (46–87 mg/kg) (Rai et al. 2015b). This was demonstrated by the early pre-breeding biofortification crosses, which heavily utilized the chosen iniadi source for Fe and Zn biofortification at ICRISAT.

Pearl millet cultivation is dominated by hybrids more than OPVs in India owing to their high yield and uniformity in plant types and maturity, enabling better harvest and crop rotations. Selection of suitable hybrid parents (B and R-lines) with high Fe and Zn is a prerequisite in developing high yielding biofortified hybrids. A study using biofortification-breeding pipelines consisting of advanced breeding lines, population progenies and hybrids showed higher genetic variations for these two

Table 9.2 Studies on genetic variability and correlation for iron and zinc contents in pearl millet

SN	Genetic materials details	Micronutrient variability		Correlation coefficient (Fe vs. Zn)	Analytical technique	References
		Iron (mg/kg)	Zinc (mg/kg)			
Germplasm collections						
1	Germplasm Accessions (Core collections) (504)	30–90	30–74	0.43**	ICP-OES and XRF	Govindaraj et al. (2020)
2	Sudanese pearl millet germplasm (225)	19.7–86.4	13.5–82.4	–	–	Bashir et al. (2014)
3	WCA pearl millet germplasm (72)	24.2–48.7	19.8–43.4	–	–	Pucher et al. (2014)
4	Iniadi germplasm accessions (297)	51–121	46–87	0.77**	ICP-OES	Rai et al. (2015b)
Hybrids						
1	52 single cross hybrids	46.9–85.0	36.4–69.9	0.65**	AAS	Velu et al. (2008a)
2	90 single cross hybrids	34.1–75.8	26.4–52.9	–	AAS	Velu et al. (2011)
3	264 single cross hybrids	30.0–80.0	31.0–70.0	0.75**	AAS	Govindaraj et al. (2013)
4	32 single cross hybrids (18 ms, 31 R and 3 control)	31–67	22–34	0.69**	ICP-OES	Kanatti et al. (2014a)
	28 single cross hybrids (10 ms, 28 R and 3 control)	44–76	31–44	0.65**	ICP-OES	
5	15 high –Fe candidate single cross hybrids	46–56	37–44	057**	ICP-OES	Rai et al. (2016)
Populations and population progenies						
1	Improved populations developed by ICRISAT and NARS partners in India and Africa (68)	42.0–79.9	27.2–50.2	0.84**	AAS	Velu et al. (2008b)
2	Iniadi population progenies (S_1) from PVGGP-6 (30)	29.9–77.2	30.7–63.0	0.82**	AAS	Gupta et al. (2009)
	Non-Iniadi population progenies (S_1) from IAC-ISC-TCP-1 (24)	26.8–48.3	28.2–50.9	0.80**		

3	Population progenies (S_0 and S_1) of ICTP 8203, JBV 3, AIMP 92901 and ICMR 312 (320)	30–133	32–92	–	–	Govindaraj et al. (2012)
4	Open pollinated varieties (OPVs) (18)	42–67	37–52	0.86**	ICP-OES	Rai et al. (2016)
Advance hybrid parental lines						
1	Advanced breeding lines, population progenies and hybrids (796)	18.0–135.0	22.0–92.0	0.49** to 0.71**	ICP, XRF, NIRS	Rai et al. (2012)
2	Recombinant inbred lines derived from ICMS 8511-S1-17-2-1-1-B-P03 X AIMP 92901-S1-183-2-2B-08 (317)	22.9–154.5	19.3–121	0.90**	ICP-OES	Mahendrakar et al. (2019)
3	Advance hybrid parental lines (B and R-lines), population and germplasm progenies (281)	35–116	21–80	0.76**	ICP-MS	Pujar et al. (2020a)

** Significant at the $p<0.01$ probablilty level

micronutrients (18–135 mg/kg Fe and 22–92 mg/kg Zn) (Rai et al. 2012). Recombinant inbred lines (RILs) developed from the cross between ICMS 8511-S1-17-2-1-1-B-P03 X AIMP 92901-S1-183-2-2B-08 showed almost sevenfold variations in Fe (22.9–154.5 mg/kg) and Zn (19.3–121 mg/kg) (Mahendrakar et al. 2019). Evaluation of advanced hybrid parental lines from ICRISAT, that is designated R- and B-lines, population progenies and germplasm progenies revealed a substantial amount of variability for Fe (35–116 mg/kg) and Zn (21–80 mg/kg) (Pujar et al. 2020a). Studies on the evaluation of single-cross hybrids (including commercial hybrids) showed a good prospect of variability ranging from 30 to 85 mg/kg for Fe and 22 to 70.0 mg/kg for Zn (Velu et al. 2008a, 2011; Govindaraj et al. 2013; Kanatti et al. 2014a; Rai et al. 2016). Studies on populations, composites and population progenies also revealed substantial variability for Fe ranging from 26.8 to 133 mg/kg and 27.2 to 92 mg/kg for Zn in pearl millet (Velu et al. 2008b; Gupta et al. 2009; Govindaraj et al. 2012; Rai et al. 2016). These high Fe and Zn cultivars and/or breeding lines can be exploited in the breeding programme and some of the better performing hybrids and open-pollinated varieties can be proposed to cultivate to a large extent.

9.4 Micronutrient Phenotyping Progress

Pearl millet breeding so far has aimed at improving grain yields either by developing varieties or hybrids through exploiting heterosis inherited by the virtue of its cross-pollination nature. The development of resistance to diseases and pests has also significantly contributed to grain yield increment. Considering the impact of nutritional traits on human health, especially Fe and Zn micronutrients that are present in abundance, pearl millet breeding has now started focusing on improving grain micronutrients and grain yield under the biofortification programme. This largely uses conventional breeding strategies that consist of high throughput phenotyping in combination with the advanced genomics approaches to bring the desired improvement in micronutrients level in the cultivars. A significant initiative has been made in this area, engaging partners from the governmental and private sectors, including the National Agricultural Research System (NARS) and the International Crops Research Institute for the Semi-Arid Tropics (ICRISAT). Additionally, to be bred into national varieties of pearl millet, the Indian Council of Agricultural Research (ICAR) has also set minimum requirements for iron (42 mg/kg) and zinc (32 mg/kg). (at the 52nd Annual Meeting of the Pearl Millet Improvement Project in 2017). This in turn would help to bring these micronutrients into the core breeding programme to focus grain micronutrients in parallel with grain yield.

The main concern in the biofortification effort aimed at increasing the Fe and Zn levels in cultivars is a robust phenotyping platform for micronutrient quantification in pearl millet grains. In pearl millet, open-pollinated (OP) seed samples are used as cost-effective and reliable Fe and Zn content estimations. Although OP-seeds are the preferable option, their aluminium (Al) content is checked to rule out any potential dust contamination (Rai et al. 2015a). Screening of large breeding lines is a huge task

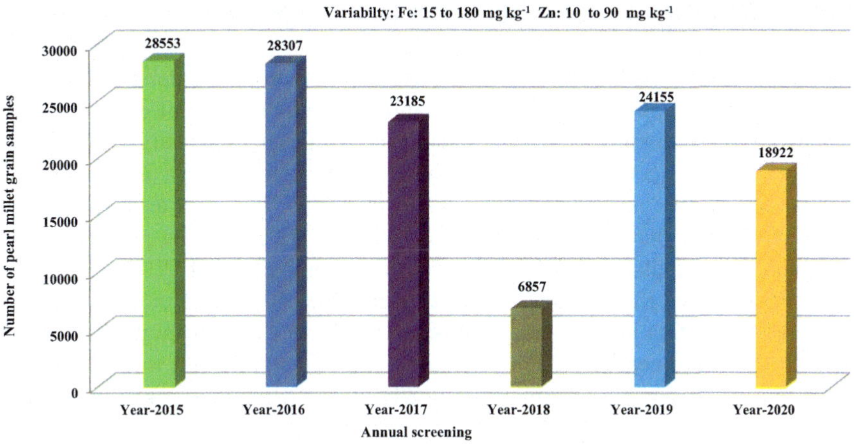

Fig. 9.1 Year-wise number of pearl millet grain samples analysed for grain iron (Fe) and zinc (Zn) content through Energy-Dispersive X-ray Fluorescence (ED-XRF) at HarvestPlus, ICRISAT, India

and requires well established, low-cost and high-throughput reliable analytical techniques. Only Atomic Absorption Spectrometry (AAS) and Inductively Coupled Plasma Optical Emission Spectrometry (ICP) were available for micronutrient quantification in the early stages of the biofortification effort. Both AAS and ICP use destructive methods of sampling for micronutrient estimates, which in turn are time-consuming and above all are a costly affair. A single sample estimate for Fe and Zn through the ICP method costs almost 18 USD and requires many days (Rai et al. 2012). For the study of plant grain samples, an Energy-Dispersive X-ray Fluorescence (XRF) system has recently been used. For the determination of grain Fe and Zn content in pearl millet, this technique uses a non-destructive method of grain sampling that has been calibrated and confirmed (Paltridge et al. 2012). Highly significant correlation ($r \geq 0.80^{**}$; $P < 0.01$) established between XRF and ICP (Govindaraj et al. 2016a) shows that the XRF is more effective than the ICP for high-throughput Fe and Zn quantification in pearl millet grains. On average, 300 samples per day can be analysed, which costs around <2.0 USD per sample. This, in turn, would help to screen large breeding lines and provide the breeder with an opportunity to discard low Fe and/or Zn lines at the early stages of generation advancement. XRF machine was established at ICRISAT, Patancheru in the year 2010 and has now been used for handling breeding lines at ICRISAT and its partner's centre. This, in turn, has helped for screening of more and more breeding lines from NARS partners who are playing a major role in the production and dissemination of biofortified cultivars. For instance, in the last 5–6 years (2015–2020), about 130,000 pearl millet samples were analysed in XRF which marks more than 21,000 samples per year (Fig. 9.1). This is highly possible and achieved with the help of the dedicated XRF labs at ICRISAT and HarvetsPlus.

9.5 Genetics of Micronutrients in Pearl Millet

Determining the best breeding strategies requires a thorough understanding of the genetics of micronutrients under consideration. The efficient use of genes in hybrid parents depends on knowledge of the abundance of genes (especially major genes), their proportional contributions, their linkages and their position in the genome. Numerous genetic experiments (via diallel mating design, line tester and North Caroline designs) have conclusively shown that the Fe and Zn in pearl millet are predominantly regulated by the additive gene action (Velu et al. 2011; Govindaraj et al. 2013, 2016b; Kanatti et al. 2014a, 2016a) (Table 9.3). Therefore, it is expected that intra-population improvement will be successful, and creating high-Fe and high-Zn hybrids would necessitate the breeding of these micronutrients into both female (A/B-lines/Seed parent) and male (R-line/Pollinator) hybrid parental lines. This claim was further supported by a small number of hybrids that had considerable variations from the mid-parental values (variations occurring in both positive and negative directions). No hybrid was found to perform better than its higher-Fe or higher-Zn parental line (Velu et al. 2011; Govindaraj et al. 2013; Kanatti et al. 2014a). Further evidence showing the grain Fe and Zn content in pearl millet is under additive genetic control included a considerably high and positive connection between the mid-parent value and the hybrid performance, predictability ratios close to unity, and a higher proportion of GCA to SCA ratio (Govindaraj et al. 2013).

A gene interaction study conducted through generation mean analysis disclosed the existence of a duplicate type of epistasis for Fe and Zn. This further suggests that it is best to avoid selecting for high Fe and Zn recombinants in early segregating generations (Boubacar Gaoh et al. 2020). Furthermore, the selection response and breeding behaviour of the hybrid-parents breeding programme are greatly influenced by knowledge of reciprocal cross differences and maternal effects. An investigation into the maternal effect on the accumulation Fe and Zn content in grains (Kanatti et al. 2018) revealed that high-Fe traits could be transferred into the elite genetic background through crossing programmes using lines with high-Fe as either a male or female parent (taking into account the seed and restorer gene pool) and selecting for elite agronomic performance with the high-Fe trait in the segregating populations. Breeding for biofortified hybrids requires information related to combining ability and heterosis for micronutrients. A highly significant positive correlation ($r = 0.89$; $P < 0.01$ for Fe and 0.92; $P < 0.01$ for Zn content) was observed between the performance per se of the lines and their general combining ability (Govindaraj et al. 2013) implies that high general combining lines can be effectively selected by selecting for their performance per se. This is not unexpected considering the predominant role of additive gene action for Fe content. Further studies are underway to develop a genomic understanding of these micronutrients.

The genotype's innate ability to transmit traits from one generation to the next, or from parents to offspring, is known as heritability. Heritability is evaluated during selection and utilized as an indicator of trait transmission. The Fe and Zn concentrations in pearl millet have demonstrated a high degree of heritability (Vagadiya et al. 2013; Vinodhana et al. 2013; Kumar et al. 2014; Govindaraj

Table 9.3 Gene action studies on grain iron and zinc contents in pearl millet

SN	Mating design	Trial size	Analytical technique	Trait	Proportion of GCA and SCA	Predominant gene action	Author
1	Diallel	10 parents = 90 F_1 (Sib-mating)	AAS	Fe and Zn	GCA > SCA	Additive	Velu et al. (2011)
2	Line × tester	8B × 9R =72 F_1 (self-pollination)	AAS	Fe and Zn	GCA > SCA	Additive	Govindaraj et al. (2013)
		16B × 12R = 192 F_1 (self-pollination)	AAS	Fe and Zn	GCA > SCA	Additive	
3	Line × tester	196 hybrids	ICP-OES	Fe and Zn	GCA > SCA	Additive	Kanatti et al. (2014a)
4	Selfing and Sib-mating	Intra-population genetic variance study in ICMR312	AAS	Fe and Zn	GCA > SCA	Additive	Govindaraj et al. (2016b)
5	NCD-1	160 full-sib progenies	ICP-OES	Fe and Zn	–	Additive	Kanatti et al. (2016a)

2011; Govindaraj et al. 2016b; Pujar et al. 2020a). The existence of high heritability signifies that simple selection and advancement would boost the genetic gain of these micronutrients.

9.6 Micronutrient Traits Association

A farmer's preference and key economic characteristics include grain yield, grain size and maturity. These characteristics must not be sacrificed in the pursuit of new traits, such as high Fe and Zn content in any crop. So, we looked into how these agronomic characteristics related to Fe and Zn contents in grains. A highly substantial and positive correlation was found between Fe and Zn content in the 68-entry population trial conducted during the 2004 rainy season and 2005 summer season ($r = 0.84$), indicating that simultaneous selection for both micronutrients might be successful. Besides this, both Fe and Zn were found highly significantly and positively correlated with grain size ($r = 0.46$–0.56), indicating that selecting for Fe and Zn does not compromise grain size and, in fact, results in larger grain size (or vice versa). None of these micronutrients were linked to flowering time. Such associations could, however, be an artefact of mostly large-seeded and relatively early maturing iniadi-based populations with higher Fe and Zn content, and non-iniadi populations with relatively smaller seed, late flowering and low Fe and Zn content. This topic was thus investigated further using S_1 progenies from four iniadi-based populations. Fe and Zn content were found to have highly significant and positive correlations in all four populations ('r' ranging from 0.66 to 0.85). In any of the populations, these micronutrients were not associated with flowering time. In addition, in terms of grain size, the correlation of both micronutrients was positive and significant in two populations but uncorrelated in the other two. Another study found a highly significant and positive correlation between Fe and Zn ($r = 0.80$–0.82) in S_1 progenies from two diverse populations (PVGGP6 with a large seed size and IAC-ISC-TCP1 with a small seed size) (Gupta et al. 2009). Except for a significant and positive correlation between Fe and time to flowering, neither of these two micronutrients was correlated with grain yield, grain size or time to flowering. Such studies, the findings of which will have a direct impact on developing early maturing and large-seeded hybrids with high yield potential and high Fe and Zn levels, will be pursued further using a variety of materials and test environments to examine a general trend of these character associations.

In pearl millet, grain Fe content is the target trait to be improved, and Zn is improved as an associated trait. This is possible because of the presence of a significantly very high positive correlation ($r = 0.43$–0.90, $P < 0.01$) between Fe and Zn (Rai et al. 2012; Govindaraj et al. 2012; Kanatti et al. 2016b; Pujar et al. 2020a). The biofortification programme focuses on improvement in micronutrient contents. In doing so, one must understand the possible impact that the improvement for Fe and Zn could leave on the other economically important traits such as grain yield. Previous research has found a moderate to high significant positive correlation between grain Fe and Zn content and 1000 grain weight in pearl millet (Velu et al.

2007, 2008a, b; Kanatti et al. 2014b; Pujar et al. 2020a) (Table 9.4). This shows that improvement for Fe and Zn can be possible without compromising grain weight. Furthermore, studies on the relationship between Fe and/or Zn and grain yield in pearl millet have revealed both positive and low to moderately negative relationships (Gupta et al. 2009; Kanatti et al. 2014b; Rai et al. 2016; Pujar et al. 2020a). Wherever the negative association is found, it is either very weak in magnitude or non-significant, indicating that improving Fe and Zn content in pearl millet would not reduce grain yield. To combine these traits in pearl millet, it would be appropriate to practice balanced selective breeding in some of the specific breeding crosses and their derived segregating populations.

9.7 Biofortified Breeding Approaches

9.7.1 Open-Pollinated Variety Breeding

The improvement of the pearl millet population began in India in the 1930s. Previously, mass selection and progeny testing were used, and several open-pollinated varieties with improved grain yield were developed. Pearl millet grain Fe and Zn content is largely under additive genetic control. As a result, population-improvement (inter- and intra-population) strategies such as mass selection, recurrent selection for general combining ability (GCA) or composite breeding would offer a great opportunity for micronutrient improvement. Figure 9.2 depicts the OPV screening and progeny performance-based selection and random mating for developing the newer version or biofortified varieties. Earlier, a study based on the intra-population improvement of two different populations (AIMP92901 and ICMR312) showed highly significant genetic variance for S_1s (single plant selections), and HS (Half sib) progenies justify the presence of large variability for Fe and Zn within the population itself and can be explored for improvement (Govindaraj et al. 2016b). Additionally, the presence of additive genetic effects and S_1's larger genetic variance compared to HS suggested that OPV may be created by purposeful selection of Fe and Zn as target attributes. The effectiveness of single plant selection among four different populations revealed a highly significant positive correlation for grain Fe ($r = 0.58$–0.75, $P < 0.01$) and Zn ($r = 0.61$–0.73, $P < 0.01$) content between S_0 (base population) and S_1 (population progenies). This finding suggests that the single plant progeny selection in pearl millet can be used successfully for the simultaneous genetic improvement of Fe and Zn micronutrients (Govindaraj 2011). It's interesting to note that there was no significant correlation between the two grain micronutrients and grain mass (1000-grain weight) indicating that improvement for Fe and Zn through a single plant section would not reduce grain mass. Three generations of progeny-based selection in an OPV (ICTP 8203) resulted in improvement of grain Fe and Zn content in a population, and one of the improved versions was found superior and designated as ICTP 8203 Fe 10^{-2}. After testing at the national level in more than 40 trials, this was found superior in Fe (71 mg/kg Fe) and yield (2.2 t/ha), hence re-released in the name of 'Dhnashakti'. This variety recorded almost 9%

Table 9.4 Studies on the association of grain Fe and Zn contents with agronomic traits 1000-grain weight (GW), grain yield (GY) and days to flowering (DF)

SN	Entries	Type of entries	Trait	GW		GY		DF		Author
1	120	Hybrids, populations, inbred lines and germplasm accessions	Fe	0.80	**	–		–		Velu et al. (2007)
			Zn	0.85	**					
2	52	Hybrids	Fe	0.34	*	–		−0.35	*	Velu et al. (2008a)
			Zn	0.35	*			−0.22		
3	259	68 Populations	Fe	0.56	**	–		−0.29		Velu et al. (2008b)
			Zn	0.46	**			−0.28		
		50 S_1 of ICTP 8203 and 47 S_1 each of CGP, GGP, PVGGP6 populations	Fe	0.25–0.36	*	–		−0.22 to −0.08		
			Zn	0.19–0.31	*			−0.31 to 0.17		
4	54	30 S_1 from PVGGP-6 and 24 S_1 from IAC-ISC-TCP-1	Fe	0.2 and 0.27		−0.02 and 0.16		−0.20 and 0.40	*	Gupta et al. (2009)
			Zn	0.14 and 0.33		−0.01 and −0.10		−0.17 and 0.26		
5	178	Hybrids	Fe	−0.15 to 0.32		−0.58** to −0.13		−0.59** to −0.05		Rai et al. (2012)
			Zn	−0.05 to 0.29		−0.32 to 0.17		−0.22 to 0.34		
6	196	Hybrids	Fe	0.42	**	−0.29	**	–		Kanatti et al. (2014a)
			Zn	0.43	**	−0.26	**	–		
7	32	Hybrids (18 ms, 31 R and 3 control)	Fe	0.16		−0.55	**	−0.47	**	Kanatti et al. (2014b)
			Zn	0.19		−0.32		−0.23		
	28	Hybrids (10 ms, 28 R and 3 control)	Fe	−0.06		−0.52	**	−0.34		
			Zn	0.19		−0.32		−0.23		

*,** Significant at the 0.05, 0.01 probability levels, respectively, *GW* 1000 grain weight, *GY* grain yield, *DF* days to flowering, *Fe* iron, *Zn* zinc

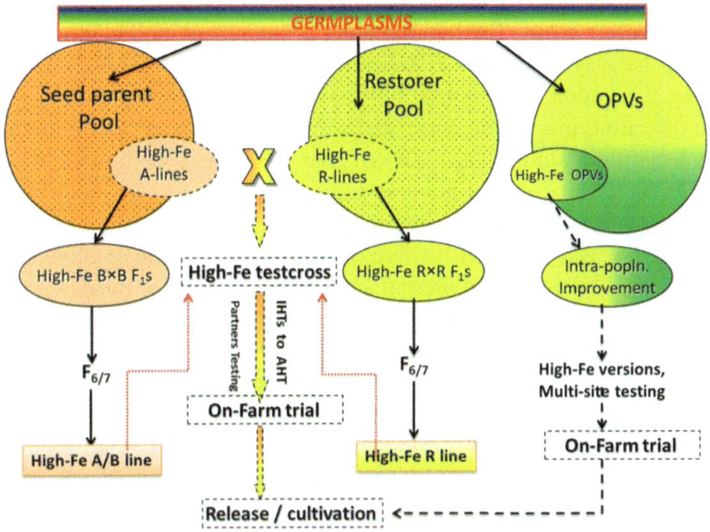

Fig. 9.2 Fast-track breeding approach followed at ICRISAT for biofortified hybrid breeding in India and OPV development in West Africa (Govindaraj et al. 2019)

higher Fe content and 11% higher grain yield than the original population (Rai et al. 2014).

9.7.2 Hybrid Breeding

High cross-pollination (>85%) (Burton 1983) and the presence of the well-established male sterile system in pearl millet have greatly aided in the development of new hybrids by allowing for the exploitation of the heterosis. Hybrids being vigorous and high yielding have become popular over the years in farmer's fields in pearl millet in India. About five million hectare of pearl millet cultivated land in India is occupied by hybrids. In contrast to grain yield, the Fe and Zn micronutrients in pearl millet, which are controlled by additive gene action, do not exhibit better parent heterosis. Therefore, both the pollinator parent and the seed parent must possess genes linked to high levels of micronutrients in grains in order to produce a high Fe and Zn biofortified hybrid. This makes it very much necessary to characterize the entire available hybrid parental lines for micronutrient content and only selected lines should be used in the hybridization programme. To demonstrate the biofortification hybrid breeding, a fast-track breeding approach was followed at ICRISAT, as shown in Fig. 9.2. Screening of all available elite hybrid parents (seed parents and restorers) for Fe and Zn content is the essential and first step in identifying elite lines/donors.

Hybrid parent breeding forms an integral part of the commercial hybrid development programme. This involves the development of high Fe and/or Zn seed parent and pollinator parent with good agronomic characteristics particularly grain yield and early flowering types. The hybrid parent development with high micronutrients depends on the magnitude of variability that exists in the base population, that is segregating generation such as F_2s, OPVs or composites and that further depends upon the direction, magnitude and pattern of inbreeding effects (Govindaraj et al. 2019). Interestingly studies have shown that inbreeding would not significantly affect, or it would marginally increase both the micronutrients (Rai et al. 2017). The primary source of high Fe and Zn in pearl millet has been iniadi germplasm (Velu et al. 2011; Rai et al. 2012; Govindaraj et al. 2013). The commercial hybrids that are present in the market at present are developed from the iniadi source. Genes introgressed into both male and female parental lines from the single germplasm source would narrow down the genetic diversity between the two parental groups (B/A and R-lines). This, in turn, would contribute to the reduction of heterosis of commercially important traits like grain yield that are predominantly controlled by non-additive (dominant and over-dominant) gene action. This requires the intervention of the modern genomics breeding approach wherein selective genes for Fe and Zn content from such similar germplasm can be transferred into parental groups without disturbing the genetic diversity of other important traits that would play a vital role in breeding for biofortification programmes. On the other hand, new sources of Fe and Zn germplasm are being investigated at ICRISAT and would undoubtedly aid in the genetic diversification of parental lines of high Fe and Zn hybrids.

9.7.3 Biofortified Pearl Millet Cultivars

Biofortification breeding in pearl millet funded by HarvestPlus has certainly helped to reduce the Fe and Zn associated malnutrition in India and African regions by developing some of the high Fe and Zn biofortified varieties and hybrids. The first mineral biofortified product of any crop cultivar issued in India, Dhanashakti (High grain Fe content), was developed together by the International Crops Research Institute for the Semi-Arid Tropics (ICRISAT) and Mahatma Phule Krishi Vidyapeeth was released in 2012 in Maharashtra and later in 2013 throughout India. 'Dhanashakti' has 71 mg/kg Fe and 40 mg/kg Zn content in grains reaching 65,000 farmers by 2015. The Dhanashakti seeds are available with Nirmal Seed Company and State Seed Corporations in Maharashtra, Telangana, Karnataka, Haryana, Uttar Pradesh and Rajasthan. ShaktiVardhak Seed Company is marketing ICMH 1201, a high-iron pearl millet hybrid developed by ICRISAT, under the name Shakti 1201 is having 75 mg/kg Fe and 40 mg/kg Zn content in grains. The grain yield of ICMH 1201 hybrid is 30% greater than that of Dhanashakti while having the same levels of Fe and Zn. Due to its increased grain yield and high iron content, it quickly spread to more than 35,000 farmers in Maharashtra and Rajasthan (Govindaraj and Rai 2016). According to estimates from feeding trials, 'Dhanashakti' and 'Shakti 1201' would meet more than the daily iron requirement of 0.84 mg for men, as well as 70% of the daily requirement of 1.65 mg for non-pregnant, non-lactating women, and 42% of the daily requirement of 2.8 mg for pregnant women. These estimates assumed 240 g/day of grain consumption at 7.5% of bioavailability. The above-mentioned grain consumption rate will also supply 80% of the daily required allowance of 12 mg of zinc.

The first wave of biofortified hybrids was released for a national level in collaboration with the agricultural universities of Maharashtra (VNMKV and MPKV) and Haryana (CCSHAU) in the year 2018 (Table 9.5). These hybrids are AHB 1200 Fe (ICMH 1202), HHB 299 (ICMH 1203) and Phule Mahashakti (DHBH 1211/MH 2078). These biofortified hybrids contain more than 70 mg/kg Fe and 35 mg/kg Zn in grains. A biofortified variety (OPV) 'Chakti' was officially released by the government of Niger in the year 2018. Chakti has the advantage of early maturity (40 days earlier than local varieties) and 30% higher grain yield than local varieties, which is also resistant to downy mildew disease. Three biofortified hybrids with high Fe and Zn content—AHB 1269 Fe, RHB 234 and RHB 233—were made available in India in 2019. The grain yield of all three hybrids was close to 3.2 t/ha, but the Fe and Zn contents in these hybrids varied from 83 to 91 and 41 to 46 mg/kg, respectively.

Given that all breeding efforts among partners (in the public and private sectors) are currently concentrated on the development of hybrids, the HarvestPlus-supported pearl millet biofortification programme in India is specifically targeted at developing hybrids. Therefore, the development of high-Fe versions of ICTP 8203 and ICMV 221 constitutes the extent of ICRISAT's involvement in OPV development under this project. The attention now switches to the production of high-yielding and high-Fe hybrids, but only as a demonstration technology, with the major strategic breeding and screening concerns having been essentially resolved. The development of high-Fe hybrids will move to the NARS in the public and private sectors after it has been successfully shown that they can be developed without jeopardizing grain yield or any other significant agronomic features. A step has already been taken towards hybrid development. In attempt to develop high-yielding hybrids with high Fe and Zn content, high-Fe lines discovered through screening of the existing hybrid parents and breeding lines are being used. The fact that these hybrids were produced using parental lines that were not particularly developed for high Fe content suggests that the likelihood of producing higher, high-Fe hybrids will increase once those parental lines are employed for hybrid production. This would necessitate including biofortification into the entire breeding effort.

9.7.4 Genomic Progress and Application Prospects

Most productivity and quality traits are usually quantitatively inherited, and this includes Fe and Zn content that is largely governed by additive gene action. Quantitative trait loci are areas of a genome that include genes linked to a specific quantitative trait (QTLs). The identification of QTLs based on only conventional phenotypic evaluation is not precise. The development of DNA markers has brought a breakthrough that helped to identify the genetic region/chromosomal region governing quantitative traits through the construction of a linkage map (Mohan

Table 9.5 Biofortified hybrids and varieties (OPVs) of pearl millet released in Africa and India under biofortification breeding programme

SN	Cultivar name	Commercial name	Hybrid/OPV	Year of release	Fe (mg/kg)	Zn (mg/kg)	Grain yield (t/ha)	Country name
1	ICTP 8203 10-2	Dhanashakti	OPV	2013	71	40	2.2	India
2	ICMH 1201	Shakti-1201	Hybrid (TLS)	NA	75	40	3.2	India
3	ICMH 1203	HHB 299	Hybrid	2017	73	41	3.3	India
4	DHBH 1211/MH 2078	Phule Mahashakti	Hybrid	2018	87	41	2.9	India
5	Chakti	Chakti	OPV	2018	60	45	1.5	Niger (Africa)
6	MH 2072	AHB 1200 Fe	Hybrid	2018	77	39	3.2	India
7	ABV 04	ABV 04	OPV	2018	70	63	2.9	India
8	MH 2185	AHB 1269 Fe	Hybrid	2019	91	43	3.3	India
9	MH 2174	RHB 234	Hybrid	2019	84	41	3.2	India
10	MH 2173	RHB 233	Hybrid	2019	83	46	3.2	India
11	MH 2179	HHB 311	Hybrid	2019	83	39	3.2	India
12	Moti Shakti	GHB1225	Hybrid	2020	72	43	3.0	India

et al. 1997). Working on QTL/linkage mapping in pearl millet has been started a few decades ago (Liu et al. 1994). Exploration of molecular markers to construct linkage maps to identify genomic regions governing Fe and Zn contents in pearl millet particularly in the interest of a biofortification programme has started recently. A linkage map was constructed with 106 RILs (F_6) from ICMB 841-P3 × 863B-P2 and two co-localized QTLs for Fe and Zn on LG3 were identified (Kumar et al. 2016). Another QTL mapping study on 317 RILs (F_6) derived from ICMS 8511-S1-17-2-1-1-B-P03 × AIMP 92901-S1-183-2-2-B-08 revealed 11 QTLs for Fe and 8 QTLs for Zn (Kumar et al. 2018). Recently, 6 QTLs for Fe (LGs 3, 5 and 7) and 10 QTLs for Zn (LGs 3, 4, 5, 6 and 7) were identified from an association mapping study in pearl millet (Anuradha et al. 2017). A GWAS study at ICRISAT consisting of 281 diverse inbred lines could able to identify 18 genome-wide marker-trait associations (MTAs) for Fe and 43 for Zn (Pujar et al. 2020b). For the Fe and Zn concentrations in pearl millet, more than 45 QTLs were found; however, the phenotyping variance ranged from 6.7 to 20.5% for Fe and from 5.7 to 26.4% for Zn (Kumar et al. 2016, 2018). The advent of high-throughput genotyping platforms (HTGP) and genomic platforms will reduce the cost of genotyping and screening a large set of breeding pipelines would be highly possible. Therefore, using these platforms, further verification and validation of these reported markers would be of great prospect for their mainstreaming biofortification breeding in national and international programmes. To date, few SNP-based diagnostic markers are identified, and recommended for screening for Fe/Zn traits and terminal drought tolerance in pearl millet (Table 9.6). These markers were developed at ICRISAT and made affordable for usage by the general people (CGIAR-EiB low-density genotyping service). With the help of this application, pearl millet drought-tolerant biofortified varieties or hybrids may be produced. On the other hand, the in-silico discovery and selection of 29 genes associated to Fe and Zn metabolism revealed tissue- and stage-specific expressions for Fe and Zn metabolism genes at various developmental stages in Fe and Zn-contradictory parents. Fe and Zn candidate genes were reported to be found in the PglZIP, PglNRAMP and PglFER gene families (Mahendrakar et al. 2020). The findings of this work can serve as the basis for the functional analysis of various Fe and Zn metabolism gene homologs and their eventual use in pearl millet molecular breeding.

9.7.5 Iron Bioavailability in Biofortified Pearl Millet

The percentage of a nutrient in a meal that can be absorbed and used by the body for metabolic activities is known as bioavailability (Welch and Graham 2004). The crop biofortification programme places a high priority on determining the bioavailability of Fe and Zn in newly developed genetically improved cultivars because only a small percentage of the accumulated minerals in edible parts are bioavailable. The levels of bioavailability of Fe and Zn in seeds and grains of staple food crops are as low as 5% and 25%, respectively (Bouis and Welch 2010). Considering the impact of bioavailability of micronutrients such as Fe and Zn in biofortification breeding

Table 9.6 Available diagnostic markers for terminal drought and Fe/Zn traits screening in pearl millet (source: https://excellenceinbreeding.org)

Trait designation	Chr	Allele 1	Alternate allele	SNP/INDEL	Gene/QTL	Donor sources	Susceptible controls	Marker origin
Terminal drought tolerance	LG2	T	C	T/C	DT QTL	PRLT2/89-33	H77/833-2	ICRISAT
Terminal drought tolerance	LG2	G	A	G/A	DT QTL	PRLT2/89-34	H77/833-2	ICRISAT
Terminal drought tolerance	LG2	C	T	C/T	DT QTL	PRLT2/89-33	H77/833-2	ICRISAT
Terminal drought tolerance	LG2	C	T	C/T	DT QTL	PRLT2/89-33	H77/833-2	ICRISAT
Terminal drought tolerance	LG2	A	G	A/G	DT QTL	PRLT2/89-33	H77/833-2	ICRISAT
Terminal drought tolerance	LG2	G	A	G/A	DT QTL	PRLT2/89-33	H77/833-2	ICRISAT
Terminal drought tolerance	LG2	G	T	G/T	DT QTL	PRLT2/89-33	H77/833-2	ICRISAT
Terminal drought tolerance	LG2	T	C	T/C	DT QTL	PRLT2/89-33	H77/833-2	ICRISAT
Terminal drought tolerance	LG2	A	G	A/G	DT QTL	PRLT2/89-33	H77/833-2	ICRISAT
Terminal drought tolerance	LG2	C	T	C/T	DT QTL	PRLT2/89-33	H77/833-2	ICRISAT
Grain Fe/Zn content	n.a.	C	T	C/T	–	863B	n.a.	ICRISAT
Grain Fe/Zn content	n.a.	A	T	A/T	–	863B	n.a.	ICRISAT
Grain Fe/Zn content	n.a.	A	G	A/G	–	863B	n.a.	ICRISAT
Grain Fe/Zn content	n.a.	A	T	A/T	–	863B	n.a.	ICRISAT

Chr chromosome, *LG* linkage groups

programmes, several studies were conducted among different groups of consumers, especially on children and pregnant women to account for the efficacy of biofortified pearl millet cultivar to provide the daily required Fe and Zn micronutrients.

Twenty Beninese women with low iron levels participated in an iron absorption trial using a composite test meal that contained millet paste made from ordinary, iron-biofortified and iron-fortified pearl millet flour (Cercamondi et al. 2013). Iron absorption was measured based on the erythrocyte incorporation of stable iron isotopes. It was found that regular-iron millet had an iron absorption rate of 7.5%, which was similar to that of iron-biofortified millet and suggested that consumption of iron-biofortified millet would increase women's iron absorption by twofold. Another iron absorption study conducted among 40 Indian (Karnataka) children aged 2 years showed that the mean iron absorption of Fe among children with iron-fortified pearl millet (0.67 ± 0.48) was three times higher than the regular millet-based food (0.23 ± 0.15 mg/day) (Kodkany et al. 2013). It was also interesting and promising to see that the mean zinc absorption among children was also higher (0.95 ± 0.47 mg/day) than the regular (0.67 ± 0.24 mg/day). A similar study was conducted among 246 children aged 12–16 years from Maharashtra. Iron biofortified pearl millet meal increased almost 1.6 times higher chances of recovering the required Fe concentration in the body (Finkelstein et al. 2015). Thus, all these studies comprehensively suggest that the consumption of high biofortified pearl millet grains would easily meet more than 80% of the required daily quantities of Fe and 100% of Zn requirements.

It is well recognized that a variety of dietary elements, including both inhibitors and enhancers of absorption, have an impact on the bioavailability of iron and zinc. The most powerful inhibitors are tannins, dietary fibre, calcium and phytic acid (PA), while organic acids are known to increase iron absorption (Sandberg 2002; Hambidge et al. 2010). For populations whose diets mostly consist of grains and legumes, phytate, a compound of phytic acid and mineral elements, lowers the concentration of nutrient elements that are bioavailable and consequently causes health issues, such as Fe and Zn deficiencies (Liu et al. 2006). These substances are typical plant metabolites, and even slight variations in their content could have a large impact on the bioavailability of micronutrients. According to a study on pearl millet Zn was chelated solely by phytates, whereas Fe was chelated by both phytates and insoluble fibres. The chelating effect of these substances was also greater than that of phytates in the event of increased fibre and tannin levels (Lestienne et al. 2005). Studies on animals have demonstrated the beneficial effects of diets rich in low-phytate maize on enhancing the use of minerals (Veum et al. 2001; Li et al. 2000). Therefore, while evaluating Fe and Zn deficiencies, the inhibitory action of phytates should be taken into consideration.

9.8 Conclusion

Climate change and ever-increasing population are driving for better energy-cum nutrition requirements through staple crops including pearl millet, particularly in India and sub-Saharan Africa. Continuously evolving on the staple crop breeding mandate is very crucial to provide the demanding traits that are required for the growing population and food markets. Such research for development-based investments will reduce hunger in all forms (including hidden hunger) across the regions. Pearl millet as such is a highly nutritious cereal compared to rice, wheat and maize but reported micronutrient baselines (42 mg/kg Fe and 32 mg/kg Zn) in commercial cultivars call for constant nutrition improvement to reduce the genetic gap between yield and nutrition over time. ICRISAT breeding materials and global germplasm collections have both demonstrated significant variation for both micronutrients in pearl millet biofortification studies on grain Fe and Zn content. Many of these sources of Fe and Zn have these levels of micronutrients and come from exceptional genetic origins. Increased use of some of the most promising sources will result in stronger genetic gains for Fe/Zn and desired features in pearl millet mainstream breeding. The breeding of high-Fe and high-Zn hybrids with high yield potential, on the other hand, will be greatly accelerated by developing collaborations with user-research groups, both in the public and commercial sector, and by extending quick and affordable XRF (rapid and non-destructive) screening facilities.

The possibility of improving the Fe and Zn concentrations of pearl millet is predicted by the lack of a significant negative correlation between these minerals and grain yield. Fe and Zn linked malnutrition in India and sub-Saharan Africa could be reduced by mainstreaming breeding priorities for Fe and Zn in pearl millet. There have only been 12 biofortified pearl millet cultivars released so far, and only a small number of them have reached farmers through the public seed system. Fair scaling of public bred cultivars through private marketing networks can hasten the spread of novel varieties to farmers with limited resources. In order to speed up the breeding process and product development for these micronutrients, new, affordable genomics methods will be deployed. In the same way that has been done so far for the non-biofortified materials, these biofortified breeding lines and hybrid parents will be regularly distributed to and used by user-research organizations (both public and private sector). As a result, nutrition will be greatly improved and the production of biofortified hybrids will become regular.

Acknowledgement Authors greatly acknowledge the support and funding of the HarvestPlus programme of the CGIAR and Agriculture for Nutrition and Health (A4NH) for pearl millet biofortification breeding and dissemination projects.

References

Anuradha N, Satyavathi CT, Bharadwaj C, Nepolean T, Sankar SM, Singh SP, Meena MC, Singhal T, Srivastava RK (2017) Deciphering genomic regions for high grain iron and zinc content using association mapping in pearl millet. Front Plant Sci 8:e00412. https://doi.org/10.3389/fpls.2017.00412

Bashir EM, Ali AM, Ali AM, Melchinger AE, Parzies HK, Haussmann BIG (2014) Characterization of Sudanese pearl millet germplasm for agro-morphological traits and grain nutritional values. Plant Gen Res 12:35–47

Boubacar Gaoh BS, Gangashetty PI, Mohammed R, Dzidzienyo DK, Tongoona P (2020) Generation mean analysis of pearl millet [*Pennisetum glaucum (L.) R. Br.*] grain iron and zinc contents and agronomic traits in West Africa. J Cereal Sci 96:103066. https://doi.org/10.1016/j.jcs.2020.103066

Bouis HE (2003) Micronutrient fortification of plants through plant breeding: can it improve nutrition in man at low cost. Proc Nutr Soc 62:403–411

Bouis HE, Welch RM (2010) Biofortification a sustainable agricultural strategy for reducing micronutrient malnutrition in the global south. Crop Sci 50:20–32

Burton GW (1983) Breeding pearl millet. Plant Breed Rev 1:162–182

Cercamondi CI, Egli IM, Mitchikpe E, Tossou F, Zeder C, Hounhouigan JD, Hurrell RF (2013) Total iron absorption by young women from iron-biofortified pearl millet composite meals is double that from regular millet meals but less than that from post-harvest iron-fortified millet meals. J Nutr Sep 143(9):1376–1382. https://doi.org/10.3945/jn.113.176826

Directorate of Millets Development (2020) Satyavathi T. C. Project Coordinator Review, 55th annual group meeting, ICAR-AICRP on Pearl millet, Angrau, Tirupati, 13–15 March, 2020

Djanaguiraman M, Perumal R, Ciampitti IA, Gupta SK, Prasad PVV (2018) Quantifying pearl millet response to high temperature stress: thresholds, sensitive stages, genetic variability and relative sensitivity of pollen and pistil. Plant Cell Environ 41:993–1007. https://doi.org/10.1111/pce.12931

FAO IFAD, UNICEF, WFP and WHO (2019) The state of food security and nutrition in the world 2019. Safeguarding against economic slowdowns and downturns. FAO, Rome

FAOSTAT (2020) FAOSTAT online database (2020). FAO, Rome

Finkelstein JL, Mehta S, Udipi SA, Ghugre PS, Luna SV, Wenger MJ, Murray-Kolb LE, Przybyszewski EM, Haas JDA (2015) Randomized trial of iron-biofortified pearl millet in school children in India. J Nutr 145(7):1576–1581. https://doi.org/10.3945/jn.114.208009

Govindaraj M (2011) Genetics of grain iron and zinc concentration in pearl millet (*Pennisetum glaucum* (L.) R. Br.). Ph.D. thesis, Tamil Nadu Agricultural University

Govindaraj M, Rai KN (2016) Breeding biofortified pearl millet cultivars with high iron density. Indian Farm 65(12):53–55

Govindaraj M, Rai KN, Shanmugasundaram P, Rao AS (2012) Efficiency of single plant selection for grain iron and zinc density in pearl millet. Eur J Plant Sci Biotechnol 6:114–117

Govindaraj M, Rai KN, Shanmugasundaram P, Dwivedi SL, Sahrawat KL, Muthaiah AR, Rao AS (2013) Combining ability and heterosis for grain iron and zinc densities in pearl millet. Crop Sci 53:507–517

Govindaraj M, Rai KN, Pfeiffer WH, Kanatti A, Shivade H (2016a) Energy-dispersive X-ray fluorescence spectrometry for cost-effective and rapid screening of pearl millet germplasm and breeding lines for grain iron and zinc density. Commun Soil Sci Plant Anal 47:2126–2134

Govindaraj M, Rai KN, Shanmugasundaram P (2016b) Intra-population genetic variance for grain iron and zinc contents and agronomic traits in pearl millet. Crop J 4:48–54

Govindaraj M, Yadav OP, Srivastava RK, Gupta SK (2019) Conventional and Molecular Breeding Approaches for Biofortification of Pearl Millet. In: Qureshi A, Dar Z, Wani S (eds) Quality Breeding in Field Crops. Springer, Cham. https://doi.org/10.1007/978-3-030-04609-5_4

Govindaraj M, Rai KN, Kanatti A, Upadhyaya HD, Shivade H, Rao AS (2020) Exploring the genetic variability and diversity of pearl millet core collection germplasm for grain nutritional traits improvement. Sci Rep 10:21177. https://doi.org/10.1038/s41598-020-77818-0

Gupta SK, Velu G, Rai KN, Sumalini K (2009) Association of grain iron and zinc content with grain yield and other traits in pearl millet (*Pennisetum glaucum* (L.) R. Br.). Crop Improv 36:4–7

Hambidge KM, Miller LV, Westcott JE, Sheng X, Krebs NF (2010) Zinc bioavailability and homeostasis. Am J Clin Nutr 91:1478S–1483S

Indian Council of Medical Research (2010) Nutrient requirements and recommended dietary allowances for Indians. A report of the expert group of the Indian Council of Medical Research. National Institute of Nutrition, Hyderabad

Kanatti A, Rai KN, Radhika K, Govindaraj M, Sahrawat KL, Rao AS (2014a) Grain iron and zinc density in pearl millet: combining ability, heterosis and association with grain yield and grain size. Springerplus 3:763

Kanatti A, Rai KN, Radhika K, Govindaraj M, Sahrawat KL, Srinivasu K, Shivade H (2014b) Relationship of grain iron and zinc content with grain yield in pearl millet hybrids. Crop Improv 41:91–96

Kanatti A, Rai KN, Radhika K, Govindaraj M (2016a) Tester effect on combining ability and its relationship with line performance per se for grain iron and zinc densities in pearl millet. Crop Sci 56:1–8

Kanatti A, Rai KN, Radhika K, Govindaraj M (2016b) Genetic architecture of open-pollinated varieties of pearl millet for grain iron and zinc densities. Indian J Genet Plant Breed 76:299–303. https://doi.org/10.5958/0975-6906.2016.00045.6

Kanatti A, Govindaraj M, Rai KN, Rao AS (2018) Maternal effect on grain iron and zinc densities in pearl millet. In: 1st National genetics congress on genetics for sustainable food, health and nutrition security, 14–16 Dec 2018. ICAR-Indian Agricultural Research Institute, New Delhi

Kodkany BS, Bellad RM, Mahantshetti NS, Westcott JE, Krebs NF, Kemp JF, Hambidge KM (2013) Biofortification of pearl millet with iron and zinc in a randomized controlled trial increases absorption of these minerals above physiologic requirements in young children. J Nutr 143(9):1489–1493. https://doi.org/10.3945/jn.113.176677. Epub 2013 Jul 10

Kumar Y, Lamba AS, Yadav HP, Kumar VD (2014) Studies on variability and character association under rainfed conditions in pearl millet (*Pennisetum glaucum* L.) hybrids. Forage Res 39(4): 175–178

Kumar S, Hash CT, Thirunavukkarasu N, Singh G, Rajaram V, Rathore A, Senapathy S, Mahendrakar MD, Yadav RS, Srivastava RK (2016) Mapping quantitative trait loci controlling high iron and zinc content in self and open pollinated grains of pearl millet [*Pennisetum glaucum* (L.) R. Br.]. Front Plant Sci 7:1636

Kumar S, Hash CT, Nepolean T, Mahendrakar MD, Satyavathi CT, Singh G, Rathore A, Yadav RS, Gupta R, Srivastava RK (2018) Mapping grain iron and zinc content quantitative trait loci in an Iniadi-derived immortal population of pearl millet. Genes 9(5):248. https://doi.org/10.3390/genes9050248

Lestienne I, Besancon P, Caporiccio B, Lullien-Pellerin V, Treche S (2005) Iron and zinc *in vitro* availability in pearl millet flours (*Pennisetum glaucum*) with varying phytate, tannin and fiber contents. J Agric Food Chem 53:3240–3247

Li YC, Ledoux DR, Veum TL, Raboy V, Ertl DS (2000) Effects of low phytic acid corn on phosphorus utilization, performance, and bone mineralization in broiler chicks. Poult Sci 79: 1444–1450

Liu CJ, Witcombe JR, Pittaway TS, Nash M, Hash CT, Busso CS, Gale MD (1994) An RFLP-based genetic map of pearl millet (Pennisetum glaucum). Theor Appl Genet 89:481–487

Liu ZH, Wang HY, Wang XE, Zhang GP, Chen PD, Liu DJ (2006) Genotypic and spike positional difference in grain phytase activity, phytate, inorganic phosphorus, iron and zinc contents in wheat (*Triticum aestivum* L). J Cereal Sci 44:212–219

Mahendrakar MD, Kumar S, Singh RB, Rathore A, Potupureddi G, Gupta R, Kavi Kishor PB, Srivastava R (2019) Genetic variability, genotype × environment interaction and correlation analysis for grain iron and zinc contents in recombinant inbred line population of pearl millet [*Pennisetum glaucum* (L). R. Br.]. Indian J Genet 79:545–551. https://doi.org/10.31742/IJGPB.79.3.3

Mahendrakar MD, Parveda M, Kishor PBK, Srivastava R (2020) Discovery and validation of candidate genes for grain iron and zinc metabolism in pearl millet [*Pennisetum glaucum* (L.) R. Br.]. Sci Rep 10:16562. https://doi.org/10.1038/s41598-020-73241-7

Mani UV, Prabhu BM, Damle SS, Mani I (1993) Glycemic index of some commonly consumed foods in Western India. Asia Pac J Clin Nutr 2:111–114

Ministry of Agriculture & Farmers Welfare (2021) Press Information Bureau, Government of India, 24 Feb. PIB, Delhi

Mohan M, Nair S, Bhagwat A, Krishna TG, Yano M, Bhatia CR, Sasaki T (1997) Genome mapping, molecular markers and marker-assisted selection in crop plants. Mol Breed 3:87–103. https://doi.org/10.1023/A:1009651919792

Muthusamy V, Hossain F, Thirunavukkarasu N, Choudhary M, Saha S et al (2014) Development of β-carotene rich maize hybrids through marker-assisted introgression of β-carotene hydroxylase allele. PLoS One 9(12):e113583. https://doi.org/10.1371/journal.pone.0113583

NIN (2003) Nutritive value of Indian foods. In: Gopalan C, Deosthale YG (eds) Nutritive value of Indian foods. National Institute of Nutrition, Hyderabad

Paltridge NG, Palmer LJ, Milham PJ, Guild GE, Stangoulis JCR (2012) Energy-dispersive X-ray fluorescence analysis of zinc and iron concentration in rice and pearl millet grain. Plant Soil 361:251–260

Parthasarathy Rao P, Birthal PS, Reddy BVS, Rai KN, Ramesh S (2006) Diagnostics of sorghum and pearl millet grains-based nutrition in India. ISMN 46:93–96

Proceedings of the 52nd Annual Pearl Millet Workshop All India Coordinated Research Project on Pearl Millet (2017). Held at: Punjab Agricultural University, Ludhiana (Punjab), 28–30 April

Pucher H, Høgh-Jensen J, Gondah CT, Hash CT, BIG H (2014) Micronutrient density and stability in west African pearl millet: potential for biofortification. Crop Sci 54:1709–1720

Pujar M, Gangaprasad S, Govindaraj M, Gangurde SS, Kanatti A, Kudapa H (2020a) Genetic variation and diversity for grain iron, zinc, protein and agronomic traits in advanced breeding lines of pearl millet [*Pennisetum glaucum* (L.) R. Br.] for biofortification breeding. Genet Resour Crop Evol 67:2009–2022. https://doi.org/10.1007/s10722-020-00956-x

Pujar M, Gangaprasad S, Govindaraj M, Gangurde SS, Kanatti A, Kudapa H (2020b) Genome-wide association study uncovers genomic regions associated with grain iron, zinc and protein content in pearl millet. Sci Rep 10:19473. https://doi.org/10.1038/s41598-020-76230-y

Rai KN, Govindaraj M, Rao AS (2012) Genetic enhancement of grain iron and zinc content in pearl millet. Qual Assur Saf Crops Foods 4:119–125

Rai KN, Patil HT, Yadav OP, Govindaraj M, Khairwal IS, Cherian B, Rajpurohit BS, Rao AS, Kulkarni MP (2014) Dhanashakti: a high-iron pearl millet variety. Indian Farm 64:32–34

Rai KN, Velu G, Govindaraj M, Wolfgang P, Aluri R (2015a) Seed set and Xenia effects on grain iron and zinc density in pearl millet. Crop Sci 55:821. https://doi.org/10.2135/cropsci2014.04.0305

Rai KN, Velu G, Govindaraj M, Upadhyaya HD, Rao AS, Shivade H, Reddy KN (2015b) Iniadi pearl millet germplasm as a valuable genetic resource for high grain iron and zinc densities. Plant Genet Resour 13:75–82

Rai KN, Yadav OP, Govindaraj M, Pfeiffer WH, Yadav HP, Rajpurohit BS, Patil HT, Kanatti A, Rathore A, Rao AS, Shivade H (2016) Grain iron and zinc densities in released and commercial cultivars of pearl millet (*Pennisetum glaucum*). Indian J Agric Sci 86:11–16

Rai KN, Govindaraj M, Kanatti A, Rao AS, Shivade H (2017) Inbreeding effects on grain iron and zinc concentrations in pearl millet. Crop Sci 57:1–8

Sandberg AS (2002) Bioavailability of mineral in legumes. Br J Nutr 88(3):S281–S285

Vagadiya KJ, Dhedhi KK, Joshi HJ (2013) Genetic variability, heritability and genetic advance of grain yield in pearl millet. Agric Sci Dig 33(3):223–225

Velu G, Rai KN, Muralidharan V, Kulkarni VN, Longvah T, Raveendran TS (2007) Prospects of breeding biofortified pearl millet with high grain iron and zinc content. Plant Breed 126:182–185

Velu G, Rai KN, Sahrawat KL, Sumalini K (2008a) Variability for grain iron and zinc contents in pearl millet hybrids. J SAT Agric Res 6:1–4

Velu G, Rai KN, Sharawat KL (2008b) Variability for grain iron and zinc content in a diverse range of pearl millet populations. Crop Improv 35(2):186–191

Velu G, Rai KN, Muralidharan V, Longvah T, Crossa J (2011) Gene effects and heterosis for grain iron and zinc density in pearl millet (*Pennisetum glaucum* (L.) R. Br). Euphytica 180:251–259

Veum TL, Ledoux DR, Raboy V, Ertl DS (2001) Low-phytic acid corn improves nutrient utilization for growing pigs. Am Soc Anim Sci 79:2873–2880

Vinodhana KN, Sumathi P, Sathya M (2013) Genetic variability and inter-relationship among morpho-economic traits of pearl millet [*Pennisetum glaucum* (L.)R. Br.,] and their implications in selection. Int J Plant Anim Environ Sci 3(2):145–149

Welch RM, Graham RD (2004) Breeding for micronutrients in staple food crops from a human nutrition perspective. J Exp Bot 55:353–364

Yadav OP, Upadhyaya HD, Reddy KN, Jukanti AK, Sushil P, Tyagi PK (2017) Genetic resources of pearl millet: status and utilization. Indian J Plant Genet Res 30(1):1–7

Yadav OP, Singh D, Dhillon B, Mohapatra T (2019) India's evergreen revolution in cereals. Curr Sci 116:1805–1808

Physiological and Molecular Bases of Drought and Heat Tolerance in Pearl Millet

10

B. Sine, Aïda Ndiaye, G. Kanfany, A. Grondin, Adama Ndiaye, A. Faye, O. Sy, O. Diack, N. A. Kane, and L. Laplaze

Abstract

Pearl millet is one of the most important sources of nutrition for millions of people in arid and semi-arid areas in Africa and Asia. Farmers have had, throughout its domestication, to select cultivars adapted to their environments. So, pearl millet appears as an interesting crop model to study drought adaptation. However,

B. Sine · A. Faye · N. A. Kane
Centre de Recherche pour l'Amélioration de l'Adaptation à la Sécheresse, Institut Sénégalais de Recherches Agricoles, Thiès, Senegal

LMI LAPSE, Institut de Recherche pour le Développement, Dakar, Senegal

A. Ndiaye · A. Ndiaye · O. Sy · O. Diack
Centre de Recherche pour l'Amélioration de l'Adaptation à la Sécheresse, Institut Sénégalais de Recherches Agricoles, Thiès, Senegal

G. Kanfany
LMI LAPSE, Institut de Recherche pour le Développement, Dakar, Senegal

Centre National de Recherche Agronomique, Institut Sénégalais de Recherches Agricoles, Bambey, Senegal

A. Grondin
Centre de Recherche pour l'Amélioration de l'Adaptation à la Sécheresse, Institut Sénégalais de Recherches Agricoles, Thiès, Senegal

LMI LAPSE, Institut de Recherche pour le Développement, Dakar, Senegal

UMR DIADE, Institut de Recherche pour le Développement and Université de Montpellier, Montpellier, France

L. Laplaze (✉)
LMI LAPSE, Institut de Recherche pour le Développement, Dakar, Senegal

UMR DIADE, Institut de Recherche pour le Développement and Université de Montpellier, Montpellier, France
e-mail: laurent.laplaze@ird.fr

V. A. Tonapi et al. (eds.), *Pearl Millet in the 21st Century*,
https://doi.org/10.1007/978-981-99-5890-0_10

current and future climatic changes pose challenges to its cropping sustainability. Drought and heat are the main factors of climate change. To accelerate pearl millet adaptation and improve its productivity to cope with climate change, its mechanisms of adaptation must be dissected. Here, we review the state of research on the physiological and molecular bases of pearl millet adaptation to drought and heat. However, pearl millet remains a neglected crop, and progress in research remains to be made.

Keywords

Pearl millet · Mechanisms of adaptation · Stress · Drought · Heat · Tolerance

10.1 Climate Change

Climate change is a major threat to agriculture and food security (Kang et al. 2009; Godfray et al. 2010; Wang et al. 2018). The most conspicuous climate changes in recent times are the increase in atmospheric temperatures due to increasing levels of greenhouse gases (Solomon et al. 2007; Stott et al. 2010; Christidis et al. 2012; Wang et al. 2018) and associated changes in the water cycle (Bates et al. 2008; Collins et al. 2013; Jung et al. 2002; Balling and Cerveny 2003; Fauchereau et al. 2003; Trenberth et al. 2007). A continued rise in global temperature is predicted if greenhouse gas (GHG) emissions continue unabated, according to the Fifth Assessment Report (AR5) of the Intergovernmental Panel on Climate Change (IPCC). The frequency of warm days, warm nights, and heat waves have increased, while the frequency of cold days and cold nights has decreased (Sillmann et al. 2013). Based on emission scenarios, the IPCC predicts a temperature increase between 0.3 and 0.7 °C by 2035 with a medium degree of confidence (Kirtman et al. 2013). In a study assessing long-term projections of climate change, Collins et al. (2013) showed that increase in global mean surface temperatures for 2081–2100, relative to 1986–2005, will likely be in the 5–95% range of CMIP5 (Coupled Model Intercomparison Project Phase 5) models; 0.3–1.7 °C (Representative Concentration Pathway 2.6 or RCP2.6), 1.1–2.6 °C (RCP4.5), 1.4–3.1 °C (RCP6.0), 2.6–4.8 °C (RCP8.5). As a result, an increase in temperature from 0.3 to 4.8 °C is expected by the end of the twenty-first century. This increase in mean temperatures is expected to be greater in the tropics and subtropics than in the mid-latitudes (high confidence; Kirtman et al. 2013).

Along with the increase in temperature, global warming is associated with changes in the intensity/uncertainties of precipitation. Future climate projections show that precipitation could increase or decrease depending on the latitude of the area. These results corroborate those of Bates et al. (2008), who reported that over the past century, precipitation has mainly increased over northern high-latitude lands, while notable decreases have occurred in recent years from 10°S to 30°N. This observed effect is expected to continue in the coming years as climate model simulations predict an increase in precipitation in high latitudes and parts of the tropics and decreases in some subtropics and lower mid-latitudes by the end of the twenty-first century (Collins et al. 2013).

Arid and semi-arid areas, especially those in Asia and Africa, are very vulnerable to climate change. Temperature and precipitation variability is expected to increase in these areas. In Africa, it is predicted that the temperature will rise faster compared to the rest of the world. This increase could exceed 2 °C by the middle of the twenty-first century and 4 °C by the end of the twenty-first century (Niang et al. 2014; Adhikari et al. 2015; Djanaguiraman et al. 2018) and will lead to reduced crop yields (Fischer et al. 2005a, b; Howden et al. 2007; Liu et al. 2016). Regarding rainfall, significant negative trends are observed in West Africa and the Sahel. The Sahelian region of West Africa experienced a decrease in rainfall between the late 1950s and the late 1980s (Dai et al. 2004). Recently, Sultan and Gaetani (2016) predicted an increase in extreme events and increased variability in precipitation from year to year. Precipitation projections for the twenty-first century are not spatially homogeneous in West Africa. As a result, there is a large dispersion in the representation of rainfall from one regional model to another, both on a seasonal and intra-seasonal scale (Salack et al. 2012b). Beyond the changes in the total amount of rain, precipitation patterns are also predicted to change in the Sahelian area. For instance, Salack et al. (2011) identified two agro-climatic zones where rainfall breaks frequently occur at the start and end of the rainy season (Salack et al. 2012a). These rainfall breaks accompanied by heat stress are likely to affect crop yield. These factors constitute a major agronomic problem that contributes to severe yield losses of up to 10% for pearl millet in arid and semi-arid areas (South Asia and Africa) (Knox et al. 2012). It is therefore urgent to better understand heat and drought stress mechanisms in pearl millet to develop new agricultural practices and varieties adapted to these stresses and ensure food and nutritional security in sub-Saharan Africa.

10.2 Effects of Drought and Heat on Pearl Millet and Annual Plant Growth and Development

Pearl millet [*Pennisetum glaucum* (L.) R. Br.] is an important food crop grown under hot and dry conditions in arid and semi-arid regions of Africa and Asia (Arya and Yadav 2009; Ullah et al. 2017). Since these areas are characterized by low and irregular rainfall, high temperatures, and low soil fertility, they constitute the main constraints for its production. These stresses often occur simultaneously, making it very difficult to separate the effects of each on plants. However, the combined stresses have a negative impact on plant growth and productivity, which is more pronounced than the individual impacts (Craufurd and Peacock 1993; Prasad et al. 2008; Dreesen et al. 2012).

10.2.1 Drought Effects

Drought stress is one of the major constraints limiting crop production worldwide (Forster et al. 2004; Khan et al. 2010). Drought can be defined as a deficit of adequate moisture necessary for normal plant growth in the complete life cycle (Zhu 2002). Plants are subject to drought when the stock of water in the soil is limited, and the vapor pressure deficit is very low (Anjum et al. 2011b). It has a considerable influence on plant growth and development. Pearl millet is sensitive to drought stress at the vegetative and reproductive stages (Vadez et al. 2012). The effects of drought stress depend on the degree, duration, stage of crop development, and tolerance level of the species. Drought stress during the growth phase affects cell division and elongation, the main processes involved in plant growth. Due to the reduction in turgor pressure, cell elongation appears to be one of the most drought-sensitive processes. Even though cell division is less sensitive than cell elongation, it can be affected by mild drought stress. Tardieu et al. (2000) reported that soil moisture deficit resulted in a reduction in the length of the cell division zone and the relative division rate. According to Alves and Setter (2004), the reduction in leaf area was caused largely by development delay and a reduction in cell division in the youngest meristematic leaves.

However, the effects of drought on the leaves depend on the intensity of the stress. A slight drought causes a reduction in the rate of expansion, number, and size of leaves, while severe stress decreases the rate of leaf elongation, which can even cause leaf growth to stop. The most common negative effect of drought stress on crop plants is the reduction in growth. This has been reported in pearl millet by many authors (Muchow 1989; Winkel et al. 1997; Aparna et al. 2014; Kholová et al. 2016). Drought reduces leaf size, stem extension, and root proliferation, disrupts plant–water relationships, and reduces water use efficiency (Farooq et al. 2009; Anjum et al. 2011b). These results agree with those of Khan et al. (2001) and Anjum et al. (2011a), who reported a significant decrease in the growth of maize under drought in terms of plant height, stem diameter, leaf area, and plant biomass.

In addition, drought has been shown to alter physiological processes (Radhouane 2008, 2013; Pinheiro and Chaves 2010; Anjum et al. 2011a; Ghatak et al. 2016). Photosynthesis, transpiration rate, stomatal conductance, water use efficiency, intrinsic water use efficiency, and intercellular CO_2 have been reduced by drought (Anjum et al. 2011a). In millet, severe drought stress leads to a reduction in photosynthesis (Ashraf et al. 2001; Golombek 2003; Radhouane 2009). This reduction has also been reported in C4 cereals such as maize (Boyer and Westgate 2004) and sorghum (Ogbaga et al. 2014). One or more steps in the photosynthetic process can be affected by water deficit, such as the diffusion of CO_2 through the stomata and into intercellular spaces (Flexas et al. 2006). For Slatyer (1973), almost all the decrease in photosynthesis must be attributed to stomatal closure. On the other hand, according to Bois (1993) and Golombek (2003), the decrease in photosynthesis of pearl millet is due jointly to stomatal resistance and non-stomatal changes. The imposition of drought stress also resulted in a decrease in chlorophyll content (Ghatak et al. 2021), which could be related to the increase in the activity of the enzyme chlorophyllase (Ashraf et al. 1994).

In terms of production, drought stress strongly affects yield during the grain-filling period (Aparna et al. 2014). In sorghum, the yield can be reduced by more than 36% and 55% when water stress occurs in the vegetative and reproductive phases respectively (Assefa et al. 2010). Stress reduces grain size and weight, resulting in a reduction in grain yield (Arya et al. 2010). Many studies corroborate this result (Barnabás et al. 2008; Alqudah et al. 2011; Aparna et al. 2014; Debieu et al. 2018). A severe reduction in panicle filling occurs under drought (Winkel et al. 2001) due to a reduction in the assimilate partitioning and activities of sucrose and starch synthesis enzymes (Farooq et al. 2009; Anjum et al. 2011b). However, according to Aparna et al. (2014), the decrease in grain yield depends more on the number than the size of the seeds. Therefore, it is due to the combined effect of a reduction in the number of panicles (productive tillers) and seeds. In pearl millet, a strong correlation was found between grain yield and grain number (Bidinger and Raju 2000). In wheat, drought stress reduced yield following tiller abortion and a lower grain number per spike (Izanloo et al. 2008). Thus, drought most often occurs during the vegetative and reproductive stages of pearl millet, causing drastic effects on growth and productivity.

10.2.2 Heat Effects

Heat stress (increase in air temperatures above the optimum) is an agricultural problem in many parts of the world (Wahid et al. 2007). It is a key determinant of crop growth and productivity (Al-Khatib and Paulsen 1999), whose adverse effects on cereals vary with the timing, duration, and intensity (sternness) of stress (Barnabás et al. 2008; Fahad et al. 2016a). In Africa and India, soil temperatures generally exceed 45 °C and sometimes reach 60 °C (Yadav et al. 2010). This explains the important place occupied by pearl millet cultivation in these areas because the optimum temperature for its development fluctuates between 33 and 34 °C (Ashraf and Hafeez 2004).

Indeed, pearl millet is the most heat-tolerant cereal crop and needs high temperatures to grow. This has been illustrated by numerous works showing that the growth of pearl millet is optimal up to a temperature of 35 °C (Arya et al. 2014), but beyond that, inhibition of the normal growth process is noted (Ashraf and Hafeez 2004; Yadav et al. 2010, 2016; Arya et al. 2014; Djanaguiraman et al. 2018).

A temperature above the optimum can delay or prevent germination. This germination delay was reported in pearl millet cultivars subjected to supra-optimal temperatures compared to controls (Khalifa and Ong 1990). In pearl millet and maize, heat stress has been reported to cause a decrease in the final percentage of germinated seeds and the germination rate (Ashraf and Hafeez 2004). Yadav et al. (2016) showed that increasing the temperature can decrease the germination rate or even inhibit germination depending on the tolerance of the genotype, which agrees with the results of Wahid et al. (2007).

In later stages, heat stress can also negatively affect the vegetative growth of plants. It induces a reduction in shoot dry mass, relative growth rate, and net assimilation rate (Ashraf and Hafeez 2004; Wahid 2007). On the roots, temperatures above optimal lead to a decrease in primary root length, number of lateral roots, and their angle of emergence (Calleja-Cabrera et al. 2020). In addition, it causes a decrease in shoots or root's water status, root's hydraulic conductivity, or leaf stomatal conductance, which only occurs when the stress is prolonged (Heckathorn et al. 2013).

Reduction in photosynthetic activities by higher temperature has been reported by numerous studies (Al-Khatib and Paulsen 1999; Crafts-Brandner and Salvucci 2000, 2002; Prasad et al. 2004; Fahad et al. 2016b; Yadav et al. 2016). This reduction in photosynthesis was attributed to the damage to chlorophyll pigments, a decline in leaf nitrogen contents, blockage of PSII reaction center and electron flow decreased quantum efficiency (Fv/Fm), and down-regulation of PSII photochemistry (Fahad et al. 2016b, 2017). Heat stress decreases the activation and activity of Rubisco (Prasad et al. 2004) by inhibiting the activity of the enzyme Rubisco activase (Crafts-Brandner and Salvucci 2000, 2002). Heat also reduces the photochemical efficiency of photosystem II (PSII), which appears to be the most heat-sensitive photosynthetic component (Al-Khatib and Paulsen 1999).

On the other hand, despite the multiple negative effects of heat stress at the vegetative stage, plants appear more sensitive at the reproductive stages (Farooq et al. 2011; Prasad et al. 2017). Recently, much work has been done on cereals to identify the growth stage(s) most sensitive to heat stress during reproductive development (Prasad and Djanaguiraman 2014; Prasad et al. 2015; Djanaguiraman et al. 2018). Two periods were identified, ranging from 10 to 12 days and 2 to 0 days before anthesis in pearl millet (Djanaguiraman et al. 2018), 8 to 6 days and 2 to 0 days before anthesis in wheat (Prasad and Djanaguiraman 2014), 10 to 5 days before anthesis and 5 days before and 5 days after anthesis in sorghum (Prasad et al. 2015). During these periods, heat stress causes maximum decreases in pollen germination percentage, seed number (Djanaguiraman et al. 2018), and floret fertility (Prasad and Djanaguiraman 2014; Prasad et al. 2015).

In pearl millet, high temperatures (>40 °C) often coincide with the flowering and grain-filling stages (Gupta et al. 2015). Heat stress during these stages leads to a decrease in grain number and weight, leading to poor crop yield and quality (Bita and Gerats 2013; Djanaguiraman et al. 2018). According to Sultan et al. (2013), raising the temperature to 6 °C would lead to a 41% reduction in pearl millet yield. Several studies have reported a decrease in the yield of pearl millet and other cereals under heat stress (Gupta et al. 2015; Prasad et al. 2017; Djanaguiraman et al. 2018; Qaseem et al. 2019; Jagadish 2020).

According to Fahad et al. (2016a, b), the decrease in pollen germination mainly results from the retention of pollen in dehiscent anthers. However, an increase in the content of reactive oxygen species and a decrease in the activity of antioxidant enzymes in pollen and pistils have been reported in millet (Djanaguiraman et al. 2018). Pistils were more sensitive than pollen grains because they had relatively higher reactive oxygen species and lower antioxidant enzyme activity. Increased production of reactive oxygen species under these conditions may be responsible for decreased germination and pollen viability in sorghum (Prasad and Djanaguiraman 2011).

Heat stress also causes a decrease in spikelet fertility due to a decrease in the production and number of pollens on the stigma (Prasad et al. 2006a) and inhibition of panicle emergence (Prasad et al. 2006b). This decrease in spikelet fertility resulted in fewer filled grains, lower grain weight per panicle, and lower harvest index. According to Gupta et al. (2015), heat stress causes reproductive sterility in pearl millet leading to a drastic reduction in grain yield. In addition, the increase in temperature promotes a higher rate of evapotranspiration, which ultimately reduces soil moisture and available water needed to fill the grains.

10.2.3 Heat and Drought Combined Effects

In arid and semi-arid areas, drought and heat stress often occur simultaneously (Shah and Paulsen 2003; Barnabás et al. 2008). This effect occurs because of a negative correlation between temperature and precipitation on inter-annual scales because dry conditions favor more sunshine and less evaporative cooling (Trenberth and Shea 2005; Zscheischler and Seneviratne 2017). However, few studies examining the impact of the combined effects of both stresses on crops, let alone pearl millet, have been carried out. The few studies carried out on this subject have shown that the combined effects on growth and productivity considerably exceed the simple effects (Savin and Nicolas 1996; Prasad et al. 2008; Dreesen et al. 2012; Qaseem et al. 2019).

Drought and heat stresses, by their intensity and duration, can influence the growth and development of plants. At the leaf scale, the distribution of the relative elongation rate was independently affected by these stresses, which had quasi-additive effects (Tardieu et al. 2000). Additionally, leaf elongation rates were positively correlated with leaf temperatures and negatively with vapor pressure deficit and pre-dawn leaf water potential (Welcker et al. 2007). This shows the strong relationship that exists between leaf elongation rates and various physiological components that can be indicative of drought and heat stress. In terms of growth, drought, and heat affect stem growth and plant height (Katerji et al. 1994; Winkel et al. 1997; Prasad et al. 2006b).

Similarly, these stresses can also impact the transition and duration of the developmental stage. Cooper et al. (2009) reported a reduction in the length of the growing period of plants in dry tropical regions. This reduction is the result of both the rapid development of the leaf canopy and an increase in the overall growth rate of the crop stimulated by heat stress. Higher temperature results in faster development, and therefore shorter, growth phase duration.

Heat stress combined with drought can cause stomatal closure leading to an increase in leaf temperature (Rizhsky et al. 2002), unlike the simple effect of heat stress, which promotes the opening of the stomata, thus leading to cooling of the leaves by transpiration. Consequently, the water status of the plant is partly linked to the temperature, which affects several parameters of the plant. According to Shah and Paulsen (2003), the interactions between stresses were pronounced, and the consequences of drought were more severe at high temperatures than at low temperatures on all physiological parameters. Barnabás et al. (2008) confirm these statements by asserting that the synchronization of the two results in even greater severity of drought stress. Thus, heat increases the intensity of the drought by causing the soil to dry faster. This effect results from the increase in the vapor pressure deficit of the air, which favors a greater demand for evapotranspiration. In addition to warming, the indirect effect of heat on water evapotranspiration from the soil can have a great impact on plants.

On the other hand, antagonistic interactions between heat and drought have been reported on net photosynthesis. In wheat, Lu and Zhang (1999) asserted that drought stress increases the tolerance of PSII to heat stress. However, many authors have argued that drought tends to dramatically exacerbate the effects of heat stress on plant growth and photosynthesis (Shah and Paulsen 2003; Xu and Zhou 2005a, b, 2006).

The effect of drought and heat stresses on root growth depends on the intensity of the stress. Moderate stress results in greater root growth due to increased distribution of carbohydrates to the roots and greater exploration of the soil caused by drought and heat stress, respectively. On the other hand, severe stress leads to a reduction in the number, length, and diameter of the roots, which becomes more important when drought and heat stress are associated (Prasad et al. 2008).

When these constraints occur during grain development, they cause significant yield losses in cereals (Chaves et al. 2003; Bai et al. 2004; Prasad et al. 2008). Knowing that starch represents a major part of the dry weight of cereals, this reduction is due to a decrease in its accumulation. On the other hand, data on the possible variation of this trait in pearl millet are limited.

10.2.3.1 Mechanisms of Adaptation to Climate Variability (Drought and Heat) in Pearl Millet

Faced with environmental constraints, plants must be able to react and adapt to increase their chance of survival. Thus, they are developing different strategies to adapt and resist drought and heat stresses. These strategies are characterized by a strong ability to set up biochemical, molecular, and physiological responses which influence various cellular processes in the plant. Drought survival mechanisms of plants are like those used to cope with heat stress (Wu et al. 2018). Escape, avoidance, and tolerance mechanisms have long been considered important strategies for drought adaptation (Chaves et al. 2003). These mechanisms have also been reported by Kooyers (2015) and Li et al. (2017).

Drought escape allows some plants to cope with stress by completing their full development cycle before the water deficit sets in the soil (Annerose 1990). Drought avoidance, on the other hand, is the ability of plants to maintain high levels of water potential in their tissues by reducing water loss or improving water uptake (Ludlow and Muchow 1990). Escape and avoidance strategies may be the most effective for survival and reproduction when drought stress is mild-tomoderate (Kooyers 2015).

Mechanisms of dehydration avoidance include morphological and functional modifications such as leaf area size, leaf rolling, stomatal conductance, and osmotic adjustment (Blum 2011; Kadioglu et al. 2012). However, when drought stress becomes severe, plants must be able to rely on tolerance strategies to avoid desiccation. Thus, drought tolerance appears to be the capacity of plants to resist water deficit while maintaining appropriate physiological activities (Xiong et al. 2006). Therefore, the response of plants to drought depends on the species, duration of the drought as well as the timing of application (Sanchez et al. 2002; Pinheiro and Chaves 2011).

Due to its drought and heat tolerance (Arya et al. 2010), pearl millet is an ideal model for studying the heat and drought resistance mechanism of cereal crops.

Leaf Rolling and Stomatal Conductance

In plants, many changes occur in the leaf, both in structure and morphology, in response to drought and heat stress. Leaf rolling is an abiotic stress avoidance mechanism (Kadioglu and Terzi 2007; Kadioglu et al. 2012). It occurs when plants are under stress and is caused by folding in the midrib of upper leaves and changes in their leaf's orientations (Kusaka et al. 2005b). This leads to a reduction in leaf temperature via a decrease in incident radiation (O'toole et al. 1979; Heckathorn and DeLucia 1991) and thus offers protection against the effects of excessive radiation (Kadioglu and Terzi 2007). Leaf rolling effectively reduces light interception, transpiration, and leaf dehydration (Kadioglu and Terzi 2007).

However, reduced transpiration has often been associated with leaf senescence which results in an efficient and rapid decrease in leaf area. It can extend the duration of soil water availability by reducing the plant's water requirements and losses. According to Wallace et al. (1993) and Soegaard and Boegh (1995), senescence constitutes the main mechanism responsible for the reduction of transpiration. However, it can be a limiting factor for the accumulation of crop biomass and hence grain yield due to its irreversible effect. Unlike senescence, the reversible nature of leaf rolling provides flexibility when drought is temporary and intermittent.

However, sorghum and pearl millet studies have shown that leaf rolling did not occur until after stomatal closure (Blum and Sullivan 1986; Ludlow and Muchow 1990). Therefore, the immediate response of plants under drought stress is the stomatal closure to prevent water loss through transpiration (Cornic and Massacci 1996; Assmann et al. 2000; Ghatak et al. 2016; Buckley 2019). Through the transpiration stream, drought induces root-to-leaf signaling promoted by soil drying, which causes stomatal closure (Farooq et al. 2009; Anjum et al. 2011b). This stomatal closure is higher and faster in tolerant genotypes (Ghatak et al. 2021) and appears to be more effective in reducing water loss than leaf rolling. It has been reported that the decrease in water use by stomata is greater pre-anthesis than post-anthesis because of the ontogenetic decline in the range of stomatal conductance (Winkel et al. 2001).

Under drought conditions, plants tend to have lower stomatal conductance, which decreases as drought stress increases (Ghatak et al. 2021). This decrease in conductance allows the plant to conserve water and maintain an adequate water status of the leaves, hence the close relationship between them. A significant correlation between stomatal conductance and leaf xylem water potential has been reported by Matsuura et al. (1996). However, a low stomatal conductance could be partly related to a difference in stomatal density. According to Slama (2002), the increase in the number of stomata per unit area could be one of the factors of resistance to water deficit in cereals if accompanied by appropriate photosynthesis activity. It can decrease water loss and increase the net uptake of CO_2, which allows the plant to maintain photosynthesis. Increased stomatal density can also affect crop yield. The variety of durum wheat with the highest yield and the largest kernels has a higher stomatal density at the beard and flag leaf (Slama 2002). On the other hand, according to Kholová et al. (2010), stomatal regulation is more important than stomatal density in regulating the loss of water in pearl millet. Nevertheless, stomatal conductance and leaf rolling have been shown to be reliable physiological indicators of drought tolerance in plants (Kadioglu and Terzi 2007).

Heat stress, often associated with a high vapor pressure deficit (VPD), also causes leaf rolling in plants (Omarova et al. 1995) and stomatal closure (Maroco et al. 1997). It causes a change in the temperature of the leaves, which can be an important factor in controlling the leaf water status under stress. The study by Kadioglu and Terzi (2007) showed that leaf rolling was linearly correlated with osmotic potential and leaf temperature. Thus, the physiological role of leaf rolling was reported as the maintenance of adaptive potential by increasing the efficiency of water metabolism in wheat flag leaves under heat stress (Sarieva et al. 2010). In fact, this is consistent with the fact that leaf rolling results in more efficient use of water during photosynthesis (Kadioglu and Terzi 2007). In addition, a reduction in leaf area was noted in pearl millet and sorghum under high VPD (Choudhary et al. 2020). According to Choudhary et al. (2020), water conservation when increasing VPD depends primarily on reduced leaf area and somewhat on transpiration restriction in these two crops. This restriction of transpiration has also been reported by Kholová et al. (2010). However, partial stomatal closure is achieved by limiting the transpiration rate under conditions of high VPD. Stomatal sensitivity to VPD was correlated with the hydraulic conductance of leaves relative to the total leaf area (Ocheltree et al. 2014). However, reducing the transpiration of pearl millet under a high vapor pressure deficit has been proposed to be beneficial for crop yield under such conditions (Kholová et al. 2010).

Root Characteristics

Pearl millet is one of the most abiotic stress-tolerant cereal crops in part due to its strong root system. Rapid root growth at depth can offer a chance for survival in harsh conditions as water uptake requests deep roots because of the quick water drainage of the sandy soils where millet is usually grown. The advantage of deeper root systems was demonstrated in pearl millet (Faye et al. 2019), sorghum (Chopart et al. 2008b), wheat (Kirkegaard et al. 2007; Christopher et al. 2008), maize (Sinclair and Muchow 2001; Hund et al. 2009), and rice (Manschadi et al. 2010; Wasson et al. 2012). In pearl millet, the root system is made up of several types of roots, namely the primary root (emerges from the seed and the mesocotyl connecting the seed and the base of the stem), crown roots (emerges from the base of the stem), lateral roots (appear on primary or crown roots), and secondary roots (ramifications of lateral roots) (Passot et al. 2016). Crown roots form most of the root system, even though the primary root characterizes the root system at the start of pearl millet growth. This primary root will regress from 1 month after sowing (within 2 months after germination) (Maiti and Bidinger 1981; Passot 2016). Passot et al. (2016) showed that the number of central metaxylem vessels constitutes the major difference between the different root types.

Different lateral root types have been reported in pearl millet, rice, and maize (Passot et al. 2016, 2018; Hochholdinger and Tuberosa 2009; Rebouillat et al. 2009). The different types of lateral roots in cereals have been identified through anatomical studies of roots, often based on traits such as root diameters and vascularization (Varney et al. 1991; Watt et al. 2008; Henry et al. 2016; Passot et al. 2016). Recently, Passot et al. (2018) were able to classify roots based on their growth rate profiles. The study found three types of lateral roots with similar characteristics in pearl millet and maize. This revealed three types of lateral roots with similar characteristics in pearl millet and maize.

Under drought stress, pearl millet root growth is reoriented toward deeper soil layers that retain more water. Several studies have argued that root growth orientation was only dependent on soil depth (Chopart and Siband 1999; Chopart et al. 2008a, b; Faye et al. 2019). This dependence differed between thick roots and fine roots. Thick root growth was horizontal in shallow soils and became more and more vertical with increased depth, unlike the growth orientation of fine roots, which was only marginally dependent on soil depth (Faye et al. 2019). This result agrees with those found in sugarcane and sorghum (Chopart et al. 2008a, b). They claimed that fine roots appeared isotropic when thick roots were horizontal near the surface and gradually became vertical in deeper horizons. According to Passot et al. (2016), the thick roots correspond to the seminal or crown roots, while the fine ones probably correspond to the different types of laterals.

Increasing water uptake is a way of avoiding stress. It takes place via the roots, and its transport from the soil to the xylem vessels uses two pathways: the apoplastic pathway and the cell-to-cell pathway, which summarizes the transcellular and the symplastic pathways (Steudle 2001). The type of path depends mainly on environmental conditions. Water flows through the apoplastic path under non-stressful conditions due to hydrostatic forces while it flows through the cell-to-cell path under stressful conditions due to osmotic forces. Aquaporins, water channels present in cell membranes, enable cell-to-cell water transport (Prado and Maurel 2013; Chaumont and Tyerman 2014). They are involved in the physiology of plant growth (Maurel et al. 2015), thus influencing the hydraulics, transpiration, and water conservation of the soil. Their importance has been demonstrated in pearl millet where they contribute up to 84% to the hydraulic conductivity of roots (Grondin et al. 2020). Interestingly, aquaporins contribution was higher in root hydraulic conductivity for a pearl millet line with lower water use efficiency (Grondin et al. 2020). Aquaporins are also well known for their response to drought stress (Alexandersson et al. 2005; Aroca et al. 2012; Grondin et al. 2016). Several types of aquaporin families have been identified with different functions. Many of them are involved in the regulation of water uptake by roots under drought conditions (Aroca et al. 2012).

On the other hand, root length is an important trait for tolerance to drought stress. Root length increased in all genotypes under drought stress (Ghatak et al. 2021). This increase in the root system has been reported by many authors (Kusaka et al. 2005a; Ghatak et al. 2016). It has been reported in cereals that a deep root system allows water uptake from deep layers of the soil in drought-stressed environments (Kondo et al. 2000; Kashiwagi et al. 2006; Manschadi et al. 2010; Wasson et al. 2012; Steele et al. 2013; Wasaya et al. 2018; Faye et al. 2019). In sorghum and pearl millet, there is a positive correlation between drought tolerance and root length (Matsuura et al. 1996), even though the roots of pearl millet were found to be longer than those of sorghum (Rostamza et al. 2013). In addition to length, root density can be a factor in drought tolerance. In a study on maize, Zhan et al. (2015) argue that reduced lateral root density improves drought tolerance. This reduction in density is associated with deeper rooting resulting in lower root length density for thick roots than for fine roots (Chopart et al. 2008b).

Therefore, fine root diameter, specific root length, specific root surface area, root angle, and root length density are considered useful traits for improving plant productivity under drought conditions (Wasaya et al. 2018). Increased root airspaces (aerenchyma) and root xylem diameter have also been linked with greater yield under drought conditions in maize (Chimungu et al. 2015) and the conservation of water resources to laid grain filling in wheat (Richards and Passioura 1989) respectively. Increased root growth can also be helpful for heat stress conditions. It allows plants to maintain their water potential despite significant transpiration. Under these conditions, plants develop strategies to restrict water loss. The reduction of transpiration in pearl millet under high vapor pressure deficit (VPD) was associated with aquaporins function (Reddy et al. 2017b). VPD-insensitive genotypes increased their transpiration rate, which may be since they used more symplastically mediated water transport (aquaporins) pathways than VPD-sensitive genotypes (Reddy et al. 2017b). Thus, water uptake in warmer soil appears to be positively correlated with aquaporin activity in wheat (Carvajal et al. 1996). In mature maize plants, increasing temperature slows lateral root growth to promote the development of long axial roots to reach water in deeper soil layers (Hund et al. 2008). Therefore, leaf and roots hydraulic conductance plays an important role in the response of plants to evaporative demand (Ocheltree et al. 2014).

Osmotic Adjustment

Osmotic adjustment is considered a major drought adaptation mechanism (Kusaka et al. 2005b; Izanloo et al. 2008; Sanders and Arndt 2012; Blum 2017). It allows the maintenance of water absorption and cell turgor pressure thanks to the accumulation of solutes. As a result, increasing the number of osmotically active substances in the cell leads to a more negative osmotic potential, which can improve the degree of cell hydration, maintaining turgor in leaf tissue and other metabolically active cells (Sanders and Arndt 2012). Osmotically active substances can be either organic solutes (amino acids, glycerol, sugars, and other low molecular weight metabolites) or inorganic ions (Na^+, K^+, Ca^{2+}, and Cl^-). Several studies have reported the important role of these solutes in tolerance to abiotic stresses (Chaves and Oliveira 2004; Ashraf and Foolad 2007; Chen and Jiang 2010; Verslues and Sharma 2010). Depending on the adjustment capacity, the types of solutes accumulated, and their relative contribution to lowering osmotic potential, the osmotic adjustment may vary between species and genotypes (Chen and Jiang 2010).

Variation in the osmotic adjustment of cultivars in response to drought has been reported in many crop plants, including pearl millet. Local varieties from more arid areas exhibited a greater capacity for osmotic adjustment (Blum and Sullivan 1986). However, the latter depends not only on the stage of development of the plant (Chimenti et al. 2006) but also on the degree and duration of the water deficit (Shangguan et al. 1999; Kusaka et al. 2005a, b; Nio et al. 2011). Kusaka et al. (2005b) studied the contribution of several solutes to the osmotic adjustment of two pearl millet cultivars, one susceptible (IP8949) and the other tolerant (IP8210) to drought stress. They reported that, for both accessions, the stem exhibited higher osmotic adjustment than the younger and expanded leaves, while their decrease in relative water content was different. In addition, an increase in the concentration of organic components (sucrose, glucose, proline, and QAC) was noted in both accessions in response to drought stress. This has also been reported in durum wheat, where drought stress increased sugar and proline concentrations and decreased nitrate levels (Bajji et al. 2001). Sugars were the main solutes that contributed to osmotic adjustment, especially in growing leaves, followed by proline and quaternary ammonium compounds (Bajji et al. 2001). However, in pearl millet, accumulation of proline was greater (more than four times) in young leaves of the tolerant genotype than in the susceptible genotype (Kusaka et al. 2005b). This increase has also been reported in maize up to 10 days after stress application and declines when stress becomes severe (Anjum et al. 2011a). A strong accumulation of proline increases the cell solute concentration, resulting in increased water potential in the tissues and decreased cellular damage. As a result, it constitutes the first response of plants exposed to drought stress and contributes to the immediate recovery of plants after stress.

In addition to proline, glycine betaine, and soluble sugars contribute to osmotic adjustment and stress adaptation in pearl millet. However, the accumulation of organic components remained lower than that of K^+ and NO_3^- (Kusaka et al. 2005b). The concentration of K^+ increased both in the cell sap of the leaves and the stems of the two accessions at the onset of stress. The accumulation of K^+ in the cell sap reached a very high level and was relative to decreasing relative water content. This proves the role of inorganic compounds in contradiction to the conclusion of Bajji et al. (2001). Fischer et al. (2005a, b) found a positive correlation between osmotic adjustment and grain yield under moisture deficit. Likewise, in a critical review examining 26 published studies in which osmotic adjustment and yield were measured under drought stress in variable genotypes from 12 crops, Blum (2017) reported a positive and significant association between osmotic adjustment and performance in 24 published studies. In pearl millet, drought tolerance was more correlated with osmotic adjustment capacity than total root length under severe drought stress (Kusaka et al. 2005a).

Like drought stress, heat stress also leads to changes in the accumulation of compatible osmolytes (Sakamoto and Murata 2002). The early synthesis of these osmolytes (proline, glycine betaine, or soluble sugars) compensates for the effect of the decrease in leaf water potential, which appears as an immediate response to heat. By maintaining the cell water balance and membrane stability and buffering the cellular redox potential, the accumulation of osmolytes regulates osmotic activities and protects cellular structures from high temperatures (Farooq et al. 2008). The role of glycine betaine in photosynthesis in plants under heat stress has been reported by Allakhverdiev et al. (2008). The activation of Rubisco is maintained by the production of glycine betaine in the chloroplasts by sequestering Rubisco activase near thylakoids and preventing its thermal inactivation (Allakhverdiev et al. 2008). However, there is a difference between species in their ability to synthesize glycine betaine under heat-stress conditions (Ashraf and Foolad 2007). In maize and sugarcane, a strong accumulation of glycine betaine has been reported in response to elevated temperatures (Quan et al. 2004; Wahid and Close 2007). At the same time, the increase in temperatures causes the accumulation of proline and soluble carbohydrates in wheat while the levels of valuable proteins are reduced (Qaseem et al. 2019). Thus, despite the lack of studies on pearl millet, work on other crops has shown the role of osmotic adjustment in tolerance to heat stress.

Transpiration Efficiency

In times of stress, plants need to develop mechanisms to conserve soil water or maximize their water use efficiency to alleviate the effect of stress. Several studies have been carried out in recent years to determine the water use efficiency of crops (Vadez et al. 2011, 2013a, 2014, 2021; Schittenhelm and Schroetter 2014; van Oosterom et al. 2021). Water use efficiency appears to be an important criterion for evaluating the water supply. It provides a simple and quick measure of how the available water can be converted into biomass and grain (Sekhon et al. 2010). Defined as the accumulation of biomass per unit of water transpired (Xin et al. 2009), transpiration efficiency can also be understood at the leaf level as the intrinsic water-use efficiency. It is the ratio of the instantaneous rates of CO_2 assimilation and transpiration at the level of the stomata (Condon et al. 2002). Thus, increased biomass or photosynthesis, decreased transpiration, or a combination of both can result in reduced water use reflecting higher water use efficiency of plants. However, Xin et al. (2009) reported that transpiration efficiency based on biomass production was strongly correlated with increased biomass accumulation rather than reduced water use.

In a recent study comparing the transpiration efficiency of C4 cereals, Vadez et al. (2021) found that transpiration efficiency in maize was higher than in millet and somewhat higher than in sorghum (Fig. 1.12). This difference between species could be explained by differences in the ability to restrict transpiration under high VPD (Vadez et al. 2014). Likewise, Choudhary et al. (2020) reported that maize conserves water by limiting transpiration during increased VPD and under higher soil moisture than sorghum and pearl millet. In addition, the transpiration efficiency seems to depend on the type of soil. It was higher in high clay than in sandy soil under a high VPD (Vadez et al. 2021), while the differences in transpiration efficiency between maize and sorghum were not visible in Alfisol and sandy soil.

However, there is variation in transpiration efficiency within species that has been demonstrated for C4 plant species (Mortlock and Hammer 2000; Xin et al. 2009; Vadez et al. 2011, 2014, 2021). Transpiration efficiency was higher under stress conditions (Mortlock and Hammer 2000). Sorghum genotypes with low internal CO_2 concentration and improved photosynthetic capacity may be a factor explaining the high transpiration efficiency in some lines (Xin et al. 2009). Under water deficit conditions, water use efficiency was also increased in sweet sorghum, while for maize, it was reduced (Zegada-Lizarazu et al. 2012). Bhattarai et al. (2020) confirm this result by stating that the water use efficiency was highest for sorghum cultivars, followed by millet and maize. According to Blum (2005), water use efficiency is often associated with drought tolerance and improved crop yields under stress conditions. Thus, genotypes with higher water use efficiency have greater survivability.

However, water extraction also seems to play an important role during grain filling. Differences in yield under terminal drought in sorghum have been reported to be determined by TE, followed by water extraction (Vadez et al. 2013b). Although the amount of water extracted by tolerant and susceptible genotypes was similar under drought stress, tolerant genotypes extracted less water before anthesis and more water after anthesis. This explains the lower yield of sensitive genotypes than that of tolerant lines under drought stress (Vadez et al. 2013a). Thus, the early conservation of water during pre-anthesis increases the yield of pearl millet during terminal drought (Kholová and Vadez 2013; Kholová et al. 2010). The drought tolerance of pearl millet is explained by the higher water use efficiency. Therefore, improving transpiration efficiency can effectively increase pearl millet yield in arid and semi-arid regions.

10.3 Molecular Basis of Stress Tolerance

Pearl millet is an important model for studying the physiological and molecular mechanisms of drought tolerance. Yet, compared to other cereals, the molecular mechanisms of drought stress tolerance in pearl millet remain elusive. Recently, genome-wide association studies (GWAS) have detected some genes associated with the domestication and differentiation of local millet varieties that have adapted to climatic conditions. These genes are related to heading date and plant height (Lakis et al. 2012), flowering time, morphological character, and yield (Saïdou et al. 2009; Vigouroux et al. 2011; Clotault et al. 2012; Diack et al. 2020), fitness under irregular climatic conditions (Ousseini et al. 2017), biomass and stay green (Debieu et al. 2018). Molecular responses involve a set of genes and signal transduction pathways that are highly regulated. The tolerance mechanisms begin with the detection of the stress, which causes a series of signal molecules transported in the leaves via the xylem.

Among the signals, abscisic acid (ABA) is a very important drought response pathway. It is a key root-to-shoot signal of drought stress (Xu et al. 2010) even though it is variable among species (Zhang et al. 2005; Jia and Zhang 2008). However, how the root cells detect the moisture state of the soil remains a mystery. Root tissues synthesize ABA upon detection of stress, which induces stomatal closure, thereby reducing transpiration and photosynthesis and allowing plant adaptation to drought conditions (De Ollas et al. 2013). ABA is also involved in regulating aquaporin activity, which contributes to the maintenance of the favorable water status of the plant (Parent et al. 2009; Reddy et al. 2017a). The ABA-dependent signaling response to stress involves different genes and transcription factors (Tuteja 2007). These genes and transcription factors are either involved in ABA biosynthesis or induced by the presence of ABA. For example, under drought conditions, strong expression of the 9-*cis*-epoxycarotenoid dioxygenase gene (NCED) is the first step in ABA biosynthesis (Qin and Zeevaart 1999; Behnam et al. 2013), provides evidence of ABA accumulation. In addition, several transcripts such as the WRKY transcription family (Jaiswal et al. 2018; Chanwala et al. 2020), PYL/PYR, PP2C SnRK2, ABRF DREB2A/2B, AREB1, RD22BP1, NAC, and MYC/MYB family (Tuteja 2007; Bhargava and Sawant 2013; Dudhate et al. 2018; Zhang et al. 2021) are known for their role in phytohormone signaling and response to abiotic stresses in plants. Furthermore, during drought stress, reactive oxygen species (ROS) accumulation increases considerably (Farooq et al. 2009; Alam et al. 2010). ROS form a natural by-product of normal oxygen metabolism and plays an important role in cell signaling. ROS include oxygen ions, free radicals, and peroxides, whose common characteristic is their ability to cause oxidative damage to proteins, DNA, and lipids (Apel and Hirt 2004). They target various organelles, including chloroplasts, mitochondria, and peroxisomes, resulting in premature leaf senescence or plant death (Ma et al. 2013). However, a versatile and cooperative antioxidant system tightly controls improved ROS production. It modulates the intracellular concentration of ROS and defines the redox status of the cell. Superoxide dismutase (SOD), catalase (CAT), glutathione reductase (GR), ascorbate peroxidase (APX), and guaiacol peroxidase (GPOX) constitute the main mechanisms of entrapment (Cruz de Carvalho 2008, Takayuki et al. 2013). An increase in the levels of these antioxidant enzymes (AOX) has been noted in pearl millet in water-deficit stress conditions (Vijayalakshmi et al. 2012). On the other hand, prolonged drought stress leads to the ineffectiveness of the antioxidant system causing cell damage and death. Many genes conferring tolerance to drought stress have been identified in plants. These genes were presented in the study by Kumar et al. (2018) and classified into two groups (Bray 1997; Yamaguchi-Shinozaki and Shinozaki 2005; Kumar et al. 2018). The first group includes functional genes encoding proteins whose catalytic activities are responsible for the protection of cells and organs against stress, regulatory genes encoding proteins necessary for signal transduction and the regulation of expression of genes. The second group comprises several genes and transcription factors responsive to drought, such as the binding gene to elements sensitive to dehydration, aquaporin, abundant proteins of late embryogenesis, and dehydrins (Farooq et al. 2009).

In pearl millet seeds, Ghatak et al. (2016) also identified heat shock proteins (HSP), molecular chaperones, storage proteins, and abundant late embryogenesis (LEA) with increased levels. These proteins help stabilize the folding and conformation of structural proteins and the functionality of enzymes (Wang et al. 2004), hence their protective function.

In recent years, proteomic studies have become increasingly important as proteins are the main drivers of all cellular events. At the proteomic level, several studies have been carried out to understand the effect of drought stress on cereals such as millet (Ghatak et al. 2016, 2021), maize (Riccardi et al. 2004), and wheat (Ford et al. 2011; Komatsu et al. 2014; Ghatak et al. 2021). In pearl millet, a shotgun proteomics approach was used to study protein signatures from different tissues under drought and control conditions (Ghatak et al. 2016). Proteins have been identified and quantified in the root, leaf, and seed tissues. However, there is a pronounced change in the proteome of stressed plants compared to control conditions. Putative drought-sensitive proteins have also been identified in the root (271 proteins), seed (159 proteins), and leaf (292 proteins). Leaf tissue showed the most significant changes, followed by roots and seeds.

A high temperature also triggers important molecular changes in plants. Many transcripts and proteins alter their expression and levels to prevent or reverse the effects of heat on proteins. Under stress, plants synthesize (induce) a set of heat shock proteins (HSPs), unlike in normal conditions where they are almost absent. Based on their molecular weight, HSPs are divided into different families with distinct functions. Numerous studies have shown the role of HSP families in thermotolerance (Gurley 2000; Queitsch et al. 2000; Sun et al. 2002; Hu et al. 2010; Reddy et al. 2010; Nitnavare et al. 2016). Heat stress transcription factors (Hsfs) are the main regulators of heat stress response gene expression (Baniwal et al. 2004; Kotak et al. 2007; Schramm et al. 2008). Since protein aggregation is irreversible, HSPs appear important in the thermo-tolerance reaction and act as molecular chaperones to prevent the denaturation or aggregation of target proteins and facilitate protein refolding (Ahuja et al. 2010; Scharf et al. 2012). In addition, they lead to an improvement in physiological parameters and membrane stability or hydration of cellular structures (Camejo et al. 2005; Ahn and Zimmerman 2006; Wahid and Close 2007). These improvements promote adequate plant growth and development under heat stress. Thus, it has been reported that the expression of genes inducing HSPs may be an important mechanism for increasing tolerance to heat stress (Wahid et al. 2007).

The plant cuticle is a protective layer made of lipids and waxes present on the surface of aerial organs. It forms a barrier against pathogen infection and limits water losses through transpiration. A similar hydrophobic barrier, the Casparian strip, is formed at the periphery of the endodermis's root vascular bundle and is thought to contribute to the limitation of water loss. Recent results suggest that these lipid-made barriers are important components of water potential and heat stress tolerance in pearl millet. Indeed, sequencing of the pearl millet genome revealed an expansion of gene families involved in the biosynthesis and transport of cut-in, suberin, and wax components of the cuticle and Casparian strip (Varshney et al. 2017; Debieu et al. 2017). Moreover, an association genetics study for tolerance to water stress at the vegetative stage in pearl millet led to the identification of four genetic loci associated with increased biomass production under early drought stress. One of these associations contained a gene encoding an enzyme (3-ketoacyl-CoA synthase or KCS) catalyzes the elongation of C24 fatty acids during wax and suberin biosynthesis (Debieu et al. 2018). Altogether, this suggests that wax and fatty acids biosynthesis could be targeted to increase water and heat stress tolerance in dryland cereals.

10.4 Breeding Pearl Millet for Drought and Heat Tolerance

Current climate models' prediction showed that more inter- and intra-annual variability in rainfall and temperature are expected in sub-Saharan Africa (Brown and Lall 2006; Sultan and Gaetani 2016). This situation will increasingly threaten pearl millet production in this part of the world, where this cereal is one of the most important sources of nutrition for more than 90 million people (Anuradha et al. 2017). Therefore, breeders should develop efficient breeding strategies to accelerate the development of improved varieties with better tolerance to drought and heat stresses. During the last decades, several attempts to improve phenotypic screening methods, study the genetic variation in breeding lines, and develop high-yielding and tolerant pearl millet varieties to heat and drought stress using conventional and molecular breeding methods have been accomplished.

Drought is considered the primary abiotic constraint for pearl millet production and is caused by the low and erratic rainfall distribution. In breeding, progress has been made in the development of screening techniques, identification of sources of tolerance, development of early maturing varieties, and identification of QTL linked to drought tolerance (Yadav et al. 2017). Screening of 21 genotypes for osmotic stress tolerance (as a proxy for drought) using PEG 6000 revealed three genotypes (TNBH 0538, TNBH 0642, and ICVM 221) tolerant to moisture stress at germination and early growth stages (Govindaraj et al. 2010). However, pearl millet grain yield is more affected by post-flowering and terminal drought stresses (Kholová and Vadez 2013). To overcome these stresses, drought escape mechanisms have been successfully exploited by targeting early maturity (Yadav and Rai 2013). It has been demonstrated that early flowering pearl millet genotypes with low biomass, few basal tillers and high harvest index can tolerate terminal drought stress (Bidinger et al. 2005). Across West African countries, several landraces characterized by earliness, high grain yield, bold grain, and compact and conical panicles have contributed to the development of pearl millet cultivars adapted to drought-prone areas, including ICTP-8203 and GB-8735 (Wilson et al. 2008). These varieties flower within 40–45 days and mature within less than 75 days, making them suitable cultivars for arid zones. Interestingly, recurrent drought in the 1980s led to selection by farmers in Niger for earlier flowering pearl millet varieties (Vigouroux et al. 2011).

Another important research strategy was using molecular markers to identify genomic regions associated with drought tolerance in pearl millet. Several major QTLs with significant effects on pearl millet grain in terminal drought stress environments were identified and successfully used in marker-assisted selection to improve drought-sensitive pearl millet lines (Yadav et al. 2002, 2004; Serraj et al. 2005; Bidinger et al. 2005). Similarly, potential QTLs for tolerance to water stress during the vegetative phase were identified (Debieu et al. 2018).

For heat tolerance, a good amount of work has been accomplished in breeding. Both field and greenhouse heat tolerance screening techniques have been developed and improved at ICRISAT for pearl millet (Gupta et al. 2015). These techniques were widely used in assessing the effect of heat on a large number of hybrid parental lines, germplasm accessions, and improved varieties across several field locations in India for four consecutive years. The field screenings led to the identification of five hybrid parental lines and a germplasm accession as new sources of resistance to heat tolerance that has been used to develop a high-yielding and heat-tolerant composite variety (Gupta et al. 2016).

10.5 Conclusion

Climate change in arid and semi-arid areas is led by heat and drought, often intermittent or terminal. Drought is due to soil water and vapor pressure deficits. The climate models predicted an increase in climate change effects in the near or long term. Heat and drought are distinctive stresses but often occur simultaneously on crops. They are studied separately or jointly to understand their effects on the plant better. Dry spells and heat stress are the factors constituting major agronomic problems that cause severe yield losses for pearl millet in arid and semi-arid areas. Their effects on pearl millet depend on intensity, phase of occurrence, and duration. However, the flowering period is more vulnerable for pearl millet. Drought and heat affect aerial and root development and physiological and molecular performance. This results in a proportional decrease in pearl millet yield.

In general, drought survival mechanisms of plants are similar to those used to cope with heat stress. Escape, avoidance and tolerance mechanisms have long been considered important strategies for drought adaptation. Due to its drought and heat tolerance, pearl millet is an ideal model for studying the heat and drought resistance mechanism of cereal crops. The avoidance mechanism is commonly used by pearl millet. It responds to water deficit with leaf rolling, senescence, and stomatal conductance, leading to reduced water loss when water soil stock becomes limiting. Also, transpiration restriction under high vapor pressure deficit reducing the transpiration of pearl millet is beneficial for crop yield under terminal drought. Increasing water uptake by root morphological and functional modifications reinforces pearl millet avoidance of heat and drought. Few studies on pearl millet mention osmotic adjustment, the leading mechanism of heat and drought tolerance. These tolerance mechanisms led to changes in the accumulation of compatible osmolytes. The water use efficiency also explains the drought tolerance of pearl millet.

The molecular mechanisms of drought stress tolerance in pearl millet remain elusive. However, gene discovery is ongoing with the detection of genes associated with domestication, genes conferring tolerance to drought stress like gene families involved in the biosynthesis and transport of antioxidant and constitutive molecules. The role of these genes has to be precise and validated to be useful for pearl millet improvement. Progress in pearl millet breeding is based essentially on drought escape mechanisms, with the development of early flowering and maturing varieties. However, molecular breeding started to use QTLs to improve terminal drought-sensitive pearl millet. Heat-tolerant varietal sources are also identified and used in pearl millet improvement.

Progress is made in understanding the molecular bases of pearl millet adaptation to drought and heat. However, pearl millet in research is still far from other major cereals and remains a neglected crop. So, it is urgent to address drought and heat adaptation mechanisms to breed pearl millet varieties for the benefit of the dry-land farmers.

References

Adhikari U, Nejadhashemi AP, Woznicki SA (2015) Climate change and eastern Africa: a review of impact on major crops. Food Energy Secur 4:110–132

Ahn Y, Zimmerman JL (2006) Introduction of the carrot HSP17. 7 into potato (*Solanum tuberosum* L.) enhances cellular membrane stability and tuberization in vitro. Plant Cell Environ 29:95–104

Ahuja I, de Vos RC, Bones AM, Hall RD (2010) Plant molecular stress responses face climate change. Trends Plant Sci 15:664–674

Alam I, Sharmin SA, Kim K-H et al (2010) Proteome analysis of soybean roots subjected to short-term drought stress. Plant and Soil 333:491–505

Alexandersson E, Fraysse L, Sjövall-Larsen S et al (2005) Whole gene family expression and drought stress regulation of aquaporins. Plant Mol Biol 59:469–484

Al-Khatib K, Paulsen GM (1999) High-temperature effects on photosynthetic processes in temperate and tropical cereals. Crop Sci 39:119–125

Allakhverdiev SI, Kreslavski VD, Klimov VV et al (2008) Heat stress: an overview of molecular responses in photosynthesis. Photosynth Res 98:541–550

Alqudah AM, Samarah NH, Mullen RE (2011) Drought stress effect on crop pollination, seed set, yield and quality. In: Lichtfocus E (ed) Alternative farming systems, biotechnology, drought stress and ecological fertilisation. Springer, Dordrecht, pp 193–213

Alves AA, Setter TL (2004) Response of cassava leaf area expansion to water deficit: cell proliferation, cell expansion and delayed development. Ann Bot 94:605–613

Anjum S, Wang L, Farooq M et al (2011a) Methyl jasmonate-induced alteration in lipid peroxidation, antioxidative defence system and yield in soybean under drought. J Agron Crop Sci 197:296–301

Anjum S, Wang L, Farooq M et al (2011b) Brassinolide application improves the drought tolerance in maize through modulation of enzymatic antioxidants and leaf gas exchange. J Agron Crop Sci 197:177–185

Annerose D (1990) Recherches sur les mécanismes physiologiques d'adaptation à la sécheresse. Application au cas de l'arachide (*Arachis hypogea* l.) cultivée au Sénégal, Thèse de doctorat, Université Paris VII

Anuradha N, Satyavathi CT, Bharadwaj C et al (2017) Deciphering genomic regions for high grain iron and zinc content using association mapping in pearl. Front Plant Sci 8:1–17

Aparna K, Hash C, Yadav R, Vadez V (2014) Seed number and 100-seed weight of pearl millet (*Pennisetum glaucum* L.) respond differently to low soil moisture in genotypes contrasting for drought tolerance. J Agron Crop Sci 200:119–131

Apel K, Hirt H (2004) Reactive oxygen species: metabolism, oxidative stress, and signal transduction. Annu Rev Plant Biol 55:373–399

Aroca R, Porcel R, Ruiz-Lozano JM (2012) Regulation of root water uptake under abiotic stress conditions. J Exp Bot 63:43–57

Arya RK, Yadav H (2009) Stability of grain yield and its contributing traits in white and grey grain hybrids of pearl millet (Pennisetum glaucum). Indian J Agric Sci 79:941–944

Arya R, Yadav H, Yadav A, Singh M (2010) Effect of environment on yield and its contributing traits in pearl millet. Forage Res 36:176–180

Arya R, Singh M, Yadav A et al (2014) Advances in pearl millet to mitigate adverse environment conditions emerged due to global warming. Forage Res 40:57–70

Ashraf MY, Azmi AR, Khan AH, Ala S (1994) Effect of water stress on total phenols, peroxidase activity and chlorophyll content in wheat [Triticum aestivum L.]. Acta Physiol Plant 16(3):185–191

Ashraf M, Foolad MR (2007) Roles of glycine betaine and proline in improving plant abiotic stress resistance. Environ Exp Bot 59:206–216

Ashraf M, Hafeez M (2004) Thermotolerance of pearl millet and maize at early growth stages: growth and nutrient relations. Biol Plant 48:81–86

Ashraf M, Ahmad A, McNeilly T (2001) Growth and photosynthetic characteristics in pearl millet under water stress and different potassium supply. Photosynthetica 39:389–394

Assefa Y, Staggenborg SA, Prasad PV (2010) Grain sorghum water requirement and responses to drought stress: a review. Crop Manage 9:1–11

Assmann SM, Snyder JA, Lee YJ (2000) ABA-deficient (aba1) and ABA-insensitive (abi1-1, abi2-1) mutants of Arabidopsis have a wild-type stomatal response to humidity. Plant Cell Environ 23:387–395

Bai Y, Han X, Wu J et al (2004) Ecosystem stability and compensatory effects in the Inner Mongolia grassland. Nature 431:181–184

Bajji M, Lutts S, Kinet J-M (2001) Water deficit effects on solute contribution to osmotic adjustment as a function of leaf ageing in three durum wheat (Triticum durum Desf.) cultivars performing differently in arid conditions. Plant Sci 160:669–681

Balling RC, Cerveny RS (2003) Compilation and discussion of trends in severe storms in the United States: Popular perception v. climate reality. Nat Hazards 29:103–112

Baniwal SK, Bharti K, Chan KY et al (2004) Heat stress response in plants: a complex game with chaperones and more than twenty heat stress transcription factors. J Biosci 29:471–487

Barnabás B, Jäger K, Fehér A (2008) The effect of drought and heat stress on reproductive processes in cereals. Plant Cell Environ 31:11–38

Bates B, Kundzewicz Z, Wu S, Palutikof J (eds) (2008) Climate change and water. Intergovernmental Panel on Climate Change Secretariat, Geneva

Behnam B, Iuchi S, Fujita M et al (2013) Characterization of the promoter region of an Arabidopsis gene for 9-cis-epoxycarotenoid dioxygenase involved in dehydration-inducible transcription. DNA Res 20:315–324. https://doi.org/10.1093/DNARES/DST012

Bhargava S, Sawant K (2013) Drought stress adaptation: metabolic adjustment and regulation of gene expression. Plant Breed 132:21–32

Bhattarai B, Singh S, West CP et al (2020) Water depletion pattern and water use efficiency of forage sorghum, pearl millet, and corn under water limiting condition. Agric Water Manag 238: 106206. https://doi.org/10.1016/j.agwat.2020.106206

Bidinger F, Raju D (2000) Response to selection for increased individual grain mass in pearl millet. Crop Sci 40:68–71

Bidinger FR, Serraj R, Rizvi SMH et al (2005) Field evaluation of drought tolerance QTL effects on phenotype and adaptation in pearl millet [*Pennisetum glaucum* (L.) R. Br.] topcross hybrids. Field Crop Res 94:14–32. https://doi.org/10.1016/j.fcr.2004.11.006

Bita C, Gerats T (2013) Plant tolerance to high temperature in a changing environment: scientific fundamentals and production of heat stress-tolerant crops. Front Plant Sci 4:273. https://doi.org/10.3389/fpls.2013.00273

Blum A (2005) Drought resistance, water-use efficiency, and yield potential—are they compatible, dissonant, or mutually exclusive? Aust J Agr Res 56:1159–1168

Blum A (2011) Plant water relations, plant stress and plant production. In: Blum A (ed) Plant breeding for water-limited environments. Springer, New York, pp 11–52

Blum A (2017) Osmotic adjustment is a prime drought stress adaptive engine in support of plant production. Plant Cell Environ 40:4–10

Blum A, Sullivan C (1986) The comparative drought resistance of landraces of sorghum and millet from dry and humid regions. Ann Bot 57:835–846

Bois J-F (1993) Effets de la contrainte hydrique sur la photosynthèse du mil. In: Hamon S (ed) Le mil en Afrique, diversité génétique et agro-physiologique: potentialités et contraintes pour l'amélioration génétique et l'agriculture. Orstom, Paris, pp 191–198

Boyer J, Westgate M (2004) Grain yields with limited water. J Exp Bot 55:2385–2394

Bray EA (1997) Plant responses to water deficit. Trends Plant Sci 2:48–54

Brown C, Lall U (2006) Water and economic development: the role of variability and a framework for resilience. Nat Resour Forum 30(4):306–317

Buckley TN (2019) How do stomata respond to water status? New Phytol 224:21–36

Calleja-Cabrera J, Boter M, Oñate-Sánchez L, Pernas M (2020) Root growth adaptation to climate change in crops. Front Plant Sci 11:544. https://doi.org/10.3389/fpls.2020.00544

Camejo D, Rodríguez P, Morales MA et al (2005) High temperature effects on photosynthetic activity of two tomato cultivars with different heat susceptibility. J Plant Physiol 162:281–289

Carvajal M, Cooke D, Clarkson D (1996) Plasma membrane fluidity and hydraulic conductance in wheat roots: interactions between root temperature and nitrate or phosphate deprivation. Plant Cell Environ 19:1110–1114

Chanwala J, Satpati S, Dixit A et al (2020) Genome-wide identification and expression analysis of WRKY transcription factors in pearl millet (Pennisetum glaucum) under dehydration and salinity stress. BMC Genomics 21:1–16. https://doi.org/10.1186/s12864-020-6622-0

Chaumont F, Tyerman SD (2014) Aquaporins: highly regulated channels controlling plant water relations. Plant Physiol 164:1600–1618

Chaves M, Oliveira M (2004) Mechanisms underlying plant resilience to water deficits: prospects for water-saving agriculture. J Exp Bot 55:2365–2384

Chaves MM, Maroco JP, Pereira JS (2003) Understanding plant responses to drought—from genes to the whole plant. Funct Plant Biol 30:239–264

Chen H, Jiang J-G (2010) Osmotic adjustment and plant adaptation to environmental changes related to drought and salinity. Environ Rev 18:309–319

Chimenti CA, Marcantonio M, Hall A (2006) Divergent selection for osmotic adjustment results in improved drought tolerance in maize (*Zea mays* L.) in both early growth and flowering phases. Field Crop Res 95:305–315

Chimungu JG, Maliro MF, Nalivata PC et al (2015) Utility of root cortical aerenchyma under water limited conditions in tropical maize (Zea mays L.). Field Crop Res 171:86–98

Chopart J-L, Siband P (1999) Development and validation of a model to describe root length density of maize from root counts on soil profiles. Plant and Soil 214:61–74

Chopart J-L, Rodrigues SR, De Azevedo MC, de Conti MC (2008a) Estimating sugarcane root length density through root mapping and orientation modelling. Plant and Soil 313:101–112

Chopart J-L, Sine B, Dao A, Muller B (2008b) Root orientation of four sorghum cultivars: application to estimate root length density from root counts in soil profiles. Plant Root 2:67–75

Choudhary S, Guha A, Kholova J et al (2020) Maize, sorghum, and pearl millet have highly contrasting species strategies to adapt to water stress and climate change-like conditions. Plant Sci 295:110297. https://doi.org/10.1016/j.plantsci.2019.110297

Christidis N, Stott PA, Jones GS et al (2012) Human activity and anomalously warm seasons in Europe. Int J Climatol 32:225–239

Christopher J, Manschadi A, Hammer G, Borrell A (2008) Developmental and physiological traits associated with high yield and stay-green phenotype in wheat. Aust J Agr Res 59:354–364

Clotault J, Thuillet A-C, Buiron M et al (2012) Evolutionary history of pearl millet (Pennisetum glaucum [L.] R. Br.) and selection on flowering genes since its domestication. Mol Biol Evol 29: 1199–1212

Collins M, Knutti R, Arblaster J et al (2013) Long-term climate change: projections, commitments and irreversibility. In: Stocker T, Quin D, Plattner G et al (eds) Climate change 2013-the physical science basis: contribution of Working Group I to the Fifth Assessment Report of the Intergovernmental Panel on Climate Change. Cambridge University Press, Cambridge, pp 1029–1136

Condon AG, Richards R, Rebetzke G, Farquhar G (2002) Improving intrinsic water-use efficiency and crop yield. Crop Sci 42:122–131

Cooper P, Rao K, Singh P et al (2009) Farming with current and future climate risk: advancing a 'Hypothesis of Hope' for rainfed agriculture in the semi-arid tropics. J SAT Agric Res 7:1–19

Cornic G, Massacci A (1996) Leaf photosynthesis under drought stress. In: Baker N (ed) Photosynthesis and the environment. Advances in photosynthesis and respiration, vol 5. Springer, Dordrecht, pp 347–366

Crafts-Brandner SJ, Salvucci ME (2000) Rubisco activase constrains the photosynthetic potential of leaves at high temperature and CO2. Proc Natl Acad Sci 97:13430–13435

Crafts-Brandner SJ, Salvucci ME (2002) Sensitivity of photosynthesis in a C4 plant, maize, to heat stress. Plant Physiol 129:1773–1780

Craufurd P, Peacock J (1993) Effect of heat and drought stress on sorghum (Sorghum bicolor). II Grain yield. Exp Agric 29:77–86

Cruz de Carvalho MH (2008) Drought stress and reactive oxygen species: production, scavenging and signaling. Plant Signal Behav 3:156–165

Dai A, Lamb PJ, Trenberth KE et al (2004) The recent Sahel drought is real. Int J Climatol 24:1323–1331

De Ollas C, Hernando B, Arbona V, Gómez-Cadenas A (2013) Jasmonic acid transient accumulation is needed for abscisic acid increase in citrus roots under drought stress conditions. Physiol Plant 147:296–306

Debieu M, Kanfany G, Laplaze L (2017) Pearl millet genome: lessons from a tough crop. Trends Plant Sci 22(11):911–913

Debieu M, Sine B, Passot S et al (2018) Response to early drought stress and identification of QTLs controlling biomass production under drought in pearl millet. PloS One 13:1–19. https://doi.org/10.1371/journal.pone.0201635

Diack O, Kanfany G, Gueye MC et al (2020) GWAS unveils features between early-and late-flowering pearl millets. BMC Genomics 21:1–11. https://doi.org/10.21203/rs.3.rs-25381/v1

Djanaguiraman M, Perumal R, Ciampitti I et al (2018) Quantifying pearl millet response to high temperature stress: thresholds, sensitive stages, genetic variability and relative sensitivity of pollen and pistil. Plant Cell Environ 41:993–1007

Dreesen FE, De Boeck HJ, Janssens IA, Nijs I (2012) Summer heat and drought extremes trigger unexpected changes in productivity of a temperate annual/biannual plant community. Environ Exp Bot 79:21–30

Dudhate A, Shinde H, Tsugama D et al (2018) Transcriptomic analysis reveals the differentially expressed genes and pathways involved in drought tolerance in pearl millet [Pennisetum glaucum (L.) R. Br]. PLoS One 13:1–14. https://doi.org/10.1371/journal.pone.0195908

Fahad S, Hussain S, Saud S et al (2016a) Exogenously applied plant growth regulators affect heat-stressed rice pollens. J Agron Crop Sci 202:139–150

Fahad S, Hussain S, Saud S et al (2016b) A combined application of biochar and phosphorus alleviates heat-induced adversities on physiological, agronomical and quality attributes of rice. Plant Physiol Biochem 103:191–198

Fahad S, Bajwa AA, Nazir U et al (2017) Crop production under drought and heat stress: plant responses and management options. Front Plant Sci 8:1147. https://doi.org/10.3389/fpls.2017.01147

Farooq M, Basra S, Wahid A et al (2008) Physiological role of exogenously applied glycine betaine to improve drought tolerance in fine grain aromatic rice (Oryza sativa L.). J Agron Crop Sci 194: 325–333

Farooq M, Wahid A, Kobayashi N et al (2009) Plant drought stress: effects, mechanisms and management. Sustain Agric 29:185–212

Farooq M, Bramley H, Palta JA, Siddique KH (2011) Heat stress in wheat during reproductive and grain-filling phases. Crit Rev Plant Sci 30:491–507

Fauchereau N, Trzaska S, Rouault M, Richard Y (2003) Rainfall variability and changes in southern Africa during the 20th century in the global warming context. Nat Hazards 29:139–154

Faye A, Sine B, Chopart J-L et al (2019) Development of a model estimating root length density from root impacts on a soil profile in pearl millet (*Pennisetum glaucum* (L.) R. Br). Application to measure root system response to water stress in field conditions. PloS One 14:e0214182. https://doi.org/10.1371/journal.pone.0214182

Fischer G, Shah MN, Tubiello F, Van Velhuizen H (2005a) Socio-economic and climate change impacts on agriculture: an integrated assessment, 1990–2080. Philos Trans R Soc B Biol Sci 360:2067–2083

Fischer R, Sayre K, Reynolds M (2005b) Osmotic adjustment in wheat in relation to grain yield under water deficit environments. Agron J 97:1062–1071

Flexas J, Ribas-Carbó M, Bota J et al (2006) Decreased Rubisco activity during water stress is not induced by decreased relative water content but related to conditions of low stomatal conductance and chloroplast CO2 concentration. New Phytol 172:73–82

Ford KL, Cassin A, Bacic AF (2011) Quantitative proteomic analysis of wheat cultivars with differing drought stress tolerance. Front Plant Sci 2:44

Forster B, Ellis R, Moir J et al (2004) Genotype and phenotype associations with drought tolerance in barley tested in North Africa. Ann Appl Biol 144:157–168

Ghatak A, Chaturvedi P, Nagler M et al (2016) Comprehensive tissue-specific proteome analysis of drought stress responses in *Pennisetum glaucum* (L.) R. Br.(Pearl millet). J Proteomics 143: 122–135

Ghatak A, Chaturvedi P, Bachmann G et al (2021) Physiological and proteomic signatures reveal mechanisms of superior drought resilience in pearl millet compared to wheat. Front Plant Sci 11: 600278

Godfray HCJ, Beddington JR, Crute IR et al (2010) Food security: the challenge of feeding 9 billion people. Science 327:812–818

Golombek S (2003) Effect of drought on gas exchange and carbon metabolism in pearl millet. Technological and Institutional Innovations for Sustainable Rural Development. Deutshe Tropentag, Göttingen, 8–10 October 2003, pp 339–348

Govindaraj M, Shanmugasundaram P, Sumathi P, Muthiah A (2010) Simple, rapid and cost effective screening method for drought resistant breeding in pearl millet. Electron J Plant Breed 1:590–599

Grondin A, Mauleon R, Vadez V, Henry A (2016) Root aquaporins contribute to whole plant water fluxes under drought stress in rice (*Oryza sativa* L.). Plant Cell Environ 39:347–365

Grondin A, Affortit P, Tranchant-Dubreuil C et al (2020) Aquaporins are main contributors to root hydraulic conductivity in pearl millet [*Pennisetum glaucum* (L) R. Br.]. PLoS One 15: e0233481. https://doi.org/10.1371/journal.pone.0233481

Gupta S, Rai K, Singh P et al (2015) Seed set variability under high temperatures during flowering period in pearl millet (*Pennisetum glaucum* L.(R.) Br.). Field Crop Res 171:41–53

Gupta S, Ameta V, Pareek S, et al. (2016) Genetic enhancement for flowering period heat tolerance in pearl millet (*Pennisetum glaucum* L.(R.) Br.). In: 7th international crop science congress, Beijing. http://oar.icrisat.org/9757/1/Page1.pdf. Accessed 15 Jul 2021

Gurley WB (2000) HSP101: a key component for the acquisition of thermotolerance in plants. Plant Cell 12:457–460
Heckathorn SA, DeLucia EH (1991) Effect of leaf rolling on gas exchange and leaf temperature of Andropogon gerardii and Spartina pectinata. Bot Gaz 152:263–268
Heckathorn SA, Giri A, Mishra S, Bista D (2013) Heat stress and roots. In: Climate change and plant abiotic stress tolerance. Wiley, Hoboken, pp 109–136
Henry S, Divol F, Bettembourg M et al (2016) Immunoprofiling of rice root cortex reveals two cortical subdomains. Front Plant Sci 6:1139. https://doi.org/10.3389/fpls.2015.01139
Hochholdinger F, Tuberosa R (2009) Genetic and genomic dissection of maize root development and architecture. Curr Opin Plant Biol 12:172–177
Howden SM, Soussana J-F, Tubiello FN et al (2007) Adapting agriculture to climate change. Proc Natl Acad Sci 104:19691–19696
Hu X, Li Y, Li C et al (2010) Characterization of small heat shock proteins associated with maize tolerance to combined drought and heat stress. J Plant Growth Regul 29:455–464
Hund A, Fracheboud Y, Soldati A, Stamp P (2008) Cold tolerance of maize seedlings as determined by root morphology and photosynthetic traits. Eur J Agron 28:178–185
Hund A, Ruta N, Liedgens M (2009) Rooting depth and water use efficiency of tropical maize inbred lines, differing in drought tolerance. Plant and Soil 318:311–325
Izanloo A, Condon AG, Langridge P et al (2008) Different mechanisms of adaptation to cyclic water stress in two South Australian bread wheat cultivars. J Exp Bot 59:3327–3346
Jagadish SK (2020) Heat stress during flowering in cereals–effects and adaptation strategies. New Phytol 226:1567–1572
Jaiswal S, Antala TJ, Mandavia M et al (2018) Transcriptomic signature of drought response in pearl millet (Pennisetum glaucum (L.) and development of web-genomic resources. Sci Rep 8: 1–16. https://doi.org/10.1038/s41598-018-21560-1
Jia W, Zhang J (2008) Stomatal movements and long-distance signaling in plants. Plant Signal Behav 3:772–777
Jung H, Choi Y, Oh J, Lim G (2002) Recent trends in temperature and precipitation over South Korea. Int J Climatol 22:1327–1337
Kadioglu A, Terzi R (2007) A dehydration avoidance mechanism: leaf rolling. Bot Rev 73:290–302
Kadioglu A, Terzi R, Saruhan N, Saglam A (2012) Current advances in the investigation of leaf rolling caused by biotic and abiotic stress factors. Plant Sci 182:42–48
Kang Y, Khan S, Ma X (2009) Climate change impacts on crop yield, crop water productivity and food security–a review. Prog Nat Sci 19:1665–1674
Kashiwagi J, Krishnamurthy L, Crouch J, Serraj R (2006) Variability of root length density and its contributions to seed yield in chickpea (Cicer arietinum L.) under terminal drought stress. Field Crop Res 95:171–181
Katerji N, Tardieu F, Bethenod O, Quetin P (1994) Behavior of maize stem diameter during drying cycles: comparison of two methods for detecting water stress. Crop Sci 34:165–169
Khalifa F, Ong C (1990) Effect of supra-optimal temperatures on germination of pearl millet (*Pennisetum glaucum* (L) R. BR.) hybrids. Ann Arid Zone 29:279–288
Khan MB, Hussain N, Iqbal M (2001) Effect of water stress on growth and yield components of maize variety YHS 202. J Res Sci 12:15–18
Khan H, Paull J, Siddique K, Stoddard F (2010) Faba bean breeding for drought-affected environments: A physiological and agronomic perspective. Field Crop Res 115:279–286
Kholová J, Vadez V (2013) Water extraction under terminal drought explains the genotypic differences in yield, not the antioxidant changes in leaves of pearl millet (Pennisetum glaucum). Funct Plant Biol 40:44–53. https://doi.org/10.1071/FP12181
Kholová J, Hash CT, Kakkera A et al (2010) Constitutive water-conserving mechanisms are correlated with the terminal drought tolerance of pearl millet [*Pennisetum glaucum* (L.) R. Br.]. J Exp Bot 61:369–377

Kholová J, Zindy P, Malayee S et al (2016) Component traits of plant water use are modulated by vapour pressure deficit in pearl millet (*Pennisetum glaucum* (L.) R. Br.). Funct Plant Biol 43: 423–437

Kirkegaard J, Lilley J, Howe G, Graham J (2007) Impact of subsoil water use on wheat yield. Aust J Agr Res 58:303–315

Kirtman B, Power SB, Adedoyin AJ et al (2013) Near-term climate change: projections and predictability. In: Stocker T, Quin D, Plattner G et al (eds) Climate change 2013: the physical science basis. Contribution of Working Group I to the Fifth Assessment Report of the Intergovernmental Panel on Climate Change. Cambridge University Press, Cambridge, pp 953–1028

Knox J, Hess T, Daccache A, Wheeler T (2012) Climate change impacts on crop productivity in Africa and South Asia. Environ Res Lett 7:034032. https://doi.org/10.1088/1748-9326/7/3/034032

Komatsu S, Kamal AH, Hossain Z (2014) Wheat proteomics: proteome modulation and abiotic stress acclimation. Front Plant Sci 5:684. https://doi.org/10.3389/fpls.2014.00684

Kondo M, Murty MV, Aragones DV (2000) Characteristics of root growth and water uptake from soil in upland rice and maize under water stress. Soil Sci Plant Nutr 46:721–732

Kooyers NJ (2015) The evolution of drought escape and avoidance in natural herbaceous populations. Plant Sci 234:155–162

Kotak S, Larkindale J, Lee U et al (2007) Complexity of the heat stress response in plants. Curr Opin Plant Biol 10:310–316

Kumar S, Sachdeva S, Bhat K, Vats S (2018) Plant responses to drought stress: physiological, biochemical and molecular basis. In: Vats S (ed) Biotic and abiotic stress tolerance in plants. Springer, India, pp 1–25

Kusaka M, Lalusin AG, Fujimura T (2005a) The maintenance of growth and turgor in pearl millet (*Pennisetum glaucum* [L.] Leeke) cultivars with different root structures and osmo-regulation under drought stress. Plant Sci 168:1–14

Kusaka M, Ohta M, Fujimura T (2005b) Contribution of inorganic components to osmotic adjustment and leaf folding for drought tolerance in pearl millet. Physiol Plant 125:474–489

Lakis G, Navascués M, Rekima S et al (2012) Evolution of neutral and flowering genes along pearl millet (Pennisetum glaucum) domestication. PloS One 7:e36642. https://doi.org/10.1371/journal.pone.0036642

Li P, Ma B, Xiong Y, Zhang W (2017) Morphological and physiological responses of different wheat genotypes to chilling stress: a cue to explain yield loss. J Sci Food Agric 97:4036–4045

Liu B, Asseng S, Müller C et al (2016) Similar estimates of temperature impacts on global wheat yield by three independent methods. Nat Clim Change 6:1130–1136

Lu C, Zhang J (1999) Effects of water stress on photosystem II photochemistry and its thermostability in wheat plants. J Exp Bot 50:1199–1206

Ludlow M, Muchow R (1990) A critical evaluation of traits for improving crop yields in water-limited environments. Adv Agron 43:107–153

Ma C, Wang Z, Kong B, Lin T (2013) Exogenous trehalose differentially modulate antioxidant defense system in wheat callus during water deficit and subsequent recovery. Plant Growth Regul 70:275–285

Maiti R, Bidinger F (1981) Growth and development of the pearl millet plant. Research Bulletin no. 6. ICRISAT, Patancheru

Manschadi A, Christopher J, Hammer G, Devoil P (2010) Experimental and modelling studies of drought-adaptive root architectural traits in wheat (*Triticum aestivum* L.). Plant Biosyst 144: 458–462

Maroco JP, Pereira JS, Chaves MM (1997) Stomatal responses to leaf-to-air vapour pressure deficit in Sahelian species. Funct Plant Biol 24:381–387

Matsuura A, Inanaga S, Sugimoto Y (1996) Mechanism of interspecific differences among four gramineous crops in growth response to soil drying. Jpn J Crop Sci 65:352–360

Maurel C, Boursiac Y, Luu D-T et al (2015) Aquaporins in plants. Physiol Rev 95:1321–1358

Mortlock MY, Hammer GL (2000) Genotype and water limitation effects on transpiration efficiency in sorghum. J Crop Prod 2:265–286
Muchow R (1989) Comparative productivity of maize, sorghum and pearl millet in a semi-arid tropical environment I. Yield potential. Field Crops Res 20:191–205
Niang I, Ruppel OC, Abdrabo MA et al (2014) Africa climate change 2014: impacts, adaptation, and vulnerability: part B: regional aspects. In: Barros V, Field C, Dokken D et al (eds) Contribution of working group II to the fifth assessment report of the Intergovernmental Panel on Climate Change. Cambridge University Press, Cambridge, pp 1199–1266
Nio S, Cawthray G, Wade L, Colmer T (2011) Pattern of solutes accumulated during leaf osmotic adjustment as related to duration of water deficit for wheat at the reproductive stage. Plant Physiol Biochem 49:1126–1137
Nitnavare RB, Yeshvekar RK, Sharma KK et al (2016) Molecular cloning, characterization and expression analysis of a heat shock protein 10 (Hsp10) from Pennisetum glaucum (L.), a C 4 cereal plant from the semi-arid tropics. Mol Biol Rep 43:861–870
O'toole J, Cruz R, Singh T (1979) Leaf rolling and transpiration. Plant Sci Lett 16:111–114
Ocheltree T, Nippert J, Prasad PV (2014) Stomatal responses to changes in vapor pressure deficit reflect tissue-specific differences in hydraulic conductance. Plant Cell Environ 37:132–139
Ogbaga CC, Stepien P, Johnson GN (2014) Sorghum (Sorghum bicolor) varieties adopt strongly contrasting strategies in response to drought. Physiol Plant 152:389–401
Omarova E, Bogdanova E, Polimbetova F (1995) Regulation of water-loss by the leaves of soft winter-wheat with different organization of leaf structure. Russ J Plant Physiol 42:383–385
van Oosterom E, Kulathunga M, Deifel K et al (2021) Dissecting and modelling the comparative adaptation to water limitation of sorghum and maize: role of transpiration efficiency, transpiration rate and height. In Silico Plants 3:diaa012
Ousseini I, Bakasso Y, Kane N et al (2017) Myosin XI is associated with fitness and adaptation to aridity in wild pearl millet. Heredity 119:88–94. https://doi.org/10.1038/hdy.2017.13
Parent B, Hachez C, Redondo E et al (2009) Drought and abscisic acid effects on aquaporin content translate into changes in hydraulic conductivity and leaf growth rate: a trans-scale approach. Plant Physiol 149:2000–2012. https://doi.org/10.1104/pp.108.130682
Passot S (2016) Exploring pearl millet root system and its outcome for drought tolerance, Thèse de doctorat, Université Montpellier
Passot S, Gnacko F, Moukouanga D et al (2016) Characterization of pearl millet root architecture and anatomy reveals three types of lateral roots. Front Plant Sci 7:829. https://doi.org/10.3389/fpls.2016.00829
Passot S, Moreno-Ortega B, Moukouanga D et al (2018) A new phenotyping pipeline reveals three types of lateral roots and a random branching pattern in two cereals. Plant Physiol 177:896–910
Pinheiro C, Chaves M (2010) Photosynthesis and drought: can we make metabolic connections from available data? J Exp Bot 62:869–882
Pinheiro C, Chaves M (2011) Photosynthesis and drought: can we make metabolic connections from available data? J Exp Bot 62:869–882
Prado K, Maurel C (2013) Regulation of leaf hydraulics: from molecular to whole plant levels. Front Plant Sci 4:255
Prasad PV, Djanaguiraman M (2011) High night temperature decreases leaf photosynthesis and pollen function in grain sorghum. Funct Plant Biol 38:993–1003
Prasad PV, Djanaguiraman M (2014) Response of floret fertility and individual grain weight of wheat to high temperature stress: sensitive stages and thresholds for temperature and duration. Funct Plant Biol 41:1261–1269
Prasad PV, Boote KJ, Vu JC, Allen LH Jr (2004) The carbohydrate metabolism enzymes sucrose-P synthase and ADG-pyrophosphorylase in phaseolus bean leaves are up-regulated at elevated growth carbon dioxide and temperature. Plant Sci 166:1565–1573
Prasad PV, Boote K, Allen L Jr et al (2006a) Species, ecotype and cultivar differences in spikelet fertility and harvest index of rice in response to high temperature stress. Field Crop Res 95:398–411

Prasad PV, Boote KJ, Allen LH Jr (2006b) Adverse high temperature effects on pollen viability, seed-set, seed yield and harvest index of grain-sorghum [*Sorghum bicolor* (L.) Moench] are more severe at elevated carbon dioxide due to higher tissue temperatures. Agric For Meteorol 139:237–251

Prasad PV, Staggenborg S, Ristic Z (2008) Impacts of drought and/or heat stress on physiological, developmental, growth, and yield processes of crop plants. In: Response of crops to limited water: Understanding and modeling water stress effects on plant growth processes, vol 1. ASA, CSSA, Madison, pp 301–355

Prasad PV, Djanaguiraman M, Perumal R, Ciampitti IA (2015) Impact of high temperature stress on floret fertility and individual grain weight of grain sorghum: sensitive stages and thresholds for temperature and duration. Front Plant Sci 6:820

Prasad PV, Bheemanahalli R, Jagadish SK (2017) Field crops and the fear of heat stress—opportunities, challenges and future directions. Field Crop Res 200:114–121

Qaseem MF, Qureshi R, Shaheen H (2019) Effects of pre-anthesis drought, heat and their combination on the growth, yield and physiology of diverse wheat (*Triticum aestivum* L.) genotypes varying in sensitivity to heat and drought stress. Sci Rep 9:1–12

Qin X, Zeevaart JA (1999) The 9-cis-epoxycarotenoid cleavage reaction is the key regulatory step of abscisic acid biosynthesis in water-stressed bean. Proc Natl Acad Sci 96:15354–15361

Quan R, Shang M, Zhang H et al (2004) Engineering of enhanced glycine betaine synthesis improves drought tolerance in maize. Plant Biotechnol J 2:477–486

Queitsch C, Hong S-W, Vierling E, Lindquist S (2000) Heat shock protein 101 plays a crucial role in thermotolerance in Arabidopsis. Plant Cell 12:479–492

Radhouane L (2008) Caractéristiques hydriques du mil (*Pennisetum glaucum* (L.) R. Br.) en présence de contraintes hydriques. C R Biol 331:206–214

Radhouane L (2009) La photosynthèse du mil (*Pennisetum glaucum* (L.) R. Br.) en présence de contrainte hydrique et saline. J Agric Environ Int Devel 103:185–200

Radhouane L (2013) Les échanges gazeux d'un écotype autochtone de mil (*Pennisetum glaucum* LR Br.) au cours d'un dessèchement hydrique. J Anim Plants Sci 2:2552–2566

Rebouillat J, Dievart A, Verdeil J-L et al (2009) Molecular genetics of rice root development. Rice 2:15–34

Reddy PS, Mallikarjuna G, Kaul T et al (2010) Molecular cloning and characterization of gene encoding for cytoplasmic Hsc70 from Pennisetum glaucum may play a protective role against abiotic stresses. Mol Genet Genomics 283:243–254

Reddy KS, Sekhar KM, Reddy AR (2017a) Genotypic variation in tolerance to drought stress is highly coordinated with hydraulic conductivity–photosynthesis interplay and aquaporin expression in field-grown mulberry (Morus spp.). Tree Physiol 37:926–937. https://doi.org/10.1093/treephys/tpx051

Reddy PS, Tharanya M, Sivasakthi K et al (2017b) Molecular cloning and expression analysis of aquaporin genes in pearl millet [*Pennisetum glaucum* (L) R. Br.] genotypes contrasting in their transpiration response to high vapour pressure deficits. Plant Sci 265:167–176

Riccardi F, Gazeau P, Jacquemot M-P et al (2004) Deciphering genetic variations of proteome responses to water deficit in maize leaves. Plant Physiol Biochem 42:1003–1011

Richards R, Passioura J (1989) A breeding program to reduce the diameter of the major xylem vessel in the seminal roots of wheat and its effect on grain yield in rain-fed environments. Aust J Agr Res 40:943–950

Rizhsky L, Liang H, Mittler R (2002) The combined effect of drought stress and heat shock on gene expression in tobacco. Plant Physiol 130:1143–1151

Rostamza M, Richards R, Watt M (2013) Response of millet and sorghum to a varying water supply around the primary and nodal roots. Ann Bot 112:439–446

Saidou AA, Mariac C, Luong V et al (2009) Association studies identify natural variation at PHYC linked to flowering time and morphological variation in pearl millet. Genetics 182(3):899–910

Sakamoto A, Murata N (2002) The role of glycine betaine in the protection of plants from stress: clues from transgenic plants. Plant Cell Environ 25:163–171

Salack S, Muller B, Gaye AT (2011) Rain-based factors of high agricultural impacts over Senegal. Part I: integration of local to sub-regional trends and variability. Theor Appl Climatol 106:1–22
Salack S, Muller B, Gaye AT et al (2012a) Analyses multi-échelles des pauses pluviométriques au Niger et au Sénégal. Sécheresse 23:3–13. https://doi.org/10.1684/sec.2012.0335
Salack S, Sultan B, Oettli P et al (2012b) Representation of rainfall in regional climate models and application to millet yield estimations in Senegal. Science and changements planétaires/ Sécheresse 23:14–23. https://doi.org/10.1684/sec.2012.0332
Sanchez A, Subudhi P, Rosenow D, Nguyen H (2002) Mapping QTLs associated with drought resistance in sorghum (*Sorghum bicolor* L. Moench). Plant Mol Biol 48:713–726
Sanders GJ, Arndt SK (2012) Osmotic adjustment under drought conditions. In: Aroca R (ed) Plant responses to drought stress: from morphological to molecular features. Springer, Berlin, pp 199–229
Sarieva G, Kenzhebaeva S, Lichtenthaler H (2010) Adaptation potential of photosynthesis in wheat cultivars with a capability of leaf rolling under high temperature conditions. Russ J Plant Physiol 57:28–36
Savin R, Nicolas ME (1996) Effects of short periods of drought and high temperature on grain growth and starch accumulation of two malting barley cultivars. Funct Plant Biol 23:201–210
Scharf K-D, Berberich T, Ebersberger I, Nover L (2012) The plant heat stress transcription factor (Hsf) family: structure, function and evolution. Biochim Biophys Acta 1819:104–119
Schittenhelm S, Schroetter S (2014) Comparison of drought tolerance of maize, sweet sorghum and sorghum-sudangrass hybrids. J Agron Crop Sci 200:46–53
Schramm F, Larkindale J, Kiehlmann E et al (2008) A cascade of transcription factor DREB2A and heat stress transcription factor HsfA3 regulates the heat stress response of Arabidopsis. Plant J 53:264–274
Sekhon H, Singh G, Sharma P, Bains T (2010) Water use efficiency under stress environments. In: Yadav S, Mc Neil D, Redden R et al (eds) Climate change and management of cool season grain legume crops. Springer, New York, pp 207–227
Serraj R, Hash CT, Rizvi SMH et al (2005) Recent advances in marker-assisted selection for drought tolerance in pearl millet. Plant Prod Sci 8:334–337. https://doi.org/10.1626/pps.8.334
Shah N, Paulsen G (2003) Interaction of drought and high temperature on photosynthesis and grain-filling of wheat. Plant Soil 257:219–226
Shangguan Z, Shao M, Dyckmans J (1999) Interaction of osmotic adjustment and photosynthesis in winter wheat under soil drought. J Plant Physiol 154:753–758
Sillmann J, Kharin VV, Zwiers F et al (2013) Climate extremes indices in the CMIP5 multimodel ensemble: Part 2. Future climate projections. J Geophys Res Atmos 118:2473–2493
Sinclair TR, Muchow RC (2001) System analysis of plant traits to increase grain yield on limited water supplies. Agron J 93:263–270
Slama A (2002) Étude comparative de la contribution des différentes parties du plant du blé dur dans la contribution du rendement en grains en irrigué et en conditions de déficit hydrique. Thèse de doctorat en biologie, faculté des sciences de Tunis
Slatyer R (1973) Effects of short periods of water stress on leaf photosynthesis. In: Slatyer R (ed) Plant response to climatic factors proceedings of theuppsala symposium. UNESCO, Paris, pp 271–276
Soegaard H, Boegh E (1995) Estimation of evapotranspiration from a millet crop in the Sahel combining sap flow, leaf area index and eddy correlation technique. J Hydrol 166:265–282
Solomon S, Manning M, Marquis M et al (eds) (2007) Climate change 2007-the physical science basis: Working group I, contribution to the fourth assessment report of the IPCC. Cambridge University Press, Cambridge
Steele K, Price A, Witcombe J et al (2013) QTLs associated with root traits increase yield in upland rice when transferred through marker assisted selection. Theor Appl Genet 126:101–108
Steudle E (2001) The cohesion-tension mechanism and the acquisition of water by plant roots. Annu Rev Plant Biol 52:847–875

Stott PA, Gillett NP, Hegerl GC et al (2010) Detection and attribution of climate change: a regional perspective. Wiley Interdiscip Rev Clim Change 1:192–211
Sultan B, Gaetani M (2016) Agriculture in West Africa in the twenty-first century: climate change and impacts scenarios, and potential for adaptation. Front Plant Sci 7:1262
Sultan B, Roudier P, Quirion P et al (2013) Assessing climate change impacts on sorghum and millet yields in the Sudanian and Sahelian savannas of West Africa. Environ Res Lett 8(1): 014040
Sun W, Van Montagu M, Verbruggen N (2002) Small heat shock proteins and stress tolerance in plants. Biochim Biophys Acta 1577:1–9
Takayuki Y, Hiroyuki K, Takashi M, Hiroshi M (2013) Simultaneous Transcriptome Analysis of Sorghum and Bipolaris sorghicola by Using RNA-seq in Combination with De Novo Transcriptome Assembly. PLoS ONE 8(4):e62460. https://doi.org/10.1371/journal.pone.0062460
Tardieu F, Reymond M, Hamard P et al (2000) Spatial distributions of expansion rate, cell division rate and cell size in maize leaves: a synthesis of the effects of soil water status, evaporative demand and temperature. J Exp Bot 51:1505–1514
Trenberth KE, Shea DJ (2005) Relationships between precipitation and surface temperature. Geophys Res Lett 32:L14703. https://doi.org/10.1029/2005GL022760
Trenberth KE, Jones PD, Ambenje P et al (2007) Observations. Surface and atmospheric climate change. In: Solomon S, Qin M, Manning Z et al (eds) Climate change 2007: the physical science basis. Cambridge University Press, Cambridge, pp 235–336
Tuteja N (2007) Abscisic acid and abiotic stress signaling. Plant Signal Behav 2:135–138
Ullah A, Ahmad A, Khaliq T, Akhtar J (2017) Recognizing production options for pearl millet in Pakistan under changing climate scenarios. J Integr Agric 16:762–773
Vadez V, Krishnamurthy L, Hash C et al (2011) Yield, transpiration efficiency, and water-use variations and their interrelationships in the sorghum reference collection. Crop Pasture Sci 62: 645–655
Vadez V, Hash T, Kholova J (2012) II. 1.5 Phenotyping pearl millet for adaptation to drought. Front Physiol 3:386
Vadez V, Kholová J, Yadav RS, Hash CT (2013a) Small temporal differences in water uptake among varieties of pearl millet (*Pennisetum glaucum* (L.) R. Br.) are critical for grain yield under terminal drought. Plant and Soil 371:447–462
Vadez V, Kholova J, Zaman-Allah M, Belko N (2013b) Water: the most important 'molecular' component of water stress tolerance research. Funct Plant Biol 40:1310–1322
Vadez V, Kholova J, Medina S et al (2014) Transpiration efficiency: new insights into an old story. J Exp Bot 65:6141–6153
Vadez V, Choudhary S, Kholová J et al (2021) Transpiration efficiency: insights from comparisons of C4 cereal species. J Exp Bot 72:5221–5234
Varney GT, Canny MJ, Mccully ME (1991) The branch roots of Zea. I. First order branches, their number, sizes and division into classes. Ann Bot 67:357–364
Varshney RK, Shi C, Thudi M et al (2017) Pearl millet genome sequence provides a resource to improve agronomic traits in arid environments. Nat Biotechnol 35:969–976
Verslues PE, Sharma S (2010) Proline metabolism and its implications for plant-environment interaction. Arabidopsis Book 8:1–23
Vigouroux Y, Cédric M, de Mita S et al (2011) Selection for earlier flowering crop associated with climatic variations in the Sahel. PloS One 6:1–9. https://doi.org/10.1371/journal.pone.0019563
Vijayalakshmi T, Varalaxmi Y, Jainender S et al (2012) Physiological and biochemical basis of water-deficit stress tolerance in pearl millet hybrid and parents. Am J Plant Sci 3(12):1730–1740
Wahid A (2007) Physiological implications of metabolite biosynthesis for net assimilation and heat-stress tolerance of sugarcane (Saccharum officinarum) sprouts. J Plant Res 120:219–228
Wahid A, Close T (2007) Expression of dehydrins under heat stress and their relationship with water relations of sugarcane leaves. Biol Plant 51:104–109

Wahid A, Gelani S, Ashraf M, Foolad MR (2007) Heat tolerance in plants: an overview. Environ Exp Bot 61:199–223
Wallace J, Lloyd C, Sivakumar M (1993) Measurements of soil, plant and total evaporation from millet in Niger. Agric For Meteorol 63:149–169
Wang W, Vinocur B, Shoseyov O, Altman A (2004) Role of plant heat-shock proteins and molecular chaperones in the abiotic stress response. Trends Plant Sci 9:244–252
Wang J, Vanga SK, Saxena R et al (2018) Effect of climate change on the yield of cereal crops: a review. Climate 6:41
Wasaya A, Zhang X, Fang Q, Yan Z (2018) Root phenotyping for drought tolerance: a review. Agronomy 8:24. https://doi.org/10.3390/agronomy8110241
Wasson AP, Richards R, Chatrath R et al (2012) Traits and selection strategies to improve root systems and water uptake in water-limited wheat crops. J Exp Bot 63:3485–3498
Watt M, Magee LJ, McCully ME (2008) Types, structure and potential for axial water flow in the deepest roots of field-grown cereals. New Phytol 178:135–146
Welcker C, Boussuge B, Bencivenni C et al (2007) Are source and sink strengths genetically linked in maize plants subjected to water deficit? A QTL study of the responses of leaf growth and of anthesis-silking interval to water deficit. J Exp Bot 58:339–349
Wilson J, Sanogo M, Nutsugah S et al (2008) Evaluation of pearl millet for yield and downy mildew resistance across seven countries in sub-Saharan Africa. Afr J Agric Res 3(5):371–378
Winkel T, Renno J-F, Payne W (1997) Effect of the timing of water deficit on growth, phenology and yield of pearl millet (*Pennisetum glaucum* (L.) R. Br.) grown in Sahelian conditions. J Exp Bot 48:1001–1009
Winkel T, Payne W, Renno J (2001) Ontogeny modifies the effects of water stress on stomatal control, leaf area duration and biomass partitioning of Pennisetum glaucum. New Phytol 149: 71–82
Wu W, Ma B, Whalen JK (2018) Enhancing rapeseed tolerance to heat and drought stresses in a changing climate: perspectives for stress adaptation from root system architecture. Adv Agron 151:87–157
Xin Z, Aiken R, Burke J (2009) Genetic diversity of transpiration efficiency in sorghum. Field Crop Res 111:74–80
Xiong L, Wang R-G, Mao G, Koczan JM (2006) Identification of drought tolerance determinants by genetic analysis of root response to drought stress and abscisic acid. Plant Physiol 142:1065–1074
Xu Z, Zhou G (2005a) Effects of water stress and nocturnal temperature on carbon allocation in the perennial grass, Leymus chinensis. Physiol Plant 123:272–280
Xu Z-Z, Zhou G-S (2005b) Effects of water stress and high nocturnal temperature on photosynthesis and nitrogen level of a perennial grass Leymus chinensis. Plant and Soil 269:131–139
Xu ZZ, Zhou GS (2006) Combined effects of water stress and high temperature on photosynthesis, nitrogen metabolism and lipid peroxidation of a perennial grass Leymus chinensis. Planta 224: 1080–1090
Xu Z, Zhou G, Shimizu H (2010) Plant responses to drought and rewatering. Plant Signal Behav 5: 649–654
Yadav RS, Hash C, Bidinger FR et al (2002) Quantitative trait loci associated with traits determining grain and stover yield in pearl millet under terminal drought stress conditions. Theor Appl Genet 104:67–83
Yadav RS, Hash CT, Bidinger FR et al (2004) Genomic regions associated with grain yield and aspects of post-flowering drought tolerance in pearl millet across stress environments and tester background. Euphytica 136:265–277
Yadav A, Narwal M, Arya R (2010) Stability studies for seedling traits with supra-optimal temperature exposure at seedling stage in pearl millet [*Pennisetum glaucum* (L.) R. Br.]. Forage Res 32:65–70
Yadav OP, Rai KN (2013) Genetic improvement of pearl millet in India. Agric Res 2:275–292

Yadav A, Arya R, Singh M et al (2016) Heat tolerance in pearl millet: a review. Forage Res 42:65–81

Yadav OP, Singh DV, Vadez V et al (2017) Improving pearl millet for drought tolerance-retrospect and prospects. Indian J Genet Plant Breed 77:464–474. https://doi.org/10.5958/0975-6906.2017.00062.1

Yamaguchi-Shinozaki K, Shinozaki K (2005) Organization of cis-acting regulatory elements in osmotic-and cold-stress-responsive promoters. Trends Plant Sci 10:88–94

Zegada-Lizarazu W, Zatta A, Monti A (2012) Water uptake efficiency and above-and belowground biomass development of sweet sorghum and maize under different water regimes. Plant Soil 351:47–60

Zhan A, Schneider H, Lynch JP (2015) Reduced lateral root branching density improves drought tolerance in maize. Plant Physiol 168:1603–1615

Zhang X, Wang T, Li C (2005) Different responses of two contrasting wheat genotypes to abscisic acid application. Biol Plant 49:613–616

Zhang A, Ji Y, Sun M et al (2021) Research on the drought tolerance mechanism of Pennisetum glaucum (L.) in the root during the seedling stage. BMC Genomics 22:1–14. https://doi.org/10.1186/S12864-021-07888-5

Zhu J-K (2002) Salt and drought stress signal transduction in plants. Annu Rev Plant Biol 53:247–273

Zscheischler J, Seneviratne SI (2017) Dependence of drivers affects risks associated with compound events. Sci Adv 3:e1700263

Forage Pearl Millet: Issues and Strategies for Genetic Improvement

11

P. Kaushal, A. K. Roy, D. R. Malaviya, N. R. Bhardwaj, and R. K. Agrawal

Abstract

Pearl Millet [*Pennisetum glaucum* (L.) R. Br.] is an important dual-purpose crop, where grain is consumed as a staple food or food supplement, and stover or green biomass is used for livestock feeding. In India, it is popularly known as Bajra or bajri. It has several desirable attributes like tolerance to heat and moisture stress, good nutritional quality, and high mineral content and hence widely cultivated in drier parts of India and the world.

Several popular fodder varieties have been developed in the past utilizing breeding tools like composites, recurrent selection, and targeted hybridization. Being highly cross-pollinated, it is difficult to maintain the genetic purity of a variety. Targeted breeding efforts have resulted in a series of new multi-cut forage pearl millet under the irrigated condition which is now becoming popular in the southern and central part of India because of its high yield, nutritive quality, and providing green fodder during the lean period in 2–3 cuts.

Exploiting the gene pool of the agamic complex of *Pennisetum*, attempts have been made with varying degrees of success to introgress several desired traits in the pearl millet cultivars to make it more adaptable and highly productive with better nutritive quality. Successful bispecific and trispecific hybrids in different combinations have been reported using the species such as *P. violaceum*, *P. squamulatum*, *P. orientale*, and *P. purpureum*. Traits like perenniality, multi-cut nature, improved quality, and apomictic components have been introgressed in pearl millet.

P. Kaushal
ICAR-Indian Grassland and Fodder Research Institute, Jhansi, India

A. K. Roy (✉) · D. R. Malaviya · N. R. Bhardwaj · R. K. Agrawal
ICAR-Indian Grassland and Fodder Research Institute, Jhansi, India

V. A. Tonapi et al. (eds.), *Pearl Millet in the 21st Century*,
https://doi.org/10.1007/978-981-99-5890-0_11

The future pearl millet improvement program should concentrate on developing biofortified forage or dual-type varieties with resistance to important diseases like downy mildew and blast. Multi-cut with improved quality will be an added advantage to make it more acceptable across different agro-climatic zones of the country.

11.1 Introduction

Pearl millet or Bajra [*Pennisetum glaucum* (L.) R. Br.] is cultivated as a dual-purpose food and fodder crop in several countries, particularly in Africa and Asia. It is cultivated as a forage crop in India, Australia, Brazil as well as Latin America and utilized as dry stover as well as green forage. It is a short duration, cross-pollinated diploid ($2n = 2x = 14$) species. Being a C4 plant, it has excellent photosynthetic efficiency and high production potential (Sanou et al. 2012). It is highly allogamous (Jauhar and Hanna 1998). Several landraces of this crop are available and represent genetically heterozygous and heterogeneous open-pollinating populations. It is a nutritionally rich crop, and several biofortified varieties have been notified in India with higher levels of minerals like iron, zinc, and calcium, as well as high levels of proteins and better amino acid profile as compared to maize or sorghum (Rai et al. 2008a, b). Pearl millet is mainly grown as a dual-purpose crop for grain and stover. It is also grown as a single-cut or multi-cut crop for green fodder. Pearl millet can withstand harsh growing conditions like low moisture, high temperature, and low soil fertility.

Pearl millet is cultivated over 30 million ha in arid and semi-arid tropical regions. The area of cultivation of this crop is estimated to be >18 million ha in Africa and >10 million ha in Asia (Raheem et al. 2021). It is used as a staple food by nearly 90 million people in the African Sahelian region and the north-western part of India mainly Rajasthan, Haryana, and Gujarat (Srivastava et al. 2020). In India, it is cultivated with average grain productivity of 12.43 q/ha in around 6.93 million ha area with production of around 8.61 million tons (Directorate of Millet Development 2020). Major pearl millet-producing states in India are Rajasthan (4.283 million tons), followed by Uttar Pradesh (1.302), Haryana (1.079), Gujarat (0.961), and Maharashtra (0.66) (Satyavathi et al. 2013, 2021). These states contribute nearly 90% of the total grain production in India.

In India, for fodder purposes, it is mainly cultivated as the rainfed rainy (*Kharif*) season single-cut crop. However, it is now gaining popularity in southern, western, and central India as a multi-cut forage crop during summer under irrigated conditions. A series of high-yielding varieties were recently released in both single-cut and multi-cut systems for green fodder. Nutritionally the fodder from pearl millet is better with low hydrocyanic acid and oxalic acid (considered as undesirable anti-quality factors). It is rich in protein and minerals like calcium and phosphorus (Gupta 1975; Arya et al. 2009).

The grain yield of bajra with improved cultural practices is nearly 3.0–3.5 q of grain and about 100 q of dry stover/ha under irrigated conditions and about 12–15 q of grain and 70–75 q of dry stover under rainfed conditions. The green fodder yield varies from 350 to 400 q/ha with 15–20% dry matter and 7–10% crude protein, and 50–55% dry matter digestibility. Under the multi-cut irrigated system, the green fodder yield, from summer to rainy season, is nearly 800 q/ha in 3–4 harvests as green fodder. Table 11.1 lists common pearl millet cultivars used for forage purposes.

11.2 Origin and Distribution

Archaeological retrievals indicate the origin and domestication of pearl millet to be the southern part of the Sahara Desert in West Africa (Manning et al. 2011). Evolutionary biology studies, including resequencing of diverse genotypes, also support it (Oumar et al. 2008; Burgarella et al. 2018). Based on geographical diversity and distribution, Harlan et al. (1975) proposed the center of origin to be a belt from western Sudan to Senegal. The theory gains strength from the fact that cultivated pearl millet, along with wild *Pennisetum* species and intermediate genotypes, is distributed in the Sahel (Burgarella et al. 2018). Sharma et al. (2021) based on studies indicated that the crop was domesticated in the West African Sahel around 4000 years ago and later spread to India and East Africa. Burgarella et al. (2018) reported that high genetic diversity in the crop is due to wild-to-crop gene flow, which leads to adaptive introgression of 15 genomic regions and diversity hotspots in the Eastern and Western Sahel.

The occurrence of high diversity at the phenotypic and genotypic levels can be attributed to the highly outcrossing nature of the crop (Satyavathi et al. 2013). Considerable variation for different characters can be observed within and between different cultivars. However, based on microsatellite markers studies in cultivated and wild Pennisetum studies, it is reported that a considerable loss of genetic diversity has already occurred (Oumar et al. 2008; Mariac et al. 2006). International efforts have been made to conserve the genetic diversity of the crop. In total, ~91% of this wild germplasm collection is conserved in 10 gene banks in 8 countries. About 4900 accessions of 56 wild *Pennisetum* species are conserved in 52 gene banks in 38 countries. The RS Paroda gene bank at ICRISAT, Patancheru, India, holds the world's largest collection of germplasm (24,373 accessions) from 51 countries, which includes several valuable crop wild relatives, cultivars, landraces, obsolete varieties, and advanced breeding lines (genebank.icrisat.org). (Reviewed in Sharma et al. 2021.)

Table 11.1 Details about important fodder pearl millet varieties—identified/notified

Variety	Parentage and method of breeding	Year	Breeding institution	Characteristics	Area of adoption
TSFB 18-1	IP 22269, ICRISAT-developed high tillering gene pool; population improvement through recurrent selection	2021	PJTSAU, Hyderabad & ICRISAT, Hyderabad	Multi-cut irrigated system during summer, palatable and digestible; high tillering, high crude protein and digestibility	Maharashtra, Madhya Pradesh, and Gujarat
BAIF Bajra-6	ICMV 1613 from ICRISAT, Hyderabad Pure line selection	2021	BAIF Urulikanchan, Pune	Multi-cut irrigated system during summer, palatable; high digestibility	Gujarat, Maharashtra, and Madhya Pradesh
BAIF Bajra-5	Selection from line ICMV 05777	2021	BAIF Urulikanchan, Pune	Multi-cut irrigated system during summer, very tall, thick stem, high digestibility	Gujarat, Maharashtra, and Madhya Pradesh
TSFB 17-7	IP 6073, a landrace from Central African Republic Population improvement through recurrent selection	2021	PJTSAU, Hyderabad & ICRISAT, Hyderabad	As rainfed in the *Kharif* season, in single cut dark green foliage, glabrous stem, maturity seed to seed (110–120 days), palatable and digestible	Tamil Nadu, Andhra Pradesh, Telangana, and Karnataka
K-25	A-962A, B-07888B R 104M Conventional hybridization. Hybrid	2021	Kanchan Ganga Seed Company Pvt. Ltd, Hyderabad	As rainfed in *Kharif* season, in single cut system	Tamil Nadu, Andhra Pradesh, Telangana, and Karnataka
ADV0061	The source material of male parent introduced South Africa EC-474901-Heterosis Breeding	2021	UPL Limited, Hyderabad	High green and dry fodder yield suitable as fodder multi-cut bajra in summer irrigated conditions and also in *Kharif* rainfed conditions in single cut system	Tamil Nadu, Andhra Pradesh, Telangana, Karnataka, Madhya Pradesh, Maharashtra, and Gujarat

HTBH 4902	PM0103A × PM0268R Heterosis breeding using A5 CMS system and short-day photosensitive pollinator line	2021	Hytech Seed India Private Limited Hyderabad	High green and dry fodder yield suitable as fodder multi-cut bajra in summer irrigated condition	Telangana, Andhra Pradesh, Karnataka, and Kerala
TSFB-15-4	Recurrent selection from IP 20485, IP20550, IP 20555, IP 20584	2019	PJTSAU, Hyderabad	Kharif season under rainfed conditions in single cut system	Telangana, Andhra Pradesh, Tamil Nadu, Puducherry, and Karnataka
TSFB-15-8	Recurrent selection form IP 22269, IP15352, HTGP	2019	PJTSAU, Hyderabad	Kharif season under rainfed conditions in single cut system	Telangana, Andhra Pradesh, Tamil Nadu, Puducherry, and Karnataka
AFB-37	Mass selection form cross between Bajra Bawal × MRB-8	2019	AAU, Anand	Single cut system in rainfed kharif season, light green foliage, thin stem, tall, high tillering and leafy	Gujarat
Moti Bajra (APFB-09-1)	Polycross population of APFB-2, Giant Bajra and BAIF Bajra 1	2016	PJTSAU, Hyderabad	Rainfed single cut kharif season and multi-cut summer season Tall, high tillering, dark green foliage	Telangana
RBB-1	Composite developed by selection of plants from RBC-2, NDFB-3, NDFB-5 and Giant Bajra	2016	SKRAU, Bikaner	Rainfed/irrigated kharif season, high yield, tolerant to dry conditions	Rajasthan
APFB-09-1	Recurrent selection from derivative of crossing between APFB-2, Giant Bajra and BAIF Bajra 1	2015	PJTSAU, Hyderabad	Timely sown, rainfed condition in *Kharif* season, single cut, better quality	UP, Odisha, Jharkhand, West Bengal, Assam, Bihar
PAC 981	Hybrid between 981F × 981M Heterosis breeding	2013	Advanta Limited, Hyderabad	Rainfed/irrigated kharif season, high yield	Punjab, Haryana, Rajasthan, Gujarat, MP, Maharashtra, UP B
AFB-3	Population improvement of local material from North Gujarat	2011	AAU, Anand	Single cut, Kharif season	Rajasthan, Punjab, Haryana, Western UP, Uttarakhand plains

(continued)

Table 11.1 (continued)

Variety	Parentage and method of breeding	Year	Breeding institution	Characteristics	Area of adoption
BAIF Bajra-1	Selection from the base population of Giant Bajra, pure line selection	2010	BAIF Urulikanchan, Pune	Rainfed single cut kharif season, high yield	Rajasthan, Punjab, Haryana, UP, Uttarakhand plains, Gujarat, MP, Maharashtra
Avika Bajra-19	Selection from Nagore, Rajasthan	2009	ICAR-IGFRI RRS Avikanagar	Rainfed single cut kharif season, high yield	Rajasthan, Punjab, Haryana, Western UP, Uttarakhand plains
Raj Bajra chari-2	Full sib selection from a random mating population of 20 crosses	1990	SKRAU, Bikaner	High yield, very popular, single cut, Kharif season	Whole Bajra growing area in the country
Giant Bajra	Derivative of cross between Australian Bajra × Local Bajra from Dhule district	1985	MPKV, Rahuri	High yield, very popular, single cut, Kharif season	Whole Bajra growing area in the country

11.3 Male Sterility System

Five cytoplasmic male sterility (CMS) systems are mainly used in the breeding programs of the crop. In earlier programs, three CMS sources viz., A1, A2, and A3, were used (Burton and Athwal 1967). Now two more sources (A4, A5) were identified which are derived from different accessions of *P. violaceum* (Marchais and Pernes 1985; Hanna 1989). *P. violaceum* being sexually compatible with pearl millet is an important source of stable male sterility-inducing cytoplasm (Rai et al. 1996).

Kaushal et al. (2004) reported the development of one male-sterile plant from a cross between pearl millet and its wild progenitor *P. violaceum*. The thermosensitive male sterility trait was found to be genetically controlled. The plant showed sterility at lower temperatures and was found to be fertile at higher temperatures (mean daily temperature >20 °C). Four to 5 days before anthesis was reported to be the most sensitive period for sterility induction.

11.4 Contribution of Crop Wild Relatives (CWR) in Pearl Millet Breeding

The genus *Pennisetum* Rich belongs to the tribe Paniceae of the subfamily Panicoideae in the family Poaceae. The large genus having ~80 to 140 species (Brunken 1977) is reported to have diverse basic chromosome numbers ($x = 5, 7, 8$, or 9) (Jauhar 1981). Similarly, wide diversity has been reported for other traits like diploid to octoploid accessions, and sexual to apomictic reproduction. The life cycle of the genus also exhibits species ranging from annual to perennial as well as biennial ones (Martel et al. 1997). The occurrence of species in recent divergent clades with basic chromosome numbers ($x = 5, 7, 8$) indicates that the trend of evolution was toward the increased size and reduced number of chromosomes from a higher basic chromosome number ($x = 9$) with a short length (Martel et al. 2004).

P. pedicellatum, P. purpureum, and *P. orientale* have fodder value in arid and semi-arid regions. Other wild species belonging to secondary and tertiary gene pools of pearl millet are also important as donors of desirable traits for pearl millet improvement (Hanna 1986). Several important traits like forage yield, quality, cytoplasmic male sterility, and stress tolerance can be transferred from wild *Pennisetum* species to cultivated pearl millet.

Cross-compatibility relationship studies among cultivated pearl millet and CWRs classified these species into primary (GP1), secondary (GP2), and tertiary (GP3) gene pools following Harlan and de-Wet (1971). The important constituents of GP1 with AA genome and $2n = 2x = 14$ include domesticated diploid species, *P. glaucum* ssp. *Glaucum,* wild progenitor, *P. violaceum* (Lam.). L. Rich., *P. mollissimum* Hochst. Normal chromosome pairing in the hybridization results in fertile hybrids in these species, and thus they can serve as parents for transferring desirable traits into pearl millet.

The secondary gene pool (GP2) includes mainly two perennial species Napier or elephant grass *P. purpureum* ($2n = 4x = 28$ with A′A′BB genome) and apomictic *P. squamulatum* Fresen. ($2n = 8x = 56$). These species have also been utilized to transfer desirable traits, but the hybrids are highly sterile.

The tertiary gene pool (GP3) includes the other cross-incompatible species. The only species having large chromosomes ($2n = 2x = 14$) in this gene pool is *P. schweinfurthii* (Hanna and Dujardin 1986). Strong incompatibility barriers exist between the members of GP3 and GP1 or GP2. Only in vitro techniques or complex hybrid bridges can be used to transfer desirable traits in cultivated pearl millet from these GP3 species. Tertiary gene pool species with $x = 9$ have better chances of hybridization with pearl millet than species with $x = 5$ (Dujardin and Hanna 1989a).

These CWRs are a source of abiotic and biotic stress-tolerant genes and alleles to enrich the genetic base of cultivated pearl millet. A few such species like *P. glaucum* ssp. *Monodii* (Maire) Br., *P. purpureum* Schumach, and *P. squamulatum* Fresen., *P. violaceum* have been used in the past to broaden the genetic base of pearl millet (Dujardin and Hanna 1989b; Jauhar and Hanna 1998; Kaushal et al. 2008; Obok et al. 2012).

Interspecific and interploidal hybridization has important contributions to the development of diversity for crop improvement, especially the production of cytogenetic stocks and new variants. However, interspecific and interploidal hybridization often suffers from incompatibility, which may operate at both pre-fertilization and post-fertilization stages.

A lot of potential still exists in transferring desirable traits from wild *Pennisetum* species to cultivated pearl millet, and successful hybrids have also been reported between pearl millet and wild species with objectives of improving gene pool, studying apomixis inheritance pattern, cytotaxonomic relationships, etc. Several such examples are reported with a fair degree of success, such as *P. purpureum* (Burton 1944), *P. dubium* (Gildenhuys and Brix 1958), *P. squamulatum* (Patil et al. 1961; Dujardin and Hanna 1983, 1985; Kaushal et al. 2007), *P. schweinfurthii* (Hanna and Dujardin 1986), *P. orientale* (Patil and Singh 1964; Zadoo and Singh 1986), *P. mezianum* (Nagesh and Subrahmanyam 1996), and *P. setaceum* (Hanna 1979). *P. pedicellatum, P. squamulatum,* were particularly considered as donors of forage traits, whereas, *P. orientale, P. squamulatum,* were considered donors for apomictic genes (reviewed in Sharma et al. 2020, 2021). *P. squamulatum* hybrids with pearl millet were utilized for characterization of a genomic region controlling apospory (Ozias-Akins et al. 1998, 2003) as they segregated for reproductive behavior (apomictic or sexual). In intervarietal and interploidal crosses involving *P. squamulatum* and backcross generations, pollen–pistil interaction studies indicated abnormalities like pollen tube arrest, coiling, bursting, and clumping of pollen tubes.

Hybridization of a new cytotype of *P. squamulatum* ($2n = 8x = 56$) (Roy et al. 2003) with tetraploid bajra resulted in two hybrids that segregated for several traits. It opened up the possibility of transferring several useful traits such as perenniality, apomixis, and tolerance to stress, to the cultivated bajra (Kaushal et al. 2007). A new octoploid perennial cytotype of *P. pedicellatum* has also been reported (Zadoo et al.

1997) which can be exploited for improving the pearl millet through bridge species. Interspecific hybrids with *P. squamulatum* ($2n = 54, 56$) with tetraploid pearl millet ($2n = 4x = 28$) (Dujardin and Hanna 1983; Kaushal et al. 2007) were reported earlier. Hybrids between pearl millet and *P. orientale* were reported when both parents were at diploid and tetraploid levels (Patil and Singh 1964; Hanna and Dujardin 1982).

Several trispecific hybrids involving pearl millet as one of the parents were also reported. A few examples include NBS hybrids involving *P. glaucum, P. squamulatum* and *P. purpureum* (Burton 1944), GOS hybrids involving *P. glaucum, P. orientale*, and *P. squamulatum* (Kaushal et al. 2010), an hybrid between *P. glaucum, P. orientale*, and *P. setaceum* (Dujardin and Hanna 1984). The GOS hybrid was unique in the sense as it represented three gene pools having three ploidy levels as well as two basic chromosome numbers belonging to two morphological sections and divergent modes of reproduction (sexual as well as apomictic), offering a model to study their interaction on morphological, agronomical and reproductive traits. These hybrids (F1GO, BC1GO, BC2GO, GOS) (Zadoo and Singh 1986; Kaushal et al. 2010) also present a ploidy series with serially added *P. glaucum* genomes to the *P. orientale* monoploid genome. The hybrids contained $2n = 44$ chromosomes (GGGSSO) representing 21, 14, and 9 chromosomes from *P. glaucum, P. squamulatum,* and *P. orientale,* respectively. Such interspecific hybrids in various combinations also help in understanding dosage-dependent gene regulation (Cervigni et al. 2008; Riddle et al. 2006).

Genomic tools and diagnostic markers have helped accelerate the breeding cycle and precision in introgression of traits in cultivated backgrounds. The "Pearl Millet ~1000 Genome Resequencing Project" (Varshney et al. 2017) resequenced 31 wild *Pennisetum* accessions and has contributed to the repository of more than 29 million genome-wide single nucleotide polymorphisms (SNPs). Mariac et al. (2006) studied 46 wild accessions over 421 cultivated lines from Niger and demonstrated greater allelic diversity and a higher number of alleles.

11.5 Fodder Quality and Nutritive Value

The term forage quality includes forage nutritive value as well as palatability. Livestock productivity and animal growth depend on the genetic makeup of the livestock and forage quality. The nutritive value of forage crops is an important parameter for animal productivity and growth, as it is required to meet the essential nutritional demands of livestock. The desirable herbage quality of pearl millet or hybrids is the juicy and succulent green biomass up to the harvesting stage. It may be 50% flowering stage for forage varieties whereas stay green for grain varieties. Stover quality of grain varieties is also important as many farmers keep the stover as animal feed.

Morpho-physiological stages and harvesting intervals are important factors influencing the nutritive quality of forage and its digestibility. In general, it is accepted that forage yields increase, and nutritive value declines with increasing maturity. Hence, more and more dairy farmers are opting for a multi-cut system during *Kharif* as well as the summer season. Forage digestibility is another component having a direct correlation with stages of growth. Therefore, the perfect harvesting regime is an important factor for utilizing the full nutritive potential of forages.

Pearl millet is utilized in various forms like grazing, green chop, hay, or silage because of its good quality, and being sufficient in protein with balanced amino acid components. It has a low content of fiber and lignin. Crude protein usually ranges from 9 to 12% depending upon the cutting stage and soil fertility condition in normally fertilized soils. It has high mineral content, especially calcium and iron. The forage is highly palatable at vegetative stages. Pearl millet stover has a low nutritive value; however, it is a valuable source of fodder for livestock, particularly during the lean period. In India, the utilization of grain as feed for livestock or poultry also holds great promise because of its high nutritive value. Significant variation in the pearl millet varieties for crude protein and NDF contents in stover were reported (Umutoni et al. 2021).

Dual-purpose (grain and forage) varieties with higher forage quality are another area of interest. Pearl millet "brown midrib" mutants contain less lignin, have more crude protein, and have higher dry matter digestibility (Hassanat 2007); hence, such mutants are used in hybridization programs to improve the quality of existing varieties.

Forage yield and quality traits are polygenic and quantitatively inherited. Pure line, backcross breeding, or composite population improvement could be used to improve these traits. Khairwal et al. (2009) suggested the recurrent and divergent selection to improve IVDMD through decreasing cell wall concentration, reducing lignin concentration, increasing ready energy, and/or increasing crude protein.

Several studies indicate the existence of considerable variability in quality parameters like dry matter, protein, fiber content, and digestibility (Jindal et al. 2009; Blümmel et al. 2007; Amodu et al. 2007). Improving tillering capacity, leafiness, and sweetness of stem through genetic modifications has been suggested to improve nutritive quality (Mathur et al. 1993). The residual sugars in the stover or accumulated in green forage are the parameters of forage quality.

QTLs for drought tolerance, stover quality contributing to the improvement of metabolizable energy, in-vitro organic matter digestibility, sugar content, and gas volume were mapped onto the linkage groups (LGs) 2, 4, and 5 (Nepolean et al. 2009). These QTLs were introgressed into four parental lines of existing hybrids. Three QTLs for dry stover yield were found across LG3, LG5, and LG6 (Nepolean et al. 2006).

Stay-green genotypes are of high significance because of less cellulose and lignin and higher protein. These also have early vigor, better heat tolerance, and high biomass and hence are a better option from a nutritive point of view (Narwal et al. 2009).

11.6 Bajra × Napier Hybrid

Napier grass and pearl millet are highly cross-compatible in both ways cross (Burton 1944; Hanna et al. 2004; Dowling et al. 2013). The hybrids are usually a combination of desirable traits from both the parents, such as vigor, drought resistance, forage quality from pearl millet and faster growth, perenniality, and better dry matter from Napier (Timbo et al. 2010). Several perennial hybrid varieties are widely cultivated in tropical and subtropical Asia, Africa, Southern Europe, and America. In India, it is cultivated in several states like Punjab, Haryana, Uttar Pradesh, Bihar, Madhya Pradesh, Orissa, Gujarat, West Bengal, and Assam, and is now being spread in other states also.

Bajra × Napier is an interspecific hybrid between bajra (*Pennisetum glaucum*) and common Napier grass (*Pennisetum purpureum*). It was first produced in South Africa, and the cultivar was named Babala Napier hybrid or Bana grass. In India, Tamil Nadu Agricultural University, Coimbatore, and Indian Agricultural Research Institute, Pusa, took the initiative in hybridization, resulting in the release of the Cumbu Napier hybrid and Pusa giant Napier, respectively. High heterosis showing high fodder yield coupled with better crude protein and digestibility are the characteristics of these hybrids. These hybrids are usually a good combination of traits like better quality, fast growth from bajra with a deep root system, high productivity, multi-cut regeneration capability, tillering, and perenniality from Napier grass. However, these hybrids are sterile and propagated vegetatively through stem cuttings or rooted slips. The interspecific hybrids are sterile triploid (AA′B genomes) with $2n = 3x = 21$ chromosomes (7 from pearl millet and 14 from Napier). These triploid hybrids are usually highly variable because of the heterozygosity of Napier grass, even if the pearl millet parent is an inbred (Hanna et al. 2004). These high-yielding tropical fodder hybrids are utilized mostly by undercut and carry systems for stall feeding. It can also be converted to good-quality silage or hay. A series of high-yielding varieties have been developed under the NARS system (Table 11.2). CO(CN)-4 and CO(BN)-5 became highly popular and revolutionized the fodder scenario in the country.

11.7 *Pennisetum* Hybrid (*P. glaucum* × *P. squamulatum*)

A new combination of interspecific hybridization was used to develop drought tolerant, perennial hybrid at ICAR- IGFRI, Jhansi, which resulted in a variety Bundel-Bajra-Squamulatum Hybrid-1 (BBSH-1) (Table 11.3). It is a multi-cut perennial hybrid. It is a derivative of the cross between *P. glaucum* Tetra-$12n = 28$ (IG99-748/INGR 09047) × *P. squamulatum* $2n = 56$ (INGR 06017/ IC546955). Plants are erect, multi-tillered with perennial growth habits, tolerant to major insects and diseases, and have good quality and digestibility (Roy et al. 2020, 2022).

Table 11.2 Details about important fodder Bajra × Napier hybrid varieties (*Pennisetum glaucum* × *P. purpureum*)

Variety	Parentage and method of breeding	Year	Breeding institution	Characteristics	Area of adoption
BNH-11	BAIF Bajra-1 × BRN	2020	BAIF, Uralikanchan, Pune	Tall, quick regeneration, soft long, broad leaves, profuse tillering, non-lodging	Punjab, Haryana, Rajasthan, UP, Gujarat, MP, Maharashtra, CG, Karnataka, Andhra Pradesh, Telangana, Kerala, TN
BNH 14	BAIF Bajra-1 × BRN	2020	BAIF, Uralikanchan, Pune	Tall, quick regeneration, dark green foliage, thin stem, soft long, narrow, semi-erect leaves, profuse tillering	Punjab, Haryana, Rajasthan, Karnataka, Andhra Pradesh, Telangana, Kerala, TN
Co-6 [TNCN-1280]	Fodder Bajra-Co-7 × FD 459	2019	TNAU, Coimbatore	Profuse tillering, tall, leafy, long and broad leaves, high crude protein and digestibility	Punjab, Haryana, Rajasthan, UP, Gujarat, MP, Maharashtra, CG
PBN 351	BAIF Bajra-1 × NGL-9	2019	PAU, Ludhiana	Hairless stem, long and broad leaves	UP, Gujarat, MP, Maharashtra, CG
PBN-342	Kale Bajra × N-22	2018	PAU, Ludhiana	Perennial, high leaf stem ratio, quick regeneration	Punjab, Haryana, UP, Rajasthan, Odisha, Karnataka, Assam, TN
Phule Gunwant	Giant Bajra × FD-473-1	2017	MPKV, Rahuri	Perennial, high leaf stem ratio, quick regeneration	Maharashtra
BNH-10	BAIF Bjara-1 × BRN-2	2015	BAIF, Uralikanchan, Pune	Perennial, very long green foliage, broad, semi-drooping leaves, soft, succulent	Punjab, Haryana, UP, Jharkhand, Odisha, Gujarat, MP, Maharashtra, Karnataka, Andhra Pradesh, Kerala, TN
CO (BN)-5 [TNCN 074]	Fodder Bajra IP 20594 × FD 437	2014	TNAU, Coimbatore	Profusely tillering, highly leafy, long and broad leaves, high yield, quick regeneratio,	All over India
DHN-6 (Sampoorna)	IPM-14188 × FD 484	2009	ICAR-IGFRI, Jhansi RRS Dharwad	Perennial, robust and erect plant with long broad and droopy leaves, quick regeneration	Karnataka
CO (CN)-4	Fodder Combu CO8 × FD 461	2008	TNAU, Coimbatore	Profusely tillering, highly leafy, high yield, quick regeneratio,	Tamil Nadu

APBN-1	IPN 12159 × Kasate-52440	1997	ANGRAU, Hyderabad	Tall plant, perennial, quick regeneration	Andhra Pradesh
Swetika-1 (IGFRI-3)	Bajra (PSB-2) × Napier	1983	ICAR-IGFRI, Jhansi	Perennial, erect growth, thin stem, quick regeneration	Andhra Pradesh, Karnataka, Kerala, HP, Assam

Note: All varieties developed through interspecific hybridization followed by clonal selection, are perennial, multi-cut and cultivated under irrigated conditions

Table 11.3 Details about fodder Bajra × squamulatum hybrid variety (*Pennisetum glaucum* × *P. squamulatum*)

Variety	Parentage and method of breeding	Year	Breeding institution	Characteristics	Area of adoption
BBSH-1	Derivative of cross between Bajra Tetra-1 (INGR09047) × *P. squamulatum* (INGR 06017 $2n = 56$)	2020	ICAR-IGFRI, Jhansi	Erect, multi-tillered plants, perennial, multi-cut, high quality, and digestibility	Punjab, Haryana, Gujarat, Maharashtra, HP, Assam

11.8 Apomixis

Apomixis is a process of producing seed without the normal sexual process of meiosis and fertilization, and it is a potent tool to fix heterozygosity and hybrid vigor. It produces true-to-type maternal seed. It is proposed as a "next-generation breeding technology" (Hand and Koltunow 2014).

Apospory, a form of apomixis, is prevalent in many *Pennisetum* species (Ozias-Akins et al. 1998). The GOS hybrid, as mentioned above, offered the first report of the existence of apomeiosis independent of parthenogenesis and polyploidy in the genus *Pennisetum* (Kaushal et al. 2010). It was further exploited to identify differentially regulated genes involved in the development of apomeiotic Embryo sac. Complete and independent expression of apomeiosis, including an apomeiotic non-parthenogenetic inter-specific hybrid between two sexual diploid species viz., *P. glaucum* and *P. orientale*, has been reported (Kaushal et al. 2010).

Expression pattern analysis of a total of 96 non-redundant transcripts indicated their involvement in floral development during apomeiotic embryo sac development (Panicum-type aposporous). These transcripts were recovered after a comparative transcriptome analysis of differentially expressed genes during embryo sac (ES) development in a sexual *P. glaucum* (genotype 81A1) and its aposporic non-parthenogenetic interspecific hybrid (BC1GO) (Sahu et al. 2012).

Detailed molecular analysis of genomic regions governing apomixis in natural apomictic systems led to the identification and characterization of key genes involved in apomictic reproduction. These include genes controlling parthenogenesis, such as ASGR-BBML (Apospory Specific Genomic Region-Baby Boom; *Pennisetum squamulatum*) (Conner et al. 2015). Independent recurrent parthenogenesis, such as lop mutants in Hieracium; PsBBML in *Pennisetum*, has been achieved experimentally (Koltunow et al. 2011; Conner et al. 2015; Mirzagadheri and Horandl 2016). By utilizing the PsASGR-BBML gene, parthenogenetic haploids in sexual crops such as pearl millet, rice, and maize were successfully developed (Conner et al. 2015, 2017), and such genes are reported to be conserved across Paniceae species (Worthington et al. 2016).

Concerted efforts have been made in the past to transfer apomixis genes from *P. orientale* into pearl millet (Dujardin and Hanna 1987; Hanna and Dujardin 1982). Similarly, *P. purpureum* ($2n = 4x = 28$) was used as a bridge species (Dujardin and Hanna 1989b, c) to transfer apomixis genes from *P. squamulatum* ($2n = 6x = 54$) into tetraploid pearl millet.

Successful demonstration of the presence of *P. squamulatum* DNA in the obligate apomictic backcross line resulting from the hybridization of *P. squamulatum* and pearl millet indicated that the genes for apomixis could be transmitted by a single chromosome (Ozias-Akins et al. 1993).

Transferring apomixis genes to cultivated pearl millet will have several advantages as desired gene combination with heterozygosity, and hybrid vigor can be fixed, and there will be no need to produce crosses every year. *P. orientale, P. squamulatum,* and *P. purpureum* have been used in various combinations to get interspecific hybrids and even trispecific hybrids by several authors (reviewed in

Kaushal et al. 2019; Sharma et al. 2021). However, till now, no success has been achieved in getting apomictic bajra, whereas several desired combinations have been obtained to get perennial plants and varieties.

11.9 Breeding Aspect

Evaluation of pearl millet germplasm for forage traits was started in the 1960s. Various morpho-agronomic parameters were studied in different institutions, and utilizing biometric tools, the heritability, genetic advance, correlation, and path coefficient were analyzed. In general, it was reported that the plant height, tiller number, and the number of leaves have a direct and positive correlation with the green fodder and dry matter yield. The estimates of variability like range, coefficient of variation, heritability, and genetic advance have been used for designing suitable selection criteria for selecting high fodder-yielding lines. The development of core and mini-core collections (Upadhyaya et al. 2011) has helped the breeder to use the desired germplasm.

ICRISAT accessions of pearl millet, including landraces and advanced breeders' line, revealed considerable variability for various components contributing to fodder yield such as plant height, number of tillers/plant, stem thickness, number of leaves, leaf length and leaf width (Khairwal et al. 2009). Global pearl millet collections were analyzed and reported by several authors, which indicated high variability for protein, mineral contents like phosphorus, calcium, and anti-nutritional factors like oxalic acid (reviewed in Kumar et al. 2012).

The earlier programs included tools like mass selection, S1 progeny selection, recurrent selection, and backcross method, as well as population breeding program, and have resulted in very good cultivars. Nowadays, selective hybridization utilizing the CMS system is more prevalent, especially in developing multi-cut high-yielding hybrids. ICRISAT germplasm is contributing significantly in recent cultivars. Although significant progress has been made, the potential for further improvements exists, especially to target biotic and abiotic stress as well as nutritional quality for both green fodder and stover (Yadav 2010; Yadav and Rai 2013).

Evaluation of landraces indicated significant differences and heterosis among accessions in total biomass, grain, and stover yields (Yadav and Bidinger 2008). These findings suggest that landraces can be exploited for developing dual-purpose varieties and the introgression of desirable traits. Landrace-based top cross hybrids may provide an alternative to the Open-pollinated variety. In general, population hybrids and composite varieties are better options as compared to open-pollinated varieties. These require a national integrated breeding program with effective sharing of germplasm and progenies. The significant positive correlation of grain and fodder yields with the major contributing traits could help in developing high-yielding dual-purpose cultivars.

Forage yield was found to have positively correlated with plant height, leaf, tiller, node number, and stem diameter (Lopez-Dominguez et al. 2001). Yadav et al. (2012) reported that the dry fodder yield was significantly and positively associated with effective tillers/plant, plant height, ear length, ear weight, and negatively with panicle emergence. A positive correlation between fodder and grain yield was reported by Reddy et al. (2003). Fodder yield and digestibility were positively correlated and showed a positive association with several agronomic traits like plant height, leaf number, and number of tillers (Navale et al. 1995; Kumar et al. 2012). Fodder yield was reported to be positively associated with days to maturity, number of productive tillers, flag leaf area, ear girth, and grain yield (Harer and Karad 1998).

Kaushal et al. (2003) reported embryo cloning in pearl millet, a tool to produce multiple plantlets from immature zygotic embryos. This has the scope to provide an opportunity for rapid clonal propagation of Pennisetum embryos in lesser time.

Serba et al. (2020) evaluated 100 pearl millet accessions, comprising 83 open-pollinated varieties (OPV), 5 composites and 1 synthetic variety, 5 hybrids, and 6 landraces at 14 environments in Burkina Faso, Mali, Niger, and Senegal for 2 years. Significant differences among genotypes and prevalence of high genotype-by-environment interaction effects were reported. Significant positive correlations of plant height, panicle length, and panicle yield with grain yield opened up the possibility for simultaneous improvement.

Upadhyaya et al. (2017) evaluated 326 pearl millet accessions, including 318 landraces selected for fodder traits from the world collection at ICRISAT genebank, India, in 3 seasons to identify promising sources for fodder yield. The major reasons for adopting landraces include adaptation to climate vagaries, better fodder yield, and stability, and abiotic and biotic stress tolerance. Accumulation of such traits in landraces might have occurred due to continuous selection by farmers for better yield (Bellon 1990; Bhatnagar 2002).

11.10 Forage Traits

Moussa et al. (2021) in a survey based on individual and focus group interviews of 508 stakeholders in Niger, from 7 major ethnic groups, prepared a forage pearl millet profile. The forage profile included the criteria of producing abundant tillers, tall nature, long and wide leaves, slender stems with many leaves, capacity to regenerate after grazing, and producing more biomass than grains quickly.

Various studies indicate that late maturity, plant height, thick stems, long and broad leaves, high tillering, and sweetness of stem are the desired traits for good fodder variety. Landraces have not been fully exploited in breeding programs (Jindal et al. 2009). Dry matter protein, fiber, and lignin, as well as in vitro and in vivo dry matter digestibility, are the indicators of forage quality (Ouendeba et al. 1996; Jindal et al. 2009).

Morpho-physiological stages and clipping intervals are two important factors, influencing the nutritive quality of forages. Variations in the nutritional composition of forage species with the advancement in growth need consideration while determining the optimum stage of harvesting without adversely affecting forage productivity. Among many factors influencing the quality of forage, an in-depth knowledge of the effect of seasonal variation on the nutritional composition of forages is a prerequisite for determining the best harvesting regimes for obtaining good quality fodder.

11.11 Genetics of Important Pearl Millet Diseases

Pearl millet is relatively less affected by diseases compared to other major cereals due to its hard nature and suitability for growing in harsh conditions. However, there are a few major diseases adversely affecting the crop at various growth stages. The two most important diseases that affect forage pearl millet and reduce forage yield, as well as quality, are Downy mildew and Blast.

11.12 Pearl Millet Downy Mildew

Downy mildew caused by *Sclerospora graminicola* (Sacc.) J. Schrot is an important disease and causes economic losses in India and Africa under favorable environmental conditions. The disease manifests in the form of pale, chlorotic broad streaks on the leaves extending from base to tip, which turn brown, and the leaves become shredded longitudinally with the progression of the disease. In severe infection, the downy fungal growth can be seen on the upper as well as the lower surface of the leaves. The ear either fails to form or, if formed, is malformed to green leafy structures, gradually becoming brown and dry with the progression of the disease. It results in an economic loss of both fodder and seed yield. Grain yield loss due to this disease is estimated to vary between approximately 20% and 40% (Singh 1995; Hash et al. 1999; Hess et al. 2002). Average forage yield losses of 33% can be caused by the disease if appropriate management practices are not applied (AICRP-FCU Annual report Part-I Kharif 2016 to 2019 2018). The development and utilization of resistance sources is the most economic method of managing the disease. A large number of germplasm accessions and breeding lines have been identified as potential resistance sources (Singh et al. 1987, 1997), which have been subsequently used in the development of disease-resistant varieties and hybrids. Studies indicate that resistance is governed by major dominant genes with both additive and non-additive gene action (Basavaraju et al. 1980, 1981; Shinde et al. 1984). Quantitative trait loci (QTL) for downy mildew resistance have been identified utilizing marker-based linkage maps (Jones et al. 1995, 2002; Azhaguvel 2001; Breese et al. 2002), and some of them have been found to be specific to different pathotypes of *Sclerospora graminicola* (Hash et al. 1999; Hash and Witcombe 2002; Jones et al. 2002). In addition, efforts were also made to pyramiding QTL for incomplete resistance from diverse sources in the commercially successful parental lines that have become susceptible to the new, more virulent pathotypes (Hash et al. 2006). This sort of approach will result in the development of stable and durable resistance against this disease.

11.13 Pearl Millet Blast

The blast disease, caused by *Magnaporthe grisea* (Herbert) Barr [anamorph: *Pyricularia grisea* (Cooke) Sacc.], has emerged as one of the most destructive diseases in the recent past mainly driven by factors such as commercialization of new hybrids with varying levels of susceptibility and climate change. It affects forage and grain yield (Sharma et al. 2013). The blast causes substantial yield losses in grain (Timper et al. 2002) and forage pearl millet (Wilson and Gates 1993). Due to its widespread occurrence in epidemic proportions in different parts of the country, the blast has now been identified as one of the major goals of pearl millet improvement programs.

The disease appears as small, roundish, elliptical, and diamond-shaped to elongated lesions with grey centers, which subsequently enlarge and become necrotic. It produces chlorosis and early senescence of young leaves (Kato 2001). These symptoms on leaf and stem appear during all stages, from seedling to flowering, giving a burnt appearance in severe outbreaks. Severely infected plants produce very few shriveled grains. Blast disease severity increases in high plant-density fields coupled with humid weather (Thakur et al. 2011).

Identification and utilization of resistance sources is an effective and economical method for management. Resistance to *Magnaporthe* blast in pearl millet tends to be dominant or partially dominant (Wilson and Hanna 1992). Resistance to *M. grisea* against Indian isolate was reported to be governed by a single dominant gene (Gupta et al. 2012). The wild sub-species, *P. glaucum* subsp. *Monodii*, has been identified as an important source for pearl millet improvement programs as genetic resistance to rust, leaf blast, and smut (Wilson and Hanna 1992) has been identified in "*monodii*" accessions. These resistance features have been incorporated into inbred lines to develop pearl millet forage (Hanna et al. 1988) and grain hybrids (Hanna and Wells 1993). Moreover, the quality genome of reference genotype Tift23D2B1-P1-P5 has been sequenced, and these genetic and genomic resources can be used to map and deployment of effective blast resistance genes as well as to search for other blast resistance loci (Sehgal et al. 2015) in pearl millet. Two major blast resistance QTLs, on LG 4 and 7 of pearl millet line 863B-P2, were identified which can be utilized (Yadav et al. 2021).

11.14 Discussion and Conclusion

Concerted efforts in the past have resulted in successful conservation and utilization of global genetic diversity in the crop. Wide diversity exists in the area of origin and place of domestication. Being a fast-growing C4 plant highly tolerant to low water conditions, it is a favorite crop for drought-prone areas globally. The high mineral content and nutritive parameters make the crop suitable as a biofortified crop and have been found crucial for the well-being of poor persons. The crop is now gaining enhanced attention for both fodder and grain purposes. The introduction of perenniality, apomixis, multi-cut nature, and regeneration potential from wild relatives has made the crop more resilient, suitable for different situations, and adaptable. Several multi-cut forage varieties have been developed in the recent past, which have extended the cultivation of crop under irrigated conditions for green and nutritious fodder, especially in central, western, and southern India. New varieties of the Bajra × Napier hybrid, although sterile and vegetatively propagated, are being adopted across the country and have the potential to become the most widely accepted fodder crop in the country. The wild relatives should be used to develop more such hybrids with biomass production potential coupled with improved nutritive quality. The crop is now facing problems with two important diseases, Downy mildew, and blast. Screening germplasm, especially the landraces and crop wild relatives, for identifying the source of resistance is the need of the hour.

The future program should concentrate on widening the genetic base of the crop, especially for abiotic and biotic stress tolerance by hybridization with crop wild relatives. Screening of progenies in different agro-climatic zones will lead to the identification of suitable varieties for several states. The biofortified varieties for grain, as well as fodder, should be identified and popularized to provide a cheap source of nutrition to the people and livestock. The lines with good regeneration potential will offer a cheap source of high-quality fodder biomass for livestock for a prolonged period.

Acknowledgments Authors are also immensely thankful to the Indian Council of Agricultural Research, New Delhi, and the Department of Biotechnology and Director, ICAR-IGFRI, Jhansi, for felicitating the research work.

References

AICRP-FCU, 2016-2019. Annual Reports: Part-I (Kharif 2016-Kharif 2019) (2018) All India Coordinated Research Project on forage crops and utilization. ICAR-IGFRI, Jhansi

Amodu JT, Adeyinka I, Kallah M, Alawa JP (2007) Evaluation of pearl millet accessions for yield and nutrient composition. J Biol Sci 7:379–383. https://doi.org/10.3923/jbs.2007.379.383

Arya RK, Yadav HP, Yadav AK, Deshraj (2009) Forage symposium on "Emerging trends in forage research and livestock production, Feb. 16-17, 2009 at CAZRI, RRS, Jaisalmer (Rajasthan) India, pp 56–59

Azhaguvel P (2001) Linkage map construction and identification of QTLs for downy mildew (*Sclerospora graminicola*) resistance in pearl millet (*Pennisetum glaucum* (L.) R. Br.). PhD thesis, Department of Agricultural Botany, Agricultural College and Research Institute, Tamil Nadu Agricultural University, Madurai, Tamil Nadu, India

Basavaraju R, Safeeulla KM, Murty BR (1980) The role of gene effects and heterosis for resistance to downy mildew in pearl millet. Indian J Genet 40:537–548

Basavaraju R, Safeeulla KM, Murty BR (1981) Genetic variance and heritability for resistance to downy mildew in pearl millet. Indian J Genet 41:137–143

Bellon MR (1990) The ethnoecology of maize production under technological change. PhD dissertation, University of California, Davis

Bhatnagar SK (2002) Measuring and monitoring genetic erosion of pearl millet landraces through participatory approach. Project report (December 1999–December 2001) submitted to IPGRI, Regional Office for South Asia, New Delhi

Blümmel M, Bidinger FR, Hash CT (2007) Management and cultivar effects on ruminant nutritional quality of pearl millet (*Pennisetum glaucum* (L.) R.Br.) stover: II. Effects of cultivar choice on stover quality and productivity. Field Crop Res 103(2):129–139

Breese WA, Hash CT, Devos KM, Howarth CJ (2002) Pearl millet genomics – an overview with respect to breeding for resistance to downy mildew. In: Leslie JF (ed) Sorghum and millets diseases. Iowa State Press, Ames, pp 243–246

Brunken JN (1977) A systematic study of *Pennisetum* sect. *Pennisetum* (Gramineae). Am J Bot 64: 161–176. https://doi.org/10.1002/j.1537-2197.1977.tb15715.x

Burgarella C, Cubry P, Kane NA, Varshney RK, Mariac C, Liu X, Shi C, Thudi M, Couderc M, Xu X, Chitikineni A, Scarcelli N, Barnaud A, Rhoné B, Dupuy C, François O, Berthouly-Salazar C, Vigouroux Y (2018) A Western Sahara centre of domestication inferred from pearl millet genomes. Nat Ecol Evol 2:1377–1380. https://doi.org/10.1038/s41559-018-0643-y

Burton GW (1944) Hybrids between Napier grass and cattail millet. J Hered 35:227–232

Burton GW, Athwal DS (1967) Two additional sources of cytoplasmic male-sterility in pearl millet and their relationships to TIFT 23A. Crop Sci 7:209–211

Cervigni GD, Paniego N, Pessino S, Selva JP, Díaz M, Spangenberg G, Echenique V (2008) Gene expression in diplosporous and sexual *Eragrostis curvula* genotypes with differing ploidy levels. Plant Mol Biol 67:11–23

Conner JA, Mookkan M, Huo H, Chae K, Ozias-Akins P (2015) A parthenogenesis gene of apomictic origin elicits embryo formation from un-fertilized eggs in a sexual plant. Proc Natl Acad Sci U S A 112:11205–11210. https://doi.org/10.1073/pnas.1505856112

Conner JA, Podio M, Ozias-Akins P (2017) Haploid embryo production in rice and maize induced by PsASGR-BBML transgenes. Plant Reprod 30:41–52. https://doi.org/10.1007/s00497-017-0298-x

Directorate of Millet Development (2020) Crop profile. http://www.aicpmip.res.in/aboutus.html

Dowling CD, Burson BL, Foster JL, Tarpley L, Jessup RW (2013) Confirmation of pearl millet-Napier grass hybrids using EST-derived simple sequence repeat (SSR) markers. Am J Plant Sci 4:1004–1012. https://doi.org/10.4236/ajps.2013.45124

Dujardin M, Hanna WW (1983) Apomictic and sexual pearl millet × *Pennisetum squamulatum* hybrids. J Hered 74:277–279. https://doi.org/10.1093/oxfordjournals.jhered.a109783

Dujardin M, Hanna WW (1984) Pseudogamous parthenogenesis and fertilization of a pearl millet x *Pennisetum orientale* derivative. J Hered 75:503–504

Dujardin M, Hanna WW (1985) Cytology and reproduction of reciprocal backcrosses between pearl millet and sexual and apomictic hybrids of pearl millet x *Pennisetum squamulatum*. Crop Sci 25: 59–62

Dujardin M, Hanna WW (1987) Inducing male fertility in crosses between pearl millet and *Pennisetum orientale* Rich. Crop Sci 27:65–68. https://doi.org/10.2135/cropsci1987.0011183X002700010017x

Dujardin M, Hanna WW (1989a) Crossability of pearl millet with wild *Pennisetum* species. Crop Sci 29:77–80. https://doi.org/10.2135/cropsci1989.0011183x0029000100

Dujardin M, Hanna WW (1989b) Developing apomictic pearl millet-characterization of a BC3 plant. J Genet Plant Breed 43:145–151

Dujardin M, Hanna WW (1989c) Fertility improvement in tetraploid pearl millet. Euphytica 42: 285–289. https://doi.org/10.1007/BF00034466

Gildenhuys P, Brix K (1958) Cytological abnormalities in *Pennisetum dubium*. Heredity 12:441–452

Gupta VP (1975) Forage Res 1:54–60. Cited in Kumar et al 2012

Gupta SK, Sharma R, Rai KN, Thakur RP (2012) Inheritance of foliar blast resistance in pearl millet (*Pennisetum glaucum* L. (R.) Br.). Plant Breed 131:217–219

Hand ML, Koltunow AMG (2014) The genetic control of apomixis: asexual seed formation. Genetics 197:441–450. https://doi.org/10.1534/genetics.114.163105

Hanna WW (1979) Interspecific hybrids between pearl millet and fountain grass. J Hered 70:425–427

Hanna WW (1986) Utilization of wild relatives of pearl millet. In: Proceedings of international pearl millet workshop, 7–11 April 1986, ICRISAT, Patancheru, India, pp 33–42
Hanna WW (1989) Characteristics and stability of a new cytoplasmic-nuclear male sterile source in pearl millet. Crop Sci 29:1457–1459. https://doi.org/10.2135/cropsci1989.0011183X002900060026x
Hanna WW, Dujardin M (1982) Apomictic interspecific hybrids between pearl millet and *Pennisetum orientale* L.C. Rich. Crop Sci 22:857–859. https://doi.org/10.2135/cropsci1982.0011183X002200040038x
Hanna WW, Dujardin M (1986) Cytogenetics of *Pennisetum schweinfurthii* Pilger and its hybrids with pearl millet. Crop Sci 26:449–453. https://doi.org/10.2135/cropsci1986.0011183X002600030001x
Hanna WW, Wells HD (1993) Registration of parental line Tift 89D2, rust resistant pearl millet. Crop Sci 33:361–362
Hanna WW, Wells HD, Burton GW, Hill GM, Monson WG (1988) Registration of 'Tifleaf 2' pearl millet. Crop Sci 28(6):1023
Hanna WW, Chaparro CJ, Mathews BW, Burns JC, Sollenberger LE, Carpenter JR (2004) Perennial *Pennisetums*. In: Moser LE, et al (ed) Warm-season (C4) grasses. Agron. Monogr. 45. ASA, CSSA, and SSSA, Madison, pp 503–535
Harer PN, Karad SR (1998) Correlation and path coefficient analysis in pearl millet (Pennisetum glaucum L.). J Maharastra Agric Univ 23:132–135. Cited in Kumar et al 2012
Harlan JR, de-Wet JMJ (1971) Toward a rational classification of cultivated plants. Taxonomy 20: 509–517. https://doi.org/10.2307/1218252
Harlan JR, de Wet JMJ, Stemler ABL (1975) Plant domestication and indigenous African agriculture. In: Harlan JR, deWet IMJ, Stemler ABL (eds) Origin of African domestication. Mouton, The Hague, pp 1–9
Hash CT, Witcombe JR (2002) Gene management and breeding for downy mildew resistance. In: Leslie JF (ed) Sorghum and millets pathology 2000. Iowa State Press, Ames, pp 27–36
Hash CT, Singh SD, Thakur RP, Talukdar BS (1999) Breeding for disease resistance. In: Khairwal IS, Rai KN, Andrews DJ, Harinarayana G (eds) Pearl millet breeding. Oxford & IBH, New Delhi, pp 337–379
Hash CT, Thakur RP, Rao VP, Bhasker Raj AG (2006) Evidence for enhanced resistance to diverse isolates of Pearl Millet Downy Mildew through gene pyramiding. Int Sorghum Millets Newslett 47:134–138
Hassanat F (2007) Evaluation of pearl millet forage. PhD dissertation, Department of Animal Science, McGill University, Montreal
Hess DE, Thakur RP, Hash CT, Sereme P, Magill CW (2002) Pearl millet downy mildew: problems and control strategies for a new millennium. In: Leslie JF (ed) Sorghum and millets diseases. Iowa State Press, Ames, pp 37–42
Jauhar PP (1981) Cytogenetics and breeding of pearl millet and related species. Alan R. Liss, New York
Jauhar PP, Hanna WW (1998) Cytogenetics and genetics of pearl millet. Adv Agron 64:1–26
Jindal Y, Yadav R, Pahuja SK, Malik V (2009) Forage symposium on "Emerging trends in forage research and livestock production, Feb. 16-17, 2009 at CAZRI, RRS, Jaisalmer, pp 74–78
Jones ES, Liu CJ, Gale MD, Hash CT, Witcombe JR (1995) Mapping quantitative trait loci for downy mildew resistance in pearl millet. Theor Appl Genet 91:448–456
Jones ES, Breese WA, Liu CJ, Singh SD, Shaw DS, Witcombe JR (2002) Mapping quantitative trait loci for resistance to downy mildew in pearl millet: field and glasshouse screens detect the same QTL. Crop Sci 42:1316–1323
Kato H (2001) Rice blast control. In: Pesticide Outlook. Japan Crop Protection Association (JCPA), Tokyo, pp 23–25
Kaushal P, Roy AK, Zadoo SN, Choubey RN (2003) Embryo cloning in *Pennisetum glaucum:* production of multiple plantlets from immature zygotic embryos. Range Manag Agrofor 24(2): 156–158

Kaushal P, Roy AK, Zadoo SN, Choubey RN (2004) Cytogenetic analysis of Thermosensitive Genic Male Sterility (TGMS) recovered from a *Pennisetum glaucum* (L.) R. Br. *P. violaceum* (Lam.) L. Rich cross. Cytologia 69(4):409–418

Kaushal P, Roy AK, Khare A, Malaviya DR, Zadoo SN, Choubey RN (2007) Crossability and characterization of interspecific hybrids between sexual *Pennisetum glaucum* (pearl millet) and a new cytotype ($2n=56$) of apomictic *P. squamulatum*. Cytologia 72:111–118. https://doi.org/10.1508/cytologia.72.111

Kaushal P, Khare A, Zadoo SN, Roy AK, Malaviya DR, Agrawal A, Choubey RN (2008) Sequential reduction of *Pennisetum squamulatum* genome complement in *P. glaucum* ($2n = 28$) × *P. squamulatum* ($2n = 56$) hybrids and their progenies revealed its octoploid status. Cytologia 73:151–158. https://doi.org/10.1508/cytologia.73.151

Kaushal P, Khare A, Siddiqui SA, Agrawal A, Paul S, Malaviya DR, Roy AK, Zadoo SN (2010) Morphological, cytological and reproductive characterization of tri-species hybrids (GOS) between *Pennisetum glaucum, P. orientale* and *P. squamulatum*. Euphytica 174:261–281

Kaushal P, Dwivedi KK, Radhakrishna A, Srivastava MK, Kumar V, Roy AK, Malaviya DR (2019) Partitioning apomixis components to understand and utilize gametophytic apomixis. Front Plant Sci 10:256

Khairwal IS, Rajpurohit BS, Kumari DS (2009) Pearl millet: a potential forage and fodder crop. In: Forage symposium on emerging trends in forage research and livestock production, 16–17 February 2009, CAZRI, RRS, Jaisalmer, pp 39–49

Koltunow AMG, Johnson SD, Rodrigues JCM, Okada T, Hu Y, Tsuchiya T et al (2011) Sexual reproduction is the default mode in apomictic *Hieracium* subgenus Pilosella, in which two dominant loci function to enable apomixis. Plant J 66:890–902. https://doi.org/10.1111/j.1365-313X.2011.04556.x

Kumar A, Arya RK, Sunil K, Dharmender K, Suresh K, Ravish P (2012) Advances in pearl millet fodder yield and quality improvement through breeding and management practices. Forage Res 38(1):1–14

Lopez-Dominguez U, Maiti RK, Wesche-Ebeling P, Ramirez RGL, Verde-Star J (2001) Agro-biological factors influencing the productivity and forage quality of some pearl millet [*Pennisetum glaucum* (L.) R. Br. Emend stuntz]. Res Crops 2:263–277

Manning K, Pelling R, Higham T, Schwenniger JL, Fuller DQ (2011) 4500-Year old domesticated pearl millet (*Pennisetum glaucum*) from the Tilemsi Valley, Mali: new insights into an alternative cereal domestication pathway. J Archaeol Sci 38:312–322. https://doi.org/10.1016/j.jas.2010.09.007

Marchais L, Pernes J (1985) Genetic divergence between wild and cultivated pearl millets (*Pennisetum typhoides*). 1. Male sterility. Zeitschrift fur Pflanzenzüchtung 95:103–112

Mariac C, Luong V, Kapran I, Mamadou A, Sagnard F, Deu M et al (2006) Diversity of wild and cultivated pearl millet accessions (*Pennisetum glaucum* [L.] Br.) in Niger assessed by microsatellites. Theor Appl Genet 114:49–58. https://doi.org/10.1007/s00122-006-0409-9

Martel E, De Nay D, Siljak-Yakovlev S, Brown S, Sarr A (1997) Genome size variation and basic chromosome number in pearl millet and fourteen related *Pennisetum* species. J Hered 88:139–143. https://doi.org/10.1093/oxfordjournals.jhered.a023072

Martel E, Poncet V, Lamy F, Siljak-Yakovlev S, Lejeune B, Sarr A (2004) Chromosome evolution of *Pennisetum* species (Poaceae): implications of ITS phylogeny. Plant Syst Evol 249:139–149. https://doi.org/10.1007/s00606-004-0191-6

Mathur PN, Appa Rao S, Sapra RL, Mengesha MH, Rana RS (1993) Catalogue: evaluation of pearl millet germplasm. Part-2. National Bureau of Plant Genetic Resources (NBPGR) and International Crops Research Institute for the Semi-Arid Tropics (ICRISAT) collaborative programme. Print-Line, p 215

Mirzagadheri G, Horandl E (2016) The evolution of meiotic sex and its alternatives. Proc Biol Sci 283:1221. https://doi.org/10.1098/rspb.2016.1221

Moussa H, Kindomihou V, Houehanou T D., Chaibou M, Souleymane O, SoumanaIdrissa, Dossou J, Sinsin B (2021) Farmers' perceptions of fodder performances of pearl millet (Pennisetum glaucum (L.) R. Br) accessions in Niger. Heliyon 7 : e07965

Nagesh CH, Subrahmanyam NC (1996) Interspecific hybridization of *Pennisetum glaucum* (L.) R. Br. with wild relatives. J Plant Biochem Biotechnol 5:1–5. https://doi.org/10.1007/BF03262970

Narwal MS, Yadav AK, Bal C (2009) Mainstay characters of dual purpose pearl millet hybrids. In: Forage symposium on "Emerging trends in forage research and livestock production, Feb. 16–17, 2009 at CAZRI, RRS, Jaisalmer, pp 50–55

Navale PA, Nimbalkar CA, Kulkarni VM, Wattammer MJ, Harinarayana G (1995) Correlations and path analysis in pearl millet. J Maharastra Agric Univ 20:43–46. Cited in Kumar et al 2012

Nepolean T, Blummel M, Bahasker Raj AG, Rajaram V, Senthilvel S, Hash CT (2006) International sorghum and millets newsletter, vol 47, pp 149–152. Cited in Kumar et al 2012

Nepolean T, Blummel M, Hash CT (2009) Forage symposium on "Emerging trends in forage research and livestock production, Feb. 16–17, 2009 at CAZRI, RRS, Jaisalmer (Rajasthan) India, pp 68–73

Obok EE, Aken'Ova ME, Iwo GA (2012) Forage potentials of interspecific hybrids between elephant grass selections and cultivated pearl millet genotypes of Nigerian origin. J Plant Breed Crop Sci 4:136–143

Ouendeba B, Hanna WW, Ejeta G, Nyquest WE, Santini JB (1996) Forage yield and digestibility of African pearl millet landraces in diallel with missing cross. Crop Sci 36:1517–1520. Cited in Kumar et al 2012

Oumar I, Mariac C, Pham JL, Vigouroux Y (2008) Phylogeny and origin of pearl millet (*Pennisetum glaucum* [L.] R. Br) as revealed by microsatellite loci. Theor Appl Genet 117: 489–497. https://doi.org/10.1007/s00122-008-0793-4

Ozias-Akins P, Lubbers EL, Hanna WW, McNay JW (1993) Transmission of the apomictic mode of reproduction in *Pennisetum*: co-inheritance of the trait and molecular markers. Theor Appl Genet 85:632–638. https://doi.org/10.1007/BF00220923

Ozias-Akins P, Roche D, Hanna WW (1998) Tight clustering and hemizygosity of apomixis-linked molecular markers in *Pennisetum squamulatum* implies genetic control of apospory by a divergent locus which may have no allelic form in sexual genotypes. Proc Natl Acad Sci U S A 95:5127–5132. https://doi.org/10.1073/pnas.95.9.5127

Ozias-Akins P, Akiyama Y, Hanna WW (2003) Molecular characterization of the genomic region linked with apomixis in Pennisetum/Cenchrus. Funct Integr Genomics 3:94–104

Patil BD, Singh A (1964) An interspecific cross in genus *Pennisetum* involving two basic numbers. Curr Sci 33:255

Patil BD, Hardas MW, Joshi AB (1961) Auto-allopolyploid nature of Pennisetum squamulatum Fresen. Nature 199:419–420

Raheem D, Dayoub M, Birech R, Nakiyemba A (2021) The contribution of cereal grains to food security and sustainability in Africa: potential application of UAV in Ghana, Nigeria, Uganda, and Namibia. Urban Sci 5:8. https://doi.org/10.3390/urbansci5010008

Rai KN, Virk DS, Harinarayan G, Rao AS (1996) Stability of male sterile sources and fertility restoration of their hybrids in pearl millet. Plant Breed 115:494–500

Rai KN, Gowda CLL, Reddy BVS, Sehgal S (2008a) The potential of sorghum and pearl millet in alternative and health food uses. Compr Rev Food Sci Food Saf 7:340–352

Rai KN, Hash CT, Singh AK, Velu G (2008b) Adaptation and quality traits of a germplasm-derived commercial seed parent of pearl millet. Plant Genet Resour Newslett 154:20–24

Reddy BVS, Reddy PS, Bindinger F, Blummel M (2003) Crop management factors influencing yield and quality of crop residues. Field Crop Res 84:57–77. Cited in Kumar et al 2012

Riddle NC, Kato A, Birchler JA (2006) Genetic variation for response to ploidy change in *Zea mays* L. Theor Appl Genet 114:101–111

Roy AK, Kaushal P, Zadoo SN, Choubey RN (2003) Identification of a new cytotype of *Pennisetum squamulatum* Fresen with 2n = 56 chromosomes. Range Manag Agrofor 24:71–73

Roy AK, Agrawal RK, Chand S, Ahmad S, Kumar RV, Mall AK, Bhardwaj NR, Mawar R, Singh DN, Kantwa SR, Faruqui SA(2020) Database of forage crop varieties 2020. ICAR-All India Coordinated Research Project on forage crops and utilization, 370 pages. ISBN 978-81-948917-2-7

Roy AK, Chakraborti M, Radhakrishna A, Dwivedi KK, Srivastava MK, Saxena S, Paul S, Aarti K, Malaviya DR, Kaushal P (2022) Alien genome mobilization and fixation utilizing an Apomixis Mediated Genome Addition (T057AMGA) in *Pennisetum* to improve domestication traits of *P. squamulatum*. Theor Appl Genet 135:2555–2575

Sahu PP, Gupta S, Malaviya DR, Roy AK, Kaushal P, Prasad M (2012) Transcriptome analysis of differentially expressed genes during embryo sac development in apomeiotic non parthenogenetic interspecific hybrid of *Pennisetum glaucum*. Mol Biotechnol 51:262–271

Sanou J, Bayala J, Paulin B, Teklehaimanot Z (2012) Photosynthesis and biomass production by millet (*Pennisetum glaucum*) and taro (*Colocasia esculenta*) grown under baobab (*Adansonia digitata*) and néré (*Parkia biglobosa*) in an agroforestry parkland system of Burkina Faso, West Africa. Exp Agric 48:283. https://doi.org/10.1017/S0014479712000014

Satyavathi CT, Tiwari S, Bharadwaj C, Rao AR, Bhat J, Singh SP (2013) Genetic diversity analysis in a novel set of restorer lines of pearl millet [*Pennisetum glaucum* (L.) R. Br] using SSR markers. Vegetos 26:72–82. https://doi.org/10.5958/j.2229-4473.26.1.011

Satyavathi CT, Ambawat S, Khandelwal V, Srivastava RK (2021) Pearl Millet: a climate-resilient nutricereal for mitigating hidden hunger and provide nutritional security. Front Plant Sci 12: 659938. https://doi.org/10.3389/fpls.2021.659938

Sehgal D, Skot L, Singh R, Srivastava RK, Das SP, Taunk J et al (2015) Exploring potential of pearl millet germplasm association panel for association mapping of drought tolerance traits. PLoS One 10(5):e0122165

Serba DD, Sy O, Sanogo MD, Issaka A, Ouedraog M, Ang IK, Drab I, Kanfany G (2020) Performance of dual-purpose pearl millet genotypes in West Africa: importance of morphology and phenology. Afr Crop Sci J 28(4):481–498

Sharma R, Upadhyaya HD, Manjunatha SV, Rai KN, Gupta SK, Thakur RP (2013) Pathogenic variation in the pearl millet blast pathogen *Magnaporthe grisea* and identification of resistance to diverse pathotypes. Plant Dis 97:189–195

Sharma S, Sharma R, Govindaraj M, Singh MR, Satyavathi CT, Srivastava RK, Gumma MK, Kilian B (2020) Harnessing wild relatives of pearl millet for germplasm enhancement: challenges and opportunities special issue: adapting agriculture to climate change: a walk on the wild side. Crop Sci 61(1):177–200

Sharma S, Sharma R, Pujar M, Yadav D, Yadav Y, Rathore A, Mahala RS, Singh I, Verma Y, Deora VS, Vaid B, Jayalekha AK, Gupta SK (2021) Use of wild *Pennisetum* species for improving biotic and abiotic stress tolerance in pearl millet. Crop Science Special issue: adapting agriculture to climate change: a walk on the wild side. Crop Sci 61:289–304. https://doi.org/10.1002/csc2.20408

Shinde RB, Patil FB, Sangave RA (1984) Resistance to downy mildew in pearl millet. J Maharashtra Agric Uni 9:337–338

Singh SD (1995) Downy mildew of pearl millet. Plant Dis 79:545–550

Singh SD, Gopinath R, Pawar MN (1987) Effects of environmental factors on asexual sporulation of *Sclerospora graminicola*. Indian Phytopathol 40:186–193

Singh SD, Wilson JP, Navi SS, Talukdar BS, Hess DE, Reddy KN (1997) Screening techniques and sources of resistance to downy mildew and rust in pearl millet. Information Bulletin no. 48. International Crops Research Institute for the Semi-Arid Tropics, Patancheru. 36 pp

Srivastava RK, Bollam S, Pujarula V, Pusuluri M, Singh RB, Potupureddi G et al (2020) Exploitation of heterosis in pearl millet: a review. Plants 9:807. https://doi.org/10.3390/plants9070807

Thakur RP, Sharma R, Rao VP (2011) Screening techniques for pearl millet diseases. Information bulletin no. 89. International Crops Research Institute for the Semi-Arid Tropics, Patancheru, p 48

Timbo AL, Davide LC, Pinto JEBP, Pireira AV (2010) Protoplast production from Napier grass and pearl millet triploid hybrids. Ciênc Agrotec 34(5):1219–1222

Timper P, Wilson JP, Johnson AW, Hanna WW (2002) Evaluation of pearl millet grain hybrids for resistance to *Meloidogyne* spp. and leaf blight caused by *Pyricularia grisea*. Plant Dis 86:909–914

Umutoni C, Bado V, Whitbread A, Ayantunde A (2021) Assessment of stovers of dual-purpose Pearl Millet varieties as feed for goats in the West African Sahel. Int J Livestock Res 11(12): 37–43. https://doi.org/10.5455/ijlr.20210712082123

Upadhyaya HD, Yadav D, Reddy KN, Gowda CLL, Singh S (2011) Development of pearl millet mini core collection for enhanced utilization of germplasm. Crop Sci 51:217–233

Upadhyaya HD, Reddy KN, Pattanashetti SK, Kumar V, Ramachandran S (2017) Identification of promising sources for fodder traits in the world collection of pearl millet at the ICRISAT genebank. Plant Genet Resour 16:1–10. https://doi.org/10.1017/S147926211700003X

Varshney RK, Shi C, Thudi M, Mariac C, Wallace J, Qi P et al (2017) Pearl millet genome sequence provides a resource to improve agronomic traits in arid environments. Nat Biotechnol 35:969–974

Wilson JP, Gates RN (1993) Forage yield losses in hybrid pearl millet due to leaf blight caused primarily by *Pyricularia grisea*. Phytopathology 83:739–743

Wilson JP, Hanna WW (1992) Disease resistance in wild *Pennisetum* species. Plant Dis 76:1171–1175

Worthington M, Heffelfinger C, Bernal D, Quintero C, Zapata YP, Perez JG et al (2016) A parthenogenesis gene candidate and evidence for segmental alloployploidy in apomictic *Brachiaria decumbens*. Genetics 203:1117–1132. https://doi.org/10.1534/genetics.116.190314

Yadav OP (2010) Drought response of pearl millet landrace-based populations and their crosses with elite composites. Field Crop Res 118:51–56. https://doi.org/10.1016/j.fcr.2010.04.005

Yadav OP, Bidinger F (2008) Dual purpose landraces of pearl millet (*Pennisetum glaucum*) as sources of high stover and grain yield for arid zone environments. Plant Genet Resour 6:73–78. https://doi.org/10.1017/S1479262108993084

Yadav OP, Rai KN (2013) Genetic improvement of pearl millet in India. Agric Res 2:275–292. https://doi.org/10.1007/s40003-013-;0089-z

Yadav AK, Narwal MS, Arya RK (2012) Int J Plant Breed Genet 6:115–128. Cited in Kumar et al 2012

Yadav OP, Gupta SK, Govindraj M, Sharma R, Varshney RK, Srivastava RK, Rathore A, Mahala RS (2021) genetic gains in pearl millet in India: insights into historic breeding strategies and future perspective. Front Plant Sci 12:654038

Zadoo SN, Singh A (1986) Recurrent addition of the *Pennisetum americanum* genome in a *P. americanum* × *P. orientale* hybrid. Plant Breed 97:187–189

Zadoo SN, Roy AK, Choubey RN (1997) Cytology of a perennial octoploid cytotype of *Pennisetum pedicellatum* Trin: a new report. Range Manag Agrofor 18(1):35–39

The Major Diseases of Pearl Millet in the Indian Sub-continent: Current Scenarios in Resistance and Management Strategies

12

Shilpi Bansal, Alexander Balamurugan,
Mallana Gowdra Mallikarjuna, Sumer Pal Singh, S. Chandra Nayaka,
and Ganesan Prakash

Abstract

Pearl millet is a highly nutritive food crop grown across the world for its grain and forage purposes. However, due to the lower production of the crop in India, it is anticipated that demand will surpass the anticipated growth. Infection by various phytopathogens is the major cause of decreased production, mainly due to downy mildew, ergot, blast, rust and smut diseases as the major contributors to yield reduction. Downy mildew is the most destructive of all, with a distinctive biology. Ergot is damaging due to its direct effect on grain and forage. The blast disease initially reported in rice is now a severe threat to pearl millet. Incidences of smut and rust are also posing a serious challenge to plants. Changes in climatic conditions and the development of new hybrids led to the origin of new virulent strains with the potential to evade the resistant mechanism. This emphasizes the need for the establishment of new management techniques, including chemical, biological and host plant resistance practices, so as to increase grain yield. The current chapter is an attempt to understand the epidemiology, aetiology and life cycle of the globally important pearl millet diseases to ensure the development of successful management techniques.

S. Bansal
Division of Plant Pathology, ICAR-Indian Agricultural Research Institute, New Delhi, India

Department of Science and Humanities, SRM Institute of Science and Technology, Delhi NCR Campus, Modinagar, Delhi, India

A. Balamurugan · G. Prakash (✉)
Division of Plant Pathology, ICAR-Indian Agricultural Research Institute, New Delhi, India

M. G. Mallikarjuna · S. P. Singh
Division of Genetics, ICAR-Indian Agricultural Research Institute, New Delhi, India

S. C. Nayaka
Department of Studies in Applied Botany and Biotechnology, University of Mysore, Mysore, India

V. A. Tonapi et al. (eds.), *Pearl Millet in the 21st Century*,
https://doi.org/10.1007/978-981-99-5890-0_12

Keywords

Pearl millet · Diseases · Pathogens · Management strategies

12.1 Introduction

Pearl millet (*Pennisetum glaucum* (L.) R. Br.), a member of the Poaceae family, is an important cereal crop extensively grown in arid and semi-arid regions of Asia and Africa. Among millets, pearl millet is grown the most across the world and covers 30 million hectares (Satyavathi et al. 2021). Globally, pearl millet is the sixth most cultivated cereal crop, while in India, it holds the third position after rice and wheat. Rajasthan is the largest producer of pearl millet in terms of area and production in India. Gujarat stands first in terms of productivity in India (Sreekala et al. 2023). The crop is gaining popularity as it can withstand harsh environmental conditions and has high nutritive value. It is climate-resilient and can counter various climatic changes better than other crops, such as it can survive in less fertile, highly alkaline soil, drought conditions, high temperatures etc. (Srivastava et al. 2022; Lin et al. 2023). Only about 8% of pearl millet cultivated area is irrigated across India, mostly cultivated as a rainfed crop. It is a powerhouse of nutrition which is also rich in micronutrients like zinc and iron (Chakraborty et al. 2022). The use of the plant is not only restricted to the food industry but also used as dry fodder and stover crop (Newman et al. 2010; Yadav et al. 2012). It occupies an area of 6.93 million ha with an average production of 8.61 million tonnes and productivity of 1243 kg ha^{-1} (Directorate of Millets Development 2020). In India, the production value of bajra (pearl millet) was 10.49 million metric tonnes in the financial year 2020–2021.

In comparison to open-pollinated varieties, single-cross hybrids of pearl millet have shown a 25–30% grain yield advantage (Rai et al. 2006). The research on Pearl millet improvement in India is carried out by the All India Coordinated Research Project on Pearl Millet (AICRP-PM), administered by the Indian Council of Agricultural Research (ICAR) through a network of 13 AICRP-PM centres and several voluntary ones. To create domestic and global demand and to provide nutritional food to the people, the Government of India proposed to United Nations to declare 2023 as the International Year of Millets (IYoM). Thus, due to the increasing importance of millets, the year 2023 is being celebrated as the International Year of Millets.

Pearl millet crop production is affected by various plant pathogens, including fungal, bacterial and viruses (Table 12.1). Among various pathogens infecting pearl millet, fungal pathogens are of great importance, which cause major yield losses in pearl millet crop production. The fungal disease includes downy mildew (*Sclerospora graminicola)*, blast (*Magnaporthe grisea*), smut (*Moesziomyces penicillariae*), ergot (*Claviceps fusiformis*) and rust (*Puccinia substriata)* diseases. The fungal disease causes severe yield losses in grain production and also increases production costs. In this chapter, we discussed the important diseases of pearl millet, the mode of infection, the life cycle and their management.

Table 12.1 Globally recognized important diseases of pearl millet caused by fungi, bacteria, viruses, nematodes and parasitic plants

S. no.	Name of the disease	Casual organisms
Fungal diseases		
1.	Downy mildew	*Sclerospora graminicola* (Sacc.) Schroet.
2.	Ergot	*Claviceps fusiformis*
3.	Pyricularia leaf spot	*Pyricularia grisea* (Cke.) Sacc
4.	Rust	*Puccinia substriata* Ell. & Barth. var. *indica* Ramachar & Cumm.
5.	Smut	*Moesziomyces penicillariae* (Bref.) Vanky
6.	Southern blight	*Sclerotium rolfsii* Sacc.
7.	Top rot	*Fusarium moniliforme* Sheldon
8.	Rhizoctonia blight	*Rhizoctonia solani* Kuhn
9.	Bipolaris leaf spot	*Bipolaris setariae* (Saw.) Shoem
10.	Cercospora leaf spot	*Cercospora penniseti* (Chupp)
11.	Curvularia leaf spot	*Curvularia penniseti* (Mitra) Boedijn
12.	Zonate leaf spot	*Gleocercospora sorghi* Bain & Edgerton
13.	Myrothecium leaf spot	*Myrothecium roridum* Tode ex. Fr
14.	Phyllosticta leaf blight	*Phyllosticta penicillariae* Speg.
15.	Seedling blight	Various fungi
16.	Head mould	Various fungi
17.	Dactuliophora leaf spot	*Dactuliophora elongate* Leaky
Bacterial diseases		
18.	Bacterial spot	*Pseudomonas syringae* van Hall
19.	Bacterial leaf streak	*Xanthomonas campestris* (Pammel) Dowson pv. *Pennamericanum*
20.	Bacterial leaf stripe	*Pseudomonas avenae* Manns
Viral diseases		
21.	Black streaked dwarf virus	*Reoviridae fijivirus*
22.	Guinea grass mosaic virus	*Potyvirus*
23.	Maize dwarf mosaic virus	*Potyvirus*
24.	Maize streak virus	*Geminivirus*
25.	Satellite panicum mosaic virus	*Panicum mosaic satellite virus*
26.	Wheat streak mosaic virus	*Potyviridae*
Parasitic flowering plant		
27.	Witchweed	*Strigaher monthica* Benth.
		Striga asiatica (L.) Kuntze
Nematode diseases		
28.	Burrowing nematode	*Radopholous similis* (Cobb) Thorne
29.	Dagger nematode	*Xiphinema americanum* Cobb
30.	Root-knot nematode	*Meloidogyne incognita* (Kofoid & White) Chitwood
		M. javonica (Treub.) Chitwood
		M. arenaria (Neal) Chitwooda
31.	Cyst nematode	*Heteroderag ambiensis* Merny&Netscher
32.	Root lesion nematode	*Pratylenchus mulchandi* Nandakumar & Khera

12.2 Fungal Diseases of Pearl Millet

12.2.1 Downy Mildew or Green Ear Disease

Sclerospora graminicola (Sacc.) Schroet is the causative agent of downy mildew, which generates deleterious effects on pearl millet plants widely grown in Asian and African countries. The first incidence of downy mildew in India has been dated back to 1907 in ill-drained lands (Butler 1907). However, the downy mildew turned into an epidemic in 1971 when HB3 hybrid variety was severely infected and resulting in a loss of about 4.6 million metric tonnes (Kumari et al. 2021). Due to the continuous cultivation of these hybrids, the incidence of epidemics continued in successive years. It has been estimated that annual loss in grain yield globally may account for up to 20%, while in India alone, this loss may rise up to 40% (Nene and Singh 1976; Singh et al. 1993). Initially, all the hybrids harbouring A1 cytoplasmic-nuclear male sterility, which was basically introduced to increase the production of pearl millet, but failed due to the recurrence of the green ear (Arya and Kumar 1976; Thakur et al. 2001). This even led to the belief that cytoplasm is responsible for the disease, but later on, this idea was discarded. In recent years the pearl millet hybrids with more than 20% downy mildew incidence have been sidelined from cultivation (Rao et al. 2005).

12.2.2 Pathogen Biology

S. graminicola, a member of the Oomycetes class, Sclerosporales order and Sclerosporaceae family, is an obligate biotrophic parasite (Thines et al. 2015). It exhibits both sexual and asexual life cycles. In the asexual or vegetative phase, the mycelium penetrates the host tissues and lies in its intercellular spaces and draws nutrition through haustoria. Haustoria are specialized structures which may either be simple or branched and digitate or globose (Weston 1929). The hyphae are systemic, multinucleate, coenocytic, highly branched and grow in a fulsome fashion and produce many zoospores on the abaxial side of a leaf. The sporulation event is promoted by photosynthate, low temperature and high humidity. Once sporulation stops, the pathogen automatically starts producing oospores and switches to sexual reproduction.

Asexual phase Sterigmata present at the sporangiophore branches are the site of the production of sporangia. The sporangiophores arise from stomata and are hypophyllous or amphigenous, short, dichotomously branched, stout, determinate and clavate, with dimensions of 15–22 × 12–21 μm. These consist of 1–6 short pedicels. Naturally, the sporangia are produced at night (1–4 am), while in artificial conditions, these are produced 6 h after infection. These are ellipsoid or broadly elliptic, hyaline, thin-walled and papillate (Jouan and Delassus 1971). Under ideal conditions of 20 °C temperature and 95–100% humidity, approximately 1.5×10^5 sporangia/cm^2 leaf can be produced in one night (Singh et al. 1993), which results in prominent and typical white 'downy' growth (Francis and Williams 1983).

Sexual spores Oospores mark the beginning of a sexual phase of the fungus and are demarcated by brownish-yellow, spherical, thick-walled structures. Once the plants seize to grow and the asexual life cycle of the pathogen is complete, the pathogens produce antheridia and oogonia, which mate to produce oospores. The size of the mature oospore is 32 μm. The characteristic feature of *S. graminicola* is the fusion of the oogonial wall with the oospore wall. Oospores can germinate via two modes, first by germ tubes and second by zoospore liberation with 28 ± 2 °C as the ambient temperature for germination.

12.2.3 Disease Symptoms

Plants grown on the infected soil express symptoms within 5–6 days of establishment and are systemic in nature. These symptoms are of two types: downy mildew and green ear. The plants infected with downy mildew undergo changes in various morphological traits viz., reduction in the height of the plant height, leaves and nodes, which ultimately affects the yield of the crop. Pearl millet develops symptoms during the initial stage itself and may die before attaining maturity. Symptoms of the disease appear initially on the second leaf, which progresses to other leaves and panicles; however, in case of high severity, symptoms appear on the first leaf itself. Chlorosis near the base of an infected leaf is the typical first sign of the illness, which is followed by sporulation, or the 'half-leaf' symptom, on the underside of the affected leaf. A prominent margin between the basal diseased region and the noneased tip region is characteristic of half-leaf symptoms. White downy growth is produced by an abundance of white asexual sporulation on the lower leaf surface of an infected chlorotic leaf when relative humidity is high (>95%) and the temperature is moderate (20–22 °C). The subsequent formation of oospores causes the leaves to turn reddish brown and eventually dry. Shredding of leaves happens in severe cases. Because of root retardation, severely infected plants (up to 60%) are typically stunted and lack panicles. Green ear symptoms appear during the panicle emergence on the ear head, with florets undergoing varying degrees of malformations, transformations and proliferation. This phenomenon is also called termed virescence. These leafy structures are capable of producing spores in some conditions. Generally, four types of malformations have been observed: (a) whole inflorescence transforms into a green leafy tuft, called a 'green ear'; (b) upper half of the ear head contains normal flowers while the lower half is proliferated; (c) no leafy structures are malformed; however, bristles become long and (d) shoots remain stunted and produce leafy tufts at the top. In latent infections, the green ear is the only manifestation of the disease. Primary infection is due to infected seeds and soil. It was observed that in infected plants, normal ears often develop into green ears when the disease outgrows them at the primary shoots, which is followed by tufting

Fig. 12.1 Different stages of downy mildew symptoms on pearl millet. (**a**) Initation of downy mildew symptoms on ear head, (**b**) half leaf symptoms, (**c**) inflorescence started transforming into green ear, (**d**) whole inflorescence transforms into a green leafy tuft, called a 'green ear', (**e**) an infected plant shows chlorosis, (**f**) stunted growth

and leaf yellowing. Leaf yellowing is an indicator of the susceptibility of the plant towards the disease (Fig. 12.1).

12.2.4 Disease Cycle

Oospores which lie in the soil initiate the infection process and may survive up to 10 years (Borchhardt 1927). Soil, wind and water act as the carriers of oospores on the seed surface. These oospores are covered by thick impermeable walls, which prevent them from desiccation before infecting the underground plant parts at the seedling stage. The oospores germinate into a germ tube which produces appressorium. Based on the plant part involved (coleorhiza, roots and underground portions of stem, radicles, lower portions of the seedling coleoptiles) and developmental stage, appressorium could be formed at three different sites: epidermal cell junction or over stomata or directly over epidermal cells. Thereafter, the pathogen extends towards the meristematic region. Since the apical meristem of the young seedlings is protected only by coleoptile, whereas the apical meristem of the older plants is covered by a number of leaf sheaths, younger seedlings are more prone to penetration by pathogens which marks the beginning of the systemic response.

Sporangia or zoospores mark the secondary infection within and among fields (Singh and Williams 1980) at the apical meristem region. On its abaxial surface under humid conditions, the systemically infected leaves produce a large number of sporangia. These sporangia release 1–12 zoospores (ICRISAT 1987) which germinate, penetrate stomata or epidermis, and induce the infection cycle. The optimal condition for zoospore to release from sporangium is 30 °C in about 2 h 40 min; however, the temperature may vary from 10 to 45 °C. Lower temperature supports zoospore infectivity for a longer time, with an average infectivity of about 4 h at 30 ° C (Singh and Gopinath 1985). Zoospores emerge through a pore produced by the release of an operculum. These zoospores swim for 30–60 min, encyst and subsequently germinate by forming a germ tube. High relative humidity, low temperature and well-proportioned rainfall substantially affect the infection and disease spread phenomenon (Jeger et al. 1998; Gupta and Singh 1999). Under appropriate environmental conditions, there is a continual production of sporangia till the time tissue necrosis or senesce is pulled off.

12.2.5 Disease Management

12.2.5.1 Cultural Control

The collection of infected plants should be removed immediately from the field to stop further spread. Selection of disease-free seeds, weeding operations, adjustment of date of sowing, planting with proper space and summer ploughing can reduce the primary source of inoculums.

12.2.5.2 Biological Control

Treatment of susceptible seeds with another fungus *Penicillium chrysogenum* resulted not only in increased plant growth but also in reducing the incidence of disease by 28%. Additionally, the treatment enhanced the expression of *CHS, LOX, POX, Pr-1*a and *CHT* genes, with the *CHS* gene showing maximum expression of 3.5-fold at 24 h and the *Pr-1a* gene showing threefold expression at 12 h (Murali et al. 2013).

12.2.5.3 Chemical Control

The peculiar disease cycle of oomycetes *S. graminicola* has made the regular fungicides ineffective, thus developing the need to quest for oomycete-specific fungicides. Metalaxyl, a systemic fungicide, has been found to be effective against the fungus. The plant parts such as leaves, stem and roots absorb metalaxyl which in turn interferes with rRNA synthesis, thereby inhibiting protein and ergosterol synthesis. The nanoparticle bio-fabricated zinc oxide was reported to reduce the downy mildew incidence by induction of disease resistance and enhance plant growth promotion activities (Nandhini et al. 2019).

12.2.5.4 Host Resistance

Adapting host resistance through breeding approaches is a cost-effective strategy to combat the disease; however, it is noted that the plant should have a minimum of 8% tolerance capacity before notification of the symptoms. The male-sterile lines and male parent lines have been identified in pearl millet as possessing a resistance mechanism against downy mildew (Prakash et al. 2014). In India, Thakur et al. (2001) identified a highly sustainable resistant line, that is, IP 18292, against all the six pathotypes prevalent in India. Singh (1995) reported several disease-resistant varieties developed through open-pollination and hybrid breeding approaches and cultivated in India and West Africa by ICRISAT. The single dominant gene (*Rsg1*) based resistance in downy mildew was demonstrated by Singh and Talukdar (1998).

12.3 Smut

Another major disease infecting pearl millet is smut disease. *Tolyposporium penicillariae* is the causal agent of smut which resides as a teleutospore in either infected soil or infected seed. The disease finds its occurrence across the globe and is widespread in India, Burkina Faso, Cameroon, Gambia, Nigeria, Pakistan, Senegal, Western Africa, the United States, Zimbabwe etc. (Wilson 2000). In India, states like Rajasthan, Haryana, Punjab and Gujarat are the drastically affected areas. The disease accounts for a 5–30% loss in grain yield, especially in commercial F1 hybrids and also in exotic early maturing cultivars in comparison to naturally pollinated varieties grains (Thakur 1989). The disease severity is known to be a factor of season, geographic distribution, developmental stage etc. (Dashora et al. 2008)

12.3.1 Pathogen Biology

Infected seeds produce black powdery masses containing teleutospores which are very compact and look like spore balls. The infected seeds varied in size between 2 and 325 × 50–175 μm in diameter. The favourable temperature for the germination of teleutospores is 30 ° C. After the germination, the extended mycelium, called pro-mycelium, is four-celled and develops further both lateral and terminal sporidia. Fungus growing on nutrient-rich agar media is composed of budding sporidia, which are hyaline, spindle-shaped and single-celled, having 8–25 μm diameter length (Dashora et al. 2008).

12.3.2 Disease Cycle

The fungus remains in a dormant phase in the seed or soil as a teleutospore serving as the primary inoculum. Under favourable environmental conditions such as optimum temperature and humidity, these resting spores germinate into mycelial clusters and produce sporidia. In order to form a dikaryotic infection hypha, two sporidia that are of compatible mating types are needed. Stigma at the flowering stage is attacked by these air-borne sporidia, which causes infection. This infection converts the ovary into sori which is a mass of spores that replaces the grain of the pearl millet. After inoculation, it takes 14 days for spore production and further 7–14 days for maturation of the sori, which is followed by another cycle of sporidia formation to initiate infection (Dashora et al. 2008).

12.3.3 Disease Symptoms

During the early stages of sori formation, its colour is bright green, which changes to brown, followed by black as it attains maturity. It is during this final stage, when the colour of sori turns black, that disease symptoms are visible in the plant. In comparison to normal grains (1–2 mm), the size of sori is 0.3–0.4 cm long and 0.2–0.3 cm wide, and it protrudes out between the glumes. The sori appear more oval or conical enlarged bodies that replace the inside material of the grain; however, the membrane remains intact. The lower portion of the ear head is the soft target of the fungus (Fig. 12.2).

12.3.4 Disease Management

12.3.4.1 Chemical Control

Despite the use of chemicals harmful to plants and the high cost involved in it, various chemicals have been tried to control the onset and breakage of the disease. Dashora and Kumar (2009) tested the efficacy of four different fungicides, namely, thiram, bavastin, mancozeb and blitox, for their ability to inhibit fungal growth. It was observed that none of the fungicides was 100% effective; however, the thiram, after third week of spray, was the most effective fungicide with 84.9% suppression at 200 μg/mL concentration, followed by bavastin. Treatment of infected seeds with sterilized sand and ethanol for 2 min by stirring method was also found beneficial in removing the infection, which could further prevent the spread of disease (Chakrabarty et al. 2011). Carboxin had always been the preferable fungicide; however, it was noticed that the application of Hexaconazole and Propiconazole was even more effective, with 97.63% and 97.43% efficacy (Meena et al. 2012).

12.3.4.2 Biological Control

Treatment of seeds with *Gliocladium virens* (6 g/kg) and raw cow/goat milk for 18 h, along with soil drenching with *G. virens* (10 g/m^2), protected the plant by 58.9%

Fig. 12.2 Smut symptoms—green and mature sori. Matured sori turn brown and rupture to release dark-brown to black sporeballs of numerous teleutospores

(Dashora et al. 2008). Other than this, some plant extracts such as neem oil, aloe vera gel, jamun and eucalyptus leaf have also been known to control fungal infection (Khanna et al. 2014).

12.3.4.3 Host Resistance

Pearl millet resistance to smut is believed to be partially or completely dominant with additive gene effects (Yadav et al. 2000). Considering the limitations posed by other control methods use of resistant cultivars is the method of choice to get rid of smut disease. Also, this is the most economical method relative to all. Initially, the smut-resistant lines were obtained from gene bank collections of Mali, Nigeria, Senegal and Zimbabwe (Murty et al. 1967), followed by Togo, Cameron, Lebanon and India representing diverse agro-ecological zones (Thakur et al. 1986, 1992). A large number of smut-resistant lines are now available, viz. GHB 719, GHB 558, PUSA 23, RHB 121, JBV 2, PB 106, ICMH 356, GHB 757, GHB 744, HHB 197, PUSA 266, GHB 538, ICTP 8203, ICMV 155, B 2301, SABURI, SHARDHA, PUSA 383 and ICMV 221, ICMB 92888, ICMB 92777, IP 19874, GHB 732, MH 1317, RAJ 171, ICMPS 900-9-3, ICMPS 1600-2-4, CZP 9802 and ICMPS 2000-5-2 (Thakur and King 1988b). Proper management of pollination further adds to better control of disease as local varieties and germplasm are more resistant to smut than hybrids (Thakur et al. 1983).

12.4 Ergot

Ergot, or sugary disease of pearl millet, is yet another destructive disease that can cause crop loss of up to 70%. It is caused by *Claviceps fusiformis*. The history of the disease goes late back to 1945, reported by Thomas et al. (1945) in India; however, the first epidemic was reported in 1956 in Maharashtra (Bhide and Hegde 1957). The occurrence of the disease is mainly restricted to the eastern hemisphere and not the western hemisphere and is present in almost all pearl millet-grown areas (Thakur and King 1988a, b). The disease symptoms are seen in India (Delhi, Haryana, Maharashtra, South Indian states, Uttarakhand), Pakistan, African countries (Gambia, Ghana, Malawi, Nigeria, Senegal, Somalia, Tanzania, Uganda etc.) (Ramakrishnan 1971; Rachie and Majmudar 1980; Rothwell 1982). F1 hybrids developed to increase grain yield suffered losses of up to 58–70% due to ergot disease (Natarajan et al. 1974). Ergot affects seed quality, viability, germination and yield, but the major harm it causes is the production of toxic alkaloids, which has detrimental effects. The grain of pearl millet is contaminated by the pathogen sclerotia that harbour three major classes of ergot alkaloids: clavinet alkaloids, ergopeptines and D-lysergic acid (Hulvová et al. 2013). St. Anthony's fire, a condition where the patient suffers from intense burning pain and cyanosis, was reported after the consumption of infected grains. Other symptoms include nausea, vomiting, hallucinations, gangrene and somnolence, the alkaloids are harmful not only to humans but also to birds and animals, which leads to tissue necrosis and loss of the tail. Agroclavine-induced toxicity prohibited mice from raising litters when fed with 2–3% sclerotia as the alkaloid affected the development of the mammary gland (Mantle 1968).

12.4.1 Pathogen Biology

Sexual reproduction Sexual reproduction in *C. purpurea* occurs by fusion between antheridia (male) and ascogonia (female) to form diploid nuclei by a phenomenon called karyogamy which is followed by meiosis resulting in a haploid state. After this process, sexual fruiting bodies will produce a cup-like structure called perithecia which contains multiple sac-like structures called asci (Agrios 2005). Each asci has eight ascospores with 2.0 μm in diameter, 60–70 μm long. Once asci attain maturity, ascospores are released from it and perithecia into the air (up to 7–15 cm in height). These ascospores thereafter are disseminated in air, windblown, or transported by insects. The disseminated ascospores, considered primary inoculum, further land on the stigma or ovary of the host and cause infection on them (Schumann and Uppala 2000).

Asexual reproduction Upon successful landing on the host flower or stigma, the ascospore triggers to germinate and produce long filamentous hyphae (mycelium) that proliferate to colonize on the ovary of the host flower. From matured mycelium, the fungus produces numerous asexual conidia, which function as secondary inoculum from palisade conidiophores, and this occurs in the yellowish colour honeydew secretion drops. Honeydews attract insects, and contaminated insects spread the secondary inoculum to healthy plants over distances and nearby plants. Further, rain splashing also disperses the conidia to nearby plants. Over time, conidia and honeydew production stop, the entire ovary is invaded by hyphae, and hyphal threads turn to become thicker and change to colour from brown-black (Agrios 2005). This structure becomes compact-mass tissue called pseudo-parenchyma, which further develops into hard, dark-coloured sclerotia that serve as an inoculum source for the next season (Schumann and Uppala 2000).

12.4.2 Disease Symptoms

The onset of the disease starts with oozing out of cream to pink or light honey-coloured mucilaginous droplets from the infected spikelet on pearl millet panicles. These droplets are rich in both macro and microconidia (asexual spores) (Bahadur et al. 2021). The severity of the infection converts the ear head to soft, sticky and black in colour, and the honeydew drips on larvae and soil. In 10–15 days, these droplets get dried and turn into dark brown-black sclerotia that protrude out from florets. Harvesting and threshing cause the mixing of sclerotia with grain which serve as the primary inoculum for the next infection cycle (Thakur and King 1988b) (Fig. 12.3).

Fig. 12.3 Ergot symptoms—honeydew and sclerotia. Both macro- and microconidia are produced in the honeydew

12.4.3 Disease Cycle

The onset of the disease begins with the release of numerous ascospores from the sclerotia available in the soil or contaminated seeds from the previous season. These ascospores, with the help of winds, get attached to stigmas of flowering pearl millet plants, germinate and infect the ovaries. The hyphae invade and grow down to the ovary axis to establish host–pathogen interaction. In this ovary, one-celled conidia are produced by the spacelial stroma, which is exuded in honeydew (Thakur et al. 1984). This honeydew acts as an insect attractant and helps in the transfer of disease. Production of honeydew goes hand in hand with sclerotia formation and maturity, which happens to be in 4–5 weeks. Vernalization of sclerotia requires a temperature in the range of 0–10 °C for 25 days (Mitchell and Cooke 1968). The ambient conditions for disease incidence and spread are overcast skies, drizzling rain with 80–85% relative humidity and 20–30 °C moderate temperature. After 4–6 days of inoculation, honeydew symptoms appear and post 15–20 days of it, fully developed sclerotia are visible. These sclerotia fall from infected panicles, get mixed in soil and seed during harvesting and threshing, and again prepare inoculum for the next crop cycle. Pollination (Thakur and Williams 1980) and length of protogyny have proved to have a critical role in ergot epidemiology (Willingale et al. 1986). Shorter protogyny favours rapid pollination, which increases seed set and reduces infection. The severity of the disease increases in case of pollen washed by rains which inhibits pollination, and as a result, cytoplasmic male-sterility increases the incidence of ergot infection (Thakur et al. 1992).

12.4.4 Disease Management

12.4.4.1 Cultural Control

Burying of sclerotia deep down the soil by ploughing method after harvesting of plant inhibits germination of sclerotia and release of ascospores, thereby reducing primary inoculum for infection. A high concentration of nitrogen elevates infection, while a high concentration of potassium lowers infection. Thus, judicious use of the two macronutrients may regulate plant infection (Thakur 1984). Early sowing in the month of July also reduces infection (Sharma 2012). Treatment of seeds with 10% NaCl solution is also an effective method as the weight of the sclerotia and seeds is different where the seed, due to high weight, will settle down and sclerotia will float in the solution (Nene and Singh 1976). Collateral hosts like *Cenchrus ciliaris* and *Panicum antidotale* serve as hosts to ergot pathogens (Singh et al. 1983; Thakur and Kanwar 1978). Removal of these hosts stops the supply of honeydew inoculum to pearl millet.

12.4.4.2 Chemical Control

Various chemicals have been tested for their efficiency in controlling ergot. Two to three spraying of ziram or a combination of copper oxychloride with zineb before the onset of panicle emergence was found effective (Sundaram 1975). Similarly, treatment of panicles with 0.1% Bavistin or 0.2% Tilt, or 0.2% Mancozeb controlled ergot incidence and spread (Singh 2000). Cuman-L at 200 ppm concentration was also found suitable (Thakur 1984). 86.4% efficiency was achieved in controlling the disease spread with a single spray of 8-Hydroxyqunoline (500 ppm) on ear heads (Kumar et al. 1989).

12.4.4.3 Biological Control

Various biocontrol agents have been tested for their potential to control ergot, such as *Fusarium semitectum*, *F. sambucinum*, *Aspergillus niger*, *Trichoderma harzianum. T. viride* and *Bacillus subtilis* (Mahadevamurthy et al. 1988; Thakur and King 1988b).

12.4.4.4 Host Resistance

The incorporation of resistant lines in pearl millet cultivation is the method of choice to control the spread of ergot disease. ICMS 7703, ICMV135, WC-C75 and ICTP 8203ICMV 155 are some of the resistant lines released in India.

12.5 Rust

Rust infection in pearl millet has been visualized in most of the continents of the world viz Asia, North and South America and Africa also. Initially, the disease was confined to the western hemisphere and Africa, but later it extended to central Brazil and other countries (de Carvalho et al. 2006). Indian states like Tamil Nadu, Karnataka, Rajasthan, Gujarat and Maharashtra showed the maximum incidence of it. The disease is caused by *Puccinia penniseti* Zimm. Witon with a characteristic feature of the production of uredo and teleutospores on a specific host and other spores on the alternate host, especially in brinjal and other Solanum species. It is more severe in the post-rainy season than during the rainy or summer season. Across India, the mean rust severity varies from 10% to 45%, where the disease mainly occurs during the post-rainy seasons, which is more favourable than the rainy or summer season. Since the disease appears late in the season, generally after completion of the grain filling stage, the loss caused by it is small, making rust disease less important. The disease holds importance in the case of multi-cut forage hybrids where rust reduces the digestible dry matter yield (Wilson et al. 1991).

12.5.1 Pathogen Biology

Rust-infected lesion on the leaf surface develops uredinial spores as shiny brown in colour, followed by dark brown. At the time of the harvesting stage or end of the pathogen life cycle, they develop black spores called telial pustules. Uredia spores are erumpent, sub-epidermal, golden yellow and echinulate with 3–4 equatorial germ pores. Telia are irregularly distributed, erumpent, sub-epidermal and paraphysate with dark brown to black colour. Uredial spores mostly develop on younger leaves or stems; however, teliospores produce on older plantlets when the life cycle of the disease begins to complete and act as the potential source for the next season (de Carvalho et al. 2006).

12.5.2 Disease Symptoms

The rust pathogen infects all stages of the crop. Lower and older leaves are most affected. Initial symptoms appear as minute, reddish-brown pustules that occur in groups that represent the circular and reddish-brown to reddish-orange coloured uredospores. These uredospores are responsible for the spread of the disease among the crop. After the ageing of pustules, they become darker in colour, ovoid and subepidermal teliospores. The leaves get wilted from top to bottom. If disease incidence is higher, more rust pustules can appear on stems, and the plants will fall down. At the severe stage of infection, death of the diseased leaf occurs. These teliospores contain hard outer layers which assist in living in the soil for a longer period (Fig. 12.4).

12.5.3 Disease Cycle

Pearl millet rust is a macrocyclic and heteroecious rust. Ramakrishnan and Soumini (1948) reported brinjal (*Solanum melangena*) as the alternate host. This rust pathogen has all five stages, on which uredial and telial stages occur on the pearl millet plant and the basidial, pycnial and aecial stages appear on the alternate host plant brinjal (de Carvalho et al. 2006). The disease cycle starts as the uredospores land on the surface of the pearl millet leaf. It produces raised brown rusty spots on the leaf. On maturity, these uredospores produce the black colour teliospores. These teliospores produce the promycelium and form the basidium, over it produces four basidiospores. These basidiospores never infect the pearl millet crop. These basidiospores infect the brinjal plant and produce minute, oily yellow spots on the upper surface of the leaf, which contains the pycniospores. These pycniospores, in fusion with the receptive hyphae, produce cup-shaped aeciospores on the lower surface of the leaf, which in turn produces repeated spores, that is, uredospores which never infect the brinjal plant. Thus, the disease cycle starts.

12.5.4 Disease Management

12.5.4.1 Cultural Practices

The disease spreads majorly through the uredospores, which are produced on the alternate host brinjal plant. Thus, stripping away these proxy hosts, such as brinjal and grassy weeds, may aid in the reduction of rust disease. Sowing from December to May results in a lesser incidence of the disease. Removal and destruction of infected plants will limit the disease's spread.

12.5.4.2 Chemical Management

Three fungicides, viz., Triadimefon, Hexaconazole and Propiconazole, have been tested for the management of pearl millet rust; Hexaconazole @ 0.1% was found to be most effective (Nagaraja and Patil 2014). Pretreatment of pearl millet with salicylic acid imparted resistance to a virulent isolate of the *Puccinia penniseti* (Crampton et al. 2009).

Fig. 12.4 Rust symptoms-infected leaves initially show pinhead chlorotic flecks, which later turn into reddish-orange, round to elliptical pustules on both surfaces

12.5.4.3 Biological Control

Expression of an antifungal protein identified from a heterologous system, that is, *Aspergillus giganteus* in pearl millet, reduces the incidence of not only rust but also downy mildew (Girgi et al. 2006). Secondary metabolites extracted from the rust mycoparasite, *Sphaerellopsis paraphysica*, were found to be rich in phenol, decane, ester and acids with potent antifungal properties. Spray of diluted crude extract of this mycoparasite on infected plants resulted in 92.67% lysis of *P. substriata* uredospores for the management of rust in pearl millet (Ashmitha et al 2020). Trichoderma spp. is yet another biocontrol agent with the potential to remove the disease (Harman 2000).

12.5.4.4 Host Resistance

Being one of the most effective methods to combat rust susceptibility, various lines which were either naturally occurring (germplasm or local landraces) or were developed by breeding approaches have been identified (Govindarajan et al. 1984; Wilson et al. 1989; Singh 1990). Sharma et al. (2009) enlisted the hybrid parental lines developed so far, which included ICMB 96222, ICMP 451-P8, ICMR 06999 and ICMP 451-P6. The phenomenon of delay in rusting is governed by genetic markers, and it has been established in various studies (Sokhi and Singh 1984; Pannu et al. 1996). Resistance developed against one pathogen sometimes provides an additional advantage by conferring resistance against another pathogen also. Lines ICML nos. 12–16 are perfect for it, which possess resistance against downy mildew and rust (Singh 1990).

12.6 Blast

The disease is caused by a filamentous, ascomycetous fungus *Magnaporthe grisea,* and considered as most important disease in India (Bansal et al. 2022). This disease was first reported by Mehta et al. (1953) at Government Research Farm, Kanpur, India. In earlier days this disease was considered as a minor disease now it is increasing at an alarming level (Thakur et al. 2009). *M. grisea* has high genomic variability from *M. oryzae*, a very important disease in rice. This fungus shows high-level host-specificity, which infects only pearl millet and barley and does not infect rice (Srinivasachary et al. 2002). Biologically, *Magnaporthe grisea* is a hemibiotroph fungus that primarily infects green tissues and subsequently develops brown lesion and further kill the remaining green tissues from programmed necrotrophic growth (Perfect and Green 2001; Münch et al. 2008). Substantial yield losses caused by *M. grisea* have been reported in pearl millet (Timper et al. 2002) earlier.

12.6.1 Pathogen Biology

The asexual conidia of *M. grisea* appear to be pyriform in shape, having three-celled septation on the matured conidium along with a small protruding appendage on the base cell. Each matured conidium measures approximately 17.5–30.8 × 5.9–8.8 μm (Mehta et al. 1953). The conidiophores of the fungus are produced in clusters from each stoma, and they are rarely solitary with 2–4 septation. The basal area of the conidiophores is swollen and tapers towards the lighter apex. This fungus initially survives in the alternate host, weeds, or infected plant debris and initiates the new infection when the pearl millet becomes available (primary inoculum). In each spindle-shaped blast lesion, the fungus develops numerous conidia (secondary inoculum) and further disseminates in the air or by insects to the next field. These conidia initiate secondary infection and extend their life cycle till the host dies eventually.

12.6.2 Disease Symptoms

Magnaporthe blast symptoms in pearl millet are identified by the spindle-shaped grey leaf spot with dark brown margin on leaves and stem. Initial blast symptoms start from the tiny specks or lesions on the leaves that broaden and then turn necrotic, with a yellow border and ultimately, the young leaves dry. Blast lesions usually develop at any aerial part of the leaf surface and extend to the outermost layer of the leaves. Young lesions initially show pale green to greyish green, later turning yellow to grey with age. Lesions are usually elliptical or spindle-shaped, approximately 0.25–0.35 × 0.15–0.25 cm (Bansal et al. 2023). Lesion centres are grey and water soaked when fresh but turn brown, often surrounded by a chlorotic halo which will turn necrotic (Kato 2001). Nayaka et al. (2021) described the various symptoms of different parts of pearl millet blast. Some of the important symptoms are lesions on glumes, rotting and discolouration of panicle, lesions and rotting of seed, stunting or resetting of the stem, internal red necrosis, mould growth on lesions, especially on stems, abnormal colours and forms of leaves, abnormal leaf fall, fungal growth on leaves, necrotic areas and rotting of leaves, in severe disease incidence damping off, die back and even death of the whole plant (Fig. 12.5).

12.6.3 Disease Cycle

Pearl millet blast disease primarily spreads through the ascospores present in the soil debris, some of the weeds, and the grassy plant, which serves as the alternate host for the pathogen. On a single lesion, the pathogen produces about 20,000 conidia which cause the secondary spread of the disease at the field level. Conidia land on the leaf and requires free water to germinate. A short germ tube emerges from matured conidia and further develops to form balloon-shaped appressoria at the tip of the germ tube. From each appressorium further develops a needle-like structure called a

Fig. 12.5 Pearl millet leaf blast disease symptoms. Varied degree of blast symptoms on pearl millet plants in field

penetrating or infection peg. This infection needle enters the cuticle layer of the host cell, which further produces secondary hyphae to release toxins. Once the toxin is released from the infected spots, then a typical spindle-shaped lesion appears on the lead, followed coalescing lesion occurs. The disease is polycyclic in nature, with a spore-to-spore cycling time period of 7 days. At the end of the cropping season, the fungus undergoes a sexual phase whereby two opposite mating types form perithecium (a specialized bulbous structure with an elongated neck) by coming into contact with one another. Each perithecium contains a specialized spore-forming structure called an ascus, with each ascus bearing eight number of ascospore, which serves as a potential inoculum source for the next cropping season.

12.6.4 Disease Management

12.6.4.1 Cultural Practices

Cultural practices include the collection and destruction of infected stubbles, field sanitation, selection of certified and disease-free seed lots, removal of weeds in the field, summer ploughing, crop rotation, proper water drainage, avoiding excess nitrogenous fertilizers and adjustment of date sowing, respectively. Excess application of nitrogen and phosphorus can increase the plants' susceptibility; hence, application of these fertilizers can be optimized (Kato 2001).

12.6.4.2 Chemical Control

For the control of blast disease of pearl millet, foliar spraying of carbendazim 0.05% (ICBR 1:3.85) or 1.0 g/L at 15 days intervals is recommended from the initiation of the disease. Melanin biosynthesis inhibitors include tricyclazole, which acts on the melanin produced on the germinating conidia (Kurahashi 2001). Apart from tricyclazole, the fungicides also recommended to manage the blast includes pyroquilon, benomyl, iprobenfos, diclocymet, carpropamid and felimzone (Kato 2001). Azoxystrobin is a strobilurin fungicide that has been shown to inhibit fungal respiration by binding to the cytochrome *b* complex III.

12.6.4.3 Biological Control

Pearl millet blast is reported to be managed effectively by using the potential fungal biocontrol agent *Trichoderma harzianum* (Gouramanis 1995). The bacterial biocontrol agents such as *Pseudomonas fluorescens*, *Bacillus pumilus* and *Bacillus subtilis* were also found to control blast pathogen by induction of disease resistance and enhancing the plant growth-promoting functions (Krishnamurthy and Gnanamanickam 1998; Yoshihiro et al. 2003). *Streptomyces* species, actinobacteria, has been known to control the proliferation of blast disease (Zarandi et al. 2009).

12.6.4.4 Host Resistance

The Indian genotypes showed that the resistance to *M. grisea*is was governed by a single dominant gene (Gupta et al. 2012). Sharma et al. (2013) made an attempt to identify the resistance sources against the various pathotypes of *M. grisea* and identified three accessions such as IP 21187, IP 11036 and IP 7846, which were resistant to all pathotypes. Blast-resistant pearl millet originated from the same accession of *P. glaucum* sp. *monodii* in which the *Rr*1 rust gene was discovered (Hanna et al. 1987). Resistance against pearl millet blast from ICMB 06222, an elite parent line reported in ICRISAT, India, to isolate Pg45 was governed by a single dominant gene (Gupta et al. 2012).

12.7 Conclusion

Despite the advancement in modern science, ensuring food security for the growing population is a challenge. The cause of the problem is various biotic and abiotic factors that restrict the production of food crops. Once considered the food of the lower class and neglected, pearl millet has now become the food of choice owing to the enormous health benefits associated with it. Similar to other food crops, there is a huge gap in the demand and supply of pearl millet, for which stress factors can be blamed primarily the biotic stress. Increasing production by large-scale cultivation is always an option, but the loss of yield caused by diseases is also inevitable. Thus, a strong and sustainable disease management strategy is required for which a proper understanding of the pathogen life cycle, its symptoms and its biology is a pre-requisite. Seed treatment with chemicals is one of the effective methods to control disease spread, but owing to chemical toxicity and its residues in harvested products, chemical management is not always recommended. In this regard, none of the methods can replace the efficacy of host-resistant lines. However, the issue of breaking the resistant mechanism by the new emerging virulent strains of pathogens needs to be addressed by incorporating novel biotechnological approaches.

References

Agrios GN (2005) Plant pathology. Elsevier, Amsterdam

Arya HC, Kumar A (1976) Diseases of bajra-A serious problem of Rajasthan desert economy. Trans Indian Soc Desert Technol Univ Cent Desert Stud 1:177–182

Ashmitha PS, Sendhilvel V, Raguchander T (2020) Metabolites of the rust mycoparasite, *Sphaerellopsis paraphysata* for management of rust in pearl millet. J Biol Control 20:94–100

Bahadur A, Khan MR, Jamil A (2021) Pearl millet (BAJRA): Downy Mildew and Ergot. In: Khan MR, Haque Z, Ahamad F (eds) Diseases of nationally important field crops. Today and Tomorrow Printers and Publishers, New Delhi, pp 203–216

Bansal S, Mallikarjuna MG, Balamurugan A, Nayaka SC, Prakash G (2022) Composition and codon usage pattern results in divergence of the zinc binuclear cluster (Zn (II) 2Cys6) sequences among ascomycetes plant pathogenic fungi. J Fungi 8(11):1134

Bansal S, Mallikarjuna MG, Reddy B, Balamurugan A, Achary VMM, Reddy MK, Prakash G (2023) Characterization and validation of hypothetical virulence factors in recently sequenced genomes of Magnaporthe species. Physiol Mol Plant Pathol 124(101):969

Bhide VP, Hegde RK (1957) Ergot on bajri [*Pennisetum typhoides* (Burm.) Stapf and Hubbard] in Bombay state. Curr Sci 26(4):116–116

Borchhardt A (1927) Operations of the phytopathological section of the Agricultural Experiment Station in the Eastern Steppe region (Ekaterinostaff Government) in the 1925, pp 3–37

Butler EJ (1907) Some diseases of cereals caused by *Sclerospora graminicola*. Thacker, Spink & Company

Chakrabarty SK, Girish GA, Anitha K, Thakur RP (2011) Eliminating smut (*Moesziomyces penicillariae*) from pearl millet seeds under transboundary movement. Indian J Plant Genet Resour 24(3):340–342

Chakraborty A, Viswanath A, Malipatil R, Semalaiyappan J, Shah P, Ronanki S et al (2022) Identification of candidate genes regulating drought tolerance in Pearl Millet. Int J Mol Sci 23(13):6907

Crampton BG, Hein I, Berger DK (2009) Salicylic acid confers resistance to a biotrophic rust pathogen, *Puccinia substriata*, in pearl millet (*Pennisetum glaucum*). Mol Plant Pathol 10 (2):291–304

Dashora K, Kumar A (2009) In-vitro evaluation of relative effectiveness of some fungicides against smut (*Tolyposporium penicillariae*) disease of pearl millet. Pestology 33(4):45–47

Dashora K, Kumar A, Bhansali RR (2008) Smut disease of pearl millet: biology and control. Adv Appl Microbiol 36:238–240

de Carvalho ADO, Soares DJ, do Carmo MGF, da Costa ACT, Pimentel C (2006) Description of the life-cycle of the pearl millet rust fungus–*Puccinia substriata* var. *penicillariae* with a proposal of reducing var. *indica* to asynonym. Mycopathologia 161:331–336

Directorate of Millets Development (2020) Available online at: http://millets.dacfw.nic.in

Francis SM, Williams RJ (1983) *Sclerospora graminicola* [Descriptions of fungi and bacteria]. CABI, Wallingford. (77), Sheet-770

Girgi M, Breese WA, Lörz H, Oldach KH (2006) Rust and downy mildew resistance in pearl millet (*Pennisetum glaucum*) mediated by heterologous expression of the *afp* gene from *Aspergillus giganteus*. Transgenic Res 15:313–324

Gouramanis GD (1995) Biological and chemical control of rice blast disease (*Pyricularia oryzae*) in Northern Greece. CIHEAM 15:61–68

Govindarajan K, Nagarajan C, Raveendran TS, Prasad MM, Shanmugam N (1984) New source of resistance to pearl millet rust disease. Madras Agric J

Gupta GK, Singh D (1999) Role of edaphic factors in the development of downy mildew (*Sclerospora graminicola*) in pearl millet. J Agric Sci 133(1):61–68

Gupta SK, Sharma R, Rai KN, Thakur RP (2012) Inheritance of foliar blast resistance in pearl millet (*Pennisetum glaucum* L. (R.) Br.). Plant Breed 131:217–219

Hanna WW, Wells HD, Burton GW (1987) Registration of pearl millet inbred parental lines, Tift 85D2A1 and Tift 85D2B1. Crop Sci 27:1324–1325

Harman GE (2000) Myths and dogmas of biocontrol: changes in the perceptions derived from research on *Trichoderma harzianum* T-22. Pl Dis 84:377–393. https://www.statista.com/statistics/1286469/india-production-of-bajra/

Hulvová H, Galuszka P, Frébortová J, Frébort I (2013) Parasitic fungus *Claviceps* as a source for biotechnological production of ergot alkaloids. Biotechnol Adv 31(1):79–89

ICRISAT (International Crops Research Institute for the Semi-Arid Tropics) (1987) ProcInternat pearl millet workshop, 7–11 April 1986. ICRISAT Center, India, Patancheru, p 36

Jeger MJ, Gilijamse E, Bock CH, Frinking HD (1998) The epidemiology, variability and control of the downy mildews of pearl millet and sorghum, with particular reference to Africa. Plant Pathol 47(5):544–569

Jouan B, Delassus M (1971) Principales maladies des misetsorghos observes au Niger (In Fr. Summaries In En. Es.). Agron Trop 26(8):830–860

Kato H (2001) Rice blast control. In: Pesticide Outlook. Japan Crop Protection Association (JCPA), Tokyo, pp 23–25

Khanna A, Raj K, Sangwan P (2014) Smut of pearl millet: current status and future prospects. Chief Editor Dr. RK Naresh, pp 123–129

Krishnamurthy K, Gnanamanickam SS (1998) Biological control of rice blast by *Pseudomonas fluorescens* strain Pf7-14: evaluation of a marker gene and formulations. Biol Control 3:158–165

Kumar A, Kumar A, Nath V (1989) On the efficacy of 8-hydroxyquinoline in pearl millet ergot control (Abst.). Indian Phytopathol 42:292

Kumari P, Godika S, Kumar S, Ghasolia RP (2021) Downey mildew of pearl millet and it's management. Planta 2:395–404

Kurahashi Y (2001) Melanin biosynthesis inhibitors (MBIs) for control of rice blast. Pestic Outlook. 12(1):32–35

Lin M, Dong Z, Zhou H, Wu G, Xu L, Ying S, Chen M (2023) Genome-wide identification and transcriptional analysis of the MYB gene family in pearl millet (Pennisetum glaucum). Int J Mol Sci 24(3):2484

Mahadevamurthy S, Prakash HS, Shetty HS (1988) Effect of some antagonists on the sclerotial germination of *Claviceps fusiformis'*. J BiolControl 2:120–122

Mantle PG (1968) Inhibition of lactation in mice following feeding with ergot sclerotia [*Claviceps fusiformis* (Loveless)] from the bulrush millet [*Pennisetum typhoides* (Staph and Hubbard)] and an alkaloid component. Proc R Soc Lond B Biol Sci 170(1021):423–434

Meena RL, Mathur AC, Majumdar VL (2012) Management of Pearl millet smut through cultural practices and fungicides. Indian Phytopathol 65(3):268–271

Mehta PR, Singh B, Mathur SC (1953) A new leaf spot disease of bajra (*Pennisetum typhoides* Staph and Hubbard) caused by a species of *Pyricularia*. Indian Phytopathol 5:140–143

Mitchell DT, Cooke RC (1968) Some effects of temperature on germination and longevity of sclerotia in Claviceps purpurea. Trans Br Mycol Soc 51(5):721–729

Münch S, Lingner U, Floss DS, Ludwig N, Sauer N, Deising HB (2008) The hemibiotrophic lifestyle of *Colletotrichum* species. J. Plant Physiol 165(1):41–51

Murali M, Sudisha J, Amruthesh KN, Ito SI, Shetty HS (2013) Rhizosphere fungus *Penicillium chrysogenum* promotes growth and induces defence-related genes and downy mildew disease resistance in pearl millet. Plant Biol 15(1):111–118

Murty BR, Upadhyay MK, Manchanda PL (1967) Classification and cataloguing of a world collection of genetic stocks of Pennisetum. Indian J Genet Plant Breed 27:313–394

Nagaraja H, Patil PV (2014) Development of integrated spray schedule for the management of pearl millet rust in Northern dry zone of Karnataka. Karnataka J Agric Sci 27(3):308–311

Nandhini M, Rajini SB, Udayashankar AC, Niranjana SR, Lund OS, Shetty HS, Prakash HS (2019) Biofabricated zinc oxide nanoparticles as an eco-friendly alternative for growth promotion and management of downy mildew of pearl millet. Crop Prot 121:103–112

Natarajan US, Raja VDG, Selvaraj S, Parambaramani C (1974) Grain loss due to ergot disease in bajra hybrids. Indian Phytopathol 27(2):254–256

Nayaka SC, Hosahatti R, Prakash G, Satyavathi CT, Sharma R (2021) Correction to: blast disease of cereal crops: evolution and adaptation in context of climate change. In: Nayaka SC, Hosahatti R, Prakash G, Satyavathi CT, Sharma R (eds) Blast disease of cereal crops. Fungal biology. Springer, Cham. https://doi.org/10.1007/978-3-030-60585-8_15

Nene YL, Singh SD (1976) Downy mildew and ergot of pearl millet. Proc Natl Acad Sci 22(3): 366–385

Newman Y, Jennings E, Vendramini J, Blount A (2010) Pearl millet (*Pennisetum glaucum*): overview and management. SSAGR-337, one of a series of the Agronomy Department, Florida Cooperative Extension Service, Institute of Food and Agricultural Sciences, University of Florida, pp 1–6

Pannu PPS, Sokhi SS, Aulakh KS (1996) Resistance in pearl millet against rust. Indian Phytopathol 49(3):243–246

Perfect SE, Green JR (2001) Infection structures of biotrophic and hemibiotrophic fungal plant pathogens. Mol. Plant Pathol 2(2):101–108

Prakash HS, Nayaka CS, Kini KR (2014) Downy Mildew disease of pearl millet and its control. In: Future challenges in crop protection against fungal pathogens. Springer, Berlin, pp 109–129

Rachie KO, Majmudar JV (1980) Pearl millet. The Pennsylvania State University Press, University Park. 307pp

Rai KN, Kulkarni VN, Thakur RP, Haussmann BI, Mgonja MA (2006) Pearl millet hybrid parents research: approaches and achievements. Gowda CLL, Rai KN, Potential Welfare Benefit of Millets Improvement Research at ICRISAT: Multi country-Economic Surplus model approach

Rao VP, Thakur RP, Rai KN, Sharma YK (2005) Downy mildew incidence on pearl millet cultivars and pathogenic variability among isolates of *Sclerospora graminicola* in Rajasthan. SAT eJournal 46:107–110

Ramakrishnan TS (1971) Diseases of millets. Indian Council of Agricultural Research, New Delhi. 152pp

Ramakrishnan TS, Soumini CK (1948) Studies on cereal rusts I. *Puccinia penniseti* Zimm. and its alternate host. Indian Phytopathol 1:97–103

Rothwell A (1982) A revised list of plant diseases occurring in Zimbabwe. Kirkia 12:233–351

Satyavathi CT, Ambawat S, Khandelwal V, Srivastava RK (2021) Pearl millet: a climate-resilient nutricereal for mitigating hidden hunger and provide nutritional security. Front Plant Sci 12:659938. https://doi.org/10.3389/fpls.2021.659938

Schumann GL, Uppala S (2000) Ergot of rye. The Plant Health Instructor. American Phytopathological Society, Saint Paul. https://doi.org/10.1094/PHI-I-2000-1016-01

Sharma R (2012) Studies on effect of showing dates on ergot severity of pearl millet. Adv Plant Sci 25:583–584

Sharma R, Thakur RP, Rai KN, Gupta SK, Rao VP, Rao AS, Kumar S (2009) Identification of rust resistance in hybrid parents and advanced breeding lines of pearl millet. J SAT Agric Res 7:1–4

Sharma R, Upadhyaya HD, Manjunatha SV, Rai KN, Gupta SK, Thakur RP (2013) Pathogenic variation in the pearl millet blast pathogen *Magnaporthe grisea* and identification of resistance to diverse pathotypes. Plant Dis 97:189–195

Singh SD (1990) Sources of resistance to downy mildew and rust in pearl millet. Plant Dis 74 (11):871–874

Singh SD (1995) Downy mildew of pearl millet. Plant Dis 79:545–550

Singh RS (2000) Plant diseases, 7th edn. Oxford and IBH, New Delhi

Singh SD, Gopinath R (1985) A seedling inoculation technique for detecting downy mildew resistance in pearl millet. Plant Dis 69(7):582–584

Singh SD, Talukdar BS (1998) Inheritance of complete resistance to pearl millet downy mildew. Plant Dis 82:791–793

Singh SD, Williams RJ (1980) The role of sporangia in the epidemiology of pearl millet downy mildew. Phytopathology 70(12):1187–1190

Singh G, Vyas KL, Bhatt BN (1983) Occurrence of pearlmillet ergot on Cenchrusciliaris Pers. in Rajasthan. Indian J Agric Sci 53(6):481–483

Singh SD, King SB, Werder J (1993) Downy mildew disease of pearl millet. Information Bulletin no. 37. International Crops Research Institute for the Semi-Arid Tropics, Patancheru

Sokhi SS, Singh BB (1984) Components of slow rusting in pearl millet infected with *Puccinia penniseti*. Indian J Mycol Plant Pathol 14(2):190–192

Sreekala AD, Anbukkani P, Singh A, Dayakar Rao B, Jha GK (2023) millet production and consumption in india: where do we stand and where do we go? Natl Acad Sci Lett. 46(1):65–70

Srinivasachary HS, Shivayogi S, Vaishali MG, Shashidhar HE, Kumar KG (2002) Genetic analysis of rice blast fungus of southern Karnataka using DNA markers and reaction of popular rice genotypes. Curr Sci 25:732–735

Srivastava RK, Yadav OP, Kaliamoorthy S, Gupta SK, Serba DD, Choudhary S et al (2022) Breeding drought-tolerant pearl millet using conventional and genomic approaches: achievements and prospects. Front Plant Sci 13:530

Sundaram NV (1975) Ergot of bajra. In: Raychoudhary SP, Verma A, Bhargava KS, Mehrotra BS (eds) Advances in mycology and plant pathology. New Delhi, India: Prof. R.N. Tandoll's birthday celebration committee, pp 155–160

Thakur DP (1984) Ergot disease of pearl millet. Rev Trop Plant Pathol 1:297–328

Thakur RP (1989) Flowering events in relation to smut susceptibility in pearl millet. Plant Pathol 38(4):557–563

Thakur DP, Kanwar ZS (1978) Ability of naturally incident *Claviceps microcephala* from *Panicum* antidotale to produce ergot symptoms in *Pennisetum typhoides*. Indian J Agric Sci 48(9): 540–542

Thakur RP, King SB (1988a) Registration of six smut resistant germplasms of pearl millet. Crop Sci 28(2):382–383

Thakur RP, King SB (1988b) Ergot disease of pearl millet. International Crops Research Institute for the Semi-Arid Tropics, Patancheru

Thakur RP, Williams RJ (1980) Pollination effects on pearl millet ergot. Phytopathology 70(2): 80–84

Thakur RP, Rao KS, Williams RJ (1983) Effects of pollination on smut development in pearl millet. Plant Pathol 32(2):141–144

Thakur RP, Rao VP, Williams RJ (1984) The morphology and disease cycle of ergot caused by *Claviceps fusiformis* in pearl millet. Phytopathology 74(2):201–205

Thakur RP, Rao KS, Williams RJ, Gupta SC, Thakur DP, Nafade SD et al (1986) Identification of stable resistance to smut in pearl millet. Plant Dis 70(1):38–41

Thakur RP, King SB, Rai KN, Rao VP (1992) Identification and utilization of smut resistance in pearl millet. International Crops Research Institute for the Semi-Arid Tropics, Patancheru

Thakur RP, Rai KN, Rao VP, Rao AS (2001) Genetic resistance in pearl millet male-sterile lines to diverse pathotypes of *Sclerospora graminicola*. Plant Dis 85:621–626

Thakur RP, Sharma R, Rai KN, Gupta SK, Rao VP (2009) Screening techniques and resistance sources for foliar blast in pearl millet. J SAT Agric Res 7:1–5

Thines M, Telle S, Choi YJ, Tan YP, Shivas RG (2015) Baobabopsis, a new genus of graminicolous downy mildews from tropical Australia, with an updated key to the genera of downy mildews. IMA Fungus 6(2):483–491

Thomas KM, Ramakrishnan TS, Srinivasan KV (1945) The occurrence of ergot in south India. Proc Indian Acad Sci B 21:93–100

Timper P, Wilson JP, Johnson AW, Hanna WW (2002) Evaluation of pearl millet grain hybrids for resistance to *Meloidogyne* spp. and leaf blight caused by *Pyricularia grisea*. Plant Dis 86:909–914

Weston WH (1929) A new Sclerospora from Australia. Phytopathology 19:1107–1115

Willingale J, Mantle PG, Thakur RP (1986) Postpollination stigmatic constriction, the basis of ergot resistance. Phytopathology 76:536–539

Wilson JP (2000) Pearl millet diseases: a compilation of information on the known pathogens of pearl millet: Pennisetum Glaucum (L.) R. Br (No. 716). US Department of Agriculture, Agricultural Research Service, Washington, DC

Wilson JP, Burton GW, Wells HD, Zongo JD, Dicko IO (1989) Leaf spot, rust, and smut resistance in pearl millet landraces from central Burkina Faso. Plant Dis 73:345–349. https://doi.org/10.1094/PD-73-0345

Wilson JP, Gates RN, Hanna WW (1991) Effect of rust on yield and digestibility of pearl millet forage. Phytopathology 81(2):233–236

Yadav HP, Prem S, Sabharwal PS (2000) Inheritance of smut resistance in pearl millet. Czech J Genet Plant Breed 36(2):45–47

Yadav OP, Rai KN, Rajpurohit BS, Hash CT, Mahala RS, Gupta SK et al (2012) Twenty-five years of pearl millet improvement in India. ICAR, New Delhi

Yoshihiro T, Mitsuo H, Hayato H, Futoshi K (2003) Biological control of rice blast disease by *Bacillus subtilis* IK-1080. Ann Phytopathol Soc Japan 69:85–93

Zarandi EH, Bonjar GH, Dehkaei FP, Moosavi SAA, Farokhi PR, Aghighi S (2009) Biological control of rice blast (*Magnaporthe oryzae*) by use of *Streptomyces sindeneusis* isolate 263 in Greenhouse. Am J Appl Sci 6:194–199

Pearl Millet Breeding for Enhancing Yield and Stability: Strategies, Achievements, and Perspectives

13

O. P. Yadav, S. K. Gupta, and P. Sanjana Reddy

Abstract

Pearl millet is cultivated under the most adverse agro-climatic conditions challenged by low and erratic rainfall, high mean temperature, high potential evaporation, and infertile and shallow soils with poor water holding capacity resulting in the huge temporal and spatial variation in its productivity and unstable production. The objective of this chapter is to review the strategies for achieving higher productivity and greater stability in India and Africa and to suggest future approaches and necessary interventions in pearl millet breeding meet the forthcoming challenges to provide higher and stable yields. The major strategies for enhancing yield included strategic use of germplasm resources and cultivar development (mostly hybrids) with targeted traits as per the requirement of production ecologies. On the other hand, the approach to augmenting stability has been improving genetic resistance to diseases, increasing tolerance to abiotic stresses, and addressing regional adaptation. Following the adoption of high-yielding, disease-resistant and drought-tolerant cultivars and crop production technology, pearl millet productivity has been consistently increasing in India. In view of increasing demand of pearl millet grain and stover in future, the higher yields are to be targeted. Pearl millet cultivation is likely to become more challenging because of predicted intense drought stress, rise in temperature, and greater disease incidences in sub-Sahara Africa and South Asia; yields must be

O. P. Yadav
ICAR, Central Arid Zone Research Institute, Jodhpur, India

S. K. Gupta (✉)
ICRISAT, Patancheru, Telangana, India
e-mail: s.gupta@cgiar.org

P. Sanjana Reddy
ICAR-Indian Institute of Millets Research, Hyderabad, India

V. A. Tonapi et al. (eds.), *Pearl Millet in the 21st Century*,
https://doi.org/10.1007/978-981-99-5890-0_13

increased at a much faster rate with greater stability in challenging agro-ecologies. Speed breeding and molecular-marker assisted breeding are going to play a very important role to enhance genetic gains in future. Heterotic grouping of hybrid parental lines would be an important strategy to increase the magnitude of heterosis on a long-term basis. Mainstreaming the bio-fortification is essentially needed to amalgamate higher productivity with nutritional traits to address both energy and micronutrient malnutrition issues.

13.1 Introduction

Food is the most basic need of humans. Providing adequate, safe, and nutritious food for the estimated human population of 10.4 billion in 2050 on a sustainable basis in the climate-change scenario is the greatest future challenge in agriculture as arable land and available water for agriculture are limited. Increasing crop production per unit area and developing crops grown with less water seem one of the most promising solutions. Pearl millet is the potential candidate crop for meeting this challenge as it is the world's sixth most important food crop after wheat, maize, rice, sorghum, and barley.

Pearl millet forms a major staple cereal in the drier regions of South Asia (SA) and Sub-Sahara Africa (SSA). It is critically important in providing food, nutrition, and livelihood security for more than 90 million farmers in these regions. A C4 plant, pearl millet has very high photosynthetic efficiency and dry matter production capacity under water-limited conditions. Therefore, it is valued not only for its nutritious grain for human consumption but also for its dry stover for ruminants in the crop-livestock farming system. In addition, pearl millet has a remarkable ability to respond to favorable growing conditions because of its short developmental stages and capacity for high growth rate, thus making it an excellent crop for the short growing season and under improved management conditions. Consequently, it is emerging as an important alternative crop for feed, food, fodder, and relay crop in Brazil, Canada, Mexico, the United States, West Asia, North Africa, and Central Asia.

Pearl millet is cultivated under adverse agro-climatic conditions where other crops like rice, maize, or sorghum fail to produce economic yields. Agro-environments of pearl millet are characterized by low and erratic rainfall, high mean temperature, high potential evaporation, and infertile and shallow soils with poor water holding capacity resulting in huge temporal and spatial variation in the crop productivity and unstable production. Pearl millet production is likely to become more challenging because of predicted intense drought stress and rise in temperature in Sub-Sahara Africa (Sultan et al. 2013) and South Asia (Rama Rao et al. 2019). On the other hand, pearl millet demand is anticipated to increase in the future because of increasing human and livestock populations in these regions and its use as healthy food and feed, in addition to its non-conventional use as raw material in the beverage industry (Rai et al. 2008; Serba et al. 2020). There is little scope to enhance production by expanding its cultivation, especially in SA. Therefore, its production must be increased, more so in challenging agro-ecologies, by increasing productivity per unit of land.

Like any other crop, the breeding programs in pearl millet have constantly been making efforts to develop high-yielding cultivars with built-in resistance to various biotic stresses and tolerance to abiotic stresses to enhance yield and stability. The breeding programs have been appropriately sensitive to embrace innovations and techniques in genomics and molecular biology to discover new traits, combine multi-stress tolerance, and improve breeding efficiency.

The intent of this chapter is to review the strategies for achieving higher productivity and greater stability in India and Africa and to assess their outcome and impact. We also suggest future approaches and necessary interventions in pearl millet breeding to meet the forthcoming challenges and enhance the speed and efficiency of cultivar development to provide higher and stable yields taking full cognizance of production constraints in pearl millet cultivation across various regions. Several innovations in seed production technologies and delivery mechanisms of improved seed helped further to harvest the benefits of better products in the target ecologies.

13.2 Production Constraints

Pearl millet is challenged by biotic and abiotic constraints due to its cultivation in diverse agro-ecologies. The nature and extent of these constraints vary widely in different regions in SA and SSA. Most of the crop growing environments are highly drought-, disease- and heat-prone. These factors, either individually or jointly, cause a huge loss in production.

Among the biotic constraints, diseases can cause losses of up to 30–40% in crop production if susceptible cultivars are grown. Both the quantity and quality of foliage and grain are adversely affected by diseases. Downy mildew (*Sclerospora graminicola*) is the most devastating disease in SA and SSA, affecting grain yield by converting panicle into the leafy structure and fodder yields by reducing total leaf surface area. Blast (*Magnaporthe grisea*) and rust (*Puccinia substriata* var. *indica*) affect severely vegetative tissues resulting in reduced stover yield and quality and indirectly decreasing grain yield. Other diseases include smut (*Moesziomyces penicillariae*) and ergot (*Claviceps fusiformis*), affect the developing grain in the panicle.

Striga remains the most serious biotic constraint in West Africa, especially in fields with low soil fertility. In the absence of effective screening techniques and confirmed resistance sources in the recent past, breeding for resistance to Striga has received negligible attention. However, now that effective screening techniques are in place (Wilson et al. 2004), work has been initiated to identify the resistance sources.

Abiotic constraints in pearl millet production include drought, heat, salinity, and nutrient stresses because cultivation is largely in rainfed production systems in the summer season on mostly sandy and sandy loam soils. The crop is most commonly cultivated in the regions receiving annual rainfall ranging from 150 to 600 mm, most of which is received from June to September. Because of rainfall's low and erratic distribution, drought is the primary abiotic constraint. Drought often comes with heat stress as the maximum air temperatures around 43 °C are common at the beginning of the rainy season. Pearl millet is becoming an important crop during the summer (February to May) in parts of India (states of Gujarat, Rajasthan, and Uttar Pradesh), where it is grown under high input and management conditions. High temperature, often more than 42 °C, persists during flowering and grain-filling stages of the crop in parts of western India. Tolerance to high temperatures is becoming an important research area in the genetic improvement of the crop.

Nutrient stress is very prominent in the pearl millet growing systems as the crop is largely cultivated on sandy and sandy loam soils that are inherently low in nitrogen (N) and phosphorous (P) contents. However, its adaptation to low nutrients has received negligible attention, assuming that this issue can be easily addressed through the external use of fertilizers. Limited studies on this aspect (Gemenet et al. 2015) have indicated the possibility of breeding nutrient-use-efficient cultivars. There is increasing interest in pearl millet nowadays to understand its adaptation to low nutrients.

13.3 Strategies for Enhancing Yield

Developing cultivars able to produce more with given resources has been a top priority since the beginning. The process is largely governed by the available technologies and key production challenges in the target environment.

The major strategies in the past for enhancing yield included hybrid development for exploiting heterosis, strategic use of germplasm resources to promote genetic diversification, prioritization of traits for different production ecologies, and tailoring plant types. On the other hand, augmenting stability has improved genetic resistance to diseases, increased tolerance to abiotic stresses, and enhanced regional adaptation.

13.3.1 Heterosis Exploitation

The outcrossing nature of pearl millet makes it amenable to various options in developing commercial cultivars that include composite, synthetic, topcross hybrid, three-way hybrid, and single cross hybrid. While composites, synthetics, and topcross hybrids can exploit heterosis partly, single cross hybrids (mainly referred to as hybrids) can exploit full heterosis.

Like maize, the heterosis for grain yield was also observed in pearl millet since the 1940s. Still, the hermaphrodite nature of flowers of small size limited the ability to exploit it fully at the commercial level. The discovery of cytoplasmic male sterility (CMS) in 1958 (Burton 1965) at Tifton, Georgia, and the availability of good fertility restorers in Indian germplasm laid the strong foundation for hybrid breeding in India in the mid-60s (Athwal 1966).

Five hybrids based on the male-sterile line Tift 23A were released between 1965 and 1969. Hybrids made a quick and significant impact on pearl millet production. However, the cultivation of a few hybrids with a narrow genetic base on a large scale led to downy mildew (DM) epidemics in the early 1970s. Such epidemics reappeared whenever a few hybrids occupied a large area year after year (Singh 1995). Recurring DM epidemics in pearl millet hybrids in India prompted intensified efforts on genetic diversification of hybrid parental lines, especially after the 1980s. This involved both cytoplasmic and nuclear diversification of parental lines.

The use of a diverse range of germplasm material from India and Africa with varied phenotypic characteristics was strategically exploited to diversify the genetic base of both seed and restorer parents (Andrews and Anand Kumar 1996; Rai et al. 2009a; Yadav et al. 2012c; Patil et al. 2020). On the other hand, searching for alternate CMS sources other than A1 (Burton 1965), A2, and A3 (Athwal 1961, 1966) led to identifying additional sources viz. A4 (Hanna 1989) and A5 (Rai 1995). Extensive characterization of these sources established instability of A2 and A3 sources, whereas A4 and A5 were more promising (Rai et al. 1996). This was followed by developing several lines based on A4 and A5. After initial difficulties in restoring these two CMS, breeding efforts were intensified for developing restorers for A4 and A5 CMS systems (Rai et al. 1996, 2009b). Understanding the genetics of A4 (Gupta et al. 2012a) and A5 CMS (Gupta et al. 2018) further helped in the utilization of these CMS sources. Thus, genetic and cytoplasmic diversification of hybrid parental lines, production of high-yielding and good combiners, and cultivar breeding for diverse production systems have been strategically undertaken. In the last four decades, hybrid breeding has received a very high priority in India, using genetically diverse parental lines targeting various production ecologies (Rai et al. 2009a; Yadav et al. 2012a).

For the successful exploitation of heterosis, traits that received high priority while breeding for the development of B- and R-lines have been yield per se (especially A-lines), lodging resistance, desirable panicle, and grain traits, complete and stable male sterility of A-lines, high levels of male fertility restoration of R-lines, good pollen shedding in B- and R-lines, flowering synchrony between A- and R-lines, and high levels of resistance to diseases and pests. Most of these traits are important as they are related to the economy of seed production. During the last 40 years of breeding hybrid parental lines, seed yield levels in production plots of hybrids have increased significantly, mainly due to the development of high-yielding parental lines and improved management practices. This is evidenced by the average seed yields of A-lines bred in the 1980s, which were 0.8–1.0 t/ha, whereas current yield levels have been reported to be around 1.5–2.0 t/ha (experts from the seed industry, Yadav et al. 2015) which resulted in considerably improved economics of hybrid seed production in pearl millet. While breeding for highly productive hybrid parental lines, the high general combining ability (GCA) of lines is very important as most studies have indicated no correlation or positive correlation between the grain yield per se of lines and their GCA. Thus, high general combiners are equally likely or even more likely to occur in lines with high grain yield than in any other yield group (Gupta et al. 2020; Gowda et al. 2006). Hence, considering the seed production economy, it would be judicious to breed for seed parents possessing high yield per se and high GCA.

Before the advent of hybrids, the genetic improvement of pearl millet heavily depended upon population improvement through mass selection and ear-to-row progeny testing to generate open-pollinating varieties (OPVs). With the establishment of the International Crops Research Institute for the Semi-Arid Tropics (ICRISAT) in the 1970s, the population improvement program saw new strides through the introgression of exotic germplasm. The utilization of African germplasm in breeding programs and improvement by recurrent selection resulted in several trait- and adaptation-specific composites and generated diverse breeding material, which was later utilized as base materials for developing high-yielding cultivars, OPVs, and hybrids.

Various recurrent selection methods (mass selection, gridded-mass selection, recurrent restricted phenotypic selection, S1, and S2 progeny selection, half-sib selection, and full-sib selection) have been used with varying successes in the genetic improvement of several trait-based composites (Rai et al. 1999). Many trait-based composites of the broad genetic base have been developed. Many of these were improved by recurrent selection, and several OPVs were developed (Khairwal et al. 2009; Yadav and Rai 2013) that had significantly contributed to their niche areas and continue to do so.

13.3.2 Strategic Use of Germplasm Resources

Genetic variability is bedrock on which the success of a genetic improvement program depends. The global pearl millet germplasm collections include more than 55,000 accessions of pearl millet (with several potential duplicates) conserved at the International Crops Research Institute for the Semi-Arid Tropics (ICRISAT) (Upadhyaya et al. 2012a, b; Hay et al. 2013) and the ICAR-National Bureau of Plant Genetic Resources (NPBGR) (Yadav et al. 2017). These include landraces, improved elite material, wild relatives, genetic stocks that are potential sources of specific phenotypic traits, resistance to diseases and insect pests, and tolerance to various abiotic stresses like drought and high temperature. To ensure unrestricted access to the world community, the FAO-designated gene bank at ICRISAT has placed its germplasm collections under the auspices of the Food and Agricultural Organization of the United Nations (FAO) in 1994. As per the agreement with the International Treaty on Plant Genetic Resources for Food and Agriculture (ITPGRFA), ICRISAT supplies germplasm free of cost to the global research community using the Standard Material Transfer Agreement (SMTA).

13.3.2.1 Targeting Specific Phenotypic Traits

The first criterion of using the germplasm remains the phenotypic traits that are subjected to manipulation in breeding programs. Characterizing germplasm for morpho-agronomic traits empowered breeders to use germplasm having desired trait or a combination of traits important in the target environments. In general, Indian pearl millet germplasm has mainly contributed to earliness, high tillering, high harvest index, and local adaptation (Yadav and Bidinger 2007), whereas African material has been a good source of high head volume, large seed size, and disease resistance (Anand Kumar and Appa Rao 1987; Rai and Anand Kumar 1994).

Strong and variable choice of phenotypic traits in different breeding programs has prompted the creation of trait-specific gene pools. Different methods have been employed for developing gene pools (Burton 1978; Singh and Jika 1988; Appa Rao et al. 1998; Yadav and Bidinger 2007). Experience in pearl millet indicated that gene pools have slightly narrowed phenotypic variation and obscured recessive traits (Burton 1978; Appa Rao et al. 1998; Upadhyaya et al. 2007a). Nonetheless, such trait-specific gene pools and populations are attractive options to most breeders as it is much easier to look for the traits in the gene pools and populations in which random-mating would have produced recombinants along with the specific trait of interest. Trait-specific gene pools and populations can be easily distributed and used as a source population for improvement through single-plant selection, recurrent selection, and mass selection.

While the use of diverse germplasm helps create additional variation, introgression from other germplasm sources or elite material is vital to diversify the breeding program's genetic base further. The most effective strategy for achieving this objective has been the hybridization of genetic materials with complementary traits (Presterl and Weltzien 2003; Yadav and Rai 2011), resulting in new gene combinations (Patil et al. 2020).

13.3.2.2 Searching for Sources of Disease Resistance

Several genetic stocks have been used as sources of disease resistance. Field evaluation of germplasm accessions and stability tests across locations in India and Africa has identified several sources of high levels of downy mildew resistance (Singh et al. 1987, 1990; Andrews et al. 1985; Mengesha et al. 1990; Thakur et al. 2001; Upadhyaya et al. 2007b; Sharma et al. 2015). Germplasm accessions obtained from India, Uganda, Nigeria, Lebanon, and Senegal yielded six promising smut-resistant lines (Thakur and King 1988; Wilson et al. 1991; Thakur et al. 1992). A screening effort involving 2752 germplasm accessions obtained from 19 countries and some unknown sources did not yield any accession with an acceptable level of resistance to ergot (Thakur et al. 1993). However, accessions with low ergot susceptibility (0–10%) and >75% selfed seed set from India, Nigeria, Togo, and Uganda were identified. Higher levels of ergot resistance were developed from the less susceptible lines by inter-mating and pedigree breeding (Thakur and King 1988). Several accessions from different African countries and India were found to show resistance to rust (Singh et al. 1987, 1990, 1997) and blast (Sharma et al. 2013). Recently, new sources of blast resistance have been developed in cultivated pearl millet backgrounds using wild *P. glaucum* subsp. *violaceum* (Sharma et al. 2020).

13.3.2.3 Searching for Adaptation Traits

Germplasm accessions and other genetic stocks are valuable for drought tolerance and escape. Earliness is important, especially for arid regions (Murty et al. 1967; Harinarayana et al. 1988; Kumar and Appa Rao 1987; Yadav et al. 2003). Salinity is another important production constraint, especially in some parts of India. Genetic variation has been reported, and specific accessions are identified as having salinity tolerance (Ashraf and Mcneilly 1992; Chopra and Chopra 1997; Kulkarni et al. 2006; Krishnamurthy et al. 2007; Upadhyaya et al. 2007b). Peacock et al. (1993) identified a few genotypes from Rajasthan for better survival under high soil surface temperatures during the initial stage of development. Large genetic variation for tolerance to heat at the reproductive stage among pearl millet breeding lines and populations has also been observed, and heat-tolerant lines have been identified (Gupta et al. 2015). Based on multi-locational screening, 5 maintainer lines, ICMB 92777, ICMB 05666, ICMB 00333, and ICMB 02333, were found to have >60% seed set when the air temperature during flowering exceeded 42 °C.

13.3.2.4 Sources of Nutritional Traits

Evaluation of 297 Iniadi germplasm accessions originating from Western Africa (Togo, eastern Ghana, southern Burkina Faso, and western Benin) has indicated wide variability for Fe and Zn density levels (Rai et al. 2008, 2014, 2016; Velu et al. 2008; Upadhyaya et al. 2007b; Govindraj et al. 2011).

13.3.3 Trait Prioritization

Most of the pearl millet breeding programs in India's public and private sectors are guided by well-characterized product profiles for different segments. However, such information may not always be available in a documented form. Product profiles are developed considering the mandate of the program, and the requirements of a wide range of stakeholders, including producers and consumers, to utilize the financial and human resources in the program efficiently. For instance, the pearl millet breeding program at ICRISAT organizes a "Consultation Meeting" biennially with crop improvement leaders from both the public and private sectors to identify the priority traits to be addressed on a short- and long-term basis. In such surveys, it was revealed that five production segments are targeted in different pearl millet breeding programs in India: (1) Rainy season: dual-purpose hybrids for A- and B-zones (better-endowed environment), (2) rainy season drought-adapted hybrids for A1-zone (drought-prone environment), (3) Summer season hybrids, (4) forage hybrids for niche areas, (5) Superior grain quality (grain Fe/Zn) hybrids. The priority traits for each of these segments are also identified in addition to high productivity, which is common across all segments. For rainy-season hybrids, traits include downy mildew resistance, lodging tolerance, and blast resistance. Cytoplasmic diversification of parental lines is also of high priority in this segment. Drought tolerance, stay-green, and early maturity remain the priority traits for drought-prone environments, while heat tolerance ranked highest in the summer hybrid market.

Multi-cut and disease resistance are identified as priority traits for forage hybrids. Overall, the survey indicated disease resistance (DM, blast, and rust) as the prioritized traits followed by lodging and heat tolerance.

13.3.4 Plant Type

The concept of plant type in crop improvement started to receive major attention with the discovery of dwarfing genes in wheat and rice. The concept is particularly relevant for the modernization of traditional agriculture, where genetic diversity for plant types could help develop improved crop varieties responsive to applications of fertilizers, irrigation, and other farm inputs. Traits linked to the plant type concept have been strategically manipulated in pearl millet. Considering the diversity in pearl millet growing environments across regions, grouping various growing regions into mega-environments is a successful strategy (Gupta et al. 2013) which involves the development and deployment of cultivars having traits conditioning adaptation to a specific mega-environment. For instance, private sector breeding programs have largely focused on relatively better-endowed environments emphasizing breeding for dual-purpose hybrids. As a result, private sector hybrids are generally taller, later in maturity, with longer panicles, and with fewer effective tillers/plant (AICPMIP, unpublished data). In contrast, in most public sector programs that target drought-prone environments, hybrids are generally shorter in height and early in maturity, with smaller panicles and a higher number of effective tillers/plant. Genetic variation for these plant-type linked traits has been utilized as per need (Yadav et al. 2004b, 2017). As a result, panicle length has increased from 16.7 to 22.0 cm, panicle diameter from 2.4 to 3.0 cm, and 1000-grain weight from 10 to 12 g in seed parents developed during the last four decades. Mean productive tillers per plant were found unchanged. However, specific lines have been bred with >35 cm long panicle length, >3.5 cm thick panicles, >5 productive tillers per plant, and a 1000-grain weight of >15 g.

The plant type concept has implications for breeding hybrid parental lines. The d_2 dwarfing gene-based shorter height is the most dominant plant type developed in seed parents breeding (Rai and Hanna 1990; Rai and Rao 1991) as it reduces the risk of lodging in high-management conditions and helps in easy detection of off-type and pollen shedders in the seed production plots.

The A-lines are bred for complete and stable male sterility, and the B-lines for profuse pollen production ability across the seasons and sites. In breeding A-lines, high grain yield potential, both as lines per se and in hybrids (i.e., combining ability), is the most important consideration. The plant type of seed parents profoundly affects the plant type of their hybrids (Yadav et al. 2000). Hence, the development of a large number of seed parents and their utilization as per desired traits in hybrids targeted for specific mega-environment has been a highly successful strategy in pearl millet (Yadav et al. 2012c).

The foremost requirement in the restorer lines is to produce highly fertile hybrids and profuse pollen that remains viable at air temperatures as high as 42–44 °C. It is desirable to breed pollinators of 150–180 cm in height, but no shorter than the A-line with built-in attributes of panicle, maturity, and tillering that will be preferred by farmers in the hybrids. Both seed parents and pollinators must have an acceptable level of lodging resistance and resistance to various diseases.

13.4 Strategies for Enhancing Stability

Pearl millet being a rainfed crop, the temporal variation in its productivity can be very high due to differences in the amount and distribution of rainfall. Similarly, diseases can have catastrophic effects on production, if the season is favorable from a disease-development point of view. Highly unstable production leads to farmers' distress, and variable income adversely affects their livelihood. Not only is higher productivity important, but reliable production is also equally essential. The latter that be achieved by enhancing the resilience to both biotic and abiotic stresses.

13.4.1 Abiotic Stress Resilience

13.4.1.1 Drought Tolerance

Drought is the most important constraint in pearl millet production and is predicted to intensify further due to future climate change requiring a greater focus on improving crop tolerance to drought. Though the level of drought tolerance of pearl millet is much higher than that of other cereals, developing cultivars with a greater degree of built-in drought tolerance is the best approach to enhance temporal and spatial stability in production. This is the most challenging task in pearl millet breeding as the nature of drought in the different target populations of crop growing environments (TPE) is highly variable in its timing, severity, and period.

Breeding for enhanced drought tolerance has tremendously benefited from a comprehensive understanding of pearl millet response to drought at various growth stages. Differential sensitivity of various growth stages to water stress helped to identify target traits in pearl millet for drought ecology for manipulation in breeding programs.

Drought stress during the early establishment phase results in seedling mortality, causing uneven crop stand and yield losses in pearl millet. This issue has been addressed by crop management under such conditions, rather than breeding. Drought stress at the vegetative stage affects pearl millet growth (Shivhare and Lata 2017) by reducing the dominance of the main shoot due to the accumulation of abscisic acid (Yadav et al. 2012a). Further drought stress during the vegetative phase delays the flowering time of the main shoot and reduces biomass, total leaf area, and specific leaf area; exposure to drought reduces biomass and total and specific leaf area. Drought at the reproductive stage causes a severe impact on yield and stability (Mahalakshmi et al. 1987; Winkel et al. 1997; Kholová and Vadez 2013). Bidinger et al. (1987) reported that terminal drought stress could cause 60% yield losses in pearl millet due to a decrease in the number of panicles per plant and reduced grain mass. The seed setting that determines the number of grains per panicle is usually less affected if terminal stress occurs after flowering.

Dissection of drought tolerance in terms of physiology, phenology, and morphology of the crop has helped in understanding the yield formation process under drought conditions (van Oosterom et al. 2003; Yadav 2011). The key physiological feature of drought tolerance of pearl millet includes greater uptake of water at critical crop growth stages (Vadez et al. 2013; Tharanya et al. 2018), conservation of water use (Kholová et al. 2010a; Vadez et al. 2013) by limiting transpiration at high atmospheric evaporative demand (Kholová et al. 2010a, b; Medina et al. 2017; Tharanya et al. 2018; Choudhary et al. 2019), dehydration avoidance (Bidinger and Witcombe 1989), growth maintenance through the stability of cellular membrane (Blum and Ebercon 1981; Vadez and Sinclair 2001; Kholová et al. 2010a, b), and radiation reflectance (Bidinger and Witcombe 1989).

Phenology plays a critical role in determining performance under drought as early flowering helps escape terminal drought conditions (Bidinger et al. 1987; Fussell et al. 1991; van Oosterom et al. 1995; Yadav et al. 2012c). Genetic variation for earliness is abundant in pearl millet (Rai et al. 1997; Rattunde et al. 1989; Yadav et al. 2017), and several sources from India and western Africa, especially Iniadi-type landraces, have been widely used as a source of earliness (Andrews and Anand Kumar 1996).

Traits related to canopy development are reportedly tightly associated with water use (Vadez et al. 2015; Tharanya et al. 2018). Early vigor that signifies rapid crop development in the initial stages has been correlated with drought tolerance (Manga and Yadav 1995), claiming that crops with faster leaf area development could intercept greater radiation and reduce water losses by evaporation. However, there are some concerns that greater transpiration from a larger leaf area will quickly exhaust soil water resources and may cause intense water shortage at growth stages more sensitive to drought (Winkel and Do 1992). High tillering is another useful morphological trait associated with tolerance to mid-season drought (Rai et al. 1997; Yadav et al. 2001, 2017). This trait has been successfully manipulated in breeding programs targeting drought environments (van Oosterom et al. 2003, 2006). The panicle harvest index (PHI) suggests the plants' ability to set and fill grains under water-limiting conditions is an indicator of drought tolerance (Bidinger et al. 1987; Fussell et al. 1991; Yadav 1994; Yadav et al. 2012c).

Several screening protocols have been developed to assess drought tolerance in target environmental conditions. Osmotic solutions with polyethylene glycol (PEG) have frequently been used in in vitro screening to induce water stress and maintain uniform water potential throughout the experimental period. The vivo screening methods have included pot screening using well-watered and water-stressed conditions (Kholová et al. 2010a, b, 2016; Choudhary et al. 2019). Screening in pots is simple and cost-effective, but it is very difficult to screen large populations. Therefore, high throughput and automated phenotyping platforms like LeasyScan, capable of screening large populations/germplasm materials for drought adaptive traits, are preferred (Vadez et al. 2015; Tharanya et al. 2018; Kar et al. 2020). The Lysimeter approach provides a bridge between field and laboratory-based research

(Vadez et al. 2013). Screening under field conditions is done by evaluating the test material through multi-locational trials conducted in drought-prone regions (Yadav et al. 2012c) or by growing pearl millet in a rain-free season under adequate water supply but withholding irrigations to expose pearl millet to drought at desired stage.

The role of adapted germplasm has been emphasized for drought breeding because the measured performance under drought stress is largely a result of adaptation to stress conditions rather than that of high potential yield. Therefore, the base material required for a successful breeding program differs for drought and more favorable environments. Traditional landraces that evolved in dry areas due to natural and human selection over centuries usually exhibit good adaptation to drought. Attempts have been made to exploit these landraces in pearl millet breeding programs in a systematic way. Cycles of mass selection in genetically heterogeneous landraces were found to increase yield considerably (Bidinger et al. 1995; Yadav and Manga 1999; Yadav and Bidinger 2007) and have also been shown as a useful germplasm source in developing drought-adapted varieties (Yadav 2004) and deriving inbred restorer lines for hybrids (Yadav et al. 2009, 2012a).

Molecular breeding is fast emerging as a supplement approach to enhance drought adaptation faster with greater precision. Efforts in this direction started with the development of molecular marker-based genetic linkage maps, followed by the identification of quantitative trait loci (QTL) determining yield under drought environments. Genome regions underlying drought tolerance-related traits have been identified and mapped (Yadav et al. 2002, 2004a; Serba and Yadav 2016). Several such genomic regions are being transferred to improve performance under water-limiting conditions (Bidinger et al. 2007; Sharma et al. 2014).

13.4.1.2 Heat Tolerance

Heat stress can have a significant effect on the pearl millet plant at two stages, either at the seeding stage or at the flowering period/reproductive stage. Soil temperatures exceeding 45 °C during the first 10 days of pearl millet sowing can lead to a poor plant stand. This trait is a complex trait due to the prevalence of multiple environmental factors like soil temperature, soil moisture, and soil composition in the field conditions, which makes it difficult to do screening for heat stress at the seedling stage; hence, very few breeding efforts have been made on this trait. However, Peacock et al. (1993) reported genetic differences in seedling survival under high soil surface temperatures, and some artificial screening techniques were reported to be effective for screening for seedling heat tolerance (Soman and Peacock 1985; Howarth et al. 1997).

Heat tolerance at the reproductive stage emerged as an important target trait, as the pearl millet crop established itself in high heat stress as a summer crop in northwestern parts of India around 2000 (Yadav and Rai 2013), and its cultivation is further expanding. High heat stress (often above 42 °C) coincides with the flowering stage, leads to the lower seed set, and reduces grain yields (Gupta et al. 2015; Djanaguiraman et al. 2018). In the last decade, flowering-period heat stress screening protocols for pearl millet were standardized for screening under both controlled environment facilities (growth chambers) and under field conditions in heat stress-prone target ecology (Gupta et al. 2015). Other investigations revealed that stigma is more heat sensitive than pollen; large genetic variation exists among breeding lines and within open-pollinating populations; the boot-leaf stage is more heat sensitive than the panicle-emergence stage, and heat tolerance behaves as a dominant trait in pearl millet (Gupta et al. 2015). Multi-locational and multi-year field screening from 2009 to 2014 in target ecology (Gujarat, western U.P., and south Rajasthan) in the summer season led to the identification of heat-tolerant breeding materials, which included several maintainer lines (ICMB 92777, ICMB 05666, ICMB 00333, ICMB 01888, ICMB 02333, and ICMB 03555), improved populations (ICMV 82132, MC 94, ICTP 8202, and MC-Bulk), and germplasm accessions (IP 19799, IP 19877, and IP 19743). These identified lines were further used to develop crosses and heat-tolerant composites, followed by selections in target ecology through the shuttle breeding approach (Gupta et al. 2016, 2019) to facilitate the pyramiding of heat tolerance in new hybrid parental lines.

13.4.2 Biotic Stresses Resistance

As compared to other cereals, pearl millet is less affected by diseases. The major diseases of current significance are caused by fungi. These include downy mildew (*Sclerospora graminicola* (Sacc.) J. Schroet), blast (*Magnaporthe grisea*), rust (*Puccinia substriata* (Ellis & Barth) var. indica), smut (*Moesziomyces penicillariae* (Bref.) Vanky), and ergot (Claviceps fusiformis Loveless). Developing and applying effective screening techniques and evaluating germplasm, breeding material, and hybrid parental lines for their reaction to various diseases have been integral components of pearl millet improvement.

13.4.2.1 Diseases Resistance

Among the diseases, downy mildew is the most important constraint, especially on those hybrids with DM-susceptible lines in their parentage or when a specific hybrid is cultivated continuously in a large area.

Downy mildew Following a good understanding of the epidemiology of DM pathogen, highly effective field (Williams et al. 1981) and greenhouse (Singh and Gopinath 1985) screening techniques have been developed which easily differentiate between resistant and susceptible progenies and have been widely utilized. Screening techniques under greenhouse conditions have been further refined (Singh et al. 1993) that are highly useful for testing breeding progenies between growing seasons (Singh 1995).

Several genetic stocks and selections from germplasm accessions and elite breeding materials have shown a high degree of stability for resistance across sites and years. The identified sources of resistance were effectively utilized in developing DM-resistant male-sterile lines and pollinators. Moreover, mutation-induced resistance was also successfully utilized. The residual variability for DM resistance could also be exploited to improve the resistance levels of susceptible material (Singh et al. 1992).

Several putative QTLs have been identified that determine a significant proportion of DM resistance in pearl millet. Resistance alleles at two DM QTLs, one each on linkage groups 1 (LG1) and 4 (LG4), were added to the male parent (H 77/833-2) of one of the most widely grown hybrids, HHB 67, through marker-assisted backcross breeding using a selection from elite parent ICMP 451 as the resistance donor (Breese et al. 2002). A more DM-resistant version of this early-maturing hybrid was released as "HHB 67 Improved" (Hash et al. 2006). To further enhance the DM resistance level of this hybrid, four new QTLs from different donor parents are being transferred to the pollen parent of HHB 67 Improved. Replacing the susceptible hybrid with its disease-resistant version, created by marker-assisted selection, has also been an effective strategy.

Smut Smut, considered a minor disease, gained considerable importance after the widespread cultivation of hybrids. Smut is most severe on the upwind borders of isolated fields, especially on the earliest flowering panicles when pollen availability is limited. It can be more serious on hybrids with more synchronous flowering and poor fertility restoration, especially when rainfall and high humidity coincide with flowering, resulting in poor anther dehiscence.

A highly effective screening technique for smut resistance involves inoculation of panicles by injecting an aqueous suspension of sporidia into the boot, covering the inoculated panicles with parchment paper selfing bags, providing high humidity (>80% RH), removing bags 2–3 weeks after inoculation, and scoring for smut severity. Significant progress has been made in incorporating smut resistance from unadapted source materials into commercially useful male-sterile lines using the pedigree-bulk breeding procedure. ICMA 88006 was the first smut-resistant male-sterile line developed using these methods. Several other smut-resistant male-sterile lines have been bred. The inherited recessive 'tr' allele that conditions trichomelessness to most above-ground plant parts, including stigmas, confers a useful degree of smut resistance (Wells et al. 1987; Wilson 1995).

Ergot Field screening techniques (Thakur and Williams 1980) for ergot disease include bagging of panicles at the boot-leaf stage with selfing bags to allow stigma emergence in a pollen-protected environment, inoculating panicles at stigma emergence and briefly opening the bags, and spraying the panicles with an aqueous conidial suspension produced from honeydew of infected panicles, providing high humidity (>80% RH), and removing bags and scoring ergot severity. Screening a large number of pearl millet inbred lines at different locations revealed no lines with satisfactory ergot resistance. Resistant lines were developed by intermating plants scored as less-susceptible and selecting resistant progenies under high disease pressure for several generations following pedigree and recurrent selection procedures. A number of lines thus developed have shown a high level of resistance across different locations in India and western Africa over several years. Lines with high levels of resistance combined with good agronomic traits have also been identified and used as donors for developing ergot-resistant varieties. About 300 ergot-resistant inbred lines and populations have been evaluated for their agronomic attributes and reaction to diseases (Thakur et al. 1993). A number of these lines and populations possess improved agronomic traits and combined resistance to ergot, smut, DM, and rust (Hash et al. 1997). Inheritance of ergot resistance is relatively complex and may involve cytoplasmic × nuclear interactions (Thakur et al. 1993). Resistance behaves as a recessive trait, and it is polygenically controlled.

Rust Rust of pearl millet is considered a minor problem because of its late appearance, generally after the grain filling stage. However, where the disease is observed as early as the seedling stage, it can cause a substantial reduction in grain and fodder yield and quality. Several concerted efforts have been made to identify resistance sources for pearl millet rust. A large number of germplasm accessions have been evaluated for their reaction to DM and rust, and many accessions with resistance to these two diseases have been identified (Singh 1995).

Blast Blast pathogen isolate of pearl millet does not cross-infect rice and finger millet and vice versa. Although it was considered a minor disease of pearl millet until the 1990s, its incidence has increased considerably in recent years. Field and greenhouse screening techniques have been developed, and sources of resistance have been identified (Sharma et al. 2013). Inheritance studies have indicated major single-gene control with resistance being dominant and a single gene effective in imparting resistance (Gupta et al. 2012b). However, three independent dominant genes for blast resistance have also been reported (Hanna and Wells 1989).

13.4.2.2 Insect-Pests Resistance

Pearl millet attracts lesser infestation of insect pests compared to other cereals. Some of the insect pests with significant economic importance include shoot fly (*Atherigona approximata*), stem borers (*Chilo partelus* in India and *Coniesta ignefusalies* in western Africa) and white grubs (*Holotrichia consanguinea*) in India, ear head worms (*Helicoverpa armigera*), gray weevil (*Myllocerus* species), and leaf roller (*Marasmia trapezalis*) (Raghvani et al. 2008). The distribution and damage of insect pests vary in different regions. Long-term monitoring revealed that no single control method is effective against any insect. It requires an integrated approach, including cultural and chemical control (Sharma and Youm 1999). The pest incidence on commercial cultivars and experimental test genotypes is closely monitored, and no breeding programs are undertaking insect resistance as the target trait.

13.4.3 Regional Adaptation

Pearl millet is cultivated across a wide range of agro-climatic conditions. A single cultivar cannot be expected to perform well under all environmental conditions (Rosielle and Hamblin 1981; Ceccarelli 1989). A cultivar planted outside its adaptation zone will suffer yield reduction due to significant genotype × location (GL) interactions. Therefore, planning breeding and testing activities require subdividing the testing environments into relatively more homogeneous groups of locations, called mega-environments, where specific genotypes can be targeted for each location.

Since pearl millet is primarily grown in marginal environments, which are stressed in a variety of ways, leading to large genotype × environment interactions, an understanding of the nature and relative contribution of genotype, locations, and GL interaction effects assumes greater importance in site selection, cultivar testing, and possible recommendation of genotypes for cultivation. Latest statistical methods such as additive main effects, multiplicative interaction genotype main effects, and genotype × environment interaction (GGE) biplots (Yan and Tinker 2006) are used for genotype × environment data analysis. Plant breeders and agronomists have found GGE biplots useful in mega-environment analysis and evaluation of test environments (Thomason and Phillips 2006).

Although the subdivision of crop growing region into several mega-environments increases the work for breeders to breed for each mega-environment separately, it also helps select genotypes for greater yield stability within relatively well-defined and homogeneous environments, increasing the efficiency of breeding programs by targeting genotypes to appropriate production areas. Furthermore, even if the breeding goal is a wide adaptation, the best strategy would be to identify several mega-environments and use the most effective test locations to select for wide adaptation (Gauch and Zobel 1997). Therefore, the pearl millet cultivation area in India has been divided into three different mega zones based on the geographic location, rainfall pattern, and local adaptation of the crop (Gupta et al. 2013).

Pearl millet hybrids bred for low rainfall mega-zone had a relatively smaller canopy than hybrids bred for higher rainfalls and differences in canopy development linked to soil water conservation (Vadez et al. 2015). It was also reported that the hybrids adapted to lower rainfall had higher water use efficiency than those from the highest rainfall zones. In addition, genotypes bred for lower rainfall zones showed lower leaf area, dry matter, thicker leaves, root development, and exudation than the ones bred for high rainfall zone. These differences in the water use response and growth showed that breeding for different agro-ecological zones also bred for different genotype strategies related to plant water use (Medina et al. 2017). Van Oosterom et al. (2003) explained the specific adaptation of local landraces to drought stress environments in Rajasthan, India, by analyzing the yield components of pearl millet and concluded that landraces were beneficial under limited water availability as they produced small-sized main shoot panicle, which favored the productive basal tillers and minimized the delay in anthesis. Tharanya et al. (2018) showed that in environments with unlimited water access, pearl millet genotypes with high tillering and biomass accumulation led to increased water use and yield. On the contrary, in severely water-limited environments, where water can be stored in the soil profile, crop production benefited from less vigorous growth with few tiller/smaller canopy and expression of stay-green phenotypes. In this regard, it is evident that the effect of genotype by environment interactions (GEI) determines the success of specific adaptation. It is also demonstrated that traits that support crop production in one environment might bring a production penalty in another (Yadav and Weltzien 2000). Therefore, the characterization of the crop environment is an important step in understanding GL interactions and achieving stability in production.

13.5 Achievements and Impact

13.5.1 Deployment of Improved Cultivars and Outcome

Breeding strategies in pearl millet have evolved comprehensively over several decades, considering the serious challenges in its cultivation across diverse production environments. A better understanding of its adaptation mechanism to stress environments, uninterrupted access to improved germplasm, and systematic breeding efforts have led to the development and release of a very large number of cultivars, both hybrids and OPVs, in India and Africa. The choice of cultivars for different production environments has become wider with time. For example, more than 125 pearl millet hybrids are in the market today compared to the limited alternative to farmers in the later part of the twentieth century. Development and strategic utilization of genetically diverse and a very large number of male-sterile lines and pollinators in India resulted in ample diversity in commercial hybrids of pearl millet that ensured containment of downy mildew which was threatening hybrid per se in pearl millet a few decades ago. No major outbreak of downy mildew has been witnessed since the mid-1980s due to the cultivation of several disease-resistant hybrids and the frequent replacement of older hybrids with new ones. Diversity also provided buffering mechanism against environmental vagaries.

The high-yielding cultivars, especially hybrids (Fig. 13.1), have been widely adopted by Indian farmers. The area under improved cultivars has increased considerably over the years. Currently, more than 75% of pearl millet area is annually planted using hybrids and a few improved OPVs in India. On the other hand, the adoption of improved cultivars in Africa has been limited.

Three factors have largely contributed to the large-scale adoption of hybrids in India. Firstly, the availability of disease-resistant hybrids in the broad maturity duration (60–90 days) has enabled farmers to choose from a wide range of available cultivars with appropriate trait combinations that they consider fit to meet their requirements in different crop production environments of various states. Secondly, there is the recovery of seed cost by farmers with even 10% higher grain yield with improved cultivars with a low seed requirement of 5 kg/ha. Thirdly, highly profitable seed production, distribution, and marketing have attracted the private sector to deliver hybrid seeds efficiently.

Fig. 13.1 A good crop of an early maturing and disease-resistant pearl millet hybrid MPMH 17 in Rajasthan, India

13.5.2 Impact of Breeding

Following the adoption of high-yielding and disease-resistant cultivars (mostly hybrids) and crop production technology, pearl millet productivity has been consistently increasing in India (Fig. 13.2). During the last 70 years, the productivity has gone up from 303 kg/ha during 1951–1955 to 1235 kg/ha during 2016–2020 that translates to an increase of more than 300% owing to the widespread use of high-yielding and disease-resistant cultivars with improved production technology. This productivity increase in pearl millet assumes a greater significance in two ways. First, more than 85% of pearl millet is grown as rainfed and often on marginal lands. Second, pearl millet research has attracted much lesser investment in infrastructure and human resources development than other major cereal crops. Despite this, the magnitude of yield improvement in pearl millet under rainfed conditions successfully demonstrated technology-led development and highlights the role of hybrid technology in raising crop productivity even in marginal drylands.

13.5.3 Realized Yield Gains in Pearl Millet Vis-à-Vis Other Cereals

A critical analysis of the genetic improvement in pearl millet has been recently done (Yadav et al. 2019) in which seven decades of breeding were divided into four phases, each phase having its uniqueness. During phase I (1950–1966), when genetic improvement largely concentrated on the enhancement of yield in locally adapted materials, the rate of productivity improvement was 4.5 kg/ha per year (Fig. 13.3). Discovery and utilization of CMS in hybrid development marked the second phase of genetic improvement (1967–1983) in which an annual increase of

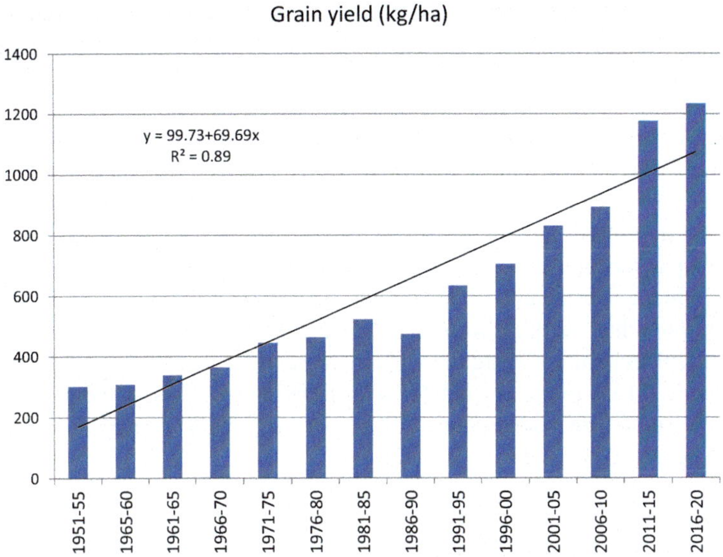

Fig. 13.2 Pearl millet productivity from 1950 to 2020. (Source: DAC, Government of India as on August 7, 2021, available at http://www.agricoop.nic.in)

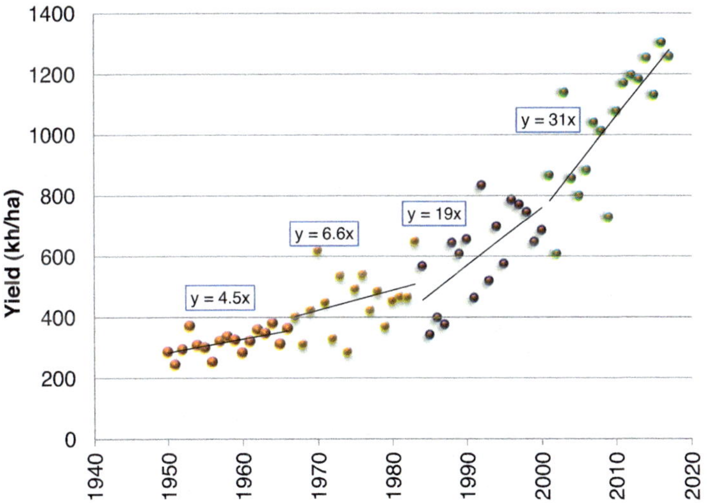

Fig. 13.3 Productivity trend of pearl millet in four phases since 1950. The figures in boxes are the rate of increase in yield (kg/ha/year) in each of the four phases. Phase I is for 1950–1966, phase II for 1967–1983, phase III for 1984–2000, and phase IV for 2001–2018

6.6 kg/ha in productivity was realized despite the large-scale cultivation of a few hybrids. Many genetically diverse CMS lines were developed and utilized in hybrid breeding during phase III (1984–2000), and the productivity increase was 19.0 kg/ha per year. During phase IV (2001–2018) when genetic improvement put much greater emphasis on the genetic diversification of hybrids and adaptation to niche areas of cultivation, the rate of improvement in grain productivity further increased to 31.1 kg/ha per year, which is 470% higher than the productivity gain achieved during the first phase.

A comparison of yield increases in pearl millet vis-à-vis the other four major cereals in India after the mid-1980s presents very interesting information. Following the adoption of high-yielding, disease-resistant, and stress-tolerant cultivars and crop management technology, there is a yield increase of 26% in sorghum, 59% in wheat, 69% in rice, 113% in maize, and 162% in pearl millet. These yield gains translate into 0.9% annual gains in sorghum, 2.0% in wheat, 2.3% in rice, 3.8% in maize, and 5.4% in pearl millet (Yadav et al. 2021).

13.6 Future Perspectives

In view of the increasing demand for pearl millet grain and stover in the future, higher yields are to be targeted and achieved. At the same time, pearl millet cultivation is likely to become more challenging because of predicted intense drought stress, rise in temperature, and greater disease incidences in SSA and SA. Therefore, its production must be increased much faster with greater stability in challenging agro-ecologies.

13.6.1 Achieving Higher Yields

Productivity is the function of technology adoption and the nature of production environments. Efforts are continually needed to develop new cultivars with a built-in capacity to produce more with a given amount of resources using the new opportunities.

Pearl millet has shown impressive genetic gains in India for the past seven decades (Yadav et al. 2019). The genetic gain needs to be further accelerated, for which several prospects exist. Genomics-assisted breeding offers a great opportunity for taking genetic gains to the next level in pearl millet by improving the precision and efficiency of the breeding program. The ~1000 genomes sequencing project has been a major milestone in pearl millet improvement (Varshney et al. 2017). This work has laid a solid foundation for carrying out trait discovery, mapping, and deploying QTLs/alleles/candidate genes linked to traits of economic interests.

Heterotic grouping of hybrid parental lines would be an important long-term strategy to increase the magnitude of heterosis. A diverse range of breeding materials has historically been used to develop either seed parents (B-lines) or restorers parents (R-lines) depending upon their specific phenotypic traits. The existence of B- and

R-lines as separate groups has been found responsible for behaving as two separate broad heterotic pools, as B × R hybrids reported significantly higher levels of heterosis than B × B or R × R hybrids (Singh et al. 2018). Recently, in a study involving 320 R- and 260 B-lines derived from pearl millet breeding programs in India, 2 B- and R-line heterotic groups were identified based on the heterotic performance and combining ability (Gupta et al. 2020). This study also indicated that distinct parental groups could be formed based on molecular markers, which can help in assigning hybrid parental lines into heterotic groups to develop high-yielding hybrids. Now, there is a need to select the appropriate testers to categorize the new hybrid parental lines or new germplasm into heterotic groups to enhance the genetic gains in pearl millet.

Targeting hybrids for niche areas would play an enormously important role in further increasing yields. A comprehensive study conducted in arid regions of India comparing 142 hybrids and 84 composites over 12 years in 94 environments for their performance reported that hybrids yielded significantly higher grain than composites, with an overall superiority of 25% (Yadav et al. 2012b). Therefore, hybrids are likely to play a much greater role than composites in enhancing pearl millet productivity in drought-prone regions, including SSA. Another consideration in SSA is the multiple sowings, especially in severely drought-prone areas. It is often argued that the composites have the edge over hybrids as they are self-perpetuating, and harvested seeds can be used to plant the next crop. However, such an option would come with a significant penalty for grain yield. Recently, a breeding program in Africa has been initiated with encouraging results in releasing hybrids for cultivation. This should also encourage the seed supply system in this region and extend the benefits of hybrids to African continents.

Dual-purpose pearl millet hybrids would be critically important for a crop-livestock-based farming system. Recent growth and future projections of aggregate food demand patterns suggest a substantial increase in the demand for meat, milk, and eggs in India by 2050. The current supply and projected demand of dry stover indicate a 23% deficit. Pearl millet, a C4 species with high photosynthetic potential and biomass production ability, has a greater role to play in arid and drier semi-arid regions to bridge the gap between demand and supply of dry fodder in dry periods. Pearl millet grain is also a potential source to supplement maize supply in the poultry and cattle feed industry.

Mainstreaming bio-fortification is essentially needed to amalgamate higher productivity with nutritional traits to address energy and micronutrient malnutrition issues. Large genetic variability has been found in the elite germplasm and breeding material to develop high-yielding cultivars with higher levels of these micronutrients. Interestingly, a negative association has not been reported between grain yield and micronutrients in pearl millet, suggesting the feasibility of combining high yield with a greater concentration of micronutrients.

Speed breeding and big-data analytics also hold a great promise. Using traditional breeding methods, developing a new crop cultivar takes about a decade or more, with 6 or 7 years spent in seasonal generational advancements to arrive at elite materials for testing and release. New environmentally controlled facilities, known as "RapidGen," have been developed to shorten the 7-year window significantly. The availability of ultra-high-speed computing protocols, crop simulations, big data analytics, the internet of things, artificial intelligence, and machine learning will provide an opportunity to run complex queries and scenario analyses, enabling researchers to focus on specialized research. Using the quantitative genetics-based crop simulation algorithms to understand and optimize existing pearl millet breeding pipelines and take measures to refine them further is the need of the hour.

13.6.2 Targeting Greater Stability

In view of climate change encountered in agriculture, the need for the climatic resilience of pearl millet is greater than ever to provide higher and stable yields.

Disease resistance would continue to play a major role in providing stability in pearl millet production. The experience in resistance breeding for downy mildew has indicated that most hybrids become susceptible in about 5–6 years of cultivation in the same area due to selection pressure in the pathogen. This situation calls for continuous monitoring of the virulence of pathogen populations through reaction on host differential, this is essential to identify resistance effective against new virulent pathotypes. Genome sequencing of pearl millet and pathogens of DM and blast will help understand the molecular basis of compatible/incompatible host × pathogen interaction and provide a greater opportunity for breeders and pathologists to control the diseases. With the availability of genomic tools, identification of QTLs determining resistance to particular diseases, and demonstration of the success of marker-assisted backcrossing, it now appears possible to stack target QTLs in the parental lines of hybrids having multiple resistance to various pathotypes of downy mildew, blast, and rust. Insect pests and parasitic weeds (Striga species), which are also important yield-reducing factors in Africa, must be tackled in future programs.

Pearl millet is the most water-use efficient major cereal under water-limiting environments and is also the heat-tolerant cereal. It is also the most salinity-tolerant cereal next only to barley. These attributes of pearl millet put it in a uniquely advantageous position to address the multiple environmental challenges, which are likely to be of increasingly serious proportion in the future. Precision phenotyping protocols for drought and heat will determine the success of achieving tolerance to multiple stresses. While genotyping has become considerably cheaper and more precise recently, precision phenotyping has been a major challenge. Full advantage of genomic resources can be taken only when quick, accurate, and cost-effective phenotypic data, including root systems, are available for genetic dissection of drought tolerance and selection of drought-resilient genotypes. The usefulness of high throughput and automated phenotyping platforms like LeasyScan has been demonstrated in screening many genotypes for drought tolerance. There exists a much greater need to enhance the capacity for drought tolerance breeding programs to generate quick and accurate data through the use of drones, near-infrared imaging, and remote sensing.

Improving the nutrient use efficiency of pearl millet certainly makes a case for future research for bringing in more stability as it is mainly cultivated on sandy and sandy-loam soils that are inherently low in their nitrogen (N) and phosphorous (P) contents. The limited research has indicated the possibility of breeding nutrient-use-efficient (NUE) cultivars of pearl millet. Looking to soil degradation and water contamination due to nitrogen leaching in the subsurface or groundwater, pearl millet can be an important source of native genes that adapt to low nutrient conditions. There is a need for a systematic study to understand the relevant traits' priority and their magnitude of variability for NUE using core breeding materials, including mini-core collections available at genebank.

Harnessing abundant genetic diversity in pearl millet would determine the future of new trait discovery. World collection of germplasm (>23,000 from 52 countries), including wild species, provides a great resource to look for new sources of economic traits, disease resistance, abiotic stress tolerance, and better nutritional quality. A small fraction of germplasm has been utilized so far primarily due to the huge number of germplasm accessions and the presence of undesired traits in the unadapted genetic background. These twin problems have been largely circumvented. The development of core and mini-core collections is prompting breeders to use the desired germplasm in broadening the genetic base of commercial cultivars, which is essential to reduce the chances of disease epidemics and mitigate the effects of climate change.

In addition to the breeding programs addressing the enhanced adaptation to prevalent abiotic stresses and greater resistance to predominant biotic stresses of various regions, a sustained increase in pearl millet productivity requires the integration of suitable cultural practices in its diverse production environments for improved cultivars. On-farm demonstrations of improved cultivars and production technologies have established that the pearl millet yields at farm levels can easily be enhanced by 20–25% by adopting suitable agro-techniques (Yadav et al. 2012b). Intensive management, including higher planting density, irrigation scheduling, and recommended use of mineral fertilizers in better-endowed areas, would play a critical role in harnessing the potential yield of improved cultivars. Widely spaced crops, integration of legumes in pearl millet-based cropping system to maintain soil fertility, and micro-dosing of nutrients are very important to enhance productivity gains in drought-prone regions further. Machine-harvestable plant type and lodging resistance are the need of the hour in reducing cultivation costs and enhancing profitability. The availability of extra-early-maturing cultivars with tolerance to multiple environmental constraints makes pearl millet an ideal cereal for multiple cropping. Improved crop management in the summer season has already demonstrated a yield potential of up to 4–5 t/ha of grain yield in 85 days of the crop, indicating enormous yield potential and management responsiveness of pearl millet.

References

Anand Kumar K, Appa Rao S (1987) Diversity and utilization of pearl millet germplasm. In: Witcombe JR, Beckerman SR (eds) Proceedings of the international pearl millet workshop, 7-11 April 1986, ICRISAT Centre, India, ICRISAT, Patancheru, pp 69–82

Andrews DJ, Anand Kumar K (1996) Use of the West African pearl millet landrace *Iniadi* in cultivar development. Plant Genet Resour Newslett 105:15–22

Andrews DJ, King SB, Witcombe JR, Singh SD, Rai KN, Thakur RP, Talukdar BS, Chavan SB, Singh P (1985) Breeding for disease resistance and yield in pearl millet. Field Crops Res. 11: 241–258

Appa Rao S, Mengesha MH, Reddy KN (1998) Development and characterization of trait-specific genepools in pearl millet. Plant Genet Resour Newslett 113:27–30

Ashraf M, Mcneilly TM (1992) The potential for exploiting variation in salinity tolerance in pearl millet *Pennisetum americanum* Leeke. Plant Breed 108:234–240

Athwal DS (1961) Recent developments in the breeding and improvement of bajra (pearl millet) in the Punjab. Madras Agric J 48:18–19

Athwal DS (1966) Current plant breeding research with special reference to Pennisetum. Indian J Genet Plant Breed 26A:73–85

Bidinger FR, Witcombe JR (1989) Evaluation of specific drought avoidance traits as selection criteria for improvement of drought resistance. In: Baker FGW (ed) Drought resistance in cereals. CAB International, Wallingford, pp 151–164

Bidinger FR, Mahalakshmi V, Rao GDP (1987) Assessment of drought resistance in pearl millet [*Pennisetum americanum* (L.) Leeke]. II. Estimation of genotype response to stress. Aust J Agric Res 38:49–59

Bidinger FR, Mahalakshmi V, Talukdar BS, Sharma RK (1995) Improvement of landrace cultivars of pearl millet for arid and semi-arid environments. Ann Arid Zone 34:105–110

Bidinger FR, Nepolean T, Hash CT, Yadav RS, Howarth CJ (2007) Quantitative trait loci for grain yield in pearl millet under variable post-flowering moisture conditions. Crop Sci 47(3):969–980

Blum A, Ebercon A (1981) Cell membrane stability as a measure of drought and heat tolerance in wheat. Crop Sci 21:43–47

Breese WA, Hash CT, Devos KM, Howarth CJ (2002) Pearl millet genomics—an overview with respect to breeding for resistance to downy mildew. In: Leslie JF (ed) Sorghum and millets pathology 2000. Iowa State Press, Ames, pp 243–246

Burton GW (1965) Pearl millet Tift 23A release. Crops Soils 17:19

Burton GW (1978) Advances in breeding a better quality forage plant. In: "Advances in hay, silage and pasture quality". Proceedings of forage and grassland conference, Raleigh, North Carolina. American Forage and Grassland Council, Lexington

Ceccarelli S (1989) Wide adaptation: how wide? Euphytica 40:197–205

Chopra NK, Chopra N (1997) Performance of pearl millet genotypes at different salinity levels in western Rajasthan. Ann Arid Zone 27:183–189

Choudhary S, Guha A, Kholova J, Pandravada A, Messina CD, Cooper M, Vadez V (2019) Maize, sorghum, and pearl millet have highly contrasting species strategies to adapt to water stress and climate change-like conditions. Plant Sci 295:110297

Djanaguiraman M, Perumal R, Ciampitti IA, Gupta SK, Prasad PVV (2018) Quantifying pearl millet response to high temperature stress: thresholds, sensitive stages, genetic variability and relative sensitivity of pollen and pistil. Plant Cell Environ 41:993–1007

Fussell LK, Bidinger FR, Bieler P (1991) Crop physiology and breeding for drought tolerance. Research and development. Field Crops Res 27:183–199

Gauch HG, Zobel RW (1997) Identifying megaenvironments and targeting genotypes. Crop Sci 37: 311–326

Gemenet DC, Hash CT, Sanogo MD, Sy O, Zangre RG, Leiser WL, Haussmann BIG (2015) Phosphorus uptake and utilization efficiency in west African pearl millet inbred lines. Field Crops Res. 171:54–66

Govindraj M, Selvi B, Rajarathinam S, Sumathi P (2011) Genetic variability and heritability of grain yield components and grain mineral concentration in India's pearl millet (*Pennisetum glaucum* (L) R. Br.) accessions. Afr J Food Agric Nutr Dev 11:4758–4771

Gowda CLL, Rai KN, Reddy BVS, Saxena KB (2006) Hybrid parents research at ICRISAT. International Crops Research Institute for the Semi-Arid Tropics, Patancheru. 216 pp

Gupta SK, Rai KN, Govindaraj M, Rao AS (2012a) Genetics of fertility restoration of the A4 cytoplasmic-nuclear male sterility system in pearl millet. Czech J Genet Plant Breed 48:87–92

Gupta SK, Sharma R, Rai KN, Thakur RP (2012b) Inheritance of foliar blast resistance in pearl millet (Pennisetum glaucum). Plant Breed 13:217–219

Gupta SK, Rathore A, Yadav OP, Rai KN, Khairwal IS, Rajpurohit BS, Das RR (2013) Identifying mega-environments and essential test locations for pearl millet cultivar selection in India. Crop Sci 53:2444–2453

Gupta SK, Rai KN, Singh P, Ameta VL, Gupta SK, Jayalekha AK, Mahala RS, Pareek S, Swami ML, Verma YS (2015) Seed set variability under high temperatures during flowering period in pearl millet (*Pennisetum glaucum* L. (R.) Br.). Field Crops Res 171:41–53

Gupta SK, Ameta VL, Pareek S, Mahala RS, Jayalekha AK, Deora VS, Verma YS, Boratkar M, Atkari D, Rai KN (2016) Genetic enhancement for flowering period heat tolerance in peart millet (Pennisetum glaucum L. (R.) Br.). In: 7th International crop science congress, August 14–19, 2016, Beijing, China. http://oar.icrisat.org/9757/1/Page1.pdf

Gupta SK, Yadav DV, Govindaraj M, Boratkar M, Kulkarni VN, Rai KN (2018) Inheritance of fertility restoration of A5 cytoplasmic-nuclear male sterility system in pearl millet [*Pennisetum glaucum* (L.) R. Br.]. Indian J Genet Plant Breed 78:228–232

Gupta SK, Ameta VL, Pareek S, Singh I, Deora VS, Verma YS, Jayalekha AK (2019) Enhancing flowering period heat tolerance in pearl millet through shuttle breeding. Presented at 3rd agriculture and climate change conference held at Budapest, Hungary, 24-26 March, 2019. https://www.elsevier.com/__data/assets/pdf_file/0010/847882/AGRI-2019-Oral-Programme_v5_DE.pdf

Gupta SK, Patil S, Rathore A, Yadav D, Sharma LD, Mungra KD, Patil HD, Gupta SK, Kumar R, Singh V, Gupta R, Varshney RK, Rai KN, Yadav OP (2020) Identification of heterotic groups in south Asian bred hybrid parents of pearl millet. Theor Appl Genet 133:873–888

Hanna WW (1989) Characteristics and stability of a new cytoplasmic-nuclear male-sterile source in pearl millet. Crop Sci 29:1457–1459

Hanna WW, Wells HD (1989) Inheritance of pyricularia leaf spot resistance in pearl millet. J Hered 80:145–147

Harinarayana G, Appa RS, Mengesha MH (1988) Prospects of utilizing genetic diversity in pearl millet. In: Paroda et al (eds) Genetic resources, Indian perspective. NBPGR, New Delhi, pp 170–182

Hash CT, Witcombe JR, Thakur RP, Bhatnagar SK, Singh SD, Wilson JP (1997) Breeding for pearl millet disease resistance. In: Proceedings of international conference on genetic improvement of sorghum and pearl millet held from 22–27 September 1996 at Lubbock, Texas, USA, pp 337–373

Hash CT, Sharma A, Kolesnikova-Allen MA, Singh SD, Thakur RP, Bhasker Raj AG, Ratnaji Rao MNV, Nijhawan DC, Beniwal CR, Sagar P, Yadav HP, Yadav YP, Srikant BSK, Khairwal IS, Howarth CJ, Cavan GP, Gale MD, Liu C, Devos KM, Breese WA, Witcombe JR (2006) Teamwork delivers biotechnology products to Indian small-holder crop-livestock producers: pearl millet hybrid "HHB 67 improved" enters seed delivery pipeline. SAT eJournal 2(1):1–3. http://www.icrisat.org/Journal/bioinformatics/v2i1/v2i1teamwork.pdf

Hay FR, Hamilton RS, Furman BJ, Upadhyaya HD, Reddy KN, Singh SK (2013) Cereals. In: Normah MN, Chin HF, Reed BM (eds) Conservation of tropical plant species. Springer, New York, pp 293–316

Howarth CJ, Pollock CJ, Peacock JM (1997) Development of laboratory-based methods for assessing seedling thermotolerance in pearl millet. New Phytol 137:129–139

Kar S, Tanaka R, Korbu LB, Kholová J, Iwata H, Durbha SS, Adinarayana J, Vadez V (2020) Automated discretization of transpiration restriction to increasing VPD features from outdoors high-throughput phenotyping data. Plant Methods 16:140

Khairwal IS, Rai KN, Yadav OP, Rajpurohit BS, Negi S (2009) Pearl millet cultivars: seeds of choice. All India Coordinated Pearl Millet Improvement Project, Mandor

Kholová J, Vadez V (2013) Water extraction under terminal drought explains the genotypic differences in yield, not the anti-oxidant changes in leaves of pearl millet (Pennisetum glaucum). Funct Plant Biol 40(1):44–53

Kholová J, Hash CT, Lava Kumar P, Yadav SR, Koèová M, Vadez V (2010a) Terminal drought tolerant pearl millet [*Pennisetum glaucum* (L.) R. Br.] have high leaf ABA and limit transpiration at high vapor pressure deficit. J Exp Bot 61:1431–1440

Kholová J, Hash CT, Kakkera A, Kočová M, Vadez V (2010b) Constitutive water-conserving mechanisms are correlated with the terminal drought tolerance of pearl millet [Pennisetum glaucum (L.) R. Br.]. J Exp Bot 61:369–377

Kholová J, Zindy P, Malayee S, Baddam R, Murugesan T, Kaliamoorthy S, Hash CT, Votrubová O, Soukup A, Kočová M, Niang M (2016) Component traits of plant water use are modulated by vapour pressure deficit in pearl millet (*Pennisetum glaucum* (L.) R. Br.). Funct Plant Biol 43: 423–437

Krishnamurthy L, Serraj R, Rai KN, Hash CT, Dakheel AJ (2007) Identification of pearl millet [*Pennisetum glaucum* (L.) R. Br.] lines tolerant to soil salinity. Euphytica 158:179–188

Kulkarni VN, Rai KN, Dakheel AJ, Ibrahim M, Hebbara M, Vadez V (2006) Pearl millet germplasm adapted to saline conditions. SAT eJournal 2:1–4

Kumar AK, Appa RS (1987) Diversity and utilization of pearl millet germplasm. In: Witcombe JR, Beckerman SR (eds) Proceedings of the international pearl millet workshop, 7–11 April 1986. ICRISAT Centre, Patancheru, pp 69–82

Mahalakshmi V, Bidinger FR, Raju DS (1987) Effect of timing of water deficit on pearl millet (*Pennisetum americanum*). Field Crop Res 15:327–339

Manga VK, Yadav OP (1995) Effect of seed size on development traits and ability to tolerate drought in pearl millet. J Arid Environ 29:169–172

Medina S, Gupta SK, Vadez V (2017) Transpiration response and growth in pearl millet parental lines and hybrids bred for contrasting rainfall environments. Front Plant Sci 8:1846. https://doi.org/10.3389/fpls.2017.01846

Mengesha MH, Appa RS, Reddy CR (1990) Characteristics and potential uses of some genetic stocks of pearl millet in the world collection. Indian J Plant Genet Resour 3:1–7

Murty BR, Upadhyaya MK, Manchanda PL (1967) Classification and cataloguing of a world collection of genetic stocks of *Pennisetum*. Indian J Genet 27:313–394

Patil KS, Gupta SK, Marathi B, Danam S, Ramesh T, Rathore A, Das RR, Dangi KS, Yadav OP (2020) African and Asian origin pearl millet populations: genetic diversity pattern and its association with yield heterosis. Crop Sci 60:3035–3048

Peacock JM, Soman P, Jayachandran R, Rani AU, Howarth CJ, Thomas A (1993) Effect of high surface soil temperature on survival in pearl millet. Exp Agric 29:215–225

Presterl T, Weltzien E (2003) Exploiting heterosis in pearl millet population breeding in arid environments. Crop Sci 43:767–776

Raghvani KL, Juneja RP, Ghelani YH, Parmar GM, Dangaria CJ (2008) Influence of abiotic factors on population fluctuations of major insect pest of pearl millet. Indian J Appl Entomol 22:48–50

Rai KN (1995) A new cytoplasmic-nuclear male sterility system in pearl millet. Plant Breed 114: 445–447

Rai KN, Anand Kumar K (1994) Pearl millet improvement at ICRISAT—an update. Int Sorghum Millets Newslett 35:1–29

Rai KN, Hanna WW (1990) Morphological characteristics of tall and dwarf pearl millet isolines. Crop Sci 30:23–25

Rai KN, Rao AS (1991) Effect of d2 dwarfing gene on grain yield and yield components in pearl millet near-isogenic lines. Euphytica 52:25–31

Rai KN, Virk DS, Harinarayana G, Rao AS (1996) Stability of male sterile sources and fertility restoration of their hybrids in pearl millet. Plant Breed 115:494–500

Rai KN, Anand Kumar K, Andrews DJ, Gupta SC, Ouendeba B (1997) Breeding pearl millet for grain yield and stability. In: Proceedings of the international conference on genetic improvement sorghum pearl millet, Lubbock, Texas, 23–27 Sept. 1996. INTSORMIL, Lincoln, pp 71–83

Rai KN, Murty DS, Andrews DJ, Bramel Cox PJ (1999) Genetic enhancement of pearl millet and sorghum for the semi-arid tropics of Asia and Africa. Genome 42:617–628

Rai KN, Hash CT, Singh AK, Velu G (2008) Adaptation and quality traits of a germplasm-derived commercial seed parent of pearl millet. Plant Genet Res Newslett 154:20–24

Rai KN, Gupta SK, Bhattachajee R, Kulkarni VN, Singh AK, Rao AS (2009a) Morphological characteristics of ICRISAT-bred pearl millet hybrid parents. International Crops Research for the Semi-Arid Tropics, Patancheru, p 176

Rai KN, Khairwal IS, Dangaria C, Singh AK, Rao AS (2009b) Seed parent breeding efficiency of three diverse cytoplasmic-nuclear male sterility systems in pearl millet. Euphytica 165:495–507

Rai KN, Patil HT, Yadav OP, Govindaraj M, Khairwal IS, Cherian B, Rajpurohit BS, Rao AS, Shivade H, Kulkarni MP (2014) Notification of crop varieties and registration of germplasm: pearl millet variety 'Dhanashakti'. Indian J Genet Plant Breed 74:405–406

Rai KN, Yadav OP, Govindaraj M, Pfeiffer WH, Yadav HP, Rajpurohit BS, Patil HT, Kanatti A, Rathore A, Rao AS, Shivade H (2016) Grain iron and zinc densities in released and commercial cultivars of pearl millet (Pennisetum glaucum). Indian J Agric Sci 86:11–16

Rama Rao CA, Raju BMK, Rao AVMS, Reddy DY, Meghana YL, Swapna N, Ravindra CGR (2019) Yield variability of sorghum and pearl millet to climate change in India. Indian J Agric Econ 74:350–362

Rattunde HF, Witcombe JR, Frey KJ (1989) Structure of variation among morphological and physiological traits in three pearl millet composites. Euphytica 43:233–244

Rosielle AA, Hamblin J (1981) Theoretical aspects of selection for yield in stress and non-stress environments. Crop Sci 21:943–946

Serba DD, Yadav RS (2016) Genomic tools in pearl millet breeding for drought tolerance: status and prospects. Front Plant Sci 7:1724

Serba DD, Yadav RS, Varshney RK, Gupta SK, Mahalingam G, Srivastava RK, Gupta R, Ramasamy P, Tessso TT (2020) Genomic designing of pearl millet: a resilient crop for arid and semi-arid environments. In: Kole C (ed) Genomic designing of climate-smart cereal crops. Springer, Cham, pp 221–286. https://doi.org/10.1007/978-3-319-93381-8_6

Sharma HC, Youm O (1999) Host plant resistance in integrated pest management. In: Khairwal IS, Rai KN, Andrews DJ, Harinarayana G (eds) Pearl millet breeding. Oxford and IBH Co, Pvt. Limited, New Delhi, pp 381–418

Sharma R, Upadhyaya HD, Manjunatha SV, Rai KN, Gupta SK, Thakur RP (2013) Pathogenic variation in the pearl millet blast pathogen *Magnaporthe grisea* and identification of resistance to diverse pathotypes. Plant Dis 97:189–195

Sharma PC, Singh D, Sehgal D, Singh G, Hash CT, Yadav RS (2014) Further evidence that a terminal drought tolerance QTL of pearl millet is associated with reduced salt uptake. Environ Exp Bot 102:48–57

Sharma R, Upadhyaya HD, Sharma S, Gate VL, Raj C (2015) Identification of new sources of resistance to multiple pathotypes of *Sclerospora graminicola* in the pearl millet mini core germplasm collection. Crop Sci 55:1–10

Sharma S, Sharma R, Pujar M, Yadav D, Yadav YP, Rathore A, Mahala RS, Singh I, Verma YS, Deora V, Vaid B, Jayalekha AK, Gupta SK (2020) Utilization of wild Pennisetum species for improving biotic and abiotic stress tolerance in pearl millet (*Pennisetum glaucum* L.). Crop Sci 61:289–304

Shivhare R, Lata C (2017) Exploration of genetic and genomic resources for abiotic and biotic stress tolerance in pearl millet. Front Plant Sci 7:2069

Singh SD (1995) Downy mildew of pearl millet. Plant Dis 79:545–550

Singh SD, Gopinath R (1985) A seedling inoculation technique for detecting downy mildew resistance in pearl millet. Plant Dis 69:582–584
Singh BB, Jika N (1988) Five pearl millet genepools in Niger. Plant Genet Resour Newslett 73:29–30
Singh SD, Andrews DJ, Rai KN (1987) Registration of ICML 11 rust resistant pearl millet germplasm. Crop Sci 27:367–368
Singh SD, King SB, Reddy PM (1990) Registration of five pearl millet germplasm sources with stable resistance to rust. Crop Sci 30:1165
Singh SD, Singh P, Andrews DJ, Talukdar BS, King SB (1992) Reselection of pearl millet cultivar utilizing residual variability for downy mildew reaction. Plant Breed 109:54–59
Singh SD, King SB, Werder J (1993) Downy mildew disease of pearl millet. Information Blletin no. 37. International Crops Research Institute for the Semi-Arid Tropics, Patancheru, p 30
Singh SD, Wilson JP, Navi SS, Talukdar BS, Hess DE, Reddy KN (1997) Screening techniques and sources of resistance to downy mildew and rust in pearl millet. Research Bulletin no. 48. International Crops Research Institute for the Semi-Arid Tropics, Patancheru. 30 p
Singh S, Gupta SK, Thudi M, Das RR, Vemula A, Garg V, Varshney RK, Rathore A, Pahuja SK, Yadav DV (2018) Genetic diversity patterns and heterosis prediction based on SSRs and SNPs in hybrid parents of pearl millet. Crop Sci 58:2379–2390
Soman P, Peacock JM (1985) A laboratory technique to screen seedling emergence of sorghum and pearl millet at high soil temperature. Exp Agric 21:335–341
Sultan B, Roudier P, Quirion P, Alhassane A, Muller B, Dingkuhn M, Ciais M, Guimberteau ST, Baron C (2013) Assessing climate change impacts on sorghum and millet yields in the Sudanian and Sahelian savannas of West Africa. Environ Res Lett 8:014040
Thakur RP, King SB (1988) Registration of four ergot resistant germplasms of pearl millet. Crop Sci 28:382
Thakur RP, Williams RJ (1980) Pollination effects on pearl millet ergot. Phytopathology 70:80–84
Thakur RP, King SB, Rai KN, Rao VP (1992) Identification and utilization of smut resistance in pearl millet. Research Bulletin no. 17. International Crops Research Institute for the Semi-Arid Tropics, Patancheru. 39 p
Thakur RP, King SB, Rai KN, Rao VP (1993) Identification and utilization of ergot resistance in pearl millet. Research Bulletin no. 17. International Crops Research Institute for the Semi-Arid Tropics, Patancheru, p 36
Thakur RP, Rai KN, Rao VP, Rao AS (2001) Genetic resistance of pearl millet male-sterile lines to diverse Indian pathotypes of *Sclerospora graminicola*. Plant Dis 85:621–626
Tharanya M, Kholova J, Sivasakthi K, Seghal D, Hash CT, Raj B, Srivastava RK, Rekha B, Thiyagarajan T, Yadav RS, Vadez V (2018) Quantitative Trait Loci (QTL) for water use and crop production traits collocate with major QTL for tolerance to water deficit in a fine mapping population of pearl millet (*Pennisetum glaucum* L. R. Br.). Theor Appl Genet 131:1509–1529
Thomason WE, Phillips SB (2006) Methods to evaluate wheat cultivar testing environments and improve cultivar selection protocols. Field Crops Res 99:87–95
Upadhyaya HD, Reddy KN, Gowda CLL (2007a) Pearl millet germplasm at ICRISAT gene bank – status and impact. SAT eJournal 3:1–5
Upadhyaya HD, Reddy KN, Gowda CLL, Ahmed MI, Singh S (2007b) Agroecological patterns of diversity in pearl millet [*Pennisetum glaucum* (L.) R. Br.] germplasm from India. Indian J Plant Genet Resour 20:178–185
Upadhyaya HD, Reddy KN, Ahmed MI, Gowda CLL (2012a) Identification of gaps in pearl millet germplasm from east and southern Africa conserved at the ICRISAT genebank. Plant Genet Resour 10:202–213
Upadhyaya HD, Reddy KN, Ahmed MI, Dronavalli N, Gowda CLL (2012b) Latitudinal variation and distribution of photoperiod and temperature sensitivity for flowering in the world collection of pearl millet germplasm at ICRISAT genebank. Plant Genet Resour 10:59–69
Vadez V, Kholová J (2013) Coping with drought: resilience versus risk. Targeting the most suitable G*E*M options by crop simulation modeling. Secheresse 24:274–281

Vadez V, Sinclair TR (2001) Leaf ureide degradation and N2 fixation tolerance to water deficit in soybean. J Exp Bot 52:153–159

Vadez V, Hash T, Kholova J (2012) 1.5 phenotyping pearl millet for adaptation to drought. Front Physiol 3:386

Vadez V, Kholová J, Yadav RS, Hash CT (2013) Small temporal differences in water uptake among varieties of pearl millet (*Pennisetum glaucum* (L.) R. Br.) are critical for grain yield under terminal drought. Plant Soil 371:447–462

Vadez V, Kholova J, Hummel G, Zhokhavets U, Gupta SK, Hash CT (2015) LeasyScan: a novel concept combining 3D imaging and lysimetry for highthroughput phenotyping of traits controlling plant water budget. J Exp Bot 66:5581–5593

van Oosterom EJ, Mahalakshmi V, Arya GK, Dave HR, Gothwal BD, Joshi AK, Joshi P, Kapoor RL, Sagar P, Saxena MBL, Singhania DL, Vyas KL (1995) Effect of yield potential, drought escape and drought tolerance on yield of pearl millet (*Pennisetum glaucum*) in different stress environments. Indian J Agric Sci 65:629–635

van Oosterom EJ, Bidinger FR, Weltzien RE (2003) A yield architecture framework to explain adaptation of pearl millet to environmental stress. Field Crops Res 80:33–56

van Oosterom EJ, Weltzien E, Yadav OP, Bidinger FR (2006) Grain yield components of pearl millet under optimum conditions can be used to identify germplasm with adaptation to arid zones. Field Crops Res. 96:407–421

Varshney RK, Shi C, Thudi M, Mariac C, Wallace J, Qi P, Zhang H, Zhao Y, Wang X, Rathore A, Srivastava RK, Chitikineni A, Fan G, Bajaj P, Punnuri S, Gupta SK, Wang H, Jiang Y, Couderc M, Katta MAVSK, Paudel DR, Mungra KD, Chen W, Harris-Shultz KR, Garg V, Desai N, Doddamani D, Kane N, Conner JA, Ghatak A, Chaturvedi P, Subramaniam S, Yadav OP, Berthouly-Salazar C, Hamiduo F, Wang J, Liang X, Clotault J, Upadhyaya HD, Cubry P, Rhoné B, Gueye MC, Sunkar R, Dupuy C, Sparvoli F, Cheng S, Mahala RS, Singh B, Yadav RS, Lyons E, Datta SK, Hash CT, Devos KM, Buckler E, Bennetzen JL, Paterson AH, Ozias-Akins P, Grando S, Wang J, Mohapatra T, Weckwerth W, Reif JC, Xi L, Vigouroux Y, Xu X (2017) Pearl millet genome sequence provides a resource to improve agronomic traits in arid environments. Nat Biotechnol 35:969–974

Velu G, Rai KN, Sahrawat KL, Sumalini K (2008) Variability for grain iron and zinc contents in pearl millet hybrids. J SAT Agric Res 6:1–4

Wells HD, Hanna WW, Burton GW (1987) Effects of inoculation and pollination on smut development in near-isogenic lines of pearl millet. Phytopathology 77:293–296

Williams RJ, Singh SD, Pawar MN (1981) An improved field screening technique for downy mildew resistance in pearl millet. Plant Dis 65:239–241

Wilson JP (1995) Mechanism associated with the tr allele contributing to reduced smut susceptibility of pearl millet. Phytopathology 85:966–969

Wilson JP, Burton GW, Zongo JD, Dicko IO (1991) Disease resistance and morphogenic traits of pearl millet landraces from South Burkina Faso. Crop Sci 31:641–645

Wilson JP, Hess DE, Hanna WW, Kumar KA, Gupta SC (2004) *Pennisetum glaucum* subsp. *monodii* accessions with Striga resistance in West Africa. Crop Prot 23:865–870

Winkel T, Do F (1992) Caracteres morphologiques et physiologiques de resistance du mil [*Pennisetum glaucum* (L.) R. Br.] a la secheresse. L'Agron Tropic 46:339–350

Winkel T, Renno JF, Payne WA (1997) Effect of the timing of water deficit on growth, phenology and yield of pearl millet (Pennisetum glaucum (L) R Br) grown in Sahelian conditions. J Exp Bot 48:1001–1009

Yadav OP (1994) Inheritance of threshing percentage in pearl millet. Euphytica 78:77–80

Yadav OP (2004) CZP 9802 – a new drought-tolerant cultivar of pearl millet. Indian Farm 54:15–17

Yadav OP (2011) Breeding dual-purpose pearl millet (*Pennisetum glaucum*) for northwestern India: understanding association of biomass and phenotypic traits. Indian J Agric Sci 81:816–820

Yadav OP, Bidinger FR (2007) Utilization, diversification and improvement of landraces for enhancing pearl millet productivity in arid environments. Ann Arid Zone 46:49–57

Yadav OP, Manga VK (1999) Diversity in pearl millet landraces of arid regions of Rajasthan. In: Faroda AS, Joshi NL, Kathju S, Kar A (eds) Management of arid ecosystem. Arid Zone Research Association of India & Scientific Publishers, Jodhpur, pp 57–60
Yadav OP, Rai KN (2011) Hybridization of Indian landraces and African elite composites of pearl millet results in biomass and stover yield improvement under arid zone conditions. Crop Sci 51: 1980–1987
Yadav OP, Rai KN (2013) Genetic improvement of pearl millet in India. Agric Res 2:275–292
Yadav OP, Weltzien RE (2000) Differential response of pearl millet landrace-based populations and high yielding varieties in contrasting environments. Ann Arid Zone 39:39–45
Yadav OP, Weltzien-Rattunde E, Bidinger FR, Mahalakshmi V (2000) Heterosis in landrace-based topcross hybrids of pearl millet across arid environments. Euphytica 112:285–295
Yadav OP, Weltzien RE, Bhandari DC (2001) Genetic variation and trait relationship in pearl millet landraces from Rajasthan. Indian J Genet 61:322–326
Yadav RS, Hash CT, Bidinger FR, Cavan GP, Howarth CT (2002) Quantitative trait loci associated with traits determining grain and stover yield in pearl millet under terminal drought stress conditions. Theor Appl Genet 104:67–83
Yadav OP, Weltzien-Rattunde E, Bidinger FR (2003) Genetic variation for drought response among landraces of pearl millet (*Pennisetum glaucum*). Indian J Genet Plant Breed 63:37–40
Yadav RS, Hash CT, Bidinger FR, Devos KM, Howarth CJ (2004a) Genomic regions associated with grain yield and aspects of post-flowering drought tolerance in pearl millet across stress environments and tester background. Euphytica 136:265–277
Yadav OP, Weltzien-Rattunde E, Bidinger FR (2004b) Genetic diversity among pearl millet landraces from North-Western India. Ann. Arid Zone 43:45–53
Yadav OP, Bidinger FR, Singh DV (2009) Utility of pearl millet landraces in breeding dual-purpose hybrids for arid zone environments of India. Euphytica 166:239–247
Yadav OP, Rai KN, Gupta SK (2012a) Pearl millet: genetic improvement for tolerance to abiotic stresses. In: Tuteja N, Gill SS, Tuteja R (eds) Improving crop resistance to abiotic stress. Wiley VCH, Weinheim, pp 261–288
Yadav OP, Rajpurohit BS, Kherwa GR, Kumar A (2012b) Prospects of enhancing pearl millet (*Pennisetum glaucum*) productivity under drought environments of North-Western India through hybrids. Indian J Genet Plant Breed 72:25–30
Yadav OP, Rai KN, Rajpurohit BS, Hash CT, Mahala RS, Gupta SK, Shetty HS, Bishnoi HR, Rathore MS, Kumar A, Sehgal S, Raghvani KL (2012c) Twenty-five years of pearl millet improvement in India. All India Coordinated Pearl Millet Improvement Project, Jodhpur. 122 pp
Yadav OP, Mahala RS, Rai KN, Gupta SK, Rajpurohit BS, Yadav HP (2015) Pearl millet seed production and processing. Indian Council of Agricultural Research (ICAR), Jodhpur
Yadav OP, Upadhyaya HD, Reddy KN, Jukanti AK, Pandey S, Tyagi RK (2017) Genetic resources of pearl millet: status and utilization. Indian J Plant Genet Resour 30:31–47
Yadav OP, Singh DV, Dhillon BS, Mohapatra T (2019) India's evergreen revolution in cereals. Curr Sci 116:1805–1808
Yadav OP, Gupta SK, Govindaraj M, Sharma R, Varshney RK, Srivastava RK, Rathore A, Mahala RS (2021) Genetic gains in pearl millet in India: insights into historic breeding strategies and future perspective. Front Plant Sci 12:645038
Yan W, Tinker NA (2006) Biplot analysis of multi-environment trial data: principles and applications. Can J Plant Sci 86:623–645

Salinity Stress in Pearl Millet: From Physiological to Molecular Responses

14

Srividhya S, R. Swarna, J. Jinu, C. Dheeraj, and H. S. Talwar

Abstract

Salinity stress is a major constraint for crop growth, development, and yield, worldwide. Salinity stress is among the key abiotic stresses that critically impede plant development, causing yield reductions ranging from 15% to 90% in major crops, under moderate-to-high soil salinity levels. Although regarded as climate resilient, very little information is available on pearl millet, regarding its spectrum of physiological to molecular responses, inherent mechanisms exhibited, yield losses, and stress mitigation strategies compared to other cereal crops. High salinity levels in the soil impact the growth and productivity of pearl millet, which is predominantly grown in several arid and semi-arid zones. Therefore, this chapter highlights the differential responses of pearl millet crop to salinity stress, and the need to evaluate for superior genotypic variability with greater stress tolerance mechanisms. This chapter also discusses different approaches that can be employed for crop improvement programs that target salt-tolerant genotypes suitable for varying agro-ecological conditions.

14.1 Introduction

Pearl millet (*Pennisetum glaucum* (L.)) is the most extensively cultivated millet crop for grain and forage. It is a staple crop for more than 90 million poor farmers, mainly grown in semi-arid and arid regions of Asia and Africa, where other cereals cannot be cultivated (Srivastava and Kumar 2015). It can survive under extreme environmental conditions and nutrient-deficient soils. These adaptive traits make pearl millet

Srividhya S · R. Swarna (✉) · J. Jinu · C. Dheeraj · H. S. Talwar
ICAR-Indian Institute of Millets Research, Hyderabad, India
e-mail: srividhya.s@icar.gov.in; swarna@millets.res.in

V. A. Tonapi et al. (eds.), *Pearl Millet in the 21st Century*,
https://doi.org/10.1007/978-981-99-5890-0_14

a vital cereal grown in adverse agroclimatic conditions where other crops fail to produce economic yields. Salinity is emerging as a serious threat for pearl millet cultivation, as it is mainly grown under rainfed production systems of arid and semiarid regions.

Salinity stress severely limits crop production and yield potential. Low precipitation, irrigation with saline water, and poor irrigation practices cause salinity stress. Salinization poses a major challenge in both arable and marginal agricultural lands (Litalien and Zeeb 2019). It is estimated that by 2050 the present area of over 833 million hectares of salt-affected soils will expand further fivefold, posing a threat to food security (FAO 2021). The soil is assumed to be saline or influenced by salinity if it has an electrical conductivity of more than 4 dS m^{-1} at 25 °C, where 4 dS m^{-1} ≈40 mM NaCl or greater. Cell dehydration, ionic toxicity, nutritional stress, osmotic stress, drought stress, and oxidative stress are some effects of salinity stress (Zhu 2002). The existing and future changes in soil fertility need climate efficient crops for the utilization of marginal lands for cultivation. Millets are considered to be climate resilient that can adapt to poor resource and hardy soils. Thus, understanding the stress tolerance mechanisms and genetic manipulation of millets, especially pearl millet, will aid further in attaining sustainable development efforts to search out enhanced crop performance on marginal and insignificant lands, which still remains undeciphered.

This chapter highlights the spectrum of physiological and molecular responses concerning salt tolerance and mitigation of salt stress in plant salinity stress tolerance, especially pearl millet. Approaches to improve salinity tolerance in pearl millet are limited, including the conventional agronomic, breeding, molecular breeding, and genomic approaches with some success. Various novel approaches have been attempted in pearl millet for enhancing yield under salt-stress environments. These include using adapted germplasm, genetic diversification of adapted landraces through introgression of suitable elite genetic material, and exploitation of heterosis to amalgamate salt tolerance and high yield. Molecular breeding is fast emerging as a supplement approach to enhance drought adaptation at a faster rate with greater precision. Molecular marker-based genetic linkage maps of pearl millet are available and genomic regions determining yield under salinity stress environments have been identified, preparing a road map for marker-assisted selection.

14.2 Physiological Basis of Salt Stress Tolerance in Pearl Millet

Millets are generally climate-contingent crops adapted to marginal and dry lands of arid and semi-arid regions, surviving low rainfall and poor soils. They show exceptional tolerance to various abiotic stresses, including salinity stress. The primary growing areas of the pearl millet crop fall in the arid and semi-arid zones of South Asia and West Africa (Blummel et al. 2003), where the soil salinity stress is one of the major constraints. Pearl millet [*Pennisetum glaucum* (L.) R. Br] is considered to be fairly tolerant to salinity, having three subtypes of C_4 photosynthetic pathway (phosphoenol pyruvate carboxylase (PEP-CK subtype)), NAD-malic enzyme subtype and NADP-malic enzyme subtype) (Wang and Shangguan 2010). Millets benefit greatly from the C_4 photosynthetic trait as their *Kranz anatomy* helps maintain higher CO_2 levels in bundle sheath cells, reduces photorespiratory loss, and the carboxylase activity of Rubisco is also enhanced. Many physiological and morphological advantages like lesser CO_2 compensation point, higher transpiration and water use efficiency, leaf venation features, and deeper root systems make these crops resilient to harsh climates. The growth and yield remain unaffected to changing CO_2 levels or elevated atmospheric CO_2 concentration, unlike C_3 cereals like rice and wheat. Pearl millet is classified as glycophytes and has an average salt tolerance threshold of 6 (ECe) (dS/m). The availability of fairly high levels of tolerance in *Pennisetum* species (Ashraf and McNeilly 1987, 1992; Dua 1989; Muscolo et al. 2003) offers a scope to integrate these type of tolerant crop species into appropriate breeding and management programs to improve the productivity of the saline soils. In the current changing climate scenario, millets are promising crops for climate resilience and sustainable productivity. Improved overall plant growth, higher biomass allocation, and partitioning are some additional benefits of C4 photosynthesis alongside WUE and NUE (Sage and Zhu 2011). Higher stomatal conductance, high photo assimilation, and low transpiration benefitted millet crops with increased resource use efficiency. Pearl millet crop adopts various escape mechanisms, avoidance and tolerance to unfavorable stress environments.

The rapid diffusion of photosynthetic metabolites is attributed to leaf vein density apart from their anatomical superiority. These C_4 grasses primarily have an efficient water distribution cascade compared to C_3 grasses by having a denser leaf venation pattern (higher network of the small longitudinal and transverse veins while keeping a constant density of large longitudinal veins) (Govinda et al. 2012). The leaf veins anatomy described by Altus and Canny (1982) has a hierarchical order and has different structures.

- The enormous longitudinal veins run from the leaf blade into the sheath. The huge longitudinal veins assist essentially in the longitudinal carriage of photosynthate outside the leaf blade.
- The little longitudinal veins fundamentally gather photosynthates from adjacent photosynthetic cells.
- The cross-over (transverse) veins and little longitudinal veins assume an essential part in the sideways carriage of photosynthates from the little to the enormous longitudinal veins.

14.3 Effect of Salinity Stress on Pearl Millet: Morpho-Physiological and Biochemical Changes

Plants grown under saline soil condition experience a significant amount of high osmotic stress, ion toxicities, and nutritional disorder leading to reduced plant productivity. Salinity may inhibit plant growth in two different ways. Primarily presence of excess salts in the soil solution minimizes the ability of the plant to extract water and leads to water deficit stress (Kumar et al. 2016; Yadav and Dagar 2016). Secondly, the entry of excessive amounts of salt in the transpiration stream will injure the cells in the transpiring leaves leading to reduced growth and development (Kumar et al. 2018). This is called the salt-specific or ion excess effect of salinity.

Salt stress affects all the major processes such as germination, growth, photosynthetic pigments and photosynthesis, water relations, and nutrient uptake (Shahzad et al. 2021), as explained below. However, high salinity impacts its growth and productivity in several arid zones (Krishnamurthy et al. 2007a, b). Yield loss under salinity stress in pearl millet grain yield was around 13–22% under salinity levels of 8–12 dS m^{-1} (Yadav et al. 2020). Studies have also reported an average reduction of 20% in shoot biomass productivity and about 40% in grain yield in pearl millet germplasm (Krishnamurthy et al. 2014; Toderich et al. 2018) under salinity. Under high saline conditions in pearl millet, a significant reduction of about 47–86% in grain yield and 51% in fodder yield was observed (Choudhary et al. 2019; Kulkarni et al. 2006; Ribadiya et al. 2018).

14.3.1 Germination and Seedling Establishment

Seed germination is the primary and vital phase in a plant's growth cycle determining yield. Salinity stress alters the imbibition of water by seeds by lowering the osmotic potential of surrounding soil water causing toxicity which changes the activities of enzymes of nucleic acid metabolism, alters protein metabolism, disturbs hormonal balance (ABA/GA), and reduces the utilization of seed reserves. Salinity affected sorghum seed germination by inducing dormancy and ion (Na^+, Cl^-) toxicity (Rajabi Dehnavi et al. 2020). Maiti et al. (2007) and Jain and Dev Sharma (2005) reported that high salinity levels decreased seed germination and seedling growth pearl millet by reducing root length, fresh/dry weights of roots, and coleoptiles. Similar findings were reported in sorghum under salinity stress by Ali et al. (2020).

14.3.2 Growth and Development Changes Affected by Salinity Stress

The first phase of the growth response results from the effect of salt outside the plant. The salt in the soil solution reduces seedling leaf growth and root growth. Neither Na^+ nor Cl^- builds up in growing tissues at concentrations that inhibit growth because meristematic tissues are primarily fed by the phloem from which salt is effectively excluded and rapidly elongating cells can accommodate the salt that arrives in the xylem within their vacuoles. The second phase of the growth response results from the toxic effect of salt inside the plant. Munns and Tester (2008) reviewed the response of a plant to salinity stress and summarized that reduction in shoot growth occurs in two phases: a rapid response to the increase in external osmotic pressure (**Osmotic Stress**) and a slower response due to the accumulation of Na^+ in leaves (**Ionic Stress**). In the osmotic phase, which starts immediately after the salt concentration around the roots increases to a threshold level, the rate of shoot growth falls significantly. This is mainly (but not entirely) due to the osmotic effect of the salt outside the roots. The osmotic stress not only has an immediate impact on growth, but also has a more significant effect on growth rates than the ionic stress. Older leaves get concentrated by the salt intake and continue transport into transpiring leaves, leading to leaf death. Excess loading of salts exceeding the vacuole capacity inhibits enzymatic activity. Alternatively, they might build up in the cell walls and dehydrate the cell. The excessive salt concentration decreased the osmotic potential of the soil that restricted the water uptake in sorghum while salinity stress declined yield in pearl millet through membrane rupture, decreased photosynthesis and dry matter partitioning from source to sink (Kumar et al. 2018; Rahim et al. 2020). Genetic differences in tolerance to salinity at both seedling and grain filling stages have been established, and screening techniques standardized. Also, the germplasm and breeding materials with a higher salinity tolerance have been identified to use in breeding programs (Yadav et al. 2012).

14.3.3 Salinity and Ionic Toxicity Effects in Plants

The appropriate ion ratios provide a tool for the physiological response of a plant in relation to its growth and development (Wang et al. 2003). Excessive soluble salts in the soil compete with the uptake and metabolism of essential mineral nutrients. Increased salt uptake induces specific ion toxicities like high Na^+, Cl^-, or sulfate (SO_2^-) that decrease the uptake of essential macronutrients (N, P, K, Ca). The imbalance in Na^+ and Cl^- ion ratios leads to many physiological disorders in plants. Accumulation of Na^+ ion interferes with the uptake of potassium (K^+) ion and disturbs stomatal regulation causing water loss, and inhibits protein synthesis, photosynthesis and inactivation of enzymes, while the Cl^- ion disturbs the chlorophyll production and causes chlorotic toxicity in pearl millet (Hanin et al. 2016; Kumar et al. 2018) and finger millet (Mbinda and Mukami 2021). Excess Na^+ accumulation leads to the deterioration of the structural and functional integrity of organelle membranes in pearl millet. Reduced shoot N content and increased K^+ and Na^+ content are usually associated with an adaptive salinity tolerance mechanism in pearl millet (Dwivedi et al. 2012). According to Krishnamurthy et al. (2007a, b), shoot biomass ratio associated with salt tolerance and shoot Na^+ concentration could be used as potential selection criteria for screening of pearl millet germplasm at the vegetative stage for salt stress.

14.3.4 Changes in Plant Water Relations

Determining the water status is an essential physiological parameter while selecting stress-tolerant genotypes (Parida and Das 2005). With the increased concentration of salts in the root medium, leaf water content gets reduced as lower soil water potential interferes with the plant's ability to extract water from soil and maintain turgor. This leads to water stress in plants, adversely affecting all the physiological and growth mechanisms like cell expansion, cell division, stomatal mechanisms, and key enzymatic activities (Hussain et al. 2010). Radhouane (2013) reported that salinity causes a reduction in the relative water content in pearl millet leaves. However, plants adjust osmotically (accumulate solutes) at low or moderate salt concentration and maintain a potential gradient for the influx of water in finger millet (Hasanuzzaman et al. 2017).

14.3.5 Salinity and Oxidative Stress

Besides the direct impact of salinity on plants, a common consequence of salinity is induction of excessive accumulation of Reactive Oxygen Species (ROS). ROS are products of normal cellular metabolism, but the balance between production and elimination is disturbed under stress conditions in cellular components of plants. Environmental perturbations like salinity stress induce the overproduction of reactive oxygen species (superoxide radical ($O_2^{\bullet -}$), hydroxyl radical ($OH^{\bullet}$), hydroperoxyl radical ($HO_2^{\bullet}$), hydrogen peroxide (H_2O_2), alkoxy radical ($RO^{\bullet}$), peroxy radical ($ROO^{\bullet}$), singlet oxygen ($^1O_2$)), all of which are cytotoxic to plants (Golldack et al. 2014), which results in peroxidation of membrane lipids, oxidation of protein, inactivation of key enzymes, DNA damage, and disruption of membrane integrity (Pooja et al. 2020). Salt stress causes physiological drought leading to stomatal closure, reduces carbon fixation, exposing chloroplasts to excessive excitation energy, which in turn increases the generation of ROS and membrane disruption (Toderich et al. 2018).

14.3.6 Photosynthetic Pigments and Photosynthesis

Reduced water potential is the main reason for reduced photosynthesis under salt stress. Accumulation of high concentrations of Na^+ and/or Cl^- in the chloroplasts inhibits photosynthesis by injuring chloroplast. Salinity stress is associated with stomatal closure leading to reduced carbon assimilation. Photosystem II (PS II) is a relatively sensitive component concerning abiotic stress, especially salt stress. A considerable decrease in the efficiency of PS II, electron transport chain, and CO_2 assimilation rate was reported under the influence of salinity. Salinity stress reduced chlorophyll fluorescence by disrupting the functionality of thylakoids in chloroplasts and oxygen-evolving complex (PS II) reducing the photosynthetic rate in many crops including pearl millet (Kalaji et al. 2011; Yadav et al. 2020). Salinity stress reduces the photosynthetic rate owing to cell membranes' dehydration, which reduces their permeability to carbon dioxide. Salt toxicity enhanced leaf senescence and changed enzymatic activity regulating photosynthesis, eventually resulting in negative feedback by reduced sink activity in finger millet (Mukami et al. 2020).

14.3.7 Nutrient Imbalance

Salinity induces multiple nutritional disorders affecting crop performance. The dietary disorders may result from the effect of salinity on nutrient availability, competitive uptake, transport, or distribution within the plant. Salinity disturbs nutrient uptake by increasing the soil pH (Zhu 2004) and disrupting the redox potential of soil solution. Decreased nitrogen uptake, for example in saline soils, occurs due to interaction between Na^+ and NH_4^+ or between Cl^- and NO_3^- that ultimately reduces the growth and yield of the crop. This reduction in NO_3^- uptake is associated with Cl^- antagonism. Phosphate concentration in crops decreases with salinity as it affects the activity of PO_3^- and phosphate concentration in soils is tightly controlled by the sorption process. Salinity increases sodium concentration in plant tissues resulting in ion imbalances of K, Ca, and Mg. Supra-optimal concentrations of sodium ions are toxic to cell metabolism, which can finally lead to growth inhibition.

14.4 Strategies of Adaptation and Tolerance of Pearl Millet to Salt Stress

Plants develop various physiological and biochemical mechanisms (as discussed above) to tolerate and survive under salinity stress. Na^+ exclusion is often a primary determinant of variability in salinity tolerance within a species, including millets like sorghum and pearl millet (Krishnamurthy et al. 2007a, b). Plant tolerance mechanisms to salinity stress are distinguished into three types: (a) tolerance to osmotic stress, (b) Na^+ exclusion, and (c) tissue tolerance. The major form of salt present in the soil is NaCl. A general schema of plant response to salinity stress is depicted in Fig. 14.1.

Maintaining ion homeostasis by ion uptake and compartmentalization is crucial for average plant growth during salt stress. Both glycophytes and halophytes cannot tolerate high salt concentrations in their cytoplasm. Hence, the excess salt is sequestered to the vacuole in older tissues thereby protecting the plant from salinity stress (Munns and Tester 2008). Biological membranes and their associated components play an integral role in maintaining ion concentration within the cytosol during the period of stress by regulating ion uptake and transport (Zhu 2002). Various membrane transporters like carrier proteins, channel proteins, antiporters, and symporters facilitate this phenomenon. The Na^+ ion that enters the cytoplasm is then transported to the vacuole via Na^+/H^+ antiporter (Krishnamurthy et al. 2014). Two types of H^+ pumps are present in the vacuolar membrane: vacuolar type H^+-ATPase (V-ATPase) and the vacuolar pyrophosphatase (V-PPase). Of these, V-ATPase is the most dominant H^+ pump present within the plant cell. Under stressed conditions the survivability of the plant depends upon the activity of V-ATPase. These antiporters are encoded by SOS salt overly sensitive signaling proteins (SOS1, SOS2 and SOS3) helping in ion homeostasis and salt tolerance (Fig. 14.2).

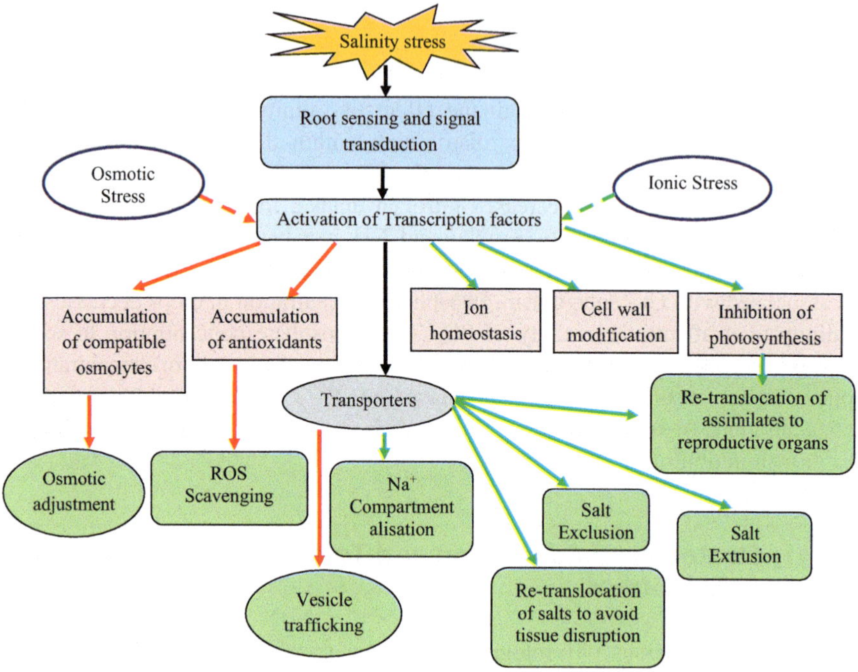

Fig. 14.1 Summary of events that take place under salt stress in plants to confer salt tolerance

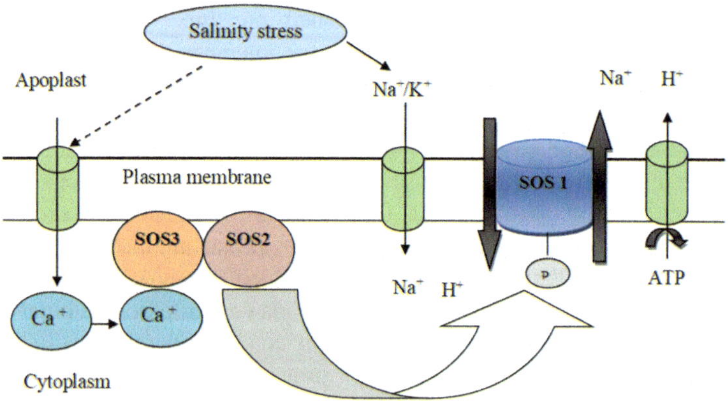

Fig. 14.2 SOS signaling pathway for ion homeostasis under salt stress

Excess Na^+ and high osmolarity are separately sensed by unknown sensors at the plasma membrane level, which then induce an increase in cytosolic ($Ca^{2+)}$. This increase is perceived by SOS3 which activates SOS2. The activated SOS3-SOS2 protein complex phosphorylates SOS1, the plasma membrane Na^+/H^+ antiporter resulting in the efflux of Na^+ ions. SOS2 can regulate NHX1 antiport activity and

V-H^+ATPase activity independently of SOS3, possibly by SOS3 like Ca^{2+} binding proteins (SCaBP) that target it to the tonoplast. Salt stress can also induce the accumulation of ABA, which using ABI1 and ABI2, can negatively regulate SOS2 or SOS1 and NHX1. Besides conferring salt tolerance, it regulates pH homeostasis and membrane vesicle trafficking.

During salinity stress, Na^+ ion due to its higher concentration in the soil competes with K^+ ion for the transporter as they both ions share the same transport mechanism, thereby limiting the uptake of K^+. Transporters located on the plasma membrane, belonging to the HKT (histidine kinase transporter) family play an essential role in salt tolerance by regulating transportation of Na^+ and K^+. A large number of genes and proteins, such as HKT and NHX encoding K^+ transporters and Na^+, K^+/H^+ antiporters, protect plants from adverse effects of salinity by maintaining K^+ homeostasis, endosomal pH regulation and preventing excess accumulation Na^+ in leaves.

14.4.1 Accumulation of Osmolytes or Compatible Solutes

Accumulation of compatible osmolytes reduces the water potential in the plant creating strong water potential gradient which helps in improving water uptake by roots under salinity stress. These compatible solutes include amino acid proline, glycine betaine, sugars, and polyols. As the accumulation of these compounds is proportional to the external osmolarity, they protect cellular structures by maintaining osmotic balance within the cell via continuous water influx (Sorahinobar et al. 2016). Role of compatible osmolytes is not limited to osmotic balance. Compatible solutes also act as low molecular weight chaperones due to their hydrophilic properties. Proline accumulation is a well-known measure adopted for alleviation of salinity stress (Sharma et al. 2019). In pearl millet intracellular proline accumulation during salinity stress not only provided tolerance towards stress but also served as an organic nitrogen reserve during stress recovery (Toderich et al. 2018; Yadav et al. 2020).

Glycine betaine is a non-toxic cellular osmolyte that helps in stress mitigation. Glycine betaine also protects the cell by protein stabilization and reduction in ROS (Veerangamallaiah et al. 2008). Hussain et al.(2010) reported increment in glycine betaine in pearl millet under NaCl treatment. Sugar alcohols are a class of polyols functioning as low molecular weight chaperones and ROS scavenging compounds. Carbohydrates such as sugars (e.g., glucose, fructose, fructans, and trehalose) and starch accumulate under salt stress involving in osmoprotection, carbon storage, and scavenging of reactive oxygen species. Osmotic adjustment is greater in millet races with smaller plants having small organs and cells. Pearl millet varieties well-adapted to saline environments showed good physiological and biochemical responses to increased salinity such as increased proline, total soluble proteins, and epicuticular wax content.

14.4.2 Polyamines

Polyamines (PA) play a crucial role in abiotic stress tolerance including salinity. Increase in the level of polyamines is correlated with stress tolerance in plants. The most common polyamines that are found within the plant system are diamine putrescine (PUT), triamine spermidine (SPD), and tetra-amine spermine (SPM). The positive effects of polyamines have been associated with the maintenance of membrane integrity, regulation of gene expression for the synthesis of osmotically active solutes, active solutes, reduction in ROS production, and controlling accumulation of Na^+ and Cl^- ion in different organs. When the seedlings of *Sorghum bicolour* treated with 0.52 mM SPM were subjected to salt stress, improvement in growth and partial increase in the activity of peroxidase and glutathione reductase enzyme was observed with a concomitant decrease in the level of membrane lipid peroxidation (Chai et al. 2010).

14.4.3 Hormonal Regulation

Production of abscisic acid (ABA) is upregulated with salinity induced soil water deficit around the root which helps in partial stomatal closure limiting transpirational water loss. The accumulation of ABA can mitigate the inhibitory effect of salinity on photosynthesis, growth, and translocation of assimilates (Cabot et al. 2009). The positive relationship between ABA accumulation and salinity tolerance is attributed to the accumulation of K^+, Ca^{2+} and compatible solutes in vacuoles of roots, which counteract with the uptake of Na^+ and Cl^-. ABA is a potential cellular signal that regulates the expression of a number of salt and water deficit responsive genes. ABA treatment in wheat induced the expression of MAPK4-like, TIP 1, and GLP 1 genes under salinity stress (Keskin et al. 2010).

Some other hormones such as salicylic acid (SA) and brassinosteroids (BR) also participate in plant salt stress responses (Yadav et al. 2020). Under salinity stress endogenous level of SA increased along with the increase in the activity of salicylic acid biosynthetic enzyme in rice seedling (Sawada et al. 2006). In an experiment conducted by Jayakannan et al. (2013), SA improved salinity tolerance in *Arabidopsis* by restoring membrane potential and preventing salt-induced K^+ loss via a guard cell outward rectifying K(+) (GORK) channel. *Arabidopsis* seedling pretreated with SA showed upregulation of H^+-ATPase activity, thereby improving K^+ retention during salt stress. SA alleviates reduction in photosynthesis under salt stress by enhancing nitrogen and sulfur assimilation and antioxidants.

14.4.4 Antioxidant Regulation

Antioxidant metabolism, including various enzymatic and non-enzymatic antioxidants, plays crucial role in detoxifying ROS induced by salinity stress. Salinity tolerance in pearl millet is positively correlated with the activity of enzymatic antioxidants such as superoxide dismutase (SOD), catalase (CAT) glutathione peroxidase (GPX), glutathione reductase (GR) ascorbate peroxidase (APX), and non-enzymatic antioxidants like ascorbate, anthocyanin, tocopherols, and phenolic compounds (Heidari and Jamshidi 2011; Lakshmi et al. 2017; Khan et al. 2020). Anthocyanin accumulation is largely reported in cereals exposed to salt stress (Mbarki et al. 2018).

14.4.4.1 Case Study on the Response of Pearl Millet to Varying Levels of Salt Stress

In a field study at ICRISAT, where pearl millet germplasms were screened under different ranges of soil salinity, breeding lines IP 19586 and HHVBC Tall were selected based on good growth and yield productivity (Toderich et al. 2018). While IP 19586 improved pearl millet germplasm line produced an average dry matter production of 27.6–35.0 t ha^{-1}, the HHVBC Tall line was superior since it registered highest value of seed production up to 4.3–4.7 t ha^{-1}. This implies that lines had different salt tolerance strategies at the seed germination stage. The germination rate gradually decreased from 0 to 400 mM NaCl, for the IP 19586 line. The HHVBC Tall line showed tolerance up to a high salinity level (300 mM NaCl), but it was sensitive to extra-high salinity (400 mM NaCl). For other pearl millet varieties, approximately 60% decreases in seed germination rates occurred with 1.5% NaCl treatment (Ali and Idris 2015). The study also revealed that controlled experiments allow for the investigation of plant resistance-related physiological and biochemical characteristics under ionic and osmotic stresses (Toderich et al. 2018). The study revealed the salt tolerance thresholds of pearl millet lines taken for investigation.

Salinity (combined effects of osmotic and ionic stresses) at 200 mM NaCl caused a significant twofold reduction in the growth parameters. Because a PEG-induced osmotic stress with a similar osmotic potential caused no decrease in these growth parameters, we believe the toxic actions of ions resulted in the growth reduction (Munns and Tester 2008). The HHVBC Tall line was more sensitive to ionic stress, having significantly higher proline content. Under extremely high salinity conditions, the most significant differences between the studied genetic lines were revealed. There was significant tissue dehydration in both lines, but especially in HHVBC Tall plants. Furthermore, the pigments significantly decreased in HHVBC Tall plants at 300 mM NaCl. For other *P. glaucum* varieties, significant reductions in pigment composition occurred at 100–200 mM NaCl (Sneha et al. 2013). The high proline and malondialdehyde (MDA) contents indicate the presence of high stress levels in HHVBC Tall plants under these conditions (Liu et al. 2014). The IP 19586 pearl millet germplasm line, which has a high biomass yield potential, is of interest as a forage resource for livestock in salt-affected agro-landscapes. Since soil salinization is negatively reflected in the grain production, the seed multiplication of the high-grain HHVBC Tall pearl millet line would guarantee stable seed production only in soils with low and medium levels of salinity. The study concluded that IP 19586 line of pearl millet should be used for maximal green biomass production under drought stress (0.6 MPa), while the HHVBC-Tall line should be used to produce high grain yields under low and medium salinity and drought (0.6 MPa) conditions. Depending on the timing and intensity of the salt stress, the harvest index varies from 0.2 to 0.5 (Munns et al. 2006). At the same time, low salinity levels may not reduce grain yield, even though the numbers of leaves, leaf area, and stover biomass are reduced. This may reflect a harvest index that increases with salinity or grain yield that does not decrease until a given ("threshold") salinity is reached. The findings from the study indicated that the dry biomass yield can be used as a screening and selection criterion for evaluating salt-tolerance behaviors among a large collection of plant accessions. Based on their relative dry biomasses (DMs), it is possible to select the optimum genotypes for each salinity level.

14.5 Molecular Characterization of Salt Stress

Salinity stress triggers various other stresses like osmotic stress, dehydration, and oxidative stress. Disruption of physiological processes such as stomatal movement and reduction in gaseous exchange, chlorophyll content, photosynthetic efficiency etc. are consequences of salinity stress (Chaudhry et al. 2020, 2021). There are manifold changes at the molecular level resulting from salinity stress. Higher salt content in the root environment leads to poor availability of water and nutrients (Ali et al. 2021). High salt environment promotes reactive oxygen species production in plants consequently promoting the production of antioxidant defense components. The signals generated because of the reactive oxygen species are sensed by various molecular components are transmitted through the cell such that plants exhibit transcriptional and post-transcriptional modifications to cope up with the stress. Salt-induced oxidative stress and subsequent damage to cellular organelles and their functions are reported in various crop/plant species.

14.5.1 Molecular Basis of Salt Tolerance in Plants

Plants cope up with cellular changes through sensors and signaling cascades. In plants, the environmental cue to high salinity is first recognized by transmembrane sensors. High sodium concentration causes imbalance in homeostasis of other ions such as K^+ and Ca^{2+} (Julkowska and Testerink 2015). It is reported that Ca^{2+} levels in cytosol are elevated within seconds of exposure to salinity (Knight et al. 1997). Ca^{2+} acts as a second messenger in stress signaling and Ca^{2+} ion concentration acts as one of the earliest steps in salt sensing (Lynch et al. 1989).

Only a few salt sensors have been identified so far in the plant kingdom and they include ROS-based and Na^+/Ca^{2+} membrane transporters (Wu et al. 2021). There are Na^+/H^+ transporters at both the plasma membrane and tonoplast that enable efflux of Na^+ ions out of the cell into the apoplastic space or to sequester them in the vacuole, respectively. The salt overly sensitive (SOS) signaling pathway is well characterized, which converts Ca^{2+} signaling to salt tolerance. SOS pathway is activated when plants are exposed to high levels of sodium chloride and the pathway helps in regulating osmotic homeostasis under high salinity conditions (Ishitani et al.

2000). When salt concentration in the soil increases, plants grown there tend to accumulate excessive sodium content which adversely affects various metabolic processes leading ultimately to poor plant growth. Optimum sodium levels are maintained in the cell by the coordinated action of several membrane bound transport proteins. *S*alt *o*verly *s*ensitive mutants were first identified in *Arabidopsis thaliana* in a genetic screen designed to identify the components of salt sensitivity/tolerance (Zhu et al. 1998). The key components of the SOS signaling pathway involved in Na^+ extrusion have been identified as SOS3, SOS2, and SOS1. Arabidopsis plants with a mutation in *SOS1* gene were highly sensitive to increased NaCl content in the growing medium and accumulated excessive Na^+ in them when compared to the wild-type plants (Shi et al. 2000). Biochemically, SOS1 was identified as a Na^+/H^+ antiporter located in the plasma membrane (Shi et al. 2000; Qiu et al. 2002) that regulates salt content by mediating Ca^{2+}-dependent microfilament re-organization (Zhao et al. 2011). Na^+/H^+ exchanger in the vacuolar membrane is also a target for the SOS pathway (Qiu et al. 2004). The expression of SOS1 is regulated by other members of the pathway namely, SOS2 and SOS3. SOS2 is a CBL-interacting protein kinase (CIPK) and SOS3, a (a Ca^{2+} sensor of the CBL family, Calcineurin B-like). The elevated levels of cytosolic Ca^{2+} stimulate Ca^{2+}-dependent protein kinase complex (SOS2-SOS3) that phosphorylates and initiates the activity of SOS1. SOS pathway genes are found conserved across species and their presence has been reported in many species such as *Brassica juncea*, wheat, barley, grapes, and sugarcane. Calcium signal that is initiated as a result of high salinity stress is sensed by SOS3 which is a calcium sensor belonging to calcineurin B-like family. SOS3 along with SOS2 forms a protein complex which activates SOS1 (Na^+/H^+ antiporter) transcription through protein phosphorylation so that SOS1 regulates efflux of excess sodium ions (Guo et al. 2009). SOS3-SOS2 complex also activates vacuolar Na^+/H^+ exchanger (NHX) making vacuole accumulate excess Na^+ ions.

Ion and nutrient transporters form an integral part of stress signaling. Ion transporters and channel proteins are most often regulated by Ca^{2+}-driven phosphorylation (Kudla et al. 2018). Various receptor kinases are implicated in salinity tolerance. FERONIA (FER) is one such receptor kinase to salt stress identified in Arabidopsis (Feng et al. 2018). *Fer* is also a member in the chain of Ca^{2+} signaling and is involved in cell wall integrity restoration and root growth. Increased calcium deposition in the cortex and epidermis of the roots stimulates Ca^{2+}calmodulin-dependent kinases and plasma membrane ATPase along with triggering ROS accumulation. High ROS accumulation and Ca^{2+} levels regulate the release of the stress hormone abscisic acid and initiate transcriptional changes (Jiang et al. 2012).

K^+ is the most abundant cation that is accumulated in cytosol. It is an important nutrient for plant growth, whereas high concentrations of sodium ions are harmful. Plants grown in saline environments will tend to have a high Na^+/K^+ ratio which is toxic to plants, and they attempt to maintain a high K^+/Na^+ concentration in the cytosol by regulating Na^+, K^+, and H^+ transporters. High affinity K+ transporters (HKTs) play significant role in maintaining a low Na^+/K^+ ratio (Hauser and Horie 2010). Na^+/K^+ transporters, hence, are important candidates in salt signaling and tolerance. Members of this group of genes have been identified from many species and *Gmhkt* gene from soybean imparted salt tolerance in transgenic tobacco plants (Chen et al. 2011).

One of the immediate effects of high salinity is the production of reactive oxygen species as discussed above. As high levels of them are toxic to the organism in multiple ways, antioxidant mechanism is activated to counter-act the effects. There are various antioxidant enzymes present in plants, some of which had proven roles in tackling salt stress. On overexpressing tobacco ascorbate peroxidase, the transgenic tobacco plants had salinity tolerance (Badawi et al. 2004). Late embryogenesis abundant protein (LEA), that are commonly associated with dehydration stress and cold stress tolerance were also found to have role in salinity stress tolerance (Checker et al. 2012; Chen et al. 2016). Glyoxalase pathway genes (*GlyI* and *GlyII*) have substantial, well-characterized roles in salinity tolerance (Singla-Pareek et al. 2003). These genes are involved in de-toxification of ROS and this pathway is ubiquitous in plants and animals. Their effectiveness has been proved in multiple crops through over-expression and transgenic development (Singla-Pareek et al. 2003; Bhomkar et al. 2008; Álvarez Viveros et al. 2013).

14.5.2 Transcription Factors in Salt Tolerance Mechanism

Transcription factors are major players in regulation of gene expression in any kind of stresses in plants. They bind to the cis-acting elements in the promoter regions of stress-responsive genes and regulate their expression. Potentially important transcription factors are generally identified by comparing the transcriptomes of stressed and non-stressed tissues. Various transcription factors responding to salt stress have been identified from many plant species and many have been functionally characterized. bZIP genes were upregulated in wheat genotypes under long-term salinity stress (Johnson et al. 2002). In *Arabidopsis*, salt stress induced the expression of AtWRKY8, and its target gene was identified as *RD29A*, a gene highly induced under various types of abiotic stress including salinity (Hu et al. 2013). NAC transcription factor also plays role in imparting salinity tolerance in crops like rice and wheat proved by overexpressing the gene (Song et al. 2011; Nakashima et al. 2007).

As the DREB1A/CBF3 gene from *Arabidopsis thaliana* was constitutively expressed in citrus and peanut, the transgenic events in both these plants showed tolerance to high salinity (Alvarez-Gerding et al. 2015; Sarkar et al. 2014). *Erianthus* DREB2 when expressed in sugarcane the resultant plants were better performers at drought and high-salinity conditions (Augustine et al. 2015). OsEREBP2 transcripts were increased after high salinity treatment and hence was a probable candidate for salt tolerance in rice. OsEREBP2 was found to bind the promoter region of a receptor-like kinase, *OsRMC* which negatively regulated salt stress response in rice (Serra et al. 2013). MYB transcription factors are also found to play roles in salinity tolerance in plants and it was proven by overexpression of OsMYB91 and OsMYB48-1 in rice which resulted in upregulation of salt-stress responsive genes

(Zhu et al. 2015; Xiong et al. 2014). To understand the functional contribution of these transcription factors, the common approaches adopted are generation of mutant and overexpression of the gene of interest in both crop species and in model crops like Arabidopsis and tobacco. A transcription factor, *SERF1* (Salt-Responsive ERF1) identified in rice showed a root-specific induction under high salt conditions and loss of this gene weakened the MAPK (mitogen-activated Protein Kinase) pathway and salt responsive transcription factors downstream of it (Schmidt et al. 2013). OSBZ8, a bZIP class of ABRE binding transcription factor, was highly expressed in salt tolerant rice genotypes than in susceptible one (Mukherjee et al. 2006). MdMYB46 transcription factor from apple was identified to enhance salt and osmotic tolerance in apple and the RNAi apple lines developed indicated that this factor increased salt tolerance (Chen et al. 2019).

14.5.3 Transgenics for Enhancing Salinity Tolerance

Cis-genic or transgenic expression of salt stress tolerance-related candidate genes characterizes them functionally and is often used to improve the trait in crop plants. SOS pathway genes alone or in combination have imparted salt tolerance in transgenic plants (Shang et al. 2012; Ma et al. 2014). Overexpression of *NHX* homologues from *Arabidopsis thaliana* was reported to improve salinity tolerance in major crops (Xue et al. 2004; Asif et al. 2011). Other Na^+/H^+ antiporters also were good candidates to impart salt tolerance. Vacuolar Na^+/H^+ antiporter from pokkali rice imparted resistance to salt stress when overexpressed (Amin et al. 2016). Transgenic expression of Na^+/H^+ antiporters from different species such as barley, Arabidopsis, wheat, and soybean showed promising results (Bayat et al. 2010; Apse et al. 1999; Chen et al. 2008, 2011). Expression of microbial genes like *mtlD* (mannitol biosynthesis), catalase (ROS neutralization), *betA* (from *E. coli*), and *hal1* (yeast salt tolerance gene) that are involved in osmo-tolerance in plants have been attempted to develop salinity tolerant genotypes (Abebe et al. 2003; Bhattacharya et al. 2004).

14.5.4 Understanding the Molecular Basis of Salinity Tolerance in Pearl Millet

Development of salinity tolerant genotypes is vital in the given scenario of climate change events, that is occuring globally. Apart from traditional plant breeding approaches, modern molecular biology tools are relevant and are to be used to harness maximum benefits. Whole genome reference for pearl millet that was developed in the recent years has provided momentum to pearl millet genomic research lately (Varshney et al. 2017). Several candidate genes relevant to salinity stress have been identified and cloned from pearl millet and many of them were functionally characterized.

A great extent of diversity towards salinity tolerance is found in various germplasms of pearl millet, at both genotypic and phenotypic levels (Liu et al. 2014). The local indigenous landraces grown in arid regions can survive better in the adverse climatic conditions and are known to be superior to the newly bred cultivars in terms of abiotic stress-tolerance. Therefore, they can represent valuable plant genetic resources and donor germplasm for providing new variations and stress adaptation traits. Some unique genes from pearl millet have already been employed to develop salt-tolerance in rice and groundnut (Santosh Rama Bhadra Rao et al. 2017). It is imperative to analyze the molecular mechanism of salt-tolerance in this critical hardy plant, in order to dig out more such proteins for developing stress-tolerance in the sensitive crop species through tools of biotechnology. Proteomics also represents an essential and complementary approach in the field of abiotic stress. Studies in various germplasm collection of pearl millet have also characterized for salinity tolerance at morphological and physio-biochemical level (Jha 2022) and at the proteomic level where salt susceptible accessions had $>25\%$ downregulation of proteins under salinity, while salt tolerant accessions had almost 50% unaltered protein (Jha 2022).

Vacuolar antiporters like Na^+/H^+ antiporter actively sequester excess Na^+ into the vacuoles using the proton motive force developed by ATPases and PPases. These transporters are thereby playing a vital role in regulation of effects of excess salinity. Rajagopal et al. (2007) reported the isolation and characterization of an isoform of vacuolar Na^+/H^+ antiporter from *Pennisetum glaucum* (PgNHX1) and through transgenic expression of the gene in *Brassica juncea,* they proved its role in salinity tolerance (Rajagopal et al. 2007). The same gene had been transformed into rice also and PgNHX1 overexpressing rice plants had high salinity tolerance. The transgenic plants had extensive root system in comparison to the wild type and they set flowers and seeds and completed their life cycle even at high salt concentrations of 150 mM NaCl. The toxicity effects of high ionic concentration during salt stress are alleviated by the combined action of several proteins including ATPases. They are proton pumps that generate the proton motive force required for ion transport, vacuolar ATPases being one among them. A subunit of Vacuolar ATPase gene was cloned from pearl millet along with its promoter and its expression pattern was studied (Tyagi et al. 2005). This gene was highly upregulated under salinity stress, high ABA and calcium and salicylic acid. Its promoter harbored DRE and ABRE elements. More isoforms of the c sub-unit of vacuolar ATPases were cloned and studied later (Tyagi et al. 2006). Bhaskaran and Savithramma (2011) co-expressed a vacuolar H^+ pyrophosphatase gene from *A. thaliana* and PgNHX1, a vacuolar Na^+/H^+ antiporter gene from pearl millet in tomato plants to enhance their salt tolerance. Transgenics with both the genes had enhanced tolerance to salinity than single gene transformed plants. The dual gene transgenic plants could survive well even at 200 mM NaCl concentrations by enhanced synthesis of chlorophyll and proline contents. These plants could survive in high salt conditions by sequestering the excess Na^+ ions into their vacuoles thereby reducing its toxic effects.

A gene encoding voltage-dependent anion channel from *Pennisetum glaucum* (*PgVDAC*) was identified from among the genes that are regulated under salt stress and was cloned and characterized (Desai et al. 2006). Expression analysis showed its upregulation under salinity, low temperature, and desiccation, while it did not respond to ABA application. Rice plants that had constitutive expression of *PgVDAC* could not survive, while mild levels of upregulation imparted adaptiveness to salinity stress. Late embryogenesis abundant (LEA) family members are known to contribute towards adaptation to desiccation and osmotic stress. A cDNA clone coding for group 7 LEA was identified in pearl millet (*PgLEA*) which when expressed in *E. coli* showed enhanced tolerance to high temperature and salinity (Reddy et al. 2012). Another late embryogenesis protein, dehydrin (*PgDHN*) cloned from pearl millet, was induced by high salinity and conferred salt tolerance to pearl millet plants by accumulating in their leaves (Shinde et al. 2018). A NAC transcription factor from pearl millet (PgNAC21) was shown to be induced by salinity stress and ABA treatment (Shinde et al. 2019). NAC proteins are generally highly induced under various abiotic stress conditions. PgNAC21 had ABA responsive elements and MYB factor binding sites in its promoter and MYB binding was proved by a yeast-hybrid assay. Arabidopsis plants overexpressing PgNAC21 had high salinity apart from many other attributes such as overexpression of stress-regulated genes such as *GSTF6* (*GLUTATHIONE S-TRANSFERASE 6*), *COR47* (*COLD-REGULATED 47*), and *RD20* (*RESPONSIVE TO DEHYDRATION 20*).

MicroRNAs are small regulatory RNA sequences of 20–22 bp length having multiple roles in organism development and functions. Shinde et al. (2020) identified 81 conserved and 14 novel miRNAs from a salinity tolerant pearl millet genotype that were salinity responsive. Of these, 30 miRNAs were upregulated and 51 were downregulated under salt stress. Among the target mRNAs of these miRNAs, 25% encoded transcription factors. Majority of the targets identified for these miRNAs were transcription factor and members belonging to ion transporters, protein kinases, detoxification proteins, and heat shock proteins which are involved in salinity tolerance mechanisms.

A comprehensive understanding of the salinity tolerance mechanism in pearl millet was brought out by comparative transcriptome analysis of contrasting genotypes for salinity tolerance, ICMB081 (susceptible) and ICMB 01222 (tolerant) (Shinde et al. 2018). Their study revealed a higher growth rate and a higher leaf sugar accumulation in the tolerant line under salt stress. The most conspicuous differentially expressed genes that were upregulated in tolerant line included genes of ubiquitin-mediated proteolysis and phenyl propanoid pathways and genes from glycolysis/gluconeogenesis pathways. The candidate genes identified in various studies are promising tools for functional characterization of salinity stress tolerance and for employing as donors for transgenics or genome editing.

14.5.5 High-Throughput Approaches for Phenotyping Salt Stress Tolerance in Pearl Millet

Developing quantifiable high-throughput phenotyping (HTP) approaches are the prerequisite to eliminate the bottlenecks in identifying the genetic gains with precision and accelerating the breeding programs. HTPs include employing aerial and ground-based platforms to phenotype the crop behavior (traits) throughout their growth stages under abiotic or biotic stresses. The HTP platforms applies the imaging techniques (satellite imaging, high pixel mobile camera, drone imaging, robot imaging), light sensors, remote sensing, mobile applications that are non-destructive, non-invasive with different levels of automation. These techniques offer wide range of advantages where data can be recorded over a large area with high resolution, are easy to operate and portable, lesser operational cost, and fewer cons like battery capacity, suitability over all areas, high initial costs, weather hindrance, and data interpretation and image quality.

Hence, an in-depth understanding of plant stress is pivotal for improving yield protection for sustainable production systems (Pessarakli 2019). Plant scientists rely on crop phenotyping for precise and reliable trait collection and utilization of genetic resources and tools to accomplish their research goals. Plant phenotyping is defined as the comprehensive assessment of complex traits of plants such as development, growth, resistance, tolerance, physiology, architecture, yield, ecology, and the elementary measurement of individual quantitative parameters that form the foundation for complex trait assessment (Li et al. 2014). Breeding programs generally aim to phenotype large populations for numerous traits throughout the crop cycle (Sandhu et al. 2021a, b). Low-throughput phenotyping has high cost, laborious, destructive, less accurate, low resolution, and environment inference. The development of high-throughput phenotyping (HTP) has largely overcome the problems crop stress phenotyping due to the automation, imaging techniques, and artificial intelligence. HTP has offered great potential for non-destructive and effective field-based plant phenotyping. Under the HTP, manual, semi-autonomous, or autonomous platforms equipped with single or multiple sensors record temporal and spatial data, resulting in storage of large amounts of data and analysis (Kaur et al. 2021; Sandhu et al. 2021b). Several HTP platforms exist and are presently employed to phenotype different abiotic stress-associated traits in various crops (Table 14.1).

These HTP platforms include several imaging platforms, techniques (satellite imagery, mobile camera, Drone imaging, robot imaging), sensors, analytical tools that include indices and formulas (Stress tolerance index, tolerance index, salt tolerance, NDVI, Green leaf index, chlorophyll index), machine and data learning. The outcome of the phenotypic data encompasses a large amount of physiological, morphological, and biochemical data to identify stress tolerant genotypes by assessing their traits under varying environments. The major challenge here is handling the data collected from different formats of platform, interpreting the spectral images, environmental interferences, which have to be effectively optimized for accurate results and cost-effective phenotyping. These approaches also need to consider the performance of crops under multi-locational trials.

Table 14.1 HTPs platforms available for phenotyping salt tolerance in pearl millet that has been utilized in other crops under abiotic stress (adapted and modified from Gill et al. 2022)

HTP platform	Crop	Morpho-physiological traits studied	References
LemnaTec 3D scanalyzer system	Rice	Salt tolerance traits	Hairmansis et al. (2014)
PHENOPSIS	Arabidopsis	Plant responses to water stress	Granier et al. (2006)
PHENODYN	Rice and maize	Soil water status (drought scenarios), leaf elongation rate, and micrometeorological variable	Sadok et al. (2007)
GROWSCREENFLUORO	Arabidopsis	Leaf growth and chlorophyll fluorescence that allow detection of stress tolerance	Jansen et al. (2009)
BreedVision	Wheat	Lodging, plant moisture content, biomass yield, or tiller density	Busemeyer et al. (2013)
RhizoTubes; RADIX	Wheat and maize	Root-related traits under non-stressed and stressed conditions	Jeudy et al. (2016)
PHENOVISION	Maize	Detection of drought stress and recovery	Asaari et al. (2019)
PhénoField	Wheat	Characterization of different abiotic stresses	Beauchêne et al. (2019)
PhenoImage	Wheat, sorghum	Plant responses to water stress	Zhu et al. (2021)
Field Scanalyzer	Wheat	Morphological traits	Virlet et al. (2016)
CropQuant	Wheat	Performance-related traits	Zhou et al. (2017)
PhenoRoots	Cotton (*Gossypium hirsutum* L.)	Root-related traits	Martins et al. (2020)
Self-propelled electric HTPP platform	Wheat (*Triticum aestivum*)	Plant height	Pérez-Ruiz et al. (2020)

14.6 Breeding Strategies for Salt Stress Tolerance in Pearl Millet

Pearl millet and its wild relatives are rated to be fairly tolerant to salinity (Ashraf and McNeilly 1987) and can be more profitably grown in saline soils. Large genotypic variation was reported to exist in pearl millet for salinity response in terms of whole plant response (Ashraf and McNeilly 1987, 1992; Dua 1989). Moreover, availability of high levels of tolerance in other species of Pennisetum (Ashraf and McNeilly 1987, 1992; Muscolo et al. 2003) and within the *P. glaucum* (Dua 1989) offers a scope for understanding the traits related to tolerance and to integrate these tolerant crop species/genotypes into appropriate management programs to improve the productivity of the saline soils. The differential responses of plants toward salinity stress rely upon their genetic make-up and the environment. Therefore, screening a large number of genotypes is essential to select the superior genotypes with greater stress tolerance. The candidate genes could be identified from those potential genotypes, and transferred to other salt-sensitive crops by plant breeding or transgenic approaches (Jha 2019). Germplasm screening for salinity stress tolerance has been performed in several plant species, namely rice, wheat, maize, and sorghum (Morton et al. 2019), but only a few candidate genes have been identified for stress tolerance, owing to the complex nature of salinity stress (Jha 2018; Lakra et al. 2018). Large genotypic variation has been observed in pearl millet toward salinity stress tolerance. A wide range of pearl millet breeding lines has been evaluated extensively for salt tolerance (Krishnamurthy et al. 2007a, b; Toderich et al. 2018). Typically, the landraces and wild relatives of a crop species exhibit genetic diversity and are known to harbor novel genes for environmental adaptation and other agronomic important traits. Therefore, these genotypes can be used as valuable genetic resources for developing abiotic stress tolerance (Hoang et al. 2016; Quan et al. 2018). Despite having a wide genetic diversity and a large germplasm collection available at the national repositories, limited reports are available for the identification and selection of superior genotypes for abiotic stress tolerance in pearl millet (Shivhare and Lata 2017).

Various traits of the plants' adaptability under salinity stress conditions have been reported from several studies. Genomic approaches and HTP phenotyping provide new insights and create a pipeline to support breeders to overcome salinity stress for crop improvement. Plants cope with salinity stress using various mechanisms, and these mechanisms can be exploited using strategies as mentioned above. Approaches like mutation breeding, wild relative exploration, Marker assisted breeding (MAB), Double haploid coupled with novel approaches like CRISPRCas9 technology and GWAS (Genome Wide Association Studies) have resulted in developing salinity tolerant crops including millets. In pearl millet, the essential genomic resources are available, although this information is limited in other millets. Pearl millet has a high level of DNA marker polymorphism between elite inbred parental lines of popular hybrids (Vadez et al. 2012). Information on the DNA markers (AFLP, RFLP, RAPD, expressed sequence tag-based (EST) markers, sequence-tagged sites (STSs), simple sequence repeat (SSRs/microsatellites), DArTs, CISP and SNP) for identification of QTLs are already available to develop salt tolerant lines and other abiotic stresses based on the molecular and biochemical bases coupled to higher crop yield. Based on this available marker information, gene linkage map has been constructed which is about 700 cM and map QTLs for salt tolerance (Singh et al. 2016; Sharma et al. 2014). Also, a final genetic map in pearl millet is about 716.7 cM with 23.23/cM overall average density of SNPs and 1.66 unique linkage bins per cM (Vadez et al. 2012).

Screening of pearl millet germplasm has resulted in the development of advanced breeding materials, improved population including OPVs, gene pools and composites, parental lines of potential hybrids, and germplasm accessions with high grain and forage yield presumably with a high degree of salinity tolerance.

The available genotypes for salinity stress tolerance in pearl millet documented and reported by Shivhare and Lata (2017) include ICMB 02111, ICMB 94555, ICMB 95333, ICMB 00888, ICMB 01222, ICMP 451, IP 3732, IP 3757, IP8210, and PRLT 2/89-33, 10876 and 10878 (Sudan), 18406 and 18570 (Namibia), and ICMV93753 and ICMV 94474 (India); 863-B, CZI 98-11, CZI 9621, HTP 94/54. These available materials can be released for cultivation after extensive validation of their yield performances at on-farm trials. "HASHAKI I," a salt tolerant pearl millet variety, has been identified in Uzbekistan in 2012 as a high-forage variety for salt-affected areas. The identified salinity-tolerant pearl millet lines should be utilized in breeding programs to develop salinity-tolerant locally adapted cultivars (both OPVs and hybrids) (Yadav et al. 2012).

With respect to the functional validation of genes and QTLs, voltage-dependent anion channel gene *VDAC* (Desai et al. 2006), formation of LEA genes (Reddy et al. 2012), *Pg*NHX1 Vacuolar Na^+/H^+ transporter gene (Verma et al. 2007), and reduced salt uptake DT-QTL at linkage group 2 (Sharma et al. 2011, 2014) are a few of them reported to confer salt tolerance. But till date, only few efforts have been aligned in the area of transcriptome analysis and transgenic approach in pearl millet crop for delineating the mechanism of abiotic stress tolerance (Choudhary et al. 2021). Hence, development of recombinant pearl millet varieties is still in nascent stage despite its economic importance and hence calls for a concerted effort for generation and evaluation of transgenic lines under various stress conditions. Also, in an effort to develop whole genome information, several organizations are taking a consensus effort to sequence the pearl millet genome, to augment breeding programs and generate ample gene resources. In regard to this, Tift 23DB2B1 has been chosen as a global pearl millet reference genotype, to develop its draft genome sequence through whole genome shotgun and bacterial artificial chromosome (BAC) sequence approaches.

14.7 Salinity Management Practices and Recent Advances for Stress Tolerance

Salinity stress affects the physiological and biochemical processes of the plant, which results in reduced seed germination, growth, and yield in many crops including pearl millet. Different management approaches are needed to mitigate the adverse effects of salinity stress to enhance the production of pearl millet. Identification of salt-tolerant genotypes along with appropriate site-specific crop management practices would increase pearl millet productivity in saline soils. Recent approaches in use of plant growth promoting rhizobacteria and advanced genetic and molecular techniques proved effective in reducing the effects of salinity stress in pearl millet.

14.7.1 Agronomic Approaches

Agronomic approaches involving the use of soil amendments (chemical and organic) will reclaim the saline soils and reduce the adverse effects of salinity on crop production. In pearl millet, application of potassium, phosphorus, gypsum, silicon & boron and organic amendments e.g., manure, biochar, compost, and crop residues have been used to improve plant growth under salinity stress. The concentration of sodium is increased in saline conditions resulting in imbalanced uptake and accumulation of essential nutrients. Increase in Na^+ content will reduce Ca^{2+} and K^+ content in saline soils. Application of K fertilizer reduces the adverse effects of salinity through its role in stomatal regulation, osmoregulation, protein synthesis, and energy status. Application of potassium in saline soil significantly increased the grain yield and resulted in decreased sodium and increased potassium content in leaves of pearl millet (Heidari and Jamshidi 2011). Application of silicon and potassium humate resulted in better growth parameters and biochemical components under saline conditions (Hassanein et al. 2017). Boron is an important element for many biochemical and physiological reactions of plants (Sezer 2014). Boron application alleviated the negative effect of salinity and improved grain yields in pearl millet (Salem 2020). Boron improved potassium concentration and maintained membrane integrity. Boron (B) application reduces the toxicity of aluminum by stabilizing integrity of the proteins, controlling the activities of antioxidant enzymes and secondary metabolites and lowering reactive oxygen species and Al concentrations (Riaz et al. 2018).

Organic amendments proved as an effective strategy for saline soil amelioration. Organic amendments improve soil chemical and physical properties. Solid waste, vermicompost, and cow dung influence soil salinity and alleviate its adverse effects on the growth of plants by changing the physico-chemical properties of soil. Application of organic amendments improved the growth performance of pearl millet in saline conditions (Diatta 2016; Araújo et al. 2022). Biochar also improved physico-chemical properties of soil, including soil cation exchange capacity, pH, water holding capacity, surface area, and soil structure under abiotic stresses (Bamminger et al. 2016). Biochar application improved potassium availability uptake and decreased sodium availability and uptake under salt stress. It enhanced the soil quality, availability and concentrations of nutrients in plant, and chlorophyll synthesis, on the other hand it reduced Na, Cl, and proline in the leaf tissue (Ding et al. 2022). It is concluded that exogenously applied organic matters such as plant residues, manure, a by-product of municipal or farming activities, etc. are an efficient and feasible way to mitigate the effects of salinity on plant growth and soil health (Meena et al. 2018).

14.7.2 Seed Priming

Seed priming is one of the most economical and easiest technique for successful crop production under salinity stress conditions. It is defined as a pre-sowing technique of treating seeds with various priming agents like water or other chemicals to enhance the germination percentage. It hastens the germination rate and improves seedling establishment in crops like maize, pearl millet, and wheat. To hasten the germination rate, seedling establishment, and crop yields, a number of priming procedures have been developed, including hydropriming (soaking seed in water), osmopriming (soaking seed in nutrients, hormones, or chemicals), and halopriming (soaking seed in salt solution). The positive effect of seed priming arises due to synthesis of certain germination-promoting substances, enhancing pre germination metabolites, early DNA replication, greater ATP availability, enzyme activation, osmotic adjustments, and membrane reorganization through restoring their original structures and reducing leakage of metabolites (Paparella et al. 2015). Primed seeds exhibit reduced photo and thermal dormancy, a wider range of germination temperatures, and a superior ability to deal with weeds and pests in addition to synchronous and quick emergence. Several research findings evidenced the role of seed priming to improve salt stress tolerance in pearl millet. Zida et al. (2017) observed the positive effect of hydro priming on germination rate under saline conditions. Many researchers reported an increase in grain yield to the tune of 13–30% by soaking pearl millet seeds for 8–16 h in water (Jidda and Anaso 2017). Khan et al. (2020) reported that seed priming with silver nanoparticles at 20 mM improved the growth attributes and antioxidant enzyme activities of pearl millet by reducing oxidative damage, and increased the salinity tolerance by reducing Na^+ uptake and maintaining the Na^+/K^+ ratio. Priming pearl millet seeds with salicylic acid @100 μM for 8 h can improve the germination (%) and seedling vigor of the genotypes under higher salinity level (Anju et al. 2019).

14.7.3 Plant Growth-Promoting Rhizobacteria (PGPR) to Ameliorate Salinity Stress

Bacteria used for promoting plant growth are termed as plant growth-promoting rhizobacteria. These are rhizospheric or endophytic bacteria that colonize the rhizosphere and enhance germination, root and shoot length, increase uptake of minerals, and yield, and enhance tolerance against drought and salinity stress (Lugtenberg and Kamilova 2009). A variety of PGPR genera such as *Azospirillum*, *Aeromonas*, *Acetobacter*, *Achromobacter*, *Pseudomonas*, *Bacillus*, *Flavobacterium*, *Chryseobacterium*, *Sinorhizobium*, *Bradyrhizobium*, etc. have been identified for maintaining the growth of various crop plants grown under high salt environment (Etesami and Maheshwari 2018). PGPR plays a major role in imparting salt tolerance to plants (Paul and Lade 2014; Srivastava and Kumar 2015; Qin et al. 2016). Application of halotolerant bacteria will be advantageous for improving the crop yields in salt affected areas. Mechanisms underlining the ability by which bacteria encourage plant growth and avert damage induced by high salt concentration in the soil include phytohormones production such as indoleacetic acid (IAA), gibberellic acid, cytokinins, and ethylene (Spaepen et al. 2009; Mishra et al. 2010); synthesis of enzyme ACC deaminase to diminish the content of ethylene in the roots of developing plants (Dey et al. 2004); synthesis of osmoprotectants (Berg et al. 2013); solubilization of minerals, like phosphorus and potassium; and modulation of antioxidant enzymes (Kohler et al. 2009). Kushwaha et al. (2020) reported that pearl millet plants treated with bacillus strain resulted in significantly higher plant growth compared to untreated seed grown under saline environment (200 mM NaCl). Kayasth et al. 2014 reported *Gordonia* sp. a salt tolerant bacterial inoculant for growth promotion of pearl millet under saline soil conditions. Khushdil et al. (2019) reported that pearl millet plants inoculated with *Aspergillusterreus* improves the NaCl tolerance in pearl millet by ameliorating the physicochemical attributes of the host plants.

14.7.4 Application of Hormones

Plant hormones, namely auxins, gibberellins, cytokinins, abscisic acid, and salicylic acid, are produced within the plant at very low content and have the potential to regulate plant development and ameliorate the effects of various biotic and abiotic stresses including salt stress. Application of hormones in pearl millet elevates osmotic adjustment to maintain turgor, improve nutrient uptake, accumulate antioxidants, and detoxify reactive oxygen species, thereby maintaining membrane and enzyme stability under stress conditions (Kaya et al. 2010). Abscisic acid (ABA) is a signaling molecule that mediates the responses to salt stress (Knight and Knight 2001; Nishiyama et al. 2011), and is considered a vital signal of salt tolerance because of its rapid biosynthesis and significant accumulation in plant cells upon exposure to salt stress conditions. The ameliorations effects on salinity stress were reported in other millet crops. Addition of abscisic acid (ABA) to the induction medium containing 200 mM NaCl improved salt tolerance of finger millet over those without ABA in association with the appearance of several ABA-responsive proteins (Uma et al. 1995). Hussain et al. (2010) and Yadav et al. (2020) reported that application of salicylic acid proved beneficial to mitigate adverse effects of salt stress by significantly improving physiological traits, biochemical traits, and ultimately improved grain yield in pearl millet. Salem (2020) reported that plant height (cm), panicle length (cm), panicles number/m^2, grain weight/panicle (g), seed index (g) and grain protein content percentage as well as biological, grain, and protein yields (kg/fed) were significantly increased by salicylic acid (SA) under saline soil conditions.

14.8 Conclusion and Future Perspectives

Though pearl millet is a crop with inbuilt capacity to withstand harsh environments and therefore is capable of cultivation in saline lands for grain and forage production, but still salinity stress acts as a significant abiotic constraint for its cultivation in several areas. There is a significant decline in the growth and yield of pearl millet crop in saline areas, which eventually affects its productivity. As compared to other cereal crops only limited information has been available on response to soil salinity in pearl millet. Efforts should be focused on collection and characterization of pearl millet germplasm so that potential sources of genetic variation for salinity tolerance could be identified and incorporated in crop improvement programs. A concerted effort is also needed to develop and upgrade phenotypic screens for abiotic stress tolerance. Unravelling the whole genome sequence will help in further crop improvement programs and ideotype development.

References

Abebe T, Guenzi AC, Martin B, Cushman JC (2003) Tolerance of mannitol-accumulating transgenic wheat to water stress and salinity. Plant Physiol 131(4):1748–1755. https://doi.org/10.1104/pp.102.003616

Ali S, Idris A (2015) Germination and seedling growth of pearl millet (*Pennisetum glaucum* L.) cultivars under salinity conditions. Int J Plant Res 1:1–5

Ali AYA, Ibrahim MEH, Zhou G, Nimir NEA, Jiao X, Zhu G, Elsiddig AMI, Suliman MSE, Elradi SBM, Yue W (2020) Exogenous jasmonic acid and humic acid increased salinity tolerance of sorghum. Agron J 112:871–884

Ali M, Afzal S, Parveen A, Kamran M, Javed MR, Abbasi GH, Malik Z, Riaz M, Ahmad S, Chattha MS et al (2021) Silicon mediated improvement in the growth and ion homeostasis by decreasing Na+ uptake in maize (*Zea mays* L.) cultivars exposed to salinity stress. Plant Physiol Biochem 158:208–218

Altus D, Canny M (1982) Loading of assimilates in wheat leaves. I. The specialization of vein types for separate activities. Funct Plant Biol 9:571

Álvarez Viveros MF, Inostroza-Blancheteau C, Timmermann T et al (2013) Overexpression of *GlyI* and *GlyII* genes in transgenic tomato (*Solanum lycopersicum* Mill.) plants confers salt tolerance by decreasing oxidative stress. Mol Biol Rep 40:3281–3290. https://doi.org/10.1007/s11033-012-2403-4

Alvarez-Gerding X, Espinoza C, Inostroza-Blancheteau C, Arce-Johnson P (2015) Molecular and physiological changes in response to salt stress in *Citrus macrophylla* W plants overexpressing *Arabidopsis CBF3/DREB1A*. Plant Physiol Biochem 92:71–80

Amin USM, Sudip B, Elias Sabrina M, Samsad R, Taslima H, Richard M, Seraj Zeba I (2016) Enhanced salt tolerance conferred by the complete 2.3 kb cDNA of the rice vacuolar Na+/H+ antiporter gene compared to 1.9 kb coding region with 5′ UTR in transgenic lines of rice. Front Plant Sci 7:14. https://doi.org/10.3389/fpls.2016.00014

Anju UL, Doddagoudar SR, Pattanashetti SK, Gowda B, Vijaykumar K (2019) Influence of seed priming on seed germination, seedling growth, peroxidase activity, proline and total soluble sugar content of pearl millet (Pennisetum glaucum L.) under salinity stress. IJCS 7(5):508–514

Apse MP et al (1999) Salt tolerance conferred by overexpression of a vacuolar Na+/H+ antiport in Arabidopsis. Science 285(5431):1256–1258

Araújo CDA, Lira JBD, Magalhães ALR, Silva TGFD, Gois GC, Andrade APD, Araújo GGLD, Campos FS (2022) Pearl millet cultivation with brackish water and organic fertilizer alters soil properties. Ciência Animal Brasileira 22

Asaari MSM, Mertens S, Dhondt S et al (2019) Analysis of hyperspectral images for detection of drought stress and recovery in maize plants in a high-throughput phenotyping platform. Comput Electron Agric 162:749–758

Ashraf M, McNeilly T (1987) Salinity effects on five cultivars/lines of pearl millet (*Pennisetum americanum* [L] Leeke). Plant Soil 103:13–19

Ashraf M, McNeilly T (1992) The potential of exploting variations in salinity tolerance in pearl millet (*Pennisetum americanum* [L] Leeke). Plant Breed 104:234–240

Asif MA, Zafar Y, Iqbal J, Iqbal MM, Rashid U, Ali GM et al (2011) Enhanced expression of *AtNHX1*, in transgenic groundnut (*Arachis hypogaea* L.) improves salt and drought tolerance. Mol Biotechnol 49(3):250–256. https://doi.org/10.1007/s12033-011-9399-1

Augustine SM, Narayan JA, Syamaladevi DP, Appunu C, Chakravarthi M, Ravichandran V, Tuteja N, Subramonian N (2015) Overexpression of *EaDREB2* and pyramiding of *EaDREB2* with the pea DNA helicase gene (*PDH45*) enhance drought and salinity tolerance in sugarcane (*Saccharum* spp. hybrid). Plant Cell Rep 34(2):247–263

Badawi GH, Kawano N, Yamauchi Y, Shimada E, Sasaki R, Kubo A, Tanaka K (2004) Over-expression of ascorbate peroxidase in tobacco chloroplasts enhances the tolerance to salt stress and water deficit. Physiol Plant 121(2):231–238

Bamminger C, Poll C, Sixt C, Högy P, Wüst D, Kandeler E et al (2016) Short-term response of soil microorganisms to biochar addition in a temperate agroecosystem under soil warming. Agric Ecosyst Environ 233:308–317

Bayat F, Shiran B, Belyaev DV, Yur'eva NO, Sobol'kova GI, Alizadeh H et al (2010) Potato plants bearing a vacuolar Na^+/H^+ antiporter *HvNHX2* from barley are characterized by improved salt tolerance. Russ J Plant Physiol 57:696–706

Beauchêne K, Leroy F, Fournier A et al (2019) Management and characterization of abiotic stress via PhénoField®, a high-throughput field phenotyping platform. Front Plant Sci 10:904

Berg G, Alavi M, Schmidt CS, Zachow C, Egamberdieva D, Kamilova F, Lugtenberg B (2013) Biocontrol and osmoprotection for plants under salinated conditions. Mol Microb Ecol Rhizosphere 1:561–573

Bhaskaran S, Savithramma DL (2011) Co-expression of *Pennisetum glaucum* vacuolar Na^+/H^+ antiporter and *Arabidopsis* H^+-pyrophosphatase enhances salt tolerance in transgenic tomato. J Exp Bot 62(15):5561–5570. https://doi.org/10.1093/jxb/err237

Bhattacharya RC, Maheswari M, Dineshkumar V, Kirti PB, Bhat SR, Chopra VL (2004) Transformation of *Brassica oleracea* var. capitata with bacterial *betA* gene enhances tolerance to salt stress. Sci Horticult 100:215–227

Bhomkar P, Upadhyay CP, Saxena M, Muthusamy A, Prakash NS, Poggin K, Hohn T, Sarin NB (2008) Salt stress alleviation in transgenic *Vigna mungo* L. Hepper (blackgram) by overexpression of the *glyoxalase I* gene using a novel Cestrum yellow leaf curling virus (*CmYLCV*) promoter. Mol Breed 22:169–181

Blummel M, Zerbini E, Reddy BVS, Hash CT, Bidinger F, Khan AA (2003) Improving the production and utilization of sorghum and pearl millet as livestock feed: progress towards dual-purpose genotypes. Field Crops Res 84:143–158

Busemeyer L, Mentrup D, Möller K, Wunder E, Alheit K, Hahn V et al (2013) BreedVision—a multi-sensor platform for non-destructive field-based phenotyping in plant breeding. Sensors 13 (3):2830–2847

Cabot C, Sibole JV, Barceló J, Poschenrieder C (2009) Abscisic acid decreases leaf Na^+ exclusion in salt-treated *Phaseolus vulgaris* L. J Plant Growth Regul 28(2):187–192

Chai YY, Jiang CD, Shi L, Shi T, Gu W (2010) Effects of exogenous spermine on sweet sorghum during germination under salinity. Biol Plant 54:145–148. https://doi.org/10.1007/s10535-010-0023-1

Chaudhry UK, Gökçe ZN, Gökçe AF (2020) Effects of salinity and drought stresses on the physio-morphological attributes of onion cultivars at bulbification stage. Int J Agric Biol 24(6): 1681–1689. https://doi.org/10.17957/IJAB/15.1611

Chaudhry UK, Gökçe ZNÖ, Gökçe AF (2021) Drought and salt stress effects on biochemical changes and gene expression of photosystem II and catalase genes in selected onion cultivars. Biologia 76:1–15. https://doi.org/10.1007/s11756-021-00827-5

Checker VG, Chhibbar AK, Khurana P (2012) Stress-inducible expression of barley *Hva1* gene in transgenic mulberry displays enhanced tolerance against drought, salinity and cold stress. Transgenic Res 21:939–957. https://doi.org/10.1007/s11248-011-9577-8

Chen LH, Zhang B, Xu ZQ (2008) Salt tolerance conferred by overexpression of *Arabidopsis* vacuolar Na^+/H^+ antiporter gene *AtNHX1* in common buckwheat (*Fagopyrum esculentum*). Transgenic Res 17:121–132

Chen H, He H, Yu D (2011) Overexpression of a novel soybean gene modulating Na^+ and K^+ transport enhances salt tolerance in transgenic tobacco plants. Physiol Plant 141(1):11–18. https://doi.org/10.1111/j.1399-3054.2010.01412.x

Chen H, Liu L, Wang L, Wang S, Cheng X (2016) *VrDREB2A*, a DREB-binding transcription factor from *Vigna radiata*, increased drought and high-salt tolerance in transgenic *Arabidopsis thaliana*. J Plant Res 129:263–273

Chen K, Song M, Guo Y, Liu L, Xue H, Dai H, Zhang Z (2019) *MdMYB46* could enhance salt and osmotic stress tolerance in apple by directly activating stress responsive signals. Plant Biotechnol J 17:2341. https://doi.org/10.1111/pbi.13151

Choudhary S, Vadez V, Tom Hash C et al (2019) Pearl millet mapping population parents: performance and selection under salt stress across environments varying in evaporative demand. Proc Natl Acad Sci India Sect B Biol Sci 89:201–211

Choudhary M, Singh A, Rakshit S (2021) Coping with low moisture stress: remembering and responding. Physiol Plant 172(2):1162–1169

Desai MK, Mishra RN, Verma D, Nair S, Sopory SK, Reddy MK (2006) Structural and functional analysis of a salt stress inducible gene encoding voltage dependent anion channel (VDAC) from pearl millet (Pennisetum glaucum). Plant Physiol Biochem 44(7–9):483–493

Dey RKKP, Pal KK, Bhatt DM, Chauhan SM (2004) Growth promotion and yield enhancement of peanut (Arachis hypogaea L.) by application of plant growth-promoting rhizobacteria. Microbiol Res 159(4):371–394

Diatta S (2016) Improving pearl millet (Pennisetum glaucum (L.) R. Br.) productivity in salt-affected soils in Senegal: a greenhouse and field investigation. Doctoral dissertation, Virginia Tech

Ding Z, Majrashi MA, Ghoneim AM, Ali EF, Eissa MA, Shal RE (2022) Irrigation and biochar effects on pearl millet and kinetics of ammonia volatilization from saline sandy soils. J Soil Sci Plant Nutr 22:1–13

Dua RP (1989) Salinity tolerance in pearl millet. Indian J Agric Res 23:9–14

Dwivedi SL, Upadhyaya HD, Senthilvel S, Tom Hash C, Fukunaga K, Diao X, Santra D, Baltensperge D, Prasad M (2012) Millets genet genomic research. In: Janick J (ed) Plant breeding reviews, vol 35. Wiley, New York, pp 247–375

Etesami H, Maheshwari DK (2018) Use of plant growth promoting rhizobacteria (PGPRs) with multiple plant growth promoting traits in stress agriculture: action mechanisms and future prospects. Ecotoxicol Environ Saf 156:225–246

FAO (2021) Global symposium on salt affected soils. https://www.fao.org/events/global-symposium-on-salt-affected-soils

Feng W, Kita D, Peaucelle A, Cartwright HN, Doan V, Duan Q, Liu MC, Maman J, Steinhorst L, Schmitz-Thom I, Yvon R, Kudla J, Wu HM, Cheung AY, Dinneny JR (2018) The FERONIA receptor kinase maintains cell-wall integrity during salt stress through Ca^{2+} signaling. Curr Biol 28(5):666–675.e5. https://doi.org/10.1016/j.cub.2018.01.023. Epub 2018 Feb 15. PMID: 29456142; PMCID: PMC5894116

Gill T, Gill SK, Saini DK, Chopra Y et al (2022) A comprehensive review of high throughput phenotyping and machine learning for plant stress phenotyping. Phenomics 2:156

Golldack D, Li C, Mohan H, Probst N (2014) Tolerance to drought and salt stress in plants: unraveling the signaling networks. Front Plant Sci 5:151

Govinda R, Shanta K, Vivek T, Jolly C, Coe R, Samart W, Quick W et al (2012) Towards a C_4 rice. Asian J Cell Biol 7:13–31

Granier C, Aguirrezabal L, Chenu K et al (2006) PHENOPSIS, an automated platform for reproducible phenotyping of plant responses to soil water deficit in Arabidopsis thaliana permitted the identification of an accession with low sensitivity to soil water deficit. New Phytol 169:623–635

Guo KM, Babourina O, Rengel Z (2009) Na^+/H^+ antiporter activity of the SOS1 gene: lifetime imaging analysis and electrophysiological studies on *Arabidopsis* seedlings. Plant Physiol 137: 155–165

Hairmansis A, Berger B, Tester M, Roy SJ (2014) Image-based phenotyping for non-destructive screening of diferent salinity tolerance traits in rice. Rice (NY) 7:16

Hanin M, Ebel C, Ngom M, Laplaze L, Masmoudi K (2016) New insights on plant salt tolerance mechanisms and their potential use for breeding. Front Plant Sci 7:1787

Hasanuzzaman M, Davies NW, Shabala L, Zhou M, Brodribb TJ, Shabala S (2017) Residual transpiration as a component of salinity stress tolerance mechanism: a case study for barley. BMC Plant Biol 17:107. https://doi.org/10.1186/s12870-017-1054

Hassanein RA, El Khawas SA, Khafaga HS, Abd El-Nabe AS, Abd Elrady AS (2017) Amelioration of drought stress on physiological performance of pearl millet (*Pennisetum americanum*) plant grown under saline condition using potassium humate and silicon source. Egypt J Exp Biol 13: 57–68

Hauser F, Horie T (2010) A conserved primary salt tolerance mechanism mediated by HKT transporters: a mechanism for sodium exclusion and maintenance of high K^+/Na^+ ratio in leaves during salinity stress. Plant Cell Environ 33:552–565

Heidari M, Jamshidi P (2011) Effects of salinity and potassium application on antioxidant enzyme activities and physiological parameters in pearl millet. Agric Sci China 10(2):228–237

Hoang TML, Tran TN, Nguyen TKT, Williams B, Wurm P, Bellairs S, Mundree S (2016) Improvement of salinity stress tolerance in rice: challenges and opportunities. Agronomy 6: 54. https://doi.org/10.1007/s43657-022-00048-z

Hu Y, Chen L, Wang H, Zhang L, Wang F, Yu D (2013) *Arabidopsis* transcription factor WRKY8 functions antagonistically with its interacting partner VQ9 to modulate salinity stress tolerance. Plant J 74:730–745

Hussain K, Nawaz K, Majeed A, Khan F, Lin F, Ghani A, Raza G, Afghan S, Zia-ul-Hussain S, Ali K, Shahazad A (2010) Alleviation of salinity effects by exogenous applications of salicylic acid in pearl millet (Pennisetum glaucum (L.) R. Br.) seedlings. Afr J Biotechnol 9(50): 8602–8607

Ishitani M, Liu J, Halfter U, Kim CS, Shi W, Zhu JK (2000) SOS3 function in plant salt tolerance requires N-myristoylation and calcium binding. Plant Cell 12:1667–1678. PMID:11006339

Jain A, Dev Sharma A (2005) The effect of phytoharmones (ABA, GA_3) on germination, osmoprotectant solution and enzymes of carbohydrates maetabolism in pearl millet. Israel J Plant Sci 12:56–60

Jansen M, Gilmer F, Biskup B et al (2009) Simultaneous phenotyping of leaf growth and chlorophyll fuorescence via GROWSCREEN FLUORO allows detection of stress tolerance in *Arabidopsis thaliana* and other rosette plants. Funct Plant Biol 36:902–914

Jayakannan M, Bose J, Babourina O (2013) Salycilic acid improves salinity tolerance in Arabidopsis by restoring membrane potential and preventing salt induced K+ loss via GORK channel. J Exp Bot 64(8):2255–2268

Jeudy C, Adrian M, Baussard C et al (2016) RhizoTubes as a new tool for high throughput imaging of plant root development and architecture: test, comparison with pot grown plants and validation. Plant Methods 12:31

Jha S (2018) Proteomics of salinity stress: opportunities and challenges. In: Ramakrishna A, Gill SS (eds) Metabolic adaptations in plants during abiotic stress. CRC Press, Boca Raton

Jha S (2019) Transgenic approaches for enhancement of salinity stress tolerance in plants. In: Singh SP et al (eds) Molecular approaches in plant biology and environmental challenges. Springer Nature, Singapore

Jha S (2022) Proteome responses of pearl millet genotypes under salinity. Plant Gene 29:100347. https://doi.org/10.1016/j.plgene.2021.100347

Jiang C, Belfield EJ, Mithani A, Visscher A, Ragoussis J, Mott R et al (2012) ROS-mediated vascular homeostatic control of root-to-shoot soil Na delivery in Arabidopsis. EMBO J 31: 4359–4370. https://doi.org/10.1038/emboj.2012.273

Jidda MB, Anaso AB (2017) Effects of crop improvement technologies on downy mildew of pearl millet [Pennisetum glaucum (L.) R. Br]. J Cereals Oilseeds 8(3):14–20

Johnson RR, Wagner RL, Verhey SD, Walker-Simmons MK (2002) The abscisic acid-responsive kinase PKABA1 interacts with a seed-specific abscisic acid response element-binding factor, TaABF, and phosphorylates TaABF peptide sequences. Plant Physiol 130(2):837–846

Julkowska MM, Testerink C (2015) Tuning plant signaling and growth to survive salt. Trends Plant Sci 20:586–594

Kalaji HM, Govindjee BK, Kocielniakd J, Zuk-Golaszewska K (2011) Effects of salt stress on photosystem II efficiency and CO2 assimilation of two Syrian barley landraces. Environ Exp Bot 73:64–72

Kaur B, Sandhu KS, Kamal R et al (2021) Omics for the improvement of abiotic, biotic, and agronomic traits in major cereal crops: applications, challenges, and prospects. Plants 10:1989

Kaya C, Tuna AL, Okant AM (2010) Effect of foliar applied kinetin and indole acetic acid on maize plants grown under saline conditions. Turk J Agric For 34(6):529–538

Kayasth M, Kumar V, Gera R (2014) Gordonia sp.: a salt tolerant bacterial inoculant for growth promotion of pearl millet under saline soil conditions. 3 Biotech 4(5):553–557

Keskin BC, Sarikaya AT, Ykel B, Memon AR (2010) Abscicic acid regulated gene expression in bread wheat (*Triticum aestivum L.*). Aust J Crop Sci 4(8):617–625

Khan I, Raza MA, Awan SA, Shah GA, Rizwan M, Ali B, Tariq R, Hassan MJ, Alyemeni MN, Brestic M, Zhang X (2020) Amelioration of salt induced toxicity in pearl millet by seed priming with silver nanoparticles (AgNPs): the oxidative damage, antioxidant enzymes and ions uptake are major determinants of salt tolerant capacity. Plant Physiol Biochem 156:221–232

Khushdil F, Jan FG, Jan G, Hamayun M, Iqbal A, Hussain A, Bibi N (2019) Salt stress alleviation in Pennisetum glaucum through secondary metabolites modulation by Aspergillus terreus. Plant Physiol Biochem 144:127–134

Knight H, Knight MR (2001) Abiotic stress signalling pathways: specificity and cross-talk. Trends Plant Sci 6:262–267

Knight H, Trewavas AJ, Knight MR (1997) Calcium signalling in *Arabidopsis thaliana* responding to drought and salinity. Plant J 12:1067–1078

Kohler J, Hernández JA, Caravaca F, Roldán A (2009) Induction of antioxidant enzymes is involved in the greater effectiveness of a PGPR versus AM fungi with respect to increasing the tolerance of lettuce to severe salt stress. Environ Exp Bot 65(2–3):245–252

Krishnamurthy L, Serraj RC, Hash T, Dakheel AJ, Reddy BVS (2007a) Screening sorghum genotypes for salinity tolerant biomass production. Euphytica 156:15–24

Krishnamurthy L, Serraj R, Rai KN, Hash CT and Dakheel AJ (2007b). Identification of pearl millet [*Pennisetum glaucum* (L.) R. Br.] lines tolerant to soil salinity. Euphytica 158: 179–188

Krishnamurthy L, Upadhyaya HD, Purushothaman R, Gowda CLL, Kashiwagi J, Dwivedi SL, Singh S, Vadez V (2014) The extent of variation in salinity tolerance of the minicore collection of finger millet germplasm. Plant Sci 227(2014):51–59

Kudla J, Becker D, Grill E, Hedrich R, Hippler M, Kummer U, Parniske M, Romeis T, Schumacher K (2018) Advances and current challenges in calcium signaling. New Phytol 218:414. https://doi.org/10.1111/nph.14966

Kulkarni VN, Rai KN, Dakheel AJ, Ibrahim M, Hebbara M, Vadez V (2006) Pearl millet germplasm adapted to saline conditions. SAT eJournal 2(1):1–4

Kumar A, Krishnamurthy SL, Lata C, Kumar P, Devi R, Kulshrestha N, Yadav RK, Sharma SK (2016) Salinity and drought induced changes in gas exchange attributes and chlorophyll fluorescence characteristics of rice (*Oryza sativa*) varieties. Ind J Agric Sci 86(6):718–726

Kumar A, Sharma SK, Lata C, Devi R, Kulshrestha N, Krishnamurthy SL, Singh K, Yadav RK (2018) Impact of water deficit (salt and drought) stress on physiological, biochemical and yield attributes on wheat (*Triticum aestivum*) varieties. Ind J Agric Sci 88(10):1624–1632

Kushwaha P, Kashyap PL, Kuppusamy P, Srivastava AK, Tiwari RK (2020) Functional characterization of endophytic bacilli from pearl millet (Pennisetum glaucum) and their possible role in multiple stress tolerance. Plant Biosyst 154(4):503–514

Lakra N, Kaur C, Anwar K, Singla-Pareek SL, Pareek A (2018) Proteomics of contrasting rice genotypes: identification of potential targets for raising crops for saline environment. Plant Cell Environ 41(5):947–969

Lakshmi TV, Varalaxmi Y, Yadav S, Maheswari M (2017) Differential response of antioxidative enzymes to various abiotic stresses in Pennisetum glaucum seedlings. Russ J Plant Physiol 64: 889–898

Li L, Zhang Q, Huang D (2014) A review of imaging techniques for plant phenotyping. Sensors 14: 20078–20111. https://doi.org/10.3390/s141120078

Litalien A, Zeeb B (2019) Curing the earth: a review of anthropogenic soil salinization and plant-based strategies for sustainable mitigation. Sci Total Environ 698:134235. https://doi.org/10.1016/j.scitotenv.2019.134235

Liu Y, Wang Q, Zhang Y, Cui J, Chen G, Xie B, Wu C, Liu H (2014) Synergistic and antagonistic effects of salinity and pH on germination in switchgrass (Panicum virgatum L.). PLoS One 9: e85282

Lugtenberg B, Kamilova F (2009) Plant-growth-promoting-rhizobacteria. Annu Rev Microbiol 63: 541–556. https://doi.org/10.1146/annurev.micro.62.081307.162918

Lynch J, Polito VS, Lauchli A (1989) Salinity stress increases cytoplasmic Ca activity in maize root protoplasts. Plant Physiol 90:1271–1274

Ma DM, Wr WX, Li HW, Jin FX, Guo LN, Wang J et al (2014) Co-expression of the Arabidopsis SOS genes enhances salt tolerance in transgenic tall fescue (Festuca arundinacea Schreb.). Protoplasma 251:219–231. PMID:24022678

Maiti RK, Singh VP, Sarkar NC, Purohit SS, Hernandez-Pinero J (2007) Physical basis of crop growth and productivity. In: Reaearch advances in pearl millet. Springer, Berlin, pp 39–67

Martins SM, de Brito GG, da Gonçalves WC et al (2020) PhenoRoots: an inexpensive non-invasive phenotyping system to assess the variability of the root system architecture. Sci Agric 77. https://doi.org/10.1590/1678-992x-2018-0420

Mbarki S, Sytar O, Zivcak M, Abdelly C, Cerda A, Brestic M (2018) Anthocyanins of coloured wheat genotypes in specific response to salt stress. Molecules 23:1518. https://doi.org/10.3390/molecules23071518

Mbinda W, Mukami A (2021) A review of recent advances and future directions in the management of salinity stress in finger millet. Front Plant Sci 12:734798. https://doi.org/10.3389/fpls.2021.734798

Meena MD, Narjary B, Sheoran P, Jat HS, Joshi PK, Chinchmalatpure AR, Yadav G, Yadav RK, Meena MK (2018) Changes of phosphorus fractions in saline soil amended with municipal solid waste compost and mineral fertilizers in a mustard-pearl millet cropping system. Catena 160:32–40

Mishra M, Kumar U, Mishra PK, Prakash V (2010) Efficiency of plant growth promoting rhizobacteria for the enhancement of Cicer arietinum L. growth and germination under salinity. Adv Biol Res 4(2):92–96

Morton MJL, Awlia M, Al-Tamimi N, Saade S, Pailles Y, Negrão S, Tester M (2019) Salt stress under the scalpel – dissecting the genetics of salt tolerance. Plant J 97:148–163

Mukami A, Ng'etich A, Syombua E, Odour R, Mbinda W (2020) Varietal differences in physiological and biochemical responses to salinity stress in six finger millet plants. Physiol Mol Biol Plants 26:1569–1582. https://doi.org/10.1007/s12298-020-00853-8

Mukherjee K, Choudhury AR, Gupta B et al (2006) An ABRE-binding factor, OSBZ8, is highly expressed in salt tolerant cultivars than in salt sensitive cultivars of indica rice. BMC Plant Biol 6(1):18

Munns R, James RA, Läuchli A (2006) Approaches to increasing the salt tolerance of wheat and other cereals. J Exp Bot 57(5):1025–1043

Munns R, Tester M (2008) Mechanisms of salinity tolerance. Annu Rev Plant Biol 59:651–681

Muscolo A, Panuccio MR, Sidari M (2003) Effect of salinity on growth, carbohydrate metabolism and nutritive properties of Kikuyu grass (*Pennisetum clandestinum* Hochst). Plant Sci 104: 1103–1110

Nakashima K, Tran L-SP, Van Nguyen D et al (2007) Functional analysis of a NAC-type transcription factor OsNAC6 involved in abiotic and biotic stress-responsive gene expression in rice. Plant J 51(4):617–630

Nishiyama R, Watanabe Y, Fujita Y, Le DT, Kojima M, Werner T, Vankova R, Yamaguchi-Shinozaki K, Shinozaki K, Kakimoto T, Sakakibara H, Schmülling T, Tran LSP (2011) Analysis of cytokinin mutants and regulation of cytokinin metabolic genes reveals important regulatory roles of cytokinins in drought, salt and abscisic acid responses, and abscisic acid biosynthesis. Plant Cell 23:2169–2183

Paparella S, Araujo SS, Rossi G, Wijayasinghe M, Carbonera D, Balestrazzi A (2015) Seed priming: state of the art and new perspectives. Plant Cell Rep 34:1281–1293

Parida AK, Das AB (2005) Salt tolerance and salinity effect on plants: a review. Ecotoxicol Environ Saf 60:324–349

Paul D, Lade H (2014) Plant-growth-promoting rhizobacteria to improve crop growth in saline soils: a review. Agron Sustain Dev 34:737–752

Pérez-Ruiz M, Prior A, Martinez-Guanter J et al (2020) Development and evaluation of a self-propelled electric platform for high-throughput field phenotyping in wheat breeding trials. Comput Electron Agric 169:105237

Pessarakli M (2019) Handbook of plant and crop stress, 4th edn. CRC Press, Taylor & Francis Group, Boca Raton

Pooja, Nandwal AS, Chand M, Kumari A, Rani B, Goel V, Kulshreshtha N (2020) Soil moisture deficit induced changes in antioxidative defense mechanism of four sugarcane varieties differing in their maturity. Ind J Agric Sci 90(3):56–61

Qin Y, Druzhinina IS, Pan X, Yuan Z (2016) Microbially mediated plant salt tolerance and microbiome-based solutions for saline agriculture. Biotechnol Adv 34(7):1245–1259

Qiu QS, Guo Y, Dietrich MA, Schumaker KS, Zhu J-K (2002) Proc Natl Acad Sci U S A 99:8436–8441

Qiu QS, Guo Y, Quintero FJ, Pardo JM, Schumaker KS, Zhu JK (2004) Regulation of vacuolar Na +/H+ exchange in Arabidopsis thaliana by the salt-overly-sensitive (SOS) pathway. J Biol Chem 279(1):207–215. https://doi.org/10.1074/jbc.M307982200. Epub 2003 Oct 21. PMID: 14570921

Quan R, Wang J, Hui J et al (2018) Improvement of salt tolerance using wild rice genes. Front Plant Sci 8:2269

Radhouane L (2013) Agronomic and physiological responses of pearl millet ecotype (*Pennisetum glaucum* (L.) R. Br.) to saline irrigation. Emir J Food Agric 25(2):109–116

Rahim Z, Gulnaz P, Gul S, Rehman K (2020) Ameliorating effect of salt stress (KCl, NaCl) on growth and germination parameters of pearl millet (*Pennisetum americanum*). Pak J Agric Res 33(4):951–956

Rajabi Dehnavi A, Zahedi M, Ludwiczak A, Cardenas Perez S, Piernik A (2020) Effect of salinity on seed germination and seedling development of sorghum (*Sorghum bicolor* (L.) Moench) Genotypes. Agronomy 10:859. https://doi.org/10.3390/agronomy10060859

Rajagopal D, Agarwal P, Tyagi W et al (2007) *Pennisetum glaucum* Na^+/H^+ antiporter confers high level of salinity tolerance in transgenic *Brassica juncea*. Mol Breed 19:137–151. https://doi.org/10.1007/s11032-006-9052-z

Reddy PS, Reddy GM, Pandey P et al (2012) Cloning and molecular characterization of a gene encoding late embryogenesis abundant protein from *Pennisetum glaucum*: protection against abiotic stresses. Mol Biol Rep 39:7163–7174. https://doi.org/10.1007/s11033-012-1548-5

Riaz M, Yan L, Wu X, Hussain S, Aziz O, Wang Y, Imran M, Jiang C (2018) Boron alleviates the aluminum toxicity in trifoliate orange by regulating antioxidant defense system and reducing root cell injury. J Environ Manag 208:149–158

Ribadiya TR, Savalia SG, Vadaliya BM, Davara MA (2018) Efect of salinity on yield, yield attributes and quality of pearl millet (*Pennisetum glaucum* L) varieties. Int J Chem Studies 6(6):878–882

Sadok W, Naudin P, Boussuge B et al (2007) Leaf growth rate per unit thermal time follows QTL-dependent daily patterns in hundreds of maize lines under naturally fluctuating conditions. Plant Cell Environ 30:135–146

Sage RF, Zhu XG (2011) Exploiting the engine of C_4 photosynthesis. J Exp Bot 62:2989–3000. https://doi.org/10.1093/jxb/err179

Salem E (2020) Cooperative effect of salicylic acid and boron on the productivity of pearl millet crop under the degraded saline soils conditions. Egypt J Agron 42(2):185–195

Sandhu KS, Patil SS, Pumphrey MO, Carter AH (2021a) Multi-trait machine and deep learning models for genomic selection using spectral information in a wheat breeding program. Plant Genome 14:e20119. https://doi.org/10.1002/TPG2.20119

Sandhu KS, Aoun M, Morris CF, Carter AH (2021b) Genomic selection for end-use quality and processing traits in soft white winter wheat breeding program with machine and deep learning models. Biology 10:689. https://doi.org/10.3390/BIOLOGY10070689

Santosh Rama Bhadra Rao T, Vijaya Naresh J, Sudhakar Reddy P, Reddy MK, Mallikarjuna G (2017) Expression of *Pennisetum glaucum* eukaryotic translational initiation factor 4A (*PgeIF4A*) confers improved drought, salinity, and oxidative stress tolerance in groundnut. Front Plant Sci 8:453. https://doi.org/10.3389/fpls.2017.00453

Sarkar T, Thankappan R, Kumar A, Mishra GP, Dobaria JR (2014) Heterologous expression of the *AtDREB1A* gene in transgenic peanut-conferred tolerance to drought and salinity stresses. PLoS One 9(12):e110507

Sawada H, Shim I-S, Usui K (2006) Induction of benzoic acid 2-hydroxylase and salicylic acid biosynthesis: modulation by salt stress in rice seedlings. Plant Sci 171:263–270. https://doi.org/10.1016/j.plantsci.2006.03.020

Schmidt R, Mieulet D, Hubberten H-M et al (2013) Salt-responsive ERF1 regulates reactive oxygen species-dependent signaling during the initial response to salt stress in rice. Plant Cell 25(6):2115–2131

Serra T, Figueiredo DD, Cordeiro AM et al (2013) OsRMC, a negative regulator of salt stress response in rice, is regulated by two AP2/ERF transcription factors. Plant Mol Biol 82:439–455

Sezer S (2014) Effect of boron fertilizer applications on the growth and B, N uptake of maize (*Zea mays* L.) under the different soils. J Food Agric Environ 12(2):1323–1327

Shahzad B, Rehman A, Tanveer M, Wang L, Park SK, Ali A (2021) Salt stress in *Brassica*: effects, tolerance mechanisms, and management. J Plant Growth Regulat 41:781–795. https://doi.org/10.1007/s00344-021-10338-x

Shang G, Li Y, Hong Z, Liu CL, Shao-Zhen HE, Liu QC (2012) Overexpression of SOS genes enhanced salt tolerance in Sweetpotato. J Integr Agric 11:378–386

Sharma PC, Sehgal D, Singh D, Singh G, Yadav RS (2011) A major terminal drought tolerance QTL of pearl millet is also associated with reduced salt uptake and enhanced growth under salt stress. Mol Breed 27(2):207–222

Sharma PC, Singh D, Sehgal D, Singh G, Hash CT, Yadav RS (2014) Further evidence that a terminal drought tolerance QTL of pearl millet is associated with reduced salt uptake. Environ Exp Bot 102:48–57

Sharma A, Shahzad B, Kumar V, Kohli SK, Sidhu GPS, Bali AS, Handa N, Kapoor D, Bhardwaj R, Zheng B (2019) Phytohormones regulate accumulation of osmolytes under abiotic stress. Biomol Ther 9(7):285. https://doi.org/10.3390/biom9070285

Shi H, Ishitani M, Kim C, Zhu J-K (2000) Proc Natl Acad Sci U S A 97:6896–6901

Shinde H, Tanaka K, Dudhate A, Tsugama D, Mine Y, Kamiya T, Gupta SK, Liu S, Takano T (2018) Comparative de novo transcriptomic profiling of the salinity stress responsiveness in contrasting pearl millet lines. Environ Exp Bot 155:619–627., ISSN 0098-8472. https://doi.org/10.1016/j.envexpbot.2018.07.008

Shinde H, Dudhate A, Tsugama D, Gupta SK, Liu S, Takano T (2019) Pearl millet stress-responsive NAC transcription factor *PgNAC21* enhances salinity stress tolerance in *Arabidopsis*. Plant Physiol Biochem 135:546–553. https://doi.org/10.1016/j.plaphy.2018.11.004

Shinde H, Dudhate A, Anand L, Tsugama D, Gupta SK, Liu S, Takano T (2020) Small RNA sequencing reveals the role of pearl millet miRNAs and their targets in salinity stress responses. South African J Bot 132:395–402. https://doi.org/10.1016/j.sajb.2020.06.011. ISSN 0254-6299

Shivhare R, Lata C (2017) Exploration of genetic and genomic resources for abiotic and biotic stress tolerance in pearl millet. Front Plant Sci 7:2069

Singh R, Singh Y, Xalaxo S, Verulkar S, Yadav N, Singh S et al (2016) From QTL to variety-harnessing the benefits of QTLs for drought, flood and salt tolerance in mega rice varieties of India through a multi-institutional network. Plant Sci 242:278–287

Singla-Pareek SL, Reddy MK, Sopory SK (2003) Genetic engineering of the glyoxalase pathway in tobacco leads to enhanced salinity tolerance. Proc Natl Acad Sci U S A 100:14672–14677

Sneha S, Anirudha R, Amit D, Subhash C (2013) Effect of salinity on seed germination, accumulation of proline and free amino acid in *Pennisetum glaucum* (L.) R. Br. Pak J Biol Sci 16:877–881. https://doi.org/10.3923/pjbs.2013.877.881

Song S-Y, Chen Y, Chen J, Dai X-Y, Zhang W-H (2011) Physiological mechanisms underlying OsNAC5-dependent tolerance of rice plants to abiotic stress. Planta 234(2):331–345

Sorahinobar M, Niknam V, Ebrahimzadeh H, Soltanloo H, Behmanesh M, Enferadi ST (2016) Central role of salicylic acid in resistance of wheat against *Fusarium graminearum*. J Plant Growth Regul 35:477–491. https://doi.org/10.1007/s00344-015-9554-1

Spaepen S, Vanderleyden J, Okon Y (2009) Plant growth-promoting actions of rhizobacteria. Adv Bot Res 51:283–320

Srivastava P, Kumar R (2015) Soil salinity: a serious environmental issue and plant growth promoting bacteria as one of the tools for its alleviation. Saudi J Biol Sci 22(2):123–131

Toderich K, Shuyakaya E, Rakhmankulova Z, Bukerev R, Khujanazarov T, Zhapaev R, Ismail S, Gupta SK, Yamanaka N, Boboev F (2018) Threshold tolerance of new genotypes of Pennisetum glaucum (L.) R. Br. to salinity and drought. Agronomy 8:230

Tyagi W, Rajagopal D, Singla-Pareek SL, Reddy MK, Sopory SK (2005) Cloning and regulation of a stress-regulated *Pennisetum glaucum* vacuolar ATPase c gene and characterization of its promoter that is expressed in shoot hairs and floral organs. Plant Cell Physiol 46(8): 1411–1422. https://doi.org/10.1093/pcp/pci154

Tyagi W, Singla-Pareek S, Nair S et al (2006) A novel isoform of ATPase c subunit from pearl millet that is differentially regulated in response to salinity and calcium. Plant Cell Rep 25:156–163. https://doi.org/10.1007/s00299-005-0055-8

Uma S, Prasad TG, Udaya Kumar M (1995) Genetic variability in recovery growth and synthesis of stress proteins in response to polyethylene glycol and salt stress in finger millet. Ann Bot 76(1): 43–49

Vadez V, Hash T, Kholova J (2012) Phenotyping pearl millet for adaptation to drought. Front Physiol 3:386

Varshney R, Shi C, Thudi M et al (2017) Pearl millet genome sequence provides a resource to improve agronomic traits in arid environments. Nat Biotechnol 35:969–976. https://doi.org/10.1038/nbt.3943

Veerangamallaiah G, Jyothsnakumari G, Thippaswamy M, Reddy PCO, Surabhi GK, Sriranganayakulu G, Mahesh Y, Rajasekhar B, Madhurarekha C, Sudhakar C (2008) Proteomic analysis of salt stress responses in foxtail millet (Setaria italic L. cv. Prasad). Plant Sci 175:631

Verma D, Singla-Pareek SL, Rajagopal D, Reddy MK, Sopory SK (2007) Functional validation of a novel isoform of Na+/H+ antiporter from Pennisetum glaucum for enhancing salinity tolerance in rice. J Biosci 32(6):21–628

Virlet N, Sabermanesh K, Sadeghi-Tehran P, Hawkesford MJ (2016) Field Scanalyzer: an automated robotic feld phenotyping platform for detailed crop monitoring. Funct Plant Biol 44:143–153

Wang K, Shangguan Z (2010) Photosynthetic characteristics and resource utilization efficiency of maize (*Zea mays* L.) and millet (*Setaria italica* L.) in a semi-arid hilly loess region in China. N Z J Crop Hortic Sci 38:247–254. https://doi.org/10.1080/01140671.2010.503987

Wang W, Vinocur B, Altman A (2003) Plant responses to drought, salinity and extreme temperatures: towards genetic engineering for stress tolerance. Planta 218:1–14

Wu H, Hill CB, Stefano G, Bose J (2021) New insights into salinity sensing, signaling and adaptation in plants. Front Plant Sci 11:1843

Xiong H, Li J, Liu P, Duan J, Zhao Y, Guo X, Li Y, Zhang H, Ali J, Li Z (2014) Overexpression of *OsMYB48-1*, a novel MYB-related transcription factor, enhances drought and salinity tolerance in rice. PLoS One 9(3):e92913

Xue ZY, Zhi DY, Xue GP, Zhang H, Zhao YX et al (2004) Enhanced salt tolerance of transgenic wheat (Tritivum aestivum L.) expressing a vacuolar Na+/H+ antiporter gene with improved grain yields in saline soils in the field and a reduced level of leaf Na+. Plant Sci 167:849–859. https://doi.org/10.1016/j.plantsci.2004.05.034

Yadav RK, Dagar JC (2016) Innovations in utilization of poor-quality water for sustainable agricultural production. In: Dagar et al (eds) Innovative saline agriculture. Springer, Berlin, pp 219–261. ISBN 978-81-322-2770-0

Yadav OP, Rai KN, Gupta SK (2012) Pearl millet: genetic improvement for tolerance to abiotic stresses. In: Improving crop productivity in sustainable agriculture. Wiley Blackwell, Hoboken, pp 261–268. ISBN 978-3-527-33242-7

Yadav T, Kumar A, Yadav RK, Yadav G, Kumar R, Kushwaha M (2020) Salicylic acid and thiourea mitigate the salinity and drought stress on physiological traits governing yield in pearl millet-wheat. Saudi J Biol Sci 27(8):2010–2017

Zhao YK, Wang T, Zhang WS, Li X (2011) SOS3 mediates lateral root development under low salt stress through regulation of auxin redistribution and maxima in Arabidopsis. New Phytol 189: 1122–1134. PMID: 21087263

Zhou J, Reynolds D, Websdale D et al (2017) CropQuant: an automated and scalable feld phenotyping platform for crop monitoring and trait measurements to facilitate breeding and digital agriculture. BioRxiv. https://doi.org/10.1101/161547

Zhu JK (2002) Salt and drought stress signal transduction in plants. Annu Rev Plant Biol 53:247–273

Zhu JK (2004) Plant salt tolerance and the SOS pathway. In: Proceedings of Italian Society of Agricultural Genetics. ISBN 88-900622-5-8

Zhu JK, Liu J, Xiong L (1998) Genetic analysis of salt tolerance in *Arabidopsis*: evidence for a critical role of potassium nutrition. Plant Cell 10:1181–1191

Zhu N, Cheng S, Liu X, Du H, Dai M, Zhou D-X, Yang W, Zhao Y (2015) The R2R3-type MYB gene *OsMYB91* has a function in coordinating plant growth and salt stress tolerance in rice. Plant Sci 236:146–156

Zhu F, Saluja M, Dharni JS et al (2021) PhenoImage: an open-source graphical user interface for plant image analysis. Plant Phenome J 4:e20015. https://doi.org/10.1002/ppj2.20015

Zida PE, Soalla WR, Paco SÃ (2017) Hydropriming of pearl millet (Pennisetum glaucum L.) in northern and Central Burkina Faso applying six hours of soaking and overnight drying of seeds. Afr J Agric Res 12(49):3441–3446

Weed and *Striga* Management in Pearl Millet Production Systems in Sub-Saharan Africa

15

Djibril Yonli, Hamidou Traore, and Boubacar A. Kountche

Abstract

Weeds were a major constraint to food crop production in sub-Saharan Africa (SSA) much more before soil poverty and drought became a problem. Among the weeds infesting pearl millet fields, *Cyperus* spp. and *Digitaria horizontalis* are the dominant species in terms of occurrence frequency and emerged plant density and are difficult to control. The depressive effect of weeds on pearl millet yield is compounded by the presence of parasitic species. *Buchnera hispida*, *Striga asiatica,* and *Striga hermonthica* are the main parasitic weeds of pearl millet, of which *S. hermonthica* is the most damaging and widespread. Control options involve cultural and herbicidal techniques. Some control methods, including cropping systems (crop rotation and intercropping), were recommended for *S. hermonthica* management. Compared to other *Striga* hosts, pearl millet has benefited from little research into the development of resistant varieties. Few control options to weed/*Striga* in pearl millet farming system have been designed, more research is needed to identify innovative weed control strategies in a participatory approach to conservation agriculture. The priority research needs would be to highlight (1) developed herbicide-resistant weed species, the existence of *S. hermonthica* races in SSA; (2) develop pearl millet varieties resistant

D. Yonli (✉) · H. Traore
Centre National de la Recherche Scientifique et Technologique (CNRST), Institut de l'Environnement et de Recherches Agricoles (INERA), Ouagadougou, Burkina Faso

B. A. Kountche
Centre National de la Recherche Scientifique et Technologique (CNRST), Institut de l'Environnement et de Recherches Agricoles (INERA), Ouagadougou, Burkina Faso

CIRAD, UMR AGAP Institut, Ouagadougou, Burkina Faso

UMR AGAP Institut, Univ Montpellier, CIRAD, INRAE, Institut Agro, Montpellier, France
e-mail: boubacar.kountche@cirad.fr

V. A. Tonapi et al. (eds.), *Pearl Millet in the 21st Century*,
https://doi.org/10.1007/978-981-99-5890-0_15

to *Striga* ecotypes; (3) the effect resulted in pearl millet roots × soil microorganisms and nutrients interactions from the rhizosphere level on *Striga* infection; and (4) *Striga* severity and aggressiveness induced by climate change.

Keywords

Pearl millet · Weed flora · Parasitic species · Control options · Future research needs

15.1 Introduction

Globally, infestation by weed flora and parasitic plants is one of the prominent sources of major agricultural crop loss, causing about \$95 billion each year to farmers worldwide (Chaudhary et al. 2018). In Sub-Saharan Africa (SSA), accounting for 50% of the world's agricultural land, only 1% of that land is suitable for long-term cultivation, attributable to the lack of fertile soil and poor land management practices (Edgerton 2009). While the data on the economic loss due to weed flora is still lacking, infestation by *Striga* spp. alone has been estimated to exceed \$10 billion annually, and heavily infested fields with both weeds and *Striga* being abandoned by farmers (Gressel et al. 2004; Hearne 2009; Atera et al. 2012). These staggering losses come at a time where pressure on pearl millet, a staple food and fodder crop, production is further aggravated by the ever-increasing population, depleting natural resources, and climate change (Padgham 2009; Lybbert and Sumner 2010). In addition, the farming systems in which pearl millet is produced are characterized by the dominance of smallholder farmers, responsible for the production of over 80% of the locally consumed food, while the majority of these farmers are resource-deficient and have very limited access to loans for on-farm investments (Fan and Rue 2020). Thus, pearl millet has been postulated to play a vital role in ensuring food, nutritional and economic security in SSA. Unfortunately, its production is declining due to various constraints, of which the main factors include agricultural pests such as weed flora and especially parasitic plants, affecting not only the potential production capacity but also the quality of the harvested product.

Within recent years, noticeable efforts have been made to lower many of these barriers. As a result, a wide range of control strategies, including manual, mechanical, cultural, chemical, biological, and genetic, have been deployed either individually or in an integrated manner (Haussmann et al. 2000). Although these methods have provided capacity to farmers to diminish the impact of the weeds on pearl millet production, their success is still limited as the tremendous weed seed bank problem has not been adequately addressed (Kountche et al. 2016).

To realize the full potential of pearl millet, allowing the crop to meet unprecedented challenges would require a transition to sustainable farming practices that support agroecological intensification-based cropping system and environmental health as well as accelerated development and dissemination of innovative approaches that will improve pearl millet resilience to the weed flora and *Striga*-

infection. This is an urgent task for food and nutritional security, given the importance of pearl millet in human diets in SSA.

This chapter aims to provide an overview of the strategies implemented so far to control weed flora and *Striga* spp. infestation in pearl millet production systems in Africa, with the prospect of highlighting what should be addressed in future research to ensure sustainable control of these invasive weeds.

15.2 Weed Management in Pearl Millet Production Systems in Sub-Saharan Africa

Several factors contribute to the heavy weed infestation in pearl millet fields. Indeed, the installation of the cropping season is spread over a long period with sporadic rains favorable to the development of weeds endowed with genotypic and phenotypic plasticity but which do not allow producers to carry out their sowing. This situation makes seedbed preparation operations more difficult for African farmers and gives a competitive advantage to weeds for the use of environmental resources such as soil nutrients and water compared to pearl millet plants. Mechanical or manual ridging, which is effective and recommended to control weeds without environmental pollution before the first sowing, is not commonly practiced in pearl millet fields. Apart from mowing shrubs and perennial grasses, the majority of pearl millet farmers do no-till seeding. As a result, pearl millet seedlings emerge in an overgrown field with weeds, leading to strong competition for nutrients (Fig. 15.1). In seedbed preparation, soil scraping or shallow plowing is usually done in millet fields. In favor of low soil moisture or regular rainfall, these cultural practices act on weediness as transplanting or a multiplication of the stands of ordinary or perennial grasses.

15.2.1 Weed Flora of Pearl Millet Cropping Systems

Similar to other major agricultural crops, weeds exert strong pressure on pearl millet plants to the point of smothering them in terms of high density coupled with high vegetative development. Farmers fear some weeds because of their survival and adaptability capability. They are considered major weeds whose capacity of nuisance in terms of high plant density, soil covering, and high cost of pearl millet production is significant (Table 15.1). The critical period of pearl millet-weed competition covered 15–42 days after the sowing (Dubey et al. 2023). In pearl millet, the literature about weed flora research is very limited, making the writing of this section difficult and justifying the lack of sufficient references.

Fig. 15.1 Condition of pearl millet plants in competition with weeds dominated by *Digitaria horizontalis*, emerged after the first (**a**) and the second (**b**) hoe weeding, i.e., around 15–30 and 45–60 days after sowing, respectively

15.2.2 Grain Yield Losses Attributed to Weeds

Notably, weeds compete with crops for nutrients, soil moisture, sunlight, and space when limiting, resulting in reduced yields, lower grain quality, and increased production costs (Chaudhary et al. 2018). Weed infestation leads to a significant yield reduction (Ahanchede and Gasquez 1995). Marnotte (1995) pointed out that in Sudan-Sahelian Africa, even before soil poverty was a constraint to farming, weeds were the major obstacle to good yields. Crops are increasingly invaded by weeds, and their density increases with each crop cycle. According to Koch et al. (1982), Deat and Bockel (1986), weeds are one of the major causes of crop failure in developing countries. Pearl millet production, including stalk biomass and grain yield, can be affected by weed infection (Fig. 15.1), but no African statistical data has been documented. In other places, uncontrolled weed infestation reduces pearl millet yield between 16% and 94%, depending on crop cultivars, nature and intensity of weeds, spacing, duration of weeds infestation, management practices, and environmental conditions (Mishra et al. 2018). In Sahelian countries, crop losses of

Table 15.1 Major weeds in pearl millet cropping systems in sub-Saharan Africa (Traore and Maillet 1992; Yonli et al. unpublished data)

Weed category	Species	Family
Broad-leaved grass	*Celosia trigyna* L.; *Celosia argentea* L.	*Amaranthaceae*
	Acanthospermum hispidum de Candole; *Ageratum conyzoides Linnaeueus* subsp. conyzoides; *Eclipta prostrata* (L.) L.	*Asteraceae*
	Ipomoea cocinosperma; I. eriocarpa R. Brown	Convulvulaceae
	Senna obtusifiolia (L.) H.S. Irwin & Barneby	*Fabaceae*
	Acalypha segetalis J. Mueller; *Euphorbia hirta* L.	*Euphorbiaceae*
	Hyptis spicigera L.; *Leucas martinicensis* (Jacquin) R. Brown	*Lamiaceae*
	Hisurta villosus; *Spermacoce* spp.	*Rubiaceae*
	Corchorus tridens L.; *Corchorus olitorus* L.	*Tiliaceae*
Grasses	*Dactyloctenuim aegyptium* (L.) *P. Beauv.*; *Digitaria horizontalis* Willdenow; *Cenchrus biflorus* Roxburgh; *Chloris Pilosa* Schumach.*;* *Cynodon dactylon* (L.) Pers.; *Echinochloa colona* (Linnaeus) Link; *Mnesithea granularis* (L.) de Koning & Sosef; *Pennisetum pedicellatum* Trin; *Setaria pallide-fusca* (Schumach.) Stapf & C.E. Hubb.	*Poaceae*
	Thelepogon elegans Roem. & Schult.; *Commelina forskalei Vahl*; *Cyanotis lanata* Benth	*Commelinaceae*
Sedges	*Cyperus esculentus* L.; *Cyperus iria* L., *Cyperus rotundus* L.; *Maricus squarrosus* (L.) C.B. Clarke; *Kyllinga squamulata* Thonn. ex Vahl; *Bulbostylis hispidulata* (Vahl) R.W. Haines	Cyperaceae
	Imperata cylindrica (L.) P. Beauv.	*Poaceae*
Parasitic plants	*Buchnera hispida* Buch.-Ham. ex D. Don.	*Scrophulariaceae*
	Striga hermonthica (Del.) Benth.; Striga asiatica (L.) Kuntze	*Orobanchaceae*

millet are aggravated by the existence of parasitic weeds. Thus, pearl millet yield is reduced, and production can only be increased by increasing the cultivated area (Ahanchede and Gasquez 1995) or, more often, by abandoning heavily infested fields in favor of new land (Déat et al. 1980). Nowadays, these palliative alternatives are compromised by the lack of land due to demographic pressure.

15.2.3 Control Options

As pearl millet is predominantly grown in rain-fed conditions in Africa, weeds deprive the crop of vital nutrients and moisture, affecting the yield accordingly. Because of wider row spacing and slow initial growth in pearl millet, weeds are more problematic during the juvenile crop growth period (Chaudhary et al. 2018). Therefore, early control is required to overcome weed incidence on millet growth and productivity. The concept of the critical period of weed competition, during which weeds have the greatest effect on crop growth, requires the implementation of a weeding operation within this period. Overall, weed management in pearl millet fields in Africa involves mainly manual, cultural, mechanical, and herbicidal techniques.

15.2.3.1 Manual Weeding

In the traditional farming systems, hand pulling and hoe weeding are the two most common practices used in pearl millet fields to control the weeds. These manual interventions were performed from the 2–3 leaf stage to the tillering stage. It is important to stress that both practices are time and energy-consuming, requiring much manpower. This might certainly explain the lack of success.

15.2.3.2 Mechanical Weeding

The most practical mechanical weeding includes the mounding and the ridging using animal attraction that can be asinine, oxen, or horse, depending on the locality. The mounding can be practiced from seedbed establishment to crop flowering while the ridging is implemented, especially from the millet flowering phase. Ridges can be tied to hold water, prevent run-off, and promote plant water use (Mason et al. 2015). Alternatively, tillage options available include shallow cultivation with a harrow (tines), ridging and mounding, tied ridging, and localized tillage to form micro-catchments termed "Zaï" (Fatondji et al. 2001; Nicou and Charreau 1985). As indirect weed control options, these practices can potentially increase water infiltration of early season rains and have little effect on crop root growth, with yield increases from 0% to 15% (Nicou and Charreau 1985).

15.2.3.3 Chemical Control

Over the last decades, weed infestation has worsened as a consequence of nutrient deficiencies in the soil, poor land management, labor unavailability, and to changes in cultivation habits as a sign of modernization of agriculture in Africa. Agrochemical-based weed control is therefore being extensively used in pearl millet production systems in Africa as a weed control strategy. Herbicidal weeding is carried out from seedbed preparation to the whole of the vegetative cycle of millet. However, despite its impressive results in weed dissemination, the major problem is that the majority of pearl millet producers use non-homologous herbicides and have not received adequate, if not any, training in herbicide storage, preparation prior to and application. Furthermore, herbicides are used throughout the cropping season, leading to a detrimental effect on the soil microbial and weed flora, seriously and negatively affecting the organic status of cultivated soils. Since crop yield is reduced under weed infestation, there is a need for developing promising and sustainable solutions to the dual challenges of achieving food security, while supporting a healthy environment.

15.3 Parasitic Weeds Management in Pearl Millet Farming Systems in SSA

In pearl millet production systems, three parasitic plants, including *Buchnera hispida* Buch. Ham ex D. Don, *Striga asiatica* (L.) Kuntze, and *Striga hermonthica* (Del.) Benth. (*Orobanchaceae* Ex *Scrophulariaceae*) are considered the most serious agricultural pests of economic importance (Tank et al. 2006). *B. hispida,* a facultative hemiparasite (Fig. 15.2), is generally found in isolated individuals throughout the cultivated area, especially on pearl millet plants. The parasite is widespread in tropical and subtropical areas of Africa, expanding from West through Central to East Africa, and occurs in more than 30 African countries. Moreover, the potential incidence of *B. hispida* on cereal crops, including pearl millet in the tropics, has been discussed by Iwoke and Okonkwo (1974).

Besides, *S. hermonthica* and *S. asiatica*, obligate root parasites, represent the two most economically important species in tropical areas of Africa (Haussmann et al. 2004; Parker 2012), attacking and irreversibly impacting production of monocotyledonous crops (Parker and Riches 1993). *S. asiatica* has been reported to attack pearl millet in Eastern Africa, especially in Ethiopia and Kenya, and very limited occurrences have been observed in Western Africa, such as Burkina Faso (Obilana and Ramaiah 1992). In *S. hermonthica*, no investigation has been initiated to highlight the extent of the parasitism of *S. asiatica* on pearl millet and other cereals.

15.3.1 *Striga* Distribution and Economic Incidence

Owing to its unparalleled ability to adapt to diverse climatic and environmental conditions, its high fecundity, and the longevity of the seed reserve in infested soils,

Fig. 15.2 Pearl millet field infested by *Buchnera hispida* Buch. Ham ex D. Don (in blue flowers). (Credit D. Yonli)

Striga has emerged as the major and persistent biotic threat to major staple food, feed, and fodder crops (Pennisi 2010, 2015; Kountche et al. 2016). *S. hermonthica* appears to be a highly out-crossing species, thus, it is expected to show greater diversity within a population than in related autogamous species (Hamrick 1982; Koyama 2000). This mode of pollination has contributed to the genetic variation in *S. hermonthica* plants and also restricted the geographical distribution of this species depending on the availability of pollinators (Berner et al. 1997; Mohamed et al. 2007). *S. hermonthica* is a notorious root hemiparasite on pearl millet in eight West African countries (Nigeria, Ghana, Burkina Faso, Niger, Tchad, Mali, Senegal, and Mauritania); in five East African countries (Sudan, Ethiopia, Yemen, Kenya, and Uganda); and in three South African countries (Angola, Tanzania, and Mozambique) (Obilana and Ramaiah 1992; Parker 2012). However, *Striga* infection on pearl millet (Fig. 15.3) is insignificant in East Central and Southern Africa (Gressel et al. 2004).

In the various farming systems of sub-Saharan Africa, *Striga* prevents farmers from achieving the expected grain and fodder yields of pearl millet, sorghum, rice, and maize; hence aggravating food, nutritional and economic insecurity of already resource-deprived smallholder farmers. *Striga* parasitism inflicts serious damage ranging from few percent (10–31%) to complete crop failure (100%) depending on the crop cultivar, degree of infestation, rainfall pattern, and soil degradation (Atera et al. 2012; Gressel et al. 2004; Wilson et al. 2000). The *Striga* problem is often associated with low economic resources, low soil fertility, marginal environments with continued crop monoculture, and newly infested areas regrettably due to various human and agricultural activities (Oswald 2005; Rodenburg et al. 2005; Parker 2009). From the economic perspective, *Striga* infestation has been estimated to be USD 7 to 10 billion loss annually in cereal production systems (Gressel et al. 2004; Hearne 2009; Westwood et al. 2012). Unfortunately, specific statistics about

Fig. 15.3 Pearl millet yield highly affected by *Striga hermonthica* (purple flowers) attacks in farmers' fields. Highly *Striga* infested pearl millet field showing serious damage on grain yield (**a**) and (**b**) illustrating a stunted and yellowed pearl millet plant due to *Striga* parasitism

losses due to *Striga* are not documented for pearl millet and are often combined with those of other cereals such as sorghum and/or maize (Gressel et al. 2004). For subsistence farmers in the arid and semi-arid regions of SSA, however, the highly undesirable consequences of these losses are a return to the top of the cycle, creating a worsening downward spiral and compromising a better horizon. A sustainable *Striga* control is thus fundamental to ensuring sustainable development and securing food, nutritional and economic security for millions of rural families in *Striga*-prone regions.

15.3.2 Farmers' Knowledge and Approaches Towards *Striga* Management

When dealing with a complex problem such as *Striga*, research efforts for guiding our responses to such threat have also employed a participatory appraisal of the parasite. This includes farmer surveys carried out in Mali (Hoffmann et al. 1997; Tom V.M. unpublished data), in Burkina Faso (Traoré and Yonli 1999; Rouamba et al. 2021), in Nigeria (Emechebe et al. 2004), and in Kenya (Atera et al. 2012) to get insights into farmers perception on *Striga* problem and on ground knowledge of endogenous *Striga* control methods used for coping with the parasite. It appears that some of the smallholders knew about *Striga* plants, but the majority are unaware of how it reproduces and are generally unable to differentiate *Striga* species (Hoffmann et al. 1997). Water runoff, animal dung, and wind have been reported as the main factors worsening *Striga* occurrence. Furthermore, the increased incidence and severity of *Striga* damage were attributed to the declining soil fertility and the continuous monoculture cropping of host crops. Indigenous *Striga* control strategies have been inventoried in Western Africa (Table 15.2) and the most widely used were hoe weeding and hand pulling, application of organic manure, crop rotations, and intercropping (Emechebe et al. 2004). These cultural practices have also been recommended by research, but unfortunately, they are not applied according to scientific standards. Indeed, the doses of fertilizers and the modes of association/rotation of crops are not appropriate in *Striga* controlling. Certain practices, namely, Sorghum—pearl millet rotation, may seem odd for research of bounty on board, while the existence of physiological Striga strains could justify it.

15.3.3 Conventional Strategies Towards *Striga* Control

The witchweed *Striga* has long been a devastating agricultural parasitic plant, jeopardizing production of major cereal crops, including pearl millet, and prompting research over the years that aimed at developing management strategies. As a result of decades of remarkable *Striga* research efforts, different approaches have been developed and deployed for combatting *Striga* (Haussmann et al. 2000; Teka 2014; Kountche et al. 2016), targeting different impacts on the parasite lifecycle such as the reduction of the seed bank, limitation of seed production and reduction/prevention of

Table 15.2 African farmers' *Striga* coping strategies

Burkina Faso (Traoré and Yonli 1999; Rouamba et al. 2021)	Kenya (Atera et al. 2012)	Nigeria (Emechebe et al. 2004)	Mali (Hoffmann et al. 1997; Tom V.M. unpublished data)
Common African farmers' practices			
(1) Additional hand pulling; (2) additional weeding/hoeing; (3) use of organic manure (animal dung, compost, farmyard manure or cotton seeds); (4) use of mineral fertilizers; (5) cereal-cowpea intercropping; (6) crop rotations (cereal-peanut, cotton-cereal, sorghum-pearl millet and tuber-cereal); (7) fallow			
1. Earthing up/ridges	1. Hand-pulling and burning	1. Earthing up/ridges	1. Use of ash
2. Anti-erosion managements	2. Herbicide seed dressing	2. Use of ash and lime	2. Use of *Parkia biglobosa* pod powder
3. Early planting		3. Spreading pearl millet chaff	
4. Use of *Parkia biglobosa* pod powder		4. Burning	
5. Cropping of tolerant host crop varieties		5. Deep ploughing	
6. Use of *Acacia gourmaensis* bark powder		6. Strip cropping	
7. Use of microplots locally referred to as 'zaï'		7. Early planting	
8. Mulching			
9. Use of herbicides			

seed dissemination to uninfected fields (Haussmann et al. 2000). Notably, control practices that affect germination and attachment of the parasite seed to the host are expected to be more effective as they can prevent parasitism before the host plant is irreversibly damaged and contribute to parasite seed bank reduction.

15.3.3.1 Cultural Methods

As the most traditional practice, cultural methods include hand weeding, tillage, and planting methods, improved soil fertility, cowpea intercropping with cereal, and rotation of cash/trap crops with cereals (Haussmann et al. 2000; Kuchinda et al. 2003; Hess and Dodo 2004; Goldwasser and Rodenburg 2013). Rotation of sesame (*Sesamum indicum* L.) with pearl millet and/or in association has been reported as a natural suicide germination strategy of *Striga* seeds (Hess and Dodo 2004). Despite its high potential to significantly contribute to *Striga* seed bank reduction, this strategy has, however, received little interest. Investigating new sesame production strategies could provide a sustainable alternative to enhance staple food crops' resilience to *Striga*. The rotation of cereals like pearl millet with false hosts such as soybean (*Glycine max* (L.) Merr), cotton (*Gossipium hisurtum* L.), and voandzou (*Voandzeia subterranea* L.), which stimulate *Striga* seed germination but do not allow its fixation (Parkinson et al. 1987), has been recommended to farmers as the technique allows the reduction of the *Striga* seed bank in infested-soil. Only after several years of implementation can the degree of *Striga* infestation be reduced to a non-damaging level. However, the lack of cultivable land does not allow farmers to rotate cereals with legumes that are not part of their staple diet. In addition, genotypic differences exist between varieties within the same false-host species, implying that research should recommend legume varieties resistant to *S. gesnerioides* (Willd.) Vatke and exhibiting suicidal germination potential to *S. hermonthica* seeds (Traore et al. 2011).

Certain fertilization techniques have been shown to be of value in the control of *Striga*. Nitrogen fertilizers applied at high rates reduce *Striga*-related production losses by increasing the vigor of the host crop (Parker 1984; Kim and Adetimirin 1997). Cechin and Press (1993) reported that nitrogen fertilizers affect host exudation, while Pieterse (1991) emphasized that they inhibit the radicle elongation of *S. hermonthica*. However, the high cost of nitrogen fertilizers makes their use difficult on the farm.

An integrated management system called the "Push–Pull" technology (PPT) has been developed by the International Centre of Insect Physiology and Ecology (ICIPE) to control *Striga hermonthica* and insect pests (Khan et al. 2008). This technology involved intercropping cereal with a repellent crop *Desmodium* (*Desmodium uncinatum* Jacq.) (push), and planting an attractive trap crop, Napier grass (*Pennisetum purpureum* Schumach or *Brachiaria*) (pull), as a border crop around this inter-crop. *S. hermonthica* is controlled by *Desmodium* and induces abortive germination of *Striga* seeds that fail to develop and attach onto the hosts' roots (Fig. 15.4) (Khan et al. 2008; Tsanuo et al. 2003). PPT was disseminated in Eastern Africa (Kenya, Mozambique, Uganda, Rwanda, Zimbabwe) and introduced

Fig. 15.4 Design of "Push–Pull" technology as integrated *Striga hermonthica* and insect pests management in field conditions

during 2018–2019 in three West African countries (Burkina Faso, Ghana, and Senegal). PPT has limited success in the Sahelian regions, the major pearl millet producing areas in Africa. Because of one cropping season a year in the Sahel, farmers must plant both weed species every year, whereas in the PPT concept, they are perennial, so their planting is done once for years of farming. In the Sahel, during the whole rain-off season (7–8 months), *Desmodium* plants thus cannot be alive, and *Brachiaria* shoots will be fed by animals in raving. Because temperature and rainfall are the key drivers in Sahelian areas in determining suitable habitat niches of *Desmodium*, research should investigate local weeds to select those that could play similar roles.

15.3.3.2 Chemical Control

The concerns over the *Striga* problem have also necessitated the deployment of herbicides, aiming to mitigate the parasite infection and impact on major agricultural cereal crops. It is noteworthy that chemical control has been reported to have a high impact rate. Two post-emergence herbicides Triclopyr, 2,4-D and Triclopyr +2,4-D, were applied to *Striga hermonthica* affecting sorghum in Burkina Faso. Herbicide applications using Triclopyr or 2,4-D (at 1 L ha^{-1} on 70 and on 85 days after sowing (DAS)) and Triclopyr +2,4-D (at 0.5 L ha^{-1} on 70 and on 85 DAS) significantly reduced the number of emerged *Striga*, *Striga* flowering and seed formation (Traore et al. 2000). However, these results are not systematically transposable to pearl millet crops. Indeed, Clopiramid, 2,4-D, dicamba, picloram, and prosulfuron, belonging to Auxin-mimic and acetolactate inhibitor class herbicides, were evaluated by Dembele et al. (2005) in sorghum and pearl millet growth in Mali as seed priming agents. Their results showed that none of the herbicides used consistently reduced *S. hermonthica* on pearl millet, whereas their reducing effect on the purple witch-weed on sorghum was significant. The most promising herbicide for pearl millet may be dicamba, which showed a slight and non-significant reduction in *Striga* densities in the field (Dembele et al. 2005). So far, the herbicides used for parasitic weeds include, for example, glyphosate, imidazolinones, glufosinate, 2,4-D, and dicamba (Eplee and Norris 1987; Aly 2007). However, it is important to stress that the output of chemical herbicides application can take various forms, including shifts in weed flora, and disturbed environmental and human health, owing to their hazardous effects. Mounting evidence highlights the many challenges posed by chemical herbicides that even the cost of *Striga* control and limitations of other control methods now necessitate further development of innovative alternative strategies.

15.3.3.3 Biological Control

Deploying the natural enemies, especially insects, bacterial and fungal antagonists, has long been considered as a potential alternative strategy to suppress *Striga* infestation. Ultimately, insect parasitoids have been reported on *S. hermonthica* plants in Eastern Africa (Greathead and Milner 1971), Nigeria (Williams and Caswell 1959), and Burkina Faso (Traore et al. 1996). However, most of these insects are polyphagous, among which we distinguish crop pests such as *Spodoptera* spp. and *Helicoverpa armigera* (Greathead 1984). Besides, bacteria have been isolated in West Africa from the rhizosphere of sorghum and maize infested by *S. hermonthica*. Two races (L1 and L2) of the bacterium *Azospirillum brasilense* from Mali (Bouillant et al. 1997), 15 isolates of the bacteria *Pseudomonas fluorescens* and *P. putida* from Nigeria (Ahonsi et al. 2002) significantly inhibited in in vitro germination of *S. hermonthica* seeds. Two other species of bacteria, namely *Bacillus subtilis* Cohn and *Pontoena agglomerans* [*Enterobacter agglomerans* (Beijerinck) comb. Nov], were isolated from *S. hermonthica* plants in Sudan (Abbasher et al. 1996). Moreover, the biological control of *Striga* through the use of the pathogenic fungus, *Fusarium oxysporium* as a mycoherbicide, has been recommended to farmers (Marley et al. 1999, 2005; Elzein and Kroschel 2004; Yonli et al. 2006; Zahran et al. 2008). To our knowledge, this control option has not been practically or extensively deployed in farmers' fields. For the effective use of natural enemies of *S. hermonthica* identified by the research programs (insects, bacteria, fungi), no bio-herbicide was proposed to the end-users.

15.3.3.4 Genetic Control

As a key strategy towards *Striga* management, genetic control through the deployment of resistant varieties has gained a marked interest since the approach is believed to offer the most cost-effective and sustainable control of the pernicious weed (Ejeta 2007; Haussmann et al. 2000; Hearne 2009; Wilson et al. 2000; Yoder and Scholes 2010). However, *Striga* resistance in pearl millet has been much more elusive than in other cereals, such as sorghum and rice. Resistance of 274 *Pennisetum glaucum* subsp. monodii accessions were evaluated, and four accessions, including PS 202, PS 637, PS 639, and PS 727, were identified to be resistant to *Striga*, providing useful sources of *Striga* resistance for improving cultivated pearl millet in West Africa (Wilson et al. 2000; Wilson et al. 2004). Hence, improving *Striga* resistance in cultivated pearl millet has historically been challenging due to the limited genetic diversity for *Striga* resistance, lack of knowledge of resistance mechanisms and their molecular genetic basis, and in-field phenotyping constraints (Kountche et al. 2016). Nevertheless, significant progress has been made toward generating resistant varieties during the last two decades. Research conducted by the International Crops Research Institute for the Semi-Arid Tropics (ICRISAT) and its collaborators resulted in the identification of donor sources in cultivated pearl millet, and the development through a field-based phenotypic recurrent selection of the first *Striga*-resistant varieties (Kountche et al. 2013). Future efforts need to focus on unraveling the pearl millet-*Striga* interplay, thereby harnessing the as-yet largely untapped genetic potential of existing pearl millet germplasm.

15.3.3.5 Integrated *Striga* Management (ISM)

Although farmers have been provided with a wide range of control options, the fact remains that none of these methods has proved to effectively subdue the *Striga* problem (Joel 2000; Oswald 2005). It has been widely agreed that an integrated approach, incorporating a variety of strategies in a wise way, could provide the most comprehensive and sustainable way to deal with *Striga* (Ejeta 2007; Marley et al. 2004; Elzein et al. 2008; Kanampiu et al. 2003). Nevertheless, it has become evident that the massive and long-lived *Striga* seed bank accumulated over the years has hindered the efficiency of the conventional strategies applied individually or in an integrated manner, leading to the never-ending *Striga* problem (Kountche et al. 2016). Integrated *Striga* management (ISM) has been promoted in several African countries by ICRISAT, the International Institute for Tropical Agriculture (IITA), along with the technical assistance of national agricultural research scientists (NARS) using the Farmer Field School (FFS), a participatory agricultural extension approach, based on "learning by discovery" (Van de Fliert 1993). The FFS learning process builds on existing knowledge and enables farmers to evaluate new and existing technologies in their own fields and to adapt new technologies to their conditions and means. FFS has been upgraded to the cluster-based farmer field school (CBFFS) system developed by IITA and NARS in West Africa (Nathaniels 2005) to perform integrated *Striga* and soil fertility management (ISSFM). ISSFM is a cropping system approach that is not focused on a single technology and involves (1) a sound knowledge of *Striga* biology and control, (2) combinations of multiple *Striga* control options, (3) adaptation of control techniques to local conditions, and (4) a long-term *Striga* reduction and soil fertility improvement.

15.3.4 Emerging Strategies Toward Ending with Striga Problem

Given the evident and limited success of current methods to effectively contain the parasite invasion, a paradigm shift in how the *Striga* problem is approached scientifically and in development terms is required to ensure sustainable and rational management of the parasite. This is possible only when the tremendous *Striga* seed bank, is significantly, if not utterly, depleted to reach the least prejudicial level to host crop production. Towards ending the pernicious *Striga* problem, efforts are now being directed in harnessing the genetic potential for steady resistance and exploring the seed germination dependency on host-released phytohormones, aiming to develop an environmentally friendly alternative *Striga* management package.

Interestingly, the past decades have seen marked developments in plant phenomics and omics—genomic, transcriptomic, proteomic, and metabolomic—approaches. An opportunity lies ahead to dissect complex, quantitative traits when both genotype and phenotype can be assessed at a high level of detail. This is especially true for *Striga* research in pearl millet, for which forward genetics studies have yielded little progress in our understanding of the genetic layout of the traits. Since the molecular mechanisms underlining *Striga* resistance are yet to be elucidated, combining different omics approaches will help in dissecting pearl millet genes associated with *Striga* resistance traits, targeting both pre- and post-attachment resistance factors. This, in turn, will foster an in-depth understanding of the link between genotype and phenotype. A direct outcome will certainly be the identification, and mapping of several quantitative trait loci (QTLs) linked to resistance genes and their flanking markers, which will pave the way for pyramiding the identified QTLs/genes into and fast-tracking the development of locally adapted farmers-preferred elite varieties.

Besides optimizing the genetic potential, the suicidal germination approach has recently received considerable attention in the fight against the witchweed *Striga*. Although the approach is not new (Eplee 1975), the strategy has emerged as one of the most powerful means to deplete the accumulated seed bank, owing to its ability to induce destructive parasite seed germination in the host absence (Kountche et al. 2016; Samejima et al. 2016; Zwanenburg et al. 2016; Kountche et al. 2019). It does not only contribute towards lowering *Striga* infestation, leading to improved host crop productivity, but has the potential to revolutionize *Striga* management. Furthermore, there have been ongoing discussions about how the recent developments in strigolactone (SL) research can be further harnessed to develop new strategies for the sustainable control of the parasite. An integrated approach combining genetic resistance and suicidal germination technologies is likely to increase the rate and efficiency and provide the next generation of cost-effective and environmentally friendly alternative *Striga* control options for the well-being of resource-limited smallholders in SSA.

15.4 Gaps and Future Research Needs

Considering the unprecedented challenges on food security with the persistent weeds problem, the ever-increasing global population, depleting natural resources, and climate change, harnessing each and every single potential means of major crop improvement and tackling all potential causes are the need of the hour. In light of their striking and damage on major staple crops production, one should ask the question about what would be one of the best steps to ending noxious weed flora and witchweed *Striga* problems in infested fields.

While the genetic potential of pearl millet is yet largely to be exploited, deployment of the new and amazing gene editing technology, such as the Clustered Regularly Interspaced Short Palindromic Repeats/CRISPR-associated (CRISPR/Cas9) system, could potentially help in improving the efficiency of conventional breeding. As more information on the molecular mechanisms underlying host plant–*Striga* interaction becomes available, it is feasible to manipulate host plant genomes by disrupting genes that contribute to parasite susceptibility.

Despite the research efforts, our knowledge of the *Striga* seed bank-level, distribution, and dynamics is still very limited, attributable to various factors such as crop rotation, intercropping, wind, variable growing seasons, and anthropogenic spread. Notably, the assessment and characterization of *Striga* infestation distribution appear to be a crucial step prior to developing effective and rationally designed interventions for increased and sustainable pearl millet productivity. Moreover, attempts to elucidate *S. hermonthica* strains have been initiated (Ramaiah 1984; Kim et al. 1994; Freitag et al. 1996; Bozkurt et al. 2015). Two clusters of *S. hermonthica* populations were distinguished in Sudan, and the millet strain slightly differed and was more specific to its host (Ali et al. 2009). However, no study successfully identified *S. hermonthica* strains as it was done for *S. gesnerioides* (Cardwell and Lane 1995; Li and Timko 2009). Indeed, in the same area, a crop variety can prove to be resistant to *S. hermonthica* in one location and susceptible to another one. Besides, "intracrop specific" describes strains reacting in a different manner to cultivars of a single host crop, whereas "intercrop specific" describes strains reacting in a different manner to different host crops (Kim et al. 1994). The genetic variability of *S. hermonthica* has not been sufficiently evaluated relative to its wide distributions (Mohamed et al. 2007). The genotypic identification of *S. hermonthica* ecotypes and their mapping at the scale of each African country seems a prerequisite for a better understanding of the variability in the parasite aggressiveness, leading to the development of sustainable resistant pearl millet varieties.

Notably, it is postulated that the root system of host plants and the microorganisms from its rhizosphere interact, consequently influencing the quality and quantity of host root exudates responsible for the stimulation of *Striga* seeds (Bouwmeester et al. 2007). It is worth noting that a deeper understanding of the interactions *Striga* × soil microorganisms × pearl millet crop will provide insights into the variation in *Striga* infestation in the various farming systems. Importantly, the grain yield loss of the host crop due to *Striga* is positively correlated with the soil fertility level (Showemimo et al. 2002). Thus, modeling the levels of soil type and fertility, *Striga* ecotype and infestation, climate variability, and the resulting yield losses appears to be another area yet to explore in order to predict the impact of *Striga* parasitism on pearl millet production. This should be integrated with generating individual data about parasitic weeds (*S. hermonthica, S. asiatica, Buchnera hispida*) incidence on pearl millet production, yet to be documented. The availability of such data will increase awareness of the real *Striga* problem among producers and policymakers.

Climate envelope models can serve as a guideline for understanding the present distribution of parasitic weeds and predicting their potential future geographic distribution in light of climate and land use change. However, forecasts across African continental ranges prove difficult due to (1) the local climate variability and adaptations, (2) the host plant/management variations, and, (3) the local varieties within target species.

References

Abbasher AA, Hess DE, Sauerborn J, Kroschel J (1996) Effect of different Fusarium spp on seed germination of: Striga hermonthica (sorghum and millet strains); S. asiatica and S. gesnerioides. In: Monero MT, Cubero JI, Berner D, Musselman LJ, Parker C (eds) Advances in parasitic plant research. Dirección General de Investigación Agraria, Seville, pp 880–887

Ahanchede A, Gasquez J (1995) Mauvaises herbes des cultures pluviales au Nord-Est du Bénin. Agriculture et Développement 7:22–29

Ahonsi MO, Berner DK, Emechebe AM, Lagoke ST (2002) Selection of rhizobacterial strains for suppression of germination of Striga hermonthica (Del.) Benth. seeds. Biol Control 24(2):143–152

Ali RAMA, El-Hussein AA, Mohamed KI, Babiker AGT (2009) Specificity and genetic relatedness among *Striga hermonthica* strains in Sudan. Life Sci Int J 3(3):1159–1166

Aly R (2007) Conventional and biotechnological approaches for control of parasitic weeds. Vitr Cell Dev Biol Plant 43:304–317

Atera EA, Itoh K, Azuma T, Ishii T (2012) Farmers' perspectives on the biotic constraint of *Striga hermonthica* and its control in western Kenya. Weed Biol Manag 12:53–62

Berner DK, Winslow MD, Awad AE, Cardwell KF, Mohan Raj DR, Kim SK (1997) *Striga* research methods – a manual, 2nd edn. International Institute of Tropical Agriculture (IITA), Ibadan

Bouillant ML, Miché L, Ouedraogo O, Alexandre G, Jacoud C, Sallé G, Bally R (1997) Inhibition of Striga seed germination associated with sorghum growth promotion by soil bacteria. Comptes Rendus de l'Académie des Sciences-Series III-Sciences de la Vie 320:159–162

Bouwmeester HJ, Roux C, Lopez-Raez JA, Becard G (2007) Rhizosphere communication of plants, parasitic plants and AM fungi. Trends Plant Sci 12:224–230

Bozkurt ML, Muth P, Parzies HK, Haussmann BIG (2015) Genetic diversity of east and west African *Striga hermonthica* populations and virulence effects on a contrasting set of sorghum cultivars. Weed Res 55:71–81

Cardwell KF, Lane JA (1995) Effect of soils, cropping system, and host phenotype on incidence and severity of *Striga gesnerioides* on cowpea in West Africa. Agric Ecosyst Environ 53:253–262

Chaudhary C, Dahiya S, Rani S, Pandey S (2018) Review and outlook of weed management in pearl millet. Int J Chem Stud 6(2):2346–2350

Cechin I, Press MC (1993) Nitrogen relations of the sorghum-Striga hermonthica host-parasite association: germination, attachment and early growth. New Phytol 124:681–687

Deat MM, Bockel N (1986) Identification des situations agricoles en Afrique sahélienne en vue d'améliorer leur adaptation à la sécheresse. CIRAD, Paris. 84 p

Deat M, Sement G, Fontenay P (1980) Premières observations sur la levée de quelques adventices tropicales. VIe Colloque International sur l'Ecologie, la Biologie et la systématique des mauvaises herbes. 7–8 Mai 1980. Montpellier, pp 321–328

Dembele B, Dembele D, Westwood JH (2005) Herbicide seed treatments for control of purple witchweed (*Striga hermonthica*) in *sorghum* and millet. Weed Technol 19:629–635

Dubey RP, Chethan CR, Choudhary VK, Mishra JS (2023) A review on weed management in millets. Indian J Weed Sci 55(2):141–148

Edgerton MD (2009) Increasing crop productivity to meet global needs for feed, food, and fuel. Plant Physiol 149(1):7–13

Ejeta G (2007) Breeding for *Striga* resistance in *sorghum*: exploitation of an intricate host-parasite biology. Crop Sci 47:S216–S227

Elzein A, Kroschel JK (2004) *Fusarium oxysporum* Foxy 2 shows potential to control both *Striga hermonthica* and *S. asiatica*. Weed Res 44:433–438

Elzein A, Kroschel J, Leth V (2006) Seed treatment technology: An attractive delivery system for controlling root parasitic weed Striga with mycoherbicide. Biocontrol Sci Technol 16(1):3–26

Emechebe AM, Ellis-Jones J, Schulz S, Chikoye D, Douthwaite B, Kurreh I, Tarawali IG, Hussaini MA, Kormawa P, Sanni A (2004) Farmers' perception of the *Striga* problem and its control in northern Nigeria. Exp Agric 40:215–232

Eplee RE (1975) Ethylene: a witchweed seed germination stimulant. Weed Sci 23:433–436

Eplee RE, Norris RS (1987) Chemical control of Striga. In: Musselman LJ (ed) Parasitic weeds in agriculture: vol. I, Striga. CRC Press, Boca Raton, Fl, pp 173–182

Fan S, Rue C (2020) The role of smallholder farms in a changing world. In: The role of smallholder farms in food and nutrition security. Springer International Publishing, Cham, pp 13–28

Fatondji D, Martius C, Vlek P (2001) Zaï - a traditional technique for land rehabilitation in Niger. ZEFnews 8:1–2

Freitag J, Hess DE, Welz HG (1996) Pearl millet and sorghum specific races of *Striga hermonthica* in Niger, pp 471–478. In: Monero M, Cubero JI, Berner D, Musselman IJ, Parker C (eds) Advances in parasitic plant research, sixth international parasitic weed symposium, April 16-18, Junta de Andalucia, 923 pp

Goldwasser Y, Rodenburg J (2013) Integrated agronomic management of parasitic weed seed banks. In: Parasitic Orobanchaceae: Parasitic mechanisms and control strategies. Springer Berlin Heidelberg, Berlin, Heidelberg, pp 393–413. Jun 9

Greathead DJ (1984) The natural enemies of Striga spp. and the prospects for their utilisation as biological control agents. In: Striga, biology and control, pp 133–160

Greathead DJ, Milner JE (1971) A survey of Striga spp.(Scrophulariaceae) and their insect natural enemies in East Africa with a discussion on the possibilities of biological control. Trop Agric 48 (2):111–124

Gressel J, Hanafi A, Head G, Marasas W, Obilana B, Ochanda J, Souissi T, Tzotzos G (2004) Major heretofore intractable biotic constraints to African food security that may be amenable to novel biotechnological solutions. Crop Prot 23:661–689

Hamrick JL (1982) Plant population genetics and evolution. Am J Bot 69:1685–1693

Haussmann BIG, Hess DE, Welz HG, Geiger HH (2000) Improved methodologies for breeding *Striga*-resistant sorghums. Field Crop Res 66:195–211

Haussmann BIG, Hess DE, Omanya GO, Folkertsma RT, Reddy BVS, Kayentao M, Welz HG, Geiger HH (2004) Genomic regions influencing resistance to the parasitic weed Striga hermonthica in two recombinant inbred populations of sorghum. Theor Appl Genet 109:1005–1016

Hearne SJ (2009) Control - the Striga conundrum. Pest Manag Sci 65:603–614

Hess DE, Dodo H (2004) Potential for sesame to contribute to integrated control of Striga hermonthica in the west African Sahel. Crop Prot 23:515–522

Hoffmann G, Diarra C, Dembele B (1997) Les espèces parasites herbacées des cultures vivriéres en Afrique : biologie, in Dact, méthodes de lutte. Agriculture et développement 13:30–51

Iwoke FIO, Okonkwo SNC (1974) Facultative hemi-parasitism in *Buchnera hispida* Harm. Ex D. Don. Ann Bot 38:993–1002

Joel DM (2000) The long-term approach to parasitic weeds control: manipulation of specific developmental mechanisms of the parasite. Crop Prot 19:753–758

Kanampiu FK, Kabambe V, Massawe C, Jasi L, Friesen D, Ransom JK, Gressel J (2003) Multi-site, multi-season field tests demonstrate that herbicide seed-coating herbicide-resistance maize controls *Striga* spp. and increases yields in several African countries. Crop Prot 22:679–706

Khan ZR, Midega CAO, Amudavi DM, Hassanali A, Pickett JA (2008) On-farm evaluation of the 'push–pull' technology for the control of stemborers and *Striga* weed on maize in western Kenya. Field Crop Res 106:224–233

Kim SK, Adetimirin VO (1997) Responses of tolerant and susceptible maize varieties to timing and rate of nitrogen under Striga hermonthica infestation. Agron J 89:38–44

Kim SK, Akintunde AY, Walker P (1994) Responses of maize, sorghum and millet host plants to infestation by *Striga hermonthica*. Crop Prot 13(8):559–567

Koch W, Beshir ME, Unterladstatter R (1982) Crop losses due to weeds. Improving weed management. FAO plant production and protection paper, Rome 6-10 September, 1982, vol 44, pp 153–165

Kountche BA, Hash CT, Dodo H, Laoualy O, Sanogo MD, Timbeli A, Vigouroux Y, This D, Nijkamp R, Haussmann BI (2013) Development of a pearl millet *Striga*-resistant genepool: response to five cycles of recurrent selection under *Striga*-infested field conditions in West Africa. Field Crop Res 154:82–90

Kountche BA, Al-Babili S, Haussmann BIG (2016) *Striga*: a persistent problem on millets. In: Biotic stress resistance in millets. Elsevier, Amsterdam, pp 173–203

Kountche BA, Jamil M, Yonli D, Nikiema MP, Blanco-Ania D, Asami T, Zwanenburg B, Al-Babili S (2019) Suicidal germination as a control strategy for *Striga hermonthica* (Del.) Benth. in smallholder farms of sub-Saharan Africa. Plants People Planet 1:107–118

Koyama ML (2000) Molecular markers for studying pathogen variability: implications for breeding for resistance to Striga hermonthica. In: Haussmann BIG, Hess DE, Koyama ML, Grivet L, Rattunde HFW, Geiger HH (eds) Breeding for Striga resistance in cereals. Margraf Verlag, Weikersheim, Germany, pp 227–254

Kuchinda NC, Kureh I, Tarfa BD, Shinggu C, Omolehin R (2003) On-farm evaluation of improved maize varieties intercropped with some legumes in the control of *Striga* in the northern Guinea Savanna of Nigeria. Crop Prot 22:533–538

Li J, Timko PJ (2009) Gene-for-gene resistance in *Striga*-cowpea associations. Science 325:1094
Lybbert T, Sumner D (2010) Agricultural technologies for climate change mitigation and adaptation in developing countries: policy options for innovation and technology diffusion. ICTSD/Food & Agricultural Trade, issue brief no. 6, 32 pp
Marley PS, Ahmed SM, Shebayan JAY, Lagoke STO (1999) Isolation of *Fusarium oxysporum* with potential for biocontrol of the witchweed (*Striga hermonthica*) in the Nigerian savanna. Biocontrol Sci Tech 9:159–163
Marley PS, Aba DA, Shebayan JAY, Musa R, Sanni A (2004) Integrated management of Striga hermonthica in sorghum using a mycoherbicide and host plant resistance in the Nigerian Sudano-Sahelian savanna. Weed Res 44:157–162
Marley PS, Kroschel J, Elzein A (2005) Host specificity of *Fusarium oxysporum* Schlecht (isolate PSM 197), a potential mycoherbicide for controlling *Striga* spp. in West Africa. Weed Res 45: 407–412
Marnotte P (1995) Utilisation des herbicides : contraintes et perspectives. Agriculture et Développement 7:12–21
Mason SC, Nouri MN, Pale S (2015) Pearl millet production practices in semi-arid West Africa: a review. Exp Agric 51(4):501–521
Mishra JS, Kumar R, Upadhyya PK, Hans H (2018) Weed management in millets. Indian Farm 68(11):77–79
Mohamed KI, Bolin JF, Musselman LJ, Peterson AT (2007) Genetic diversity of Striga and implications for control and modeling future distributions. In: Integrating new technologies for striga control: towards ending the witch-hunt. World Scientific Publishing, Singapore, pp 71–84
Nathaniels NQR (2005) Cowpea, farmer field schools and farmer-to-farmer extension: a Benin case study. AgREN Network Paper 148, p 15
Nicou R, Charreau C (1985) Soil tillage and water conservation in semi-arid West Africa. In: Ohm HW, Magy JG (eds) Appropriate technologies for farmers in semi-arid West Africa. Purdue University, West Lafayette, pp 9–32
Obilana AT, Ramaiah KV (1992) Striga (witchweeds) in sorghum and millet: knowledge and future research needs. In: de Milliano WAJ, Frederiksen RA, Bengston GD (eds) Sorghum and millets diseases: a second world review. International Crops Research Institute for the Semi-Arid Tropics (CP 741), Patancheru, pp 187–201
Oswald A (2005) *Striga* control – technologies and their dissemination. Crop Prot 24:333–342
Padgham J (2009) Agricultural development under a changing climate: opportunities and challenges for adaptation. Agriculture and Rural Development & Environment Departments, Joint Departmental discussion paper-issue 1, 168 pp
Parker C (1984) The influence of Striga species on sorghum under varying nitrogen fertilization. In: Proceedings of the Third International Symposium on Parasitic Weeds. International Center for Agricultural Research in the Dry Areas, pp 90–98
Parker C (2009) Observations on the current status of *Orobanche* and *Striga* problems worldwide. Pest Manag Sci 65:453–459
Parker C (2012) Parasitic weeds: a world challenge. Weed Sci 60:269–276
Parker C, Riches CR (1993) Parasitic weeds of the world: biology and control. Caslefield Press, Kettering. 332 pp
Parkinson V, Efron Y, Bello L, Dashiel K (1987) Trap crops as a cultural measure in Striga control in Africa. FAO Plant Prot Bull 35:51–54
Pennisi E (2010) Armed and dangerous. Science 327:804–805
Pennisi E (2015) How crop-killing witchweed senses its victims, vol 350. American Association for the Advancement of Science, New York, p 146
Pieterse AH (1991) The effect of nitrogen on the germination of seeds of Striga hermonthica and Orobanche crenata. In: Progress in Orobanche research. Proceedings of the International Workshop on Orobanche Research, Obermarchtal, Germany, pp 115–124

Ramaiah KV (1984) Physiological specialization of *Striga hermonthica* and crop speficity. In: Third international symposium on parasitic weeds, Allepo, Syria, pp 58–65

Rodenburg J, Bastiaans L, Weltzien E, Hess DE (2005) How can field selection for *Striga* resistance and tolerance in sorghum be improved? Field Crop Res 93:34–50

Rouamba A, Shimelis H, Drabo I, Laing M, Gangashetty P, Mathew I, Mrema E, Shayanowako AIT (2021) Constraints to pearl millet (Pennisetum glaucum) production and farmers' approaches to Striga hermonthica management in Burkina Faso. Sustainability 13:8460

Samejima H, Babiker AG, Takikawa H, Sasaki M, Sugimoto Y (2016) Practicality of the suicidal germination approach for controlling *Striga hermonthica*. Pest Manag Sci 72:2035–2042

Showemimo FA, Kimbeng CA, Alabi SO (2002) Genotype response of sorghum cultivars to nitrogen fertilization in the control of *Striga hermonthica*. Crop Prot 21:867–870

Tank DC, Beardsley PM, Kelchner SA, Olmstead RG (2006) Review of the systematics of Scrophulariaceae and their current disposition. Aust Syst Bot 19:289–307

Teka HB (2014) Advance research on *Striga* control: a review. Afr J Plant Sci 8:492–506

Traore H, Maillet J (1992) Flore adventice des cultures céréalières annuelles du Burkina Faso. Weed Res 32:279–293

Traoré H, Yonli D (1999) Study of farmers' perception of the *Striga* problem and inventory of indigenous *Striga* control methods in Burkina Faso. In: Kroschel J, Mercier-Quarshie H, Sauerborn J, Margra Verlag W (eds) Advances in parasitic weed control at on-farm level, vol I. Joint action to control *Striga* in Africa, Germany, pp 197–205

Traore D, Vincent C, Stewart RK (1996) Association and synchrony of Smicronys guineanus and Smicronys umbrinus (Coleoptera: Curculionidea) and the parasitic weed Striga hermonthica (Del.) Benth. (Scrophulariaceae). Biol Control 3:307–315

Traore H, Ouedraogo O, Salle G (2000) Contrôle du Striga hermonthica (Del.) Benth.par la combinaison de la variété, du 2,4-D ou de l'arrachage au Burkina Faso. Etudes et recherches sahéliennes N° 4–5:86–93

Traore H, Yonli D, Diallo D, Sereme P (2011) Suicidal germination of Striga hermonthica (Del.) Benth. by cotton, cowpea and groundnut Genotypes in Burkina Faso. Int J Agric Res 6(1):49–57

Tsanuo MK, Hassanali A, Hooper AM, Khan Z, Kaberia F, Pickett JA, Wadhams LJ (2003) Isoflavanones from the allelopathic aqueous root exudate of Desmodium uncinatum. Phytochemistry 64(1):265–273

Van de Fliert E (1993) Integrated pest management: farmer field schools generate sustainable practices. Wageningen Agricultural University Papers, Wageningen, The Netherlands, pp 3–93

Westwood JH, dePamphilis CW, Das M, Fernandez-Aparicio M, Honaas LA, Timko MP, Wafula EK, Wickett NJ, Yoder JI (2012) The parasitic plant genome project: new tools for understanding the biology of *Orobanche* and *Striga*. Weed Sci 60:295–306

Williams CN, Caswell GH (1959) An insect attacking Striga. Nature 184:1668–1668

Wilson JP, Hess DE, Hanna WW (2000) Resistance to *Striga hermonthica* in wild accessions of the primary gene pool of *Pennisetum glaucum*. Phytopathology 90:1169–1172

Wilson JP, Hess DE, Hanna WW, Kumar KA, Guptae SC (2004) *Pennisetum glaucum* subsp. monodii accessions with *Striga* resistance in West Africa. Crop Prot 23:865–870

Yoder JI, Scholes JD (2010) Host plant resistance to parasitic weeds; recent progress and bottlenecks. Curr Opin Plant Biol 13:478–484

Yonli D, Traore H, Hess DE, Sankara P, Sereme P (2006) Effect of growth medium, Striga seed burial distance and depth on efficacy of Fusarium isolates to control *Striga hermonthica* in Burkina Faso. Weed Res 46:73–81

Zahran E, Kohlschmid E, Sauerborn J, Abbasher AA, Muller-Stover D (2008) Does an alication as seed coating stabilize the efficacy of biological control agents of *Striga hermonthica* under field conditions? J Plant Dis Prot S 21:467–471

Zwanenburg B, Mwakaboko AS, Kannan C (2016) Suicidal germination for parasitic weed control. Pest Manag Sci 72:2016–2025

Crop Simulation Models for Climate Change Adaptation in Pearl Millet

16

R. Swarna, S. Srividhya, C. Dheeraj, B. Bhargavi, C. Deepika, and B. Bhaskar

Abstract

Globally, climate change has a negative impact on food security, particularly in developing nations where a growing population is battling food and nutritional insecurity. It is anticipated that there will be a 10% increase of dry land areas worldwide, along with increased climate variability and extreme weather events (drought and heat stress). Pearl millet is a resilient and climate-smart nutricereal, ideal for areas vulnerable to drought and heat stress. It is a water saving, drought tolerant, and climate change complaint crop. In spite of these unique features, the yields of pearl millet are very low due to lack of optimum production practices under changing climate conditions. A better understanding on the impact of climate change on crop productivity and possible adaptation strategies to mitigate these effects is needed to increase its productivity. Many pearl millet crop simulation models are in use to simulate crop growth and development in response to management decisions. However, recent advances in modeling enabled to simulate the effect of climate change on crop yields and predict adaptation strategies to mitigate these adverse effects. These models assist the researchers, stakeholders. and policy makers to address and adapt to the current and future climate changes.

R. Swarna (✉) · S. Srividhya · C. Dheeraj · C. Deepika
ICAR - Indian Institute of Millets Research, Hyderabad, India
e-mail: swarna@millets.res.in

B. Bhargavi
ICAR – Central Research Institute for Dry land Agriculture, Hyderabad, India

B. Bhaskar
ICAR - National Bureau of Plant Genetic Resources, Hyderabad, India

V. A. Tonapi et al. (eds.), *Pearl Millet in the 21st Century*,
https://doi.org/10.1007/978-981-99-5890-0_16

Keywords

Pearl millet · Crop model · Climate change · Adaptation · Productivity

16.1 Introduction

Indian agriculture is greatly impacted by global climate change, which is already underway as seen by the increased frequency of climate change-related events. According to the Intergovernmental Panel on Climate Change (IPCC), the temperature increase by the end of this century is anticipated to be between 1.8 and 4.0 °C. Depending on the scenario of future development, the IPCC forecasts a temperature rise of 0.88 to 3.16 °C by 2050 and 1.56 to 5.44 °C by 2080 for the Indian region (South Asia) (IPCC 2007). Despite the positive impacts of higher CO_2 concentration on a number of crops, accompanying temperature rise and increased rainfall variability would have a significant impact on food production. It is estimated that India's crop production could decline by 10–40% by 2100 AD (IPCC 2007). The dynamics of diseases and pests will change substantially. Reduction in the productivity of crops signifies the main threat to food security, amidst the burgeoning world's population. The population is supposed to grow to about 9 billion in 2050 and food requirement are expected to escalate by about 85% (Fig. 16.1).

Indian agriculture mainly depends on monsoon as lion's share of cropped area is under rainfed farming. For good crop production, a successful monsoon, or an adequate volume and even distribution of rainfall along with optimum humidity

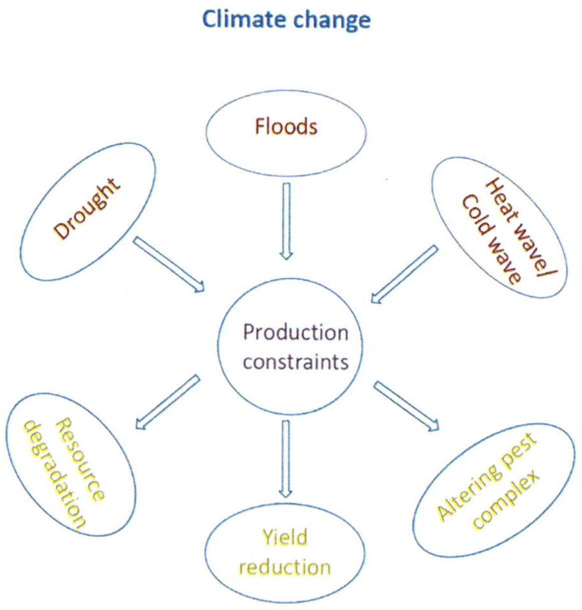

Fig. 16.1 Impact of climate change on crops

and temperature, is always critical. The monsoon is already influenced by a number of factors, such as the occurrence of El Nino and La Nina, deforestation, pollution, and other anthropogenic activities, but now, global climate change is an additional component that is causing Indian agriculture to face abnormal weather conditions. In this context, pearl millet assumes greater importance due to its unique features of drought tolerance and resilience to adverse environmental conditions. It is an important staple food crop grown in arid and semi-arid parts of Asia and Africa. In India, it is mostly cultivated in marginal soils under rainfed conditions and occupies an area of 6.93 million hectare with a total production of 8.61 million tones and productivity of 1243 kg ha^{-1} (Directorate of Millet Development 2020). High temperatures and low and high erratic rainfall are the major constraints for millet production. The optimum temperatures for pearl millet growth are 33 °C day and 28 °C night. Rise in temperature above 33 °C/28 °C decreases the grain yield by reducing the length of growing period and decreases the grain yield (Ahmed et al. 2019). The future projections of climate change showed that crops yield would be reduced due to the rise in temperature in arid and semi-arid regions. Increase in CO_2 concentration has a beneficial effect in growth and yield of the crop. High concentration of CO_2 partially minimized the negative impacts of rising temperature (Mitchell et al. 1995). The response of climate indicators such as temperatures, precipitation, and CO_2 on crops can be assessed using crop models. There is dire need to develop suitable adaptations strategies and assess the impact of climate change impacts on crops.

Crop simulation models are convenient tools for the assessment of impact of climate change on crops and suggesting suitable climate adaptation strategies (Hoogenboom et al. 2019). They simulate the crop development, growth, and yield by the interaction of soil, plant, and atmospheric variables such as temperature and precipitation (Jones et al. 2003). Various crop models have been used for assessing the impact of climate change on pearl millet. This book chapter highlights the importance of crop models in climate change assessment and various adaptation strategies for increasing the productivity of pearl millet under future climate related risks.

16.2 Pearl Millet Responses to Climate Change

Climate change is one of the biggest challenges to the world in present times. It is defined as significant changes in the average values of meteorological elements, such as precipitation and temperature, for which averages have been computed over a long period (World Meteorological Organization 1992). The world's food and nutritional security is threatened by climate change. The greenhouse effect causes the temperature to rise when greenhouse gas emissions in the atmosphere rise. The average world temperature is expected to climb by 2 °C by 2100, which would result in significant global economic losses. The concentration of CO_2, which makes up a significant portion of greenhouse gases, has increased plant growth and productivity because of increased photosynthesis. However, rising temperatures also result in higher pest infestation, altered weed flora, shorter crop duration, and increased crop respiration and evapotranspiration rates. The past few decades indicate that significant changes in climate at a global level were the result of enhanced human activities that altered the composition of the global atmosphere (IPCC 2007). Since 1750, there has been a 150%, 40%, and 20% increase in the concentration of greenhouse gases such as methane (CH_4), carbon dioxide (CO_2), and nitrous oxide (N_2O), correspondingly (IPCC 2014). The amount of carbon dioxide emissions, which make up the majority of greenhouse gases, increased from 22.15 billion metric tons in 1990 to 36.14 billion metric tons in 2014.

The increase in greenhouse gases has an impact on the atmosphere's growing temperature. These infrared-active gases, primarily carbon dioxide (CO_2), ozone (O_3), and water vapor (H_2O), absorb thermal radiations emitted by the earth's atmosphere and surface, warming the planet and causing the greenhouse effect. However, as a result of global warming, extremes in temperature and precipitation are more likely to be seen in the near future. Extreme precipitation phenomena, either heavy rainfall or drought, are dependent on a region's geography. The increased intensity and frequency of precipitation also impacts soil erosion. Anomalies in precipitation have a negative impact on agriculture, especially in underdeveloped countries. It has a big impact on farmland acreage in addition to crop production. Evidence points to dry anomalies as the cause of the about 9% rate of agricultural expansion in the developing world over the past 20 years, as farmers increase the acreage to make up for production losses (Zaveri et al. 2020). The rise in temperature is likely to reduce the yield of many crops by reducing their duration. The aggregate production of wheat, rice, and maize is expected to decrease if both the temperate and tropical regions experience a warming of 2 °C (Challinor et al. 2014). Climate change in general has more impact on tropical regions, as tropical crops remain closer to their high-temperature optima, and thereby experience high-temperature stress during elevated levels of temperature. Climate change has been predicted to cause more frequent and severe droughts in the near future across the majority of the world's regions, with an increase in the drought affected areas from 15.4 to 44.0% by 2100.

Major crops' yields in drought-affected areas are predicted to drop by more than 50% by 2050 and by over 90% by 2100. Pearl millet [Pennisetum glaucum (L.) R. Br.] is a primary food grain and fodder crop cultivated on more than 31.2 million ha (FAO 2019) throughout the dry and semi-arid tropical regions of Asia, Africa, and Latin America. Pearl millet is well adapted to areas with high temperatures, low soil fertility, and drought.

Effective energy production mechanism (C_4 photosynthesis) has helped pearl millet in rapid generation of biomass and adapt to suboptimal and semi-arid all over the world. It plays a prominent role in the integrated agricultural and livestock particularly in rainfed areas due to its drought hardiness and tolerance to high temperature (Arya et al. 2010). India is the single largest producer of pearl millet in the world both in terms of area (6.93 million ha) and production (8.61 Tg) (Directorate of Millet Development 2020).

Pearl millet cultivation in arid and semi-arid tropical regions is likely to be affected by changing climatic factors, such as reduced precipitation, prolonged dry spells, and intense heat waves. Recently, pearl millet cultivation is being extended during the summer (February–May) season under irrigated condition in parts of Gujarat, Rajasthan, and Uttar Pradesh. In these areas, the maximum air temperatures vary from 40 to 48 °C, and sometimes it may exceed 50 °C during the month of May. This high temperature usually coincides with flowering to grain-filling stage of the crop and has adverse affects on the crop yield potential. The effects of terminal drought stress associated with heat stress have induced yield losses of between 55 and 67% (Krishnamurthy et al. 2011). India's production of pearl millet has decreased by 6.2% as a result of drought stress (Priyaadharshini et al. 2019). Though pearl millet is a hardy crop, the rapidly changing climate will have adverse effects on its production which signifies the need to understand the impact of abiotic stress on pearl millet. Various morphological and physiological responses of pearl millet to changing climate induced heat and drought stress are discussed below.

16.2.1 Responses of Pearl Millet to Heat and Drought Stress

16.2.1.1 Morphological and Phenological Responses

Excessive radiation and high temperatures in tropical areas are significant barriers to plant growth and development. High temperatures may cause scorching of the twigs and leaves along with visual symptoms of sunburn, leaves senescence, growth inhibition, and discoloration of leaves. Elevated temperatures can reduce the germination potential of the seeds and, thus, results in poor germination and stand establishment in sorghum (Tandzi et al. 2019). Adverse effects of high temperature vary with the timing, duration, and severity of the heat stress (Fahad et al. 2016). High temperature stress reduced number of spikes and number of florets per plant in pearl millet (Djanaguiraman et al. 2018). Seed set in sorghum was also negatively affected under high temperatures (Fahad et al. 2016). In pearl millet, reproductive stages of plants were relatively more sensitive to higher temperatures compared to vegetative stages (Prasad 2013). Inside the floret, anthers and pollens were more susceptible to heat stress than ovules. High temperatures induce floret sterility with reduced anther dehiscence, poor shedding of pollens, poor germination of pollen grains on the stigma, decreased elongation of pollen tubes, and reduced pollen germination. High temperature during flowering stage decreases pollen germination, pollen tube growth, and decreases fertilization in pearl millet (Rao and Patil 2015).

Growth and net assimilation rate were adversely affected in pearl millet under heat and drought stress (Debieu et al. 2018). The initial effect of drought on the plants is the poor germination and impaired seedling establishment. Zhang et al. (2021) reported negative impacts of drought stress on germination, seedling growth, root and shoot dry weight, hypocotyl length, and vegetative growth in pearl millet. Drought limits the process of cell growth mainly due to the loss of turgor (Taiz and Zeiger 2006). Water limited conditions result in impaired cell elongation because of the poor water flow from xylem to the nearby cells (Asthir 2015). Reduced turgor pressure and slow rate of photosynthesis under drought conditions mainly limit the leaf expansion. Begg and Burton (1971) observed that most of the pearl millet genotypes reached anthesis sooner at the higher temperatures which shortened the grain filling duration leading to yield reduction. Though pearl millet is relatively heat tolerant than other major cereal crops, very high temperature (≥40 °C) at reproductive stages (boot-leaf stage and stigma emergence–anthesis stage) shows detrimental effects that may severely impact seed set and grain yield as a whole (Gupta et al. 2015).

16.2.1.2 Water and Nutrient Relations

Plant water status is of pivotal importance under changing temperature conditions. Temperature increase proves lethal under limited supply of water (Machado and Paulsen 2001). Unfortunately, heat stress commonly coincides with the water scarcity under field condition. Leaf tissue water contents diminish rapidly on exposure to high temperatures. Heat stress also has negative impact on the root conductance. Generally, water loss under heat stress is more during day time mainly due to increased rate of transpiration, ultimately impairing certain important physiological processes in millets (Ajithkumar and Panneerselvam 2014). Heat stress also reduces the number, mass, and growth of the roots which ultimately limit the supply of water and nutrients to the above ground parts of the plant. Activity of the major enzymes like nitrate reductase involved in the nutrient metabolism can also be significantly reduced under high temperature stress.

Drought stress disturbs all these factors of water relations in plants; however, stomatal conductance is the most affected (Farooq et al. 2009). A significant reduction in the leaf water potential and transpiration rate was observed under the drought conditions which ultimately raised the leaf and canopy temperature in pearl millet (Kholova et al. 2010a, b). Another important physiological regulation which gets adversely affected under water stress is water use efficiency which is the ratio of the dry matter accumulated to the water consumed. Efficient cultivars of pearl millet higher water use efficiency under drought and heat stress. Maintaining higher leaf relative water content was reported to contribute drought tolerance in pearl millet cultivars by Meena et al. (2017).

Drought stress greatly impacts the nutrient relations of the plants. Many important nutrients including, nitrogen, silicon, magnesium, and calcium are up taken by roots along with water. The drought conditions limit the movement of these nutrients via diffusion and mass which leads to retarded plant growth. Plants increase the length and surface area of roots and change their architecture in order to capture the less mobile nutrients (Lynch and Brown 2001). The soil moisture deficit at times reduces the growth of the roots and, hence, reducing the uptake of the less mobile nutrients such as phosphorus (Garg 2003). Root–microbe interactions also play an important role in nutrient relations of a plant. The composition and activity of soil microbial colonies are negatively affected by soil water deficit which eventually disturbs the plant nutrient relations (Schimel et al. 2007). Mostly under drought conditions, N uptake is increased, P uptake is declined, and potassium remains unaffected. Overall, drought and heat stresses affect nutrient cycling, uptake and availability to plants by hampering different physiological functions of plants. However, nutrient relations become more complicated due to interactive effects of different nutrients on each other and overall plant physiology.

16.2.1.3 Photosynthesis

One of the key physiological phenomena affected by the drought and heat stress in plants is photosynthesis (Farooq et al. 2009). Reduced leaf expansion, incorrect operation of the photosynthetic apparatus, and leaf ageing are the key factors affecting photosynthesis. The main sources of damage as a result of the high temperature stress are the C metabolism occurring in the stroma and the light-dependent chemical processes occurring in the thylakoid. Increased leaf temperature and photon flux density have an impact on thermo-tolerance adjustment of the PSII. Under higher temperatures, the oxygen evolving complex is also vulnerable to significant damage, which could lead to an unbalanced flow of electrons to the PSII acceptor site. The D_1 and D_2 proteins are also subjected to denaturation under higher temperature. Photosynthesis is limited due to disruption in electron transport chain and the reduced RuBP regeneration capacity (De Ronde et al. 2004).

Arya et al. (2014) reported photosynthesis rate maximizing up to temperatures of above 35^0C and after that it inhibited normal growth in pearl millet. Exposure to high temperature usually results in reduction in chlorophyll biosynthesis (Dutta et al. 2009). Reduced accumulation of the chlorophyll in the plants may be due to either decreased biosynthesis of the chlorophyll or due to its increased degradation or combined effect of both under high temperature stress. The chlorophyll biosynthesis inhibition under high temperature stress is actually attributed to the deactivation of various enzymes like activity of 5-aminolevulinate dehydratase, an important enzyme in the pyrrole biosynthesis pathway (Mohanty et al. 2006). Meena et al. (2021) reported that heat stress reduced the Soil Plant Analysis Development (SPAD) values of pearl millet genotypes indicating loss and degradation of chlorophyll.

Adenosine diphosphate-glucose pyrophosphorylase, sucrose phosphate synthase, and invertase are a few of the crucial enzymes whose activity is reduced under high temperature stress, which has a significant impact on the starch and sucrose production. Many plant species experience a decline in net photosynthesis due to reduction in the activation state of the CO_2 binding enzyme Rubisco. Although the catalytic activity of Rubisco increases with temperature, photorespiration is increased and the net photosynthetic rate is constrained by its poor affinity toward CO_2 and ability to interact with O_2 (Crafts-Brandner and Salvucci 2002).

The first and foremost response of almost all the plants to moisture stress is stomatal closure in order to avoid the water loss through transpiration. Stomatal closure checks CO_2 intake which leads to oxidative damage and inhibition of C assimilation. Impaired functioning of the Rubisco becomes the major factor affecting photosynthesis (Bota et al. 2004). The activity of the enzymes involved in photosynthesis and other plant processes may be severely impacted by ion toxicity as a result of increased solute levels in the cytoplasm. Biosynthesis of the Rubisco is reduced due to the decrease in its small subunits. Rubisco is rendered inactive when inhibitors like 2-carboxyaribinitol 1-phosphate bind to the enzyme's catalytic site. According to reports, the primary reasons limiting photosynthesis during drought are decreased phosphorylation and poor ATP generation.

Under drought conditions, maintaining higher chlorophyll content was observed as drought tolerance trait in pearl millet genotypes (Meena et al. 2021). The ratio between variable fluorescence (F_v) and maximum fluorescence (F_m), or F_v/F_m, reflects the maximum quantum efficiency of PSII and is one of the most heats affected parameters. A decline in F_v/F_m is frequently observed when plants are subjected to abiotic stress, including heat (Sharma et al. 2015).

16.2.1.4 Assimilate Partitioning

Under heat stress, source and sink activities are reduced, which have a significant impact on growth and, ultimately, economic yield. Improvement in the mobilization efficiency of assimilates from leaves and other plant parts can be a key strategy for improving grain filling. The balance of assimilates is disturbed by drought as most of them are moved to the roots to increase water uptake. The export of assimilates from source to sink generally depends upon the rate of photosynthesis and the sucrose concentration in leaves. Drought impairs the process of photosynthesis and decreases the sucrose content which ultimately reduce the rate of export from source to sink (Kim et al. 2000). Drought also limits the ability of the sink to utilize the incoming assimilates efficiently. Moreover, the activity of acid invertase is negatively affected under moisture stress which disrupts the phloem loading and unloading. Connection between grain development and transfer of assimilates from the leaves has been reported as the central adaptation mechanism in pearl millet to terminal drought stress (Winkel and Do 1992). Yadav et al. (2004) reported that in pearl millet, co-mapping of the harvest index and panicle harvest index with grain yield revealed better drought tolerance by greater partitioning of dry matter from stover to grain.

16.2.1.5 Oxidative Stress and Membrane Damage

Oxidative damage is usually a subsequent stage of most of the abiotic stresses in plants. Exposure of plants to drought stresses initially causes oxidative damage by the formation of reactive oxygen species (ROS). These ROS pose serious threat to the cell functioning by damaging lipids and proteins. The ROS are mainly produced in the chloroplast; however, reaction of oxygen with the components of electron transport chain in mitochondria also results in the generation of ROS.

Drought and heat stress occurring independently or combined have detrimental effects on plasma membranes and lead to their destabilization in various plants (ElBasyoni et al. 2017). Under drought and heat conditions, tolerant sorghum genotypes are able to retain cell membrane stability (CMS), decrease, and delay lipid peroxidation (Nxele et al. 2017). Cell membrane stability has been adopted as a tool in screening for drought and heat resistance in sorghum genotypes (Hemantaranjan et al. 2014). Due to their high thermostability, dehydrins and late embryogenesis abundant (LEA) proteins, which are produced in response to heat and drought stress, are crucial for stabilizing cell membrane structures in resistant sorghum genotypes. Increased malondialdehyde (MDA) content represents membrane peroxidation and drought tolerant pearl millet genotypes were screened and selected based on lower MDA contents (Meena et al. 2017).

16.2.1.6 Yield

Yield is basically the intricate integration of the various physiological processes. High temperature shocks during the reproductive phase can cause substantial reduction in the yield. Heat stress has a severe impact on grain quality as it reduces the starch and protein contents substantially. Gupta et al. (2015) reported a significant decline in the grain weight and total number of grains in pearl millet under elevated temperatures.

Physiological processes are negatively affected by the drought stress. The negative impacts of drought on the yield mainly depend upon the severity of the stress and the plant growth stage. Significant yield losses have been reported in major field crops due to drought stress. The drought induced at the pre anthesis stage shortened the time to anthesis while after anthesis reduced the period of grain filling. The process of the grain filling is controlled by four major enzymes, i.e., sucrose synthase, starch synthase, starch branching enzyme, and adenosine diphosphate glucose pyrophosphorylase (Taiz and Zeiger 2006). A decreased activity of these enzymes has been reported under the drought conditions which have a negative impact on the yield of major cereal crops. The exposure of plants to drought stress at the flowering may result in complete sterility in pearl millet, which is usually due to the disturbed assimilate movement to the developing ear (Yadav et al. 2004). Drought-induced reduction in the yield might be due to various factors such as decreased rate of photosynthesis, disturbed assimilate partitioning (Farooq et al. 2009), or poor flag leaf development, or floret abortion. Arya et al. (2010) reported stress at early stages of plant growth drastically reduced number of tillers and also stimulated early flowering, resulting in reduction in plant's normal size, which ultimately produced small ear heads, thus causing drastic reduction in grain yield along with fodder yield in pearl millet (Fig. 16.2).

16.2.2 Mechanisms Associated with Stress Tolerance

Plants are constantly under pressure from environmental stresses, and they tolerate or resist stress by various adaption and acclimation mechanisms. Plants have evolved complex physio-biochemical and molecular strategies to neutralize the effects of abiotic stress.

16.2.2.1 Adaptations to Drought Stress

Plants modify their phenology, morphology, and physiology at the cellular and molecular levels to adapt to drought. The drought-induced inhibition of growth as well as yield reductions can be attributed to adverse effects on plant functions and processes, particularly plant water uptake, water use efficiency, and the partitioning of biomass to grain.

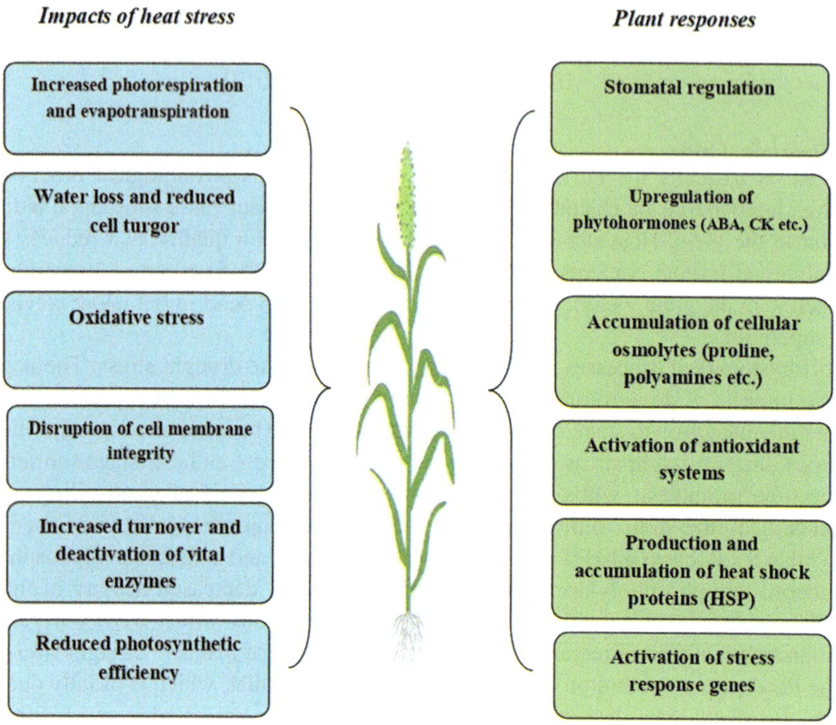

Fig. 16.2 Plant responses to heat stress

Escape Mechanisms

Plants typically adapt to droughts by shortening crop development cycles or speeding up phenological phases, a strategy known as phenological flexibility (Richards 2006). This feature has been utilized to develop short-duration drought-tolerant crop cultivars. However, severe and prolonged droughts can significantly lower the grain yields of crops drastically. In plants that utilize escape mechanisms, seed or pollen germination usually occurs before acute water shortage. However, plants with growth plasticity seem to grow slower in the dry season with few flowers but have more seeds in the normal season. Early flowering is an important attribute of pearl millet for drought escape and a key trait for the breeding process (Vadez et al. 2012).

Dehydration Avoidance

The plant adjusts to droughts by lowering water loss and getting more water through root uptake. Adaptive traits are used to set the background of a "low transpiration rate in water-saving plants" and an "osmotic adjustment in water-spender plants" to prevent dehydration. Plants mostly rely on leaf relative water content, osmotic adjustments, and root architecture to enhance the yield under drought stress (Terletskaya et al. 2020).

One way to reduce water loss through transpiration is to reduce the size of the leaves. Sequential water shortages have been shown to cause leaf shedding, which is observed to occur from the oldest to the youngest leaves. Stomatal closure also indicates drought tolerance because in response to drought stress, stomata are closed, and transpiration is decreased. Water loss is significantly influenced by the stomatal movement, stomatal density, and the resistance of plants to transpiration. Remarkable variations in the stomatal functions of various plants during droughts have been identified (Muktadir et al. 2020). Adaptations to droughts can reduce the stomata size or the number of stomata. Drought-tolerant plants exhibit adaptive root properties, including long roots, high densities of roots, and intense rooting systems. Plants selectively produce and extend their roots towards the wet part of the rhizosphere due to specific genes associated with this process. Denser roots can absorb greater amounts of water because of the more extensive root system. In pearl millet, water-saving mechanisms like transpiration efficiency, restricted transpiration during high atmospheric demand, and leaf expansion at a lower threshold of soil moisture can sustain photosynthesis for continued carbon supply to the grains during the critical period (Vadez et al. 2013; Choudhary et al. 2019).

When irrigation was withheld in the late vegetative stage, the drought-adapted pearl millet genotype PRLT-2/89-33 showed less water extraction before flowering and conserved soil moisture was hypothesized to support grain filling, whereas the drought-sensitive genotype H77/833-2 extracted more water before flowering, and depleted soil moisture could not support grain filling (Tharanya et al. 2017).

Limited transpiration under increasing atmospheric vapor pressure deficit (VPD) was identified as one of the prominent water-saving mechanisms in pearl millet and genotypic variations were reported, which helped in breeding stress tolerance (Kholova et al. 2010b). The limited transpiration trait in pearl millet operates by hydraulic regulation and partial closure of stomata contributing to water conservation. Drought-adapted genotypes have lower transpiration rates at high atmospheric evaporative demand than terminal drought-sensitive lines (Kholova et al. 2010a).

Osmoregulation

The cellular dehydration of tissues occurs when plants are exposed to extreme temperatures and droughts. The plant cell produces osmolytes such as sugars, proteins, nucleic acids, and amino acids to protect from dehydration (Sattar et al. 2020). An osmotic adjustment is the process of solute accumulation mechanisms in plant cells when the water potential is limited, which helps maintain the turgor. The accumulation of osmotic substances is controlled by the intricate cellular processes involved in water flux and osmotic adjustment during abiotic stress conditions. Sugars are critical biomolecules involved in various crucial physio-chemical mechanisms, from seed germination to senescence in cereals and grain legumes. Sugars have variety of functions such as osmolyte biosynthesis, maintaining membrane integrity, growth, and differentiation. Proline is a compatible osmolyte that guards the cellular machinery from oxidative damage and preserves the equilibrium of photosynthesis. Meena et al. (2017) reported that increased accumulation of proline imparted drought tolerance in pearl millet genotypes. Amino acid-derived compounds such as Glycine, Betaine and polyamines also contribute to abiotic stress tolerance in various plants, including rice. The accumulation of these osmolytes in the cytosol is an essential stress response to adjust the osmotic equilibrium in the plants under abiotic stress.

Antioxidant System

Reactive oxygen species (ROS) such as singlet oxygen, hydrogen peroxide, superoxide, and hydroxyl radicals are involved in numerous cellular functions. When plants are exposed to stress, ROS which exist at a low level under normal conditions tend to increase. At high levels, ROS are toxic to cells, while the same molecule at low concentrations can function as a signal transducer that activates a local and systemic plant defense response against stress (Nadarajah 2020). After being exposed to any stress, chloroplasts, peroxisomes, endoplasmic reticulum (ER), mitochondria, and apoplasts may rapidly produce ROS, which are harmful to the plant if not reduced or scavenged. Rezayian et al. (2020) reported that NO stimulates the antioxidant system and osmotic adjustment in the soybean under drought stress. According to Biswas et al. (2019), ROS and reactive carbonyl species are a feed-forward loop in auxin signaling and are crucial for the development of lateral roots.

The plant possesses antioxidant machinery with the enzymatic and non-enzymatic components to mediate redox signaling and ROS homeostasis linked to acclimation responses to abiotic stressors. In response to stress, several antioxidative enzymes are produced by the plant. Superoxide dismutase (SOD), catalase (CAT), and peroxidases (POX) are among the enzymatic components of antioxidant systems that regulate the homeostasis of ROS within organisms, as reported in wheat, sorghum, and pearl millet (Djanaguiraman et al. 2018). Ascorbic acids, tocopherol, flavonoids, glutathione, and carotenoids are examples of non-enzymatic antioxidants that effectively mitigate oxidative damage by reducing ROS activity or by collaborating with enzymatic players to achieve efficient antioxidant activity through the utilization of H_2O_2. The ascorbate–glutathione pathway comprises of AsA, GSH, and four enzymes, viz. ascorbate peroxidase, monodehydroascorbate reductase, dehydro ascorbate reductase, and glutathione reductase, which play vital roles in detoxifying ROS and ultimately mitigate oxidative damage in pearl millet under abiotic stress (Divya et al. 2019).

Phytohormones

Plant hormones play pivotal roles in controlling responses to several internal and external stimuli. Abscisic acid (ABA) is the key hormone considered to regulate the response of plants to abiotic stresses. Increased levels of endogenous ABA under drought stress conditions have been reported in many plant species including grain crops like sorghum (Yang et al. 2017). The level of ABA is also influenced by heat stress. ABA accumulates in stressed plants, interrupts their photosynthesis, and stimulates stomatal closure to minimize water loss through transpiration. ABA induces the expression of various genes which are essential for stress and tolerance reactions, such as osmoprotective synthesis enzymes. Under drought conditions, ABA is synthesized in the roots and exported to shoots causing stomatal closure.

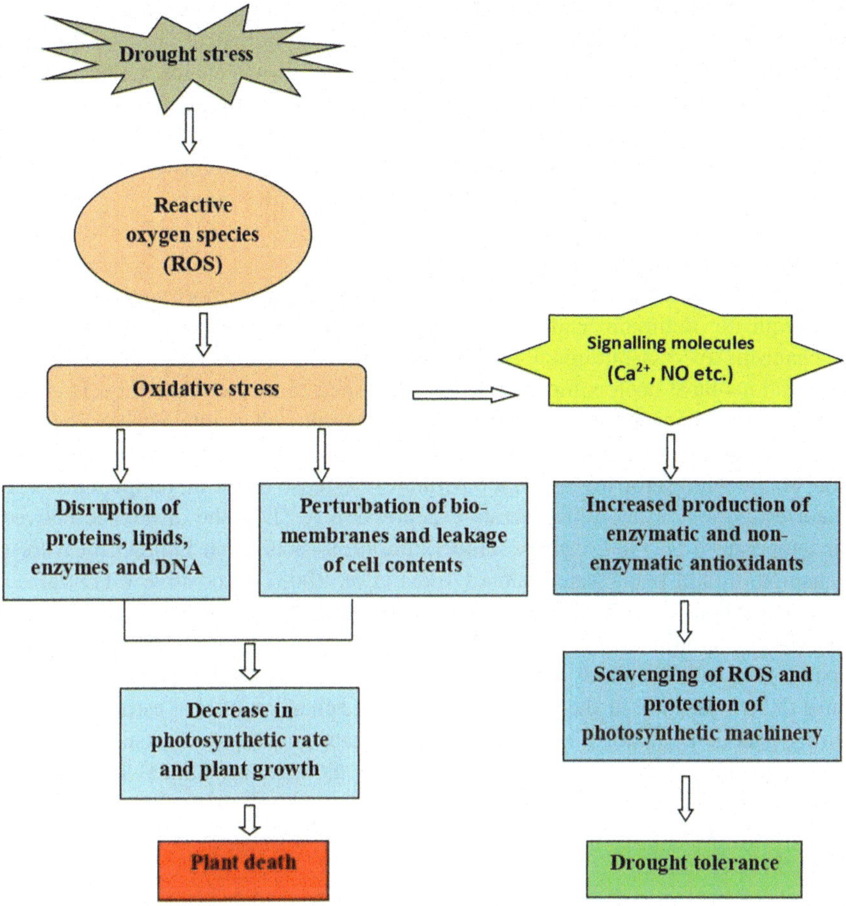

Fig. 16.3 Plant response and mechanisms under drought stress

Cytokinin (CK) postpones the premature senescence of leaves and death during drought stress and promotes adaptive traits that help enhance grain yields. The increase in endogenous levels of CK through the expression of the CK biosynthesis gene isopentenyl transferase (*IPT*) delays cell senescence caused by droughts and improves crop yields. In addition to controlling root growth and branching, CK inhibits the primary root growth and branching under drought stress which help in effective water uptake (Werner et al. 2010). Jasmonic acid also plays a vital role in abiotic stress tolerance, mainly in drought stress in plants. Kholova et al. (2010a, b) reported that under water stress conditions, leaf ABA content was recorded higher in drought tolerant genotypes of pearl millet and improved transpiration efficiency was observed (Fig. 16.3).

Adaptations to High Temperature Stress

Plants adapt to high temperature stress through morphological and physiological adjustments. Mechanisms may vary across the crop growth stages. Critical growth stages such as anthesis and grain filling are highly sensitive to above optimum temperatures. Some adaptation mechanisms to cope with high temperature stress include canopy cooling through transpiration, the involvement of heat shock proteins, antioxidant system, and the regulation of the phenology.

Transpirational Cooling

To cope with elevated ambient temperatures, plants transpire more water to maintain the requisite optimum and cooler canopies for physiological function. The mechanisms of transpirational cooling in response to high temperatures and its implications have been comprehensively illustrated (Sadok et al. 2021). This is an avoidance mechanism that allows the plant to function and maintain cooler canopies. However, this requires the availability or access to soil water resources and irrigation. At the whole crop level, leaf temperatures decrease below air temperature when water evaporates. Canopy temperature depression (CTD)—the difference between air temperature (T_a) and canopy temperature (T_c)—acts as an indirect measure of transpiration and plant water status (Araus et al. 2003). A positive CTD value is observed when the canopy is cooler than the air ($\text{CTD} = T_a - T_c$). To maintain canopy temperature in a metabolically comfortable range, plants transpire through open stomata. When plants are stressed, they temporarily close their stomata, which raise the temperature of the canopy. CTD is a mechanism of heat escape and has a strong genetic correlation with yield. Heat-tolerant genotypes of wheat had higher CTD than heat-sensitive genotypes, indicating their greater ability to maintain a cooler canopy environment (Gare et al. 2018).

Heat Shock Proteins

Heat shock proteins (HSPs), also known as stress-induced proteins, are produced in response to high temperature stress. Stressed plants typically produce less normal proteins and up-regulate HSP-related genes. About 20 HSPs have been found in plants, and the diversification of these proteins reflects the adaptation or tolerance to heat stress. The overall purpose of HSPs is to act as molecular chaperones that control the folding and aggregation of proteins. Under high temperature stress, these proteins act as chaperones to prevent the irreversible aggregation of other proteins and to promote protein refolding. The HSPs protect cells from damage and make them easier to recover after returning to normal growth conditions. The changing responses and expressions of the HSPs vary in different phases of development. HSP90 showed an increased expression under heat stress in rice (Hu et al. 2009) and wheat (Mahfouze et al. 2019).

Stay Green

Leaf senescence is a maturity stage in the life cycle of plants where leaves undergo chlorosis as a result of diverse factors such as aging, biotic, and abiotic stresses. Post-flowering drought and heat stress results in premature leaf senescence, abortion, and consequently reduced grain yield. Genotypes that possess an integrated drought adaptation trait known as stay green that delays leaf senescence are able to tolerate post-flowering drought and heat stress in sorghum (Kumar et al. 2017). Stay green expression is affected by flowering time, sink strength, and the environment. According to Van Oosterom et al. (1996), time of onset of senescence, rate of senescence, and Green Leaf Area Duration (GLAD) are independently inherited. Functional stay green character is integrated with delayed leaf senescence and photosynthetic capacity during grain filling gained in sorghum (Krupa et al. 2017).

16.3 Crop Simulation Models

Changes in phenological events can have a significant impact on how crops perform under changing climate. Crop modeling is essential for evaluating crop productivity and crop management based on crop phenology (Kadiyala et al. 2015). A phenological model can also estimate the timing of different phenological events, such as flowering, and the duration of phenological phases, such as the grain filling phase. Changes in the timing of phenological events have important implications for scientific research. Scientists use phenological data in computer models that project future climate scenarios and the projected impacts of such changes on the environment. Scientists are also interested in how phenology can inform the monitoring of drought conditions and the assessment of risk related to adverse climate changes affecting growth and productivity of plants (Jones et al. 2017; Rader et al. 2009; Challinor et al. 2013).

The use of quantitative descriptions of plant phenological processes for predicting the plant growth and development influenced by environmental changes and input management is the rationale behind Crop Simulation Modelling (CSM). CSM is a key tool in extrapolating the impact of future climate change on crop performance and ultimately the productivity. CSMs can be employed under varying agroclimatic zones, field and controlled experiments, soils, management regimes, crops, and climate change scenarios. The impact of the individual climate change components and the combined effect of climate change scenarios on crop production and externalities have been explored with such models. To counteract negative impacts of climate change and to capture some of the new opportunities of these changes, management options have been investigated. Crop simulation models can be used as decision making tool to provide crop management decisions under variousclimate change scenarios (Slafer 2003; Hodson and White 2010). Crop modeling, has two main approaches (1) use of dissected parts of plants, like leaf canopy (Leuning et al. 1995) or root structure (Dupuy et al. 2010), (2) second approach integrates models for different physiological traits to simulate the growth of the whole plant. Whole-crop computer simulation models use underlying physiology and environmental variables to calculate plant growth, or more specifically, yield and dry matter production. Both directions are increasingly coupled with breeding and molecular approaches to facilitate crop modeling for a better output. CSM has been developed either for single crop or multi- crops considering single trait (parameter) or multi-

Table 16.1 Crop parameters and climatic factors employed in crop simulation models

Parameters/key processes	Climate variables	Module
Phenological stages (anthesis, maturity, pod set, seed set, tuber initiation)	Temperature, day length	Crop module
Crop growth (aboveground biomass, leaves, stems, pods, seeds or grain yield, tubers, roots, harvest Index)	Solar radiation, temperature	Crop module
Potential and actual evapotranspiration, crop transpiration or water uptake, soil evaporation, soil physical and chemical parameters	Solar radiation, temperature, rainfall, irrigation	Soil module
Nutrient uptake (mostly nitrogen)	Nutrient supply	Management
Stresses (water deficit, nutrient stress, low or high temperature stress)	Rainfall, temperature, nutrient supply	Weather and climate module

parameters (multiple models) for gaining climate change impacts at different scenarios, locations, crops, and participants to assess the model outputs. Table 16.1 shows some of the common parameters and factors used for crop simulation models and Figs 16.4 depicts the various components and processes of crop simulation models.

The parameters considered for CM are usually determined by traits (parameter adjustment) and comparison with observed data from field trials that combine both morpho-physiological, genetic, and environmental variables to predict phenotype (Hoogenboom et al. 2004; Messina et al. 2006). This approach has several advantages over conventional methods of investigating a single trait without considering the impact of environment. Researchers now dissect model parameters into genetic factors to make cultivar-specific models. The first step is, by separating the big effect and small effect QTLs into different parts (parameters) of the eco-physiological model, for the target trait (e.g., higher grain yield or drought tolerance) which can be detected by molecular breeding strategies (Letort et al. 2008). In this situation, the eco-physiological model is like a gene network linking all of the relevant genes into a complete picture. Some QTLs may have a small effect statistically, but that does not mean the QTL is unimportant to the target trait, even though we neglect these QTLs in almost all circumstances and turn to others that can explain more variance. When we examine the underlying traits (leaf elongation rate, leaf angle, sink allocation, biomass, etc.) one by one, more QTLs might be discovered. Second step, because of G*E* interaction, some QTLs identified as beneficial to the desired trait might have the opposite effect in another location (Chenu et al. 2009). Third step, under the influence of pleiotropy, selecting for one trait may cause an adverse effect if that trait is coupled to another trait. All these issues can be reflected in a well-verified and standardized crop model. Thus, an effective crop model bridges the genotype-phenotype gap to predict a genotypic trait of any cultivar under recommended management options and various climate change scenarios (Reymond et al. 2003).

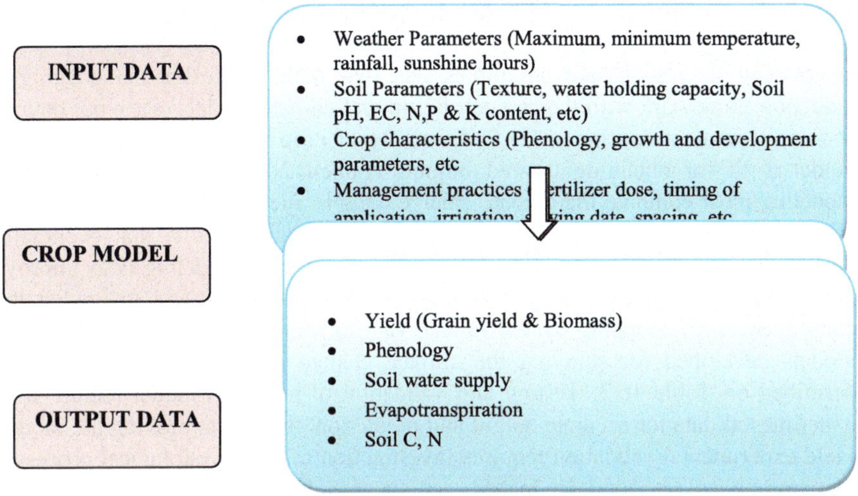

Fig. 16.4 Generic layout of a crop model and crop processes

Furthermore, CSM can be used to characterize environments based on crop performance data by connecting Geographic Information System (GIS) systems and crop models (Chapman et al. 2000). However, the outcomes from the simulated model have to be evaluated critically, require further ground-truthing via extensive field research, and need to be tested with data sets involving interactive environmental effects (Ziska and Bunce 2007) (Müller et al. 2017). For example, while using CSM for physiological traits, elevated CO_2 levels lead to stomata closure, reduction in transpiration and less canopy cooling depicting that the interaction between CO_2 and temperature is therefore important (Ziska and Bunce 2007), but it is often not considered in crop models. As the frequency of extreme high temperatures (>35 °C) during the growing season will increase, the interactions with elevated CO_2 need to be understood and considered (Attri and Rathore 2003). Similarly, factors affecting yield and grain quality are often affected by climate (Wu et al. 2004; Yang et al. 2007) but are not, simulated by crop models (e.g., protein composition, oil content), Simulation studies on the impact of climate change will require a better understanding of the physiology of yield quality and its incorporation into crop models. Case studies to explain the importance of these variables include a study reported by Fisher (2008) and Sinclair and Jamieson (2006) where physiological processes such as sink–source relationships of grain yield, correctly or incorrectly represented in crop models, could be differently affected by elevated CO_2 and increased temperature (Triboi et al. 2006; Ziska and Bunce 2007; Jifon and Wolfe 2005). Wheat grain yield has been reported to be reduced under elevated CO_2 in sink-manipulated shoots, implying that a high source: sink ratio may result in a down-regulation of photosynthetic capacity that more than offsets the direct stimulating effect of elevated CO_2 (Uddling et al. 2008).

16.3.1 Crop Models for Climate Change

Developing a crop-climate ensembles has two main processes (1) creating a modeling framework with the choice of crop and climate models (set input data or choice of parameterizations), (2) and running of crop models justifying that the model is fit for obtaining desired outputs (Lobell 2014). Hence, a simulation modeling must combine the climate change complexities with crops physiological parameters along with soil-plant-atmospheric continuum factors that eventually govern the crop performance. The major purpose of crop modeling is to improve precise understanding of crop, their physiological processes, and to predict the consequences of cropping system under varying conditions. Some of the crop models developed for studying the climate change impact on productivity are described on Table 16.2. Testing and validation of any crop model require field experimental data for accurate output and prediction, thereby improving the model. Field experimental validation requires investigation of key physiological processes like crop phenology, total dry matter accumulation, harvest index, economic yield, water and nutrient use, adaptation traits under stress environments. Hence, the structure of any crop model can be classified into crop modules, crop agronomy and management, soil module, breeding, and climate change modules. The main crop modules are linked to all the other modules (Fig. 16.5).

16.3.2 Simulation of Climate Change Impacts and Adaptation in Pearl Millet

Impact of climate change on crop productivity needs to be assessed for the development of future adaptation strategies. Crop simulation models are effective tools to quantify the impact of climate change on productivity of pearl millet and identify potential adaptation measures to ameliorate those adverse effects. Many researchers quantified the impact of current and future climate changes on crop yields (Asseng et al. 2015) and evaluate different crop management strategies (Araya et al. 2020, 2021). Singh et al. (2017) simulated the productivity of pearl millet with CSM-CERES-pearl millet model using 30 years of observed weather data and identified the major production constraints under changing climate scenarios. Adaptation measures like selection of longer duration cultivars along with drought and heat tolerance traits resulted in higher yields of pearl millet under future climate change scenarios in some parts of India and West Africa. Pearl millet and sorghum performed well under future climate scenarios. Adoption of suitable crop management strategies (i.e., N fertilization, planting date, and irrigation) positively affected sorghum and pearl millet yield, and thus considered as suitable adaptation options to reduce risks under the projected changing climate in Senegal (Araya et al. 2022).

Yadav et al. (2013) calibrated and validated CERES-millet model to simulate the impact of projected climate change on pearl millet yield in Gujarat. The *kharif* pearl millet is more vulnerable to climate change as compared to summer peal millet. Increment in yields to the tune of 4 to 9.5% was observed by adoption of various

Table 16.2 Crop simulation models available for improving production and productivity under varying conditions

Crop model	Crop	Target trait(s)	References
ARCWHEAT1	Wheat	To simulate wheat growth using crop physiological functions	Porter (1984), Weir et al. (1984)
CERES-WHEAT		To simulate the effects of cultivar, planting density, weather, soil water, and nitrogen on crop growth, development, and yield	Ritchie et al. (1985)
WINTER WHEAT			Baker et al. (1985)
SUCROS-wheat			Laar et al. (1992)
Wheat-grow model		Grain quality simulations	Pan et al. (2006)
CropSyst		N and water management tool for growth and yield	Stöckle et al. (2003)
SIRIUS		Prediction of biomass from intercepted photosynthetically active radiation (PAR) and grain growth	Jamieson et al. (1998)
Rice-Grow model	Rice (calibrated and validated for 18 popular rice varieties in 15 locations throughout Asia)	Simulation of plant architecture	Zhu et al. (2009)
ORYZAv3 improved from ORYZA 1, ORYZA W and CERES-rice		Simulates growth and development of rice including water, C, and N balance in lowland, upland, and aerobic rice ecosystems. It works in potential, water-limited, nitrogen-limited, and NxW-limited conditions	Kropff et al. (1994), Confalonieri and Bocchi (2005)
DSSAT	Annual crops including wheat, rice, maize, and various grain legumes, forage legumes, and grasses	Simulation of crop growth and development, soil and crop water, and nutrient dynamics; also includes genotype-specific parameters (GSPs)	Jones et al. (2003)
APSIM	Diverse crops	Generic model for crop management, water balance, climate impacts, cropping systems, species interactions, land use studies, soil impacts (erosion, acidity and nitrate leaching), and crop adaptation/breeding	Keating et al. (2003)
CERES-MAIZE	Maize	Optimization of maize production through manipulation of plant dates	Jones and Kiniry (1986)

(continued)

Table 16.2 (continued)

Crop model	Crop	Target trait(s)	References
STICS	Wheat, maize	Generic model for simulating crops and their water and nitrogen balances	Brisson et al. (2003)
CERES-MILLET	Pearl millet	To simulate crop growth in response to climate, soil, genotypes, and management across locations	Ritchie and Alagarswamy (1989)
BAJRAWAT		To simulate growth & development in Bajra	Rao et al. (1999)
CERES-SORGHUM	Sorghum	Simulating the crop growth based on aboveground biomass, harvest index, evapo-transpiration, and soil water	Alagarswamy and Ritchie (1989)
APSIM	Barley, canola, chickpea, cotton, cowpea, maize, millet, mung bean, peanut, pigeon pea, sorghum, soybean, sunflower, wheat, and sugarcane	For simulating growth, development and yield of crops, pastures and forests and their interactions with the soil	Keating et al. (2003)
EPIC	Several crops	To study the relationship between soil erosion and crop productivity	Kiniry et al. (1995)
SUCROS	Many crops including wheat, cotton, sugar beet, cocoa	Simulates both potential and water limited growth of a crop	Goudriaan and Van Laar (1994), van Laar et al. (1997)
DAISY	General model	Mechanistic simulation model for water, nitrogen, carbon, and pesticides in the bioactive zone near the soil surface (roughly top of canopy to the bottom of the (historic) root zone)	Hansen et al. (1990)
SALUS	General model	To simulate continuous crop, soil, water and nutrient conditions under different management strategies for multiple years	Basso et al. (2001)
InfoCrop	Rice, wheat, maize, sorghum, pearl millet, mustard, soybean, chick	Simulates the effects of weather, soil and crop management on various	Aggarwal et al. (2006)

(continued)

Table 16.2 (continued)

Crop model	Crop	Target trait(s)	References
	pea, pigeon pea, potato, and cotton	growth and yield parameters, nitrogen uptake, greenhouse gas emissions, soil water and nitrogen balance, and climate change studies	
AQUACROP	Herbaceous crops	Crop-water productivity model; simulates yield response to water of herbaceous crops and is particularly suited to address conditions where water is a key limiting factor in crop production	Steduto et al. (2009)
RZWQM2	General model	Root zone water quality model for simulating the fluxes of water, nutrients, and chemicals in the root zone	Ma et al. (2007)
HERMES	General model	Model to describe plant growth and water and nitrogen dynamics in the soil-plant system	Kersebaum (2007)
SOYGRO	Soybean	Simulates growth, development and yield of a crop growing on a uniform area of land under prescribed or simulated management; generic photosynthesis process-oriented model	Sau et al. (1999), Wilkerson et al. (1983)
CHIKPGRO	Chickpea	Based on extensive growth analysis data under changing climate conditions	Robertson et al. (2002), Soltani et al. (2006)
QCANE	Sugarcane	Simulate growth and sugar accumulation in sugarcane	Keating et al. (1999)
PNUTGRO	Peanut	For predicting phenological development, light interception, canopy growth, dry matter production, pod and seed yields	Boote et al. (1989)
GOSSYM	Cotton	Predicts crop growth and soil responses to environmental stresses	Milroy et al. (2004)

(continued)

Table 16.2 (continued)

Crop model	Crop	Target trait(s)	References
WOFOST	General model	For analyzing the growth and production of field crops under a wide range of weather and soil conditions	Hijmans et al. (1994)
CENTURY	General model	To simulate C and nutrient dynamics for different types of ecosystems including grasslands, agricultural lands, forests composed of a soil organic matter/ decomposition submodel, a water budget model, a grassland/crop submodel, a forest production submodel	Parton et al. (1988, 1993)

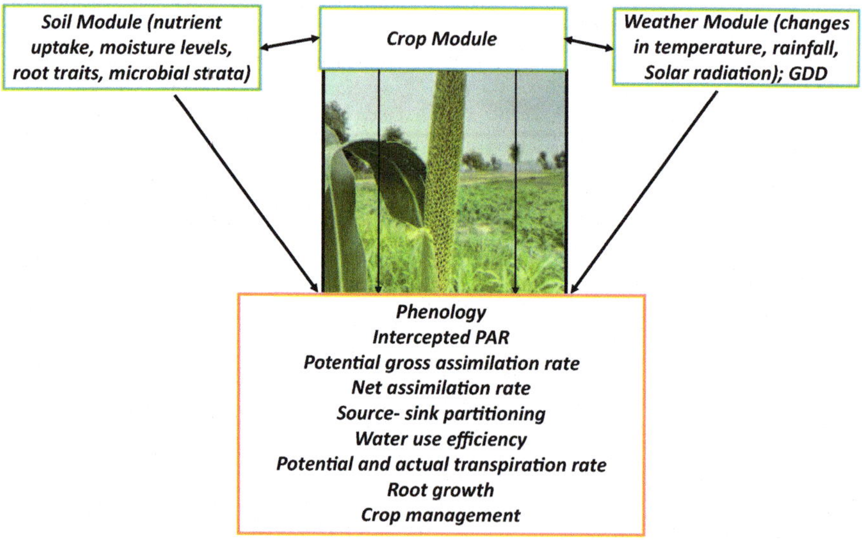

Fig. 16.5 Key processes to be incorporated for maximizing productivity of pearl millet

climate change adaptation strategies like early transplanting, varietal change, better water, and nutrient management practices. CSM-CERES-Millet is a useful tool in assessing the impact of climate change on pearl millet. Ullah et al. 2019 quantified the impact of climate change on pearl millet in arid and semi-arid environments of

Punjab, Pakistan under various climate change scenarios. The yield reduction in pearl millet is upto the tune of 7–10% under RCPs 4.5 and 8.5 in Faisalabad and 10–13% in Layyah, respectively.

16.4 Conclusion

Crop simulation models play a major role in many applications such as precision agriculture, on-farm management, and regional assessments of the impact of climate variability and change. They evaluate the quantitative effects of climate change on climate-plant interaction. Crop models are the primary tools available to assess the site-specific effects of climate change on pearl millet growth, development, and yield. Developing proactive mitigation and adaptation plans using climate data for managing inter-annual variability within the agricultural communities and the organizations that interface with them in the agricultural sector would assist in building resilience due to long-term climate changes. Crop models would facilitate a move towards a complete assessment of uncertainty in crop yield simulations. It is recommended that an integrated assessment modeling framework should be adopted that generate useful information for the farmer, planners, and policy makers. Further, these crop models assist the plant breeders in evaluating new promising plant traits of pearl millet for adapting to climate change at the selected locations and other similar environments.

References

Aggarwal PK, Kalra N, Chander S, Pathak H (2006) InfoCrop: a dynamic simulation model for the assessment of crop yields, losses due to pests, and environmental impact of agro-ecosystems in tropical environments. I. Model description. Agric Syst 89(1):1–25

Ahmed I, Ullah A, Ur Rahman MH, Ahmad B, Wajid SA, Ahmad A, Ahmed S, Hassain S (2019) Climate change impacts and adaptation strategies for agronomic crops. In: Climate change and agriculture. Springer, Cham, pp 1–14

Ajithkumar IP, Panneerselvam R (2014) ROS scavenging system, osmotic maintenance, pigment and growth status of Panicum Sumatrense under drought stress. Cell Biochem Biophys 68(3): 587–595

Alagarswamy G, Ritchie JT (1989) Predicting crop phenology. CRC Press, Boca Raton

Araus JL, Bort J, Steduto P, Villegas D, Royo C (2003) Breeding cereals for Mediterranean conditions: ecophysiological clues for biotechnology application. Ann Appl Biol 142:129–141

Araya A, Prasad PVV, Gowda PH, Djanaguiraman M, Kassa AH (2020) Potential impacts of climate change factors and agronomic adaptation strategies on wheat yields in central highlands of Ethiopia. Clim Chang 159(3):461–479

Araya A, Prasad PVV, Gowda PH, Djanaguiramana M, Gebretsadkan Y (2021) Modeling the effects of crop management on food barley production under a midcentury changing climate in northern Ethiopia. Clim Risk Manag 32:100308

Araya A, Jha PK, Zambreski Z, Faye A, Ciampitti IA, Min D, Gowda PH, Singh U, Prasad PVV (2022) Evaluating crop management options for sorghum, pearl millet and peanut to minimize risk under the projected midcentury climate scenario for different locations in Senegal. Clim Risk Manag 36:100436

Arya RK, Yadav HP, Yadav AK, Singh MK (2010) Effect of environment on yield and its contributing traits in pearl millet. Forage Res 36:176–180

Arya RK, Singh MK, Yadav AK, Kumar A, Kumar S (2014) Advances in pearl millet to mitigate adverse environment conditions emerged due to global warming. Forage Res 40(2):57–70. http://forageresearch.in

Asseng S, Zhu Y, Wang E, Zhang W (2015) Crop modeling for climate change impact and adaptation. In: Crop physiology. Academic, New York, pp 505–546

Asthir B (2015) Mechanisms of heat tolerance in crop plants. Biol Plant 59:620–628

Attri SD, Rathore LS (2003) Simulation of impact of projected climate change on wheat in India. Int J Climatol 23:693–705

Baker DN, Whisler FD, Parton WJ et al (1985) The development of winter wheat: a physical physiological process model. Agric Res Serv 38:176–187

Basso B, Ritchie JT, Pierce FJ, Braga RP, Jones JW (2001) Spatial validation of crop models for precision agriculture. Agric Syst 68(2):97–112

Begg JE, Burton GW (1971) Comparative study of five genotypes of pearl millet under a range of photoperiods and temperatures. Crop Sci 11:803–805

Biswas MS, Fukaki H, Mori IC, Nakahara K, Mano J (2019) Reactive oxygen species and reactive carbonyl species constitute a feed-forward loop in auxin signaling for lateral root formation. Plant J 100:536–548

Boote KJ, et al (1989) User's guide, Florida Agric. Experiment J No. 8420, University of Florida, USA

Bota J, Flexas J, Medrano H (2004) Is photosynthesis limited by decreased Rubisco activity and RuBP content under progressive water stress? New Phytol 162:671–681

Brisson N, Gary C, Justes E et al (2003) An overview of the crop model STiCS. Eur J Agron 18: 309–332

Challinor AJ, Stafford MS, Thornton P (2013) Use of agro-climate ensembles for quantifying uncertainty and informing adaptation. Agric For Meteorol 170:2–7

Challinor AJ, Watson J, Lobell DB, Howden SM, Smith DR, Chhetri N (2014) A meta-analysis of crop yield under climate change and adaptation. Nat Clim Chang 4:287–291

Chapman SC, Cooper M, Hammer GL, Butler DG (2000) Genotype by environment interactions affecting grain sorghum. II. Frequencies of different seasonal patterns of drought stress are related to location effects on hybrid yields. Aust J Agric Res 51(2):209–221

Chenu K, Chapman SC, Tardieu F, McLean G, Welcker C, Hammer GL (2009) Simulating the yield impacts of organ-level quantitative trait loci associated with drought response in maize: a "gene-to-phenotype" modeling approach. Genetics 183(4):1507–1523

Choudhary S, Guha A, Kholova J, Pandravada A, Messina CD, Cooper M et al (2019) Maize, sorghum, and pearl millet have highly contrasting species strategies to adapt to water stress and climate change-like conditions. Plant Sci 295:110297

Confalonieri R, Bocchi S (2005) Evaluation of CropSyst for simulating the yield of flooded rice in northern Italy. Eur J Agron 23:315–326

Crafts-Brandner SJ, Salvucci ME (2002) Sensitivity of photosynthesis in maize, a C_4 plant to heat stress. Plant Physiol 129:1773–1780

De Ronde JAD, Cress WA, Kruger GHJ, Strasser RJ, Staden JV (2004) Photosynthetic response of transgenic soybean plants containing an Arabidopsis P5CR gene, during heat and drought stress. J Plant Physiol 61:1211–1244

Debieu M, Sine B, Passot S, Grondin A, Akata E et al (2018) Response to early drought stress and identification of QTLs controlling biomass production under drought in pearl millet. PLoS One 13(10):e0201635

Directorate of Millet Development (2020) Crop profile. Available from http://www.aicpmip.res.in/aboutus.html

Divya, K., Kavi Kishor, P. B., Mathur, P. B., Singam, P., Sharma, K.K., Vadez, V. And Reddy, P.S. (2019). Isolation and functional characterization of three abiotic stress inducible (Apx, Dhn and Hsc 70) promoters from pearl millet (Pennisetum glaucum L.). Mol Biol Rep 46:6039-6052

Djanaguiraman M, Perumal R, Ciampitti IA, Gupta SK, Prasad PVV (2018) Quantifying pearl millet response to high temperature stress: threshold, sensitive stages, genetic variability and relative sensitivity of pollen and pistil. Plant Cell Environ 41:993–1007

Dupuy L, Gregory PJ, Bengough AG (2010) Root growth models: towards a new generation of continuous approaches. J Exp Bot 61(8):2131–2143

Dutta S, Mohanty S, Tripathy BC (2009) Role of temperature stress on chloroplast biogenesis and protein import in pea. Plant Physiol 150:1050–1061

ElBasyoni I, Saadalla M, Baenziger S, Bockelman H, Morsy S (2017) Cell membrane stability and association mapping for drought and heat tolerance in a worldwide wheat collection. Sustainability 9(9):1606

Fahad S, Hussain S, Saud S, Khan F, Hassan S, Amanullah et al (2016) Exogenously applied plant growth regulators affect heat-stressed rice pollens. J Agron Crop Sci 202:139–150. https://doi.org/10.1111/jac.12148

FAO (2019) Food and Agriculture Organization of the United Nations, Rome

Farooq, M., Wahid, A., Kobayashi, N., Fujita, D. And Basra, S.M.A. (2009). Plant drought stress: effects, mechanisms and management. Agron Sustain Dev 29, 185–212.

Fisher RA (2008) The importance of grain or kernel number in wheat: a reply to Sinclair and Jamieson. Field Crop Res 105:15–21

Gare S, Wagh RS, Ingle AU, Soni N (2018) Effect of temperature on stem reserve mobilization for grain development in wheat. J Agric Res Technol 43:334–341

Garg BK (2003) Nutrient uptake and management under drought: nutrient-moisture interaction. Curr Agric 27:1–8

Goudriaan J, Van Laar HH (1994) Modelling potential crop growth processes. Kluwer Academic Publishers, Dordrecht, p 233

Gupta SK, Rai KN, Singh P, Ameta VL, Gupta SK, Jayalekha AK, Verma YS (2015) Seed set variability under high temperatures during flowering period in pearl millet (Pennisetum glaucum L. (R.) Br.). Field Crop Res 171:41–53

Hansen S, Jensen HE, Nielsen NE, Svendsen H (1990) DAISY: soil plant atmosphere system model. NPO Report No. A 10. The National Agency for Environmental Protection, Copenhagen, p 272

Hemantaranjan A, Bhanu N, Sigh MN, Yadak DK, Patel PK, Singh R, Katiyar D (2014) Heat stress responses and themotolerance. Adv Plants Agric Res 1(3):62–70

Hijmans RJ, Guiking-Lens IM, van Diepen CA (1994) WOFOST 6.0; User's guide for the WOFOST 6.0 crop growth simulation model. Technical Document 12. Wageningen, DLO Winand Staring Centre

Hodson D, White J (2010) GIS and crop simulation modelling applications in climate change research. Climate change and crop production. CABI Publishers, Wallingford, pp 245–262

Hoogenboom G, White JW, Messina CD (2004) From genome to crop: Integration through simulation modeling. Field Crop Res 90(1):145–163

Hoogenboom G, Porter CH, Boote KJ, Shelia V, Wilkens PW, Singh U, White JW, Asseng S, Lizaso JI, Moreno LP, Pavan W (2019) The DSSAT crop modeling ecosystem. In: Advances in crop modelling for a sustainable agriculture. Burleigh Dodds Science Publishing, Cambridge, pp 173–216

Hu W, Hu G, Han B (2009) Genome-wide survey and expression profiling of heat shock proteins and heat shock factors revealed overlapped and stress specific response under abiotic stresses in rice. Plant Sci 176:583–590

IPCC (2007) Climate change 2007: impacts, adaptation and vulnerability. In: Working Group II Contribution to the Fourth Assessment report of the Intergovernmental Panel on Climate Change. Cambridge University Press, Cambridge

IPCC (2014) Climate change 2014: synthesis report. In: Pachauri RK, Meyer LA (eds) Contribution of Working Groups I, II and III to the Fifth Assessment report of the Intergovernmental Panel on Climate Change. IPCC, Geneva, p 151

Jamieson PD, Semenov MA, Brooking IR, Francis GS (1998) Sirius: a mechanistic model of wheat response to environmental variation. Eur J Agron 8:161–179

Jifon JL, Wolfe DW (2005) High temperature-induced sink limitation alters growth and photosynthetic acclimation to elevated CO2 in bean (Phaseolus vulgaris L.). J Am Soc Hortic Sci 130: 515–520

Jones CA, Kiniry JR (eds) (1986) CERES-Maize: a simulation model of maize growth and development. Texas A&M University Press, College Station, p 194

Jones JW, Hoogenboom G, Porter CH, Boote KJ, Batchelor WD, Hunt LA, Wilkens PW, Singh U, Gijsman AJ, Ritchie JT (2003) The DSSAT cropping system model. Eur J Agron 18(3-4): 235–265

Jones JW, Antle JM, Bruno B, Boote KJ, Conant RT, Foster I, Godfray HC et al (2017) Brief history of agricultural systems modeling. Agric Syst 155:240–254

Kadiyala MDM, Nedumaran S, Singh P, Chukka S, Irshad MA, Bantilan MCS (2015) An integrated crop model and GIS decision support system for assisting agronomic decision making under climate change. Sci Total Environ 521:123–134

Keating BA, Robertson MJ, Muchow RC, Huth NI (1999) Modelling sugarcane production systems. I. Development and performance of the sugarcane module. Field Crop Res 61:253–271

Keating BA, Carberry PS, Hammer GL, Probert ME, Robertson MJ, Holzworth D, Huth NI, Hargreaves JNG, Meinke H, Hochman Z, McLean G, Verburg K, Snow V, Dimes JP, Silburn M, Wang E, Brown S, Bristow KL, Asseng S, Chapman S, McCown RL, Freebairn DM, Smith CJ (2003) An overview of APSIM, a model designed for farming systems simulation. Eur J Agron 18:267–288

Kersebaum K (2007) Modeling nitrogen dynamics in soil-crop systems with HERMES. Nutr Cycl Agroecosyst 77(1):39–52

Kholova J, Harsh CT, Lava Kumar P, Yadav RS, Kocova M, Vadez V (2010a) Terminal drought-tolerant pearl millet Pennisetum glaucum (L.) R. Br. have high leaf ABA and limit transpiration at high vapour pressure deficit. J Exp Bot 61(5):1431–1440

Kholova J, Hash CT, Kakkera A, Kocova M, Vadez V (2010b) Consecutive water-conserving mechanisms are correlated with the terminal drought tolerance of pearl millet [Pennisetum glaucum (L.) R. Br.]. J Exp Bot 61:369–377. https://doi.org/10.1093/jxb/erp314

Kim JY, Mahe A, Brangeon J, Prioul JL (2000) A maize vacuolur invertase, IVR2, is induced by water stress. Organ/tissue specificity and diurnal modulation of expression. Plant Physiol 124: 71–74

Kiniry J et al (1995) EPIC model parameters for cereal, oilseed and forage crops in the northern great – plains region. Can J Plant Sci 75(3):679–688

Krishnamurthy L, Zaman-Allah M, Purushothaman R, Irshad M, Vadez V (2011) Plant biomass productivity under abiotic stresses in SAT agriculture. In: Matovic MD (ed) Biomass—detection, production and usage. IntechOpen, London, pp 247–264

Kropff MJ, et al (1994) SARP Res. Proceedings. IRRI, Philippines, p 110

Krupa KN, Dalwai N, Shashidhar HE, Harikumar KM, Manoj Kumar HB, Bharani S, Turaidar V (2017) Mechanisms of drought tolerance in sorghum: a review. Int J Pure Appl Biosci 5(4): 221–237

Kumar M, Govindasamy V, Rane J, Choudhary RL, Raina S, George P, Aher LK, Sigh NP (2017) Canopy temperature depression (CTD) and canopy greenness associated with variation in seed yield of soybean genotypes in semi arid environments. S Afr J Bot 13:230–238

Laar HH, Goudriaan J, Keulen H (1992) Simulation of crop growth for potential and water-limited production situations (as applied to spring wheat). Simulation reports CABO-TT 27. CABO-DLO, Wageningen, p 78

Letort V, Mahe P, Cournede PH, De Reffye P, Courtois B (2008) Quantitative genetics and functional-structural plant growth models: simulation of quantitative trait loci detection for model parameters and application to potential yield optimization. Ann Bot 101(8):1243–1254

Leuning R, Kelliher F, Pury D, Schulze ED (1995) Leaf nitrogen, photosynthesis, conductance and transpiration: scaling from leaves to canopies. Plant Cell Environ 18(10):1183–1200

Lobell DB (2014) Climate change adaptation in crop production: Beware of illusions. Glob Food Sec 3:72–76

Lynch JP, Brown KM (2001) Topsoil foraging-an architectural adaptation of plants to low phosphorous availability. Plant Soil 237:225–237

Ma L et al (2007) RZWQM simulated effects of crop rotation, tillage, and controlled drainage on crop yield and nitrate-N loss in drain flow. Geoderma 140(3):260–271

Machado S, Paulsen GM (2001) Combined effects of drought and high temperature on water relations of wheat and sorghum. Plant Soil 233:179–187

Mahfouze HA, Ramadan WA, Mahfouze SA (2019) Gene expression and single nucleotide polymorphisms (SNP) marker of heat shock protein (HSP) gene in wheat (Triticum aestivum L.). Plant Arch 19:2225–2232

Meena RP, Harinder Vishwakarma H, Singh MP, Satyavathi T, Gaikwad K, Padaria JC (2017) Physiological and biochemical plant responses to drought stress in Pennisetum Glaucum (L.) R. Br Indian Res J Genet Biotech 9(3):369–381

Meena RC, Ram M, Ambawat S, Satyavathi T, Khandelwal V, Meena RB, Kumar M, Bishnoi JP, Mungra KD (2021) Screening of pearl millet genotypes for high temperature and drought tolerance based on morpho – physiological characters. Int J Environ Climate Change 11(9): 74–80

Messina C, Jones J, Boote K, Vallejos C (2006) A gene-based model to simulate soybean development and yield responses to environment. Crop Sci 46(1):456–466

Milroy SP, Bange MP, Hearn AB (2004) Row configuration in rainfed cotton systems: modification of the OZCOT simulation model. Agric Syst 82:1–16

Mitchell JF, Johns TC, Gregory JM, Tett SFB (1995) Climate response to increasing levels of greenhouse gases and sulphate aerosols. Nature 376:501–504

Mohanty S, Baishna BG, Tripathy C (2006) Light and dark modulation of chlorophyll biosynthetic genes in response to temperature. Planta 224:692–699

Muktadir MA, Adhikari KN, Merchant A, Belachew KY, Vandenberg A, Stoddard FL, Khazaei H (2020) Physiological and biochemical basis of faba bean breeding for drought adaptation—a review. Agronomy 10:1345

Müller C, Elliott J, Chryssanthacopoulos J, Arneth A, Balkovic J, Ciais P, Deryng D, Folberth C, Glotter M, Hoek S, Iizumi T, Izaurralde R, Jones C, Khabarov N, Lawrence P, Liu W, Olin S, Pugh T, Ray D, Reddy A, Rosenzweig C, Ruane A, Sakurai G, Schmid E, Skalsky R, Song C, Wang X, De Wit A, Yang H (2017) Global gridded crop model evaluation: benchmarking, skills, deficiencies and implications. Geosci Model Dev 10:1403–1422

Nadarajah KK (2020) Ros homeostasis in abiotic stress tolerance in plants. Int J Mol Sci 21:5208

Nxele X, Klein A, Ndimba BK (2017) Drought and salinity stress alters ROS accumulation, water retention and osmolyte content in sorghum plants. S Afr J Bot 108:261–266

Pan J, Zhu Y, Jiang D, Dai TB, Li YX, Cao WX (2006) Modeling plant nitrogen uptake and grain nitrogen accumulation in wheat. Field Crop Res 97:322–336

Parton WJ, Stewart JWB, Cole CV (1988) Dynamics of C, N, P and S in grassland soils – a model. Biogeochemistry 5(1):109–131

Parton WJ et al (1993) Glob Biogeochem Cycles 7:785–809

Porter JR (1984) A model of canopy development in winter wheat. J Agric Sci 102:383–392

Prasad V (2013) Importance of millet in global food and nutritional security in changing climates. In: Proc. Global consultation on millets promotion for health and nutritional security, December 18-20, 2013. Compodium of papers and abstracts. Hyderabad, Directorate of Sorghum Research, p 6

Priyaadharshini M, Sritharan N, Senthil A, Marimuthu S (2019) Physiological studies on effect of chitosan nanoemulsion in pearl millet under drought condition. J Pharmacogn Phytochem 8: 3304–3307

Rader M, Kirshen P, Roncoli C, Hoogenboom G, Ouattara F (2009) Agricultural risk decision support system for resource-poor farmers in Burkina Faso, West Africa. J Water Resour Plan Manag 135:323–333

Rao SS, Patil JV (2015) Physiology of pearlmillet. In: Tonapi VA, Patil JV (eds) Millets – ensuring climate resilience and nutritional security. Daya Publishing House, New Delhi, pp 361–377
Rao BB et al (1999) Proceeding of National workshop on dynamic crop simulation modelling for Agrometeorological advisory services, New Delhi, p 147–155
Reymond M, Muller B, Leonardi A, Charcosset A, Tardieu F (2003) Combining quantitative trait loci analysis and an ecophysiological model to analyze the genetic variability of the responses of maize leaf growth to temperature and water deficit. Plant Physiol 131(2):664–675
Rezayian M, Ebrahimzadeh H, Niknam V (2020) Nitric oxide stimulates antioxidant system and osmotic adjustment in soybean under drought stress. J Soil Sci Plant Nutr 20:1122–1132
Richards RA (2006) Physiological traits used in the breeding of new cultivars for water-scarce environments. Agric Water Manag 80:197–211
Ritchie JT, Alagarswamy G (1989) Modelling the growth and development of sorghum and pearl millet. ICRISAT, Patancheru, pp 24–26
Ritchie JT, Godwin DC, Otter-Nacke S (1985) CERES-Wheat. A simulation model of wheat growth and development. University Press, Texas A & M. College Station
Robertson MJ, Carberry PS, Huth NI, Turpin JE, Probert ME, Poulton PL, Bell M, Wright GC, Yeates SJ, Brinsmead RB (2002) Simulation of growth and development of diverse legume species in APSIM. Aust J Agric Res 53:429–446
Sadok W, Lopez JR, Smith KP (2021) Transpiration increases under high-temperature stress: potential mechanisms, trade-offs and prospects for crop resilience in a warming world. Plant Cell Environ 44:2102–2116
Sattar A, Sher A, Ijaz M, Ul-Allah S, Rizwan MS, Hussain M, Jabran K, Cheema MA (2020) Terminal drought and heat stress alter physiological and biochemical attributes in flag leaf of bread wheat. PLoS ONE 18(4):e0284070
Sau F, Boote KJ, Ruíz-Nogueira B (1999) evaluation and improvement of CrOpGrO-soybean model for a cool environment in Galicia, northwest Spain. Field Crop Res 61:273–291
Schimel J, Balser TC, Wallenstein M (2007) Microbial stress response physiology and its implications for ecosystem function. Ecology 88:1386–1394
Sharma DK, Andersen SB, Ottosen CO, Rosenqvist E (2015) Wheat cultivars selected for high Fv/Fm under heat stress maintain high photosynthesis, total chlorophyll, stomatal conductance, transpiration and dry matter. Physiol Plant 153:284–298
Sinclair TR, Jamieson PD (2006) Grain number, wheat yield, and bottling beer: an analysis. Field Crop Res 98:60–67
Singh P, Boote KJ, Kadiyala MDM, Nedumaran S, Gupta SK, Srinivas K, Bantilan MCS (2017) An assessment of yield gains under climate change due to genetic modification of pearl millet. Sci Total Environ 601:1226–1237
Slafer GA (2003) Genetic basis of yield as viewed from a crop physiologist's perspective. Ann Appl Biol 142(2):117–128
Soltani A, Robertson MJ, Rahemi-Karizaki A, Poorreza J, Zarei H (2006) Modelling biomass accumulation and partitioning in chickpea (Cicer arietinum L.). Z Acker Pflanzenbau 192:379–389
Steduto P, Hsiao TC, Raes D, Fereres E (2009) AquaCrop-the FAO crop model to simulate yield response to water: I. Concepts and underlying principles. Agron J 101(3):426–437
Stöckle CO, Donatelli M, Nelson R (2003) CropSyst, a cropping systems simulation model. Eur J Agron 18:289–307
Taiz L, Zeiger E (2006) Plant physiology, 4th edn. Sinauer Associates Inc Publishers, Sunderland
Tandzi LN, Bradler G, Mutengwa C (2019) Morphological responses of maize to drought, heat and combined stresses at seedling stage. J Biol Sci 19(1):7–16
Terletskaya NV, Lee TE, Altayeva NA, Kudrina NO, Blavachinskaya IV, Erezhetova U (2020) Some mechanisms modulating the root growth of various wheat species under osmotic-stress conditions. Plants 9:1545

Tharanya M, Kholova J, Sivasakthi K, Thirunalasundari T, Vadez V (2017) Pearl millet. In: Sinclair TR (ed) Water-conservation traits to increase crop yields in water-deficit environments: case studies. Springer, Berlin, pp 73–83

Triboi E, Martre P, Girousse C, Ravel C, Triboi-Blondel AM (2006) Unravelling environmental and genetic relationships between grain yield and nitrogen concentration for wheat. Eur J Agron 25: 108–118

Uddling J, Gelang-Alfredsson J, Karlsson PE, Sellden G, Pleijel H (2008) Source-sink balance of wheat determines responsiveness of grain production to increased [CO2] and water supply. Agric Ecosyst Environ 127:215–222

Ullah A, Ahmad I, Ahmad A, Khaliq T, Saeed U, Habib-ur-Rahman M, Hussain J, Ullah S, Hoogenboom G (2019) Assessing climate change impacts on pearl millet under arid and semi-arid environments using CSM-CERES-Millet model. Environ Sci Pollut Res 26(7): 6745–6757

Vadez V, Hash T, Kholova J (2012) II. 1.5 Phenotyping pearl millet for adaptation to drought. Front Physiol 3:386

Vadez V, Kholová J, Yadav RS, Hash CT (2013) Small temporal differences in water uptake among varieties of pearl millet (Pennisetum glaucum (L.) R Br) are critical for grain yield under terminal drought. Plant Soil 371:447–462

van Laar HH, Goudriaan J, van Keulen H (1997) Simulation of crop growth for potential and water-limited production situations. As applied to spring wheat. In: Quantitative approaches in systems analysis. Simulation reports. Wageningen, ABO-DLO/TPE-WAU, p 78

Van Oosterom EJ, Jayachandran R, Bidinger FR (1996) Diallel analysis of the stay green trait and its components in sorghum. Crop Sci 36(3):549–555

Weir AH, Bragg PL, Porter JR, Rayner JH (1984) A winter wheat crop simulation model without water or nutrient limitations. J Agric Sci 102:371–382

Werner T, Nehnevajova E, Köllmer I, Novák O, Stmad M, Krämer U, Schmülling T (2010) Root-specific reduction of cytokinin causes enhanced root growth, drought tolerance, and leaf mineral enrichment in Arabidopsis and tobacco. Plant Cell 22:3905–3920

Wilkerson GG et al (1983) Trans ASAE 26:63–73

Winkel T, Do F (1992) Caracteres morphologiques et physiologiques de resistance du mil [Pennisetum glaucum (L.) R. Br.] a la secheresse. Agron Trop 46:339–350

World Meteorological Organization (1992) International meteorological vocabulary, 2nd edn. WMO, Geneva

Wu DX, Wang GX, Bai YF, Liao JX (2004) Effects of elevated CO2 concentration on growth, water use, yield and grain quality of wheat under two soil water levels. Agric Ecosyst Environ 104:493–507

Yadav RS, Harsh CT, Bidinger FR, Devos KM, Howarth CJ (2004) Genomic regions associated with grain yield and aspects of post flowering drought tolerance in pearl millet across environments and tester background. Euphytica 136:265–277

Yadav SB, Patel HR, Lunagaria MM, Parmar PK, Chaudhari NJ, Karande BI, Pandey V (2013) Impact assessment of projected climate change on pearl millet in Gujarat. In: Shete BT (ed) Proc. of National Seminar on Climate change impacts on water resources systems. Parul Institute of Engg. & Tech, Vadodara, pp 35–38

Yang L, Wang Y, Dong G, Gu H, Huang J, Zhu J, Hongjian Y, Liu G, Han Y (2007) The impact of free-air CO2 enrichment (FACE) and nitrogen supply on grain quality of rice. Field Crop Res 102:128–140

Yang Z, Wang Y, Wei X, Zhao X, Wang B, Sui N (2017) Transcription profiles of genes related to hormonal regulations under salt stress in sweet sorghum. Plant Mol Biol Report 35:586–599

Zaveri E, Russ J, Damania R (2020) Rainfall anomalies are a significant driver of cropland expansion. Proc Natl Acad Sci U S A 117:10225–10233

Zhang, A., Ji, Y., Sun, M., Lin, C., Zhou, P., Ren, J., Luo, D., Wang, X., Ma, C., Zhang, X., Feng, G., Nie, G. And Huang, L. (2021). Research on the drought tolerance mechanism of Pennisetum glaucum (L.) in the root during seedling stage. BMC Genomics 22(1): 568
Zhu Y et al (2009) Modeling leaf shape dynamics in rice. NJAS Wageningen J Life Sci 57(1):73–81
Ziska LH, Bunce JA (2007) Predicting the impact of changing CO2 on crop yields: some thoughts on food. New Phytol 175:607–617

17 Modern Crop Management Practices for Pearl Millet Cultivation in Semi-Arid Africa

Dougbedji Fatondji, Hakeem A. Ajeigbe, and Ali Ibrahim

Abstract

Pearl millet is the most important cereal crop in the semi-arid and arid regions of West Africa, supplying a large proportion of the dietary energy for the human population, nutritious fodder for livestock, and stalk for cooking fuel. This review focuses on modern agronomic practices for increasing millet productivity in semi-arid Africa. We showed that potential opportunities exist to improve millet productivity and nutritional quality while reducing adverse climate effects on millet production in the semi-arid areas of Africa. Millet-legume intercropping, agroforestry, crop rotation systems, fertilizer micro-dosing, and water harvesting techniques, along with improved millet varieties with resilience to various adverse climate change effects, have been identified as key interventions usually promoted for enhancing pearl millet productivity in semi-arid Africa. The benefits and their associated limitations have been discussed, and potential options for improvement have been suggested for sustainable pearl millet productivity intensification in the semi-arid area of Africa.

D. Fatondji (✉)
International Crops Research Institute for the Semi-Arid Tropics (ICRISAT), West and Central Africa, Bamako, Mali
e-mail: d.fatondji@cgiar.org

H. A. Ajeigbe
International Crops Research Institute for the Semi-Arid Tropics (ICRISAT), West and Central Africa, Kano, Nigeria
e-mail: h.ajeigbe@cgiar.org

A. Ibrahim
Africa Rice Center c/o IITA, Abuja, Nigeria

Faculté d'Agronomie, Université Abdou Moumouni de Niamey, Niamey, Niger
e-mail: i.ali@cgiar.org

V. A. Tonapi et al. (eds.), *Pearl Millet in the 21st Century*,
https://doi.org/10.1007/978-981-99-5890-0_17

Keywords

Pearl millet · Production constraints · Cropping systems · Sustainable intensification

17.1 Introduction

The term "millet" refers to various grass crops whose seeds are harvested for human consumption or animal feed and belong to five genera: Panicum, Setaria, *Echinochloa, Pennisetum,* and *Paspalum*, all the Paniceae tribe; Eleusine, of the Chlorideae tribe; and Eragrostis, of the Festuceae tribe. Pearl or cattail millet (*Pennisetum glaucum*) is one of the most important cultivated millet species, with a height ranging from 1.5 to 4.0 m, vigorous growth, and exceptional grain and fodder-yielding potential. It is one of the most important dual-purpose crops and a staple food for millions of people worldwide in arid and semi-arid environments (Ajeigbe et al. 2020a, b). *Pennisetum glaucum* is the most widely cultivated millet (Azare et al. 2020), with an estimated global area of production of 24.2 million ha with approximately 45% of the world's production in West Africa, where it is important in 17 countries (FAO 2014). It is ranked as the world's sixth most important cereal crop. It has been grown in Africa and the Indian subcontinent since prehistoric times. The center of diversity, and suggested area of domestication, for the crop is in the Sahel zone of West Africa. Recent botanical archaeology research has confirmed the presence of domesticated pearl millet in northern Mali's Sahel zone between 2500 and 2000 BC (Azare et al. 2020). It is grown in semi-arid to arid zones where soils have sandy textures, low organic matter, and nutrient levels, limited and erratic rainfall, high soil and air temperatures, and the growing season is short and varies greatly between years (Stephen et al. 2015). Pearl millet is typically a short-day plant that responds to climatic variables, but some varieties are day neutral. It is not cultivated during the cold dry season due to its general sensitivity to low temperatures, particularly during the seedling and flowering stages. However, with the release of day-length neutral varieties, millet production can be promoted during the post-cold dry season. Grain maturation requires high daytime temperatures. Under favorable conditions, millet germinates well at soil temperatures ranging from 23 to 30 °C and emerges in 2–4 days. Although the crop is grown in areas where rainfall ranges from 200 to 1500 mm, it is mostly grown in areas where rainfall ranges from 250 to 700 mm. Early maturing cultivars are generally used in low-rainfall areas. Despite its drought resistance, pearl millet requires even rainfall throughout the growing season. Crop failure is caused by excessive rain during flowering. Pearl millet, like most plants, thrives in light, well-drained, loamy soils. Millet is more tolerant of poor, infertile soil than rice, maize, and sorghum. It performs poorly in clay soils and is unable to withstand waterlogging. It tolerates acidic (even low pH 4–5) and aluminum-rich subsoils (Ajeigbe et al. 2020a, b).

17.2 Overview of Millet Production Systems in Semi-Arid Africa

Pearl millet is the most important cereal crop in the semi-arid and arid regions of West Africa, supplying a large proportion of the dietary energy for the human population, nutritious fodder for livestock, and stalk for cooking fuel. The crop is generally grown under traditional low-input extensive agricultural systems. In the traditional systems, farmers hardly use purchased inputs, including seeds and fertilizers (Ajeigbe et al. 2005). In semi-arid Africa, pearl millet is most grown in intercropping systems with cowpea (Reddy et al. 1992), and occasionally in agro-forestry systems (Reij and Smaling 2008). Plant population is generally low (less than 9000 hill/ha) due to a wide planting distance which in some systems can be as wide as 1.5 × 1.5 m either intentionally planted or as a result of poor-quality seeds, seedling death through drought or a combination of the above (Ajeigbe et al. 2019). The yield and productivities from these systems are very low averaging less than 1 t/ha (Singh and Ajeigbe 2007). Pearl millet can also be grown as a stand-alone crop or intercropped with groundnut, sorghum, or maize (Ajeigbe et al. 2020a, b). Several other production systems of pearl millet exist in Semi-Arid Africa. This includes but not limited to agro-forestry, crop rotation, and bush fallow systems. In many cases, farmers are not able to produce enough grains for home consumption talk less of an excess for the markets. Any future intensification would include the use of inputs including machinery, either tractor or bull-drawn implements.

17.2.1 Intercropping Systems

Traditionally, the farmers in the savannas of West Africa practice intercropping of legumes (cowpea and groundnut) with cereals (maize millet and/or sorghum). The farmer practices different systems to suit the locations and resources at hand. Millet is planted first and may be followed by sorghum, groundnut, and cowpea in defined and undefined row arrangements and spacing. These diversified systems are to satisfy different dietary requirements and to make efficient use of labor and other resources but also as risk management strategies to avoid total crop failure in emergency cases. The systems have historically been extensive with low use of non-labor inputs. The crop is normally grown at a low density ranging from 10,000 to 20,000 plants/ha. Application rates of organic matter are low, chemical fertilizer use is negligible, and mechanical tillage before planting is practiced only in some areas. It has been noted that in intercropping millet with cowpea in the semi-arid zone, millet yield is only reduced if cowpea is planted simultaneously with millet (Ntare 1990; Ntare and Williams 1992).

17.2.2 Agroforestry System

In West Africa, pearl millet is commonly grown alongside trees, particularly *Faidherbia albida* (Del.) A. Chev (Garrity et al. 2010). *F. albida* is a multipurpose, deep-rooted, leguminous tree with reverse phenology, meaning it has leaves during the dry season but drops them during the rainy season. Pearl millet growth and yields are greater under the *F. albida* trees, with reports of a 36–169% increase (Mokgolodi et al. 2011) attributed to higher soil nutrient levels, higher water availability, improvement in microclimate, and better soil physical properties (Kho et al. 2001). Gnankambary et al. (2008) showed that N and P applications accelerated the rate of *F. albida* litter decomposition and nutrient release, thereby stimulating late-season growth under trees. According to Mokgolodi et al. (2011), a dense stand of *F. albida* per hectare added 50 t/ha of manure to the soil. Payne et al. (1998) found that soil fertility and fine-particle content decreased as one moved away from the tree's center, and that pearl millet yielded more near the tree's canopy edge. They proposed using differences in soil water, nutrient levels, and temperature between the tree's center and perimeter to diversify cropping systems and increase grain yields. Alley cropping of millet in between rows of moringa (*Moringa oliefera*) is also found in locations in Niger. Given the need for the development of research-based solutions to produce higher yields from crops for the increasingly growing human and ruminant animal population, and the global call for massive planting of trees as part of the solutions to reduce the effect of climate change and global warming, increased awareness to farmers to adopt pearl millet production in stands of *F. Albida* is necessary. *F. Albida* is the best bet candidate for a tree-crop-livestock system, especially with millet in the semi-arid and arid regions of Africa.

17.2.3 Crop Rotation Systems

In semi-arid areas of Africa, crop rotation of pearl millet with other crops has long been encouraged to boost yields, encourage more effective nutrient cycling, increase infestation with beneficial arbuscular mycorrhizae, reduce pest infestations (Abdou et al. 2012), increase productivity, stability, and sustainability, and improve soil (Buerkert et al. 2001, 2002; Kadi et al. 1990). Pearl millet has been long grown in rotation largely with leguminous crops such as cowpea and groundnuts, and in some instances with Bambara groundnut and sorghum, and to a lesser extent with soybean in many regions within the Semi-Arid Africa. As a practice that can provide long-term sustainability in crop production, particularly in the nutrient depleted areas of Semi-Arid Africa, millet should not be continuously grown on a single land exceeding one cropping season and should not be rotated with another cereal crop because of their nutrient-exuding rather than replenishing nature, and as such, the practice of rotation with leguminous crops should continue. The decision of the crops in the rotation is complex and farmers must weigh on domestic needs, monetary gains or crop market value, input availability, and soil fertility considerations.

17.2.4 Bush Fallow Systems

Bush fallow has traditionally been used in semi-arid African cropping systems to naturally improve soil organic matter and nutrient levels at a low cost. Traditionally, these systems produce crops for 3–5 years, followed by 5–10 years of bush fallow to replenish soils. Due to increased population growth, soil degradation, and economic development, bush fallow cannot currently meet crop nutrient needs and is used by only 2% of pearl millet producers in West Africa (Schlecht and Buerkert 2004; Schlecht et al. 2006).

17.3 Challenges Facing Millet Production

Pearl millet, like other crops, is faced with production challenges in Semi-Arid Africa as well as Asia. Some of the threats to its production are abiotic in nature, while others are biotic in nature. The present low yields of pearl millet in the dryland of West Africa could be broadly attributed to a lack of inputs, including improved varieties, and fertilizer, as well as low planting density in smallholder farmers' fields. Global warming is likely to have an impact on global agriculture. According to recent research, major commodity production has decreased because of global warming since 1980 (Lobell et al. 2011). Given current warming trends in Sub-Saharan Africa, cereal production could drop by up to 20% by mid-century (Schlenker and Lobell 2010). Stress is defined as a condition that limits a plant's ability to reach its full potential for growth, development, and reproduction; extreme stress kills the plant. Plants rarely grow in ideal conditions, so they are constantly stressed. For pearl millet growing in marginal environments, this is certainly the case. In this chapter, we will address some of the abiotic and biotic challenges affecting millet production and do a short analysis of the improvement of the crop under those challenges.

17.3.1 Abiotic Challenges

Among the major abiotic challenges affecting millet production are high temperature, rainfall distribution in time and space, and soil physical and chemical characteristics.

17.3.1.1 Temperature

Among the environmental factors affecting millet germination, growth, and productivity is the temperature at which plant sensitivity varies throughout its development stages. However, the seedling phase tends to be particularly vulnerable. Even though millet is grown and is adapted to arid and semi-arid tropics, the growing season in this zone is characterized by high temperatures leading to a hot, dry seedbed environment during crop establishment, inducing soil surface temperatures often greater than 55 °C (Peacock et al. 1993). High air temperature is also recorded. In

Niger, for instance, Sivakumar et al. (1984) observed mean maximum in July, the main planting period, air temperatures ranging from 35 to 40 °C in millet growing areas. This high-temperature results in a water deficit occurring in the seedbed environment. In the sandy soils of many pearl millet growing environments, moisture depletion to less than 2% often occurs three to four days after rainfall (Peacock et al. 1993). During seed germination soil moisture depletion induced by high temperature appears as the major cause of poor germination and death of seedlings. Under such conditions, the roots have to grow rapidly to access water. Seedling emergence is the second stage affected by soil and air temperature. Under high temperature-induced conditions, the shoot often must penetrate a soil surface crust, and once emerged, it is exposed to extreme temperatures. After emergence, seedling survival is also strongly affected by high soil and environmental temperature, which leads to diminishing soil water availability. At low water potentials, cell elongation is reduced, and roots may not be able to grow quickly enough to escape the rapidly drying surface layers of soil and penetrate deeper moisture-containing layers. Long-term survival in drying soil is dependent on the establishment and growth of nodal roots. In fact, pearl millet develops a single primary root upon seed germination, and the first adventitious or nodal roots appear at the shoot base after the seedling has developed two to three leaves. Reduced soil surface moisture can prevent nodal root formation (Harris 1996). To counteract this adverse effect of high temperature and water deficit, deep sowing may be the option. However, in this case, seedlings take a longer time to emerge. The above said indicates that, although the effects of heat and drought are often examined separately in the field, these stresses frequently occur concurrently with others.

Fast seedling growth and consequent early seedling establishment is one strategy to escape a stressful environment, particularly as the conditions for seedling establishment become increasingly less optimal with time after sowing. Stomph (1990) and Peacock et al. (1993) observed that post-emergence seedling death due to abiotic stress under field conditions is primarily caused by the prevalent high soil surface temperatures, at least in the first ten days following sowing, and only after that does water deficit start to take effect. All the above said indicates that temperature is among the main factor determining the rate of plant growth.

17.3.1.2 Rainfall Distribution in Time and Space

Farmer uncertainty about crop yields is increasing as a result of poor rainfall and variability, which are putting semiarid rainfed agricultural production under strain (Graef and Haigis 2001; Bayu et al. 2005). Climate change is projected to increase this variability (Nielson 1999). In general, the irregular occurrence of rainfall, rather than low annual rainfall, is a critical constraint for pearl millet production (Sivakumar and Wallace 1991). Silungwe et al. (2019) demonstrated that seasonal rainfall could vary significantly for neighboring fields separated by less than 200 m, negatively impacting pearl millet yield. Given the unpredictable rainfall conditions exacerbated by climate variability, several strategies have been developed to assist farmers in dealing with rainfall variability in semiarid areas. Land restoration using locally adapted integrated soil and water management practices has been promoted

(Nielson 1999; Zougmoré et al. 2014; Somé et al. 2000; Fatondji 2002; Silungwe et al. 2019). Furthermore, intercropping cereals and legume crops are used not only to restore soil fertility but also to ensure harvesting of the early maturing crop (legume) even if the rainy season ends early (Osman et al. 2011). Crop improvement strategies that favor climate variability, on the other hand, have given farmers opportunities to use early-maturing and drought-tolerant crop varieties to deal with rainfall variability (Ouedraogo et al. 2010). Besides, crop management that combines soil management practices and the use of improved varieties has also been used as a viable option to buffer or cope with risks associated with rainfall variability, thus making crop production in harsh and unpredictable climatic environments smarter (Bielders and Gérard 2015; Sanou et al. 2016).

17.3.1.3 Soil Physical Characteristics

Millet is grown in sandy soils with low fertility and water-holding capacity in the dryland, particularly in the Sahel. These soils are deficient in organic matter and prone to surface crusting and erosion. They are managed with few external inputs, in which case the decline in soil organic levels is an unavoidable result of clearing and cultivation, which causes other hazards (Ahn 1970; Pieri 1989; Feller and Beare 1997). Cultivation clears the soil, and when exposed to heavy rains, the almost bare soil suffers serious structural deterioration (Casenave and Valentin 1992). Water drop impact disrupts finer particles, including organic matter from sand, clogging soil pores. As a result, the Sahelian sandy soils are extremely susceptible to surface crusting (Ambouta et al. 1996). Soil surface crusting reduces infiltration rate, causes runoff, and thus water erosion, making cultivation in dry-prone areas difficult (Valentin 1986). Transport of soil particles by water and wind contributes to the removal of the soft arable soil layer, leaving the sub-soil bare, which when submitted to the alternate effect of rain drops sun leads to further formation of a lateritic crusted layer characterized by the existence of a hard pan that limits water infiltration and seed emergence. Under such conditions, high surface run-off is observed which worsens the soil condition as further degradation occurs. However, due to high population growth and consequent limitation in available arable land, farmers are constrained to use the degraded crusted soil for agricultural production.

17.3.1.4 Soil Chemical Characteristics (Nutrient Deficiency—N, P, K)

Soil nutrient deficiencies, especially in P and N, have been reported among the major constraints of pearl millet (Scott-Wendt et al. 1988). As a result, research in mineral fertilizer strategies for increasing pearl millet yields has focused on meeting its N and P requirements, which are the two most limiting nutrients on weakly buffered African soils (Bationo et al. 2003). Phosphorus is important in increasing pearl millet productivity. According to Fofana et al. (2008), the application of phosphorus fertilizer alone led to steady and substantial pearl millet grain and straw yield increases. Even though several studies have found that P availability is critical for millet production on weakly buffered, acidic sandy Sahelian soils (Manu et al. 1991; Bationo et al. 1992), the highest grain yield was obtained at the highest N and P application rates. According to Michels and Bielders (2006), pearl millet yield tripled after phosphorus addition and increased by a factor of 13.5 when additional nitrogen was applied to eroded sandy soil in the Sahel. This implies that N and P are both important factors in millet production, with P being the most limiting factor. Millet only responded to N fertilization when P fertilizer was applied. Bationo et al. (2003) conducted a study in Niger that demonstrated that traditional millet-cowpea rotation in the Sahel does not increase millet yields unless N and P fertilizers are added, and they proposed a cropping system that incorporates millet-legume rotation and a mixture of N and P fertilization as an appropriate alternative for restoring soil fertility on degraded soils. The Sahel region has a very good record of the effects of mineral fertilizer, both positive and negative, on millet yields (Bationo and Mokwunye 1991; Bationo et al. 2003, 2011; Muehlig-Versen et al. 2003). Because of significant potassium inputs from Harmattan dust, potassium (K) is not considered a limiting nutrient for pearl millet productivity in semi-arid Africa, particularly in the Sahel (Herrmann et al. 1996). On the other hand, Rebafka et al. (1994) reported large responses to K fertilizer in the absence of crop residue application. Furthermore, Voortman et al. (2004) emphasized the significance of K in explaining the spatial variability of millet growth in the Sahel. Herrmann et al. (1996) found that potassium deficiency in the soil is a limiting factor for optimal pearl millet production with long-term application of crop residues and mineral fertilizers in the Sahel. This indicates that the potassium supply from atmospheric deposition and crop residue/ compost application was insufficient to meet millet potassium requirements for higher grain yield formation. However, the use of mineral fertilizer to meet pearl millet N, P, and K demands is still very low in pearl millet-based cropping systems. According to the African Fertilizer Summit (2006), 55% of Sub-Saharan African countries consume less than 8 kg/ha of fertilizer. As a result, there is a need to increase the availability of nutrients (N, P, and K) to increase pearl millet productivity in Africa.

17.3.1.5 Biotic Challenges

The overall grain yield of pearl millet in current traditional systems is low (less than 1 t/ha) due to a complex of biotic and abiotic factors. Insect pests, parasitic flowering plants, and viral, fungal, and bacterial diseases are examples of biotic factors. Diseases caused by living organisms can spread from plant to plant or via vectors. The most common biotic challenges for pearl millet are downy mildew, Cercospora leaf spot, and striga infestation.

Pearl Millet Downy Mildew (*Sclerospora graminicola*)

The initial symptom downy mildew develops chlorosis at the base of a leaf, with subsequent leaves on the plant becoming increasingly chlorotic, with the chlorosis always extending from the base to the tips of the leaves. The underside of chlorotic leaves develops a white powdery or downy growth of sporangia and the spores that they bear. There is a clear distinction between diseased and healthy leaf tissue at the base of the leaf. Inflorescences of infected tillers develop the characteristic green-ear symptoms where the grain is replaced by whip-like vegetative structures. Inflorescences can be partially or fully converted to green ear, depending on the time in inflorescence development at which the infection occurred. A partial green-ear always begins at the base of the inflorescence. Plants infected at an early stage of development before tillering begins can be killed. Infected plants should be rouged when found in the field. Pearl millet downy mildew can be effectively controlled using metalaxyl-based fungicide seed treatments. Application rates are as low as 2 g a.i. per kg seed; using specially formulated seed-treatment products can provide complete control of downy mildew even with highly susceptible hybrids. Downy mildew remains a major biotic constraint to pearl millet production, particularly in western Africa. Control of downy mildew in pearl millet is a strategic priority for West African agriculture. Yield losses of up to 50% have been reported to occur in West Africa due to downy mildew (ICRISAT 2009), and management of the disease has become a priority in research. Intensive research on pearl millet downy mildew has been conducted by scientists at ICRISAT and in several national programs during the last three decades. Significant advances have been made in understanding the biology of the pathogen, and the epidemiology of the disease, and in developing artificial screening techniques, identifying sources of resistance, breeding for disease resistance, and developing alternative control measures.

Cercospora Leaf Spot (*Cercospora penniseti*)

Cercospora leaf spot is characterized by small dark lesions on leaves which are usually oval but may be oblong to rectangular; centers of lesions are gray to tan in color with visible black dots; lesions may be covered in spores during wet weather; lesions may also be present on the stems and are slightly longer than those on the leaves. Disease emergence occurs when high temperatures coincide with periods of high humidity. Avoid planting millet varieties that are highly susceptible to *Cercospora fungi*; control weeds in the field; rotate crops and practice good sanitation; no chemical control is needed to treat the disease.

Pearl Millet Ergot (*Claviceps fusiformis*)

Ergot is a disease of pearl millet grain-bearing heads (inflorescences). The first sign of ergot infection in pearl millet is the appearance of viscous creamy-pink exudations, called honeydew on the flowering heads. Subsequently, hard brown spiky structures are formed, called sclerotia. Ergot reduces grain yields and constitutes a health hazard to humans and animals that consume pearl millet products contaminated with the alkaloid-containing sclerotia of the causal fungus. Ergot can be managed in pearl millet, using host plant resistance, pollen management, avoidance, and sanitation. Host plant resistance to ergot can be developed using appropriate inoculation, hybridization, and selection techniques. Sources of host plant resistance are developed and made available by public- and private-sector seed companies. Avoidance of ergot is achieved through the selection of cultivars that flower during periods of fine dry weather so that pollen wash and pathogen germination are minimized. Sanitation measures involve the removal and burning of infected inflorescences as soon as ergot symptoms are seen. When this is done at the honeydew stage, it will reduce secondary spread within the crop. When done at the sclerotia stage, it reduces the inoculum bank for subsequent seasons' infections as well as the dangers of poisonous alkaloids getting into food and feed products.

Rust (*Puccinia substriata*)
Rust is indicated by small yellow or white raised spots on upper and lower leaf surfaces; spots tend to be more numerous on the lower leaf surface, and spots enlarge and develop into red-brown pustules, which may be surrounded by a yellow halo (Wilson et al. 1995). Spores can be spread via wind and survive in crop debris in the soil. Plant resistant varieties of millet if rust is a persistent problem; avoid the use of overhead irrigation; do not plant millet near eggplant or other Solanaceous crops; there are no recommended chemical controls for rust.

Smut (*Tolyposporium penicillariae* Bref)
Smut is a disease of pearl millet grain-bearing heads (inflorescences) that can significantly reduce grain yields, wherever the crop is grown. Infected florets develop as plump sacs or sori, initially green in color and then turning dark brown or black as the sori mature. The sori are filled with black spores of the causal fungus, hence the name smut. Smut is best managed using resistant cultivars

Striga (*Striga hermonthica*)
The parasitic weed *Striga* is present throughout West and Central Africa, and farmers need to monitor and control this potentially devastating pest. The biology and genetics of the parasite and the host-parasite interactions are now understood in more detail (Ejeta 2007). Farmers' experiences with the integration of control techniques indicate the need to combine at least three or more control options. Approaches that contribute to *Striga* control include: (1) enhancement of millet growth, e.g., fertilization with optimum timing and placement; (2) reduction in the number of *Striga* seeds in the soil by a) cultivating a false host (trap crop) that induces suicidal *Striga* germination, b) applying compost that increases soil microbial activity which in turn reduces the viability of *Striga* seeds, and c) cultivating a resistant variety; (3) suppression of *Striga* growth by practices such as sowing spreading cowpea that smothers the *Striga* plants; and (4) reduction in *Striga* seed production and dispersal by late weeding and particularly by hand-pulling of *Striga* plants. Early sowing of the millet crop may also help reduce *Striga* parasitism as *Striga* seeds require two to three weeks of moist "conditioning" before they can germinate (Ejeta 2007) and attach to host roots.

Nutrient-deficient soil conditions contribute to both more *Striga* germination and enhanced *Striga* growth and seed production with reduced shading by poorly developed host canopy. Compost applications showed advantages over inorganic fertilizers for integrated soil-fertility and *Striga* management for both sorghum growth and reduction in *Striga* seed viability in on-farm experimentation (Van Mourik 2007). It also applies to millet. Farmers' adoption of *Striga* control practices also depends on obtaining some benefits in the first year of application, even if benefits accrue over many years, since sorghum growers operate under severe resource constraints. The use of spreading cowpea is an example of a particularly attractive practice, as it impedes *Striga* and produces valuable hay.

17.4 Improving Millet Production Under Challenging Conditions

The increase in human and livestock population and competition for land for agriculture, recreation, industries, and other uses implies that the agricultural systems must be intensified to provide food, feed, and raw materials for the human and livestock population and industrial uses, respectively. The availability of improved millet varieties that also responds positively to added input (Ajeigbe et al. 2019) means there are scopes to increase the productivity of the millet systems through improved genotypes, use of inputs and appropriate agronomic systems. Pearl millet is the only cereal that reliably provides grain and fodder under dryland conditions in shallow and sandy soils with low fertility and water-holding capacity. It can also be grown in areas where rainfall (200–600 mm) is insufficient for maize and sorghum. Millet is said to perform better on marginal lands than other cereals (Tabo 1995). Millet will typically produce reasonable yields even with little rainfall. Furthermore, despite the persistence of various biotic stresses such as fungal, viral, and bacterial infections, pearl millet is regarded as an excellent model for biotic stress due to its ability to withstand harsh environmental conditions (Shivhare and Lata 2017).

17.4.1 Management Practices to Improve Millet Production

To address the challenges described above, various management practices have been developed to improve millet productivity in semi-arid Africa. These management practices include soil fertility management practices, cropping systems options, and water-harvesting technologies. The following subsections highlight achievements in these management practices.

17.4.1.1 Soil Fertility Management

The soil fertility management practices for improving pearl millet productivity were generally focused on combining the use of organic and inorganic fertilizers, micro-dosing fertilizer application strategy, crop residue management, manure management through corralling, and integrated use of harvesting technology and nutrient. Crop performance and cropping patterns in semi-arid environments are determined by rainfall and soil fertility (Pandey et al. 2001). Though pearl millet thrives and can be grown in low rainfall areas (200–250 mm), making it one of the most reliable cereal crops in the Arid and Semi Arid Tropics, the insufficient supply of plant nutrients use by farmers limits the efficient use of limited rainfall, resulting in unstable millet productivity (Pandey et al. 2001). Ajeigbe et al. (2020a, b) found that increasing N-fertilizer levels increased millet water use efficiency and confirmed optimal applications of 60 kg N/ha at Minjibir (rainfall 600–800 mm/annum) and 80 kg N/ha in Gambawa (rainfall 400–500 mm/annum), while maintaining 30 kg P_2O_5/ha and 30 kg K_2O/ha in both locations for optimum millet grain and stover yields. However, due to the high cost and scarcity of inorganic fertilizer for small-holder farmers, ICRISAT and its partners developed fertilizer microdosing technology and found it suitable for small-holder farmers in Africa's semi-arid and arid regions.

17.4.1.2 Micro-Dosing Application of Organic and Inorganic Fertilizer

Farmers' use of mineral fertilizers to sustain millet and sorghum production in Africa remains very low and unappealing due to their high cost (Abdoulaye and Sanders 2005; Bationo et al. 2003). ICRISAT and partners have developed fertilizer micro-dosing technology to improve fertilizer efficiency and encourage smallholder farmers to increase the on-farm application of mineral fertilizer. This technology involves applying a small amount of mineral fertilizer (one-third of the recommended rate) to the target crop seeds on the planting hill at the time of sowing or a few weeks later (Hayashi et al. 2008; ICRISAT 2009). Fertilizer micro-dosing was first tested on millet and sorghum in the Sahel (Burkina Faso, Mali, and Niger) and then promoted in other African countries on other crops (e.g., maize). Fertilizer micro-dosing is commonly regarded as a first step toward the sustainable intensification of agricultural systems in arid regions (Aune and Bationo 2008; Twomlow et al. 2010). This technology has shown promising results in increasing millet yields in the Sahel (Table 17.1). The application of small amounts of fertilizer provides farmers with a good return on investment and low financial risk (Aune et al. 2007; Hayashi et al. 2008; Bielders and Gérard 2015). Fertilizer micro-dosing tests in Niger, Mali, Burkina Faso, and Sudan resulted in a greater average grain yield of millet and sorghum (12–172%) compared with control plots (without fertilizer) (Table 17.1). However, compared with recommended levels, micro-dosing produces lower yields and is less profitable (Muehlig-Versen et al. 2003; Ibrahim et al. 2016a). Farmers would be wise to target micro-dosing preferentially to low-fertility fields or parts of fields, which is consistent with their traditional soil fertility management strategies (Bielders and Gérard 2015). Later authors have also shown that micro-dosing helps to mitigate the negative effects of late sowing in semi-arid and arid areas. The beneficial effects of fertilizer micro-dosing on yield increase result from improved nutrient utilization due to the early development of many lateral roots within the topsoil (Ibrahim et al. 2016b). Other studies have shown that combining fertilizer micro-dosing with an organic amendment increases millet yield more than a single application of fertilizer micro-dosing (Ibrahim et al. 2015a, 2016a). Although fertilizer micro-dosing has been shown to have positive effects in enhancing crop yield, economic return, and climatic risks, this technology has some limitations which need to be resolved. Fertilizer micro-dosing increases the risk of soil nutrient mining. Several studies show that fertilizer micro-dosing enhances crop yields but increases the risk of soil nutrient depletion in low-input cropping systems (Ibrahim et al. 2016a; Nziguheba et al. 2016; Tovihoudji et al. 2017). Several suggestions have been made to overcome nutrient mining associated with fertilizer technology

Table 17.1 Yield gains due to fertilizer micro-dosing in different African countries

References	Country	Crop	Research approach	Treatments	Yield (kg/ha)	% yield increase compared to control
Muehlig-Versen et al. (2003)	Niger	Millet	On-station	Control	587	
				Fertilizer broadcasting (13 kg P/ha)	1189	103
				Fertilizer microdosing (3 kg P/ha)	855	46
				Fertilizer microdosing (5 kg P/ha)	967	65
				Fertilizer microdosing (7 kg P/ha)	1046	78
Manyame (2006)	Niger	Millet	On-farm	Control	406	
				Fertilizer microdosing (2 g DAP/hill)	642	58
				Fertilizer microdosing (2g DAP/hill + urea (1 g/hill)	664	64
Hayashi et al. (2008)	Niger	Millet	On-station	Control	362	
				Fertilizer microdosing (6 g NPK/hill)	746	106
Aune and Ousman (2011)	Sudan	Millet	On-station	Control	332	
				Fertilizer microdosing (3 g NPK/hill)	436	31
				Fertilizer microdosing (6 g NPK/hill)	434	31
				Fertilizer microdosing (9 g NPK/hill)	488	47
Bagayoko et al. (2000)	Burkina Faso	Millet	On-farm	Control	237	
				Fertilizer microdosing	545	130
	Niger	Millet	On-farm	Control	328	
				Fertilizer microdosing	577	76
	Mali	Millet	On-farm	Control	1054	
				Fertilizer microdosing	1444	37

(continued)

Table 17.1 (continued)

References	Country	Crop	Research approach	Treatments	Yield (kg/ha)	% yield increase compared to control
Ibrahim et al. (2014)	Niger	Millet	On-station	Fertilizer microdosing applied at 5 cm (control)	1005	
				Fertilizer microdosing applied at 10 cm	1217	21
Bielders and Gérard (2015)	Niger	Millet	On-farm	Control	242	
				Fertilizer microdosing (2 g DAP/hill)	312	29
				Fertilizer microdosing (6 g NPK/hill)	305	26
				Fertilizer microdosing (2 g DAP/hill + 1 g urea/hill)	340	40
Ibrahim et al. (2015a)	Niger	Millet	On-station	Fertilizer microdosing (control)	555	
				Fertilizer microdosing + Manure (hill placement)	1027	85
Ibrahim et al. (2015b)	Niger	Millet	On-station	Fertilizer microdosing (control)	826	
				Fertilizer microdosing + Manure	1172	42
				Fertilizer microdosing + millet mulch	1056	28
				Fertilizer microdosing + *Acacia tumida* mulch	1157	40
Ibrahim et al. (2016a)	Niger	Millet	On-station	Control	357	
				Fertilizer microdosing (6 g NPK/hill)	498	139
				Fertilizer microdosing (2 g DAP/hill)	615	172

(continued)

Table 17.1 (continued)

References	Country	Crop	Research approach	Treatments	Yield (kg/ha)	% yield increase compared to control
Ibrahim et al. (2016b)	Niger	Millet	On-station	Control	510	
				Fertilizer broadcasting (200 kg NPK/ha)	1028	101
				Fertilizer microdosing (6 g NPK/hill)	766	50
				Fertilizer microdosing (2 g DAP/hill)	910	78
Aune et al. (2007)	Mali	Sorghum	On-farm	Control	685	
				Fertilizer microdosing (3 g NPK/hill)	966	41
				Fertilizer microdosing (6 g NPK/hill)	1235	80
Aune and Ousman (2011)		Sorghum	On-station	Control	410	
				Fertilizer microdosing (3 g NPK/hill)	617	50
				Fertilizer microdosing (6 g NPK/hill)	692	69
				Fertilizer microdosing (9 g NPK/hill)	860	110
Abdalla et al. (2015)	Sudan	Sorghum	On-farm	Control	582	
				Fertilizer microdosing (3 g NPK/hill)	650	12
				Fertilizer microdosing (6 g NPK/hill)	831	43
				Fertilizer microdosing (9 g NPK/hill)	1074	85
Coulibaly et al. (2019)	Mali	Sorghum	On-farm	Control	1218	
				Fertilizer microdosing	1639	35

% Yield increase compared to control as relative deviation of treatment from control: = (Yield in treatment/Yield in control) − 1) × 100

through combining micro-dosing with organic amendments, e.g., the use of organic manure or compost (Ibrahim et al. 2016a; Tonitto and Ricker-Gilbert 2016), or complementing it with sustainable practices such as intercropping with nitrogen-fixing legumes or retaining crop residues (Ibrahim et al. 2016a). Furthermore, the adoption of micro-dosing at a large scale requires favorable and supportive institutional arrangements, which would include access to credit, subsidizing fertilizer, and making it available in small packages to facilitate farmers' access to it by smallholder farmers (Mwinuka et al. 2017; Tabo et al. 2007). Fertilizer micro-dosing is currently limited to cereal in mono-cropping systems and has not been tested for intercropping or mixed-cropping systems. Further studies are therefore needed to determine the different rates of fertilizer micro-dosing in diversified cropping systems and the contribution of fertilizer micro-dosing, for example, in increasing mycorrhizal infection for enhancing plant growth and nutrient uptake in low-input millet-based cropping systems for reducing soil nutrient mining particularly phosphorus which is limiting in most African soils.

17.4.1.3 Crop Residues Mulching

Crop residue management is an important production practice that has the potential to reduce runoff and increase water infiltration into the soil by capturing aeolian material with higher nutrient levels (Michels et al. 1995), minimize soil crusting effect on crop emergence, enhance nutrient recycling and crust remediation through termite and microbial decomposition (Stroosnijder et al. 2001; Buerkert et al. 2000; Muehlig-Versen et al. 1997; Klaij and Hoogmoed 1993), lower soil temperature and reduce soil evaporation. Residue mulching also had the potential to increase biological N fixation through legume rotation or intercropping (Rebafka et al. 1994).

Several studies reported by Mason et al. (2014) show yield increases from leaving residue on the soil surface. In a long-term trial in Segou Mali (Coulibaly et al. 2000), crop residue incorporation consistently increased grain and stover yields (Tables 17.2 and 17.3). The seven-year mean pearl millet grain yield was 180 kg/ha/year (12%) higher in plots with crop residues incorporated than in plots with residues removed. The mean pearl millet stover yield in plots with crop residue incorporated was 211–247 kg/ha/year (15–18%) higher than in plots with residues removed or retained on the surface.

Table 17.2 Effect of crop residue management on continuous pearl millet grain yields

	1991	1992	1993	1994	1995	1996	1997	Mean
Crop management	Grain yield (kg/ha)							
Removed	1266	1472	1392	1322	1861	1820	1154	1469
Surface	1403	1687	1095	1424	1800	1896	1297	1515
Incorporated	1413	1750	1596	1429	1907	1981	1468	1649
LSD (0.05)	NS	NS	375	NS	NS	NS	NS	NS[a]
C.V. (%)	22	28	16	23	15	18	18	22

NS not significant

Source: Coulibaly et al. (2000)

[a] Significant at $P = 0.07$; LSD (0.05) = 184

Table 17.3 Effect of crop residue management on continuous pearl millet stover yields

	1991	1992	1993	1994	1995	1996	1997	Mean
Crop management	Stover yield (kg ha^{-1})							
Removed	3174	2604	3296	3825	4557	4150	1302	3273
Surface	3703	3066	2116	3845	4598	4028	1303	3237
Incorporated	2848	3662	3296	3805	4313	4359	2116	3484
LSD (0.05)	NS	NS	NS	NS	NS	NS	NS	NS
C.V. (%)	28	34	28	19	17	27	42	27

NS not significant
Source: Coulibaly et al. (2000)

This could be due to increased microbial activity, a faster rate of crop residue decomposition, and lower soil bulk density, all of which promote root proliferation and penetration. Most semi-arid African soils have low water-holding capacities and organic matter. Although agronomic studies show that leaving (or incorporating) crop residues is important for soil maintenance and pearl millet yield (Mason et al. 2014), residues are commonly removed from fields for a variety of reasons, including livestock feed and fuel (Buerkert et al. 2002). Farmers typically apply crop residues to low-producing micro-sites within fields to combat wind erosion, and economic studies have demonstrated that this is an agronomically and economically sound practice (Buerkert et al. 2002). Despite significant economic incentives to remove crop residues from fields, retaining some crop residues in the field is critical to ensuring sustainable crop and soil management practices, particularly given the current land-use intensification for crop production to increase global food production.

17.4.1.4 Millet Production and Animal Corralling

In the semi-arid zone, farmers often grow pearl millet as a continuous crop. However, soils in these zones are predominantly sandy with very low organic carbon content (0.2%). To sustain production under such conditions, replenishing the soil organic pool is necessary. In smallholder farmers' field, this can be done either through crop residue or organic manure application. However, the crop residue available at the farm level is in limited quantity for its application as mulch or fertilizer because it is used for other purposes (Baidu-Forson 1995). Organic manure is the other form of bringing organic matter to the soil. This calls for agriculture and livestock integration, a system that is applied in the Sahel either through transport to the field or animal corralling. Animal corralling consists of living the cattle or small ruminants over the night and during the day on fields that are not cultivated or after harvest for cropped fields for the livestock to feed on crop residues and weeds and stay in the field if there is sufficient food. When this is no longer the case, the shepherd leads the flock to the pasture and returns to the field at night. Like this, animal dejection and urine are left to fertilize the field resulting in the application rate of 10–20 tons ha^{-1} of dry matter depending on the number of animals and the number of corralling days. Considering that a bovine can give, on average, 1.5 kg of feces (Ayantunde et al. 2001) and shed up to 1.26 l of urine per night of corralling (Mamane 2002), and that 50–80% of the N excreted by a bovine is found in the urine and the urine is also rich in K, it is a system which favors incorporation of an important quantity of organic nutrient in the soil. In addition, this application of manure also mechanically protects the surface of the soil during the first months of the growing season, when the soil is poorly covered by culture, and traps wind sands during storms (De Rouw and Rajot 2004). The positive effect of this system on millet production is reported in many studies, which can last up to four years Gandah et al. (2003). Michels and Bielders (2006) have reported increases in grain yields of 188% following the application of farmyard manure (10,000 kg DM manure + 1250 l urine ha^{-1}) compared to mulched plots (2000 kg CR ha^{-1}). Dutordoir (2011), reporting research conducted in the Fakara in Niger, observed 179% millet straw and 168% of grain increase on newly corralled plots compared to non-amended ones. They also observed that the difference between the yield obtained in the newly corralled plot and that of the one-year old corralled plot was, on average higher than the difference existing between the 1-year old and 2-years old corralled plots, which they explain by the rapid disappearance of the urine and the exponential rate of decomposition of the easily mineralized part of OM brought to the soil. In fact, Powell et al. (1996) observed that approximately 50% of the N excreted by animals is present in the urine but can be lost rapidly through leaching, denitrification, and especially volatilization of NH_3 if not used by plants. Also, urine increases the availability of soil P (Powell and Valentin 1997). The P contained in manure mineralizes more quickly in the soil than the P of plant biomass. This rapid mineralization provides the P needed to boost root growth (De Rouw and Rajot 2004) and increase the number of productive tillers in millet, as observed by (Michels and Bielders 2006). All this is expressed in the high yield increase observed in the first year of corralling. However, the yields continue to decrease over the year but are still higher than the non-amended plots as organic matter decomposition rates decrease with increasing time because the remaining components are more loaded with lignin and polyphenols and are more difficult to mineralize (Esse et al. 2001). Hence, residues of organic matter on the soil break down more slowly after the first year and the mineral intakes available for plants decrease, which explains the gradual drop in yield observed over the years.

The corralling system is appropriate to sustain millet production under the poor soil fertility conditions of the Sahel; however, for sustainable production, the process needs to be repeated at least every 3 years. However, combining this practice with a small dose of inorganic fertilizer using the micro-dosing technology can extend this effect as we observed that the lower the fertility level of the soil, the higher is the effect of the inorganic fertilizer micro-dosing. Also, the positive effect of the combination of inorganic fertilizers with the organic ones is demonstrated through many research works.

Corralling has proved to be advantageous in terms of increasing millet yield; however, most farmers in the Sahel are poor and cannot afford to own the required number of animals for corralling a large plot in their fields. However, the system may be applied in limited dimension by targeting specific area in the field where fertility depletion is observed, which is often practiced by the famers.

17.4.1.5 Millet Performance Underwater Harvesting Technologies (Water and Nutrient Interaction)

The scarcity of cropland combined with unreliable annual rainfall due to climate change endangers the food and nutrition security needs of the semi-arid region's growing population. To improve subsistence farmers' agricultural productivity under these conditions, excess water must be stored to mitigate the negative effects of intra-seasonal dry spells (Molden et al. 2003). Several indigenous rainwater harvesting technologies are widely used by smallholder farmers in semi-arid regions, particularly in the Sahel, to collect, concentrate, store, and surface runoff for agricultural purposes. The planting basins known as Zai pits are the most common water harvesting techniques used to rehabilitate degraded soils and collect rainfall in the Sahel (Roose et al. 1993). Zaï is a small basin of varying size, typically 15–30 cm in diameter and 10–20 cm deep, used to collect runoff water (Bouzou and Dan 2004). During the dry season, this water harvesting technique is usually dug and filled with an organic amendment such as animal manure or compost (Roose et al. 1999; Fatondji et al. 2006; Amede et al. 2011). After the first rain, three to four sorghum or millet seeds are sown per pit. Zaï pits have been used successfully in the Sahel for many years and have become more popular in East and Southern Africa (Zougmoré et al. 2014; Malley et al. 2004). Farmers find Zaï extremely attractive because it reduces the risk of crop failure and provides more options for bringing into cultivation land previously not suitable for cultivation (Roose et al. 1999; Amede et al. 2011; Zougmoré et al. 2014). In Burkina Faso, the development of Zaï pits has led to an expansion of farm size with associated crop yield increment (Kabore and Reij 2004). This technique has been advocated to improve nutrient and water use efficiency, thereby increasing crop yield in a short time while restoring lands in the long term (Fatondji et al. 2006; Lahmar et al. 2012). The application of organic amendments is needed to improve the effectiveness of the zai technique for enhancing crop productivity. Compared with organic manure application on millet performance in Niger, Wildemeersch et al. (2015) showed that zaï and demi-lunes along with manure application produce significantly higher yields due to increased soil moisture levels throughout the season (Table 17.4). In another development, Wildemeersch et al. (2015) showed that after three growing seasons, soil organic

Table 17.4 Millet responses to water harvesting techniques in Niger

Treatment	Grain yield (kg/ha)	Straw yield (kg/ha)	Empty plant pockets (%)
Control	0	0	40a
Control + manure	149a	1198a	13ab
Scarification + manure	246b	–	7b
Demi-lunes + manure	249b	2188ab	12b
Zaï + manure	651c	3000b	14ab

Source: Wildemeersch et al. (2015)

carbon increased significantly from ±2.5 to ± 5 g/kg when manure was added to the water harvesting technique. In Burkina Faso, Zai pits can increase yields by up to 1.5 t/ha in years when rainfall is quite good (Roose et al. 1999). In Niger, Fatondji et al. (2006) reported that zai improved millet water use efficiency by a factor of about 2 and improved nutrient uptake in the range of 43–64% for N, 50–87% for P, and 58–66% for K as all nutrients are placed close to the plant and within reach of plant roots. However, other studies observed poor synchrony between nutrient release and pearl millet nutrient uptake, which could increase the risk of nutrient losses by leaching (Fatondji et al. 2009). In a study initiated by Fatondji and Ibrahim (2018) to evaluate the effect of varying the time of application of the manure in the Zai pit on millet production, they observed that application of manure prior to planting increased millet yields and enhanced nutrient utilization efficiency from decomposing manure under Zai pit by improving the synchrony between nutrients released and millet nutrient requirement. Hence, the quantities of N and P absorbed by millet at the tillering stage represent 61, 52, and 33% of N released and 15, 12, and 15% of P released at the same time when manure was applied before planting, at planting, and 15 days (about 2 weeks) after planting, respectively, and application of manure before planting increased millet grain yield by 16 and 20% compared to manure applied at planting and 15 days after planting, respectively. Ibrahim and Fatondji (2020) showed that split application of manure (half before planting and half at planting) enhances nutrient synchronicity, thereby increasing nutrient and rainwater utilization efficiencies under Zai pits. This water harvesting technique by creating wet conditions has a significant impact on soil biological properties. Wildemeersch et al. (2015) observed a significantly elevated nematode count with water harvesting techniques in Niger. This technique has been advocated to improve nutrient and water use efficiency, thereby increasing crop yield in a short time while restoring lands in the long term (Lahmar et al. 2012). Although the positive impact of the water harvesting technique on crop productivity is remarkable, it is worth noting that digging Zai pits is labor-intensive. Kabore and Reij (2004) reported that at least 450 h are needed for digging Zai pits on an area of 1 ha, and 150 h are needed for applying fertilizer in the dug pits. The high labor requirements make this technology unattractive for resource-constrained smallholder farmers. There is, therefore, a need to develop supportive policies aimed at facilitating access to credit for farmers for better adoption of zai techniques in the semi-arid regions.

17.5 Cropping System

17.5.1 Integrated System for Millet Production DEF (Dryland Eco-Farm)

Soil erosion (wind and water), low soil fertility (mining agriculture), low income (vicious circle of low crop yield), low water use efficiency, insufficient supply of animal feed (low biomass production and poor pasture), and poor distribution of the labor force (farmers are busy only 4 months of the year) are some of the constraints to agricultural production in Africa, particularly in the dry land zone.

Research work by ICRISAT, IFDC, IITA, and national research institutions has led to the development of technologies that have proven to be efficient in addressing individual constraints. But further work in the dryland has shown that to address these constraints, a multicomponent system is required to increase the productivity of the arable land (Mando 1997).

The Dryland Eco-Farm (DEF) is an innovation developed at ICRISAT in collaboration with NARES partners. The system combines the use of live hedges of *Acacia species*, earth bunds that turn into micro-catchments or "half-moons," high-value trees such as the domesticated *Ziziphus Mauritania* planted inside the "half-moons," annual crops (millet or sorghum, cowpea groundnut and hibiscus sabdarifa), each of them planted in half of third of the field and rotated each year. The objectives are manifold. (1) The live hedge of acacia species not only protects against wind erosion but also provides mulch for field application; (2) improved ziziphus trees (Sahel apple) produce fruits for human consumption; and (3) shrub leaves as mulch improve soil fertility and protect it from water and wind erosion. (4) Rotation improves soil fertility and system sustainability; (5) annual crops provide food for human consumption as well as residue for animal feed. In particular, the hibiscus provides income through the sale of the dry calice. Studies conducted by Fatondji et al. (2011) have shown that this system produces a millet grain yield of 1 t/ha, similar to the production of plots outside the system with fertilizer NPK applied at the rate of 100 kg ha^{-1}. As far as cowpea total biomass yield was concerned, it was 2–3 times higher on plots within the system than on plots outside the system (1500 kg ha^{-1} vs 500 kg ha^{-1}) (Fig. 17.1). As revealed by the same study, the incorporation of the acacia species shrubs into the system sustained soil fertility after three years of cropping, in contrast with plots outside the system where soil fertility levels had dropped significantly. In fact, P Bay 1 has dropped from 20.2 mg/kg to 18.7 mg/kg in the system, whereas it was 20.2 to 10 mg/kg outside the system. The same trend was observed with total nitrogen.

17.5.2 Tree–Crop Association

Agroforestry parkland is a major component of the integrated cereal-livestock and tree (CLT) system, where trees and shrubs play key role by providing many useful products and services. Trees and shrubs play various socio-economic as well as ecological and cultural roles in many rural communities in the West African Sahel (Nikiema 2005). Trees and shrubs kept and managed by farmers in their farmland are either food plants like Karite (*Vitellaria paradoxa*), fodder trees like *Faidherbia albida*, or trees for other uses like fuel wood from which Sahelian people depend on energy. Trees from the agricultural land in mixed farming systems such as the crop-livestock-tree systems (CLT) can contribute up to 10–30% of cattle diet (Nikiema 2005). They conserve and restore degraded lands, improve soil fertility, and increase crop yield. The positive effect of integrating crops and trees has been demonstrated

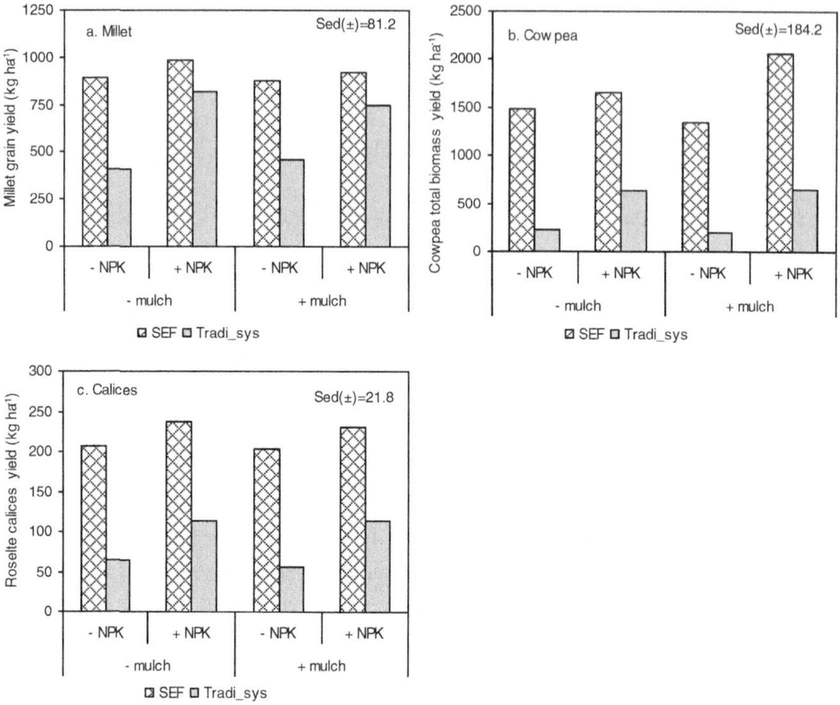

Fig. 17.1 Effect of the dryland eco-farm system on millet grain, cowpea total biomass and roselle calices yield; mean over 3 years 2004–2006; Sadore ICRISAT research station. Sed is standard error of difference between means (Adapted from Fatondji et al. 2011)

in many research works (Wezel and Böcker 1999; Wezel 2000; Wezel and Haigis 2002). In a study conducted to investigate the effect of five tree species (*Annona senegalensis*, *Balanites aegyptiaca*, *Faidherbia albida*, *Guiera senegalensis*, and *Piliostigma reticulatum*) on soil fertility and crop yield under different fertility management practices in semi-arid Niger, Diallo et al. (2021) found that tree species improved soil fertility and crop yield differently. For all soil nutrients studied, *F. albida* improved the soil quality far better than the other four tree species. The highest millet yield was recorded under *P. reticulatum* and *F. albida* for both grain (1011 and 1005 kg ha^{-1}, respectively) and straw yields (3662 and 3786 kg ha^{-1}, respectively). They observed that millet yield was two- to fourfold higher under direct tree crowns compared to close neighborhoods or treeless outside. They found the combined tree species and fertilizer effect to be additive, and the positive effect could be extended to the outside crown. However, the combined effect of mineral fertilizer with tree species on yield was better than that of manure. This is in contrast with the findings of Fatondji (personal data), who observed that millet planted in Acacia species shrubs parkland, produced 1.6 times higher millet total biomass on strips (harvested bands) that are 2 m away from the trunk of the shrub, compared to

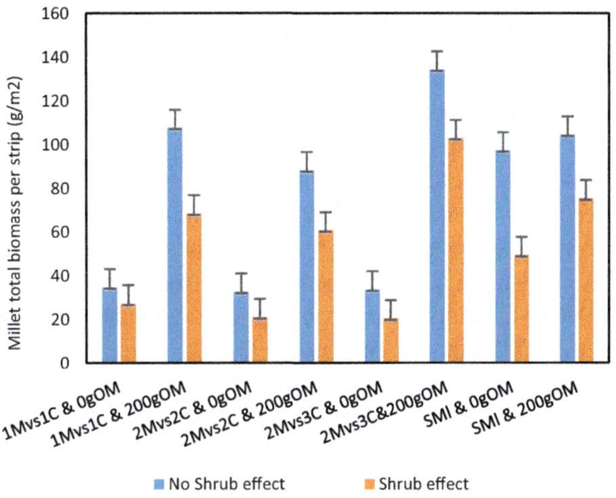

Fig. 17.2 Effect of Acacia shrub husbandry on total biomass of millet grown in strips or sole crop (Unpublished data)

strips closer on average to the shrub. Application of 200 g manure per hill led to 2.2 times greater total biomass produced on the strips that are 2 m away from the trunk, whereas 2.6 times higher biomass was obtained on the strips that are less than 2 m around the shrub when applied with the same rate of manure. This is an indication that nutrient addition compensated slightly for the suppressive effect of the shrub on millet biomass production. A comparison contract from the data indicates that millet and cowpea cropping in alternate strips reduced total biomass by 174 g/m^2 2 m away from the trunk and above and by 74 g/m^2 close to the trunk of the shrub (Fig. 17.2). The effect of shade, in this case, may have been predominant over the tree-induced fertility.

17.5.3 Millet Leguminous Crops Intercropping

This is a common practice widely used to maximize the return from the most limiting production factors, reduce risk, and capitalize on the beneficial effects of legumes on other crops (Bationo et al. 2011). Traditional pearl millet intercropping systems in Niger, for example, involve hill planting of seeds of tall, late-maturing pearl millet varieties after the first 10–20 mm of rain of the growing season (Reddy et al. 1990), and sometimes before the first rain (Reddy et al. 1990). Photoperiod-sensitive, indeterminate cowpea varieties are planted 2–6 weeks later, depending on the completion of pearl millet planting. Pearl millet plant populations are typically low (approximately 5000 hills/ha or 15,000 plants/ha), and the cowpea population is commonly 1000–5000 plants/ha (Ntare and Williams 1992) but varies depending upon cowpea growth habits and climatic conditions. In many instances, sorghum and groundnut are intercropped together, giving a complex intercropping system. Stands of high-value cash crops (*Habiscus* sp.) are occasionally planted along the border or on arable crop-missing hills. Pearl millet matures before cowpea, which is dependent upon residual soil water or late-season rains to produce reasonable yields. In general, grain legume yields are very low (Singh and Emechebe 1998) and are influenced by the degree of competition for active photosynthetic radiation (PAR) and water from pearl millet (Ndunguru and Williams 1993). In this system, pearl millet dominates cowpeas for sunlight, water, and nutrients, and the goal of farmers is a full production of pearl millet grain and stover, while cowpea grain or stover production is of secondary importance. It has been established through research that the pearl millet/cowpea and pearl millet/groundnut intercrop system productivity can be increased by choosing appropriate pearl millet (Reddy et al. 1990) and cowpea cultivars (Ntare 1990); selecting suitable plating date, populations, and spacing; and application of fertilizer (Ntare and Bationo 1992). The strip intercropping system (2 rows of millet to 4 rows of legumes (Ajeigbe et al. 2005, 2010; Singh and Ajeigbe 2007)) and its variants are being promoted across the countries of West Africa, especially Nigeria and Niger.

Pearl millet intercropping systems are site-specific, and production practices vary depending on soil, climate, inputs, and crop variety availability. Most of the available published recommendations for intercropping of pearl millet in West Africa are largely made for the Republic of Niger and are becoming obsolete due to changes in the environment (weather, climate, and soil) and varieties of the millet and legumes which have recently been developed, requiring an update in the recommendations with newly developed research-based solutions for new improved varieties and other production practices. Recent research and publications (Ajeigbe et al. 2019, 2020a, b) by ICRISAT in Nigeria have led to new recommendations of good agronomic practice, including varieties and plant populations.

17.6 Opportunities for Increased Millet Productivity

Pearl millet has recently received a lot of attention as a replacement for maize and sorghum due to its ability to produce grains in a variety of soils and harsh production environments (Dewey et al. 2012). Ajeigbe et al. (2019) found that increasing plant density increased grain and stover yield in grain pearl millet. A variety of technologies have been developed to boost pearl millet productivity in semiarid regions. The development of fertilizer micro-dosing and its combination with organic amendments such as animal manure, crop, or trees residues has offered a good opportunity to increase millet yield in the low-input millet-based cropping systems (ICRISAT 2009; Aune and Bationo 2008; Twomlow et al. 2010; Ibrahim et al. 2014). Millet varieties with resilience to various adverse climate change effects and having high nutritional values can also provide a significant contribution to enhancing millet productivity and nutritional security. For instance, the identification of millet top-cross hybrids is a potential opportunity to sustainably increase pearl millet yields and provide maximal benefits to farmers in the West Africa semiarid environment (Sattler and Haussmann 2020). However, using improved millet varieties would not provide greater yield unless they are accompanied by good crop management practices. Research should therefore focus on the development of appropriate agronomic management options for increased millet productivity in semi-arid areas.

17.7 General Analysis and Conclusion

Pearl millet is the dryland's "flagship crop," accounting for more than 49% of the population's food requirements. However, as previously stated, millet productivity and production face several challenges. Given the growing population, efforts must be made to help it express its genetic potential. Over the years, research has produced technologies that, if implemented, could counteract these challenges. This effort has been developed in various millet production domains, which contributes to improving its productivity. Despite research and development efforts to increase millet productivity, there is still a significant millet yield gap (the difference between potential yield and farmers' yield), indicating that more efforts are required to increase millet yield and production. To increase yield, millet production constraints must be addressed, and improved management options must be considered. As shown in our analysis, nutrient depletion through crop and residue removal and soil degradation (erosion) is one of the major threats to millet production (Bationo et al. 2011). Millet production without soil nutrient application, which is a widespread practice on smallholder farms in Africa, leads to high rates of soil nutrient depletion (Sanchez et al. 1997). Furthermore, the blanket application of fertilizer recommendations is widely acknowledged in Africa, though smallholder millet farmers hardly use these recommendations. Due to financial constraints, farmers often apply reduced amounts of fertilizer (ICRISAT 2009), leading to inefficient and unprofitable millet production. There is a need for locally adjusted and balanced fertilization recommendations suitable for a range of millet production environments. This would require a thorough understanding of the specific system. In addition, farmers need access to credit to be able to follow up on adaptable nutrient recommendations for improved millet productivity. The development of subsidies, improved agro-dealer networks, and re-packaging of fertilizers into smaller portions would improve fertilizer use by resource-poor farmers.

Water scarcity exacerbated by the changing climate is a challenge to millet production in semiarid environments. Rainwater harvesting techniques must be promoted to reduce water stress in millet production. However, lack of awareness and the labor involved make such practices less likely to be adopted. As a result, more incentive policies for improved credit access to resource-constrained farmers are needed to increase the adoption of water harvesting technologies for increased millet yield under water stress conditions. Poor disease and weed management have a negative impact on millet production in semiarid environments. For example, on smallholder farms, labor is scarce, making manual weeding unsuitable for improved millet production. The labor situation has been exacerbated by rural-urban or rural-rural migration for non-agricultural employment. Interventions such as small-scale mechanization and labor-saving technologies such as the use of pesticides and herbicides are required to reduce labor demands. However, no single weed control method provides an effective solution to all weed species colonizing millet farms, implying that an integrated approach to weed control in millet farms is required. Weed control using high-yielding, weed-competitive millet varieties can be investigated as a potential option for reducing yield losses caused by weed infestations. The use of improved millet varieties accounts for a substantial proportion of the yield gains obtained in the past decade (Singh et al. 2017). Improved modern varieties have higher yield potential than traditional varieties and a higher response to fertilizer application. However, there has been slow progress in the adoption of improved millet varieties as farmers still grow traditional landraces with low responsiveness to inputs. Recent adoption and impact surveys in Nigeria reveal only 18% of adoption (Vabi et al. 2020). This is a challenge to millet stakeholders. There are major constraints in the dissemination and adoption of improved varieties as they do often not possess some of the farmers' preferred qualities that traditional cultivars do. Though breeding for higher yields has been a potential objective for many years, current challenges faced by farmers require breeding programs to include many other attributes such as weed competitiveness, disease and parasitic weed resistance, and grain quality. Also, the resource use efficiency of millet varieties is becoming increasingly important, given the level of degradation on productive soils, and the cost of inputs required to produce reasonable millet yields. The seed system needs to be developed to encourage the development and release of millet hybrids.

This review demonstrated that there are opportunities to improve millet productivity and nutritional quality while reducing adverse climate effects on millet production in Africa. Millet-legume intercropping, agroforestry, crop rotation systems, fertilizer micro-dosing, and water harvesting techniques, as well as improved millet varieties resistant to a variety of adverse climate change effects, have been identified as key interventions commonly promoted for increasing pearl millet productivity in semi-arid Africa. The benefits and their associated limitations have been discussed and potential options for improvement have been proposed for sustainable pearl millet production intensification in the Semi-Arid area of Africa.

References

Abdalla E, Osman A, Maki M, Nur F, Ali S, Aune JB (2015) The response of sorghum, groundnut, sesame, and cowpea to seed priming and fertilizer micro-dosing in South Kordofan State, Sudan. Agronomy 5(4):476–490. https://doi.org/10.3390/agronomy/5040476

Abdou A, Koala S, Bationo A (2012) Long-term soil fertility trials in Niger, West Africa. In: Bationo B, Waswa B, Kihara J, Adolwa I, Vanlauwe B, Saidou K (eds) Lessons learned from long-term soil fertility management experiments in Africa. Springer, Dordrecht, pp 105–120

Abdoulaye T, Sanders JH (2005) Stages and determinants of fertilizer use in semi-arid African agriculture: the Niger experience. Agric Econ 32:167–179

Ahn PM (1970) West African soils. Oxford University Press, London

Ajeigbe HA, Singh BB, Oseni TO (2005) Cowpea-cereal Intercrop productivity in the Sudan savanna Zone of Nigeria as affected by planting pattern, crop variety and pest management. Afr Crop Sci J 13(4):269–279

Ajeigbe HA, Singh BB, Ezeaku IE, Adeosun JO (2010) On-farm evaluation of improved cowpea-cereals cropping systems for crop-livestock farmers: cereals-cowpea systems in Sudan Savanna zone of Nigeria. Afr J Agric Res 5(17):2297–2304. Available online http://www.academicjournals.org/AJAR

Ajeigbe HA, Akinseye FM, Kunihya A, Abdullahi AI, Kamara AY (2019) Response of pearl millet (Pennisetum glaucum, L.) to plant population in the semi-arid environments of Nigeria. Neth J Agric Sci 7(1):13–22

Ajeigbe HA, Ignatius IA, Abubakar HI, Folorunso MA, Tukur A (2020a) Handbook on improved pearl millet production practices in North Eastern Nigeria. Springer, Cham, p 14

Ajeigbe HA, Angarawai II, Akinseye FM, Inuwa AH, AbdulAzeez T, Vabi MB (2020b) Handbook on improved agronomy practices of sorghum production in North East Nigeria. USAID Feed the Future Nigeria Integrated Agriculture Activity. ICRISAT, Patancheru

Ambouta JMK, Valentin C, Laverdière MR (1996) Jachères et croutes d'érosion au Sahel. Sécheresse 7:269–275

Amede T, Menza M, Awlachew SB (2011) Zai improves nutrient and water productivity in the Ethiopian highlands. Exp Agric 47:7–20. https://doi.org/10.1017/S0014479710000803

Aune JB, Bationo A (2008) Agricultural intensification in the Sahel—the ladderapproach. Agric Syst 98:119–125

Aune JB, Ousman A (2011) Effect of seed priming and micro-dosing of fertilizer on sorghum and pearl millet in Western Sudan. Exp Agric 47:419–430

Aune JB, Doumbia M, Berthe A (2007) Microfertilizing sorghum and pearl millet in Mali: agronomic, economic and social feasibility. Outlook Agric 36:199–203. https://doi.org/10.5367/000000007781891504

Ayantunde AA, Fernández-Rivera S, Hierneaux PHY, van Keulen H, Udo HMJ (2001) Effect of timing and duration of grazing of growing cattle in the West African Sahel on diet selection, faecal output, eating time, forage intake and live weight changes. Anim Sci 72:117–128

Azare IM, Dantata IJ, Abdullahi MS, Adebayo AA, Aliyu M (2020) Effects of climate change on pearl millet (Pennisetum glaucum [L. R. Br.]) production in Nigeria. J Appl Sci Environ Manag 24(1):157–162

Bagayoko M, Alvey S, Neumann G, Buerkert A (2000) Root-induced increases in soil pH and nutrient availability to field-grown cereals and legumes on acid sandy soils of Sudano-Sahelian West Africa. Plant Soil 225:117–127

Baidu-Forson J (1995) Determinants of the availability of adequate millet stover for mulching in the Sahel. J Sustain Agric 5:101–116

Bationo A, Mokwunye AU (1991) Alleviating soil fertility constraints to increased crop production in West Africa: the experience in the Sahel. Fertilizer Res 29:95–115. https://doi.org/10.1007/BF01048992

Bationo A, Christianson C, Baethgen W, Mokwunye A (1992) A farm-level evaluation of nitrogen and phosphorus fertilizer use and planting density for pearl millet production in Niger. Fertil Res 31:175–184

Bationo A, Mokwunye U, Vlek PL, Koala S, Shapiro BI (2003) Soil fertility management for sustainable land use in the West African Sudano-Sahelian zone. In: Gichuri MP, Bationo A, Bekunda MA, Goma HC, Mafongoya PL, Mugendi DN, Murwuira HK, Nandwa SM, Nyathi P, Swift MJ (eds) Soil fertility management in Africa: a regional perspective. Academy Science Publishers, Seacaucus, pp 253–292

Bationo A, Kimetu B, Vanlauwe B, Bagayoko M, Koala S, Mokwunye AU (2011) Comparative analysis of the current and potential role of legumes in integrated soil fertility management in West and Central Africa. In: Bationo A, Waswa B, Okeyo JM, Maina F, Kihara J, Mokwunye U (eds) Fighting poverty in sub-Saharan Africa: the multiple roles of legumes in integrated soil fertility management. Springer, Dordrecht, pp 117–150

Bayu W, Rethman NFG, Hammes PS (2005) The role of animal manure in sustainable soil fertility management in sub-Saharan Africa: a review. J Sustain Agric 25:113–136

Bielders CL, Gérard B (2015) Millet response to microdose fertilization in south–western Niger: effect of antecedent fertility management and environmental factors. Field Crop Res 171:165–175. https://doi.org/10.1016/j.fcr.2014.10.008

Bouzou MI, Dan LN (2004) Le tassa: une technique de conservation des eaux et des sols bien adaptee aux conditions physiques et socio-economiques des glacis des regions semi-arides (Niger). Rev Geogr Alp 92:61–70

Buerkert A, Bationo A, Dossa K (2000) Mechanisms of residue mulch-induced cereal growth increases in West Africa. Soil Sci Soc Am J 64:346–358

Buerkert A, Bationo A, Piepho HP (2001) Efficient phosphorus application strategies for increased crop production in sub-Saharan West Africa. Field Crop Res 72:1–15

Buerkert A, Piepho HP, Bationo A (2002) Multi-site time-trend analysis of soil fertility management effects on crop production in Sub-Saharan West Africa. Exp Agric 38:163–183

Casenave A, Valentin C (1992) A runoff capability classification system based on surface features criteria in the arid and semi-arid areas of West Africa. J Hydrol 130:231–249

Coulibaly A, Bagayoko M, Traore S, Mason SC (2000) Effect of crop residue management and cropping system on pearl millet and cowpea yield. Afr Crop Sci J 8(4):1–8

Coulibaly A, Woumou K, Aune JB (2019) Sustainable intensification of sorghum and pearl millet production by seed priming, seed treatment and fertilizer microdosing under different rainfall regimes in Mali. Agronomy 9(10):664

Dutordoir CD (2011) Impact of fertility management practices on millet yields in Fakara (Niger). PhD thesis. Universite Catholique de Louvain. Published 27 December 2011. http://www.secheresse.info/spip.php?article15124

De Rouw A, Rajot J-L (2004) Soil organic matter, surface crusting and erosion in Sahelian farming systems based on manuring or fallowing. Agric Ecosyst Environ 104:263–276

Dewey L, Hanna WG, David B, Jeffery PW (2012) Farmers guide to pearl millet for grain in northern Nigeria, pp 24

Diallo MB, Akponikpe PBI, Tougiani A, Dougbedji F, Euloge KA (2021) Combined effect of trees and soil fertility management practices on millet yields in the Sahel. Agrofor Syst. https://doi.org/10.1007/s10457-021-00624-8

Ejeta G (2007) Breeding for resistance in sorghum: exploitation of an intricate host–parasite biology. Crop Sci 47:216. https://doi.org/10.2135/cropsci2007.04.0011IPBS

Esse PC, Buerkert A, Hiernaux P, Assa A (2001) Decomposition of and nutrient release from ruminant manure on acid sandy soils in the Sahelian zone of Niger, West Africa. Agric Ecosyst Environ 83:55–63

FAO (2014) FAOSTAT data base. Available http://faostat.fao.org/site/5676/DesktopDefault.aspx

Fatondji D (2002) Organic amendment decomposition, nutrient release and nutrient uptake by millet (Pennisetum glaucum L. R. Br) in a traditional land rehabilitation technique (zai) in the Sahel, 2002. PhD dissertation. Ecology and development Series No. 1 2002

Fatondji D, Ibrahim A (2018) Improving nutrient use efficiency from decomposing manure and millet yield under Plinthosols in Niger. Nutr Cycl Agroecosyst 110(3):485–499

Fatondji D, Martius C, Bielders C, Vlek PLG, Bationo A, Gerard B (2006) Effect of planting technique and amendment type on pearl millet yield, nutrient uptake, and water use on degraded land in Niger. Nutr Cycl Agroecosyst 76:203–217

Fatondji D, Martius C, Zougmore R, Vlek PL, Bielders C, Koala S (2009) Decomposition of organic amendment and nutrient release under the zai technique in the Sahel. Nutr Cycl Agroecosyst 85:225–239

Fatondji D, Pasternak D, Nikiema A, Senbeto D, Woltering L, Ndjeunga J, Abdoussalam S (2011) The dryland eco-farm: a potential solution to the main constraints of rain-fed agriculture in the semi-arid tropics of Africa. In: Bationo A et al (eds) Innovations as key to the green revolution in Africa. Springer, Cham, p 1115. https://doi.org/10.1007/978-90-481-2543-2_114

Feller C, Beare MH (1997) Physical control of soil organic matter dynamics in the tropics. Geoderma 79:69–116

Fofana B, Wopereis M, Bationo A, Breman H, Mando A (2008) Millet nutrient use efficiency as affected by natural soil fertility, mineral fertilizer use and rainfall in the West African Sahel. Nutr Cycl Agroecosyst 81:25–36

Gandah M, Bouma J, Brouwer J, Hiernaux P, Van Duivenbooden N (2003) Strategies to optimize allocation of limited nutrients to sandy soils of the Sahel: a case study from Niger, West Africa. Agric Ecosyst Environ 94:311–319

Garrity DP, Akinnifesi K, Ajayi OC, Weldesemayat SG, Mowo JG, Kalinganire A, Larwanou M, Bayala J (2010) Evergreen agriculture: a robust approach to sustainable food security in Africa. Food Secur 2:197–214

Gnankambary Z, Bayala J, Malmer A, Nyberg G, Hien V (2008) Decomposition and nutrient release from mixed plant litters of contrasting quality in an agroforestry parkland in the south-Sudanese zone of West Africa. Nutr Cycl Agroecosyst 82:1–13

Graef F, Haigis J (2001) Spatial and temporal rainfall variability in the Sahel and its effects on farmers' management strategies. J Arid Environ 48:221–231

Harris D (1996) The effects of manure, genotype, seed priming, depth and date of sowing on the emergence and early growth of Sorghum bicolor (L.) Moench in semi-arid Botswana. Soil Tillage Res 40(1–2):73–88. https://doi.org/10.1016/S0167-1987(96)80007-9

Hayashi K, Abdoulaye T, Gerard B, Bationo A (2008) Evaluation of application timing in fertilizer micro-dosing technology on millet production in Niger, West Africa. Nutr Cycl Agroecosyst 80 (3):257–265. ISSN 1385-1314

Herrmann L, Jahn R, Stahr K (1996) Identification and quantification of dust additions in perisaharan soils. In: The impact of desert dust across the Mediterranean. Springer, Cham, pp 173–182

Ibrahim A, Fatondji D (2020) Optimizing manure management for enhancing millet yields and resources use efficiency under rehabilitated plinthosols in the Sahel. J Arid Environ 181:104225. https://doi.org/10.1016/j.jaridenv.2020.104225

Ibrahim A, Pasternak D, Fatondji D (2014) Impact of depth of placement of mineral fertilizer micro-dosing on growth, yield and partial nutrient balance in pearl millet cropping system in the Sahel. J Agric Sci 153(8):1412–1421

Ibrahim A, Abaidoo RC, Fatondji D, Opoku A (2015a) Hill placement of manure and fertilizer micro-dosing improves yield and water use efficiency in the Sahelian low input millet-based cropping system. Field Crop Res 180:29–36

Ibrahim A, Abaidoo RC, Fatondji D, Opoku A (2015b) Integrated use of fertilizer micro-dosing and Acacia tumida mulching increases millet yield and water use efficiency in Sahelian semi-arid environment. Nutr Cycl Agroecosyst 103(3):375–388

Ibrahim A, Abaidoo RC, Fatondji D, Opoku A (2016a) Fertilizer microdosing increases crop yield in the Sahelian low-input cropping system: a success with a shadow. Soil Sci Plant Nutr 62:277–288. https://doi.org/10.1080/00380768.2016.1194169

Ibrahim A, Abaidoo RC, Fatondji D, Opoku A (2016b) Determinants of fertilizer microdosing-induced yield increment of pearl millet on an acid sandy soil. Exp Agric 52:562–578. https://doi.org/10.1017/S0014479715000241

ICRISAT (2009) Fertilizer microdosing–boosting production in unproductive lands. http://www.icrisat.org/impacts/impact-stories/icrisat-is-fertilizer-microdosing

Kabore D, Reij C (2004) The emergence and spreading of an improved traditional soil and water conservation practice in Burkina Faso. Environment and Production Technology Division discussion paper 114. http://www.ifpri.org/event/

Kadi M, Lowenberg-Deboer J, Reddy KC, Abdoulay B (1990) Sustainable millet cowpea technologies for semi-arid Niger. Indian J Dryland Agric Res Dev 4:95–98

Kho RM, Yacouba B, Yayé M, Katkoré B, Moussa A, Iktam A, Mayaki A (2001) Separating the effects of trees on crops: the case of Faidherbia albida and millet in Niger. Agrofor Syst 52:219–238

Klaij MC, Hoogmoed WB (1993) Soil management for crop production in the West African Sahel. 2. Emergence, establishment and yield of pearl millet. Soil Tillage Res 25(4):301–305

Lahmar R, Bationo BA, Dan LN, Guero Y, Tittonell P (2012) Tailoring conservation agriculture technologies to West Africa semi-arid zones: building on traditional local practices for soil restoration. Field Crop Res 132:158–167

Lobell DB, Schlenker W, Costa-Roberts J (2011) Climate trends and global crop production since 1980. Science. https://doi.org/10.1126/science.1204531

Malley ZJU, Kayombo B, Willcocks TJ, Mtakwa PW (2004) Ngoro: an indigenous, sustainable and profitable soil, water and nutrient conservation system in Tanzania for sloping land. Soil Tillage Res 77:47–58. https://doi.org/10.1016/j.still.2003.10.003

Mamane BM (2002) Test d'une méthode pour amélioration de l'efficacité de la fumure par parcage du bétail. Mémoire du fin de cycle. Faculté d'Agronomie, Université de Niamey

Mando A (1997) Effect of termites and mulch on the physical rehabilitation of structurally crusted soils in the Sahel. Land Degrad Dev 8:269–278

Manu A, Bationo A, Geiger S (1991) Fertility status of selected millet producing soils of West Africa with emphasis on phosphorus. Soil Sci 152:315–320

Manyame C (2006) On-farm yield and water use response of pearl millet to different management practices in Niger. Ph.D. Diss., Texas A&M University, Texas

Mason SC, Ouattara K, Taonda SBJ, Palé S, Sohoro A, Kaboré D (2014) Soil and cropping system research in semi-arid West Africa as related to the potential for conservation agriculture. Int J Agric Sustain 13:120

Michels K, Bielders CL (2006) Pearl millet growth on an erosion-affected soil in the Sahel. Exp Agric 42:1–17

Michels K, Sivakumar MvK, Allison BE (1995) Wind-erosion control using crop residue: effects on soil flux and soil properties. Field Crop Res 40(2):101–110

Mokgolodi NC, Setshogo MP, Ling-ling S, Yu-jun L, Chao MA (2011) Achieving food and nutritional security through agroforestry: a case of Faidherbia albida in Sub-Saharan Africa. For Stud China 13:123–131

Molden D, Murray-Rust H, Sakthivadivel R (2003) A water-productivity framework for understanding and action. In: Kijne JW, Barker R, Molden D (eds) Water productivity in agriculture: limits and opportunities for improvements. Comprehensive assessment of water management in agriculture series 1. CAB International, Wallingford

Muehlig-Versen B, Buerkert A, Bationo A, Marshcner H (1997) Crop residue and phosphorus management in millet based cropping systems on sandy soils of the Sahel. In: Renard G, Neef A, Becker K, von Oppen M (eds) Soil fertility in West African land use systems. Margraf Verlag, Weikersheim, pp 31–40

Muehlig-Versen B, Buerkert A, Bationo A, Roeheld V (2003) Phosphorus placement on acid arenosol of the west African Sahel. Exp Agric 39:307–325

Mwinuka L, Mutabazi KD, Graef F, Sieber S, Makindara J, Kimaro A, Uckert G (2017) Simulated willingness of farmers to adopt fertilizer micro-dosing and rainwater harvesting technologies in semi-arid and sub-humid farming systems in Tanzania. Food Secur 9:1237–1253

Ndunguru BJ, Williams JH (1993) The impact of varying levels of competition from pearl millet on the yields of groundnut cultivars. Exp Agric 29:29–37

Nielson H (1999) Cordons pierreux in Burkina Faso: sustainable development or stones for bread? Geografisk Tidsskrift Dan J Geogr 2:105–112

Nikiema A (2005) Agroforestry parkland species diversity: uses and management in semi-arid West Africa (Burkina Faso). PhD thesis, Wageningen University, p 102

Ntare BR (1990) Intercropping morphologically different cowpea with pearl millet in a short season environment in the Sahel. Exp Agric 26:41–47

Ntare BR, Bationo A (1992) Effects of phosphorus on yield of cowpea cultivars intercropped with pearl millet on psammentic paleustalf in Niger. Fertil Res 32:143–147

Ntare BR, Williams JH (1992) Response of cowpea cultivars to planting pattern and date of sowing in intercrops with pearl millet in Niger. Exp Agric 28:41–48

Nziguheba G, Zingore S, Kihara J, Merckx R, Njoroge S, Otinga A, Vandamme E, Vanlauwe B (2016) Phosphorus in smallholder farming systems of sub-Saharan Africa: implications for agricultural intensification. Nutr Cycl Agroecosyst 104:321–340. https://doi.org/10.1007/s10705-015-9729y

Osman AN, Ræbild A, Christiansen JL, Bayala J (2011) Performance of cowpea (Vigna unguiculata) and pearl millet (Pennisetum glaucum) intercropped under Parkia biglobosa in an agroforestry system in Burkina Faso. Afr J Agric Res 6(4):882–891

Ouedraogo M, Dembélé Y, Leopold S (2010) Perception et stratégies d'adaptation aux changements des précipitations : cas des paysans du Burkina Faso. Sciences et changements planétaires / secheresse 21:87–96. https://doi.org/10.12895/jaeid.20121.43

Pandey RK, Maranville JW, Bako Y (2001) Nitrogen fertilizer response and use efficiency for three cereal crops in Niger. Commun Soil Sci Plant Anal 32(9-10):1465–1482

Payne WA, Williams JH, Maï Moussa KA, Stern RD (1998) Crop diversification in the Sahel through use of environmental changes near Faidherbia albida (Del.) A. Chev Crop Sci 38:1585–1591

Peacock JM, Soman P, Jayachandran R, Rani AU, Howarth CJ, Thomas A (1993) Effects of high soil surface temperature on seedling survival in pearl millet. Exp Agric 29:215–225

Pieri C (1989) Fertilité des terres de savane. Bilan de trente ans de recherche et de développement agricoles au sud du Sahara. Ministère de la Coopération/Cirad, Paris

Powell J, Valentin C (1997) Effects of livestock on soil fertility in West Africa. In: Renard G, Neef A, Becker K, von Oppen M (eds) Soil fertility management in West African land use systems. Margraf Verlag, Weikersheim, pp 319–334

Powell JM, Fernández-Rivera S, Hiernaux P, Turner MD (1996) Nutrient cycling in integrated rangeland/cropland systems of the Sahel. Agric Syst 32(2/3):143–170

Rebafka FP, Hebel A, Bationo A, Stahr K, Marschner H (1994) Short- and long-term effects of crop residues and of phosphorus fertilization on pearl millet yield on an acid sandy soil in Niger, West Africa. Field Crop Res 36:113–124

Reddy KC, van der Ploeg J, Maga I (1990) Genotype effects in millet/cowpea intercropping in the semi-arid tropics of Niger. Exp Agric 26:387–396

Reddy KC, Visser P, Buckner P (1992) Pearl millet and cowpea yields in sole and intercrop systems, and their after-effects on soil and crop productivity. Field Crop Res 28:315–326

Reij CP, Smaling EMA (2008) Analyzing successes in agriculture and land management in Sub-Saharan Africa: Is macro-level gloom obscuring positive micro-level change? Land Use Policy 25:410–420

Roose E, Kabore V, Guenat C (1993) Lezai: fonctionnement, limites et amelioration d'une pratique traditionnelle africaine de rehabilitation de la vegetation et de la productivite des terres degradees en region soudano-sahelienne (Burkina Faso). Cahiers-ORSTOM. Pédologie 28: 159–173

Roose E, Kabore V, Guenat C (1999) Zai practice: a West African traditional rehabilitation system for semiarid degraded lands, a case study in Burkina Faso. Arid Soil Res Rehabil 13:343–355

Sanchez PA, Shepherd KD, Soule MJ, Place FM, Buresh RJ, Izac AMN, Mokwunye AU, Kwesiga RK, Ndiritu CG, Woomer PL (1997) Soil fertility replenishment in Africa: an investment in natural resource capital. In: Buresh RJ, Sanchez PA, Calhoun F (eds) Replenishing soil fertility in Africa. Soil Science Society of America, Madison, pp 1–46

Sanou J, Bationo BA, Barry S, Nabie LD, Bayala J, Zougmore R (2016) Combining soil fertilization, cropping systems and improved varieties to minimize climate risks on farming productivity in northern region of Burkina Faso. Agric Food Secur 5(1):1–12

Sattler FT, Haussmann BI (2020) A unified strategy for West African pearl millet hybrid and heterotic group development. Crop Sci 60(1):1–13

Schlecht E, Buerkert A (2004) Organic inputs on millet fields in western Niger: the implications of farmers' practices for sustainable agricultural production. Geoderma 121:271–289

Schlecht E, Buerkert A, Tielkes E, Bationo A (2006) A critical analysis of challenges and opportunities for soil fertility restoration in Sudano-Sahelian West Africa. Nutr Cycl Agroecosyst 76:109–136

Schlenker W, Lobell DB (2010) Robust negative impacts of climate change on African agriculture. Environ Res Lett 5:014010. https://doi.org/10.1088/1748-9326/5/1/014010
Scott-Wendt J, Chase RG, Hossner LR (1988) 2. Soil chemical variability in sandy ustalfs in semi-arid Niger, West Africa. Soil Sci 145(6):414
Shivhare R, Lata C (2017) Exploration of genetic and genomic resources for abiotic and biotic stress tolerance in pearl millet. Front Plant Sci 7:2069
Silungwe FR, Graef F, Bellingrath-Kimura SD, Tumbo SD, Kahimba FC, Lana MA (2019) The management strategies of pearl millet farmers to cope with seasonal rainfall variability in a semi-arid agroclimate. Agronomy 9(7):400
Singh BB, Ajeigbe HA (2007) Improved cowpea-cereals-based cropping systems for household food security and poverty reduction in West Africa. J Crop Improv 19:157–172
Singh BB, Emechebe AM (1998) Increasing productivity of millet-cowpea intercropping systems. In: Emechebe AM, Ikwelle MC, Ajayi O, Aminu-Kano M, Anaso AB (eds) Pearl millet in Nigerian agriculture. Ramadan Press, Maiduguri, pp 70–75
Singh P, Boote K, Kadiyala M, Nedumaran S, Gupta S, Srinivas K, Bantilan M (2017) An assessment of yield gains under climate change due to genetic modification of pearl millet. Sci Total Environ 601:1226–1237
Sivakumar M, Wallace J (1991) Soil water balance in the Sudano-Sahelian zone: need, relevance and objectives of the workshop. Soil water balance in the Sudano-Sahelian zone. Proceedings of the Niamey Workshop, February 1991
Sivakumar MVK, Huda AKS, Virmani SM (1984) Physical environment of sorghum and millet-growing areas in South Asia. In: Agrometeorology of sorghum and millet in the semi-arid tropics: proceedings of the international symposium. ICRISAT Center, Patancheru, pp 63–83
Somé L, Kambou F, Traoré S, Ouédraogo B (2000) Techniques de conservation des eaux et des sols dans la moitié nord du Burkina Faso. Sécheresse 11(4):267–274
Stephen CM, Nouri M, Siébou P (2015) Pearl millet production practices in semi-arid West Africa: a review. Exp Agric 51(4):501–521
Stomph TJ (1990) Seedling establishment in pearl millet [Pennisetum glaucum (L.) R.Br.]: The influence of genotype, physiological seed quality, soil temperature and soil water. Doctoral dissertation, University of Reading, p 210
Stroosnijder L, Ridder N, Kiepe P (2001) Nutrient balances in Sahelian villages. Adv Geoecol 33: 213–235
Tabo R (1995) Performances of sorghum and millet varieties under varying combinations of animal manures and chemical fertilizer. In: ICRISAT and collaborative programs. West and Central Africa Annual Report 1995, p 53–54
Tabo R, Bationo A, Gerard B (2007) Improving cereal productivity and farmers' income using a strategic application of fertilizers in West Africa. In: Bationo A, Kihara J, Kimetu J (eds) Advances in integrated soil fertility management in sub-Saharan Africa: challenges and opportunities. Springer, Dordrecht, pp 201–208
Tonitto C, Ricker-Gilbert JE (2016) Nutrient management in African sorghum cropping systems: applying meta-analysis to assess yield and profitability. Agron Sustain Dev 36:10. https://doi.org/10.1007/s13593-015-0336-8
Tovihoudji PG, Akponikpè PI, Agbossou EK et al (2017) Fertilizer microdosing enhances maize yields but may exacerbate nutrient mining in maize cropping systems in northern Benin. Field Crop Res 213:130–142. https://doi.org/10.1016/j.fcr.2017.08.003
Twomlow S, Rohrbach D, Dimes J, Rusike J, Mupangwa W, Ncube B, Hove L, Moyo M, Mashingaidze N, Mahposa P (2010) Microdosing as a pathway to Africa's Green Revolution: evidence from broad-scale on-farm trials. Nutr Cycl Agroecosyst 8:3–15. https://doi.org/10.1007/s10705-008-9200-4
Vabi MB, Abdulqudus IA, Angarawai II, Adogba DS, Kamara AY, Ajeigbe HA, Ojiewo C (2020) Adoption and welfare impact of pearl millet Technologies in Nigeria. International Crop Research Institute for the Semi-Arid Tropics (ICRISAT), Patancheru. Technical Working Document

Valentin C (1986) Effects of soil moisture and kinetic energy on the mechanical resistance of surface crusts. In: Callebaut C, Gabriels D, de Boodt M (eds) Assessment of soil surface sealing and crusting. University of Ghent, Belgium, pp 367–369

Van Mourik TA (2007) Striga hermonthica seed bank dynamics: process quantification and modelling Van Mourik, T.A. – [S.l.: s.n.]. Ill. PhD thesis Wageningen University. - With ref. - With summaries in English, French and Dutch. ISBN: 978-90-8504-692-9

Voortman RL, Brouwer J, Albersen PJ (2004) Characterization of spatial soil variability and its effect on millet yield on Sudano-Sahelian coversands in SW Niger. Geoderma 121:65–82

Wezel A (2000) Scattered shrubs in pearl millet fields in semiarid Niger: effect on millet production. Agrofor Syst 48:219–228

Wezel A, Böcker R (1999) Mulching with branches of an indigenous shrub (*Guiera senegalensis*) and yield of millet in semi-arid Niger. Soil Tillage Res 50:341–344

Wezel A, Haigis J (2002) Fallow cultivation system and farmers' resource management in Niger, West Africa. Land Degrad Dev 13:221–231

Wildemeersch JCJ, Garba M, Sabiou M, Steven S, Cornelis W (2015) The effect of water and soil conservation (WSC) on the soil chemical, biological and physical quality of a plinthosol in Niger. Land Degrad Dev 73:773–783

Wilson JP, Hanna WW, Gascho G, Wilson GM (1995) Pearl millet grain loss from rust infection. In: Teare ID (ed) Proceedings of the first National Grain Pearl millet. University of Georgia and USDA, Tifton, pp 54–56

Zougmoré R, Jalloh A, Tioro A (2014) Climate-smart soil water and nutrient management options in semiarid West Africa: a review of evidence and analysis of stone bunds and zaï techniques. Agric Food Secur 3:16. https://doi.org/10.1186/2048-7010-3-16

Modern Crop Management Practices for Pearl Millet Cultivation in Asia

18

B. Gangaiah and A. K. Yadav

Abstract

Pearl millet, as a rainfed crop of Asia, is subjected to varying levels of moisture stress that upon receiving low external inputs of fertilizers, pesticides including herbicides has low productivity than its potential. Scope lies in pushing up the yields through the adoption of new emerging agronomic practices holistically on a system basis. Conservation agriculture (zero tillage + residues retention + crop rotations) in pearl millet cultivation areas on cropping system mode has enormous potential to do away with conventional tillage. Rainwater harvest during excess rainfall events, its effective storage in sub (surface) structures (surface structures lining, anti-evaporates use), and use for protective irrigation by pressurized irrigation methods (drip/sprinkler) at critical stages could completely offset the moisture stress-induced pearl millet yield penalties. Scope lies for saline groundwater-based drip irrigation adoption in pearl millet and a selection of cultivars and amendments comes in handy in such situations. Integrated crop nutrition (liquid bio-fertilizers, manures, fertilizers) especially manuring, and effective weed management through herbicides (Tembotrione, atrazine) could boost crop water supplies besides augmenting the nutrient supplies to the crop. Foliar nutrition is desired under severe water stress. Use of bio-fortified stress-tolerant hybrids ds in moderate precipitation (>400 mm) and irrigated areas and high yielding open pollinated varieties in water-stressed regions (<400 mm) and their timely sowing leading to optimum plant stand maintenance (gap filling by

B. Gangaiah (✉)
ICAR-Indian Institute of Millets Research, Hyderabad, Telangana, India
e-mail: gangaiah@millets.res.in

A. K. Yadav
Bajra Section, Department of Genetics & Plant Breeding, CCS Haryana Agricultural University, Hisar, Haryana, India

V. A. Tonapi et al. (eds.), *Pearl Millet in the 21st Century*,
https://doi.org/10.1007/978-981-99-5890-0_18

seed or transplanting) are pre-requisites for taking advantage from above production technologies. Combine harvesting and organic pearl millet production could boost profits by reducing the costs and premium produce price realizations respectively.

Keywords

Pearl millet · Asia · India · Moisture conservation · Tillage · Integrated nutrition · Organic production · Weed management · Mechanization

Abbreviations

₹	Indian rupee
°C	Degree centigrade
a.i.	Active ingredient
CA	Conservation agriculture
cm	Centi meter
CR	Crop residue
CT	Conventional tillage
DAS	Days after sowing
dS	Deci Siemen
EC	Electrical conductivity
FYM	Farm yard manure
g	Gram
ha	Hectare
HYV	High yielding variety
IW/CPE	Irrigation water/cumulative pan evaporation
kg	Kilo gram
LC	Leaf compost
m	Million
Mg	Mega gram (10^9 g = ton)
mm	Milli meter
nm	Nano meter
ppm	Parts per million
PSB	Phosphate solubilizing bacteria
RDF	Recommend dose of fertilizer
SAR	Sodium absorption ratio
SC	Soluble Concentrate
SOC	Soil organic carbon
t	Ton
ZT	Zero tillage

18.1 Introduction

Pearl millet (*Pennisetum glaucum* (L.) R. Br.) is cultivated by 27 Asian countries on an 11.63 million hectares (m ha) area producing 15.737 m tonnes (m t) during 2020 as per FAO statistics. India + China accounts for 83.5 + 7.74 (91.2%) of area and 79.4 +14.62 (94%) of Asia's production, while Pakistan, Nepal and Myanmar are the other Asian countries with >1% share in area (production), i.e. 3.0 (1.69), 2.26 (2.03) and 1.31 (0.93%) respectively. Of the pearl millet produced in Asia, 245,238 t is exported, while 136,430 t are imported during 2020. Pearl millet is a predominantly rainfed (monsoon dependent) season (June–September) crop of India (Rai et al. 2015), and based on annual rainfall received and soil type, its area is grouped into three zones, i.e. A_1 zone (<400 mm rainfall receiving North-West zone with sandy soils), A zone (north and central India with >400 mm rainfall, sandy loam soils) and B zone (peninsular India with >400 mm rainfall and heavy soils). However, winter (November–February) cultivation of pearl millet in the states of Maharashtra and Gujarat and summer (February–May) irrigated cultivation in the states of Gujarat, Rajasthan and Uttar Pradesh with high levels of fertilizer inputs (Mula et al. 2009) is gaining popularity in recent times. This is evident from the fact that the pearl millet crop area under irrigation has gone up from 3.45% (1950–1951) to 9.4% during 2018–2019. In contrast to India, the majority of pearl millet production in Pakistan comes from irrigated conditions (64%) and Punjab province alone accounts for 96% of the total 0.44 m ha area (Wing 2014). Pearl millet crops grown in the A_1 zone experience very high temperatures from sowing to panicle initiation stages and the sandy soil with low water holding capacity and calcareous nature indicates a Thermo and xerophilic environment. In such situations, pearl millet was found to be a better crop compared to sorghum (Bhuja 2009) as it can produce grains with water (precipitation) as low as 200 mm (Bidinger and Hash 2004). Further, few additional fertilizer nutrients are required for the pearl millet crop that is often met through manures (in livestock dense areas) and/or legume cover/intercrops (Delate 2013). The crop's ability to survive on saline soils, use seawater for irrigation and accumulate nitrates in the lower parts of the stalks (not grain) under drought (USDA 2016) makes it an inevitable choice for farmers of the arid and semi-arid world especially in the least developed and developing countries of Asia and Africa. Owing to its production and consumption by the poorest people in the poorest countries, it is called as 'Poor man's cereal crop' (Alam et al. 2010), its richness in antioxidants, bioactive compounds, fibre and minerals (Fe, Zn, Ca, Mg, P), vitamins, lipids and high-quality protein (Uppal et al. 2015) ensures balanced nutrition of consumers. For coping with the various biotic and abiotic challenges of the pearl millet agro-ecosystem, various production technologies have been validated under the aegis of the Indian Council of Agricultural Research (ICAR), State Agricultural Universities (SAU's) and International Crops Research Institute for Semi-Arid Tropics (ICRISAT), Patancheru, India and such promising technologies were successfully transferred to stakeholder farmers through line departments of agriculture. Information on technologies evolved was available to some extent from Pakistan, while little information is available from other Asian countries (China, Myanmar,

Nepal). The developments in pearl millet production technology in Asia over time are discussed below.

18.1.1 Crop Cultivars

Selection of pearl millet cultivar (variety/hybrid) suitable for a region is the foremost decision to be made by the farmer as it decides the productivity, of course, that is moderated by soil quality, moisture/water availability, input supply (fertilizer, pesticide) and other management practices (weeding, moisture conservation). For self-saved seeds as in the case of landraces and open-pollinated varieties (OPVs), cultivar selection is a non-monetary decision. OPVs seed prices in Indian markets are around ₹30/kg. However, in the case of hybrids, whose price is 10–15 times that of OPVs, and as the seed needs to be purchased every year, cultivar selection becomes more of a monetary decision for hybrids. The higher price of hybrid seeds and associated profit margins have paved the way for the entry of the private seed industry and is a major pearl millet seed producer and supplier in India. Higher yields and associated profits have made cultivators resort to increased use of hybrids in India since 1965 and in Pakistan in the recent past.

Local landraces and open-pollinated varieties (OPVs) are popular with farmers up to 1990s in India. High-yielding OPV 'ICTP-8203' developed by ICRISAT, Patancheru, India with good agronomic traits of shorter plant height, early maturity, compact panicles, good exertion, no tip sterility and high levels of tolerance to downy mildew and released in 1988 for cultivation in Maharashtra state of India (Rai et al. 2004) rapidly spread to 50% of pearl millet area (0.8 million ha) within a short time. The development of new cultivars including hybrids subsequently has reduced the area under 'ICTP-8203' by 62.5% relegating it to dry and low fertile soils of Maharashtra (Rai et al. 2004). However, its improved version 'Dhanashakti', with high iron content and yield (9 and 11%) has promoted its cultivation in the Central and Southern parts of India (Rai et al. 2015). Pearl millet is tolerant to salinity and in salinity tolerance, it stands next to barley only (Yadav et al. 2012a) and the wide differences in salinity tolerance of pearl millet varieties give scope for selection. Studies in Haryana, India (Makarana et al. 2019), have indicated that 'AVKB-19' has a 16.3% higher grain yield than 'ICMV-15111' (1.66 t/ha) under saline conditions (pH: 8.3; EC: 6.69 dS/m). Thus, pearl millet production in risky areas of Rajasthan and Maharashtra states of India is dependent on OPVs and landraces even now.

Development and release of pearl millet hybrids in India started with the release of HB-1 in 1965 and hybrids' gradual prominence led to the complete replacement of OPVs in favourable environments and partial substitution in unfavourable environments. Shorter duration hybrids developed like HHB-67 have given scope for not only escaping from the terminal drought stress but also aiding in timely sowing of succeeding winter crops like mustard and chickpea (Rai et al. 2004). Inducing downy mildew tolerance in HHB-67 through marked assisted selection in 2005 has furthered its yield superiority over other cultivars. More and more biotic and abiotic stress tolerant hybrids are developed in recent times and thus have enhanced their use by farmers. To address zinc and iron malnutrition concerns in India, a decision was taken to release only bio-fortified cultivars, i.e. with a minimum of 42 and 32 ppm Fe and Zn concentration in grain (AICRPPM (All India Coordinated Research Project on Pearl Millet) 2021) and 'ICMH-1202' and 'ICMH-1203' are the first two bio-fortified pearl millet hybrids notified in 2018 and released (Govindaraj et al. 2020) and many are in pipeline. Here also, hybrids have predominance to varieties. As per records, of the 176 pearl millet cultivars released between 1986 and 2020, 84.7% are hybrids and are reported to have 25–30% higher yields over OPVs (Yadav et al. 2012b). In India, between 1970 and 2019, pearl millet acreage declined by 48%, but on account of doubling the productivity (103% increase), overall production increased by 11% and this productivity increase could be ascribed mainly to high-yielding varieties and partly to improved management practices.

In Pakistan, few high-yielding varieties (Ullah et al. 2017) have been developed (PARC-MS-1, PARC-MS-2, PARC-MS-3, PARC-MS-4; PARC-MS-5; 18-BY; Sahiwal, Cholistani Bajra and Barani Bajra) and a significant area has been diverted in recent times towards heat and drought tolerant hybrids from Pioneer Pakistan Seed Limited (86M33, 86M52, 86M64, 86M66, 86M86 and 86M88). Studies across six locations in Pakistan have indicated an 11.5–44% mean improvement in grain yield of pearl millet with the use of HYV over the local check and hybrid millet-7777 has 28.5% higher grain yields (Munawwar et al. 2014).

Cultivar selection of component crops is important for the successful intercropping of pearl millet. For higher system yields and land use efficiency of pearl millet intercropping with cluster bean, a prominent intercropping system of arid zone of North-Western India, medium-statured and early-maturing pearl millet cultivar 'HHB 67' was found promising to the tall and late-maturing pearl millet 'MH 179'; however, cluster bean cultivars did not differ in their performance (Yadav and Yadav 2001). Similarly, early maturing and dwarf pearl millet varieties were found promising for intercropping with cowpea, and spreading cowpea genotypes are found better performers in intercropping with pearl millet (Reddy et al. 1990). However, for intercropping with pigeon pea in the Deccan plateau and parts of central India, a combination of early maturing, tall pearl millet types (IVS-A71) and late spreading pigeon peas (PS-4) was found promising (Rao and Willey 1983). In the pearl millet and groundnut intercropping system, pearl millet roots were found to spread into groundnut rows and inhibit pods/plant and yield (Gregory and Reddy 1982) and soil microorganisms count (Hussainy et al. 2019) and thus scope lies for selecting vertically distributed rooting.

For the cropping system too, the selection of pearl millet genotypes to vacate the land for timely sowing of succeeding winter crops (wheat, mustard, chickpea) assumes huge prominence as winter crops (wheat) often grown with protective irrigation are more contributors to the system productivity and profits. Short-duration 'HHB67' pearl millet was found suitable for timely sowing of succeeding winter crops like mustard and chickpea (Rai et al. 2004).

The ever-increasing labour costs coupled with decreasing manpower in the recent past have paved the way for the mechanization of pearl millet cultivation especially harvesting for which combine harvesters are increasingly used. For this to succeed, uniform maturity of cultivars is a prerequisite. The use of high-yielding cultivar seeds without impurities assumes importance and this is more ensured with hybrids than with OPVs and landraces due to no admixtures during seed production.

18.1.2 Sowing Time

Sowing time is the most crucial non-monetary input of pearl millet productivity more so under rainfed conditions as it aids in taking full advantage of the monsoon rains. Further, sowing time also influences the temperature and radiation regimes available to the crop before flowering (photo-climate) of the crop considerably (Rao and Saxton 1995). Timely sowing results in better germination, optimum plant stand establishment, better growth, and development cumulatively aid in best pearl millet crop yield realizations. Under rainfed conditions, the time of sowing is in flux and gets altered as per the onset of monsoon during that year. Though flexibility lies in the sowing of pearl millet crops under irrigated conditions (before and even after the onset of monsoons), it's timely sowing during the rainy season assumes importance as it decides the planting of succeeding crops in the system. Though pearl millet germinates well at 23–30 °C soil temperature, the germination gets hampered at temperatures above 45 °C (Soman et al. 1981). In the Marathwada region of Maharashtra, dry sowing before the onset of the first monsoon rains is recommended. Studies have indicated that early planting with the onset of monsoon is effective in controlling shoot fly and ergot incidence in pearl millet (Thakur 1984). Gap filling should be done within one week of crop sowing by seeds; however, transplanting seedlings 2–3 weeks after sowing is more promising, provided the moisture favours it.

Studies have indicated that *kharif* pearl millet sowing should be done in the first fortnight of July in North and central India with the onset of monsoon. The normal onset of monsoon in the arid zone of Rajasthan is a bit early (late June to early July). Studies at Gwalior, Madhya Pradesh (Bhavna et al. 2020) have indicated that delayed sowing beyond 20 July, i.e. to 30 July and 9 August resulted in 12.1 and 33.3% reduction in pearl millet grain yields under rainfed conditions. In cropping system mode, delayed pearl millet sowing results in delayed sowing of succeeding crops leading to reduced system productivity.

Though pearl millet remains the best choice for sowing in dry lands even under delayed conditions, however, direct seeding beyond the optimum time of sowing results in poor plant stand, poor grain setting and thus low grain yields. Transplanting of 30 days old seedlings owing to more tillers production and better ear development was found to have higher grain yields (Jan et al. 2015; Biswas 2020) and thus a better alternative to direct seeding under delayed sowing conditions. Though it is more labour demanding and needs short-term planning for nursery etc. during monsoon season, however, during the post-rainy season, transplanting is quite possible and promising to direct seeding (with a short crop life cycle) as it is grown under irrigated conditions only. For *rabi* pearl millet in Tamil Nadu state, the first fortnight of October was found appropriate sowing time while for summer pearl millet sowing in Zone B, the last week of January to 1st week of February was ideal.

18.1.3 Plant Population, Spacing, and Seed Rate

For potential yield realization from pearl millet crop, optimum plant stand maintenance remains a pre-requisite. However, on account of its ability to produce more than one productive tiller per plant, considerable variations in plant population ranging from 50,000–150,000 plants/ha (Anand Reddy and Rao 1971) can buffer yield variations. However, this tillering ability was found inadequate to compensate for the yield losses due to uneven distribution of plants, i.e. both higher and lower population per unit area (Carberry et al. 1985) and thus uniform plant stand is desired (Soman et al. 1987). Plant population requirements of pearl millet vary with cultivar, its duration, seed size, sowing time, planting method and growing conditions.

For arid-Western plains of Rajasthan, Haryana and Kutch of Gujarat (A_1 zone with <450 mm rainfall), pearl millet planting in wider rows (60 cm) maintaining a plant population of 1.00–1.25 lakh/ha is recommended, while in areas having >450 mm rainfall (A and B zones), narrow row spacing (45 cm) is recommended and with an intra row spacing of 10–15 cm, 1.75–2.00 lakh plants/ha is obtained. In Pakistan also, 45 × 15 cm spacing proved promising for pearl millet (Ullah et al. 2020). Depending on seed size, a seed rate of 4–5 kg/ha will provide the required plant stand of pearl millet. Pearl millet crop should be sown at shallow depths (2–3 cm deep) for quick and successful emergence of seedlings. Studies have indicated that speed of germination and seedling emergence of pearl millet were positively related to seed size (Gardner 1980; Lawan et al. 1985) so bold seeds are good for successful crop establishment.

18.1.4 Land Preparation (Tillage)

Tillage is the first operation of any arable crop culture and traditionally includes primary (mouldboard plough), secondary (blade harrows) and tertiary operations (levelling). Though good tilth is a prerequisite for the greater proliferation of roots in heavy soils and pearl millet yields (Doty et al. 1975), excess tillage in light textured sandy soils was found to disintegrate clods and expose the soil to erosion and on the contrary reduced tillage was found to compact the soils low in organic matter content leading to hampered crop performance (Gupta et al. 1983) despite of its economy in time, labour and energy. Deep tillage was found to increase the profile soil moisture storage of coarse loam soil by 8 (1993) to 21 mm (1994) over 198- and 161-mm storage under minimum tillage (Saxena et al. 1997). The high soil moisture storage with deep tillage in sandy loam soil was ascribed to disturbance of the plough pan layer that is preventing vertical soil moisture movement (Kooistra et al. 1984). Laser-assisted land levelling equipment fitted with a drag scrapper for smoothening the land surface (within ±2 cm of its average micro-elevation) was found promising to conventional land levelling (an iron scraper attached to the tractor was moved on the land surface in a visual elevation level) in light and medium textured soils where pearl millet is rotated with mustard/chickpea/lentil/field pea (Singh et al. 2021a).

Tillage studies on pearl millet (2000–2008) under the aegis of All India Coordinated Research Project for Dryland Agriculture (AICRPDA), ICAR-CRIDA (Central Research Institute for Dryland Agriculture), Hyderabad, have indicated differential impacts depending on soil type and annual precipitation. In an inceptisol (Agra, Uttar Pradesh), conventional tillage, CT (summer ploughing + 3-disc harrowing's + 2 intercultural operations at 20 and 40 days after sowing, i.e. DAS) proved promising to low tillage systems. However, in Vertisol (Solapur, Maharashtra) and Aridisol (Hisar, Haryana), low tillage, i.e. exclusion of 1 ploughing (Solapur) and 1 intercultural operation at 45 days after sowing (Hisar) did not cause significant yield loss over CT, on the contrary, have better rainwater use efficiency, profits and sustainable yield indices (Sankar et al. 2012). A significant improvement in system yields (11.2%), water productivity (16.7%), infiltration rat and e, and available nutrients with reduced cost of production (14.67%) with the adoption of conservation agriculture (CA), i.e. zero tillage (ZT) with residue retention after five-pearl millet-mustard cropping cycles over CT (7.2 t/ha, 0.96 kg grain/m^3, and ₹57,556) indicates the utility of CA (Singh et al. 2021a, b). Improved soil quality under ZT due to crop residue or Leucaena mulch in other important rainfed pearl millet–chickpea/mustard/wheat cropping systems (Singh et al. 2018) further gives scope for CA. Conventional tillage in pearl millet followed by zero tillage in *rabi* crops (wheat/cluster bean) is also gaining acceptance in India (Coventry et al. 2015).

18.1.5 Manures and Fertilizers

Pearl millet crop is a prolific nutrient remover from the soil as evident from the fact that every ton of grain produced removes 31.8–17.4–61.3 kg/ha N-P_2O_5-K_2O besides 40–170–20–8 g/ha of Zn-Fe-Mn-Cu (Tandon 2004). Sulphur removal is estimated at 3.33 kg/ton of grain produced (Kumar and Kumar 2021). However, pearl millet-grown soils are poor in fertility on account of harsh climates (hot and arid). Recent soil health card data of over 30.3 million samples of India (as of 12-July 2021) also showed the moderately alkaline nature of soils (71.4%) with 82.2, 20, 15, 25.82, 33.75, 31.34, 12.52, 22.62 and 4.79% samples showing deficiency for N, P, K, S, Zn, Fe, Mn, B and Cu (Soil Health Card Dashboard (dac.gov.in)). Despite of the above wide deficiencies, the pearl millet crop has only N and P fertilizer recommendations in India and the same holds true for other Asian countries too indicating likely inhibition of crop yields due to imbalanced fertilization. Studies have indicated that fertilizers use in pearl millet (FAO 2005) is low in general (21.9–5.5–0.8 kg/ha N-P_2O_5-K_2O) and in rainfed situations in particular (18.4–4.8–0.6 kg/ha N-P_2O_5-K_2O) as compared to irrigated one (62.2–13.9–3.4

kg/ha N-P_2O_5-K_2O). The use of manures, bio-fertilizers and to some extent NP fertilizers together is supporting the crop nutrient needs partially resulting in low productivity. Thus, enormous scope lies for enhancing the pearl millet yields by matching the fertilizer applications with the soil fertility/quality, crop nutrient requirements and most importantly with soil moisture availability.

18.1.5.1 Manures

Manuring through farm yard manure (FYM) and to a little extent with crop residues (CR) is the predominant way of pearl millet crop nutrition in arid and semi-arid rainfed production systems by farmers. Soil organic carbon is the core of soil fertility and its management too, that, however, is difficult to achieve in the tropical world. The untimely and heavy rainfall events in the recent past under the influence of changing climate are conducive to the leaching loss of applied fertilizers/mineralized nutrients, especially nitrogen (N). Crop residues owing to their wide C: N ratio (>50: 1) upon application to soil immobilize the mineralized fraction of nutrients present in the soil or applied through fertilizers. FYM, on the other hand with a narrow C: N ratio (<30:1) releases nutrients slowly over prolonged periods. Thus, CR and FYM applications were found promising in preventing leaching losses of nutrients to a major extent in medium-textured soils. However, studies on coarse loamy soils (85, 8.1 and 5.5% sand, silt and clay) of Jodhpur, Rajasthan, India, reveal that pearl millet crop residues (~22 cm height of stem + roots at 2.5 t/ha) left in the field after harvest in mono-cropped regions were found to have a slow decomposition (10–15%) during the turnover periods (CAZRI 1993) gets rapidly decomposed with receipt of rains coinciding with the next crop growth and the phytotoxic substances releases into the rhizosphere (Saxena et al. 1997) were found to have inhibitive effects on yield (Saxena et al. 1997) and thus residue recycling in pearl millet monoculture regions is not a good proposition. Narrowing the C: N ratio of residues by their conversion into compost could be effective in overcoming both the immobilization and phytotoxicity effects. Not only the equal effectiveness of leaf compost (LC) as that of FYM (both at 10 t/ha) in terms of pearl millet grain yields (3.11 t/ha) but also 10.7% higher yields over the recommended dose of fertilizers (Bana et al. 2016) proves the utility of compost technology. The superiority of manure (LC and FYM) applied crop as compared to chemical fertilizers was ascribed to a balanced supply of macro and micronutrients, improved biochemical properties of soil including water holding capacity (Choudhary and Suri 2014). Enrichment of grain with Fe, Zn and protein (agronomic fortification) is the additional advantage of manuring. An improved rainwater absorption due to FYM application owing to its amorphous nature (Joshi 1987) boosting the pearl millet grain yields by 6.1% over control (Saxena et al. 1997) also depicts the soil water storage enhancement functions of manures. However, in light textured and very low organic matter containing sandy soils, leaching losses of nutrients released from decomposing organic manures (Brouwer and Powell 1998; Suzuki et al. 2017) are a matter of concern. Keeping all the concerns aside, there is a need to build/maintain soil organic carbon (SOC) pools for maintaining/improving soil productivity in pearl millet regions. A long-term field study (18 years) in an Entisol on pearl millet-cluster bean-castor sequence at Dryland Research Station, Sardar Krushi Nagar, Gujarat, India, under the AICRPDA revealed a net loss of 12 and 4.4 Mg C ha^{-1} in control (no manure and fertilizers) and CR + FYM applied plots, while integrated use of chemical fertilizers along with FYM was found to have low SOC depletion and to maintain the SOC, a C input of 3.3 Mg C ha^{-1} $year^{-1}$ was needed in the location. The results have further revealed that for every Mg increase in profile SOC stock, system yields increased by 0.46 Mg, i.e. 0.17, 0.14 and 0.15 Mg/ha of pearl millet, cluster bean, and castor (Srinivasarao et al. 2014). A similar need for the integrated use of FYM and fertilizers for enhanced SOC, available nutrients and crop productivity of pearl millet-wheat cropping system in an inceptisol was reported by Moharana et al. (2012).

18.1.5.2 Bio-fertilizers

Bio-fertilizers remain the next choice of pearl millet crop nutrition in dry lands after manures owing to their inexpensive nature despite of showing less visible impacts on yields. The importance of bio-fertilizers also emanates from the fact that they by way of substituting nitrogen chemical fertilizers, reducing ammonia volatilization (Fan et al. 2011; Jadon et al. 2018) and nitrate leaching (Wang et al. 2015) losses and thus reduces soil, water and air pollution (Singh et al. 2018) and enhances N use efficacy. Hence, Aggarwal and Venkateswarlu (1989) stressed upon giving prime importance to the use of manures and bio-fertilizers in pearl millet crop nutrition while limiting the role of chemical fertilizers to balance the crop nutrient needs. Among the bio-fertilizers, nitrogen-fixing bacteria (*Azospirillum* and *Azotobacter*), phosphate solubilizing/mobilizing organisms are more commonly used in pearl millet nutrition. However, in recent times, potassium and zinc solubilizing organisms' use is also gaining prominence. Bio-fertilizers are used either singly or in combination (consortia) as seed inoculants or soil application in direct seeded conditions. Seedling root dipping is followed in transplanted conditions.

The beneficial effect of *Azospirillum* and *Azotobacter* inoculation to pearl millet seeds was ascribed to not only atmospheric nitrogen fixation (Dart and Wani 1982) but also to their phytohormones (Indole acetic acid, cytokinin, and Gibberellin) production (Vikhe 2014; Tien et al. 1979; Gautam 2000) which promotes root functions and finally plant growth. Further, *Azotobacter* has the ability to induce resistance to soil-borne pathogens (Ponmurugan et al. 2012) and tolerate and survive better in a water-limited environment (Sivapriya and Priya 2017) owing to its dormant cell covered by a two-layered capsule; the exine and intine (Espín 2016) and thus has a wider scope for use. Depending on the edapho-climate of a location, pearl millet yields enhancements due to *Azotobacter* and *Azospirillum* inoculations by 11–12% were reported (Wani 1990).

Bio-fertilizer inoculants with species of *Pseudomonas*, *Micrococcus*, *Bacillus*, *Flavobacterium*, *Penicillium*, *Fusarium*, *Sclerotium*, *Aspergillus* etc. for bio-phosphorus conversion are used (Reddy et al. 2020; Singh et al. 2018) for enhanced pearl millet crop performance.

Utility of potassium solubilizing bacteria (KSB) in pearl millet K nutrition in pot culture studies (Rekha et al. 2018) owing to their organic acids production that dissolves K bearing minerals rapidly, chelates silicon ions and results in the release of K ions into the soil for crop uptake (Bennett et al. 1998).

Effective solubilization of soil zinc due to *Bacillus cereus* (EPP5 strain) inoculation pearl millet crop along with potassium and phosphorous solubilization and endophytic activity against *Rhizoctonia solani*, *Sclerotium rolfsii*, and *Fusarium solani* fungal pathogens (Kushwaha et al. 2020) shows the multiple roles of Zn bio-fertilizers.

Beneficial effect of co-inoculation of nitrogen-fixing (*Azotobacter chroococcum*, *Azospirillum lipoferum*, and *Acetobacter diazotrophicus*) and phosphate solubilizing microorganisms (*Pseudomonas fluorescens*) diazotrophic bacteria on growth, yield, and profits (Saxena et al. 1997; Gautam 2000; Singh et al. 2018) calls for consortia-based bio-fertilization.

18.1.5.3 Liquid-Based Bio-fertilizers

Traditional carrier and/or powder-based bio-fertilizers of *Azospirillum* and phosphate solubilizing/mobilizing organisms owing to their short shelf life, low total viable microbial load at the time of use, contamination, temperature sensitiveness etc. (Yadav and Chandra 2014) have low field efficacy. Hence, liquid formulations have evolved with the ability to address not only the above concerns but also have the scope for the inclusion of additional stress tolerance abilities (Hegde 2008; Brar et al. 2012). Further, their use as seed spray brings in additional advantages. Field studies at Coimbatore, India, have indicated enhanced germination and seedling vigour with the use of liquid bio-fertilizer (*Azospirillum* or phosphobacteria alone or as a consortium) soaked pearl millet seeds for 18 h (Raja and Anandham 2020). Further, yield enhancements in pearl millet crop and its succeeding crops of wheat/mustard in cropping system with liquid *Azotobacter* and PSB bio-fertilizer formulation (Azo-phosphoteeka') use (Gera et al. 2020) call for their extensive use. Studies (Savita 2017) on biomix (*Azotobacter* + *Azospirillum* + PSB) have proven the role of not only seed inoculation but also of foliar spray of *Azotobacter* even when used along with fertilizers (Table 18.1) and thus liquid-based bio-fertilizers are made a part of the package of practices of pearl millet cultivation in India.

Though manures and bio fertilizers could meet the low nutrient demanding traditional varieties, the advent of high-yielding varieties especially hybrids has paved the way for greater and inevitable use of fertilizers in pearl millet crop and cropping system.

18.1.5.4 Chemical Fertilizers

Fertilization is the most crucial factor of pearl millet productivity in inherently low fertile and moisture-stressed production scenarios. Extensive research under the All India Coordinated Pearl Millet Improvement Project (AICPMIP) for different eco-regions for the *kharif* Pearl Millet crop has resulted in the development of the following fertilizer recommendations (Table 18.2).

Table 18.1 Pearl millet biomass and grain yield as influenced by biomix and foliar spray of *Azatobacter* with and without fertilizers

Treatment	Yield (t ha^{-1})	
	Grain	Biomass
No biofertilizer control	1.86	6.56
Seed treatment with Biomix	2.24	8.01
Foliar spray of *Azotobacter*[a] at 15 DAS	2.14	7.75
Foliar spray of *Azotobacter* at 30 DAS	2.18	7.63
RDF (40–20 kg ha^{-1} N-P_2O_5)	2.99	10.26
75% RDF (30–15 kg ha^{-1} N-P_2O_5)	2.54	8.88
RDF + seed treatment with Biomix	3.34	11.49
RDF + foliar spray of *Azotobacter* at 15 DAS	3.11	10.74
RDF + foliar spray of *Azotobacter* at 30 DAS	3.07	10.50
75% RDF + seed treatment with Biomix	2.88	10.23
75% RDF + foliar spray of *Azotobacter* at 15 DAS	2.70	9.79
75% RDF + foliar spray of Azotobacter at 30 DAS	2.71	9.58
CD ($P = 0.05\%$)	0.335	0.819

[a] Isolate *JFS5* at 10^8 cfu ml^{-1}

Table 18.2 Fertilizer recommendations for pearl millet in three different eco-systems

Zone	Precipitation (mm)	States	Soils and temperature	Fertilizer dose (kg/ha N-P_2O_5)
Rainfed cop				
A1	<400	Rajasthan, Gujarat, Haryana	Sandy soils; high temperature	40-20
A	>400	Rajasthan, Gujarat, Haryana other than A1 zone an all northern states	Sandy loam soils	60-30
B	>400	Maharashtra, Karnataka, Tamil Nadu, Andhra Pradesh etc. Southern states	Heavy soils, mild temperature	60-30
Irrigated crop				
		Gujarat		120-60
Intercropping system				
B Zone		Pearl millet + pigeon pea / soybean (2:1)		40-30 (*Azotobacter* + PSB inoculation)
Cropping system				
B Zone (Dhule, Maharashtra)		Pearl millet–soybean	50% RDN + 50% RDN as vermicompost	50% reduction in dose to soybean

Macronutrients

In A_1 zone (Rajasthan, Gujarat, and Haryana states receiving <400 mm precipitation with light sandy soils and high temperatures), 40–20 kg/ha N-P_2O_5 was recommended. For zone A (rest of Rajasthan, Gujarat, Haryana other than A_1 zone and all Northern states of India growing pearl millet, receiving >400 mm precipitation and have irrigation facilities, sandy loam soils) and zone B (Maharashtra, Karnataka, Tamil Nadu, Andhra Pradesh etc. Southern states with >400 mm precipitation, heavy soils with mild temperatures), 150% of A_1 zone fertilizer dose (60–30 kg/ha N-P_2O_5) is recommended.

Nitrogen application in two equal splits, i.e. half as basal and remaining half as top dressed 3–4 weeks after sowing is found promising. However, response to top dressed N depends on the availability of sufficient soil moisture. Under prolonged dry spells, top dressing of N to soil is not useful; hence, 2% urea foliar spray is recommended. Studies at Bikaner, Rajasthan, India, have indicated that application of 75% of recommended N (50% basal + 25% top-dressing at 25 DAS) along with 1.5% NPK (18: 18: 18) foliar spray at 35 DAS resulted in maximum yields and profits (Rundla et al. 2018). Further, when excess rains are received during the vegetative phase, an additional 20 kg N/ha application over recommended dose was found desirable.

Studies at Anand, Gujarat, India (Jakhar et al. 2011), have indicated 120 kg N/ha (50% as basal + 25% top dressing at 30 days after sowing (DAS) + 25% foliar spray at 60 DAS) as the optimum dose for summer irrigated pearl millet productivity. A higher phosphorous dose (60 kg/ha) is applied for irrigated summer crops compared to rainfed crops (Thakor et al. 2018).

In dry lands of western Maharashtra, the application of 50 and 20 kg K_2O fertilizers to pearl millet and chickpea (Pawar et al. 2021) and 75 kg K_2O to pearl millet in pearl millet-safflower (More et al. 2020) has enhanced the prospects of cropping system in an otherwise monocropped area.

A nitrogen fertilizer economy of 40 kg/ha in succeeding pearl millet (De and Gautam 1987) after chickpea has been reported possibly due to sparing effect coupled with biological nitrogen fixation. Similar positive impacts of summer groundnut (Patel et al. 1994) groundnut and cowpea (Giri and De 1980) on nitrogen nutrition of succeeding pearl millet were reported. The positive (80–135 kg/ha) soil N balances (crop N–grain N) after chickpea explains the beneficial effect of legume (chickpea) on succeeding cereal, i.e. pearl millet as compared to cereal-cereal rotation (Turfin et al. 2002). The need for application of NPK fertilizers to each component crop-based soil test using site-specific nutrient management for pearl millet–wheat/mustard rotations proves the non-residual effects of fertilizers (Dwivedi et al. 2009). Phosphorous fertilization of legumes is found promising in legume-pearl millet rotations as it aids in better root growth, nodulation, and nitrogen fixation. Studies on fenugreek–pearl millet rotation (Kumawat et al. 2021) have indicated that 60-5 kg/ha P_2O_5-Zn applied to winter season fenugreek (a legume) have marked residual effects on pearl millet. The N benefit from the legume component to the pearl millet in intercropping is minimal and it is from the sparing effects rather than direct transfer.

Micronutrients

Studies have indicated the widespread deficiency of zinc (Zn) in Asian soils varying from 33% in China to 50% in India (Frossard et al. 2000) and the highest of 86% in Pakistan (Kausar et al. 1976). Zinc has emerged as the most commonly deficient in pearl millet grown soils of India, while in salt-affected soils besides Zn, iron (Fe) deficiencies were also reported (Meena et al. 2018). Zinc deficiency is getting aggravated further under the impact of the green revolution (Dar 2004).

Iron (Fe) deficiency in crop plants emanates from low availability than its low soil content as Fe is the fourth most abundant element in the geosphere after oxygen, silicon and aluminium. Antagonistic effects of copper, zinc, manganese and phosphorus on iron uptake of plants and its low availability for uptake by plants in alkaline soil conditions (Frossard et al. 2000) explain its deficiency in crops.

Further, high soil iron status results in poor response to Fe fertilization. However, in practice, improving crop yields through Fe fertilization was found to be difficult. For example, Fe application to soils as ferrous sulphate ($FeSO_4$) has resulted in limited increases in crop yields (Frossard et al. 2000). Though the application of iron to soil as chelates was found more effective, the high costs involved prohibit their application on low-value staple crops (Savithri et al. 1999) like pearl millet. Thus, the foliar application of iron compounds with modest positive effects on cereal yields has been followed (Frossard et al. 2000; Rengel et al. 1999).

Looking at the importance of Fe and Zn in human nutrition, minimal grain Fe and Zn contents (42 and 32 ppm) for release of new pearl millet cultivars were fixed in 2018 and to achieve such levels of zinc and iron in grain, the development of bi-fortified varieties and fertilization assumes great importance.

When pearl millet cultivated soils are efficient in zinc, application of 25 kg/ha $ZnSO_4$ is recommended. To correct the zinc deficiency in standing crops, foliar spray of 0.2% $ZnSO_4$ at tillering and the pre-flowering stage is recommended from studies of AICRP-PM.

Soil application of Zinc / Ferrous sulphate before the final ploughing once in 3–5 years is recommended in mono-cropping areas while every year in cereal-based cropping systems. For deficiency corrections in a standing crop, foliar application (0.5%) is done; however, soil application gives prolonged solutions than foliar application. Foliar spray of 1.50 kg $ZnSO_4{\cdot}7H_2O$ on millet in soils with low zinc content was found to improve the grain yield, protein, lysine acid and soluble sugar contents by 5.54, 10.83, 58.7 and 69.9% respectively over no foliar spray control (Zong et al. 2011). Nano ZnO (18.5 nm) foliar spray (10 mg L^{-1}) after 2 weeks of germination of pearl millet (HHB-67 cultivar) at Jodhpur, Rajasthan, has revealed that on account of its greater penetration in the plant cell through the stomatal opening and natural nanopore enhancing the plant cell metabolic activities have led to 37.7 and 20.5% improvement in grain yield over no Zn (1065 kg/ha) and ZnO spray (Tarafdar et al. 2014). However, the high price of nano ZnO fertilizers at present remains a limitation to its wider adoption. This technology has great scope in the coming days.

Intercropping of pearl millet with cluster bean and organic manures (FYM, leaf compost (LC), LC + cow dung) have resulted in markedly higher Zn and Fe and Zn contents in pearl millet grain (Bana et al. 2016) that indicates the scope for agronomic bio-fortification.

18.1.6 Moisture Conservation and Irrigation

Pearl millet is a crop of rainfed areas depending on the quantity of rainfall received and its distribution within the crop season is subjected to varying levels of moisture stress and accordingly yield is limited. In India, the A_1 zone comprising of western parts of Rajasthan, Gujarat, and Haryana receives <400 mm low rainfall and thus is a frequent drought impacted area. Studies by Rao and Saxton (1995) at Jodhpur, Rajasthan, (A_1 region) for the period of 1971–1989 indicated that 89% of yield variations can be explained by water stress index (WSI) $= \sum 1 - (T/\mathrm{Tp}) \times \mathrm{SUS}$; T and Tp are the computed actual and potential transpiration in mm/day, SUS is a seasonally dependent weighting factor for grain yield susceptibility) and WSI has negative relation with yield ($y = -45.38$ WSI + 526.18; $r = -0.9427$). Water stress in arid areas of pearl millet crops gets aggravated by the low moisture retention capacity of coarse-textured soils and high evaporation losses under the influence of high temperatures. Thus, any management practice (s) that conserve the soil moisture by way of either enhancing the soil moisture storage or prolonging the moisture supply to the crop on account of reduced evaporation losses will augment crop productivity.

Studies have indicated that a shift from age-old conventional tillage (CT) practices to zero tillage (ZT) is promising for moisture conservation in drylands (Choudhary et al. 2017). When a shift from CT to ZT is not possible, adoption of modified land configuration systems, i.e. ridge and furrow method, was found promising for moisture conservation that is reflected in 22.21% enhanced pearl millet grain yield (Kanwar et al. 2015) over flat beds.

Mulching is seen as a way to conserve soil moisture with the additional benefit of erosion control that is associated with the run of water and heavy winds. Studies at Kanpur, India, on pearl millet revealed that residue mulch at 4 t/ha (Choudhary et al. 2017) combined with one weeding and hoeing at 25 days after sowing (DAS) was found to enhance the grain yield over weeding and hoeing and ridging and furrowing by 24.06 and 15.97% while using 3.57 and 1.48% less water with 28.8 and 11.8% enhanced water use efficiency (Rajput et al. 2022) over weeding and hoeing and ridging and furrowing (Table 18.3). In recent times, plastic mulching has been resorted for maximum curtailing of soil evaporation losses and thus conserve moisture and on this conserved moisture pearl millet grain yields were enhanced by 48.82 per cent (Kanwar et al. 2015); however, their high costs and short life prevent its full potential exploitation.

Combining zero tillage (Choudhary et al. 2017) with residue mulch, i.e. conservation agriculture (CA), was found more promising on account of added benefits of essential nutrients supply to crop and reduced cost of cultivation (Choudhary et al. 2016) besides moisture conservation over ZT or mulching alone.

Table 18.3 Impacts of moisture conservation practices on water use and productivity of pearl millet (mean of 2018 and 2019)

Moisture conservation practice	Dry weight of roots/ plant (g)	Total water use (mm)	Grain yield (t/ha)	Stover yield (t/ha)	Water use efficiency (kg grain/ha mm water)	Net returns (₹/ha)
Weeding and hoeing at 25 DAS	19.2	336.0	1.459	5.122	4.34	16,083
Ridging and furrowing between crop rows at 25 DAS	22.3	328.8	1.692	5.506	5.00	20,312
Weeding and hoeing at 25 DAS + residue mulch at 4 t/ha	24.4	324.0	1.810	5.801	5.59	22,062

Studies on pearl millet intercropped with pigeon pea (8:2) at Gwalior, Madhya Pradesh, India (Rajput and Bhadouriya 2019), have indicated that repeated intercultivation (3 times), maize residue mulching (5 t/ha) and *Glyricidia* green leaf manuring (10 t/ha) have enhanced the pearl millet (pigeon pea) grain (seed) yields by 12.4, 22.5 and 26.8% (34.9, 61.3 and 84.9) over control (3125 and 212 kg/ha) respectively. Farmyard manure (FYM) addition to soil owing to its amorphous nature (Joshi 1987) was also found to enhance the soil water storage and thus pearl millet grain yields by 6.1% over no FYM (Saxena et al. 1997).

Moisture-absorbing polymers like 'Hydrogel' with the ability to absorb moisture 400 times of its weight and release up to 95% under stress and its usability for 2–5 years before biodegradability (Kalhapure et al. 2016) have been explored in alleviating drought stress in drylands. Studies have indicated that coating 'Pusa hydrogel' to pearl millet seeds at 20 g/kg seed as effective moisture conservation practice (Singh 2012). Combined use of Pusa hydrogel with stover mulch was found to bring enhanced nutrient uptake besides moisture conservation (Yadav et al. 2021). Foliar application of plant bio-regulators like salicylic acid and thiourea was found to alleviate the moisture and salinity stress in pearl millet on account of enhanced proline production and N and S nutrition (Yadav et al. 2020).

Pre-soaking of seeds with growth regulators, i.e. cycocel (CCC), B-9 and Phosphon-D, Malic hydrazide (Yadava and Patil 1980; Bishnoi et al. 1994) and KNO_3 (Gupta et al. 2022; Kanwar et al. 2015) and foliar application of kinetin and CCC (Kumari 2017), has been found to induce drought hardiness in plants by enhancing relative water content and thus enhancing grain yields of pearl millet. Anti-transpirants (phenyl mercuric acetate, 320 ppm) spray was also found to improve the water functions (by reducing plant transpiration losses) and yields in pear millet (Ansari et al. 2012).

18.1.6.1 Rainwater Harvesting

The skewed precipitation distribution issues of rainfed agriculture could be effectively addressed by capturing the runoff generated from excess rainfall events during the crop season and its storage (farm ponds) for subsequent protective irrigation at critical stages of the crop growth for moisture and thus has immense potential to stabilize rainfed crops productivity including pearl millet (Rao et al. 2010). For watershed-based rainwater harvesting, catchment area treatment by land smoothening and compaction, application of soil dispersing (NaCl, $NaHCO_3$), and infiltration reducing chemicals (clay, wax, asphalt, bitumen, and bentonite) enhance the runoff in to storage structures. Crop water balance analysis for 43 pearl millet grown districts of India that accounted for 85% of acreage (3.921 m ha) has the potential to generate 374,664 ha-meter of runoff of which 359,991 ha-meter (1.818 m ha) can be harvested. From the harvested water, 100 mm protective irrigation can be given to 1.37/0.837 m ha area during normal/drought seasons (Rao et al. 2009). In arid zone, efficient underground runoff water storage structures, Tanka and Khadin have been evolved (Khan and Venkateswarlu 1993; Khan 1998) and used for providing life-saving irrigation to crops in Rajasthan and Gujarat (Narain and Goyal 2005). In other parts of India (A and B Zones), surface water storage structures of farm ponds are found promising (Rao et al. 2019). The lining of farm ponds (seepage control) and evaporation control methods (Rao et al. 2017) were found to enhance the efficient storage of harvested water for irrigation use.

18.1.6.2 Irrigation

The water requirement of pearl millet is quite low and varies with climate of a location, season etc. DSSAT model studies have estimated the water requirement as 372–412 mm in Gujarat (Pandey et al. 2008) which could be met from *kharif* precipitation itself. However, the skewed distribution of precipitation during the *kharif* season makes it difficult to achieve. Further, pearl millet is increasingly cultivated during post-rainy seasons (winter as well as summer seasons) in recent times. The above conditions call for protective and need-based irrigations respectively for higher pearl millet productivity. The cultivation of pearl millet under assured irrigation in India is increasing and has gone up from 0.311 m ha, i.e. 3.45% of the total area in 1950–1951 to 0.668 m ha, i.e. 9.40% during 2018–2019.

Studies have indicated that after germination, tillering, ear head emergence (heading), and grain filling are the most critical crop stages for irrigation in pearl millet (Kumar and Kumar 2021). Generally, pearl millet crop is flood irrigated when water is unlimited. Application of 40 mm water at 1.0 IW/CPE ratio was found promising for summer season pearl millet in Gujarat (Khafi et al. 2011). However, drip irrigation at 150% pan evaporation including fertigation (with 125% N) was promising (Alagudurai and Muthukrishnan 2014) and may be resorted to under limited irrigation water available situations.

Pearl millet can tolerate irrigation with saline waters and has lesser yield penalties, but sodium absorption ratio (SAR) was found to reduce infiltration rate (Singh et al. 1992). Studies at Nain village, Panipat district, Haryana, India (Makarana et al. 2019), have indicated that saline water irrigated pearl millet (Electrical conductivity (EC) 3, 6, and 9 dS/m) has 10.1, 20.7, and 37.3% reduced grain yields over control (2.17 t/ha). The relative yield (RY) of pearl millet under saline irrigation water can be estimated by the equation of RY = 133.1 − EC_{iw} (5.58−0.0026 R) − SAR_{iw} (0.0025 R + 0.43T) − 0.03 $(SAR_{iw})^2$; $R^2 = 0.79$ (SAR: Sodium adsorption ratio; iw = irrigation water; *R*: rain fall (mm); *T*: time period after initiation of irrigation in years), gypsum application, leaching of built up salts with good quality water irrigation (Singh 2012) and drip irrigation (Singh et al. 2019) were found to reduce the adverse impacts of saline water irrigation. Solar-powered irrigation systems (both good and saline waters) have been developed with no carbon footprint and are used for pearl millet better profits (Singh et al. 2021b; Poonia et al. 2018).

18.1.7 Weed Management

Weeds are the major biotic pressures of pearl millet cultivation more so under rainfed conditions owing to their quick and high water extraction and use potential. During the rainy season, the intermittent and continuous rains may often result in the emergence of more than a flush of weeds as compared to a single flush of weeds in the winter and summer seasons. Further, proper land preparation also results in less weed menace in the winter and summer seasons. Slow initial growth of pearl millet crop coupled with its wide row sowing (45–60 cm) provides free space for germination and growth of weeds and thus severe weed competition is created. Complex weed flora are seen in pearl millet crop and the general weed flora reported across India are given in Table 18.4. These weeds compete with the crop for both natural (light, space) and applied resources (fertilizers, moisture). Competition by weeds for applied resources is for moisture in low rainfall areas (<400 mm) and for nutrients in high rainfall areas (>400 mm rain). The yield penalties due to weeds are more in low rainfall areas as weeds deplete soil moisture rapidly and crop producing low biomass may produce poor and sometimes no yields also. In an uncontrolled (weedy check) plot, weeds and pearl millet crop removed 47.7–31.2–39.9 and 105.5–24.9–117.5 kg/ha N-P-K nutrients from the soil (Ramesh et al. 2019). Weeds by way of acting as alternate hosts and providing shelter for various pests encourages the diseases to spread (Marwat et al. 2005) and thus hamper crop performance.

Studies have indicated that in pearl millet, 15–30 days after sowing (DAS) period are critical period of crop-weed competition (Labrada et al. 1994) and the crop needs to be kept without weeds during this period and as the crop attains a height of 50 cm and weeds growing thereafter will not hamper the crop performance anymore. Depending on the number of flushes, density, duration of crop-weed competition, and management practices followed, weeds were found to incite a yield penalty of 16 (Rao et al. 2014) to 94% (Balyan et al. 1993) in pearl millet crop besides hampering the produce quality (Bahadur et al. 2015). Hence, effective weed management options are used and are evolved continuously.

Hand weeding and interculture (by the animal, power tiller, and tractor-drawn) remain the most commonly followed methods of weed control in pearl millet; they often have low efficacy on account of failure to provide weed control at critical

Table 18.4 Major weed flora of Pearl millet crop in India

Broad-leaved	Grasses	Sedges	Parasitic weeds
Ageratum conyzoides, Alhagi camelorum, Ageratum conyzoides, Amaranthus viridis, A. spinosus, Borreria articularis, Commelina benghalensis, Celosia argentea, Corchorus olitorius, Corchorus aestuans, Canabis sativa, Chrozophora rottleri, Cleome viscose, Convolvulus arvensis, Croton bonplandianum, Digera arvensis, Euphorbia geniculata, E. hirta, Eclipta alba, Fumaria parviflora, Ipomoea purpurea, Leucas aspera, Oxalis latiifolia, Parthenium hysterophorus, Phyllanthus niruri, Sonchus arvensis, Trianthema portulacastrum, Tridax procumbens	*Brachiaria ramose, Brachiaria eruciformis, Brachiaria ramose, Cynodon dactylon, Dactyloctenium aegyptium, Digitaria sanguinalis, Digitaria marginata, Digitaria ciliaris, Echinochloa Colona, E. crus-galli, Eleusine aegypticum, Ischene despaire, Paspalum paspaloides, Panicum dichotomiflorum, P. repens, Saccharum spontaneum, Sorghum halepense*	*Cyperus rotundus, C. esculentus, C. iria, Eragrostis major, Fimbristylis diphylla*	*Striga asiatica*

periods in view of untimely rains during the rainy season. Hand weeding twice at 15 and 30 DAS (Ramesh et al. 2019) was found to reduce N-P-K nutrient removal of weeds by 62.7–74.7–63.9% with a concomitant increase in crop nutrient uptake (46.1–49.4–46.6% N-P-K). To derive greater benefit from applied fertilizers, a top dressing of nitrogen is advocated after hand weeding. Selective application of nitrogen to crops through foliar spray or band application to crop rows may aid in overcoming the above rule of topdressing N after hand weeding. In broadcast sown crops as in Pakistan, there is no alternative to hand weeding (www.PAKISSAN.com). However, depleting manpower and draught animals in recent times in Asia for weeding coupled with the ever-escalating labour wages are hindering their continued use as weed management tools.

Intercropping is an effective weed management tool in pearl millet and accordingly suitable intercropping systems with better weed management and yielding abilities were developed across India and in other Asian countries. Pearl millet intercropping with black gram and green gram (1:1) was found to smother the weeds by 52.0 and 39.3% leading to higher weed control efficiency (63.5 and 65.8%) that ultimately enhanced pearl millet equivalent yields by 68.7 and 57.7%

as compared to sole pearl millet crop yields of 2633 kg/ha (Mathukia et al. 2015). Better weed control, reduced NP uptake by weeds, and higher system yields with pearl millet + mothbean/cluster bean (1:1 ratio) as compared to sole pearl millet (Kiroriwal and Yadav 2013) and pearl millet intercropping with groundnut intercropping in 1:3 ratio (Shetty and Rao 1981) and pigeon pea (Rao and Shetty 1976) establish the role of intercropping in weed management.

Further, studies have indicated the utility of paddy straw mulch (2 t/ha), stale seed-bed, and soil solarization technologies as weed management solutions in summer pearl millet (Choudhary et al. 2016). However, for effective and economical weed management reliance on herbicides has become inevitable in pearl millet cultivation across India.

Studies have indicated that pre-plant (7 days before sowing) application of saflufenacil at 50 g a.i./ha (Reddy et al. 2014) and fluchloralin at 1.0 kg a.i./ha (Virkar et al. 2007) is effective for weed management in pearl millet. Atrazine has been the most commonly used pre-emergence herbicide in the sole pearl millet crop. The crop's ability to degrade atrazine at the rhizosphere level alone or in combination with soil microorganisms (Singh et al. 2004; Zhang et al. 2014) makes it the most widely used herbicide. However, for intercropped pearl millet with other crops (especially legumes), pendimethalin at 1.50 kg a.i./ha (Pre-em) is recommended (Shinde et al. 2003) in place of atrazine. Oxadiazone at 1.00 kg a.i./ha (Ram et al. 2005) and Oxyfluorfen at 0.2 kg a.i./ha (Deshveer and Deshveer 2005) were also found promising as pre-emergence herbicides. Integration of pre-emergence herbicide application with hand weeding at 30–40 DAS was found to provide effective weed management in pearl millet; however, in recent times, post-emergence herbicides are preferred to hand weeding. For early post-emergence application, atrazine at 0.50 kg a.i./ha (Munde et al. 2013) and post-emergence application (25–30 DAS), 2,4-D Na salt at 0.5 kg a.i./ha (Mehetre et al. 2021) proved promising. A new post-emergence herbicide Tembotrione 42% SC (90–120 g a.i./ha) applied at the 3–4 leaf stage has been recommended for weed management in pearl millet from 3 years coordinated study (AICRPPM (All India Coordinated Research Project on Pearl Millet) 2021). Studies have indicated that Tembotrione at 80 g a.i/ha at 10–15 DAS + hand weeding at 30 DAS could take care of all weed problems in pearl millet that includes the exclusion of pre-emergence herbicide application (Chaudhary et al. 2022).

For control of parasitic weed Striga, hand pulling of Striga weeds along with roots and 2,4-D post-emergence application on Striga leaves were recommended. In severely Striga infested regions, non-pearl millet crop cultivation after every 2 years is recommended. Crop rotation with trap crops (cotton, sunflower, groundnut, castor, Dolichos bean, linseed etc.) that stimulate Striga seed germination without getting parasitized is also recommended for Striga control. Studies have indicated the allelopathic nature of pearl millet plants on weeds (Radhouane 2014) and thus can be used as a bio-control agent (Radhouane 2012; Malik et al. 2019).

18.1.8 Cropping Systems

Pearl millet being a crop of rainfed situations, its production in a farm/region varies as a function of rainfall received in a season and sometimes crop failures are seen. Further, climate change has resulted in the simultaneous occurrence of floods and droughts in the same place. In such situations, location-specific mixed and intercropping systems are evolved to ensure livelihood security for farm families. In favourable rainfall regions and regions with irrigation facilities, pearl millet is rotated with other crops in annual cropping systems. Pearl millet is also a component of many agroforestry systems in water stressed regions.

Over time, especially in the late twentieth century, intercropping has surpassed mixed cropping both in popularity and in the extent of adoption. Non-cereals especially legumes owing to their biological N fixation coupled with P mobilization in low fertile soils remain the most preferred mixed and intercrops of pearl millet. Mixed cropping of pearl millet with sesame in North India (Gangaiah 2008) and legumes like black gram, green gram (Singh 1995), and cluster bean (Yadav and Yadav 2000) in the arid zone of Rajasthan are most common. In the Kumaun Himalayan Bhabhar region, paddy + pearl millet + horse gram + cucurbits mixed cropping is followed (Padalia et al. 2018). An improvement to mixed cropping, i.e. sowing of binary mixtures containing blended seeds of component crops having no distinguishable lines is widely practised particularly by small landholders in the Indo-Pak subcontinent and Africa (Akhtar et al. 2013). Studies in Pakistan (Iqbal et al. 2019) have indicated that binary cropping of pearl millet with cowpea/soybean/ cluster bean (50: 50 seed mixture) is promising with land equivalent ratio (LER) values (1.44–1.48) as compared to sole crops.

Under excess rainfall conditions leading to submergence, hypoxic conditions are created and, in such situations, the pearl millet crop suffers adversely due to non-existent oxygen supply to roots that impacts transpiration, photosynthesis, and finally biomass production. In such a situation, its mixed culture with rice with the ability to maintain oxygen supplies absorbed by shoots to the apical meristem of roots through its aerenchyma and sustain respiration and photosynthesis functions (Iijima et al. 2016) and drought alleviative effects are also reported for the same system (Yamane et al. 2018) provided there is no inter plant competition, have been developed.

To ensure higher production and monetary gains from pearl millet cultivation under aberrant weather conditions, intercropping with various crops is recommended. A number of pearl millet crop rows vary from 1 to 7 when pigeon pea is a component intercrop. However, for groundnut intercropping after every 1–6 rows of ground nut, 1 pearl millet row was introduced. Pearl millet and mung bean intercropping (1:7 row ratio) with 4.04 t/ha and ₹36,380 pearl millet equivalent yield and net income was found to be 2.83 and 2.28 times more productive and profitable (Ram and Meena 2015) than sole pearl millet but were at par with sole mung bean (3.90 t/ha, and ₹34,353) in the arid region of Rajasthan (Fig. 18.1). Pigeon pea + pearl millet (1:3) intercropping was recommended under scarce rainfall (dryland) conditions in Maharashtra (Kathmale et al. 2014). Intercropped pearl millet with

Fig. 18.1 Intercropping of pearl millet with moong bean in paired rows (30: 60 cm)

Table 18.5 State-wise intercropping systems of pearl millet in India and Pakistan

State	Intercropping system
Rajasthan (India), North West Frontier Province (NWFP) of Pakistan	Pearl millet + cluster bean/moth bean/ sesame
Haryana, Gujarat, Uttar Pradesh (India)	Pearl millet + green gram/sesame/castor
Madhya Pradesh	Pearl millet + black gram/soybean
Maharashtra	Pearl millet + moth bean/pigeon pea
Karnataka, Andhra Pradesh, Telangana	Pearl millet + pigeon pea/ground nut/Castor
Tamil Nadu	Pearl millet + cowpea/sunflower
NWFP, Baluchistan (Pakistan)	Pearl millet + melon/cluster bean/mung bean/sesame

sesame, moong, and melon in the plains of Baluchistan and with guar and melon in North West Frontier Province (NWFP) of Pakistan is common (www.pakissan.com/english/allabout/crop/millets.html). Promising pearl millet-based intercropping systems of India and Pakistan are given in Table 18.5.

Pearl millet is also a component of many cropping systems and some of the most important crop rotations are pearl millet-wheat-green gram, pearl millet-wheat-pearl millet, pearl millet-barley/gram/pea/wheat/berseem, pearl millet-potato-mung bean or black gram, pearl millet-potato-wheat, pearl millet-mung bean or black gram, pearl millet-rape-wheat, pearl millet-berseem, pearl millet-*toria*-wheat etc. of these systems. Pearl millet-wheat is a popular cropping system in semi-arid Agro-Eco Region IV of India, particularly in states of South-West Uttar Pradesh, Rajasthan, Haryana, and Gujarat, covering 2.3 million ha area and is the third extensive cropping system of India. Pearl millet-wheat cropping system is also found promising for scarce Western Maharashtra region, i.e. Satara district (Phaltan and Khandala blocks) with medium deep black soils (Chaudhari et al. 2019) and is prominent in Pakistan also. Pearl millet–mustard (*Brassica juncea*) cropping system is practised in 1.0 m ha in arid and semi-arid regions of North, West, and central India (Rajasthan, Uttar Pradesh, and Madhya Pradesh) on undulated light to medium textured soils with limited irrigation conditions (Yadav et al. 2011). In Punjab, Haryana, Western dry region of Rajasthan (400 mm rain), Western Plateau and Hills Region of Maharashtra, Madhya Pradesh, Rajasthan with 680–1040 mm rainfall (India), and in parts of Pakistan, the pearl millet-lentil system is popular (Sekhon et al. 2007). In assured rainfall (Eastern Uttar Pradesh, Bihar, West Bengal) and irrigation available areas (Punjab, Haryana, western Uttar Pradesh) of the Indian subcontinent, chickpea is rotated with millet (Saxena 1987). Under rainfed uplands of the Arid ecosystem too, pearl millet–chickpea rotation is followed (Ahlawat et al. 2003). The introduction of summer crops sorghum, millet, and sesame around 6000 years ago enabled chickpeas to reappear as a significant crop overcoming its *Ascochyta* blight disease associated with winter rainfall successfully (Abbo et al. 2003). Zone-wise, pear millet-based cropping sequence includes pearl millet-moth bean/cluster bean (A1 zone) and pearl millet-mung bean/soybean (zone A and B). In Myanmar, upland cropping systems, green gram, groundnut, pigeon pea, and cow-pea are commonly grown in sequence with pearl millet. Pearl millet–seed spice cropping systems are emerging in North-West India and Pearl millet–fenugreek system is one and out of 0.21 m ha, 57% is in Rajasthan alone.

Pearl millet is an important component of agro-forestry systems, especially in the Arid and Semi-arid regions of Punjab and Haryana. In the lower hill zone of Punjab (Kandi areas), agri-silviculture system of pearl millet with *Acacia nilotica/catechu*, *Anogeissus latifolia*, *Butea monosperma*, *Dalbergia sissoo*, *Grewia optiva*, *Zizyphus* sp. trees is popular. Similarly, the majority of farmers grow trees species of *Morus*, *Leucaena*, *Ficus*, and *Cyclobalanopsis* with cereal crops (maize, wheat, millet), lentils, and vegetables as an agroforestry practice in Nepal and *Leucaena* as a hedge tree of maize or millet are prominent in middle hills (Amatya et al. 2018).

18.1.9 New Trends in Pearl Millet Cultivation: Organic Production and Mechanical Harvesting

There has been increasing emphasis on organic millet in general and pearl millet in particular in light of consumers' averseness to eat chemically grown foods on account of adverse health consequences. Inherently low external inputs (fertilizer and pesticides) used in pearl millet cultivation make organic cultivation quite easy and feasible. Studies at Durgapura, Jaipur, Rajasthan, India have indicated that the supply of N through vermicompost or FYM + decomposed organic waste + *Azotobacter* + Phosphate solubilizing bacteria (75:25) could meet the nutritional demands and aid in superior yields and income as compared to unfertilized control (Sharma et al. 2012). The role of various non-chemical agronomic operations for effective weed management including poor growth of fast-growing weeds due to the slow release of nutrients from organic sources indicates the organic weed management practices (Bhadu and Kaswan 2022). Similarly, pest management solutions are available for organic pear millet production.

Fig. 18.2 Rain water inundated pearl millet after harvest in farmer's field at Mahendragadh, Haryana, India in September, 2022

The reduced availability of manpower and their ever-escalating wages in India have paved the way for mechanization of all farm operations and the pace of mechanization is rapid in recent times. As per FICCI (Federation of Indian Chamber of Commerce and Industry) (2015), 20–30% of pearl millet area is under mechanical harvesting and threshing. Manually harvested crop is getting inundated in water on heavy rainfall events in recent times under the influence of changing climate and rain fall pattern (Fig. 18.2). Such water-logged crop after manual harvest upon threshing will have poor grain quality, marketability and thus command poor price. In this context also, combine harvester use assumes importance. If combine harvester use is not available, at least mechanical threshing is advised. Studies on pearl millet have indicated the utility of threshers as evident from high threshing efficiency (96.8%) and low energy consumption (8 Mega J/ton of input crop). The cost of operation of a thresher was only ₹85/ton grain as compared to ₹312.5 and 1650/ton in threshing by animal trampling and manual beating (Kamble et al. 2003).

18.2 Future Research and Conclusion

The concerted efforts of the research organizations have led to the development of simple scientific solutions to the abiotic (and biotic) stresses pertaining to the pearl millet agro-eco-regions of Asia, especially India and Pakistan. These technologies when superimposed over high-yielding nutrient-dense cultivars (especially hybrids) could boost the pearl millet productivity near potential (sustainable) yields that can sustain the livelihoods of its stakeholder farmers. However, the following challenges needs to be addressed through research.

- Pearl millet is becoming increasingly a non-food crop in Asia, i.e. as an animal feed, biofuel etc. producing crop. Increasing non-food use can be judged from the increasing use of hybrids in lieu of traditional varieties. As a feed crop, it has to compete with maize, though the agro-regions of pearl millet (rainfed) and maize (irrigated) cultivation are different. Biofuel (ethanol) production technology from grains is emerging; however, its industrialization is still at nascent stage that needs to be evolved.
- The cultivation practices of pearl millet to a major extent are either manual or animal power dependent even today in Asia that are not only inefficient but over time are becoming costlier and are eroding the profits. Mechanization needs to be brought in lieu of manual and animal power operations. Use of precision seed drills especially under zero till conditions, manure / fertilizer spreaders, inter-row weeders etc. needs to be adopted by farmer's.
- Fertilization of pearl millet needs to be made need based and should go beyond the primary nutrients (NPK) application. Small quantities of secondary nutrients application like sulphur and micronutrients of zinc, iron and boron in their deficit soils could bring rich yield dividends. This is evident from the Bhoochetana scheme works of ICRISAT, Patancheru, India in the states of Karnataka (2009–2016) and Odisha (2018–continued) in which need based and soil fertility fertilization has given marked yield gains. However, there is need to develop soil fertility data base for pearl millet cultivated regions.
- Development of nutrient use efficient (N, P) cultivars should be given attention as the soils are inherently poor in fertility and farmers are incapable of applying required fertilizers.
- Breeding cultivars suitable for cropping systems (intercropping and crop rotations) for better performance needs to be addressed.
- Climate change is putting enormous challenges to pearl millet cultivation. Though it is tolerant to moisture, salinity etc. stresses, untimely excess rain fall is creating hypoxic situation and thus the need for suitable interventions to overcome are immediately required.
- Cultivation of pearl millet in dry lands with life-saving irrigation provided through run off water supplied through drip and sprinkler irrigation methods needs adoption.
- Weeds are becoming more competitive under changing climate and in this situation, new potent post-emergent herbicide molecules (other than atrazine)

introduction needs due attention. Non-genetically modified herbicide tolerant cultivars development in pearl millet (as done for sorghum for imazethapyr) also merits attention.

References

Abbo S, Shtienberg D, Lichtenzveig J et al (2003) The chickpea, summer cropping, and a new model for pulse domestication in the ancient near east. Q Rev Biol 78(4):435–448

Aggarwal RK, Venkateswarlu J (1989) Long term effect of manures and fertilizers on important cropping systems of arid region. Fertiliser News 34:67–70

Ahlawat IPS, Masood A, Shivakumar BG (2003) Cropping systems res. in chickpea. In: Masood A, Shiv Kumar, Singh NB (ed.) Chickpea research in India. Indian Institute of Pulse Res (IIPR), Kanpur, p 119–131.

AICRPPM (All India Coordinated Research Project on Pearl Millet) (2021) Proceedings of 56th Online Annual Group Meeting of ICAR-AICRPPM, p 71

Akhtar MF, Ahmad AU, Zamir MSI et al (2013) Agro-qualitative studies on forage sorghum (*Sorghum bicolor* L.) sown alone and in mixture with forage legumes. Pak J Sci 65:179–185. http://agris.fao.org/agris-search/search.do?recordID=PK2014001130

Alagudurai S, Muthukrishnan P (2014) Effect of crop geometry, irrigation regimes and nitrogen fertigation on fodder quality of Bajra Napier hybrid grass. Trends in Biosciences 7(19):3092–3096

Alam MK, Ogata Y, Sako Y et al (2010) Intermediary metabolism of plasma acetic acid, glucose and protein in sheep fed a rice Straw-based diet. Asian-Aust J Anim Sci 23(10):1333–1339

Amatya SM, Cedamon E, Nuberg I (2018) Agroforestry systems and practices in Nepal. Rampur: Faculty of Forestry, Agriculture and Forestry University (AFU). p 108. https://www.iufro.org/download/file/29095/1317/Agroforestry_Systems_and_Practices_in_Nepal_2018_pdf/

Anand Reddy K, Rao RS (1971) Effect of nitrogen and spacing on the growth and yield of bajra (*Pennisetum typhyoides*). Indian J Agric Sci 5:105–108

Ansari MA, Rana KS, Rana DS et al (2012) Effect of an anti-transpirant as growth suppressant and nutrient management on growth, productivity and quality of pearlmillet (*Pennisetum glaucum*)-pigeonpea (*Cajanus cajan*) intercropping system under rainfed conditions. Indian J Agron 57(4):343–348

Bahadur S, Verma SK, Prasad SK et al (2015) Eco-friendly weed management for sustainable crop production-A review. Journal Crop and Weed 11(1):181–189

Bana RS, Pooniya V, Choudhary AK et al (2016) Influence of organic nutrient sources and moisture management on productivity, biofortification and soil health in pearlmillet (*Pennisetum glaucum*) + clusterbean (*Cyamopsis tetragonaloba*) intercropping system of semi-arid India. Indian J Agric Sci 86(11):1418–1425

Balyan RS, Kumar S, Malik RK et al (1993) Post-emergence efficacy of atrazine in controlling weeds in pearlmillet. Ind J Weed Sci 25(1and2):7–11

Bennett PC, Choi WJ, Rogera JR (1998) Microbial destruction of feldspars. Min Manage 8:149–150

Bhadu K, Kaswan B (2022) Organic weed management an important component of organic farming: a review. Pharm Innov 11(7):1655–1660

Bhavna T, Sasode DS, Bhadoria SS et al (2020) Response of varieties and dates of sowing on growth and yield of Pearl millet under arid zone. J Pharmacogn Phytochem 9(1):988–990

Bhuja P (2009). Identifikasi dan Karakterisasi Tampilan Agronomis Jewawut Lokal Nusa Tenggara Timur: Upaya Merevitalisasi Keragaman Ketersediaan Pangan Nasional. http://www.goole.co.id

Bidinger FR, Hash T (2004) Pearl millet. In: Nguyen HT, Blum A (eds) Physiology and biotechnology integration for plant breeding. Marcel Dekker, New York, pp 225–270

Bishnoi N, Karwasra R, Gupta SN et al (1994) Influence of growth retardants on water relations and yield of bajra under rain fed conditions. Indian J Plant Physiol 37(2):73–75

Biswas S (2020) Prospects and constraints of transplanted maize, wheat, sorghum and pearl millet: a review. Int J Environ Clim Change 10(5):24–43

Brar SK, Sarma SJ, Chaabouni E (2012) Shelf-life of biofertilizers: an accord between formulations and genetics. J Biofertil Biopest 3(5):e109

Brouwer J, Powell JM (1998) Increasing nutrient use efficiency in West-African agriculture: the impact of microtopography on nutrient leaching from cattle and sheep manure. Agri Ecosys Environ 71(1–3):229–239

Carberry PS, Campbell LC, Bidinger FR (1985) The growth and development of pearl millet affected by plant population. Field Crop Res II:193–205

CAZRI (1993) Annual progress report on Indo-US project on enhancing fertilizer use efficiency in conjunction with residue management in dry land crops and cropping systems in low rainfall conditions. Central Arid Zone Research Institute (CAZRI), Jodhpur, pp 1–24

Chaudhari PM, Kumbhar JS, Surve US et al (2019) On-farm crop response to plant nutrients in predominant cropping systems. J Pharmacog Phytochem 8(5):1365–1367

Chaudhary C, Hooda VS, Isha KS et al (2022) Impact of Integrated weed management practices on growth parameters, yield attributes and yield of pearl millet [*Pennisetum glaucum* L. Br. Emend. Stuntz.]. Pharm Innov 11(4):687–692

Choudhary AK, Suri VK (2014) Integrated nutrient management technology for direct-seeded upland rice (*Oryza sativa*) in northwestern Himalayas. Commun Soil Sci Plant Anal 45(6): 777–784

Choudhary GL, Rana KS, Rana DS et al (2016) Moisture management and zinc fortification impacts on economics, quality and nutrient uptake of pearl millet (*Pennisetum glaucum* L.) under rainfed conditions. Indian J Agric Sci 86(1):71–77

Choudhary M, Rana KS, Bana RS et al (2017) Energy budgeting and carbon footprint of pearl millet–mustard cropping system under conventional and conservation agriculture in rainfed semi-arid agro-ecosystem. Energy 141:1052–1058

Coventry DR, Poswal RS, Yadav A et al (2015) A comparison of farming practices and performance for wheat production in Haryana, India. Agric Syst 137:139–153

Dar WD (2004) Macro-benefits from micronutrients for grey to green revolution in agriculture. Presented at IFA International Symposium on Micronutrients, New Delhi, 21–23 February 2004, p 5. http://www.sustainablecropnutrition.com/ifa/content/view/full/8913

Dart PJ, Wani SP (1982) Non-symbiotic nitrogen fixation and soil fertility. In: Non-symbiotic nitrogen fixation and organic matter in the tropics. Transactions of the 12th International Congress of Soil Science, New Delhi, India

De R, Gautam RC (1987) Management practices to increase and stabilize pearl millet production in India. In: Witcombe JR (ed) Proceedings of the International Pearl millet Workshop. ICRISAT, Hyderabad, pp 247–253

Delate K (2013). Pearl millet. Alternative agriculture, Iowa State Extension and Outreach. https://www.extension.iastate.edu/alternativeag/cropproduction/pearlmillet.htm

Deshveer CL, Deshveer LS (2005) Weed Management in pearl millet (*Pennisetum glaucum*) with Special Reference to *Trianthema portulacastrum*. Indian J Weed Sci 37(3-4):212–215

Doty CW, Campbell RB, Reicosky DC (1975) Crop response to cheiseling and irrigation in soils with a compact A-2 horizon. Trans Am Soc Agric Eng 18:668–672

Dwivedi BS, Singh D, Tiwari KN et al (2009) On-farm evaluation of SSNM in pearl millet-based cropping systems on alluvial soils. Better Crops 3(1):25–27

Espín G (2016) Genes Involved in the Formation of desiccation�resistant cysts in azotobacter vinelandii. Stress and environmental regulation of gene expression and adaptation in bacteria. https://doi.org/10.1002/9781119004813.ch67

Fan XH, Li YC, Alva AK (2011) Effects of temperature and soil type on ammonia volatilization from slow release nitrogen fertilizers. Commun Soil Sci Plant Anal 42(10):1111–1122

FAO (2005) Fertilizer use by crop in India. Food and Agriculture Organization of United Nations, Rome. www.fao.org/3/a0257e/A0257E01.htm

FICCI (Federation of Indian Chamber of Commerce and Industry) (2015) Transforming agriculture through mechanization: a knowledge paper on Indian Farm Equipment Sector. Grant Thorton India, New Delhi. http://ficci.in/spdocument/20682/agrimach

Frossard E, Bucher M, Mächler F et al (2000) Potential for increasing the content and bioavailability of Fe, Zn and Ca in plants for human nutrition. J Sci Food Agric 80:861–879

Gangaiah B (2008) Agronomy-Kharif crops: National Sci. Digital Library, NIScPR, CSIR, New Delhi. http://nsdl.niscpr.res.in

Gardner J (1980) The effect of seed size and density on field emergence and yield of pearl millet (*Pennisetum americanum* (L.) K. Schum.]. M.S. Thesis. Agronomy Department, Kansas State University, Manhattan, Kansas State

Gautam RC (2000) Effect of organic manure, bio-fertilizer and inorganic fertilizers on growth, yield and quality of rainfed pearl millet. Ann Agric Res 21(4):459–464

Gera R, Kumar A, Mohanty SR (2020) Liquid biofertilizer azo phosphoteeka. Technical Bulletin No AINP SBB/HAU/2020/02. ICAR All India Network Project on Soil Biodiversity Biofertilizer (AINP SBB), Department of Microbiology, CCSHAU, Hisar, Haryana. p 1–4

Giri G, De R (1980) Effect of preceding grain legumes on the nitrogen uptake and growth of dry land pearl millet. Plant Soil 56(3):458–465

Govindaraj M, Virk PS, Kannati A et al (2020) Biofortified pearl millet cultivars offer potential solution to tackle malnutrition in India. In: Kang MS (ed) Quantitative genetics, genomics and plant breeding. CAB International, Oxon, pp 385–396

Gregory PJ, Reddy MS (1982) Root growth in an intercrop of pearl millet/groundnut. Field Crop Res 5:241–252

Gupta JP, Aggarwal RK, Gupta GN et al (1983) Effect of continuous application of farm yard manure and urea on soil properties and production of pearl millet in western Rajasthan. Indian J Agric Sci 53:53–56

Gupta NK, Gupta S, Singh J et al (2022) On-farm hydro and nutri-priming increases yield of rainfed pearl millet through physio-biochemical adjustments and anti-oxidative defense mechanism. PLoS ONE 17(6):e0265325

Hegde SV (2008) Liquid biofertilizers in Indian agriculture. Biofert Newsl 2008:17–22

Hussainy SAH, Brindavathy R, Vaidyanathan R (2019) Influence of intercropping and different levels of irrigation on the performance and nutrient dynamics of groundnut (*Arachis hypogaea* L.). Res Crops 20:296–300

Iijima M, Awala SK, Watanabe Y et al (2016) Mixed cropping has the potential to enhance flood tolerance of drought-adapted grain crops. J Plant Physiol 192:21–25

Iqbal MA, Hamid A, Hussain I et al (2019) Competitive indices in cereal and legume mixtures in a South Asian environment. Agron J 111(1):242–249

Jadon P, Selladurai R, Yadav SS et al (2018) Volatilization and leaching losses of nitrogen from different coated urea fertilizers. J Soil Sci Plant Nutr 18(4):1036–1047

Jakhar GR, Golada SL, Sadhu AC (2011) Influence of levels and time of application of nitrogen on growth, yield and nitrogen uptake by pearl millet during summer. Madras Agric J 98(10/12): 347–349

Jan A, Khan I, Ali S et al (2015) Sowing dates and sowing methods influenced on growth yield and yield components of pearl millet under rain fed conditions. J Environ Earth Sci 5(1):105–109

Joshi NL (1987) Seedling emergence and yield or pearl millet on naturally occurring crusted arid soils in relation to sowing and cultural methods. Soil Tillage Res 10:103–112

Kalhapure A, Kumar R, Singh VP et al (2016) Hydrogels: a boon for increasing agricultural productivity in water-stressed environment. Curr Sci 111(11):1773–1779

Kamble HG, Srivastava AP, Panwar JS (2003) Development and evaluation of a pearl-millet thresher. J Agric Eng 40(1):18–25

Kanwar S, Sharma S, Karwasara PK et al (2015) Effect of moisture conservation practices and seed hardening on pearl millet. [*Pennisetum glaucum* (L.)] under rain-fed conditions. Res Environ Life Sci 8(1):126–128

Kathmale DK, Dhadge SM, Satpute NR et al (2014) Evaluation of pigeonpea (*Cajanus cajan* L.) based intercropping systems under semi-arid vertisol in scarcity zone of Maharashtra. Indian J Dryland Agric Res Dev 29(1):27–34

Kausar MA, Chaudhry FM, Rashid A et al (1976) Micronutrient availability to cereals from calcareous soils. I. Comparative Zn and Cu deficiency and their mutual interaction in rice and wheat. Plant Soil 45:397–410

Khafi HR, Mehta AC, Bunsa BD et al (2011) Response of summer pearl millet (*Pennisetum glaucum* L.) to irrigation scheduling. Crop Res 41(1-3):28–30

Khan MA (1998) Design package and guide lines for Khadin constructions. Report submitted to Government of Rajasthan. p 20

Khan MA, Venkateswarlu J (1993) Rainwater harvesting for increasing water availability in arid Rajasthan. J Inst Eng 6:18–26

Kiroriwal A, Yadav RS (2013) Effect of intercropping systems on intercrops and weeds. Int J Agric Food Sci Tech 4:643–646

Kooistra MJ, Bouma J, Boersma OH et al (1984) Physical and morphological characterization of undisturbed and disturbed plow plans in a sandy loam soil. Soil Tillage Res 4:405–417

Kumar P, Kumar A (2021) Water management for improving pearl millet production under irrigated environment: a review. Agric Rev 42(2):225–229

Kumari S (2017) Effect of Kinetin (6-FAP) and Cycocel (CCC) on growth, metabolism and cellular organelles in pearl millet (*Pennisetum glaucum*) under water stress. Int J Curr Microbiol App Sci 6(8):2711–2719

Kumawat S, Jinger D, Yadav GN (2021) Application of phosphorus and zinc improved the nutrient dynamics, productivity, and profitability of fenugreek (Trigonella foenum-graecum)-pearl millet (*Pennisetum glaucum*) cropping system. Indian J Agron 66(4):488–494

Kushwaha P, Kashyap PL, Srivastava AK et al (2020) Plant growth promoting and antifungal activity in endophytic Bacillus strains from pearl millet (*Pennisetum glaucum*). Braz J Microbiol 51(1):229–241

Labrada R, Caseley JC, Parker C (eds) (1994) Weed management for developing countries (Vol. 120). Food & Agriculture Organization.

Lawan M, Barnett FL, Khaleeq B et al (1985) Seed density and seed size of pearl millet as related to field emergence and several seed and seedling traits. Agron J 77(4):567–571

Makarana G, Kumar A, Yadav RK et al (2019) Effect of saline water irrigations on physiological, biochemical and yield attributes of dual purpose pearl millet (*Pennisetum glaucum*) varieties. Indian J Agric Sci 89(4):624–633

Malik MWI, Hussain I, Baloch MS et al (2019) *Pennisetum glaucum* suppresses growth of some weed species. Pak J Weed Sci Res 25(4):337–347

Marwat KB, Hussain Z, Saeed M et al (2005) Chemical weed management in wheat at higher altitudes. Pak J Weed Sci Res 11(3–4):102–107

Mathukia RK, Mathukia PR, Polara AM (2015) Intercropping and weed management in pearlmillet (*Pennisetum glaucum*) under rainfed condition. Agric Sci Digest A Res J 35(2):138–141

Meena BL, Kumar P, Kumar A et al (2018) Influence of zinc and iron fertilization on growth and yields of pearl millet and mustard under salt affected soils. In: Biswas DR, Bandyopadhyay KK (eds) Abstracts: national seminar on developments in soil science. Indian Society of Soil Science, New Delhi, p 74

Mehetre S, Pawar P, Pawar R et al (2021) Effect of Pre and post emergence herbicides application on economics of pearl millet (*Pennisetum glaucum* L.). Int J Plant Soil Sci 33(24):180–185

Moharana PC, Sharma BM, Biswas DR et al (2012) Long-term effect of nutrient management on soil fertility and soil organic carbon pools under a 6-year-old pearl millet–wheat cropping system in an Inceptisol of subtropical India. Field Crop Res 136:32–41

More NB, Amrutsagar VM, Pawar AB et al (2020) Potash fertilization for enhancing the productivity of pearl millet-safflower sequence under dryland condition. Int J Chem Stud 8:2958–2963

Mula RP, Rai KN, Dangaria CJ et al (2009) Pearl millet as a post-rainy cool season crop: case studies from Gujarat and Maharashtra, India. J SAT Agric Res 7:1–7

Munawwar MH, Yousaf M, Fateh J et al (2014) Participatory pearl millet varietal evaluation and selection in Pakistan. Sarhad J Agric 30(1):1–6

Munde SD, Patel JC, Ali S et al (2013) Weed control study in kharif pearl millet (Pennisetum glaucum). Bioinfolet. 10(2A): 464–468

Narain P, Goyal RK (2005) Rainwater harvesting for increasing water productivity in arid zones. J Water Manage 13(2):132–136

Padalia K, Bargali SS, Bargali K et al (2018) Socio-economic analysis based on energy input and output of mixed cropping systems of Bhabhar Region (Shiwalik Range of Kumaun Himalaya, India). Curr Agric Res J 6(2):123

Pandey V, Patel VJ, Vadodaria RP et al (2008) Irrigation water requirement and production potentials of major crops over Narmada canal command area in Gujarat. J Agrometeorol 10: 314–320

Patel TD, Patel JR, Patel JB (1994) Nutrient-use pattern and balance sheet of major nutrient under summer groundnut - rainy season pearl millet crop sequence in middle Gujarat. Indian J Agron 39(4):532–538

Pawar A, Amrutsagar VM, More NB et al (2021) Potassium management for enhancing the productivity of pearl millet-chickpea sequence on drylands of Maharashtra. Indian J Fert 17(10):1050–1057

Ponmurugan K, Sankaranarayanan A, Al-Dharbi NA (2012) Biological activities of plant growth promoting *Azotobacter sp.* isolated from vegetable crops rhizosphere soils. J Pure Appl Microbiol 6(4):1–10

Poonia S, Jain D, Santra P et al (2018) Use of solar energy in agricultural production and processing. Indian Farm 68(09):104–107

Radhouane L (2012) Allelopathic effects of pearl millet (*Pennisetum glaucum* L.) R.Br.] on seed germination and seedling growth. 1st Biotech. World Cong, UAE, Dubai, p 14–15

Radhouane L (2014) Allelopathic effects of pearl millet (*Pennisetum glaucum*) seeds on seedling growth of three cereals. Int J Innov Sci Res 6:18–24

Rai KN, Gaur PM, Hash CT et al (2004) Development of crop cultivars for increased and stable production in dry lands of the semi-arid tropics. J Arid Land Stud 14S:69–72

Rai KN, Gupta SK, Govindaraj M et al (2015) Pearl millet improvement for enhanced productivity – strategies and impact. Indian Farm 65:19–22

Raja K, Anandham R (2020) Infusing microbial consortia on seed quality and seedling vigour in pearlmillet [*Pennisetum glaucum* (L.) R. Br.]. Microbiol Res J Int 30(4):43–55

Rajput RL, Bhadouriya NS (2019) Effect of moisture conservation practices on growth and yield of pearl millet+ pigeonpea intercropping system. Legum Res 42(4):547–549

Rajput PK, Kumar Y, Katiyar AK et al (2022) Response of pearl millet cultivars to different moisture conservation practices under rainfed condition on light textured soil of central Uttar Pradesh, India. Plant Arch 22(1):105–108

Ram K, Meena R (2015) Evaluation of pearl millet and mungbean intercropping systems in arid region of Rajasthan (India). Bangladesh J Bot 43(3):367–370

Ram B, Chaudhary GR, Jat AS et al (2005) Effect of integrated weed management and intercropping systems on growth and yield of pearl millet. Indian J Agron 50(3):254–258

Ramesh N, Kalamani M, Baradhan G et al (2019) Influence of weed management practices on nutrient uptake and productivity of hybrid Pearl Millet under different herbicides application. Plant Arch 19(2):2893–2898

Rao AS, Saxton KE (1995) Analysis of soil water and water stress for pearl millet in an Indian arid region using the SPAW Model. J Arid Environ 29(2):155–167

Rao MR, Shetty SVR (1976) Some biological aspects of intercropping systems on crop-weed balance. Indian J Weed Sci 8(1):32–43

Rao MR, Willey RW (1983) Effects of genotype in cereal/pigeonpea intercropping on the alfisols of the semi-arid tropics of India. Exp Agric 19(1):67–78

Rao KV, Venkateswarlu B, Vittal KPR et al (2009) Water harvesting potential assessment in rainfed regions of India. Rainwater Harvesting and Reuse through Farm Ponds. In: Proceedings of National Workshop cum-brain storming on rainwater harvesting and reuse through farm ponds: experiences, issues and strategies, p 67

Rao KV, Venkateswarlu B, Sahrawath KL et al (2010) Proceedings of National Workshop cum-brain storming on rainwater harvesting and reuse through farm ponds: experiences, issues and strategies. Hyderabad, ICAR-CRIDA, p 242

Rao AN, Wani SP, Ladha JK (2014) Weed management Res. in India - an analysis of the past and outlook for future. In: Souvenir (1989-2014). DWR Publication No. 18. Directorate of Weed Research, Jabalpur, p 1–26

Rao CS, Rejani R, Rao CR et al (2017) Farm ponds for climate-resilient rainfed agriculture. Curr Sci 112:471–477

Rao CR, Rao KV, Raju BMK et al (2019) Levels and determinants of economic viability of rainwater harvesting farm ponds. Indian J Agric Econ 74(4):539–551

Reddy KC, van der Ploeg J, Maga I (1990) Genotype effects in millet/cowpea intercropping in the semi-arid tropics of Niger. Exp Agric 26:387–396

Reddy SS, Stahlman PW, Geier PW et al (2014) Tolerance of foxtail, proso and pearl millets to saflufenacil. Crop Prot 57:57–62

Reddy GC, Goyal RK, Puranik S et al (2020) Biofertilizers toward sustainable agricultural development. In: Varma A, Tripathi S, Prasad R (eds) Plant microbe symbiosis. Springer, Cham, pp 115–128

Rekha DLM, Lakshmipathy R, Gopal GA (2018) Effect of Integrated use of biofertilizers, chemical fertilizers and farmyard manure on soil health parameters of pearl millet (*Pennisetum glaucum* L.). J Soil Sci Plant Health 2:2. https://doi.org/10.4172/JSPH.1000111

Rengel Z, Batten GD, Crowley DE (1999) Agronomic approaches for improving the micro nutrient density in edible portions of field crops. Field Crop Res 60:27–40

Rundla S, Bairwa RC, Choudhary R et al (2018) Effect of foliar supplementation of N, P and K on nutrient dynamics and productivity of pearl millet (*Pennisetum glaucum*). Indian J Agron 63(4): 528–531

Sankar GM, Mishra PK, Sharma KL et al (2012) Efficient tillage and nutrient practices for sustainable pearl millet productivity in different soil and agro-climatic conditions. Exp Agric 48(1):1–20

Savita (2017) Response of pearl millet to Rhizosphere bioferilizers and foliar application of isolate under rainfed conditions. M.Sc. Thesis, CCS Haryana Agricultural University, Hisar

Savithri P, Perumal R, Nagarajan R (1999) Soil and crop management technologies for enhancing rice production under micronutrient constraints. Nutr Cycl Agroecosyst 53:83–92

Saxena MC (1987) Agronomy of chickpea. In: Saxena MC, Singh KB (eds) The chickpea. CAB International, Wallingford, pp 207–232

Saxena A, Singh DV, Joshi NL (1997) Effects of tillage and cropping systems on soil moisture balance and pearl millet yield. J Agron Crop Sci 178:251–257

Sekhon HS, Singh G, Ram H (2007) Lentil based cropping systems. In: Yadav SS, McNeil DL, Stevenson PC (eds) Lentil: an ancient crop for modern times. Springer, Dordrecht, pp 107–126

Sharma PK, Sudesh K, Chaudhary GR (2012) Effect of organic nutrient management on productivity and soil fertility status in pearl millet (*Pennisetum glaucum*). Res Crops 13(2):503–506

Shetty SVR, Rao AN (1981) Weed-management studies in sorghum/pigeon pea and pearl millet/ groundnut intercrop systems: some observations. In: Proceedings of the International Workshop on Intercropping. ICRISAT, Patancheru, pp 238–248

Shinde SH, Pawar VS, Suryawanshi GB et al (2003) Integrated weed management studies in pigeonpea + pearl millet intercropping (2:2) system. Indian J Weed Sci 35:90–92

Singh R (1995) Mixed cropping in pearl millet (*Pennisetum glaucum*) in arid zone of Rajasthan. Indian J Agric Sci 65:276–278

Singh H (2012) Effect of hydrogel on growth, yield and water use efficiency in pearl millet (*Pennisetum glaucum*) production. Forage Res 38(1):27–28

Singh RB, Minhas PS, Chauhan CPS et al (1992) Effect of high salinity and SAR waters on salinization, sodication and yields of pearl millet and wheat. Agric Water Manage 21(1-2): 93–105

Singh N, Megharaj M, Kookana RS et al (2004) Atrazine and simazine degradation in pennisetum rhizosphere. Chemosphere 56:257–263

Singh G, Bhattacharyya R, Das TK et al (2018) Crop rotation and residue management effects on soil enzyme activities, glomalin and aggregate stability under zero tillage in the Indo-Gangetic Plains. Soil Tillage Res 184:291–300

Singh R, Singh AK, Yadav SR et al (2019) Effect of saline water and fertility levels on pearl millet-psyllium crop sequence under drip irrigation in arid region of Rajasthan. J Soil Salinity Water Qual 11(1):56–62

Singh YP, Tomar SS, Singh S et al (2021a) Impact of community based solar lift irrigation system on productivity and economic profitability in rain fed conditions. In: Madhu M, Jakhar P, Adhikary PP, Kumar G, Jyothiprava Dash C, Beer K, Sahoo DC, Mondal D, Mishra PK (eds) Land and water resource conservation: issues, options and experiences. International Books & Periodicals Supply Service, Delhi, pp 243–257

Singh YP, Tomar SS, Singh S et al (2021b) Influence of tillage and relay cropping of berseem in mustard–pearl millet system on soil quality, productivity of crops and water. Agric Res 212: 105069

Sivapriya SL, Priya PR (2017) Selection of hyper exopolysaccharide producing and cyst forming Azotobacter isolates for better survival under stress conditions. Int J Curr Microbiol App Sci 6(6):2310–2320

Soman P, Bidinger FR, Peacock JM et al (1981) Seeding establishment: a preliminary survey taken in Aurepally during kharif 1981. ICRIST internal report. ICRISAT, Patancheru

Soman P, Jayachandran R, Bidinger FR (1987) Uneven variation in plant to plant spacing in pearl millet. Agron J 79(5):891–895

Srinivasarao CH, Venkateswarlu B, Lal R et al (2014) Long term manuring and fertilizer effects on depletion of soil organic carbon stocks under pearl millet cluster bean castor rotation in Western India. Land Degrad Dev 25(2):173–183

Suzuki K, Matsunaga R, Hayashi K et al (2017) Effects of long-term application of mineral and organic fertilizers on dynamics of nitrogen pools in the sandy soil of the Sahel region, Niger. Agri Ecosys Environ 242:76–88

Tandon HLS (2004) Fertilizers in Indian agriculture from 20th to 21st century. Fertiliser Development and Consultation Organisation

Tarafdar J, Ramesh R, Himanshu M et al (2014) Development of zinc nanofertilizer to enhance crop production in pearl millet (*Pennisetum americanum*). Agric Res 3:257–264

Thakor KP, Usadadia VP, Savani NG et al (2018) Effect of irrigation schedule and nitrogen management on productivity, profitability of summer pearl millet grown under clay soils of south Gujarat. Int J Agric Innov Res 6(4):10–11

Thakur DP (1984) Ergot disease of pearl millet. Rev Trop Plant Pathol 1:297–328

Tien T, Gaskins MH, Hubbell DM (1979) Plant growth substances produced by Azospirillum brrasilense and their effect on the growth of pearl millet (*Pannicum americanum* L.). Appl Environ Microbiol 37:1016–1024

Turfin JE, Herridge F, Robertson MJ (2002) Nitrogen fixation and soil nitrate interactions in field-grown chickpea (*Cicer arietinum*) and faba bean (*Vicia faba*). Austr J Agric Res 53:599–608

Ullah A, Ahmad A, Khaliq T et al (2017) Recognizing production options for pearl millet in Pakistan under changing climate scenarios. J Integr Agric 16(4):762–773

Ullah A, Ahmad I, Rahman MH et al (2020) Optimizing management options through empirical modeling to improve pearl millet production for semi-arid and arid regions of Punjab, Pakistan. Sustainability 12(18):7715

Uppal RK, Wani SP, Garg KK et al (2015) Balanced nutrition increases yield of pearl millet under drought. Field Crop Res 177:86–97

USDA (2016) Pearl millet hybrids for gain. Available at https://www.ars.usda.gov/southeast-area/tifton-ga/crop-genetics-and-breeding-research/docs/pearl-millethybrids-for-grain/. Accessed 29 April 2020

Vikhe PS (2014) Azotobacter species as a natural plant hormone synthesizer. Res J Recent Sci 2277:2502

Virkar KM, Patil HM, Khairnar AV et al (2007) Integrated weed management in pearl millet-pigeonpea intercropping system. J Farm Syst Res Dev 13(2):245–247

Wang H, Gao J, Li X et al (2015) Nitrate accumulation and leaching in surface and groundwater based on simulated rainfall experiments. PLoS One 10(8):e0136274

Wani SP (1990) Inoculation with associative nitrogen-fixing bacteria: Role in cereal grain production improvement. Ind J Microbiol 30(4):363–393

Wing EAS (2014) Finance Division, Government of Pakistan, Islamabad. Highlights of Pakistan Economic Survey

Yadav AK, Chandra K (2014) Mass production and quality control of microbial inoculants. Proc Indian Natl Sci Acad 80:483–489

Yadav RS, Yadav OP (2000) Differential competitive ability and growth habit of pearl millet and cluster bean cultivars in a mixed cropping system in the arid zone of India. J Agron Crop Sci 185(1):67–71

Yadav RS, Yadav OP (2001) The performance of cultivars of pearl millet and cluster bean under sole cropping and intercropping systems in arid zone conditions in India. Exp Agric 37(2): 231–240

Yadav OP, Rai KN, Khairwal IS et al (2011) Breeding pearl millet for arid zone of north western India: constraints, opportunities and approaches. All India coordinated pearl millet improvement project, Jodhpur, p 28

Yadav OP, Rai KN, Rajpurohit BS et al (2012a) Twenty-five years of pearl millet improvement in India. All India Coordinated Pearl Millet Improvement Project, Jodhpur, p 122

Yadav OP, Rajpurohit BS, Kherwa GR et al (2012b) Prospects of enhancing pearl millet (Pennisetum glaucum) productivity under drought environments of north-western India through hybrids. Indian J Genet Plant Breed 72:25–30

Yadav T, Kumar A, Yadav RK et al (2020) Salicylic acid and thiourea mitigate the salinity and drought stress on physiological traits governing yield in pearl millet-wheat. Saudi J Biol Sci 27(8):2010–2017

Yadav SK, Dudwal BL, Yadav VL et al (2021) Effect of moisture conservation practices and zinc fertilization on nutrient status and quality of pearl millet *Pennisetum glaucum* (L.) under rainfed Condition. Indian J Agric Res 55(5):629–633

Yadava RBR, Patil BD (1980) Effect of growth retardants on plant growth, yield and seed carbohydrates of bajra. Forage Res 6:175–179

Yamane K, Araki C, Watanabe Y et al (2018) Close mixed planting with pearl millet improves drought tolerance in rice by the increased access to deep water. Plant Soil 423(1):397–410

Zhang Y, Ge S, Jiang M et al (2014) Combined bioremediation of atrazine-contaminated soil by pennisetum and Arthrobacter sp. Strain DNS10. Environ Sci Pollut Res Int 21:6234–6238

Zong X, Wang H, Song Z et al (2011) Foliar Zn fertilization impacts on yield and quality in pearl millet (*Pennisetum glaucum*). Front Agric China 5:552–555

Hybrid Seed Generation System Management to Ensure the Seed Quality in Pearl Millet

19

N. Kannababu, B. V. Venkatesh Bhat, Sooganna, K. Raghunath, R. Venkateswarlu, Jinu Jacob, V. M. Malathi, A. Srikanth, T. Nepolean, and Vilas A. Tonapi

Abstract

In pearl millet, a number of hybrids with higher yield, variable maturity, tolerance to various diseases, and insect pests were developed and released for general cultivation across India. Increase in area under cultivation with superior genetic and productive hybrids eventually necessitated the production of large quantities of seed with high-quality standards. Quality seed production is the prerequisite to ensure the true to type of any hybrid in order to realize its genetic potential at farmer's field. Pearl millet hybrid seed production requires a high level of competence to ensure the seed quality in terms of genetic purity, seed germination, vigor, and health. Both public and private sectors made significant progress during last 30 years in seed production and distribution of pearl millet; however, the major market share of private sector is higher due to their excellence and expertise in marketing hybrid cultivars. High-quality seed can be obtained only from a genetically pure crop that depends upon the genetic constitution of nucleus and breeder seed followed by foundation and certified seed. Therefore, effective implementation of seed generation system with scientific methods of seed crop management only can ensure the hybrid seed quality of pearl millet.

N. Kannababu · B. V. Venkatesh Bhat · Sooganna · K. Raghunath · R. Venkateswarlu · J. Jacob · V. M. Malathi · A. Srikanth · T. Nepolean
ICAR-Indian Institute of Millets Research, Hyderabad, TS, India

V. A. Tonapi (✉)
Indian Institute of Millets Research, Hyderabad, TS, India
e-mail: director.millets@icar.gov.in; tonapi@millets.res.in

V. A. Tonapi et al. (eds.), *Pearl Millet in the 21st Century*,
https://doi.org/10.1007/978-981-99-5890-0_19

Abbreviations

°C	Degree centigrade
AICPMIP	All India Coordinated Pearl millet Improvement Program
A-line	Male sterile line
B-line	Maintainer line
cm	Centimeter
CMS	Cytoplasmic-nucleus male sterility
DAS	Days after sowing
DUS	Distinctiveness, uniformity and stability
h	Hour
ha	Hectare
ICAR	Indian Council of Agriculture Research
ICRISAT	International Crops Research Institute for Semi-arid Tropics
km	Kilometer
m	Meter
NSC	National Seed Corporation
R-line	Restorer line
SAU	State Agriculture Universities
SFCI	State Farms Corporation of India
SMR	Seed multiplication ratio
SRR	Seed replacement rate
SSC	State seeds corporation

19.1 Introduction

Pearl millet (*Pennisetum glaucum* R. Br.) provides livelihood and nutritional security for >90 million people and widely adapted in semiarid and arid zones of South Asia and sub-Saharan Africa (Serba et al. 2020). In India, it occupies 6.93 m ha with the production of 8.61 mt and productivity of 1243 kg/ha and is ranked fourth in importance among food crops (Directorate of Millets Development 2020). In addition, it is a stable source of grain as well as fodder in arid and semiarid regions of the country. Nutrient composition in its grain consists of 10–12% protein, 65–70% carbohydrates, and an ample amount of crude fiber and mineral matter. Whereas grain production is the main purpose of pearl millet cultivation in India, its stover (dry stalk) for animal fodder is an important secondary product in subsistence agriculture. Pearl millet also provides high-quality green forage. High-quality beer and sprit are also produced from pearl millet grain. It constitutes the third most important cereal after rice and wheat.

In pearl millet, protogyny is the flowering pattern where the early emergence of stigma takes place in the flower than dehiscence of anthers that enable cross-pollination. This type of flowering does not require any emasculation practices for cross-pollination; however, it can be realized only under synchronous flowering of parents. The possibility of single-cross hybrid development was realized due to the introduction of cytoplasmic-nuclear male sterility (CMS) in pearl millet that resulted towards development of male steriles (A-lines), maintainers (B-lines), and restores (R-lines) (Athwal 1996). In India, the pearl millet hybrid development program aimed towards meeting the requirements of various zones particularly addressing the diseases and production constraints specific to those regions. Improved cultivars, mostly hybrids cover >70% area of pearl millet cultivation in India. The A- and B-zones of India with >400 mm annual rainfall and with two separate agro-ecologies are mostly occupied by the hybrids of pearl millet (Gupta et al. 2013). High-quality seed is the prerequisite to establish a uniform crop stand that facilitates production seeds with high genetic purity. The seed with highest genetic purity combined with physical purity, healthiness, vigor, and viability traits determines crop's productivity and therefore seed remains as key input in modern agriculture. Strong seed production planning should be in place considering the influence of many factors on seed quality. Both public and private sectors made significant progress during last 30 years in seed production and distribution of pearl millet. however, the major market share of by private sector is higher due to their excellence and expertise in marketing due to their high involvement in developing hybrid cultivars. Availability of high yielding hybrids with disease resistance and different maturity groups ranging from 60-90 days resulted in development of effective contractual hybrid seed production of pearl millet in India. The Indian Council of Agricultural Research institutes and the State Agricultural Universities to meet the countries requirement release improved pearl millet varieties and hybrids. Through the efforts of all India coordinated crop improvement program of pearl millet, a number of hybrids with higher yield, variable maturity, tolerance to various diseases, and insect pests have been developed for general cultivation across the country. Increase in area under cultivation with improved varieties with superior genetic and productive hybrids eventually demands the production of good quality seed in required quantities. Pure seed can be obtained only from a genetically pure crop that depends upon the genetic constitution of nucleus and breeder seed followed by foundation and certified seed.

19.2 Flowering, Pollination, and Genetic Mechanism of Hybrid Seed Production

19.2.1 Flowering

Pearl millet inflorescence is panicle or false spike, which consists of spikelets surrounding the central axis in involucre and generally, the panicle length varies from 20 to 25 cm with a circumference of 7–9 cm. The common shapes of the panicle are either cylindrical or conical however may vary considerably. The central rachis bears fascicles on rachillae that have short hair. The compactness of the panicle depends on fascicles density and rachillae length. The fascicle contains spikelets surrounded by a wall of bristles (i.e., involucre). The prolongation of the fascicle axis determines the length of bristles. The spikelet usually contains two florets and sometimes two to four florets. The upper floret is bisexual or hermaphrodite, while the lower one is staminate. Due to protogynous flowering pattern, the styles start protruding 2–3 days after panicle emergence. Stylar protrusion starts initially in upper middle portion and proceeds both in upward and downward direction of the panicle. In the hermaphrodite flowers, stigma emerges before anthers become receptive to the external pollen from other plants. The emergence of stigma takes 2–3 days and remains receptive for next 2–3 days.

19.2.2 Pollination

Protogyny is the device for natural cross-pollination in pearl millet with about 80% out crossing. However, a high degree of self-pollination (31%) was also reported in this crop because the side tillers come in flowering after main culm. This promotes self-pollination between main culm and side tiller panicles due to protogynous nature. Maximum flowering takes place at 10 pm and mid night in Indian conditions. The time of panicles exertion from boot leaf varies in inbreeds and hybrids. Spike emergence takes 6 days from sheath with maximum on 4 to 5th day (Bhatnagar and Kumar 1960). Pollen shedding in a panicle takes place for about 3 days. The anthesis occurs between 8 am and 2 pm with a peak at about 10 am. High humidity and low temperature slow anther emergence, while low humidity and high temperature speed up anthesis. The major cross-pollinating agent is the wind; however, insects also may contribute to some extent. After the emergence, the stigma remains receptive for about 2–3 days; however, it varies with genotype and environment. Protogyny and the early emergence of stigma and late dehiscence of anthers enable cross-pollination without resorting to emasculation, but asynchronous flowering prevents its full realization. Once the protogynous stage reaches, the entire panicle can be dusted with the desired viable pollen with no or very little chances of self-pollination. Being highly cross-pollinated, pearl millet seed crop requires more distance more than 1000 m isolation from any other pearl millet crops.

19.2.3 Genetic Mechanism of Hybrid Seed Production

In pearl millet, the pure seed of single-cross-hybrids can be produced on large scale with the genetic mechanism using cytoplasmic genic male sterility system comprising of A-line (male sterile), B-line (maintainer), and R-line (restorer). The interaction of the sterility inducing factors in cytoplasm with genetic factors in the nucleus results in cytoplasmic genic male sterility system. The source of male fertile plants may be the germplasm or genetic stocks or varieties. By crossing the male fertile plants with a male sterile plant, the potentiality of a male and female (A-line) parents can be judged with close observations on respective hybrids in a nursery. The selected plants can be bagged before anthesis and later after few weeks the seed set percentage can be observed on each plant. Three types of hybrids can be observed in the crosses with male sterile (A-line) viz., (1) hybrids with no seed set, (2) hybrids with 100% seed set, and (3) hybrids with partial seed set. The first category hybrids indicate that the corresponding male parents unable to restore the fertility and the male sterility is maintained. These are the non-restorers/maintainers but potential new male sterile (A-lines). The second category hybrids with complete seed set indicate the corresponding pollen parent is a potential restorer (R-line) that can restore completely the male sterility, which can be used to produce the hybrids. The third category hybrids with partial seed set indicate that the corresponding male parents are neither stable restorers (R-line) nor maintainers (B-line), which can be rejected. The second category hybrids with total seed set are closely evaluated for various agronomic traits, hybrid vigor, biotic and abiotic resistance etc. in comparison with local checks, and the identified hybrids can be advanced for further studies.

19.3 Seed Generation System and Production Chain of Pearl Millet Hybrids

Being highly cross-pollinated, pearl millet seed crop requires more distance of about 1 km isolation from any other pearl millet crop. The genetic purity of the resulting seed must reflect all the specified traits in Distinctiveness, Uniformity and Stability (DUS) guidelines (revised) for the conduct of test on pearl millet (Protection of Plant Varieties and Farmers' Rights Authority (PPV and FRA) Government of India 2019). In pearl millet, the single cross-hybrids are produced in large scale by using the genetic mechanism of cytoplasmic genic male sterility system that comprises male sterile (A-line), and its corresponding maintainer (B-line) and fertility restorer (R-line). The seed is multiplied through a system of seed generations, namely nucleus, breeder, foundation, and certified seed classes. Breeder seed class is the progeny of nucleus seed and foundation seed is the progeny of breeder seed. The certification is not required for breeder seed production but requires through a formal monitoring system of production program. While planning the seed production of individual seed class, the demand forecasting is very important for total certified seed requirement. Forecasting of the demand can be done based on seed multiplication ratio, seed replacement rate, and additional seed requirement. The nucleus seed is produced in small quantities under the direct supervision of originating or sponsored breeder of the institute and this class does not require any formal certification system. The nucleus seed is the initial handful quantity of pure seed to produce breeder, foundation and certified seeds. In case of hybrid seed production, the male sterile, maintainer, and restorer lines are multiplied through breeder and foundation seed classes in order to meet the requirement of hybrid seed production (certified seed class).

19.3.1 Nucleus Seed Production of Pearl Millet Parental Lines (A-, B-, and R-Lines)

The public sector organizations such as AICPMIP centers, SAUs, ICRISAT, and some private companies produce the nucleus and breeder seed. The responsibility of supply and safe storage of nucleus and breeder seed of particular hybrid usually lies with the organization that bred and release it. Production of nucleus and breeder seed of parental lines is a prior requirement for a hybrid seed production. Maintenance breeding involves nucleus seed production and seed increases of cytoplasmic male-sterile line (A-line), maintainer line (B-line) and restorer line (R-line). Personal attention of the breeder is most important while producing the nucleus seed class as it provides the basis for all other classes of seed thus requires to be in purest form (Andrews and Harinarayana 1984). It is essential to store nucleus seed stock of sufficient quantity to meet breeder seed demand for 5 years and retaining adequate quantity in cold storage for long term as backup stock. Timely purification measures are followed if required by closely monitoring these nucleus stock seed for any genetic changes that are happening during subsequent period of storage. Andrews and Harinarayana (1984) have suggested to restrict the area selected for nucleus seed production without exceeding 0.2 ha to enable the close monitoring by the breeder and other staff concerned to evaluate individual plants at different stages of crop growth. Breeder seed can be multiplied in one lot to meet the requirement of foundation seed for 5 years (Chopra 1982). However, it depends on the availability of isolation plots of desired size and medium-term cold storage facilities. The basic seed, i.e., maintained by breeder, is a prerequisite for starting nucleus seed production of parental lines. Generally, off-season is ideal for raising the seed plots to ensure homogeneous and uniform crops aiming at highest genetic purity of seeds. Best crop management under the direct supervision of breeder is required to maintain highest genetic purity. Ear-to-row progenies (A-line and B-line) are raised in pairs at side-by-side rows. The multiplication of A- and B-lines should be done in alternate rows (approx. 5 m length) by hand pollination with proper isolation, as the seed requirement is low. An account of steps involved in production of nucleus seed of single cross-hybrids, i.e., B-line (Maintainer), A-line (Male Sterile), and R-line (Restorer), is presented in Tables 19.1, 19.2, and 19.3; Figs. 19.1 and 19.2.

19.3.2 Breeder Seed Production of Pearl Millet Parental Lines (A-, B-, and R-Lines)

Proper understanding is necessary on important practices like selection criteria for area of seed production, crop growth and environmental influence, soil fertility, irrigation and plant population, and seed processing and storage requirements for economic and efficient quality seed production. The breeder seed plot of A-line is raised in isolation by raising alternately four rows of A-line and two rows of B-line. About 4–8 border rows of B-line are raised around the plot for adequate availability of pollen to A-line rows. Strict rogueing is essential in all rows more critically during

Table 19.1 Nucleus seed production of maintainer line (B-line) in pearl millet

Season	Step-I	Step-II	Step-III	Step-IV
Season I	Raise the B-line plants in an area of 0.05 ha and select about 1000 self-pollinated plants at flowering	Identify 200 self-pollinated plants that are confirming to specified DUS traits of B-line		
Season II	Raise plant to row progeny of self-pollinated plants in 2 replications Retain some remnant seed	Closely observe progeny rows are for all specified DUS traits Reject the rows not confirming to these specified traits	Identify about 25% best progeny rows	Bulk the seed of selected best lines that were retained as remnant
Season III	Raise in isolation the bulked seed of remnant seed.	Bulk the seed after the harvest of all the plants	It becomes nucleus seed bulk of B-line	

Table 19.2 Nucleus seed production of male sterile line (A-line) in pearl millet

Season	Step-I	Step-II	Step-III	Step-IV
Season I	Raise in isolation the alternate rows of A- and B-lines B-line seed are obtained after rejuvenation scheme of B-lines	Make paired crosses of about 200–250 between A- and B-plants confirming specified DUS traits	Label the paired crosses of A- and B-lines as A1 × B1, A2 × B2 etc. Harvest the seed of each pair and store separately	
Season II	Raise in isolation the pairs of A-line and B-line in alternate rows Retain remnant seed portion	Critically observe the pairs of A- and B-lines for all specified DUS traits If pollen shedders in A-line are noticed, reject the A-line progenies along with their corresponding B-lines	Identify uniform pairs of A- and B-lines which confirm to DUS traits	Bulk the remnant seed of the A-lines of selected pairs that becomes nucleus seed bulk of A-line

and after flowering stages (Table 19.4). At flowering, the pollen shedders if any in A-line rows should be rogued out. All the plants should confirm the distinct, uniform, and stable (DUS) traits of the cultivar as specified in its release proposal. Using the nucleus seed stock as source, the breeder seed production plot of B-lines is raised separately in an area of about 0.1 ha strictly under isolation (1000 m). At all growth stages more particularly during and after flowering the rogueing operations

Table 19.3 Nucleus seed production of restorer line (R-line) in pearl millet

Season	Step-I	Step-II	Step-III	Step-IV
Season I	Raise in 0.05 ha the R-line plants Self-pollinate about 1000 plants confirming to specified DUS traits t flowering	Identify about 200 self-pollinated plants based on plant and seed traits as specified in DUS test guidelines		
Season II	Raise plant to row progeny of self-pollinated plants in 2 replications Retain remnant seed portion.	Observe the progeny rows both quantitative and qualitative traits as per DUS test guidelines, and reject those lines not confirming Evaluate the lines for yield, agronomic, and economic traits	Select about 30-50% of best progeny rows based on evaluations for various traits If adequate number of progeny rows confirming true to type are not obtained, repeat the self-pollination for one or more generations Test these self-pollinated seed again for fertility restoration ability	Bulk the remnant seed of best lines identified
Season III	Raise in isolation the bulked seed of remnant seed	Bulk the seed of all the plants after the harvest	This becomes the nucleus seed bulk of R-line	

should be strictly followed to avoid any contamination (Table 19.4). Random mating in an isolation plot or sib-mating by hand is used for maintenance of R-lines. The breeder seed plot of R-lines is raised by bulk planting in isolation with minimum distance of 1000 m in an area of 0.1–0.2 ha aiming to maintain about 3000–5000 plants. For R-line breeder seed production also, the nucleus seed stock is the source seed. Rogueing for off types should be strictly followed at all the stages of crop growth (Table 19.4). After the harvest, the bulked seed of selected plants requires thorough processing and storage in safe conditions.

19.3.2.1 Selection of Site and Season

The breeder seed production of parental lines should aim for highest quality in terms of genetic purity, physical purity, and seed vigor. The nucleus seed is the source for production of breeder seed of A-line. The A- and B-lines are planted side by side in isolation of more than 1000 m from any other pearl millet crop. Planning of breeder seed production should aim at meeting major objectives of increasing crop productivity and timely availability of required quantities of quality seed adopting certain special practices. The preceding crop should not have been sown with pearl millet in the field selected for raising breeder seed crop. The seed production plot should have fertile and well-levelled land with good drainage and irrigation system as pearl millet is sensitive to water logging. Deep soils are not preferable for growing pearl millet

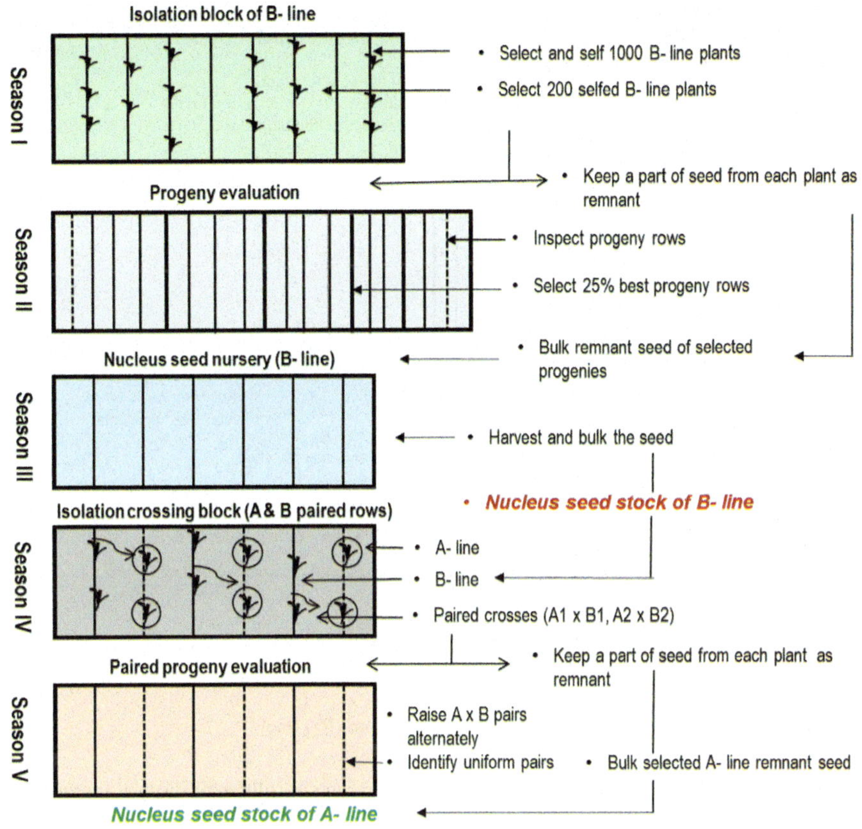

Fig. 19.1 Nucleus seed production of pearl millet maintainer (B-) line and male sterile (A-) line

seed crop during rainy season. Though pearl millet is a rainy-season crop, it can grow successfully during dry season when provided with good irrigation facilities. However, the minimum temperatures should not fall below 12 °C and the maximum temperature does not exceed 35 °C. In India, dry-season is ideal for seed production for better crop expression, seed setting, and seed quality due to the advantage of long sunshine hours, less lodging, less washing of pollen, and less incidence of disease and pest. For raising nucleus or breeder seed crops, off-season is ideal due to likely availability of required isolations, and the off-season facilitates special seed production practices like rogueing, which lead towards production seeds with high genetic purity, health, luster, vigor, and longevity.

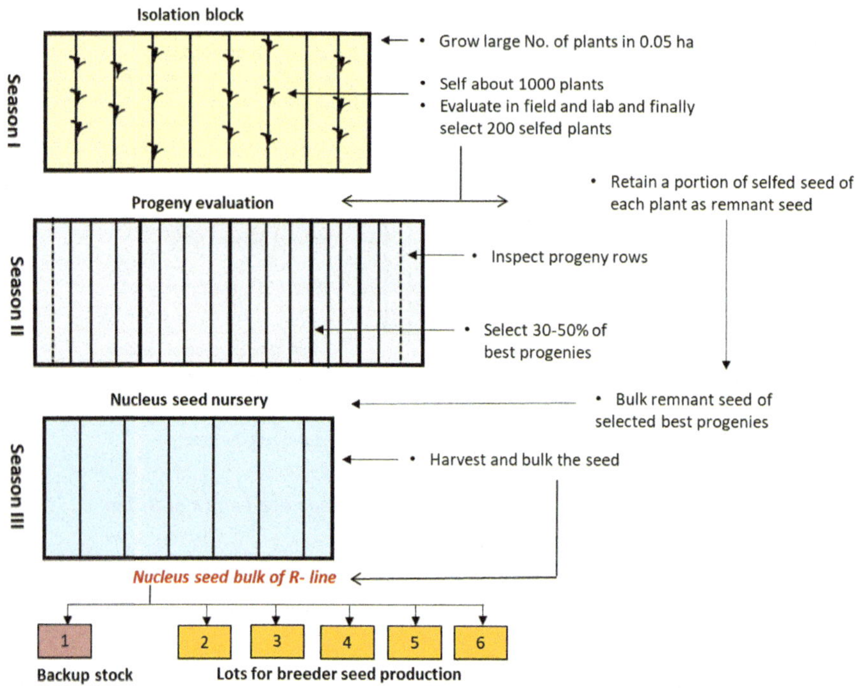

Fig. 19.2 Nucleus seed production of pearl millet restore (R-) line

Table 19.4 Different growth stages to identify the deviants for different distinct, uniform and stable traits specified for A-, B-, and R-lines of a hybrid

Stage of crop	Characters specifically observed for deviants
18 days after sowing (DAS)	Early vigor
30–35 DAS	Tillering, plant height and other traits
Boot leaf stage	Early flowering
50% flowering	Late flowering, panicle exertion and panicle type
Maturity stage	Seed type and grain yield

19.3.2.2 Requirement of Isolation Distance

The risk of genetic contamination in pearl millet is high due to its protogynous flowering behavior. Determination of isolation distance required for seed production plots of pearl millet continues to be major concern (Chopra 1982). The recommended isolation for nucleus seed production is at least 2 km, and for breeder-seed production is more than 1 km (Andrews and Harinarayana 1984). The A- and B-lines can be produced in the same seed production plot, or in different plots maintaining an isolation distance of at least 5 m. While planning isolation distance, important factors need to be considered such as wind direction and velocity, humidity, windbreaks, and plots size that is the source to spread the load of pollen. Seed village concept is the usual practice in India that can be best adapted to resolve the problem of isolation distance (Chopra 1982). All farmers of a village cultivate same variety by adopting seed village concept that facilitates seed certification convenient and cost-effective.

19.3.2.3 Sowing Method

In pearl millet, direct sowing is the usual method of sowing being practiced for all classes of seed. Transplanting of 18-to-20-day old seedlings is also a preferred method of sowing in certain areas due to several advantages over direct sowing such as saving expenditure on thinning, weeding, and irrigation at initial seedling stage. Further, transplanting method saves 30–40% of the seed quantity and establishes better plant stand leading to higher seed yield compared to direct sowing. The recommended inter-row spacing is 60–75 cm and an intra-row spacing is 15–20 cm for seed plots of nucleus and breeder seed. It is highly desirable to maintain the wider spacing and lower plant density in order to have better crop expression and to facilitate special seed production practices like rogueing of off-types and deviant phenotypes that is important operation to produce genetically pure seed. Sowing of A-line first and then the B-line is desirable irrespective of method of sowing in order to avoid the errors during sowing.

19.3.2.4 Planting Ratio of A- and B-Lines

The planting ratio depends on the pollen producing efficiency of B-lines. The ratio of four lines of A- and two lines of B- is recommended; however, the female (A-line) rows can be extended to six rows if the B-line has more ability of pollen production. The seed-production during rainy season is more prone to risk of pollen wash and ergot (a panicle disease) that is a case for closer ratio of A- and B-lines compared to seed production during dry-season. Raising 4–8 border rows of the B-line of a seed production plot can aid the supply of additional pollen leading to higher seed set in the A-line rows. In addition, these lines serve as trap for pollen from outside sources. Proper labeling of A-line rows and B-line rows is desirable to maintain in a seed production plot as per the record of sowing plan to facilitate easy identification until the end of harvest.

19.3.2.5 Synchronization of A- and B-Lines

If duration of stigma receptivity in A-line is more than 3 days, the problem of synchrony does not arise as generally there may be a gap of 2 days between A- and B-lines for flowering. Foliar spray of urea at primordial initiation stage can also advance the late flowering parent thus aiding the synchrony in flowering. Some seed growers of Andhra Pradesh state in India follow the practice of pulling out the main panicles of the A-line plants that encourages more tillering, thus aiding the synchronization of flowering between A- and B-lines further leading to higher seed yield.

19.3.2.6 Rogueing

Rogueing is very important operation during seed production to maintain the genetic purity of A-and B-lines that requires removal of pollen shedding plants from A-line and off-type plants from both A- and B-lines. Pollen shedders of cytoplasmic origin are identical to the maintainer line, and these plants can be checked thoroughly during flowering and seed setting period and rogued out immediately before harvest. In pearl millet, it was reported that majority of pollen shedders occur due to mutations in the cytoplasm (Burton 1977). Morning time is ideal to carry out rogueing of pollen shedders and off-types in the A-line as flowering starts in the night. Cooper and Burton (1965) reported that the pollen viability loss was <50% when stored at 26.7 °C. Off-type plants may arise through outcrossing or through mechanical mixtures. Rogueing of the nucleus and breeder seed plots is important to ensure quality seed production in pearl millet. The rogueing is desirable at three stages of crop growth: (1) before flowering, (2) at flowering, and (3) pre-harvest stage. Stringent rogueing is required in seed production plots for nucleus and breeder seed due to highly cross-pollinating behavior that creates small changes in various plants to converge and cause genetic shifts in the parental lines after recombination.

19.3.2.7 Harvesting, Threshing, and Processing

Always, B-line should be harvested first and labeled properly with all details both inside and outside the bag. The panicles of B-line should not be left any more in the field before starting harvest of A-line. After harvest, the A-line panicles should be heaped, sorted, dried to 12% of moisture content and packed in bags with proper labeling with all the details. The B-lines and A-lines should be threshed and processed separately to remove unwanted inert matter and achieve uniformity through size grading. Before packing, the seed should be dried using a clean drier to desirable moisture content of 8% (maximum) for packing in vapor-proof containers for safe storage.

19.3.2.8 Forecasting Land and Seed Requirements

It is important to forecast and plan the hybrid seed production. The annual certified seed requirement forecasting is dependent on seed multiplication ratio (SMR), seed replacement rate (SRR), and additional seed requirement. The available stock of particular class seed is the base factor to determine the estimation of quantity to produce subsequently. Apart from seed requirement, forecasting land requirement for different seed classes and planning accordingly is the basic step before starting the hybrid seed production. Based on seed rate, planting row ratio and seed yields of different seed classes, namely certified, foundation, and breeder seed, the required quantity of seed and area to produce the same can be estimated. Further, the sale projections of particular hybrid also need to be considered before calculating the estimations for foundation and breeder seed production.

19.3.3 Foundation Seed Production

Breeder seed is the source for sowing of foundation seed production aimed to produce genetically and physically pure seed meeting the requirement of Indian Minimum Seed Certification Standards (2013). The National Seed Corporation (NSC), State Farms Corporation of India (SFCI), State Seed Corporation (SSCs), and some private seed companies are involved in production of parental lines as foundation seed, and hybrid seed as certified seed. These A-line and R-line seeds require grow-out tests in small plots before supplying to certified seed producers. The foundation seed shall be the source for certified seed production of a hybrid.

19.4 Pearl Millet Hybrid Seed Production

The certified hybrid seed production of pearl millet requires raising of A- (male sterile) and R-(restorer) lines side by side in definite ratios for meeting the field and seed standards by adopting isolation distance to realize genetic purity (Indian Minimum Seed Certification Standards 2013). Various public and private organizations such as NSC, SSCs, SFCI, and private seed companies and seed growers are involved in certified seed production of pearl millet on large scale. However, the success of any hybrid seed production depends on agro-ecological conditions in seed production area and temperature, photoperiod, wind direction air velocity, rainfall, relative humidity, and soil chemical reaction, etc. Temperature is most important factor that influences at flowering stages of male sterile (A-) and restorer (R-) lines during pearl millet hybrid seed production. Under poor storage conditions, pearl millet hybrid seeds loose germination percentage. Storing the seed under controlled conditions is costly and it increases seed price of hybrid seeds. The seed producers in south India opt to take up the hybrid seed production during post-rainy season. Sometimes, due to lack of facilities like irrigation and due to unfavorable weather conditions, some seed producers particularly at north-west regions of India do not prefer post-rainy season. While undertaking the commercial hybrid seed production, one should consider the important factors like isolation distance, location, field conditions, irrigation facility, history of previous crops, planting row ratio of A- and R-lines, wind direction and velocity, border lines of R-line, synchronization of flowering in A- and R-lines, crop management practices, rogueing operations, and certification of seed.

19.4.1 Influence of Agro-Climatic Factors on Flowering and Seed Set

Under varying agro-climatic conditions, the final reproductive success can be judged by percentage of seed set that resulted due to stigma receptivity, pollen fertility, pollen germination, and pollen quantity traits. Some genotypes of pearl millet fail to flower when day length is >12 h, and many genotypes require long photoperiod with day length of 16 h. Due to short day conditions, early flowering occurs as reported by Burton (1965). Optimum temperature of 32 °C and appropriate photoperiod shorten the days required from sowing to flowering according to Helmars and Burton (1972). Sometimes drought also plays a role by either delaying the flowering or hastening which varies with genotype. High temperature combined with drought affects the pollen viability, stigma receptivity, and seed setting traits leading to poor seed yield and/or even complete failure of hybrid seed production.

19.4.2 Location and Field Requirement

In India, the actual pearl millet cultivation areas of rainy season are distantly located from the areas suitable for hybrid seed production in post-rainy or pre-monsoon seasons. These suitable areas have adequate facilities like irrigation, processing, and favorable ecological conditions, and hence the public and private seed industry prefer these areas ideal for better seed quality management. However, the cost of seed transport is the concern for marketing the hybrid seed at pearl millet growing areas in a short span between two seasons. In Gujarat state of India, some local seed producers opt for seed production of pearl millet during rainy season wherein the seed transport cost is low. Nevertheless, this scenario requires adequate storage facilities under controlled conditions to keep the seed viable until next rainy season. However, the seed produced in rainy season is used for sowing in the post-rainy season in some pockets of Gujarat state.

19.4.3 Isolation Distance, Cropping History, and Field Requirement

Pearl millet seed production plot requires an isolation distance of >1 km as it is highly cross-pollinated crop. The pollen can travel more than 1 km under moderate wind velocity of >30 km towards the wind direction. The wind direction in India generally changes particularly during the seed production. Physical barriers like tall, dense vegetation, hedges, and buildings may help in regulating the pollen flow to some extent. In order to overcome the isolation problems as required, the seed producers of pearl millet hybrids prefer seed village concept, where many seed growers agree to produce the same hybrid in that village or a cluster of villages that require careful operations to meet required standards for certification. Prior knowledge on cropping history is essential before selecting a field for hybrid seed production as it facilitates to plan on nutrient and irrigation requirement. If pearl millet is cultivated during previous season, such field is unsuitable as it produces volunteer plants that are unwanted. The field should be of high-quality free from water logging, light sandy soil with all facilities of irrigation, drainage, and preferably of uniform length and width.

19.4.4 Planting Row Ratio and Border Rows

The planting ratio of A- and R-line rows for hybrid seed plot is primary based on pollen production potential of the R-line and the prevailing relative humidity, wind direction, etc. The ratio with 4 rows of A-line and 2 rows of R-line favors better seed set under diverse environments in various regions in India; however, this ratio can range from 2:1 to 7:1 as per the prevailing conditions. Raising of border rows of male parent (R-line) around the hybrid seed plot can enhance the supply of pollen and ensures adequate seed set on A-lines. Further, it also prevents to some extent the entry of foreign pollen and thus reduces the chances of contamination.

19.4.5 Synchronization of Flowering in Male and Female Rows

The pollen shedding of male lines and stigma receptivity of female lines should coincide for effective crossing that is possible only when there is proper synchrony in flowering of male and female rows. Synchronization or nicking of male and female rows is essential to ensure better seed set on A-line. The time gap between male and female rows observed at flower primordia stage can be manipulated by adopting special agronomic practices that ensure synchrony in flowering between male and female parents. Some of these practices are (1) sowing male and female lines on different dates, (2) raising nurseries by sowing in different dates and transplanting both parents at same time in main field which is the usual practice by seed producers of pearl millet in Andhra Pradesh state of India. If the difference in the flowering period is more than 10–15 days, then staggered sowing of parental line, which are early in flowering, is needed. Staggered sowing is only possible under assured irrigated facilities; staggered sowing may be managed either by the dibbling or transplanting. To enhance the growth of lines lagging behind, extra-nitrogen fertilization or foliar spray may be used.

19.4.6 Rogueing

Rogueing is very important practice of hybrid seed production that must be started from vegetative stage to avoid contamination due to volunteer plants and diseased plants or any doubtful deviants in all rows with regular observation. The characters specified for parents as per distinctiveness, uniformity, and stability (DUS) traits should be observed at various crop growth stage in order to identify the deviants and off-type plants. The seed parent plant in each row should be thoroughly checked for deviants and rogueing should be performed at least once a day early in the morning particularly from ear head emergence stage. Pollen shedders if any found in rows of seed parent should be rogued out immediately. By observing the differences in height, early or late flowering, the out crosses can be removed. The rogueing operations should continue until harvesting and the diseased ears with ergot or downy mildew should be rogued out. Following points need special attention at the time of rogueing. Harvesting of R-line should be done first and late the harvest of A-line or hybrid.

19.4.7 Field Inspection

For the production of foundation and certified seed classes, a minimum of four field inspections are recommended. The first inspection shall be within 30 days after sowing to verify the isolation requirement, outcrosses, volunteer plants, planting errors, and diseases like downy mildew. The second and third inspection shall be made during flowering to identify the off-types, pollen shedders, and diseased plants (downy mildew green ear). The fourth inspection shall be made at maturity and prior to harvesting to check for the downy mildew green ear, ergot, grain smut, and to verify the true to type of planter (Indian Minimum Seed Certification Standards 2013).

19.5 Conclusion

In pearl millet, the introduction of cytoplasmic-nucleus male sterility (CMS) resulted in development of male sterile (A-), maintainer (B-), and restores (R-) lines that lead towards hybrid development programs addressing the diseases and production constraints specific to various zones. As a result, the released hybrids cover >70% area of pearl millet cultivation in India, particularly the A- and B-zones of India with >400 mm annual rainfall and with two separate agro-ecologies. During last three decades, both public and private sectors successfully organized the seed production and distribution of pearl millet hybrids. In pearl millet, the increase in area with superior genetic and productive hybrids always demands the production and availability of high-quality seed with assured genetic purity in required quantities. The realization of this is possible only by effective implementation of seed generation system through the channel of nucleus, breeder, foundation, and certified seed production programs with prescribed methods of scientific seed crop management. These scientific methods require time-to-time optimization of hybrid specific technologies in overcoming seed production constraints under varying environmental conditions.

References

Andrews DJ, Harinarayana G (1984) Procedures for the seed production of pearl millet varieties. Information Bulletin no. 16. International Crops Research Institute for the Semi-Arid Tropics, Patancheru

Athwal DS (1996) Current plant breeding research with special reference to pennisetum. Indian J Genet 26A:73–85

Bhatnagar MP, Kumar K (1960) Anthesis studies in Rajasthan bajra (Pennisetum typhoides). Indian J Agric Sci 30:185–195

Burton GW (1965) Photoperiodism in pearl millet, Pennisetum typhoides. Crop Sci 5:333–335

Burton GW (1977) Fertility sterility maintainer mutants in cytoplasmic male sterile pearl millet. Crop Sci 17:635–637

Chopra KR (1982) Technical guideline for sorghum and pearl millet seed production. Food and Agriculture Organization of the United Nations, Rome, p 110

Cooper RB, Burton GW (1965) Effect of pollen storage and hour of pollination on seed set in pearl millet Pennisetum typhoides. Crop Sci 5:18–30

Directorate of Millets Development (2020). Available https://www.millets.dacfw.nic.in

Gupta SK, Rathore A, Yadav OP, Rai KN, Khairwal IS, Rajpurohit BS, Das RR (2013) Identifying mega-environments and essential test locations for pearl millet cultivar selection in India. Crop Sci 53:2444–2453

Helmars H, Burton GW (1972) Photoperiod and temperature manipulation induces early anthesis in pearl millet. Crop Sci 12:198–200

Indian Minimum Seed Certification Standards (2013) The Central Seed Certification Board, Department of Agriculture and Cooperation, Ministry of Agriculture, Government of India, New Delhi

Protection of Plant Varieties and Farmers' Rights Authority (PPV & FRA) Government of India (2019) Distinctiveness Uniformity and stability guidelines (revised) for the conduct of test on pearl millet (Pennisetum glaucum (L.) R. Br.). http://plantauthority.gov.in/crop-guidelines.htm

Serba DD, Yadav RS, Varshney RK, Gupta SK, Mahalingam G, Srivastava RK et al (2020) Genomic designing of pearl millet: a resilient crop for arid and semi-arid environments. In: Kole C (ed) Genomic designing of climate-smart cereal crops. Springer, Cham, pp 221–286. https://doi.org/10.1007/978-3-319-93381-86

Traditional Varieties of Pearl Millet and Food Diversity

20

M. Elangovan, Karnam Venkatesh, and B. Sangappa

Keywords

Pearl millet · Traditional varieties · Food diversity

20.1 Introduction

The most extensively cultivated kind of millet is the pearl millet (*Cenchrus americanus*, also known as *Pennisetum glaucum*, also known as "Bajra" in Hindi, "Sajje" in Kannada, "Kambu" in Tamil, and "Gero" in Hausa). Since ancient times, it has been grown in the dry and semi-arid parts of Africa and the Indian subcontinent, which are characterized by marginal soils (Tako et al. 2015). The Sahel region of West Africa is the region with the greatest diversity and is thought to have domesticated the crop (Manning et al. 2011). Recent archaeobotanical studies have demonstrated that domesticated pearl millet was present in northern Mali's Sahel region between 2500 and 2000 BC. Its $2n = 14$ chromosomal number is present. It belongs to the family Poaceae.

Pearl millet is the major staple crop to meet the nutritional requirements of millions of people and especially children living in Africa and India (Pucher et al. 2014). Pearl millet can grow and yield optimally in some of the regions with poorest of the soils and driest and hottest weather conditions in Africa and India (Senthilvel et al. 2008). Pearl millet grains are nutritionally rich especially by having essential amino and fatty acids contents, essential minerals, and vitamins (Manwaring et al. 2016). The grains are also used to prepare alcoholic beverages in few regions of the globe (Loumerem et al. 2008). Additionally, researchers have reported pearl millet to

M. Elangovan (✉) · K. Venkatesh · B. Sangappa
Indian Institute of Millet Research, Hyderabad, India
e-mail: elangovan@millets.res.in

V. A. Tonapi et al. (eds.), *Pearl Millet in the 21st Century*,
https://doi.org/10.1007/978-981-99-5890-0_20

contain special oils and phenolics with antibacterial and anti-oxidative properties (Marmouzi et al. 2018).

Pearl millet being cultivated as a staple crop from ancient times has a major role in fulfilling the needs of local farmer communities. From ancient times pearl millet grains are usually used in the preparation of many food dishes such as *chapatti* (unleavened flat bread), fermented breads (*kisra*, *injera*, *dosa*, etc.), porridge, dumplings, biscuits, snacks, malt, and opaque beer (Upadhyaya et al. 2012). Therefore, there exists a lot of traditional varieties and local recipes of pearl millet, which have evolved from ancient times in western India. In this chapter we discuss in detail about the local diversity present in pearl millet germplasm and their community conservation efforts along with traditional varieties and popular food recipes.

20.2 Community Conservation of Pearl Millet

Biodiversity or germplasm diversity plays a critical role in finding solutions to challenges faced by agriculture such as scarcity of resources, incidence of pest, and disease problems. Farmers themselves have understood the importance of conserving plant biodiversity long back to meet their local requirements. The conservation of crop biodiversity also supports achievement of UN declared zero hunger challenge of "100% access to adequate food all year round" (Kumar et al. 2015). Further, the conservation of crop diversity by farmer families may help in addressing the problem on malnutrition (Kesavan and Swaminathan 2014). Understanding the importance of conservation of crop diversity at farmer level, even the Indian government started Protection of Plant Varieties and Farmers' Rights Authority (PPV&FRA) and National Biodiversity Authority (NBA) and is now awarding farmers involved in conservation of crop diversity with Genome Saviour Cash Award, and also has started National Gene Fund for sustainable management of Plant Genetic Resources, Registration for Farmers Varieties (Kumar et al. 2015).

One of the examples of community-level conservation of millets in India include, conservation of millet crop varieties by *Malayali* community and other groups of people along the western belt of Tamil Nadu in Kolli Hills as they are completely dependent on millets for their livelihood. The conservation strategies followed by communities included gathering of seed and their multiplication followed by distribution among farmers and relatives through family and village level exchange through traditional seed storage banks built using traditional methods using locally available material (Jayashree and Aram 2020). Farmer communities follow a set of time-bound rituals containing festivals and folk songs formulated around agricultural practices and seasons that help them chose the best crop variety and conserve the crop diversity. In almost all tribal region, crop diversity is conserved through few specific custodian farmer families who have a heritage of storing seeds of ancient varieties. They store the seeds using traditional methods such as using earthen clay pots.

20.3 Traditional Varieties of Pearl Millet and Food Security

Sulkhaniya Bajra—ancient Pearl millet cultivar from Rajasthan: Shri Hanumana Ram Jhuriya from the village Sulkhaniya in the arid region of Rajasthan, where the annual maximum temperature touches 50 °C, has been continuously conserving the variety through selection method and on the name of his village the variety has been given the name "*Sulkhaniya.*"

Sulkhaniya bajra is a traditional variety with drought tolerance, higher grain and fodder yield, and tastier and nutritious grains. It is favored by cattle and used in many households as medical aid due to its medicinal properties.

Sulkhaniya bajra is popularly gown in sandy loam soils of the arid and semi-arid regions. It is sown during May–June (pre-monsoon period) and very early variety maturing in around 90 days, which helps it to escape severe summer. The average grain yield is around 25 quintals per hectare with a biomass of 40 quintals per hectare. The variety is tolerant to water-deficit condition and can be grown even in longer dry periods. Productivity is not at all affected by the adverse dry condition. In fact, it becomes neutral or equal to that of normally grown cultivars.

It has good production potential under rainfed conditions. It has compact and medium thick ear heads. It is early maturing and gives a palatable and nutritious fodder. It goes well in market and is quite tolerant to diseases under field condition. It can also be used further in research and development as breeding material.

Through multi-location trials, it has been observed the variety responds better than other local land races. Farmers in all the locations showed a preference for this variety due to its qualities and better production.

For this great activity, Shri Hanumana Ram Jhuriya was awarded at the 9th National Biennial Competition, held during Festival of Innovation (FOIN) 2017, at Rashtrapati Bhawan, New Delhi, in March 2017. The seeds of the variety can be procured directly from the farmer at his address: Vill: Sulkhaniya, Teh: Ratangarh, Distt: Churu, 331022, Rajasthan.

20.3.1 Local Landraces and Their Popular Vernacular Names of Pearl Millet in India as per National Genebank (NGB), ICAR-National Bureau of Plant Genetic Resources-New Delhi

Bajra or pearl millet is known by several vernacular names in different parts of India. The accessions with the vernacular name *Karauli* are of maximum number with 78 lines followed by *Chadi* (61), *Barmer* (28), *Mathuravelllai* (25), etc. The list of pearl millet local landraces available at NGB, ICAR-NBPGR, is presented in Table 20.1.

20.3.2 Popular Vernacular Names of Pearl Millet Among the Accessions Conserved at Millets Genebank, ICAR-Indian Institute of Millets Research, Hyderabad

Millet Genebank (MGB) at ICAR-IIMR (Indian Institute of Millet Research), Hyderabad, has a collection of 126 pearl millet accessions in addition to the lines (5644 accessions) augmented from Dr. RS Paroda Genebank, International Crops Research Institute for the Semi-Arid Tropics (1134), Dr. Ramaiah Genebank, Tamil Nadu Agricultural University (2021), All India Coordinated Research Project on Pearl millet (2260), Rajmata Vijayaraje Scindia Krishi Vishwa Vidyalaya (217), and other sources (Elangovan 2021). The vernacular names recorded while collecting these accessions are Nattu cumbu (33 accessions), Khandlu (11), Bajra (11), Cumbu

Table 20.1 Traditional variety/Landraces of bajra or pearl millet accessions conserved at National Genebank at ICAR-NBPGR, New Delhi

Traditional variety/landrace	No. of accessions
Andhra Pradesh	
Bajra	51
Sodha	48
Sajjalu	22
Khandlu	12
Pitta ganti	10
Punasa ganti	8
Ganti	7
Peddha sodha	6
Ghantilu	5
Ghantlu	4
Gantelu	4
Khantalu	4
Naatu sodha	4
Ganttu	3
Javari sodha	3
Nattu cumbu	3
Poltaganti	2
Pedda ganti	2
Ghandlu	2
Ganti gantulu	2
Gantlu	2
Pedda	2
Mahudo	2
Sodha sajjalu	2
Pittaganti	1
Punasanganti	1
Javaro sodha	1
Pala ganti	1
Pedda gante	1
Paecha korra	1
Ganelu	1
Natu	1
Petha gantulu	1
Pitla gentelu	1
Punasa Ghantilu	1
Gantilu kurrai	1
Babapuri	1
Gantellu	1
Soddha	1
Palagantu	1
Mettu Dhanyam	1

(continued)

Table 20.1 (continued)

Traditional variety/landrace	No. of accessions
Peddina sodha	1
Pedha sodha	1
Gontia/Gantalu	1
Konda podu gantulu	1
Poduganti	1
Konda gantelu	1
Ganttulu	1
Bochu sodha	1
Piltaganti	1
Petldha Sodna/Kammi sodha	1
Kariganti	1
Pitta ghantilu	1
Gantalu	1
Edagaru sodha	1
Munaga	1
Arunachal Pradesh	1
Bihar	
Katua bajra	1
Delhi	
Bajra	1
Gujarat	
Bajra	32
Local cultivar	15
Bajri	9
Cultivar stain	4
Rajka bajri	2
Lujka bajri	1
Anjar	1
Porbandar	1
Mundra	1
Somnath	1
Amreli	1
Haryana	
Cultivar stain	14
Bajra	9
Himachal Pradesh	
Bajra	3
Jammu and Kashmir	
Bajra	5
Jharkhand	1
Karnataka	
Sajje	10
Javari sajje	2

(continued)

Table 20.1 (continued)

Traditional variety/landrace	No. of accessions
Telasanga local-1	2
Javari saije	1
Male parent B-5220R	1
Hola sajja	1
Lokapur local-2	1
Ramadurg local-1	1
Lokapur local-1	1
Inchal local-1	1
Chalmitanda local	1
Dsajje	1
Jumnal local-2	1
Jumnal local-1	1
Bidar local-1	1
Kumbarpeth local	1
Kerala	1
Madhya Pradesh	
Bajra	29
Desi bajra	4
Bajri	3
Deshi type	2
Pitta ganti	1
Salhar	1
Deshi bajra	1
Chota bajra	1
Maharashtra	
Bajri	11
Bajra	9
Gawran bajra	6
Deshi	3
Gawrani	3
Gaorani	2
Bajra local	1
Adchi	1
Bajiri	1
Awned bajra	1
Sajji	1
Sajgura	1
Kocin	1
Gaveti	1
Bairi	1
Rahuri	1
Awnedless bajra	1
Nasik	1

(continued)

Table 20.1 (continued)

Traditional variety/landrace	No. of accessions
Adichi	1
Derei siomithan	1
Banti	1
Kokhblwwal	1
Odisha	
Bajra	4
Ghantia	2
Gantia	1
Kurui	1
Koya	1
Jandri	1
Punjab	
Nilokheri	2
Lohar local	1
Sanodar local	1
Bangaon local	1
Rajasthan	
Bajra	193
Cultivar stain	39
Local cultivar stain	28
Bajra	25
Chadi	23
Bajara	16
Barmer	12
Karauli	11
Desi bajra	6
Local cultivar	5
Desi bajri	4
Rijaka bajra	3
Sulkhania	3
Bajri	3
Jowar	1
Avika bajra chari	1
Guar	1
Bajara desi	1
Tamil Nadu	
Nattu cumbu	23
Arisi cumbu	13
Cumbu	11
Kambu	10
Podi	10
Mathuravellai	9
Bajra	4

(continued)

Table 20.1 (continued)

Traditional variety/landrace	No. of accessions
Arisi	4
Kattu	4
Periya	4
Kullan	4
Manavari	4
Kaatu cumbu	3
Eravai	3
Naatu cumbu	2
Aelumoly	2
Oomi	2
Nattu	2
Kullan cumbu	2
Vellai	2
Mochai	2
Pottu kumbu	1
Pottu cumbu	1
Mathura vellai	1
Siru cumbu	1
Manavri	1
Pottu kambu	1
Pokku	1
Potti cumbu	1
Manavai	1
Kambaneri cumt	1
Mahuravellai	1
Kuruva	1
Cumbu CO 3	1
Velling	1
Chinna cumbu	1
Periya cumbu	1
Moonumoli	1
Kombu	1
Pokku cumbu	1
Umi	1
Vivasaya cumbu	1
Oosi	1
Mathuravelai	1
Telangana	
Pedda sajja	4
Sajja	2
Chalaka sajja	1
Deshi sajja	1
Uttar Pradesh	

(continued)

Table 20.1 (continued)

Traditional variety/landrace	No. of accessions
Bajra	22
Bajri	5
Bazara deshi	2
Badaun 17	1
Uttarakhand	
Bajra	6
Desi bajra	1

Table 20.2 Traditional variety/landraces of bajra or pearl millet accessions conserved in Millets Genebank at ICAR-IIMR, Hyderabad

Traditional variety/landrace	No. of accessions
Nattu cumbu	33
Khandlu	11
Bajra	11
Cumbu	9
Sajjalu	7
Deshi	7
Khantalu	5
Gauvrani	3
Thevana cumbu	3
Pottu cumbu	2
Periya cumbu	1
Kambu	1
Star chandra	1
Chinna cumbu	1
Bajri	1
Petta khantalu	1
Potti cumbu	1
Dadar	1
Kulla cumbu	1
Siru cumbu	1
Kullan cumbu	1
Kurui	1
Vivasaya cumbu	1
Koan	1
Koya	1
Grand total	106

(9), and Sajjalu (7). The list of pearl millet local landraces available at Millet Gene Bank, ICAR-IIMR, is presented in Table 20.2.

20.4 Health Benefits of Pearl Millet

Aids in Weight Loss Bajra is composed of complex carbs that are slowly absorbed by the body. This makes you feel satiated and prevents over-eating. The portion control reduces the risk of weight gain. Foods with a low-calorie density help you feel full but for fewer calories. Foods with a calorie density greater than 2.3 are generally considered high. Bajra has a calorie density of 1.2. Thus, foods like bajra with a low-calorie density may aid weight loss.

Manage Diabetes "Bajre ki roti" is good for diabetics. Bajra is high in fiber and gets digested slowly. There is no sudden spike in glucose levels. Hence, it is perfect to keep blood sugar levels in check. Moreover, bajra is also a good source of magnesium, which is linked to lowering the risk of diabetes. Foods that are high in fiber, especially cereal fibers like bajra, have also been associated with improved outcomes in the management of type 2 diabetes and other chronic diseases.

What is more, millet has a lower glycemic index (GI) than some refined grain products like white rice and white bread. Plus, some emerging research in animals and humans has found that millet proteins may contribute to improved blood sugar levels. On average, most types of millet have a GI value of 43–68. Foods with a GI value of 55 or below are typically considered to be low. The GI is a measure of how much certain foods affect blood sugar levels. Foods that are lower on the glycemic index are usually better choices for people with diabetes. In some cases, glycemic load (GL) may be a better measure of how a food affects blood sugar levels. GL differs from GI by also considering the typical serving size of a food. A GL of 10 or below is considered low, while a GL of 20 or more is high.

One study noted that millet flakes have a GL of 9.2, meaning they have a low GL. That said, some of the research that supports these claims did not use bajra specifically, and the use of both GI and GL in diabetes management is controversial. Therefore, more research is needed to understand exactly how millet affects blood sugar levels.

Promotes Good Gut Health The insoluble fiber content in bajra acts as a prebiotic that regulates your digestive health. The insoluble fiber also improves bowel movement, keeping digestive issues like constipation at bay.

It is Gluten Free Bajra is a boon for people who are allergic to gluten or follow a gluten-free diet. So, it is always a good idea to use bajra as an alternative to the regular wheat flour.

Can protect Against Certain Cancers The phytochemicals and polyphenols present in bajra have shown to be anti carcinogenic and antitumor in nature. Hence, bajra might prevent the growth of cancer cells and protect you against certain cancers.

Keeps the Heart Healthy Bajra is rich in omega-3 fats, magnesium, potassium, fiber, and other nutrients that help lower blood pressure. Magnesium, known for lowering blood pressure, reduces the risk of any heart attack or stroke. In addition, potassium is a good vasodilator that helps in maintaining optimum blood pressure.

Millets Are Rich in Phytochemicals The flavonoids tricin, luteolin, and acacetin are the most prevalent polyphenols found in bajra. These have been demonstrated to have anti-carcinogenic and antitumor properties, and they might even aid in slowing the spread of breast cancer cells. In addition to acting as antioxidants in our bodies, flavonoids also support a stronger immune system. Antioxidants are also thought to have a beneficial impact on aging and general health, and they guard against metabolic syndrome.

Rich in Omega-3 Fats When compared to other cereals, bajra is a better source of beneficial omega-3 fats.

Omega-3 fatty acids are recognized to be cardio protective and have been linked to lower blood pressure, triglycerides, slowed artery plaque growth, and regular heart rhythm.

Iron and Phosphorus-Rich Bajra has a modest supply of both iron and phosphorus. Iron is a crucial mineral for energy, memory, and cognitive functioning. If you have anemia (iron deficiency), you are aware that low iron levels can make you feel weak and weary because they prevent oxygen from reaching your tissues and cause carbon dioxide to build up. Foods high in iron are crucial for your health.

Contains Nutrients that May Support Healthy Hair, Skin, and Nails You may have heard that bajra is good for your hair, but the millet itself has not been studied as a hair treatment. However, bajra is a good source of many nutrients known for contributing to healthy hair, skin, and nails, including protein, vitamin B6, iron, folate, and zinc.

Use it conventionally as porridge, khichri, or roti. It is a fantastic winter cereal because of its excellent nutritional profile. Bajra can be included into many south Indian meals like uttapam or served as a salad. Even better, it can be included in the classic dosa or idly batter to boost the nutritional value of those dishes. To benefit from this wonderful traditional food, include it in one meal at least twice a week.

Nutritional Value of Pearl Millet (in Grams) per 100 G Energy (361), Carbohydrate (67), Protein (12), Fats (5), Minerals (2), Fiber (1), Calcium (42), Phosphorous (296), and Iron (8).

20.5 How to Cook Pearl Millet (Bajra)?

Bajra is a ubiquitous grain that acts as substitute to paddy, oats, quinoa, and other grain products in many cuisines. To prepare bajra, simply bring 1 cup (170 g) of millet and 2 cups (473 mL) of water and then allow it to cook for 15 min. This method should produce a light, fluffy grain.

To make Bajra as porridge, we can add up to one additional cup (237 mL) of water, dairy, or broth. We can also toast the dry millet for a few minutes before adding the liquid to bring out a rich, nutty flavor in the grain.

Prior to cooking, bajra may be soaked for hours or even days in water or a *Lactobacillus*-rich dairy like buttermilk or kefir. Fermenting millet and millet flour is common in Africa and Asia. It affects not only its flavor and taste but also likely its nutrient content. One study found that pearl millet flour that was fermented and frozen for 2 days had a 30% increase in levels of some phenolic compounds. Phenolic compounds are chemicals in plants that help your body respond to aging, inflammation, and chronic disease. While research on the topic is limited, some studies suggest that soaking or sprouting millet prior to consumption, as well as how the grain was initially processed, influences the accessibility of some of its nutrients, such as iron, zinc, calcium, and antioxidants. Bajra is commonly ground into fine flour that can be used to make roti and other types of flat bread. Yet, bajra flour is not limited to flatbreads alone. It can also be used to make cakes and pasta or as a replacement for other types of flour in many recipes.

Another way to enjoy bajra is puffed snack (same as popcorn). You can buy pre-puffed millet snacks or pop millet at home on your own. Puffed bajra can be eaten alone or used to make sweet or savory snack bars.

To pop millet, put one cup (170 g) of bajra to a dry frying pan. The millet should be heated on a medium temperature and let to sit. Once the mixture gets golden-brown, give it a quick stir before letting more grains puff up and explode.

20.5.1 Food Products from Bajra

Bajra rotis, bajra biscuits, khakras, nimkis, emolina, rawa, atta flour, bajra porridge, multigrain health drink, bajra idly mix, bajra cookies, khichdi, bajra dhebda, mini nachni, potato bajra pan cakes, and many more.

20.6 Pearl Millet Recipes

20.6.1 Millet Sushi Rolls

About 45–50 min are required for the preparation of vegan millet sushi with broccoli cream and roasted root vegetables.

Ingredients for roasted root vegetables:

- Oil—1/2 tablespoon
- Salt, 1/4 teaspoon
- 100 g slices of carrot, turnip, celery, and beet

Ingredients for broccoli cream:

- 50 g of peeled broccoli florets and sunflower seeds
- Cold water 180 mL
- Soya sauce 1 teaspoon

Millets:

- Uncooked millet, 250 g
- Water, 1.2 L
- Salt, 1/2 teaspoon
- Five nori sheets and tamari sauce are also required for serving.

How to Prepare:

- Bake your vegetables first. When the oven is preheated to 400 °F, vegetable strips should be put on a covered baking sheet with a thin layer of oil. After salting, the veggies should be roasted for 35 min, or until just beginning to brown. Let the baked veggies cool after removing the baking sheet from the oven.
- The millet should now be boiled in a mixture of water and salt. At medium heat, this mixture should boil for around 15 min. Set it aside once it softens and let it to cool.
- After 15 min of boiling, drain the water from the sunflower seeds and broccoli florets before rinsing them under cold water. Blend the cooked broccoli, sunflower seeds, cold water, vinegar, soy sauce, and oil in a food processor. After blending the mixture for a few minutes, it should be smooth.
- Collect your sushi-making tools, such as a rolling mat and nori sheets. Place a little amount of millet on a plate first, and then add some roasted vegetables and broccoli cream.
- Lightly wet the nori sheet's top with water before firmly rolling it into a sushi roll. After performing the same action for five times, slice the sushi into pieces using a sharp, moist knife. Taste the sushi made with millet rice and serve it with the remaining broccoli cream and tamari sauce.

20.6.2 Rajasthani Bajra Khichdi

Ingredients:

- Bajra
- Turmeric powder
- Moong daal
- Cumin

- Ghee
- Asafetida
- Ghee

How to prepare:
- Mix bajra, moong daal, and add water. Put all these in a pressure cooker.
- Allow it to be cooked. Then remove the lid.
- Take some ghee on pan. Once it gets hot, add cumin, asafetida, and turmeric powder.
- Once it starts crackling, add the cooked bajra and moong daal in it.
- Stir it and cook it for some time. The bajra khichdi will be ready.

20.6.3 Bajra Raab

This unique winter beverage is produced in Gujarat and Rajasthan. During the winter, bajra or pearl millet flour is frequently consumed since it is so healthy and restorative. Raab is a very thin beverage produced from millet flours that can be thickened by boiling it for a long time to resemble pearl millet porridge.

Ingredients (1 cup is equal to 240 mL; 1 tbsp is 15 mL; 1 tsp is 5 mL)

- Ghee—2 teaspoon
- Ajwain seeds, 1 teaspoon
- 4 tablespoons of pearl millet or bajra flour
- 1 tablespoon grated or powdered jaggery
- 1/2 tsp salt
- 1 teaspoon of dried ginger powder
- Two cups of water
- 1 tablespoon chopped nuts (optional)

How to prepare:
- Ghee is heated in a tiny saucepan.
- Add the ajwain seeds to the boiling ghee and let them sputter.
- Add the pearl millet flour (bajra flour) and toast it in the ghee for 2–3 min. You will begin to notice the flavor of bajra being roasted in ghee.
- Add water, salt, and ginger powder to the jaggery. Mix thoroughly to ensure that the jaggery is completely dissolved in the water and that there are no lumps of millet flour.
- It should be brought to a boil before cooking for an additional 5 min on low heat.
- The raab is complete. Add some chopped nuts (I used almonds and walnuts) on top after emptying into serving glasses. Serve hot.

20.6.4 Rajasthani Bajre Ki Khatti Raabdi

Ingredients:

- Bajra flour (1/2 cup)
- Buttermilk (3 cups)
- Cumin seeds (Jeera), for garnishing
- Salt to taste

How to prepare:

- Combine buttermilk and bajra flour in a bowl and add salt to make the Rajasthani Bajre Ki Khatti Raabdi Recipe. No lumps should remain in the liquid after mixing.
- Place a cover on it, and set it on side for a couple of hours. Usually, this mixture should be mixed in the mornings and cook it for lunch.
- Place it in a saucepan and cook it for 25–30 min on a low burner while stirring continuously, preferably with a wooden ladle. As it cooks, the mixture gets thickened.
- When the mixture becomes semi-liquid or pourable, stop cooking.
- Use raw cumin seeds as a garnish. After chilling, Raab/Raabdi will become even more thick.
- Serve Rajasthani Bajre Ki Khatti Raabdi Recipe with Bajre Ki Roti and Ker-Sangri Ka Saar for a wholesome weekday lunch. You can also serve it with Rajasthani Gatte Ki Sabzi, Phulka, and Masala Khichia.

References

Elangovan M (2021) Two decades of millets genetic resources management (2000-2020). ICAR-- Indian Institute of Millets Research (IIMR), Hyderabad, p 407

Jayashree B, Aram I (2020) Conservation of millets: the role of community leaders in Kolli Hills, South India

Kesavan P, Swaminathan M (2014) 2014 International year of family farming: a boost to evergreen revolution. Curr Sci 2014:1970–1974

Kumar NA, Nambi VA, Rani MG, King EIO, Chaudhury SS, Mishra S (2015) Community agro biodiversity conservation continuum: an integrated approach to achieve food and nutrition security. Curr Sci 2015:474–487

Loumerem M, Van Damme P, Reheul D, Behaeghe T (2008) Collection and evaluation of pearl millet (Pennisetum glaucum) germplasm from the arid regions of Tunisia. Genet Resour Crop Evol 55(7):1017–1028. https://doi.org/10.1007/s10722-008-9309-y

Manning K, Pelling R, Higham T, Schwenniger J-L, Fuller DQ (2011) 4500-years old domesticated pearl millet (Pennisetum glaucum) from the Tilemsi Valley, Mali: new insights into an alternative cereal domestication pathway. J Archaeol Sci 38(2):312–322. https://doi.org/10.1016/j.jas.2010.09.007

Manwaring HR, Bligh HFJ, Yadav R (2016) The challenges and opportunities associated with biofortification of pearl millet (Pennisetum glaucum) with elevated levels of grain iron and zinc. Front Plant Sci 7:e01944. https://doi.org/10.3389/fpls.2016.01944

Marmouzi I, Ali K, Harhar H, Gharby S, Sayah K, El Madani N, Cherrah Y, Faouzi MEA (2018) Functional composition, antibacterial and antioxidative properties of oil and phenolics from

Moroccan Pennisetum glaucum seeds. J Saudi Soc Agric Sci 17(3):229–234. https://doi.org/10.1016/j.jssas.2016.04.007

Pucher A, Høgh-Jensen H, Gondah J, Hash CT, Haussmann BIG (2014) Micronutrient density and stability in West African pearl millet-potential for biofortification. Crop Sci 54(4):1709–1720. https://doi.org/10.2135/cropsci2013.11.0744

Senthilvel S, Jayashree B, Mahalakshmi V, Kumar PS, Nakka S, Nepolean T, Hash C (2008) Development and mapping of simple sequence repeat markers for pearl millet from data mining of expressed sequence tags. BMC Plant Biol 8(1):119. https://doi.org/10.1186/1471-2229-8-119

Tako E, Reed SM, Budiman J, Hart JJ, Glahn RP (2015) Higher iron pearl millet (Pennisetum glaucum L.) provides more absorbable iron that is limited by increased polyphenolic content. Nutr J 14(1):11. https://doi.org/10.1186/1475-2891-14-11

Upadhyaya HD, Reddy KN, Irshad Ahmed M, Gowda CLL (2012) Identification of gaps in pearl millet germplasm from East and Southern Africa conserved at the ICRISAT genebank. Plant Genet Resour 10(3):202–213. https://doi.org/10.1017/S1479262112000275

Enhancing Shelf Life of Pearl Millet Flour 21

P. G. Padmaja, R. Venkateswarlu, S. P. Singh, and Vilas A. Tonapi

Abstract

Pearl millet grains are used for food purpose in Asia and Africa. The whole grains are consumed either as grits or in the form of flour and its products. These grains are rich in fat, vitamins, and minerals. However, the shelf life of milled flour is limited due to high fat content and active lipolytic enzymes that come in contact with their substrates during milling. Moreover, moisture and oxygen availability further accelerate oxidative rancidity, rendering flour unacceptable for consumption after 10–15 days. Various processing techniques are applied either to physically separate different fractions of the grain or to inactivate the lipolytic enzymes in order to enhance shelf life. Decortication, pearling, and defatting are used to remove lipid-rich aleurone layer and germ from grains, while extrusion, fermentation, and malting are used to reduce lipid content and lipolytic enzyme activity. Thermal treatment is a common method used to inactivate lipolytic enzymes before milling. Apart from conventional techniques like dry heat and hydrothermal treatment, novel thermal processing techniques like microwave and infrared heating were also experimented with. Non-thermal processing methods like gamma irradiation and high-pressure processing are effective for controlling insect infestation and microbial load, respectively. Processed flour stored in laminated packaging and the modified atmosphere was found to be effective in extending shelf life. Efforts to develop low rancid genotypes have focused mainly on screening germplasm, inbred lines, varieties, and hybrids. Processing technologies in combination with targeted breeding through marker-assisted

P. G. Padmaja (✉) · R. Venkateswarlu · V. A. Tonapi
ICAR - Indian Institute of Millets Research, Hyderabad, India
e-mail: padmaja@millets.res.in

S. P. Singh
ICAR - Indian Agricultural Research Institute, New Delhi, India

V. A. Tonapi et al. (eds.), *Pearl Millet in the 21st Century*,
https://doi.org/10.1007/978-981-99-5890-0_21

selection and genome editing are discussed in this chapter, which might help in developing low rancid pearl millet varieties with better keeping quality of flour.

Keywords

Pearl millet · Shelf life · Rancidity · Processing technologies · Packaging

Abbreviations

%	Percentage
°C	Degree Celsius
μ	Micron
γ	Gamma
BHT	Butylated hydroxytoluene
cm	Centimeter
CMS	Cytoplasmic male sterile
DNA	Deoxyribonucleic acid
FFA	Free fatty acid
h	Hour
ha	Hectare
HCl	Hydrochloric acid
HDPE	High-density polyethylene
HPP	High-pressure processing
kg	Kilogram
kGy	Kilo Gray
LDPE	Low-density polyethylene
mEq	Milli equivalent
min	Minutes
MPa	Mega Pascal
O_2	Oxygen
pH	Hydrogen ion concentration
PV	Peroxide value
RNA	Ribonucleic acid
rpm	Rotations per minute
s	Seconds
W	Watt

21.1 Introduction

Pearl millet (*Pennisetum glaucum* R. Br.) is cultivated in arid and semiarid regions of Africa and Asia. More than 90 million people are dependent on pearl millet for livelihood and nutritional security (Serba et al. 2020). In India, it occupies 6.93

million ha with an average production of 8.61 million tons and productivity of 1243 kg/ha (Directorate of Millets Development 2020). Pearl millet is consumed as whole grain and is rich in dietary fiber, vitamins, and minerals (Duodu and Dowell 2019; Kumar et al. 2018; Gopalan et al. 2007; Abdulrahman and Omoniyi 2016; Chethan and Malleshi 2007). Flavonoids available in small quantities impart antioxidant activity, which is considered important for immune function in humans (Izadi et al. 2012). Pearl millet is gluten free and its glycemic index is relatively lower than other cereals. This property is helpful for diabetes and celiac disease patients (El Khoury et al. 2018).

Despite its advantages as a staple food crop, processing and short shelf life of flour are major challenges in pearl millet (Rani et al. 2018). Improper post-harvest conditions lead to deterioration of the grain quality (Rowley 1984; Simone et al. 1994). Insect pests damage stored grains and reduce the grain's weight, which further lead to poor seed germination (Malek and Parveen 1989; Santos et al. 1990). The relative proportion of germ is high compared to other cereals and millets. Moreover, separating the germ from the endosperm is difficult due to the small size of the grain. Pearl millet is rich in lipids compared to other cereals. It has a naked caryopsis and the inner structure comprises pericarp, germ, and endosperm. The germ and the bran layers are rich in triglycerides containing unsaturated fatty acids (Fig. 21.1). Grinding of grains into flour exposes lipids to lipolytic enzymes and atmospheric oxygen. Lipases release free fatty acids (FFA) from triglycerides, which increases the flour's acidity and also imparts rancid off-flavor. These unsaturated free fatty acids are easy targets for lipoxygenase and peroxidase compared to esterified fatty acids in triglycerides (Manley and Mayer 2012). These enzymes degrade unsaturated fatty acids to peroxides which initiate a chain reaction of peroxidation (Zhang and Hamaker 2005). Concomitant availability of oxygen accelerates oxidative changes producing volatile compounds which impart rancidity and also affect the quality of processed products leading to reduced consumer

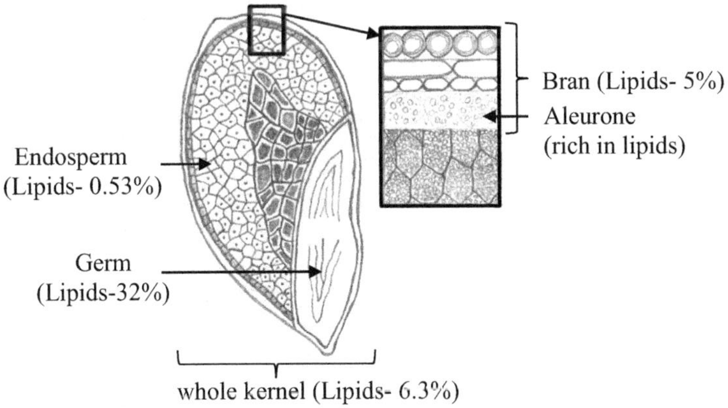

Fig. 21.1 Lipid content distribution in different portions of pearl millet grain (adapted from Abdelrahman et al. 1984; and McDonough and Rooney 1989)

acceptance. The changes also affect the pasting properties of flour (Zhang and Hamaker 2005).

21.2 Shelf Life of Pearl Millet Flour

Grains must be dehulled (pearling) or milled to flour to enable for food purposes. Milling disturbs the compartmental structure of tissues and cells, and brings enzymes into contact with their substrates. Also, milled flour becomes an easy target for attack by microbes and oxygen in the presence of moisture. Under ambient conditions, whole grain flour rich in lipid content becomes unsuitable for human consumption at 10–15 days of storage (Tiwari et al. 2014; Yadav et al. 2012a, b). The shelf life of flour is also greatly influenced by moisture content and storage conditions. Pearl millet flour contains 6–13% moisture which further increases during storage (Abioye et al. 2018; Akinola et al. 2017; Gull et al. 2016). Moisture content and temperature provide suitable conditions for altering the acidity of the products (Goyal et al. 2017). It elevates lipolytic and proteolytic activities which lead to FFA (Butt et al. 2004) and peroxide production due to hydrolytic rancidity (Varsha and Narayanan 2017). These changes result in unacceptable sensory properties (Fig. 21.2). Moisture content is also directly related to microbial growth and the production of toxins leading to food poisoning (Mustapha et al. 2014).

Flour quality depends on different processing techniques, storage conditions, packaging material, and cultivars.

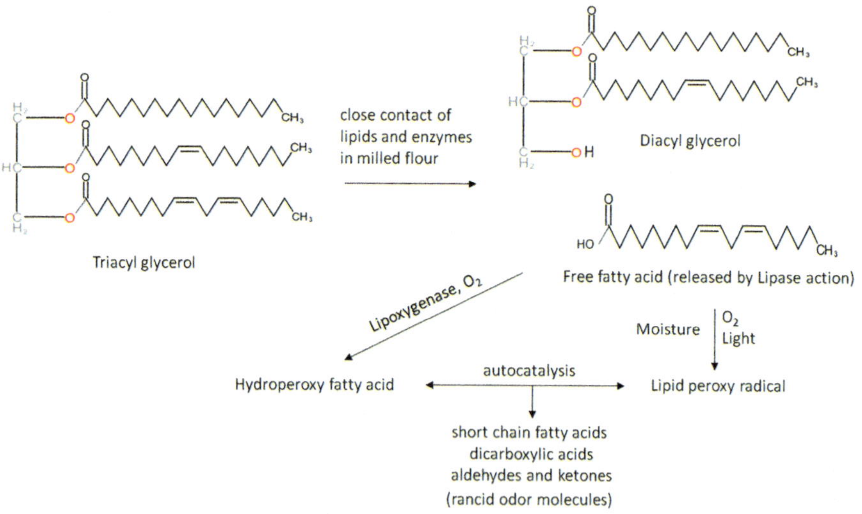

Fig. 21.2 Schematic of enzymatic and oxidative rancidity development in flour

21.3 Processing Techniques for Shelf-Life Extension of Pearl Millet Flour

The shelf life of flour can be extended by treating the grain or flour using various processing techniques (Tables 21.1 and 21.2). These techniques aim to separate the lipid-rich bran and germ, to inactivate the lipolytic enzymes and to control bacterial and fungal growth.

Various processing techniques applied to improve shelf life of pearl millet flour include mechanical processing, fermentation, malting, addition of preservatives, thermal treatments (dry heat, microwave processing, infrared heating), and gamma irradiation.

Table 21.1 Processing techniques to improve shelf life in pearl millet flour

Processing techniques	References
Milling	Carnovale and Quaglia (1973)
Removal of lipid-bearing germ	Abdelrahman et al. (1983)
Thermal treatment	Kapoor and Kapoor (1990)
Dry heat treatment	Kachare and Chavan (1992)
Hot water blanching	Chavan and Kachare (1994)
Acid-soaking	Aggarwal (1992), Chavan and Kachare (1994)
Malting	Rekha (1997)

Table 21.2 Effect of processing techniques on the storage stability of pearl millet

Processing	Effect on storage	Reference
Decortication	Loss in enzymes, fat, phytate, and phenolics	Varriano-Marston and Hoseney (1983), Abdelrahman et al. (1983), Jain and Bal (1997), Nantanga (2006)
Decortication	Decrease in lipid content	Hama et al. (2011)
Decortication	Decrease in FFA	Tiwari et al. (2014)
Decortication	Inhibition of the lipase activity	Ofosu et al. (2020)
Pearling	More shelf life	Prabhat (2015)
Defatting of the flour	No change in fat acidity and peroxide values (PV)	Kapoor and Kapoor (1990)
Fermentation	Decrease in lipid content	Chinenye et al. (2017)
Germination/ malting	Reduction in crude fat	Suma and Urooj (2014)
Germination/ malting	FFA below 1% up to 8th day of storage	Tiwari et al. (2014)
Germination/ malting	Improved digestibility Functional properties of flour are changed	Akinola et al. (2017)
Germination/ malting	66.3% reduction of FFA	Baranwal (2017)

21.3.1 Mechanical Operations

21.3.1.1 Decortication

The physical separation of pericarp and germ before storage and consumption improves sensory and cooking quality (Liu et al. 2012). Abrasive decortication causes a substantial reduction in dietary fiber, vitamins, minerals, total phenolics, and antioxidant capacity (Hama et al. 2011). This is due to the unequal distribution of nutrients in different portions of the grain. Most of the starch is concentrated in the endosperm, while pericarp and the germ contain dietary fiber, lipids, minerals, and phytic acid. Reduction in phytic acid, polyphenols, and tannins significantly improved the protein digestibility (ElShazali et al. 2011; Chandrasekara et al. 2012; Krishnan et al. 2012). Decortication reduces crude lipid content in the pericarp available for oxidation, thereby controlling the development of rancidity and contributing to better storage properties (Babiker et al. 2018).

21.3.1.2 Milling and Sieving

Bran fraction contains pericarp, seed coat, and aleurone layer. During milling, the bran is removed. The pericarp and germ layers of millets are rich in lipid content and lipolytic enzymes. Due to the larger germ size pearl millet contains higher amounts of lipids rich in unsaturated fatty acids (Sandhu and Punia 2018). Dry milling or pearling improves the shelf life of flour by removing the pericarp from the grain. Pearled flour had improved color with better overall acceptability. Flour ground at 75 rpm and packed in low-density polyethylene (LDPE) bags was superior in quality to the flour ground at 115 rpm and 400 rpm (Sidhu et al. 2016).

21.3.1.3 Fermentation

Fermentation brings about many biochemical changes that lead to better storage properties (Singh et al. 2015). Better shelf life was achieved when fermented with lactic acid bacteria (LAB) and yeast compared to unfermented products (Belz et al. 2019; Dallagnol et al. 2015; Rinaldi et al. 2017). Increasing titrable acidity and reduction in pH of the flour were observed with increase in fermentation time (Onyango et al. 2013). This makes it unsuitable for spoilage microorganisms to survive, thereby prolonging the shelf life. The microbial lipolytic enzymes break-down fatty acids and glycerol during fermentation, thereby reducing the fat content in blended flours of pearl millet and soybean (Ojokoh and Bello 2014). Fermentation of millet foods with probiotic strains improves the products color, flavor, and acceptability of the product.

21.3.1.4 Malting

Malting is controlled germination of grains to enhance enzyme activity (Amadou et al. 2011). Germinated grains are dried and roasted before milling into flour (malt). Malting enhances the nutritional and functional properties of the flour. Germination of pearl millet in moistened jute sack decreased flour pH and subsequent increase in titrable acidity (Owheruo et al. 2019).

21.3.1.5 Preservatives

Substances are added to food products to arrest microbial growth and other chemical changes responsible for spoilage (Carocho et al. 2014).

Antioxidants

The addition of butylated hydroxytoluene (BHT) to the pearl millet flour and storage for 90 days at room temperature increased the FFA value from 0.4 to 1.5% oleic acid, compared to the control value of 3.2% (Abdalgader et al. 2019). It also reduced the peroxide and fat acidity values. Natural antioxidants like ascorbic acid inhibited rancidity development by reacting with peroxide intermediates (Carocho et al. 2014).

Essential Oils

Essential oils from aromatic plants, *viz.* conehead thyme, rosemary, lemongrass, citronella, mugwort, cinnamon, and basil, and their formulations are used as preservatives (Dwivedy et al. 2016). Essential oil from *Gaultheria fragrantissima* Wall hindered aflatoxin B_1 production by *Aspergillus flavus* during storage of pearl millet. This treatment also protected millet grains from storage molds (Kumar et al. 2019).

21.3.1.6 Acid Treatment

Acid treatment denatures the enzymes by reducing the pH of the solution. Pearl millet grains treated with 0.2N HCl reduced the FFA of the flour to 73.9% at 25–30 °C after 24 h of treatment (Bhati et al. 2016).

21.3.2 Thermal Treatments

Heat increases the kinetic energy of the molecules and disrupts the non-covalent bonds in proteins. This damages the protein's secondary and tertiary structure producing randomly shaped molecules without biological activity. Enzymes being proteins with a defined three-dimensional structure lose their catalytic property to affect biochemical reactions. Heat treatment inactivates lipase enzyme at high temperatures, thereby decreasing the FFA content in grain flours (Gili et al. 2018; Mazaheri et al. 2019; Ruge et al. 2012; Xu et al. 2016). However, the level of heat treatment should be optimized to preserve the organoleptic properties of the flour for consumer acceptance. Different methods of heat treatment are dry heat, hydrothermal, microwave, and infrared treatments (Table 21.3).

21.3.2.1 Dry Heat Treatment

Dry heat treatment is the most commonly adapted processing technique wherein heat is applied directly to the material or through the air. Effective control of lipid degradation during storage was achieved by dry heat treatment of grains for 120 min (Kadlag et al. 1995). The rate of increase in lipase activity and FFA was minimal in heat-treated grains compared to the fourfold increase in control samples

Table 21.3 Effect of thermal treatments on storage stability of pearl millet

Treatment	Effect on storage	References
Dry heat treatment	Enzymes are inactivated at 100 °C for 2 h	Pruthi (1981), Kapoor and Kapoor (1990), Patel and Parameswaran (1992), Chavan and Kachare (1994), Kadlag et al. (1995), Arora et al. (2002)
Dry heat treatment	Treatment at 100 °C for 120 min reduced FFA by 63.58%	Bhati et al. (2016)
Blanching	Arresting lipase, lipoxygenase, peroxidase and polyphenol oxidase	Rani et al. (2018)
Hydrothermal treatment	Lipase activity decreased	Yadav et al. (2012a, b)
Hydrothermal treatment	11.5% reduction in FFA when treated with boiling water for 15 min and drying at 60 °C for 2 h	Jalgaonkar et al. (2016)
Hydrothermal treatment	Blanching for 90 s decreased FFA by 53.8%	Bhati et al. (2016)
Hydrothermal treatment	23.7% reduction in total plate count when cooked at 90–95 °C for 120 min	Akinola et al. (2017)
Hydrothermal treatment	Inactivation of lipase	Mohajerkhorasani et al. (2019)
Microwave	900 W treatment for 100 s reduced lipase activity by 92.9%	Yadav et al. (2012a, b)
Microwave	Inactivation of lipase	Mohajerkhorasani et al. (2019)
Ohmic heating	Decrease in water solubility and absorption indices	Dias-Martins et al. (2019)

of pearl millet flour. In addition, the flour from heat treated pearl millet grains and chapati were organoleptically acceptable (Arora et al. 2002).

21.3.2.2 Hydrothermal Processing

Hydrothermal processing or parboiling hardens the grains and improves milled grain recovery. Apart from inactivation of lipolytic enzymes and microbes, it also alters the physio-chemical properties of the flour. Although hydrothermal treatment extended the shelf-life of whole pearl millet flour (Nantanga et al. 2008), alterations in the appearance and physio-chemical properties of the products might reduce consumer acceptance.

21.3.2.3 Microwave Processing

Microwaves are non-ionizing radiations that can heat molecules having dipole moment (Sruthi et al. 2021). Food products with water produce heat in a microwave field. Indirectly this heating inactivates enzymes and, breaks down ribosomal RNA by rotation and alignment of the molecules (Banik et al. 2003). Microwave treatment reduces lipase activity and extends the flour's shelf life. However, the level of inactivation depends on temperature, moisture content, and duration of microwave treatment.

21.3.2.4 Infrared Heating

Both the surface of the grain and the inner layers are heated in infrared treatment. It is faster compared to other heating techniques (Swaminathan et al. 2015). It effectively inactivates lipase, and reduces the FFA in the sample (Li et al. 2016), thereby prolonging the storage-life and acceptability of food products. This treatment improved the physio-chemical properties with minimal alterations in the nutrient content (Barden and Decker 2016).

21.3.3 Non-thermal Processing Techniques

Non-thermal processing techniques are aimed at the denaturation of macromolecules like DNA and proteins. These techniques will not increase the temperature of the product compared to thermal techniques. Hence, heat labile micronutrients like vitamins and flavor compounds are preserved. Non-thermal techniques used for the treatment of pearl millet are

- Gamma irradiation
- High-pressure processing (HPP)

21.3.3.1 Gamma Irradiation

Gamma rays are an ionizing part of the electromagnetic radiation which ionizes molecules by knocking out electrons from the outer shells of atoms. It is used to increase food product's shelf life (Table 21.4) and safety (Cassares et al. 2020). This technology is applied for the disinfestation of food grains and packaged cereal products like *atta* (flour), *suji* (semolina), and so on. Heat treatment in combination with gamma radiation reduced microbial load significantly than the individual treatments (Kumar et al. 2018).

21.3.3.2 High-Pressure Processing (HPP)

High-pressure processing or high hydrostatic pressure processing applies very high pressure (200–800 MPa) to food products to inactivate microorganisms and enzymes

Table 21.4 Effect of non- thermal processing techniques on storage of pearl millet

Treatment	Effect on storage	Reference
Gamma irradiation	Reduced fungal counts and FFA significantly Decreased tannin and phytic acid	Mahmoud et al. (2016)
Heat treatment at 150–170 °C for 1.5 min at 300 rpm followed by γ-irradiation	Fungal count reduction by 90.56% when heat treatment combined with irradiation at 2.5 kGy Shelf life up to 60 days	Dikkala et al. (2018)

(Kaushik et al. 2016). High pressure selectively denatures macromolecules without affecting small molecules like vitamins. Moreover, the heating of the product is minimal during treatment. High-pressure soaking removes outer bran layers of grain (hydrophobic lipid layer), thereby increasing the keeping quality of flour (Zhu et al. 2016). HPP application can prevent flour spoilage by reducing microbial load without affecting nutritional qualities and functional properties.

21.4 Shelf Life of Value-Added Products

The shelf life of value-added products depends on the ingredients and processes used in the preparation. Wide range of products, viz. composite flours, instant mixes, flakes, puffs, semolina, bakery products (cakes, cookies, biscuits), extruded products, that are ready to eat (hot extrusion) and ready to cook (cold extrusion) such as vermicelli, pasta, macaroni, noodles, spaghetti, and so on can be prepared using pearl millet flour. These value-added products can help to increase the market for millet and millet- based products (Deshpande and Nishad 2021). Value-added products have different shelf life depending on the processing technology used (Table 21.5).

Table 21.5 Shelf life of value-added products of pearl millet

Products	Treatment	Shelf life	Reference
Bread	Baking	2–4 days	Nami et al. (2019)
Extruded products	Extrusion cooking	6 months shelf life	Sumathi et al. (2007)
Upma dry mix	Hydrothermal treatments (soaking up to moisture 30 ± 2%, steaming 1.05 kg cm^{-2}, 20 min)	No lipase activity after steaming 6 months stable at ambient conditions (20–35 °C) in poly ethylene pouches (75 μ)	Balasubramanian et al. (2014)
Pearl millet-based kheer mix	Nitrogen flushing	396 days at 8 °C and 288 days at 25 °C	Bunkar et al. (2014)
Pearl millet -based halwa dry mix	Soaking, steaming, pearling, pulverizing	6 months stable	Yadav et al. (2011)

21.5 Storage Periods/Structures

Pearl millet grains require safe storage before processing into flour. Grains stored in rumbu, clay pots, or polypropylene bags were less damaged compared to underground pit storage. Grains stored for a longer duration incurred more damage by insects than those stored for a relatively short period (Lale and Yusuf 2000). Grains with lesser grain borer and flour beetle infestations were higher when stored in jute bags and tin containers than in polythene bags. Insect population significantly increased after 3, 6, and 12 months of storage. The grains stored in cold storage were free from insect infestation in all the storage materials tested. Qualitative and quantitative changes were minimum when stored in polythene bags, while maximum deterioration occurred in jute bags (Mali and Satyavir 2005).

21.6 Packaging Material

Packaging material protects the products from external contamination and allows for convenient handling. Moreover, it restricts the ingress of moisture, light, and oxygen into the product. Whole grain flour of pearl millet stored in gunny bags, earthen pots, tin cans, and polythene bags became rancid on 6, 7, 8, and 10 days and inedible on 11, 12, 13, and 14 days, respectively (Chaudhary and Kapoor 1984). Flour stored in polythene bags had a better shelf life. As reported in cereals and pulses, the moisture content of the flour stored in the above materials was highly variable (Patel and Parameswaran 1992). The peroxide value of the flour stored in polythene bags increased up to 5.1 mEq peroxide per kg of fat after 62 h of storage and then decreased, whereas the value remained constant when stored in cotton bags (Kased et al. 1984). Tin containers and high-density polyethylene (HDPE) bags are suitable for storing of flour, while cotton bags are not suitable (Bhatt et al. 2017). Also, flour packed in HDPE bags and stored at refrigerated temperature had slow development of rancidity compared to storage at ambient temperature (Varsha and Narayanan 2017). Storage of flour in a sealed HDPE package prevents the generation of free radicals and the proliferation of microbes, thereby prolonging the shelf life (Ogori et al. 2013). Development of rancidity can be delayed in pearl millet flour when packed in HDPE pouches and stored at refrigeration temperature (Varsha and Narayanan 2017). The microbial quality of low temperature extruded pearl millet snacks is better in aluminum laminated pouches than compared to HDPE by effectively restricting the influence of moisture, light, and air. The total storage life of the extruded snack was extended when stored in aluminum laminate (Adesina 2021).

Packaging in polythene bags is suitable for modification of the atmosphere surrounding the product before sealing. Application of vacuum and flushing with an inert gas like nitrogen are the most common methods used. Pearl millet-based weaning food had consumer acceptance for up to 6 months when stored under vacuum, while it was limited to 4 months when stored under atmospheric air.

Weaning food under vacuum packaging had lower SPC counts at 6 months of storage (Marvesh Kumar et al. 2015).

21.7 Genetics and Breeding Approaches for Addressing Rancidity Problems in Pearl Millet

Post-harvest processing techniques available for extending the shelf life of pearl millet flour are expensive, localized, and alters the sensory properties. An alternative strategy to produce grains with a better shelf life is through conventional plant breeding, genomics, and genome editing to tackle the rancidity of pearl millet flour. Among them, genetics and breeding approaches are the most feasible, economical, and sustainable. Although different organizations are working in this direction, the substantial improvement could not be achieved as most of the studies were confined to only screening of genotypes.

Evaluation of 56 genotypes, including forty hybrids, four open-pollinated varieties, and twelve parental lines of different hybrids for two rancidity parameters namely acid value (enzymatic rancidity) and peroxide value (oxidative rancidity), has shown genetic variability for the development of rancidity in pearl millet flour. Thirteen genotypes, namely 95222-B-48, PM-15-NHB-1717, GHB 719, 9444, 89111-B-46, Super Boss, PM-19-Nandi-66, RHB-219, PM-17-Bio-13, PM-18-12KM-80, PM-16-JKBH-1294, GHB-538, and Raj-171 were found to have low rancidity (Datta Mazumdar et al. 2016). Similarly, genetic variation for fat, total phenol content, development of fat acidity, and activities of peroxidase, lipoxygenase, and polyphenol oxidase was observed among a set of pearl millet hybrids, Cytoplasmic Male Sterile (CMS) lines, inbred lines, and white composites. Among the hybrids, HHB 197 and among inbreds, HBL 11 was identified to have low rancidity (Goyal and Chugh 2017). A rancidity matrix was developed using 93 diverse genotypes of pearl millet to identify low, medium, and high rancid groups (Goswami et al. 2020). Out of these, 22 genotypes, *viz.* Gadhwal ki Dhani-1, Kaveri Super Boss, HHB 311, Chadi Bajri, Chanana Bajra-2, Dhodhsar local, Damodara Bajri, Pusa Composite 443, Gadhwal kin Dhani-3, GHB 732, Dedha Bajri, RHB 121, HHB 67 (Improved), Super Boss, 86M82, AHB 1200, KBH 108, GHB 558, RHB 223, HHB 146, Pusa purple 1, and RHB 177, were found to have low rancidity. Using multivariate techniques, HBL0828-01 was identified as a low rancid genotype (Bunty et al. 2020).

Recently, polymorphisms were observed in two lipase genes of contrasting inbreds which were directly involved in the degradation of triacylglycerols into free fatty acids. This polymorphism can be utilized in marker-assisted breeding to develop hybrids with better flour shelf life (Aher et al. 2022). Targeted editing of lipolytic enzymes using novel genome editing technologies might help in development of low rancid pearl millet genotypes.

21.8 Conclusion

Primary processing techniques control the lipolytic enzymes by either physical separation of grain fractions or by denaturation of enzymes, thereby prolonging the shelf life of millets. However, it is associated with the loss of micronutrients in the grain. Nutritional value of millet grains can be improved through fermentation and germination. Thermal treatments, both dry heat and hydrothermal, inactivate lipase. The majority of the existing research is focused on common traditionally practiced methods. Advanced research is needed to enhance shelf life of flour and value-added products with acceptable nutritional and sensory properties.

Limited efforts have been made to improve the inherent quality of the grain for low rancidity. Plant breeding and genomic approaches for shelf-life improvement in pearl millet are still in the infancy stage. Some efforts have been made to screen pearl millet hybrids, varieties, parental lines (inbreds), and germplasms for different rancidity indicators and low rancid genotypes were identified. A large number of germplasms needs to be screened for the identification of low and high rancid pearl millet lines. Then genetics of rancidity has to be worked out. This will open avenues for using genomics assisted breeding to unravel the rancidity problem. High lipase activity along with high fat content and high unsaturated fatty acids are the factors expected to be associated with rancidity in pearl millet flour. Genes responsible for high lipase activity and fat content can be modified using genome editing tools to minimize rancidity in flour. Targeted breeding strategies needs to be followed to develop hybrids/varieties with low rancidity. A holistic approach incorporating the above methodologies might help in development of pearl millet genotypes with enhanced shelf life.

References

Abdalgader MAA, Ashraf SA, Awadelkareem AM (2019) Effect of natural and synthetic antioxidant on shelf-life of different sudanese *pennisetum glaucum* L. Flours. Biosci Biotechnol Res Commun 12(3):652–657

Abdelrahman A, Hoseney RC, Varriano-Marston E (1983) Milling process to produce low fat grits from pearl millet. Cereal Chem 60:189–191

Abdelrahman A, Hoseney RC, Varriano-Marston E (1984) The proportions and chemical compositions of hand-dissected anatomical parts of pearl millet. J Cereal Sci 2:127–133

Abdulrahman WF, Omoniyi AO (2016) Proximate analysis and mineral compositions of different cereals. J Adv Food Sci Technol 3(2):50–55

Abioye VF, Ogunlakin GO, Taiwo G (2018) Effect of germination on anti-oxidant activity, total phenols, flavonoids and anti-nutritional content of finger millet flour. J Food Process Technol 9(2):719

Adesina A (2021) Effect of packaging material and storage conditions on the microbial quality of pearl millet extruded snack. J Food Process Technol 12:857

Aggarwal AJ (1992) Processing of pearl millet for its more effective utilisation. Ph.D. Thesis. CCSHAU. Hisar, India

Aher RR, Reddy PS, Bhunia RK, Flyckt KS et al (2022) Loss of function mutations in novel triacylglycerol lipase genes are associated with low rancidity in pearl millet flour. BioRxiv. https://doi.org/10.1101/2022.04.02.486827

Akinola SA, Badejo AA, Osundahunsi OF, Edema MO (2017) Effect of preprocessing techniques on pearl millet flour and changes in technological properties. Int J Food Sci Technol 52(4): 992–999

Amadou I, Gbadamosi OS, Le GW (2011) Millet-based traditional processed foods and beverages-a review. Cereal Foods World 56(3):115–121

Arora P, Sehgal S, Kawatra A (2002) The role of dry heat treatment in improving the shelf life of pearl millet flour. Nutr Health 16:331–336

Babiker E, Abdelseed B, Hassan H, Adiamo O (2018) Effect of decortication methods on the chemical composition, antinutrients, Ca, P and Fe contents of two pearl millet cultivars during storage. World J Sci Technol Sustainable Dev 15(3):278–286

Balasubramanian S, Yadav DN, Kaur J, Anand T (2014) Development and shelf-life evaluation of pearl millet based upma dry mix. J Food Sci Technol 51(6):1110–1117

Banik S, Bandyopadhyay S, Ganguly S (2003) Bioeffects of microwave - a brief review. Bioresour Technol 87(2):155–159

Baranwal D (2017) Malting: an indigenous technology used for improving the nutritional quality of grains: a review. Asian J Dairy Food Res 36:179–183

Barden L, Decker EA (2016) Lipid oxidation in low-moisture food: a review. Crit Rev Food Sci Nutr 56(15):2467–2482

Belz MCE, Axel C, Arendt EK, Lynch KM, Brosnan B, Sheehan EM et al (2019) Improvement of taste and shelf-life of yeasted low-salt bread containing functional sourdoughs using Lactobacillus amylovorus DSM 19280 and Weisellacibaria MG1. Int J Food Microbiol 302:69–79

Bhati D, Bhatnagar V, Acharya V (2016) Effect of pre-milling processing techniques on pearl millet grains with special reference to iron availability. Asian J Dairy Food Res 35(1):76–80

Bhatt VM, Dabhi MN, Rathod PJ (2017) Changes in the moisture content, free FFA and malondialdehyde of bajra flour during storage. Adv Food Sci Eng 1(2):68–73

Bunkar DS, Jha A, Mahajan A, Unnikrishnan VS (2014) Kinetics of changes in shelf life parameters during storage of pearl millet based kheer mix and development of a shelf life prediction model. J Food Sci Technol 51(12):3740–3748

Bunty S, Chugh L, Singh VK, Shekhar C, Tanwar N (2020) Characterization of rancidity indicators in selected pearl millet genotypes by multivariate analysis. Plant Arch 20:229–235

Butt MS, Nasir M, Akhtar S, Sharif MK (2004) Effect of moisture and packaging on the shelf life of wheat flour. Internet J Food Saf 4:1–6

Carnovale E, Quaglia GB (1973) Influence of temperature and humidity controlled preservation on the chemical composition of milling products from millet. Ann Technol Agric 22:371

Carocho M, Barreiro MF, Morales P, Ferreira ICFR (2014) Adding molecules to food, pros and cons: a review on synthetic and natural food additives. Compr Rev Food Sci Food Saf 34(4):377–399

Cassares M, Sakotani NL, Kunigk L, Vasquez P, Jurkiewicz C (2020) Effect of gamma irradiation on shelf life extension of fresh pasta. Radiat Phys Chem 174:108940

Chandrasekara A, Naczk M, Shahidi F (2012) Effect of processing on the antioxidant activity of millet grains. Food Chem 133:1–9

Chaudhary P, Kapoor AC (1984) Changes in the nutritional value of pearl millet flour during storage. J Sci Food Agric 35(11):1219–1224

Chavan JK, Kachare DP (1994) Effects of seed treatment on lipolytic deterioration of pearl millet flour during storage. J Food Sci Technol 31:80–81

Chethan S, Malleshi NG (2007) Finger millet polyphenols: optimization of extraction and the effect of pH on their stability. Food Chem 105(2):862–870

Chinenye OE, Ayodeji OA, Baba AJ (2017) Effect of fermentation (natural and starter) on the physicochemical, anti-nutritional and proximate composition of pearl millet used for flour production. Am J Biosci Bioeng 5(1):12–16

Dallagnol AM, Pescuma M, Rollán G, Torino MI, de Valdez GF (2015) Optimization of lactic ferment with quinoa flour as bio-preservative alternative for packed bread. Appl Microbiol Biotechnol 99(9):3839–3849

Datta Mazumdar S, Gupta SK, Banerjee R, Gite S, Durgalla P, Bagade P (2016) Determination of variability in rancidity profile of select commercial Pearl millet varieties/hybrids. DC 24. Poster presented in CGIAR Research Program on Dryland Cereals Review Meeting held at Hyderabad, India, 5-6 October 2016. International Crops Research Institute for the Semi-Arid Tropics, Patancheru, Telengana, India. http://drylandcereals.cgiar.org/index.php/determination-ofvariability-in-rancidity-profile-of-select-commercial-pearl-milletvarietieshybrids/

Deshpande SD, Nishad PK (2021) Technology for millet value-added products. In: Kumar A et al (eds) Millets and millet technology. Springer, Cham. https://doi.org/10.1007/978-981-16-0676-2_14

Dias-Martins AM, Cappato LP, daCostaMattos M, Rodrigues FN, Pacheco S, Carvalho CW (2019) Impacts of ohmic heating on decorticated and whole pearl millet grains compared to open-pan cooking. J Cereal Sci 85:120–129

Dikkala PK, Hymavathi TV, Roberts P, Sujatha M (2018) Effect of heat treatment and gamma irradiation on the total bacterial count of selected millet grains (Jowar, Bajra and Foxtail). Int J Curr Microbiol App Sci 7(2):1293–1300

Directorate of Millets Development (2020) Available online http://millets.dacfw.nic.in

Duodu KG, Dowell FE (2019) Sorghum and millets: quality management systems. Sorghum and millets. AACC International Press, Washington, pp 421–442

Dwivedy AK, Kedia A, Kumar M, Dubey NK (2016) Essential oils of traditionally used aromatic plants as green shelf-life enhancers for herbal raw materials from microbial contamination and oxidative deterioration. Curr Sci 110(2):143–145

El Khoury D, Balfour-Ducharme S, Joye IJ (2018) A review on the gluten-free diet: technological and nutritional challenges. Nutrients 10(10):1410

ElShazali AM, Nahid AA, Salma HA, Elfadil EB (2011) Effect of radiation process on antinutrients, protein digestibility and sensory quality of pearl millet flour during processing and storage. Int Food Res J 18(4):1401–1407

Gili RD, Penci MC, Torrez Irigoyen MR, Giner SA, Ribotta PD (2018) Effect of wheat germ heat treatment by fluidised bed on the kinetics of lipase inactivation. Food Bioprocess Technol 11(5): 1002–1011

Gopalan C, Sastri BVR, Balasubramanian SC, Rao BSN, Deosthale YG, Pant KC (2007) Nutritive value of Indian foods. National Institute of Nutrition, Indian Council of Medical Research, Hyderabad

Goswami S, Asrani P, Ansheef Ali TP et al (2020) Rancidity matrix: development of biochemical indicators for analysing the keeping quality of pearl millet four. Food Anal Methods 13:2147–2164. https://doi.org/10.1007/s12161-020-01831-2

Goyal P, Chugh LK (2017) Shelf life determinants and enzyme activities of pearl millet: a comparison of changes in stored flour of hybrids, CMS lines, inbreds and composites. J Food Sci Technol 54(10):3161–3169

Goyal P, Chugh LK, Berwal MK (2017) Storage effects on flour quality of commonly consumed cereals. J Appl Nat Sci 9(1):551–555

Gull A, Prasad K, Kumar P (2016) Evaluation of functional, antinutritional, pasting and microstructural properties of Millet flours. J Food Measure Charact 10(1):96–102

Hama F, Icard-Vernière C, Guyot JP, Picq C, Diawara B, Mouquet-Rivier C (2011) Changes in micro- and macronutrient composition of pearl millet and white sorghum during in field versus laboratory decortications. J Cereal Sci 54(3):425–433

Izadi Z, Nasirpour A, Izadi M, Izadi T (2012) Reducing blood cholesterol by a healthy diet. Int Food Res J 19(1):29–37

Jain RK, Bal S (1997) Properties of pearl millet. J Agric Eng Res 66:85–91

Jalgaonkar K, Jha SK, Sharma DK (2016) Effect of thermal treatments on the storage life of pearl millet (Pennisetum glaucum) flour. Indian J Agric Sci 86(6):762–767

Kachare DP, Chavan JK (1992) Effects of seeds treatment on the changes in fat acidity of pearl millet meal during storage. Indian J Agric Biochem 5(1-2):15–20

Kadlag RV, Chavan JK, Kachare DP (1995) Effects of seed treatments and storage on the changes in lipids of pearl millet meal. Plant Food Hum Nutr 47:279–285

Kapoor R, Kapoor AC (1990) Effect of different treatments on keeping quality of pearl millet flour. Food Chem 35:277–286

Kased I, Hoseney RC, Varriano-Marston E (1984) Factors affecting rancidity in ground pearl millet (*Pennisetum americanum* L. Leeke). Cereal Chem 61(2):187–192

Kaushik N, Srinivasa Rao P, Mishra HN (2016) Process optimization for thermal-assisted high pressure processing of mango (*Mangifera indica* L.) pulp using response surface methodology. LWT Food Sci Technol 69:372–381

Krishnan R, Dharmaraj U, Malleshi NG (2012) Influence of decortication, popping and malting on bioaccessibility of calcium, iron and zinc in finger millet. LWT Food Sci Technol 48:169–174

Kumar DP, Hymavathi TV, Roberts P, Sujatha M (2018) Effect of heat treatment and gamma irradiation on the total bacterial count of selected millet grains (jowar, bajra and foxtail). Int J Curr Microbiol App Sci 7(2):1293–1300

Kumar M, Dwivedy AK, Sarma P, Dkhar MS, Kayang H, Raghuwanshi R et al (2019) Chemically characterised *Artemisia nilagirica* (Clarke) Pamp. essential oil as a safe plant-based preservative and shelf-life enhancer of millets against fungal and aflatoxin contamination and lipid peroxidation. Plant Biosyst 154(3):269–276

Lale NES, Yusuf BA (2000) Insect pests infesting stored pearl millet Pennisetum Glaucum (L.) R. Br. in northeastern Nigeria and their damage potential. Cereal Res Commun 28(1/2):181–186

Li B, Zhao L, Chen H, Sun D, Deng B, Li J, Wang F (2016) Inactivation of lipase and lipoxygenase of wheat germ with temperature-controlled short wave infrared radiation and its effect on storage stability and quality of wheat germ oil. PLoS One 11(12):e0167330

Liu J, Tang X, Zhang Y, Zhao W (2012) Determination of the volatile composition in brown millet, milled millet and millet bran by gas chromatography/mass spectrometry. Molecules 17:2271–2282

Mahmoud NS, Awad SH, Madani RMA, Osman FA, Elmamoun K, Hassan AB (2016) Effect of γ radiation processing on fungal growth and quality characteristics of millet grains. Food Sci Nutr 4(3):342–347

Malek M, Parveen B (1989) Effect of insects infestation on the weight loss and viability of stored BE paddy. Bangladesh J Zool 17(1):83–85

Mali PC, Satyavir (2005) Effect of storage environment and storage material on qualitative and quantitative losses in pearl millet. Ann Arid Zone 44(2):177–184

Manley C, Mayer J (2012) Lipase. In: Mayer J, Donnelly TM (eds) Clinical veterinary advisor: birds and exotic pets. Elsevier, Amsterdam. https://doi.org/10.1016/B978-1-4160-3969-3.00427-3

Marvesh Kumar S, Sharma V, Arora S, Singh AK, Goyal A, Lal D (2015) Effect of storage conditions on sensory and microbial characteristics of developed Pearl millet based weaning food. Indian J Dairy Sci 68:5

Mazaheri Y, Torbati M, Azadmard-Damirchi S, Savage GP (2019) Effect of roasting and microwave pre-treatments of Nigella sativa L. seeds on lipase activity and the quality of the oil. Food Chem 274:480–486

McDonough CM, Rooney LW (1989) Structural characteristics of *Pennisetum americanum* (pearl millet) using scanning electron and fluorescence microscopy. Food Microstruct 8:137–149

Mohajerkhorasani S, Alami M, Kashaninejad M, Shahiritabarestani H (2019) Increasing the shelf-life of millet flour by using heat-moisture and microwave treatments. Iran J Food Sci Technol 16(86):83–93

Mustapha MB, Bousselmi M, Jerbi T, Bettaïeb NB, Fattouch S (2014) Gamma radiation effects on microbiological, physico-chemical and antioxidant properties of Tunisian millet (*Pennisetum Glaucum* LR Br.). Food Chem 154:230–237

Nami Y, Gharekhani M, Aalami M, Hejazi MA (2019) Lactobacillus-fermented sour doughs improve the quality of gluten-free bread made from pearl millet flour. J Food Sci Technol 56(9):4057–4067

Nantanga KKM (2006) Lipid stabilization and partial precooking of pearl millet by thermal treatments (12-105) MSc (Agric.) Thesis in Food Science and Technology: Department of Food Science, Faculty of Natural and Agricultural Sciences, University of 19 Pretoria. http://repository.up.ac.za/bitstream/handle/2263/26680/dissertation.pdf?sequence=1

Nantanga KKM, Seetharaman K, Kock HL, Taylor JRN (2008) Thermal treatments to partially pre-cook and improve the shelf life of whole pearl millet flour. J Sci Food Agric 88(1):892–899

Ofosu FK, Elahi F, Daliri EBM, Yeon S, Ham HJ, Kim J, Han S, Oh D (2020) Flavonoids in decorticated sorghum grains exert antioxidant, antidiabetic and antiobesity activities. Molecules 25(12):2854. https://doi.org/10.3390/molecules25122854

Ogori AF, Jatua MK, Adamu LAUM (2013) Effect of semi-densed polyethylene storage on organoleptic and chemical characteristics of flours from soaked, malted and their blend of millet grains (*Pennesitum glacum*). Am J Biol Chem Pharm Sci 1(7):81–88

Ojokoh A, Bello B (2014) Effect of fermentation on nutrient and anti-nutrient composition of millet (*Pennisetum glaucum*) and soyabean (*Glycine max*) blend flours. J Life Sci 8(27):3566–3570

Onyango CA, Ochanda SO, Mwasaru MA, Ochieng JK, Mathooko FM, Kinyuru JN (2013) Effects of malting and fermentation on anti-nutrient reduction and protein digestibility of red sorghum, white sorghum and pearl millet. J Food Res 2(1):41

Owheruo JO, Ifesan BOT, Kolawole AO (2019) Physico-chemical properties of malted finger millet (*Eleusine coracana*) and pearl millet (*Pennisetum glaucum*). Food Sci Nutr 7(2):476–482

Patel KV, Parameswaran M (1992) Effects of heat treatment on lipid degradation in bajra flour during storage. J Food Sci Technol 29(1):51–52

Prabhat KM (2015) Shelf-life enhancement of pearl millet flour. J Food Process Technol 6:10

Pruthi TD (1981) Free fatty acid changes during storage of bajra (*Pennisetum typhoideum*) flour. J Food Sci Technol 18:257–258

Rani S, Singh R, Sehrawat R, Kaur BP, Upadhyay A (2018) Pearl millet processing: a review. Nutr Food Sci 48:30–44. https://doi.org/10.1108/NFS-04-2017-0070

Rekha (1997) Efficacy of processing techniques in the utilisation of pearl millet for value-added products. M.Sc. Thesis. CCSHAU, Hisar, India

Rinaldi M, Paciulli M, Caligiani A, Scazzina F, Chiavaro E (2017) Sour dough fermentation and chestnut flour in gluten-free bread: a shelf-life evaluation. Food Chem 224:144–152

Rowley J (1984) An assessment of losses during handling and storage of millet in Mali. Trop Stored Prod Inf 47:21–33

Ruge C, Changzhong R, Zaigui L (2012) The effects of different inactivation treatments on the storage properties and sensory quality of naked oat. Food Bioprocess Technol 5(5):1853–1859

Sandhu KS, Punia S (2018) Pearl millet: flour and starch properties. Technologies in food processing. Apple Academic Press, Palm Bay, pp 309–334

Santos JP, Maia JDG, Cruz I (1990) Damage to germination of seed corn caused by maize weevil (*Sitophilus zeamais*) and Angoumois grain moth (*Sitotroga cerealella*). Pesq Agrop Brasileira 25(12):1687–1692

Serba DD, Yadav RS, Varshney RK, Gupta SK, Mahalingam G, Srivastava RK et al (2020) Genomic designing of pearl millet: a resilient crop for arid and semi-arid environments. In: Kole C (ed) Genomic designing of climate-smart cereal crops. Springer, Cham, pp 221–286. https://doi.org/10.1007/978-3-319-93381-86

Sidhu GK, Sukeeti S, Neha (2016) Influence of milling speed on proximate composition of pearl millet flour during storage. Int J Food Ferment Technol 6(1):49–56

Simone M, Failde V, Garcia M, Panadero C, De Simone M (1994) Factors associated with mechanical damage of dry beans with conventional threshing. Riv Ing Agraria 25(4):209–217

Singh AK, Rehal J, Kaur A, Jyot G (2015) Enhancement of attributes of cereals by germination and fermentation: a review. Crit Rev Food Sci Nutr 55(11):1575–1589

Sruthi NU, Premjit Y, Pandiselvam R, Kothakota A, Ramesh SV (2021) An overview of conventional and emerging techniques of roasting: effect on food bioactive signatures. Food Chem 348:129088

Suma PF, Urooj A (2014) Influence of germination on bioaccessible iron and calcium in pearl millet (*Pennisetum typhoideum*). J Food Sci Technol 51(5):976–981

Sumathi A, Ushakumari SR, Malleshi NG (2007) Physico-chemical characteristics, nutritional quality and shelf-life of pearl millet based extrusion cooked supplementary foods. Int J Food Sci Nutr 58(5):350–362

Swaminathan I, Guha M, Hunglur UH, Rao DB (2015) Optimization of infrared heating conditions of sorghum flour using central composite design. Food Sci Biotechnol 24(5):1667–1671

Tiwari A, Jha SK, Pal RK, Sethi S, Krishan L (2014) Effect of pre-milling treatments on storage stability of pearl millet flour. J Food Process Preserv 38(3):1215–1223

Varriano-Marston E, Hoseney RC (1983) Barriers to increased utilisation of pearl millet in developing countries. Cereal Foods World 28:392–395

Varsha R, Narayanan A (2017) Storage stability of biofortified pearl millet flour. Int J Agric Innov Res 5(5):709–713

Xu B, Wang LK, Miao WJ, Wu QF, Liu YX, Sun Y et al (2016) Thermal versus microwave inactivation kinetics of lipase and lipoxygenase from wheat germ. J Food Process Eng 39(3): 247–255

Yadav DN, Balasubramanian S, Kaur J, Anand T, Singh AK (2011) Optimization and shelf-life evaluation of pearl millet based halwa dry mix. J Food Sci Eng 1(4):313

Yadav DN, Kaur J, Anand T, Singh AK (2012a) Storage stability and pasting properties of hydrothermally treated pearl millet flour. Int J Food Sci Technol 47(12):2532–2537

Yadav DN, Anand T, Kaur J, Singh AK (2012b) Improved storage stability of pearl millet flour through microwave treatment. Agric Res 1(4):399–404

Zhang G, Hamaker BR (2005) Sorghum (*Sorghum bicolor* L. Moench) flour pasting properties influenced by free fatty acids and protein. Cereal Chem 82(5):534–540

Zhu SM, Hu FF, Ramaswamy HS, Yu Y, Yu L, Zhang QT (2016) Effect of high pressure treatment and degree of milling on gelatinization and structural properties of brown rice. Food Bioprocess Technol 9(11):1844–1853

Biofuel Opportunities in Pearl Millet

22

A. V. Umakanth, P. Sanjana Reddy, Subburamu Karthikeyan, Desikan Ramesh, Geetha Thanuja Kalyanasundaram, and Sivakumar Subbarayan

Abstract

Global warming and associated environmental pollution, and volatile crude oil prices have boosted major consumers across the world to sharply focus on green fuels. Global biofuel consumption is expected to rise further, mainly driven by higher blending targets in developing countries. Bioethanol from biomass sources is one of the important fuels that can be used as a fossil fuels' substitute for road transport vehicles. Millets including pearl millet can be used as an efficient source for bioethanol production. Pearl millet is a crop suitable for cultivation in marginal soils of arid and semi-arid tracts where no other cereal crop is preferred. The use of pearl millet is increasing in poultry and animal feed, as fodder, and as roofing material and in brewing industry. Among various lignocellulosic biomasses available for cellulosic ethanol production, pearl millet serves as one of the ideal feedstocks for bioethanol production. This chapter reviews the attributes that make pearl millet a potential biofuel and bioenergy crop and the ways and means for promoting the crop as an efficient feedstock for biofuel production.

A. V. Umakanth (✉) · P. S. Reddy
ICAR–Indian Institute of Millets Research, Hyderabad, India
e-mail: umakanth@millets.res.in

S. Karthikeyan · D. Ramesh · G. T. Kalyanasundaram · S. Subbarayan
Tamil Nadu Agricultural University, Coimbatore, India

V. A. Tonapi et al. (eds.), *Pearl Millet in the 21st Century*,
https://doi.org/10.1007/978-981-99-5890-0_22

22.1 Introduction

Exploration of renewable energy sources has become imperative as greenhouse gas (GHG) emissions and fossil fuel depletion are currently viewed as the major challenges of global industrial development (Niculescu et al. 2019). Biofuels produced from biomass being replenishable and environmentally friendly are increasingly becoming alternative sources of energy globally. Biofuel production is divided into different generations such as the first-generation, which uses starch or sugars from different crops, while the second-generation (2G) biofuel production is based on agricultural residues, dedicated bioenergy crops, forest trees, grasses, biomass leftovers, etc. The third-generation biofuel is from algae, which promises a higher bioethanol yield per hectare. Sugary substrates are expensive but their conversion to ethanol is at a low cost. Cellulosic materials are cheaper with higher price for conversion to ethanol. However, starchy substrates are economical (Reddy et al. 2005). It is anticipated that in the near future, the second-generation bioethanol might fully replace the first generation (Vohra et al. 2014). The second generation (2G) of biofuels can be generated by using the non-food parts of plants, such as cell walls, composed of structural polysaccharides, such as cellulose and hemicelluloses. Second-generation biofuels are considered to be advantageous over the first generation of biofuels because of availability of a wide range of plant biomass at affordable costs to the biofuel industry, absence of competition with the food production, higher energy production potential, and sustainable CO_2 balance.

In the present-day context, search for the most sustainable and high-yielding plant sources is one of the most important issues in the production of the second-generation bioethanol. Major energy crops that are available and can serve as efficient 2G feedstocks are Switchgrass, *Miscanthus*, high biomass or energy sorghum, pearl millet, and crop residues like paddy straw, wheat straw, corn stover, corn cobs, etc.

Pearl millet (*Pennisetum glaucum* L. R. Br.) has been identified as a potential feedstock to supplement maize (*Zea mays L.*) to produce ethanol in the Southeastern United States (Wilson et al. 2006a, b), and growing pearl millet for ethanol production gives more financial returns compared to corn (Wang et al. 2006). It is a crop suitable for cultivation in marginal soils of arid and semi-arid tracts (Fig. 22.1) where no other cereal crop is preferred. Being a C_4 species, it is bestowed with a very high photosynthetic efficiency leading to increased dry matter accumulation in plants. It is cultivated in 30 million ha in more than 30 countries supporting >100 million people, most of them belonging to marginal farmers category. Among the millets grown in India, 75% of total area is occupied with pearl millet. Under marginal and harsh climatic and soil conditions, pearl millet has higher yielding ability than any other crop with limited inputs. Its calorific value is 360 kcal/100 g. Pearl millet grain is a rich source of protein and amino acids and hence can be utilized as human food. Most of the grain is used as animal and poultry feed, for potable alcohol, and processed food. The biomass is a valuable livestock feed in India and Africa. In dry land areas, the crop residues are used for fodder, as fuel for cooking, and as building material. In countries like the United States, pearl millet is grown as a summer forage

Fig. 22.1 Pearl millet under highly arid conditions

crop and seed for the bird feed industry and wildlife. The use of pearl millet is increasing in brewing industry, in poultry and animal feed, as fodder, and as roofing material.

Most of the bioethanol in India is produced using sugar molasses alone. However, sorghum and pearl millet have been found to be viable alternatives in a study conducted among farmers in Madhya Pradesh, India. Estimates also suggest greater returns from millets than maize utilizing 40% less energy in processing. Millets also offer a significant cost advantage over maize as a feedstock for bioethanol production. In Haryana state of India, eight ethanol distilleries were established. Among these, six plants are using broken rice or pearl millet grain. In India, broken rice is used as a viable feedstock for alcohol industry followed by pearl millet grain (lower grade grain quality) and sorghum. The lignocellulosic biomasses, namely oats, barley, wheat, and pearl millet grains, have also been tested for saccharification and fermentation by many authors (Wu et al. 2006). The choice of the plant species depends on its easy availability, ability to thrive on poor and marginal lands, low input requirement, high biomass yields per unit area, appropriate biomass quality, and higher yield of renewable energy. These characters are best exemplified in pearl millet and hence it could be a potential feedstock for biofuel production in areas that are inhospitable for other crops.

Sweet sorghum and sweet pearl millet also qualify as energy crops as juice present in the stalks can be utilized to produce bioethanol while the bagasse can be a good source of animal feed.

Not many reports exist for sweet pearl millet. Appa Rao et al. (1982) mentioned about sweet-stalk types of pearl millet that were collected during a germplasm collecting expedition in Coimbatore and Madurai, Tamil Nadu, India. The local farmers cut the crop before flowering for fodder and the ratoon crop provides some grain and straw. When planted at the International Crops Research Institute for the Semi-Arid Tropics (ICRISAT), Hyderabad, these types flowered late and grew relatively tall in the rainy season in comparison with their growth during post-rainy season. The sweet-stalk types are characterized by long narrow leaf blades, profuse nodal tillering with asynchronous maturity, short thin spikes, and very small grains as compared with the usual types. The sweet-stalk types could be easily identified by chewing them at the dough stage. At maturity, they contained more than twice the amount of soluble sugars than the normal types. Sweet pearl millet (*Pennisetum glaucum* (L.) R. Br.) and sweet sorghum (*Sorghum bicolor* (L.) Moench) are hybrids developed to contain more fermentable sugars than standard forage types for use in ethanol production. These hybrids can have a dual purpose, as the press residues (bagasse) from the ethanol industry are suitable as animal feed (Dos Passos Bernardes et al. 2016).

22.1.1 Current Fuel Statistics

Global biofuel production was 10 billion liters (L) in 2018 to reach a record of 154 billion L. Double the growth of 2017, this 7% year-on-year increase was the highest in five years. Ethanol output expanded by 6.6 billion L, mostly owing to significantly higher production in Brazil, and combined biodiesel and hydrotreated vegetable oil (HVO) output increased by 3.6 billion L with over half of this growth in the United States. Biofuel output is anticipated to increase 25% during 2019–2024, to reach 190 billion L, with the forecast revised upward from last year owing to better market prospects in Brazil, the United States, and especially the People's Republic of China, which boasts the largest increase for any country (Renewables 2019. Analysis and forecast to 2024 IEA Publication; p. 107).

In India, the current levels of blending ethanol with petrol have reached more than 10% and an indicative target of 20% blending of ethanol in petrol and 5% blending of biodiesel in diesel is to be achieved by 2025. Biofuels have huge potential to help the country reach the blending goals, and adopting biofuels as an alternative source of energy can significantly improve farmers' income, generate employment opportunities, reduce imports, and augment waste to wealth creation. The government promotes the production and adoption of biofuels through several programs such as the "National Policy on Biofuels, 2018," which concentrates on improving the current levels of production of ethanol and biodiesel by increasing domestic production from domestic feedstocks, setting up of second-generation (2G) biorefineries, development of new feedstocks for biofuels, etc. Sugar crops such as sugar cane, sugar beet, and sweet sorghum; materials that have starch such as corn, cassava, rotten potatoes, and algae; or cellulosic materials such as bagasse, waste wood, and agricultural/forestry residues are identified as feedstocks to produce biofuels under this policy. The future of lignocellulosic fuels depends on our ability to produce large volumes of inexpensive feedstocks without threatening food security or the environment. There is every need to utilize feedstocks such as pearl millet, which can be grown with least inputs in harshest of environments and reap higher biomass, which can be utilized to produce biofuels.

22.2 Taxonomy, Botanical Description, and Reproductive Biology

22.2.1 Taxonomy

Pearl millet belongs to genus Pennisetum, family Poaceae, subfamily Panicoideae, tribe Paniceae, sub-tribe Panicinae, Section Penicillaria, and genus *Pennisetum*. The genus *Pennisetum* includes two distinct species. One of the species, *Pennisetum purpureum* Schumach, is a tetraploid with chromosome number $2n = 28$. It is perennial. The other one is the cultivated species *Pennisetum glaucum* (L.) Leeke, with chromosome number $2n = 14$. Based on seed morphological traits, the cultivated pearl millet is divided into four basic races, viz., *typhoides*, *nigritarum*, *globosum*, and *leonis*. The intermediate races are the combination of any two basic races.

22.2.2 Morphological Description

Pearl millet root system is comprised of primary, secondary, and brace roots. The radicle emerging from the germinating seeds gives rise to primary roots. The secondary roots develop from the nodes of the primary roots. The third type of roots is the brace roots that develop from the stem nodes nearer to the soil surface to provide anchorage. The plumule gives rise to the shoot that consists of main plant and the tillers. The main plant is only observed until third to fifth week after sowing depending on the genotype and the environment. The tillers appear later from the lower node. These are the primary tillers that may mature at the same time as main plant (synchronous) or have differential maturity period (non-synchronous). They may attain same height as main plant (uniform) or have different heights giving the population a non-uniform appearance. Sometimes secondary tillers may develop from the axillary buds at the upper nodes, which may be due to poor mobilization of photosynthates to the sink. The leaves appear in two vertical rows, consisting of leaf blade and leaf sheath. The pearl millet inflorescence is termed as spike/panicle. The panicle consists of a central rachis enclosed with fascicles. Each fascicle consists of one or more spikelets enclosed by a whorl of 70–80 bristles. On an average, 1600 spikelets may be present in the inflorescence of the main plant. Each spikelet consists of two florets of which the lower floret is staminate and the upper floret is hermaphrodite. The androecium pearl millet consists of three anthers while the gynoecium consists of a single ovary with two styles and a feathery stigma (Rachie and Majmudar 1980).

22.2.3 Pollination Behavior

Panicle emerges in about 4–6 days from the sheath. Flowering behavior depends on the genetic constitution of the plant. In most of the genotypes, panicle emerges from the boot followed by opening of florets. However, in certain genotypes, exertion of style may start well before emergence of the panicle from the boot leaf. The flowering starts in the upper portion of the panicle above the center and then proceeds in either direction and reaches peak by the third day. Since the stigmas emerge first as compared to anthers, the flowering pattern in pearl millet is termed as protogynous. The stigma remains receptive for 12–16 h (Rachie and Majmudar 1980). The anthers emerge after completion of the emergence of stigmas beginning in the hermaphrodite florets followed by the staminate florets. Though anthesis occurs throughout the day, the active period is between 20:00 and 02:00 h (Sundararaj and Thulasidas 1980). Another emergence follows the same pattern as stigma emergence on the panicle. The anthesis takes 1 week's time on the main plant under irrigation. The total anthesis process covering tillers takes 3 weeks to complete. Under rainfed conditions, the anthesis on main plant may be delayed by over 5 days and it may continue on the tillers till seed formation (Chalam and Venkateswarlu 1965). The stigmas dry up in 24 h after pollination and the seed-set can be observed in a week after pollination and fertilization (Burton and Powell 1968). The physiological maturity of the grain is identified by the appearance of a black spot just above the hilar region of the grain (Fussell and Pearson 1978).

22.3 Genetics of Biofuel Traits

Pearl millet, in lieu of its cross-pollinated nature, has wide variability for phenotypic traits in populations grown in diverse agro-ecologies. Virk (1988) had studied the populations from diverse ecologies. In the African populations, high genetic variability was observed for crop maturity, grain as well as fodder yields, the leaf number and stem thickness being more variable among the parameters contributing toward fodder yield. The Indian varieties had more genetic variability for grain-yield-contributing traits such as crop duration, length of the panicle, panicle weight, and grain size. Several studies found no relation between geographical diversity and genetic diversity (Andre Savery and Prasad 1995). It was also found that the genetic stocks from Africa and America were closer compared to Asia and Australia. This can also be due to the movement of pearl millet lines from Africa to America. According to Suthamathi and Stephen Dorairaj (1995), advances can be made for forage pearl millet improvement through simple selection as high heritability (h^2) coupled with high genetic advance (GA) was observed for plant height, stem weight, green fodder yield, leaf weight, and leaf/stem ratio. However, maximum GA was observed only for productive tillers for grain-yield-contributing traits.

Virk (1988) also summarized the trait associations in the four ecological groups: Indian varieties, Indian inbreds, African varieties, and American inbreds. The association of plant height (a measure of biomass) with grain yield was low in

Indian varieties, high in Indian inbreds, and stronger in African varieties and American inbreds. Late flowering and maturity were associated with high grain yield in African varieties and American inbreds, but their poor association between these two traits was observed in Indian varieties and Indian inbreds. Virk (1988) also found that panicle weight, productive tiller number, flag leaf area, harvest index, and grain weight had high direct positive effects and concluded that for improvement of grain yield, selection should be focused on the traits plant height, number of tillers, and grain size, while for improving fodder yield, greater focus should be on leaf number during selection followed by leaf size, plant height, flowering time, and number of tillers. Under saline soil conditions, number of tillers, plant height, and seedling emergence rate had an impact on green fodder yield (Patil and Jadhav 1992).

Studies conducted at a later period in high-biomass pearl millet also confirmed the inheritance contributed by additive gene action as deduced from high heritability with high genetic advance as percent of mean for green fodder yield per plant and dry matter yield per plant indicating simple selection as a desirable breeding methodology. The traits, viz., plant height, leaf breadth, and dry matter yield per plant, can be targeted during selection process for increasing the green fodder yield (Thomas and Ougham 2014).

22.4 Genomics

Cellulosic ethanol attains importance with abundant availability of low-cost feedstock of C_4 grasses including pearl millet, which is amenable to poor soil and risky crop growth conditions in tropical environments. The lignocellulosic materials represent the most abundant carbon resource and provide scope for large-scale biofuel production (Farrell et al. 2006). The lignocellulosic material comprises of structural polysaccharides of celluloses and hemicelluloses that are interlinked with phenolic polymer lignin that gives strength to cell wall and resistance to stress environments (Himmel et al. 2007; Zhao et al. 2012). The lignin content is an important factor that determines quality of feedstock and is associated with pretreatment cost for industrial application in ethanol production.

Presence of considerable variation has been identified (Blummel et al. 2003) for cell wall digestibility in animal digestive system, which indirectly indicates the lignin content. Pearl millet mapping population (ICMB 841 × 863B) was assessed for the genetic variability of stover quality traits and stover yield. The lignin content was measured as in vitro organic matter digestibility (IVOMD) for assessment of feed value for livestock. The higher digestibility of stover reflects the reduced or lowered content of cell wall lignin. For in vitro organic matter digestibility (IVOMD), quantitative trait loci (QTLs) were mapped in the ICMB 841 × 863B derived pearl millet mapping population (Nepolean et al. 2006) in the marker's interval between Xpsm761 and Xbm3RA1c for LG1, Xpsmp2066 and Xpsm3 for LG2, and Xpsm514 and XtstRA6c for LG6. Similarly, markers were identified for dry stover yield parameters between the marker's interval of Xctm10 and Xpsm174 and Xpsmp2064 and Xpsm318 at LG3 in the above mapping population. The low lignin content minimizes the pretreatment process in the separation of celluloses or hemicelluloses from lignocellulosic complexes and considerably reduces the cost and time for the preparation of fermentation process of conversion of cellulose into sugar compounds in the ethanol production. Since pearl millet grows well in all kinds of tropical environments, the early drought stress impact on growth and development affects subsequent growth stages, which ultimately reduces biomass and harvest index.

Debieu et al. (2018) studied early drought stress and identified QTLs controlling biomass production under drought in pearl millet. Marker-trait associations (MTAs) were established using pearl millet inbred lines comprising a panel of 188 inbred lines phenotyped for 11 agro-morphological traits including biomass traits, viz., stalk length, number of leaves, aerial biomass (g), and related grain yield traits under 2 conditions, viz., early drought stress and well-watered for 2 years. The Genome Wide Association Studies (GWAS) was performed with kinship matrix using the R package GAPIT (Genome Association and Prediction Integrated Tool). For most of the traits, a large range of variation was detected, with the coefficients of variation (CV) varying from 8.2% for stay-green under well-watered conditions to 98.6% for total grain weight under drought stress conditions. The drought stress led to a strong and very significant reduction in plant height, aerial biomass, and grain traits in both years under drought stress compared to well-watered conditions. Traits like panicle length, stalk length, and number of leaves were moderately reduced under drought stress.

Stay-green trait (Stg) showed significantly increased expression under drought stress in both years. Early stress was found to reduce the growth and aerial biomass in later stages for which variability for response to early drought stress was widely prevalent in the pearl millet accessions. The identification of QTLs/genes involved in early drought tolerance through phenotype/genotype association analysis in the donor lines could be source for exploration and improving genotypes for drought stress for better growth and aerial biomass production.

Stay-green is an important agronomical trait that can contribute to higher yield production under drought stress condition (Harris et al. 2007; Thomas and Ougham 2014). The genetic association studies of a panel of 175 inbred lines (Debieu et al. 2018) evaluated for both genetic and phenotypic data, particularly for stay-green, revealed four marker-trait associations (MTAs) with the single nucleotide polymorphism (SNP) markers for the stay-green trait on chromosome 6 with 12% of phenotypic variation in the panel and these marker-trait associations were found to be associated with two different polymorphisms located at the same position. The phenotypic difference of 25% was observed between above two different polymorphisms. These polymorphic positions mapped on LG6 of the reference genome were related to SNPs/stay-green association. These polymorphic loci found within the gene locus *Pgl_GLEAN_10013220* are responsible for uroporphyrin-III C-methyltransferase (UPM), an enzyme involved in siroheme biosynthetic pathway.

22.5 Breeding for High Biomass

22.5.1 Breeding Objectives

1. Development of inbreds/hybrids possessing profuse tillering and faster growth
2. Adaptability of genotypes for round-the-year cultivation to supply lignocellulosic feedstock
3. Introgression of QTLs for drought stress tolerance and biomass accumulation
4. Introgression of genes for development of low lignin traits through genetic engineering and gene silencing/editing and mutational approaches
5. Gene pyramiding for high biomass, low lignin, higher dry matter digestibility

22.5.2 Breeding Methodology

Pearl millet being a highly cross-pollinated crop, hybrids are the better option for exploitation of heterosis. Taking advantage of protogynous nature and high tillering ability, crossing can be easily taken up for executing breeding plans. The short life cycle (75–85 days) makes it amenable for handling multiple generations in a year. The crop can be sown as well as transplanted with equal efficiency depending on the availability of irrigation. Broadly, two methodologies are followed for genetic improvement in pearl millet. Recurrent selection is practiced for intra- and inter-population improvement like incorporation of desirable characters in populations, generation of new variability, and development of heterotic pools. The pedigree breeding is practiced for development of fixed lines or inbreds that are used in the hybrid programs. The inbred lines can be combined to form synthetic varieties and composite varieties that can be developed by maintaining mixture of genotypes in bulk. There is a need for constant improvement in populations to meet the changing research needs, viz., resurgence of pests and diseases, prioritizing nutrient traits, stabilizing yields under changing climates. The improvement in populations depends on the base material used and recurrent-selection procedures used. Pearl millet is amenable to different recurrent-selection schemes available such as phenotypic recurrent selection, genotypic recurrent selection (based on progeny performance), intra-population improvement methods, inter-population improvement methods, or general combining ability based or specific combining ability based recurrent-selection schemes (Allard 1960). The choice of a recurrent-selection scheme depends on the objective of the program and the availability of resources (Singh et al. 1988). The populations serve as continuous sources of variability to be exploited in hybrid breeding programs. Deriving single crosses from populations is the most common form of pedigree breeding for deriving hybrid parents (Andrews et al. 1997).

22.5.3 Breeding for High Biomass

For sustainable fuel production from cellulosic feedstock, significant breeding efforts should target biomass improvement. Biomass has the positive influence on grain and forage yield. There have been efforts to improve forage yield in pearl millet. It has been found that the tiller number and stem girth were positively related to plant height, thereby indirectly contributing toward forage yield. Most of the biomass-related traits are under the control of both additive and non-additive genes and show Mendelian inheritance. Recurrent and divergent selection has been proposed to improve in vitro drymatter digestibility (IVDMD). Multi-cut forage varieties are preferred by farmers to ensure feed security and there is an increasing effort among researchers in this direction. Apart from this, considerable progress has been made to breed pearl millet, Napier grass hybrids. Once planted, it can give eight to ten cuts in a year and if managed properly, it can provide fodder for 5–6 years (Malhi et al. 2020). Napier grass (*Pennisetum purpureum* (K.) Schum.) is a high-biomass species and can be crossed with grain pearl millet (*Pennisetum glaucum* (L.) R. Br.) to generate hybrids that are sterile triploids with very high biomass. This is suitable as a perennial forage and for biofuels. Though the hybrids are sterile, hybrid seed can be produced from seeded parents. The hybrids have the ability to efficiently utilize water and nutrients and can be grown for biomass production on more than 445 million hectares across the globe (Campbell et al. 2005) (Fig. 22.2).

Fig. 22.2 High-biomass pearl millet

22.6 Brown Midrib Pearl Millet

One of the major impediments of converting biomass to biofuels is the cost for removal of lignin apart from the high cost of enzymes used for saccharification. Genetic manipulation of biomass through introduction of brown midrib (*bmr*) genes conferring low lignin into high-biomass pearl millet cultivars would result in feedstocks with higher fermentable sugars from the lignocellulosic biomass. Purdue University research showed 50% higher yield of the fermentable sugars from the stover of certain sorghum bmr lines after enzymatic hydrolysis (Vermerris 2011). Brown midrib mutants have been isolated in maize (*Zea mays*), sorghum (*Sorghum bicolor*), and pearl millet (*Pennisetum glaucum*), arising by either spontaneous or chemical mutagenesis. The characteristic brown coloration of the leaf midveins and internodes is associated with reduced lignin content and altered lignin composition.

Only three pearl millet brown midrib mutants have been described in the literature; the first mutant came from a diethyl sulfate mutagenized population (Cherney et al. 1988), the second was a spontaneous mutant from germplasm isolated in Zimbabwe (Gupta 1995), and the third mutant arose spontaneously in Tifton, GA (Degenhart et al. 1995). The three mutants are allelic (Gupta 1995; Degenhart et al. 1995).

Efforts must be made to develop pearl millet breeding lines in brown midrib background for their utility in lignocellulosic biofuel production.

22.7 Grain-Based Ethanol

Having higher starch content, valued coproducts, and improved rate of fermentation, pearl millet has the scope for becoming an economically superior feedstock for ethanol production. Fermentation experiments were conducted to assess the biological feasibility of using pearl millet as a feedstock. Wu et al. (2006) carried out ethanol fermentation in pearl millet in flasks on rotary shaker and 5-L bioreactor. The final ethanol yield ranged from 8.71 to 16.8% v/v at dry matter concentration of 20–35% and ethanol conversion efficiency between 90 and 95.6%. Bioreactor gave better conversion than rotary shaker. By following a process similar to corn, Wilson et al. (2006a, b) demonstrated that pearl millet had 4% less conversion efficiency but fermented quickly and reached 85% conversion 12 h earlier to corn. Grinding rate of pearl millet is 53% faster and requires 40% less energy to grind than maize (Dozier et al. 2005). Even with a 10% premium on the cost of pearl millet, the net cost of ethanol production is $0.016 per L less than production using maize. Total net profit from a facility using pearl millet as the sole feedstock was $25,175,000 per year compared to $23,758,000 for maize feedstocks, a $1.4 million advantage (Wilson et al. 2007). Higher protein content endowed in pearl millet contributes to less ethanol. However, the higher distillers dried grains with soluble (DDGS) in it yield higher economic return. Therefore, pearl millets could be a potential feedstock for fuel ethanol production in areas too dry to grow corn and grain sorghum.

In Brazil, there is increasing interest to broaden ethanol feedstocks, possibly including pearl millet, as a high-biomass-yielding crop (De Assis et al. 2018). Pearl millet and sweet sorghum produce high yields of biomass with readily soluble sugars that could also be used for ethanol production (Pereira Filho et al. 2013).

22.8 Crop Residues and Their Utilization

22.8.1 Pearl Millet Wastes

Pearl Millet is a low-input-requiring, high-biomass-yielding crop tolerant to insects, diseases, and adverse weather conditions. These traits qualify this crop as an energy crop intended for biofuel production. The straw, stover, and husk are the three types of by-products generated from pearl millet processing and most of them are wastes (Fig. 22.3). In general, a package of operations is followed for millet processing to get the edible endosperm. Firstly, the pearl millet ear heads are cut from the crops by manual harvesting using a sickle or small hand knives after attaining grain maturity (FAO 2001). After harvesting the ear heads, the pearl millet stems are dried in the field for a week before being cut and stacked (Chapke et al. 2020).

For safe and long-term storage, the moisture content of grains must be low (7–9%) (Jain and Bal 1997; Chapke et al. 2020). For this purpose, the drying step is a vital process followed in millet processing. Secondly, ear heads are dried under sun or solar dryers for excess moisture removal from grains. Thirdly, the pearl millet

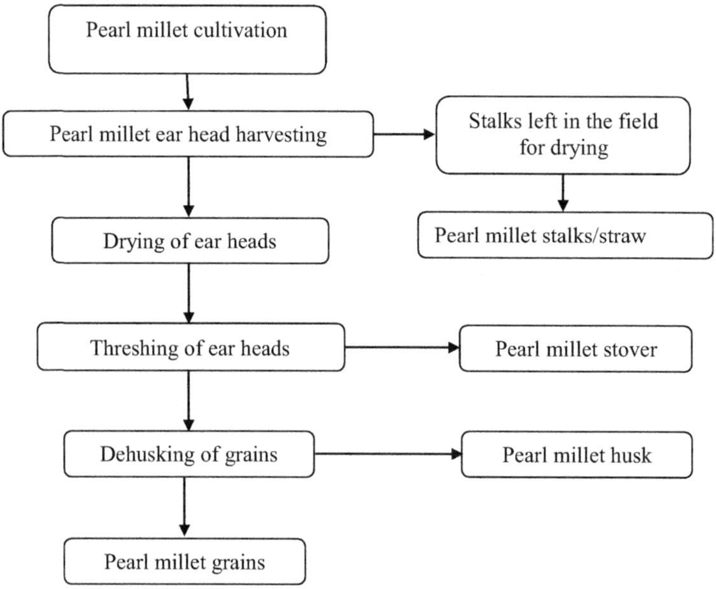

Fig. 22.3 Pearl millet processing and its by-products production

grains are separated from ear heads by manual method/pearl millet threshers. In the case of manual threshing, the ear heads are filled in gunny sacks and beaten on the floors, resulting in grain separation due to impact force (FAO 2001). During the thresher operation, the pearl millet stover and grains are separated and collected from their respective outlets provided in the threshers. Fourthly, dehusking of grains is done by manual or mechanical dehuller to remove its husk. The husk contains phenolic compounds, which have potential as a functional component in nutraceutical and food products. Finally, processed grains are filled in the bags and stored. However, pearl millet with higher oil content results in poor storability (Jain and Bal 1997).

Although pearl millet is grown in many parts of the world, mechanical intervention in millet processing is minimum. Manual processing of millets entails a great deal of drudgery in operation. Compared to other crops, the harvesting and processing machinery for pearl millet is not well developed, and research is still in its early stages.

22.8.2 Wastes Production Potential and Their Current Uses

The value of residue to product ratio (RPR) for millet straw is 2, and the potential availability of straw for India estimated for 2030 is 24.7 million metric tons (Sandra 1997; Purohit and Dhar 2018). The green fodder productivity for the pearl millet crop ranged from 20 to 35 tons per ha (Chadha 2010). The annual pearl stalks production and surplus residue availability for India was 10.93 and 1.61 million metric tons (Pathak and Chandel 2017). The RPR values for pearl millet straw, cob/stover, and husk were 1.4, 0.33, and 0.3, respectively (Sandra 1997; Cardoen et al. 2015). The annual productions of pearl millet straw, stover, and husk for India were about 10,174, 2398, and 2071 kilotons (Cardoen et al. 2015).

Generally, the pearl millet stalk includes the sheath, leaves, and core of the plant. The percentage of cellulose, hemicellulose, lignin, and ash present in the core were 52, 25, 10, and 9.4%, respectively (Yadav et al. 2019). The higher heating value of pearl millet stover, husk, and stalk were 17.39, 17.48, and 18.16 MJ/kg, respectively (Sandra 1997). The pearl millet grains are used for food, whereas the stalk is used as fodder for cattle or mostly disposed off as agricultural wastes (Yadav et al. 2019). Suttie (2000) stated that the strong stems of bulrush millet are used in traditional buildings, screens, and grain stores, and long straw is used as thatch and straws. At the same time, the stovers are used as fuel or dry fodder.

22.8.3 Briquettes

The pearl millet wastes can be directly used as fuel in the traditional or improved cook stoves. These wastes are used as fuel due to their calorific value, and it is possible to produce good-quality solid biofuels from these wastes by applying briquetting technology. A briquetting machine is used to produce a solid biofuel

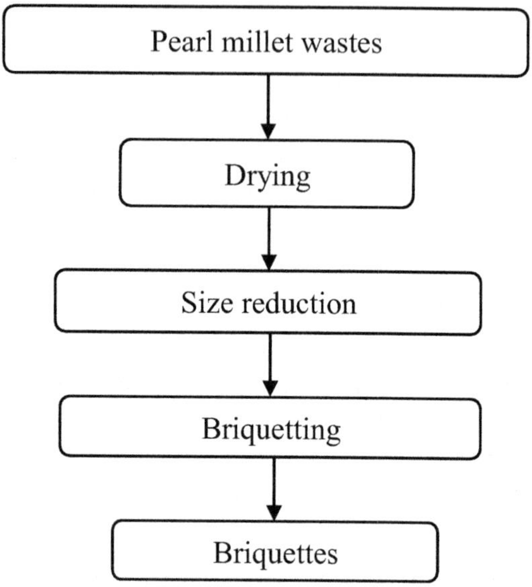

Fig. 22.4 Process flow for briquettes production

from powdered form of biomass feedstocks. The pearl millet wastes should undergo several processing operations in the manufacturing of briquettes (Fig. 22.4). An attempt was to make cylindrical biomass briquettes using sorghum panicle–pearl millet combination with the addition of cassava starch as binder, and hydraulic piston press type briquetting machine was used for this purpose (Velusamy et al. 2021). They reported that the briquettes made with a combination of 80% sorghum panicle and 20% pearl millet stover had a higher calorific value (26.87 MJ/kg) and better-quality parameters than briquettes produced with other combinations.

Millet husks are produced in huge quantities, underutilized for biofuel production, which is renewable, biodegradable, light in weight, and available at a lower price (Rosa et al. 2009; Abba et al. 2017). An investigation was conducted by Kuhe et al. (2021) to assess the feasibility of pearl millet husk briquettes to replace conventional cooking fuel (fuelwood) in the rural areas of the northern Nigeria region. For this study, they produced briquettes from pearl millet husk using Gum Arabic as a binding agent, and these briquettes have a calorific value of 15.27 MJ/kg. They reported that millet husk briquettes could save a massive quantity of fuelwood in the study area, saving about 2.257 million USD.

22.8.4 Biogas

Anaerobic digestion (AD) is a biochemical conversion process that uses a consortium of microbial activities to decompose organic wastes in an oxygen-free environment. The biogas and biodigestate are the outcomes of the AD process. The best examples for this category are livestock manures, organic fractions of domestic and

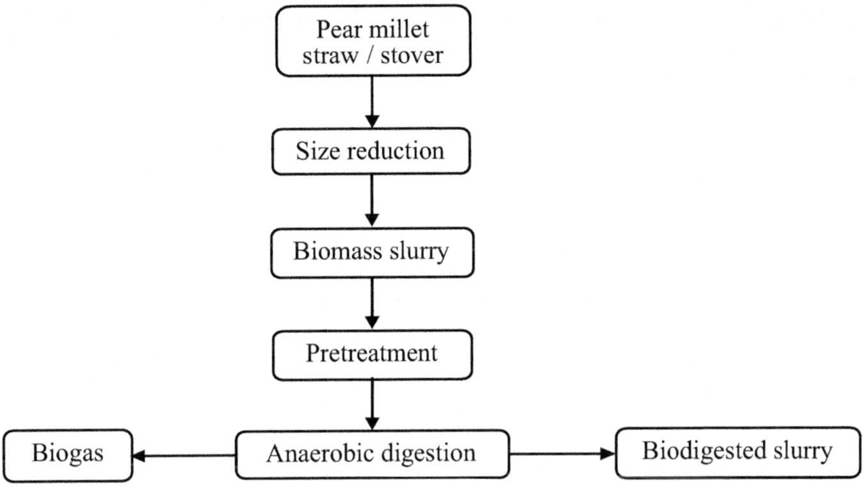

Fig. 22.5 Process flow for biogas production from pearl millet wastes

municipal solid wastes, and agro-processing industrial wastewaters. Agro-crop residues and other lignocellulosic feedstocks are also used as biogas feedstocks.

Furthermore, most of these feedstocks are not meeting the optimal requirements for better biogas yield, viz., anaerobic digestion such as C/N ratio and organic loading. However, higher lignin content in these wastes creates resistance to microbial activity for biodegradation, resulting in lower biogas yield (Negi et al. 2020; Kumar et al. 2018). Feedstock pretreatment is required to enhance microbial activities by removing lignin from biomass materials (Xu et al. 2020). Thus, it results in increased biogas production as compared to unpretreated biomass feedstocks. The process involved in converting pearl millet wastes into biogas and biodigested slurry is shown in Fig. 22.5. The pretreatment is an optional process in this case, and the inclusion of the pretreatment process can enhance biogas yield.

Some of the biogas feedstocks without pretreatment are tried for biogas production. To maintain the optimal total solid (TS) loading for better biogas yield, water is added to an equivalent or less amount to the feedstock, which is desired by TS of feedstocks. Adding water to the feedstocks is becoming an issue in the water scarcity areas. Furthermore, this makes the beneficiaries adopt the biogas technology in these areas. The solid-state anaerobic digestion (SSAD) concept offers a solution to overcome the water requirement for the biogas plant. In this concept, a lesser quantity of water is added to the biogas feedstocks. In other words, high total solid loading is practiced for the specially designed biogas plant. The performance of the SSAD reactor was conducted for two different temperatures with four different total solid (TS) content of pearl millet straw (Paritosh et al. 2019). For this study, the maximum methane for a straw with 20 and 25% TS loading was recorded at 35 °C (124.1 ± 7 L/kg volatile solids (VS)) and 55 °C (162.4 ± 9 L/kg volatile solids (VS)), respectively. Paritosh et al. (2020) experimented on SSAD of pearl millet straw with/without alkali catalyst for a retention period of 45 days. They found that alkali treatment improved the biogas yield, which showed that cumulative biogas yield was 1.09–1.65 times higher than control.

The biological pretreatment of pearl millet straw using *Chaetomium globosporum*) was attempted without or with chitinolytic bacteria (CB) for biogas production. The results showed that the biogas yield from pretreated pearl millet straw was 543 mL/gram volatile solids (gVS), which was 46% higher as compared with the raw straw sample (371 mL/gram volatile solids (gVS)). Furthermore, the biogas yield from CB added to fungi pretreated straw samples were 57 and 11% higher than raw and fungal pretreated straw samples, respectively (Yadav and Vivekanand 2021). Pearl millet straw was pretreated using *Curvularia lunata* and showed a 28% increase in biogas production than the untreated sample (Yadav and Vivekanand 2020).

22.8.5 Bio-Oil and Biochar

Pyrolysis produces either solid product/biofuel or liquid biofuels by decomposing biomass materials under an inert atmosphere and heat. The solid product generated in this process is known as biochar or charcoal, whereas the liquid product is referred to as biocrude or bio-oil. Most of the lignocellulosic biomass feedstocks are attempted for the pyrolysis process for the targeted final product. Bio-oil was produced from pearl millet using a fast pyrolysis process, and its yield was 48% (Laougé et al. 2020). These authors optimized process conditions for the pearl millet straw to get maximum bio-oil yield under reaction temperature of 400 °C and a 200 mL/min nitrogen flow rate, and particle size as 1.5 mm. Another interesting study was conducted by Bera et al. (2018) to produce biochar from pearl millet stover, and the average mass yield was found as 38% at a reaction temperature of 400 °C. They suggested that stover biochar was more suitable for ameliorating soil acidity due to its higher pH values, calcium carbonate equivalent, and carbonate content.

22.8.6 Bioethanol

Pearl millet is widely used as a forage and cover crop in many parts of the globe. The grains harvested are prone to several fungal diseases. Further, the storage of harvested grains strictly needs safe and scientific techniques to maintain the original quality and prevent pests and diseases. However, there are diversified market options available upon processing of either straw or grains. Among various lignocellulosic biomass available, pearl millet serves as an ideal feedstock for bioethanol production due to the following reasons: (1) high biomass with minimal or no fertilization and irrigation, and (2) can be grown in the soil even with low fertility and high temperature; the steps involved are depicted in Fig. 22.6.

Gohel and Duan (2012) have attempted the ethanol production from pearl millet involving liquefaction followed by simultaneous saccharification and fermentation (SSF) aided with yeast fermentation.

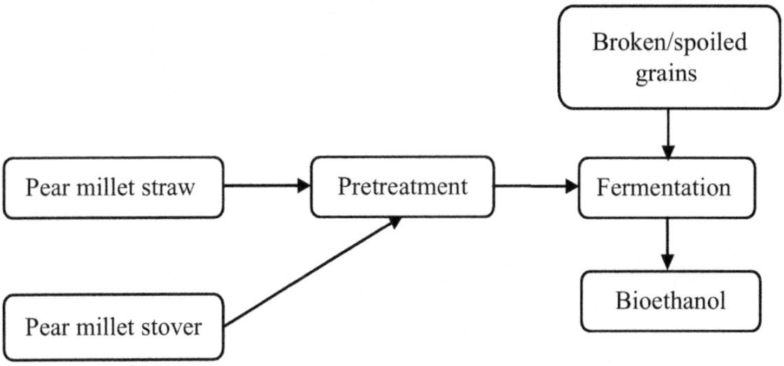

Fig. 22.6 Process flow for bioethanol production from pearl millet wastes

Table 22.1 Pretreatment method for ethanol production from pearl millet

Biomass type	Pretreatment methods	Result/product yield	References
Stover	Biological method *Penicillium* sp. AKB-24 and *Aspergillus nidulans* AKB-25	Enhanced hydrolysis	Kumar et al. (2016a)
	Alkali pretreatment	Increased fermentable sugars	Kumar et al. (2016b)
Straw	Chemical pretreatment 16% orthophosphoric acid	Increased sugar release	Maheshwari and Karthikeyan (2020)
	Biological treatment *Chaetomium globosporum*	Improved lignin removal	Yadav and Vivekanand (2019)
	Alkali catalyst $Ca(OH)_2$	Reduction of lignin and hemicellulose	Jinisha et al. (2020)
Hay	Alkaline pretreatment (2.0% sulfuric acid)	Xylan solubilization and lignin reduction	Chen et al. (2007)
Grains	Grinding + heat treatment	Faster fermentation rate	Wilson et al. (2007)

22.8.7 Pretreatment Methods for Ethanol Production

Pretreatment of the agricultural residue aims to lower the crystallinity of cellulose and enhance the accessibility to the enzymes (Geetha Thanuja et al. 2021). The pretreatment of pearl millet stover at 3% alkali dose was found to yield 65.85% celluloses, 19.35% hemicelluloses, and 9.78% lignins with substantial reduction in extractives (92.5%) and proteins (84.1%). Conversion of biomass into reducing sugars increased with increasing alkali doses up to 3% and the yield of reducing sugars was 53.13% at a hydrolysis time of 72 h (Kumar et al. 2016a).

The pretreatment techniques used for pearl millet are presented in Table 22.1. Fermentation studies revealed that pearl millet mashes containing 20, 25, 30, and 35% dry mass yielded 9, 11, 13–14, and 16–17% (v/v) of bioethanol, with fermentation efficiencies ranging from 90.0 to 95.6% (Wu et al. 2006). Estimated bioethanol from one ton of bajra stover is 201 l (Pathak and Chandel 2017).

22.8.8 Biorefinery Technologies

Biorefinery is a centralized facility used to convert the maximum percentage of biomass feedstocks into targeted final products with a generation of meager wastes in the processes. In other words, biorefinery utilizes all the parts of biomass feedstocks using appropriate biomass conversion technologies for efficient conversion into valuable products (Yoganandham et al. 2020). The selection of conversion process used in the biorefinery industries depends on (1) feedstock types and their composition, (2) targeted final products, (3) availability of viable conversion technologies, and (4) cost-effectiveness and marketability of the final products. Based on types of renewable feedstocks used, biorefineries are categorized into seven types (Kumar et al. 2020). The biorefinery concept has already been commercially launched in a few industries, such as paper and pulp, biofuel, and food (Hingsamer and Jungmeier 2019). Generally, wastes generated from pearl millet processing come under the category of lignocellulosic feedstocks. The biorefinery concept can be applied to these wastes for maximum utilization as raw material to generate multiple products, viz., bioenergy, biofuels, and biochemicals.

22.9 Challenges and Perspectives

Because of the wide variety of biomass feedstocks available globally or in India and their seasonal availability, biomass to biofuel conversion technologies should be region-specific (Kaushik and Singh 2016). The cost of raw materials plays an essential role in fixing the biofuel price. To avoid fluctuation in raw material price due to transportation cost and seasonal availability, a minimum support price for the agro-wastes may be fixed for the successful operation of biofuel/biorefinery industries. Furthermore, an excellent supply chain and logistic support for biomass transport should be available to reduce transportation costs by incorporating drying, pre-processing, and pelletization facilities at the point of material procurement. According to Passoth and Sandgren (2019), identification and improvement of microbial strains, efficient up/downstream technologies, and high-value chemical production are the efficient ways to exploit the straw biomass for multiple product production such as biofuels, biochemicals, and food. A research team of different disciplines such as agronomy, biomass conversion, process engineering, chemistry, conversion technology, genetic engineering, microbial fermentation, economics, and environmental science may be collaborated to offer solutions to the issues mentioned above (Ghosh 2016). This year, the United Nations General Assembly adopted a Resolution declaring 2023 the International Year of Millets, as proposed by India to the Food and Agriculture Organization (FAO). Millets, which include pearl millet, possess immense potential in our battles against climate change and poverty, and provide food, nutrition, fodder, and livelihood security. Being hardy, it can withstand extreme temperatures, floods, and droughts, and also help mitigate the effects of climate change through its low carbon footprint and hence may be exploited for biofuel production.

References

Abba HA, Zahari IN, Sapuan SM, Leman Z (2017) Characterization of millet (Pennisetum glaucum) husk fiber (MHF) and its use as filler for high-density polyethylene (HDPE) composites. BioRes 12(4):9287–9301

Allard RW (1960) Principles of plant breeding. Wiley, New York

Andre Savery, M. And Prasad, M.N. 1995. Genetic diversity in pearl millet, Pennisetum typhoides. Madras Agric J, 82: 8-11.

Andrews DJ, Ejeta G, Gilbert M, et al (1997) Breeding hybrid parents. Proceedings of an International Conference on the Genetic Improvement of Sorghum and Pearl Millet, held at Lubbock, Texas, September 22–27, 1996. International Sorghum and Millet Research (INTSORMIL) – International Crops Research Institute for the Semi-arid Tropics (ICRISAT), pp 173–187

Appa Rao S, Mengesha MH, Subramanian V (1982) Collection and preliminary evaluation of sweet-stalk pearl millet (Pennisetum). Econ Bot 36(3):286–290

Bera T, Purakayastha TJ, Patra AK, Datta SC (2018) Comparative analysis of physicochemical, nutrient, and spectral properties of agricultural residue biochars as influenced by pyrolysis temperatures. J Mater Cycl Waste Manage 20:1115–1127. https://doi.org/10.1007/s10163-017-0675-4

Blummel M, Zerbini E, Reddy BVS, Hash CT, Bidinger F, Khan AA (2003) Improving the production and utilization of sorghum and pearl millet as livestock feed: progress towards dual-purpose genotypes. Field Crop Res 84:143–158

Burton GW, Powell JB (1968) Pearl millet breeding and cytogenetics. Adv Agron 20:49–89

Campbell SG, Hoyle NP, Ashe MP (2005) Dynamic cycling of Eif2 through a large eIF2B-containing cytoplasmic body: implications for translation control. J Cell Biol 170:925–934

Cardoen D, Joshi P, Diels L, Sarma PM, Pant D (2015) Agriculture biomass in India: part 1. Estimation and characterization. Resour Conserv Recycl 102:39–48

Chadha KL (2010) Handbook of agriculture. Forage crops and grasses. New Delhi, ICAR, p 1356

Chalam GV, Venkateswarlu J (1965) Introduction to agricultural botany in India, vol I. Asia Publishing House, Bombay

Chapke RR, Shyam Prasad G, Das IK, Hariprasanna K, Singode A, Kanthi Sri BS, Tonapi VA (2020) Latest millet production and processing technologies. Hyderabad, ICAR-Indian Institute of Millets Research, p 82

Chen Y, Sharma-Shivappa RR, Keshwani D, Chen C (2007) Potential of agricultural residues and hay for bioethanol production. Appl Biochem Biotechnol 142(3):276–290

Cherney JH, Axtell JD, Hassen MM, Anliker KS (1988) Forage quality characterization of a chemically-induced brown-midrib mutant in pearl-millet. Crop Sci 28:783–787

De Assis RL, de Freitas RS, Mason SC (2018) Pearl millet production practices in Brazil: a review. Exp Agric 2018(54):699–718

Debieu M, Sine B, Passot S, Grondin A, Akata E, Gangashetty P, Vadez V, Gantet P, Foncéka D, Cournac L, Hash CT, Kane NA, Vigouroux Y, Laplaze L (2018) Response to early drought stress and identification of QTLs controlling biomass production under drought in pearl millet. PLoS One 13(10):e0201635. https://doi.org/10.1371/journal.pone.0201635

Degenhart NR, Werner BK, Burton GW (1995) Forage yield and quality of a brown mid-rib mutant in pearl-millet. Crop Sci 35:986–988

Dos Passos Bernardes A, Tremblay GF, Bélanger G, Seguin P, Brégard A, Vanasse A (2016) Bagasse silage from sweet pearl millet and sweet sorghum as influenced by harvest dates and

delays between biomass chopping and pressing. Bioenergy Res 9:88–97. https://doi.org/10.1007/s12155-015-9666-2
Dozier WA, Hanna W, Behnke K (2005) Grinding and pelleting responses of pearl millet-based diets. J Appl Poult Res 14:269–274
FAO (2001) Millet: post-harvest operations. Available from www.fao.org/3/a-av009e.pdf. Accessed 21 October 2021
Farrell AE, Plevin RJ, Turner BT, Jones AD, O'Hare M, Kammen DM (2006) Ethanol can contribute to energy and environmental goals. Science 311:506–550
Fussell LK, Pearson CJ (1978) Course of grain development and its relation to black region appearance in Pennisetum americanum. Field Crop Res 1:21–31
Geetha Thanuja K, Marimuthu S, Ramesh D, Karthikeyan S (2021) Paddy straw-based circular economy for sustainable waste management. In: Baskar C et al (eds) Handbook of solid waste management. Springer, Cham, pp 1–28
Ghosh SK (2016) Potential of economic utilization of biomass waste in India: implications towards SDGs. In: Seventh regional 3R forum in Asia and the Pacific, vol. 2. 2016
Gohel V, Duan G (2012) Conventional process for ethanol production from Indian broken rice and pearl millet. Bioprocess Biosyst Eng 35(8):1297–1308
Gupta SC (1995) Inheritance and allelic study of brown midrib trait in pearl-millet. J Hered 86:301–303
Harris K, Subudhi PK, Borrell A, Jordan D, Rosenow D, Nguyen H, Klein P, Klein R, Mullet J (2007) Sorghum stay-green QTL individually reduce post-flowering drought-induced leaf senescence. J Exp Bot 58(2):327–338. https://doi.org/10.1093/jxb/erl225
Himmel ME, Ding S-Y, Johnson DK, Adney WS, Nimlos MR, Brady JW (2007) Biomass recalcitrance: engineering plants and enzymes for biofuels production. Science 315:804–807
Hingsamer M, Jungmeier G (2019) Biorefineries. In: The role of bioenergy in the bioeconomy. Academic, New York, pp 179–222
Jain RK, Bal S (1997) Properties of pearl millet. J Agric Eng Res 66:85–91
Jinisha BJ, Jasmine Linda MP, Ramesh D, Kennedy ZJ, Karthikeyan S (2020) Cost-effective method for pretreatment of pearl millet biomass using catalytic downflow liquid contact reactor. Madras Agric J 2020:25–29
Kaushik N, Singh S (2016) Opportunities for bio fuels-Indian scenario. Procedia Environ Sci 35:847–852
Kuhe A, Terhemba AV, Iortyer H (2021) Biomass valorization for energy applications: a preliminary study on millet husk. Heliyon 7(8):e07802
Kumar A, Dutt D, Gautam A (2016a) Pretreatment and enzymatic hydrolysis of pearl millet stover by multi-enzymes from Aspergillus Nidulans akb-25. Cellul Chem Technol 50(7-8):781–790
Kumar A, Dutt D, Gautam A (2016b) Production of crude enzyme from Aspergillus nidulans AKB-25 using black gram residue as the substrate and its industrial applications. J Genet Eng Biotechnol 14(1):107–118
Kumar S, Paritosh K, Pareek N, Chawade A, Vivekanand V (2018) De-construction of major Indian cereal crop residues through chemical pretreatment for improved biogas production: an overview. Renew Sust Energ Rev 90:160–170
Kumar PS, Saravanan A, Jayasree R, Jeevanantham S (2020) Food industry waste biorefineries. In: Refining biomass residues for sustainable energy and bioproducts. Academic, New York, pp 407–426
Laougé ZB, Çığgın AS, Merdun H (2020) Optimization and characterization of bio-oil from fast pyrolysis of Pearl Millet and Sida cordifolia L. by using response surface methodology. Fuel 274:117842
Maheshwari P, Karthikeyan S (2020) A catalytic hydrothermal pretreatment process for sugar recovery from pearl millet biomass for Biofuel production. Madras Agric J 107:9602
Malhi GS, Sharma K, Kaur M, Gupta G (2020) Bajra Napier hybrid: an ideal crop for round the year green fodder production. Agric Food e-Newsl 2(9):907–908

Negi S, Mandpe A, Hussain A, Kumar S (2020) Collegial effect of maggots larvae and garbage enzyme in rapid composting of food waste with wheat straw or biomass waste. J Clean Prod 258:120854
Nepolean T, Blummel M, Bhasker Raj AG, Rajaram V, Senthilvel S, Hash CT (2006) QTLs controlling yield and stover quality traits in pearl millet. ICRISAT 2:1
Niculescu R, Clenci A, Iorga-Sim V (2019) Review on the use of diesel biodiesel-alcohol blends in compression ignition engines. Energies 12(7):1194. https://doi.org/10.3390/en12071194
Paritosh K, Pareek N, Chawade A, Vivekanand V (2019) Prioritization of solid concentration and temperature for solid-state anaerobic digestion of pearl millet straw employing multi-criteria assessment tool. Sci Rep 9(1):1–11
Paritosh K, Balan V, Vijay VK, Vivekanand V (2020) Simultaneous alkaline treatment of pearl millet straw for enhanced solid-state anaerobic digestion: experimental investigation and energy analysis. J Clean Prod 252:119798
Passoth V, Sandgren M (2019) Biofuel production from straw hydrolysates: current achievements and perspectives. Appl Microbiol Biotechnol 103(13):5105–5116
Pathak BS, Chandel AK (2017) Feedstock transportation, agricultural processing, logistic from farm to bio-refinery: recent developments, mechanization, and cost analysis. In: Sustainable biofuels development. Springer, Cham, pp 207–221
Patil FB, Jadhav SD (1992) Correlation and path analysis in pearl millet times Napier hybrids. J Maharashtra Agric Univ 17(2):199
Pereira Filho IA, Parrella RAC, Moreira JAA, May A, Souza VF, Cruz JC (2013) Avaliação de cultivares de sorgo sacarino [Sorghum bicolor (L.) Moench] em diferentes densidades de semeadura visando a características importantes na produção de etanol. Rev Bras Milho Sorgo 12(2):118–127
Purohit P, Dhar S (2018) Lignocellulosic biofuels in India: current perspectives, potential issues and future prospects. Aims Energy 6(3):453–486
Rachie KO, Majmudar JV (1980) Pearl millet. Pennsylvania State University Press, University Park
Reddy LVA, Reddy OVS, Basappa SC (2005) Potentiality of amylolytic yeasts for direct fermentation of starchy substrates to ethanol. Indian J Microbiol 45:1–15
Rosa SML, Santos EF, Ferreira CA, Nachgell SMB (2009) Studies on the properties of rice-husk-filled-PP composites – effect of maleate PP. Mater Res 12(3):333–338. https://doi.org/10.1590/S1516-14392009000300014
Sandra B (1997) Estimating biomass and biomass change of tropical forests: a primer vol 134 Food and Agriculture Organization of the United Nations A Forest Resources. Assessment publication, p 55
Singh P, Rai KN, Witcombe JR, Andrews DJ (1988) Population breeding methods in pearl millet improvement (Pennisetum americanum). Agron Trop 43:185–193
Sundararaj DD, Thulasidas G (1980) Botany of field crops. The MacMillan Company India Ltd., New Delhi, p 508
Suthamathi P, Stephen Dorairaj M (1995) Variability, heritability and genetic advance in fodder pearl millet. Madras Agric J 82:238–240
Suttie JM (2000) Hay and straw conservation: for small-scale farming and pastoral conditions (No. 29). Chapter IX Dry crop residues. Food & Agriculture Organization, Rome
Thomas H, Ougham H (2014) The stay-green trait. J Exp Bot 65:3889–3900
Velusamy S, Subbaiyan A, Thangam RS (2021) Combustion characteristics of briquette fuels from sorghum panicle–pearl millets using cassava starch binder. Environ Sci Pollut Res 28(17): 21471–21485
Vermerris W (2011) Survey of genomics approaches to improve bioenergy traits in maize, sorghum, and sugarcane. J Integr Plant Biol 53:105–119
Virk DS (1988) Biometrical analysis in pearl millet – a review. Crop Improv 15:1–29
Vohra M, Manwar J, Manmode R, Padgilwar S, Patil S (2014) Bioethanol production: feedstock and current technologies. J Environ Chem Eng 2014(2):573–584

Wang D, Wu X, Bean S, Wilson JP (2006) Ethanol production from pearl millet by using Saccharomyces cerevisiae. Cereal Chem 83:127–131

Wilson JP, Jurjevic Z, Hanna WW, Wilson DM, Potter TL, Coy AE (2006a) Host-specific variation in infection by toxigenic fungi and contamination by mycotoxins in pearl millet and corn. Mycopathologia 161:101–107

Wilson JP, McAloon AJ, Yee WC, McKinney J, Wang D, Bean SR (2006b) Biological and economic feasibility of pearl millet as a feed stock for ethanol production. In: 6th new crops symposium. Association for the Advancement of Industrial Crops, San Diego, p 47

Wilson JP, McAloon AJ, Yee W, McKinney J, Wang D, Bean SR (2007) Biological and economic feasibility of pearl millet as a feedstock for ethanol production. Issues New Crops New Uses 2007:56–59

Wu X, Wang D, Bean SR, Wilson JP (2006) Ethanol production from pearl millet using Saccharomyces cerevisiae. Cereal Chem 83:127–131

Xu H, Li Y, Hua D, Zhao Y, Mu H, Chen H, Chen G (2020) Enhancing the anaerobic digestion of corn stover by chemical pretreatment with the black liquor from the paper industry. Bioresour Technol 306:123090

Yadav M, Vivekanand V (2019) Chaetomium globosporum: a novel laccase producing fungus for improving the hydrolyzability of lignocellulosic biomass. Heliyon 5(3):e01353

Yadav M, Vivekanand V (2020) Biological treatment of lignocellulosic biomass by Curvularia lunata for biogas production. Bioresour Technol 306:123151

Yadav M, Vivekanand V (2021) Combined fungal and bacterial pretreatment of wheat and pearl millet straw for biogas production–A study from batch to continuous stirred tank reactors. Bioresour Technol 321:124523

Yadav M, Rengasamy RS, Gupta D (2019) Characterization of pearl millet (Pennisetum glaucum) waste. Carbohydr Polym 212:160–168

Yoganandham ST, Sathyamoorthy G, Renuka RR (2020) Emerging extraction techniques: hydrothermal processing. In: Sustainable seaweed technologies. Elsevier, Amsterdam, pp 191–205

Zhao X, Zhang L, Liu D (2012) Biomass recalcitrance. Part I: the chemical compositions and physical structures affecting the enzymatic hydrolysis of lignocellulose. Biofuels Bioprod Biorefin 6:465–482

An Ecosystem Approach to Promoting Pearl Millet: Balancing Demand and Supply 23

Ashok Dalwai

This paper seeks to highlight the role of pearl millet in promoting a food system that addresses multiple expectations from agriculture. These include food and nutrition, climate resilience and livelihood and income, which implies taking care of the needs of various stakeholders including farmer-producers and consumers. It deserves appreciation that no single millet can be discussed without reference to the millets in general. Hence, the issues are examined under two broad sections, wide Sect. 23.1—Millets, and Sect. 23.2—Pearl Millet, before arriving at conclusion and recommendation in Sect. 23.3.

23.1 Millets and Their Importance in Food Security

Millets among the important cereals grown in the tropics including India are also known as coarse cereals in many countries. In most countries including India where they are raised, millets are generally cultivated at subsistence level under poorly endowed production environment, and as a mixed- or inter-crop. They are traditionally consumed as a staple food and drink across many countries. As food, millets are nutritionally equivalent or superior to most cereals, containing high levels of nutrients. They are rich in various amino acids essential for good health, and are a unique source of provitamin A and micronutrients (Zn, Fe and Cu).

Millets, generally being thermophilic (thriving at relatively higher temperatures) and xerophilic (can reproduce with limited water input) (Saxena et al. 2018), are good in mitigating the effects of climate change on account of their low carbon

A. Dalwai (✉)
Ministry of Agriculture and Farmers Welfare, Government of India, New Delhi, India

National Rainfed Area Authority (NRAA), New Delhi, India
e-mail: ashok.dalwai@gov.in

V. A. Tonapi et al. (eds.), *Pearl Millet in the 21st Century*,
https://doi.org/10.1007/978-981-99-5890-0_23

footprint of 3218 kg equivalent of carbon dioxide per hectare (ha). The more popular cereals, wheat and rice, with 3968 kg and 3401 kg equivalent of carbon dioxide, respectively, compare poorly in this regard.

Notwithstanding several of their advantages, production of millets in many parts of the world has been on the decline over the decades. The Green Revolution that has helped most parts of the world to achieve food security is anchored around a few cereals, with wheat and rice occupying pivotal space, and to the neglect of millets. The technological advancements linked to select crops combined with robust policy support by the governments have narrowed both production and consumption patterns of the globe from a much more diverse norm of yester years. This has resulted in a more vulnerable and risky production system, and nutritionally compromised consumption pattern. The outcome though called 'food security' seems to be only 'carbohydrate security', when the need is nutrition security for the people across the continents. This is apparent from the high degree of malnourishment across several countries. In India too, both production and consumption status are no exception—major share of the food basket coming from just two cereals, wheat and rice. In the year 2020–2021, for example, of the total food grain production of 308 million tons, wheat and rice alone accounted for 231 million tons, and the rest came from maize (31.51 million tons), millets (19.64 million tons) and pulses (25.72 million tons). Indian population suffers from high level of malnourishment. Given that food and health are positively correlated, the need for a nutrition-led production in India cannot be over-emphasised.

23.1.1 Millet Production: Global Trend

Globally, the food production system has transitioned from a wide diversity to a narrow base since the adoption of the Green Revolution technology, concentrated around the dwarf varieties of wheat and rice. Biodiversity has suffered at both farm and the larger ecosystem levels, impacting agricultural system negatively.

Worldwide, one of the visible impacts has been the plummeting of area under millets over the last few decades, restraining a higher level of production that was possible. The output of all the millets together has been hovering around 30 million hectares in the last decade. The area under millets at 43.40 million ha in 1961, increased up to 45.11 million ha during the 1970s but began to experience a negative trend thereafter due to number of production constraints and policy challenges.

The production trends showed a positive growth climbing even up to 35 million tons in 2003, only to commence a slide back thenceforth to equal the starting level of about 28 million tons. The productivity trends demonstrate an increase from 600 kg ha^{-1} in 1961 to 965 kg ha^{-1} in 2008, and then a decline to about 860 kg ha^{-1} in 2018–2019.

Africa is the only region where millet production shows a substantive rise, growing from 8 million tons to over 13.7 million tons between 1979–1981 and 2019. Most of the increase in production, however, occurred during the first half of the 1980s, which has since been sustained by area expansion, mainly in the Sahel

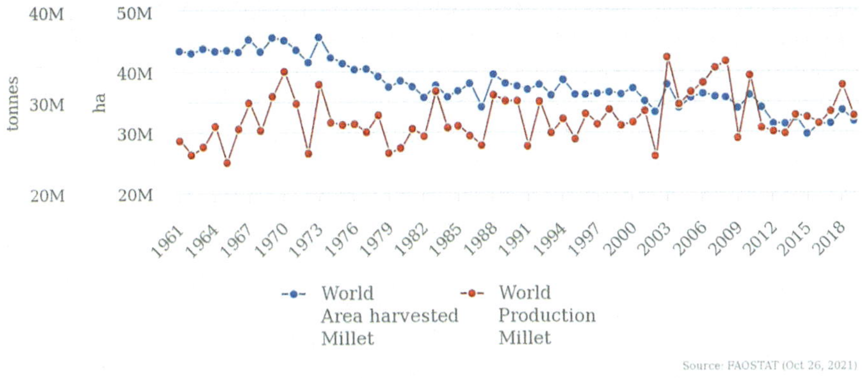

Fig. 23.1 Production/yield quantities of millet in the world during 1961–2019. (Source: FAO 2021)

and to a smaller extent in other countries. For many African countries, millet yields have remained stagnant or fallen, partly because much of the expansion has been into areas with poor soils and low, erratic rainfall.

In Asia, India has experienced growth in production despite a steep fall in the area under millets, thanks to technology-led productivity increases. This has helped the country to hold on to the production level of the year 1985–1986 (17.6 million tons) even in 2019–2020 (17.27 million tons), since when backward march of area under millets began from that year high of 32.30 million ha to a low level of 13.82 million ha in the year 2019–2020 (Fig. 23.1).

In Asia, millet is restricted almost exclusively to two countries, India and China, although Myanmar, Nepal and Pakistan also produce small quantities. The Asian nations have seen a major shift of area under millets to other crops including the horticultural crops, and more prominently since the mid-1990s. However, part of this decline was compensated for, by yield increases from 0.82 tons per ha in 2000 to 1.29 tons per ha in 2009. During the past three decades, yields have roughly doubled in China—where they are now among the highest in the world. In case of India, there has been increase too but just by more than half, as seen from the per ha productivity increasing from 0.75 tons to 1.25 tons in 2019–2020. This progress is essentially a result of successful breeding research, and the widespread dissemination of pearl millet hybrids in India and improved open-pollinated foxtail millet varieties in China.

As of now, the millets as a group are subject to a range of limitations that include decline in area coverage, high yield gaps, low prioritisation in research agenda and consequent lack of technology breakthrough. Also, public and private investments in millets are limited to seed development and production, depriving them of a broader policy incentive as available to cereals like wheat and rice. International prices for millets are highly volatile, determined largely by supply volumes, and are usually unrelated to those of other major coarse grains (now renamed as nutria-cereals in India), such as maize, sorghum or barley. Owing to their nutritional content, any improvement/development in cultivation, storage, availability, price and processing technology for millets could significantly contribute to the food and nutritional security of India's population (Michael Raj and Shanmugam 2013). Further, these millets can diversify the food basket, which is at present very narrow because of preponderance of major cereals like rice and wheat in terms of both production and distribution under the government's public distribution system (PDS).

23.1.1.1 Pearl Millet: Global Trends

In respect of pearl millet, one of the three major millets, Table 23.1 below shows the data on various parameters for countries that grow this crop majorly.

As seen from Table 23.1, pearl millet is mostly grown in India and African nations, where climate is harsh indicating the survival potential of this millet. While Niger leads the world in terms of area under pearl millet, it is India that stands only next but produces the highest quantum because of its relative yield advantage.

Table 23.1 Production of pearl millet in major millet-producing countries during 2016

S. no.	Country	Area ('000 ha)	Production ('000 tons)	Yield (kg/ha)	Percent contribution to production (%)
1	India	7129	10,280.0	1442	44.5
2	Niger	7230	3886.1	537	16.8
3	Mali	2040	1806.6	886	7.8
4	Nigeria	1736	1468.7	846	6.4
5	Sudan	3007	1449.0	482	6.3
6	Burkina Faso	1245	1056.9	849	4.6
7	Chad	1225	725.7	593	3.1
8	Senegal	858	612.6	714	2.7
9	Pakistan	492	318.3	647	1.4
10	Tanzania	341	312.4	917	1.4
11	Guinea	189	194.0	1028	0.8
12	Ghana	162	159.0	980	0.7
13	Gambia	117	101.9	871	0.4
14	Cameroon	72	100.0	1398	0.4
15	Yemen	125	86.7	691	0.4
16	Côte d'Ivoire	67	54.5	815	0.2
17	Kenya	88	54.0	611	0.2
18	Namibia	228	44.6	195	0.2
19	Angola	187	42.0	224	0.2
20	Congo	61	41.0	669	0.2
	Total	27,161	23,092	850	

Source: Report of the Committee on Doubling Farmers' Income (DFI) (2018)

23.1.2 Millet Production: Trends in India

India, with its vast tracts of semi-arid tropics and dominance of rainfed systems of agricultural production in the country, has for long been home to large number of diverse millets. The millets of India encompass three major millets, five minor millets, two lesser-known millets, one extinct millet and two pseudo-millets. Of these, all the three major millets, all five minor millets and brown top millet, one of the two lesser-known millets, are the principal nine millets raised in different parts of India. Further, sorghum (jowar), pearl millet (bajra) and finger millet (ragi) are considered as major millets because of their higher ratio of area and production relative to other millets.

Since the mid-1950s, the area under millets has undergone a sharp and steady decline (figures in million ha): 36.34 (1955–1956), 36.91 (1965–1966), 34.96 (1975–1976), 32.30 (1985–1986), 24.09 (1995–1996) and 20.84 (2005–2006). The production in million tons for the comparable years stood at 14.07 (1955–1956), 14.21 (1965–1966), 19.96 (1975–1976), 17.59 (1985–1986), 17.99 (1995–1996) and 18.14 (2005–2006). It is clear that the production has been sustained despite a visible decline in area on account of higher productivity realised through better technology.

As per data offered by FAO (which varies from the Government of India data compiled by the Directorate of Economics and Statistics [DES], Ministry of Agriculture and Farmers Welfare [FW]), India is the largest global producer with a global market share of 36.1% in millets (Fig. 23.2). The millets area coverage in India plummeted from 18.65 million ha in 1961 to 11.0 million ha in 2008 and further to 8.4 m ha in 2019. Thanks to the modern cultivation technology and high-yielding varieties (HYVs), the production has held on due to increase in yield from 0.41 tons per ha to 1.2 tons per ha in 2019 (Food and Agriculture Organization of the United Nations (FAO) 2019) (Fig. 23.3).

Once again, as per DES, Ministry of Agriculture and FW data, the area under millets at 16.99 million ha in the year 2011–2012 and 14.99 million ha in 2015–2016 reduced to 13.82 million ha in the year 2019–2020. The total production in million

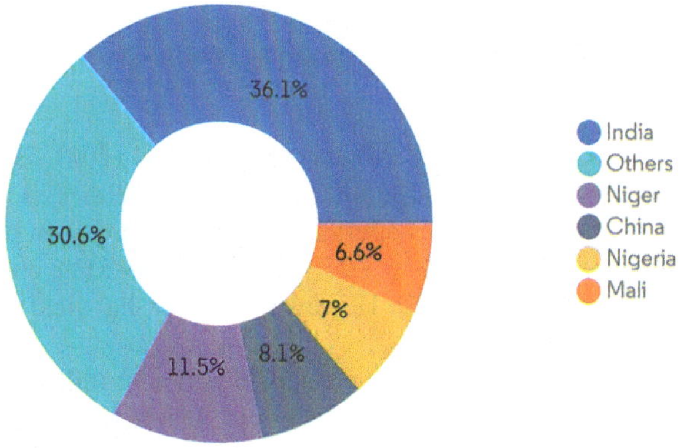

Fig. 23.2 Millet: market share (%) in global production, 2019. (*Source*: Food and Agriculture Organization of the United Nations (FAO) 2021)

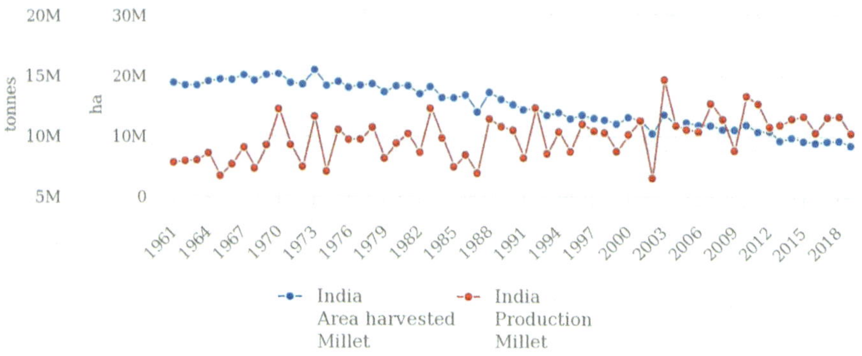

Fig. 23.3 Production/yield quantities of millets in India during 1961–2019. (*Source*: FAO 2019)

Table 23.2 India: total area under millet cultivation—trend (1995–1996 to 2019–2020)

	Area (lakh hectares)				
Year	Jowar	Bajra	Ragi	Small millets	Total millets
1955–1956	173.6	113.4	23.07	53.35	363.42
1965–1966	176.8	119.7	26.96	45.64	369.10
1975–1976	160.9	115.7	26.30	46.72	349.62
1985–1986	161	106.5	24.01	31.55	323.06
1995–1996	113.3	93.2	17.74	16.62	240.86
2005–2006	86.67	95.81	15.34	10.64	208.46
2010–2011	73.82	96.12	12.86	8.00	190.80
2011–2012	62.45	87.77	11.76	7.99	169.96
2012–2013	62.14	72.97	11.31	7.54	153.97
2013–2014	57.93	78.11	11.94	6.82	154.80
2014–2015	61.61	73.18	12.08	5.90	152.77
2015–2016	60.77	71.29	11.38	6.50	149.94
2016–2017	56.24	74.59	10.16	6.19	147.18
2017–2018	50.24	74.81	11.94	5.46	142.46
2018–2019	40.93	71.05	8.91	4.54	125.43
2019–2020	48.24	75.43	10.04	4.58	138.29

Source: Directorate of Economics and Statistics (n.d.)

tons for the comparable years stood at 18.64 (2011–2012), 14.52 (2015–2016) and 17.26 (2019–2020). The total output saw a further rise to 19.64 million tons in the year 2020–2021. This upgrade came from increase in per ha productivity from 0.97 ton/ha in 2015–2016 to 1.25 ton/ha in 2019–2020, and its further increase to 1.37 ton/ha in 2020–2021. However, the per ha yield was 1.09 tons in the year 2011–2012.

The data comparing the status of different millets for the period 1995–1996 to 2019–2020, with respect to their area, production and productivity, may be seen in

Table 23.3 India: total production of millets – trend (1995–1996 to 2019–2020)

	Production (lakh tons)				
Year	Jowar	Bajra	Ragi	Small millets	Total millets
1955–1956	67.26	34.3	18.46	20.7	140.72
1965–1966	75.81	37.52	13.27	15.55	142.15
1975–1976	95.04	57.36	27.97	19.24	199.61
1985–1986	101.97	36.64	25.18	12.17	175.96
1995–1996	93.27	53.81	25.01	7.79	179.88
2005–2006	76.3	76.84	23.54	4.72	181.40
2010–2011	69.67	103.70	21.93	4.42	199.72
2011–2012	59.79	102.76	19.29	4.52	186.36
2012–2013	52.81	87.42	15.74	4.36	160.34
2013–2014	55.42	92.50	19.83	4.30	172.05
2014–2015	54.45	91.84	20.61	3.86	170.76
2015–2016	42.38	80.67	18.22	3.91	145.17
2016–2017	45.68	97.30	13.85	4.42	161.25
2017–2018	48.03	92.09	19.85	4.39	164.36
2018–2019	34.75	86.64	12.39	3.33	137.11
2019–2020	47.72	103.63	17.55	3.71	172.61

Source: Directorate of Economics and Statistics (n.d.)

Table 23.4 India: productivity of millets—trend (1995–1996 to 2019–2020)

	Productivity (kg/ha)				
Year	Jowar	Bajra	Ragi	Small millets	Total millets
1955–1956	387.00	302.00	800.00	388.00	387.21
1965–1966	429.00	314.00	492.00	341.00	385.13
1975–1976	591.00	496.00	1064.00	412.00	570.93
1985–1986	633.00	344.00	1049.00	386.00	544.67
1995–1996	823.00	577.00	1410.00	469.00	746.82
2005–2006	880.00	802.00	1534.00	443.00	870.19
2010–2011	943.84	1078.81	1705.38	552.50	1046.76
2011–2012	957.42	1170.83	1640.81	565.28	1096.47
2012–2013	849.88	1197.96	1392.03	577.72	1041.35
2013–2014	956.57	1184.28	1661.26	630.09	1111.41
2014–2015	883.78	1255.03	1705.97	654.47	1117.78
2015–2016	697.38	1131.59	1600.62	601.51	968.23
2016–2017	812.15	1304.53	1363.15	713.84	1095.57
2017–2018	956.00	1231.03	1662.27	803.60	1153.79
2018–2019	849.06	1219.44	1390.34	733.88	1093.15
2019–2020	989.29	1373.86	1747.27	809.01	1248.12

Source: Directorate of Economics and Statistics (n.d.)

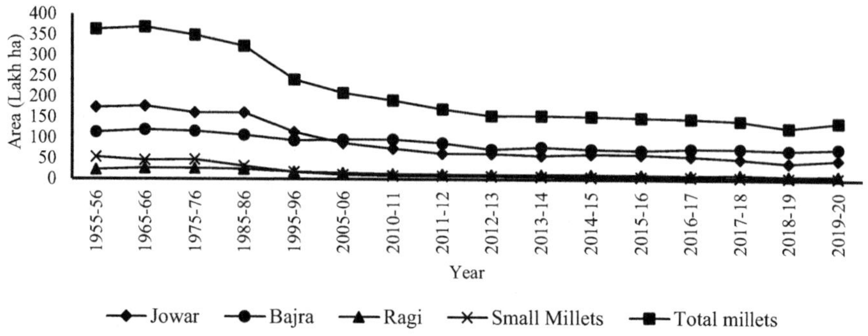

Fig. 23.4 India: total area under millet cultivation (lakh ha)—trend (1955–1956 to 2019–2020)

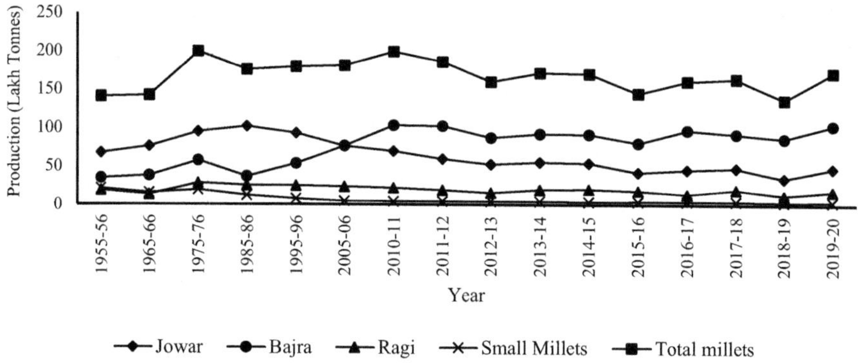

Fig. 23.5 India: total production of millets (lakh tons)—trend (1955–1956 to 2019–2020)

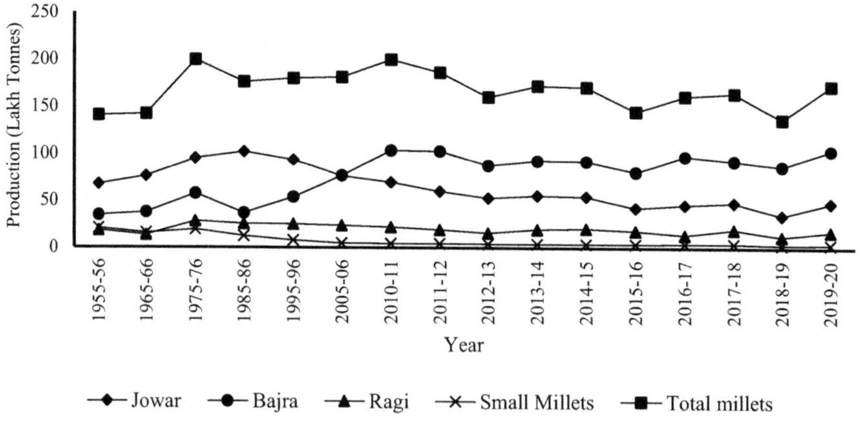

Fig. 23.6 India: productivity of millets (kg ha^{-1})—trend (1955–1956 to 2019–2020)

Tables 23.2, 23.3, and 23.4, respectively. Also may be seen the comparable data for the comparable period in the line diagrams vide Figs. 23.4, 23.5, and 23.6.

Major millet-producing states in India are Karnataka, Andhra Pradesh, Tamil Nadu, Maharashtra, Odisha, Madhya Pradesh, Rajasthan and Uttarakhand. Among these states, maximum area under millets can be seen in Rajasthan (5 million ha, of which 87% is under pearl millet), Maharashtra (4 million ha, of which 75% is under sorghum) and Karnataka (2 million ha, of which 54% is under sorghum and 32% is under finger millet) (Stanly and Shanmugam 2013).

The Doubling Farmers Income (DFI) Committee reports that the total area under millets declined at a compound annual growth rate (CAGR) of 5.4% per annum during the period of 2010–2011 to 2014–2015. During the same period, the total millet production dipped at an annual rate of 4%. On account of marginal increase in the productivity at CAGR of 1.2%, the yearly reduction in production was lower than the loss of area under millets.

23.1.3 Pearl Millet: Trends in India

As reported by the DFI Committee, pearl millet too has faced a declining trend with regard to both area and production at a CAGR of 6% and 3.5%, respectively. Though the productivity registered a positive trend of CAGR of 3.1%, it was not high enough to neutralise the major loss in area.

A more detailed status of pearl millet in terms of its area, production and productivity may be appreciated from the Tables 23.2, 23.3, and 23.4 and Figs. 23.4, 23.5, and 23.6. As these tables and figures depict, pearl millet has gained greater status by 2019–2020 compared to all other millets, including sorghum. And, this status in 2019–2020 compared to that in the year 1955–1956 is healthier in respect of all three parameters—area, production and productivity (except for ragi)—indicating the improving relevance of pearl millet even when millets as a group have suffered decline in area.

23.1.4 Millets: Decreasing Ratio of Consumption

Over the last two decades of the twenty-first century, the share of millets as a food staple has reduced in the country due to various factors affecting the availability and accessibility, as also affordability. With increasing output of wheat and rice in the country triggered by improved technology, and supported by price incentive and food distribution policy of the government, the preference for millets as a staple has suffered many a disadvantage. In terms of perception too, the large section of the society including in the rural areas has come to consider millets as inferior to major millets like wheat and rice. In result, rising incomes and growing urbanisation have incentivised higher output of wheat and rice facilitated by both higher productivity and area expansion. The expansion of area has come from diversification of that under millets. Currently, reduced demand for consumption has led to diversion of more than 50% of the millet production into alternative uses. While the use of various agri-produce for non-food processing may be good from the perspective of expanding the functional market and associated demand, it is important in case of millets to stimulate higher demand for consumption as staples in view of their enviable nutritional value.

23.1.5 Millet Production in India: Constraints

Millet production in the developing world, including in India, generally happens in poorly endowed areas that suffer from poor soil fertility, low and erratic rainfall, high temperatures, widespread *Striga* infestation, downy mildew disease and loss of grain to birds.

With increasing area under dependable sources of irrigation, improved technology and new varieties with shorter duration, the cropping intensity has been on the rise with concurrent shortening of fallow periods and accelerated decline in soil fertility. This process has pushed millet production into more marginal lands. The impact is evident in the declining yields of millet in the major producing areas over the past 15 years (e.g. western Rajasthan in India).

In addition, millets are cultivated on small and fragmented production units, which are often inter-cropped (usually with legumes and sometimes with sorghum or maize). Unreliable precipitation is a disincentive to use of inputs such as chemical fertiliser, pesticides and farm machinery; and this is further compounded by limited commercial demand. In a way a vicious cycle is created to the disadvantage of millets.

Crop improvement in millets is a challenge compared to other cereals because of their morphology, and the nature of the environment in which they are cultivated. It is also true that the national millet improvement programmes came into attention much later and remained weaker than those for many other crops.

Budgetary support for millet breeding research has been relatively lower than for other cereals. Crops like wheat and rice have dedicated research facilities, whereas millets including pearl millet have come to have one such research institute only recently, when the Indian Council of Agricultural Research (ICAR) upgraded its Indian Institute of Sorghum Research, Hyderabad, into Indian Institute of Millets Research (IIMR), Hyderabad, with a mandate to cover all the millets including sorghum. Further, there has been almost nil experience with research on millet breeding in developed countries that could be transferred to developing countries, as has been the case vis-à-vis wheat, rice and maize. Among the various millets, only pearl millet and to a small extent finger millet have so far been researched at the international level.

Breeding programmes for millets have traditionally targeted the relatively better environments, although even these environments are not as favourable as those reserved for most other crops. Hybrid grain cultivars have been developed for pearl millet in India and the United States, but perform best in areas where rainfall is reliable. The breeders find it hard to identify dual-purpose grain/stover combinations in poorly endowed regions with low and erratic rainfall patterns, and that are suitable across a range of production environments. This challenge has been addressed by the breeders by focusing on developing open-pollinated varieties that give stable grain and straw yields and suit the prevailing rainfall pattern, in preference to aiming at potential yield maximisation under more favourable conditions.

Further, even when new technologies for crop and resource management have been developed, adoption has been poor, on account of poor extension and risk-aversion of farmers cultivating in harsh environments. Since the returns on investments in millets are lower than those in other crops and non-farm enterprises, both research and development (R&D) and other activities do not attract proportionate attention from either public or private sectors.

There are several other challenges to promotion of millets that fall within the technological, research, post-harvest and policy context. These are examined in a later section that deals with promoting pearl millet.

23.1.6 Food and Health

Food and health cannot be isolated from each other. Food is integral to health, and a nutritious food that meets the prescribed levels of macro- and micronutrients of an individual is essential for a healthy living. In this context, it would be appropriate to examine the definitions of food security. According to a currently accepted definition (Food and Agriculture Organization of the United Nations (FAO) 2019), 'Food Security' is achieved when it is ensured that,

> All people, at all times, have physical, social and economic access to sufficient, safe and nutritious food which meets their dietary needs and food preferences for an active and healthy life

Food is here defined as any substance that people eat and drink to maintain life and growth. As a result, safe and clean water is an essential part of food commodities.

As per World Food Summit 'Food security' exists when,

> All people, at all times, have physical and economic access to sufficient, safe and nutritious food that meets their dietary needs and food preferences for an active and healthy life.

The *four important dimensions* of food security that emerge from these widely accepted definitions include:

- Food availability
- Food access
- Food utilisation
- Food stability

23.1.6.1 Hunger and Food System

Though there has been rapid growth in agriculture with its focus on food production, more than 820 million people in the world today are still hungry, underscoring the immense challenge of achieving the 'Zero Hunger' as targeted under Sustainable Development Goals (SDG) 2030. It is almost all sub-regions of Africa and, to a lesser extent, Latin America and Western Asia that are experiencing growing hunger. It is also highly disturbing that about 2 billion people in the world are vulnerable to moderate or severe food insecurity. This is apart from the hidden hunger that many suffer from, due to poor nutrition intake. The lack of regular access to nutritious and sufficient food is what is resulting in risk of malnutrition and poor health. Though this is a problem mostly in low- and middle-income countries, around 8% of the population in Northern America and Europe also suffer from moderate or severe food insecurity. This signifies the importance of creating awareness around healthy eating habits, even when people have adequate purchasing power. There is also gender dimension to food insecurity, since it is seen to be slightly higher among the women in comparison to men.

India has a lot to catch up with, as is obvious from the health and nutrition indicators. The country will need to do much better than what the current trends indicate, so as to meet the 2030 SDG target to halve the number of stunted children, and the 2025 World Health Assembly target to reduce the prevalence of low birth weight by 30%.

In respect of hunger and malnutrition, the prevalence of undernourishment (PoU) in India decreased from 22.2% in 2002–2004 to 14.8% in 2015–2017. The percentage of stunting in children fell from 54.2 to 38.4 between the years 2000 and 2015. However, this is still a high prevalence, compared to a global average of 23.2% in the year 2015. This points to the need for an ecosystem approach to food and nutrition security, which is possible when 'Food System' becomes the fulcrum, and not just food production. As per the concept and framework on 'Sustainable food systems' developed by the FAO,

> Food system (FS) encompasses the entire range of actors and their inter-linked value-adding activities involved in the production, aggregation, processing, distribution, consumption and disposal of food products that originate from agriculture, forestry and fisheries, and parts of broader economic, societal and natural environments in which they are embedded.

Further, FAO also defines sustainable food system (SFS) as,

> a food system that delivers food security and nutrition for all in such a way that the economic, social and environmental bases to generate food security and nutrition for future generations are not compromised.

23.1.7 Current Challenges to, and Demand from, Agriculture

Since the beginning of the settled agriculture 12,000 years ago, it has been the basis of human civilisation. Through the millennia, it has been called upon to meet societal demands for food, fodder and fibre. From the mid-twentieth century, when the nations began to see higher population growth rates adding to their net population, food security has been a major demand on the agriculture sector. In response, the new production technology that has been extractive of the nutrients in nature combined with additional area brought under the plough has caused compromise of production environment. Some of its deleterious impacts include loss of organic carbon, soil erosion, increasing extent of problematic soils, decreasing ground water level and loss of biodiversity. The threat to sustainable production environment is exacerbated by the climate change. With the world population expected to reach 9 billion by 2050, the demand from the agriculture sector is getting more complex, entailing a multipronged approach to find a solution.

In India, the challenge is similar given (1) its high population density, estimated to increase further to 1.5 billion by 2050, and (2) shrinking arable land extent due to its diversion for non-agricultural activities. Though the country has succeeded in generating required quantum of cereals, there have arisen new challenges to the existing production system. The basket of the three new concerns—nutrition for the consumers, incomes for the farmers and sustainability of the ecology—is a challenge to the agriculture sector, demanding a paradigm shift in the way it is practiced. These issues are discussed in brief in the sub-sections that follow.

23.1.7.1 Nutrition for the Consumer

Balanced nutrition is necessary for good health, general well-being and as a first defence against disease. Nutritional problems caused by an inadequate diet can manifest in many ways, including perpetuating inter-generational cycle of malnutrition. While young children and women are the most vulnerable, the challenges of malnutrition—under nutrition, micronutrient deficiencies and obesity—exist in all countries and cut across socio-economic classes. However, the causes will differ between poor and richer nations.

Emerging challenges, which include climate change, environmental sustainability and rapid technological shifts, are impacting food systems, engendering concerns over meeting nutritional requirements for a growing population. Hence the need for adopting nutrition-sensitive agriculture, which seeks to ensure the sustainable production of a variety of affordable, nutritious, culturally appropriate and safe foods in adequate quantity and appropriate quality to meet the dietary requirements of all.

India presently faces the paradox of surplus production in certain segments like cereals, but there exists deficit of protein-rich commodities like pulses and oilseeds needed to provide edible oils. Within the cereals, the nutritionally rich millets have been experiencing a short shrift resulting in almost a stagnant level of output as seen from the discussions in the preceding sections. This imbalance is manifest in high degree of malnourishment and resultant stunted and wasted growth among children, anaemia among women and lower expectancy of life at birth in general.

23.1.7.2 Sustainability in Production

Modern agricultural practices that encompass intensive tillage, intensive use of water and use of agro-chemicals have resulted in extraction of soil nutrients and water beyond carrying capacity, even as adequate food has been generated.

In India's food basket—the Indo-Gangetic Plains—the marginal rates of return on investments in agriculture are decreasing. One now needs to use more inputs to realise one unit of incremental growth. This is not a sustainable system. Simultaneously, the rainfed regions that count at 52% of the country's total arable extent have not been fully brought into science-led production system. The contemporary need is for a more horizontally spread and diverse agricultural production system for long-term sustainability. Biological diversity across the country needs to be nurtured. In this context ecologically synchronous cropping systems like millets in arid and semi-arid zones assume importance.

23.1.7.3 Support to Livelihood Systems

For developing countries including India, where a majority of families, in both the farm and non-farm sectors, derive their primary income from agriculture, livelihood issues assume critical importance. Livelihood is defined as adequate stock and flow of food and cash with an individual or a family to meet its basic needs. This implies secured ownership of, or access to, resources and income-earning activities, including reserves and assets to negotiate risks, ease shocks and meet contingencies.

There are four principal ways by which rural households can secure their livelihoods. These are: (1) Production-based livelihood—majority of the small and marginal farmers gain livelihoods by cultivating their small parcels of land. (2) Labour-based livelihood—most of the small and marginal landholders and landless rural households earn their livelihoods by engaging themselves as labour in farm and non-farm activities undertaken by private and public sector agencies. (3) Exchange- or market-based livelihood—hereunder the rural households producing surplus food and non-food agricultural products or non-farm goods sell them in the market and earn their livelihoods. (4) Transfer-based entitlements—it is the welfare support from government and various social organisations that sustains the livelihoods of the households that are short of income-earning asset or have no able-bodied person to work. Some examples include Government's welfare, social security and food assistance programmes.

The agricultural production system must ensure creation of jobs and conducive environment for the farmers to earn remunerative returns on their investments.

23.1.7.4 Income for Farmers

Agriculture employs the largest number of people for livelihood in India and is an important source of income. While India's agriculture has registered good growth, the farmers have not benefitted proportionately in terms of income. The percentage of rural households generating primary income from agriculture has gone up from 54 in 2013 to 58 in 2019 as per National Sample Survey Office's (NSSO's) Situation Assessment Survey (SAS). The income of agricultural households increased in nominal terms by 59% between 2012–2013 and 2019 (from Rs. 6426 to

Rs. 10,219). The Consumer Price Index (rural) increased by 35% (from 105 to 142) during the same period, suggesting a real income increase of 24% to agricultural households.

The agricultural sector needs to adopt an income approach that enables the farmers to generate adequate income from the agricultural enterprise. This pre-requisites adoption of a strategy that aims at (1) higher output through higher productivity, (2) lower cost of production, and (3) realisation of higher real remunerative prices on the marketed surpluses.

In promoting millets, and pearl millet in particular, suitability will need to be examined from these three standpoints besides resilience to risk and sustainability of production. When a crop or activity is tested on the anvil of these parameters, inter-crop yield advantages alone will not determine the choice of a crop. Under such a condition, a resilient crop like pearl millet may be found to be more feasible even though its yield level may not compare to be the best.

23.2 Pearl Millet

Pearl millet, *Pennisetum glaucum (L.)* commonly known as bajra in India, is probably the most drought-tolerant of all the cereals. Belonging to the grass family Gramineae, pearl millet can regenerate and produce new basal tillers to compensate for losses caused by drought or other unfavourable conditions. Thanks to its resilience, pearl millet is a better performer than other cereals under poorly endowed production conditions characterised by poor soil, moisture stress and higher heat. However, it responds very well when grown under favourable environmental conditions and good field management practices. It is grown almost entirely as a rainfed crop in the semi-arid regions of Southern Africa where the mean annual rainfall ranges from 200 to 800 mm. It is raised in India too under arid and semi-arid regions, where the rainfall is scanty and erratic, rendering such areas highly vulnerable to drought situations.

23.2.1 Origin of Pearl Millet

Pearl millet came to be domesticated about 4000–5000 years ago on the southern margins of the Saharan central highlands. It then got widely distributed across the semi-arid tropical areas of northern Africa, in the Arabian Peninsula and Asia. It is believed that pearl millet reached Southern Africa by 900–800 BCE. The primary centres of diversity for pearl millet are in Africa where cross-fertile wild species exist.

23.2.2 Importance of Pearl Millet

Pearl millet enjoys several advantages, and deserves promotion when weighed in terms of its climate resilience, nutrition content and livelihood options. Livelihood options offer scope for income generation for farmers and other stakeholders even under poorly endowed natural resource situations. These aspects are examined in the sub-sections that follow.

23.2.3 Climate Resilience

Climate change is already affecting agro-ecosystems and threatening food security by reducing crop productivity and increasing harvest uncertainty. Various studies show that the negative impact of climate change will be more severe in tropical countries like India. An increase in mean temperature of 1–2 °C is estimated to cause a decline in rice yield by about 0.75 ton/ha in efficient zones and 0.06 ton/ha in coastal regions. Likewise, in case of wheat the impact of 0.5 °C increase in winter temperature is expected to reduce yield by 0.45 ton/ha. The food security being a function of plant, animal and aquatic world is expected to suffer challenge, since climate change will have a bearing on both the terrestrial and aquatic life. All the agricultural sub-sectors that contribute to food security, namely agronomic crops, horticultural crops, dairy and livestock and fisheries, will stand to suffer in different ways and influence negatively the food and nutrition security of the world.

With expected increase in the number of weather extreme episodes due to increasing global warming in the future, one will see more of heat waves and intense rainfall bringing deleterious impact on food production and resultant food insecurity. The farming systems dependent on monsoons will therefore stand to face higher challenges demanding adoption of climate resilient agricultural practices. Large parts of the world that will face greater challenge to food production include the arid and semi-arid tracts of Asia and sub-Saharan Africa. Agriculture in the latter region is already experiencing yield reduction of major crops by up to 10% due to climate change since the beginning of this century.

Pennisetum glaucum is a crop that is highly resilient to conditions of heat and aridity. This can become a strategic crop for poorly endowed regions of the world, as global warming increases and the world gets drier and hotter. Among the major cereals, pearl millet is the one that is able to tolerate extremes of heat and drought (Lost Crops of Africa, 1996). Pearl millet has the potential to emerge as the best option in mankind's quest for 'life support' under situations of harsh environment.

23.2.4 Nutritionally Rich

Pearl millet grain among various cereals is one of the most nutritious with a cafeteria of proteins, minerals, vitamins, fats etc. Its protein content (11%) is not only high but also of exceptionally good quality. The lysine content is reported to be 3.68 mg/g of

Table 23.5 Nutritional comparison of pearl millet with sorghum, rice and wheat (in 100 g grains)

	Crop			
Contents	Pearl millet	Sorghum	Rice	Wheat
Carbohydrates (g)	61.80	67.70	78.20	64.70
Protein (g)	10.90	9.90	7.90	10.60
Fat (g)	5.43	1.73	0.520	1.47
Energy (kcal)	347.00	334.00	356.00	321.00
Dietary fibre (g)	11.50	10.20	02.80	11.20
Calcium (mg)	27.40	27.60	07.50	39.40
Phosphorus (mg)	289.00	274.00	96.00	315.00
Magnesium (mg)	124.00	133.00	19.00	125.00
Zinc (mg)	2.70	1.90	1.20	2.80
Fe (mg)	6.40	3.90	0.60	3.90
Thiamine (mg)	0.25	0.35	0.05	0.46
Riboflavin (mg)	0.20	0.14	0.05	0.15
Niacin (mg)	0.90	2.10	1.70	2.70
Folic acid (μg)	36.10	39.40	9.32	30.10

Source: NIN (2018)

protein as compared to 2.24 for wheat, 3.36 for rice, 3.0 for maize and 3.2 for sorghum. The carbohydrate content is quite high (67–72%), and starch is the major constituent. The high fat content (4–8%) compared to that of other cereals, makes pearl millet highly acceptable and palatable. While the ash content varies from 1.5 to 2.7%, it is rich in phosphorous too with 270–390 ppm. The vitamins are well represented in the crop, particularly thiamine, riboflavin and nicotinic acid.

In the semi-arid tropics and drought-prone regions of Asia and Africa, pearl millet is considered as an important source of energy after the major cereals such as wheat, rice, maize and sorghum, thanks to its rich and diverse nutritional status—carbohydrates, vitamins, minerals, essential amino acids (except for lysine and threonine), proteins, digestible fat, dietary fibre, quality protein, anti-oxidants etc. Further, pearl millet is also rich in unsaturated fatty acids (75%) and phytic acid, which are considered to be useful in lowering cholesterol, and phytate, which in turn reduces cancer risk.

Being gluten-free, it is extremely useful for people suffering from celiac diseases that make them allergic to the gluten content of wheat and other cereals. Those suffering from diseases like diabetes, obesity, diabetic heart disease, atherosclerosis and metabolic diseases are best advised to consume pearl millet due to its health beneficial properties (Kumar et al. 2020). It is also called the 'Powerhouse of Nutrition' as it is rich in essential nutrients—both macro- and micronutrients—in good quantity and quality. These include iron, zinc, magnesium, calcium, phosphorous, copper, manganese, riboflavin and folic acid. As seen from Table 23.5, pearl millet compares very well with, and even fares better than, sorghum and rice on certain parameters.

Though nutritionally rich, its consumption is limited to very few specific regions of the world because of the poor shelf life of the flour and the development of rancidity or off-odour on storage (Rani et al. 2018). Rancidity is caused by oxidative/hydrolytic enzymes such as lipase and lipoxygenase (LOX), where they hydrolyse the triacylglycerol (TAG) to diacylglycerols, glycerol, mono-glycerol and free fatty acids (Manley and Mayer 2012). Additionally, its perception as a poor man's food is also one of the reasons for poor use.

23.2.5 Income Returns and Livelihood Options: Many an Opportunity

Bajra is a labour-intensive crop and thereby generates large number of jobs at the production stage. Post harvest, it needs to be processed for consumption, and therefore provides scope for jobs. Being rich in several micro- and macronutrients, pearl millet can be processed high up the value chain for production of bio-nutrients. As high-end processing is taken up, the scope for creating additional jobs gets higher.

Income returns and livelihood opportunities are closely linked. It would therefore be useful to examine the income returns with respect to millets in general and pearl millet in particular. The Inter-Ministerial Committee on Doubling Farmers Income (DFI) constituted by the Government of India (GoI) in April 2016 estimated the returns from millet cultivation, by using the cost of cultivation (CoC) data of the millets for the year 2013–2014. In respect of small millets, for which CoC data were not available, the same was approximated from the primary survey undertaken by the Indian Institute of Millets Research (IIMR) in Madhya Pradesh and Uttarakhand during the year 2016–2017 on kodo, little millet and barnyard millet. The gross returns earned by the sorghum, pearl millet and finger millet farmers during the year 2013–2014 based on CoC data were taken as current year income and target set to be 2.5 times the base year income to resolve the time lag from 2014–2015 to 2016–2017. The computation also assumed price growth to contribute 25% of the total income and yield growth to contribute 20%. It also assumed that income growth to the extent of 45% would come from value addition and higher yields. Table 23.6 below includes the income and cost estimates of millet-growing farmers including that of bajra.

Table 23.7 below projects the annual productivity gains across various millets including pearl millet for doubling farmers' income by 2022–2023 (as against the base year) based on computation by the DFI Committee.

It may be noted that yield increase is a critical component of any strategy to realise higher production of millets, including that of pearl millet. As seen from Table 23.7, all the millets with the exception of sorghum (−0.44) have shown positive growth rates over the two decades (1996–1997 to 2014–2015). Pearl millet has registered a CAGR of 2.62% over this period. The same Table 23.7 also shows millet-wise targeted growth of yields over the period of 2017–2018 to 2022–2023.

Table 23.6 Income and cost estimates of millet-growing farmers

S. no.	Item		Jowar	Bajra	Ragi	Small millets[a]
1	Cost of cultivation (Rs./hectare)	A1	16,395.5	16,659.61	18,488.96	3470.13
2.i	Value of main product (Rs./hectare)		25,692.79	27,933.6	35,376.95	20,547.12
2.ii	Value of by-product (Rs./hectare)		7344.418	6439.684	5229.35	5150.613
3	Gross return		33,037.21	34,373.28	40,606.3	25,697.74
4	Net return over A1		16,641.71	17,713.67	22,117.34	22,227.61
5	Current (2014–2015) yield level (qtl./ha)		8.84	12.55	17.05	6.5
6	Current price level (Rs./qtl.)[b] (2016–2017)		1625	1330	1725	2220
7	Gross income projected in 2022 (2.5 times for jowar, bajra and ragi)		82,593.02	85,933.21	10,1515.8	51,395.47
8	Contribution of yield growth[b] 20%		16,518.6	17,186.64	20,303.15	10,279.09
9	Contribution of price growth[b] 25%		20,648.26	21,483.3	25,378.94	12,848.87

Source: DFI Committee Estimates based on data compiled from Directorate of Economics and Statistics (n.d.)

[a] Estimates of ICAR-IIMR-based field survey of kodo, little and barnyard millets from the states of Madhya Pradesh and Uttarakhand

[b] Current price level is minimum support price (MSP) during 2016–2017 for major millets. For small millets it is the price realised by sample farmers. Cost A1 is used; imputed value of farmer family labour (FL) and land rent if leased is additional

This emphasises the importance of yield in realising higher magnitude of livelihood options and income for the farmers growing pearl millet and other millets.

23.2.6 Promoting Pearl Millet

As already emphasised, millets are mostly grown as rainfed crops whose system is described by low productivity, low investment, poor infrastructure, poor technology and low marketable surpluses. Such a situation is not supportive of adequate incomes for the farmers. The DFI Committee has recommended that doubling of farmers' income (2022–2023 vis-à-vis 2015–2016) will depend upon four sources of income growth in respect of millets, and it summarises it as follows (Fig. 23.7):

Each of these is a value-added activity that integrates into a millet supply chain. This analysis is applicable to the longer-term approach to income-centric agriculture. Pearl millet, one of these millets, is also influenced by the same factors, and these are discussed briefly below.

Table 23.7 Annual productivity gains across millets for doubling farmers' income

S. no.	Crops	Current level		Yield growth targets					
		Yield (kg/ha)	CAGR 1996–1997 to 2014–2015	2017–2018 (% change over 2015–2016)	2018–2019 (annual increase %)	2019–2020 (annual increase %)	2020–2021 (annual increase %)	2021–2022 (annual increase %)	2022–2023 (annual increase %)
1	Sorghum	884	–0.44%	900 (1.83)	950 (5.55)	1000 (5.25)	1050 (5.00)	1150 (9.5)	1200 (4.35)
2	Pearl millet	1255	2.62%	1275 (1.59)	1300 (1.96)	1350 (3.84)	1400 (3.70)	1425 (1.78)	1450 (1.75)
3	Finger millet	1706	1.47%	1750 (2.58)	1775 (1.42)	1800 (1.40)	1850 (2.77)	1900 (2.70)	1925 (1.32)
4	Small millets	654	2.04%	750 (14.59)	775 (3.33)	800 (3.22)	820 (2.5)	850 (3.65)	875 (2.94)

Source: DFI Committee estimates based on data compiled from Directorate of Economics and Statistics (n.d.)

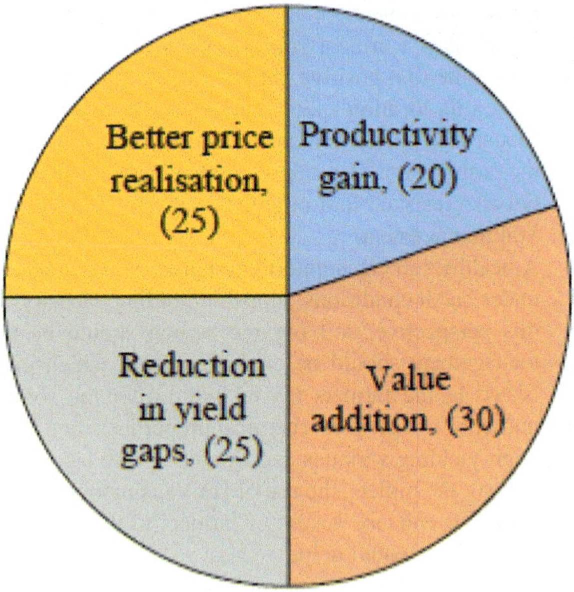

Fig. 23.7 Component-wise contribution towards doubling millet farmers' income. (*Source*: Report of the Committee on Doubling Farmers' Income (DFI) 2018)

23.2.7 Technological Interventions

1. Bridging the yield gaps to upgrade yield levels:
   This implies comparison of existing yields vis-à-vis the yield levels under front-line demonstrations (FLDs). Generally, there is a steep drop at the farmer's field level when compared to the FLD achievement due to poor agronomic practices. It is possible to bridge this gap to some extent, if not fully, in case of pearl millet, which is grown under much harsher environment.
2. Increasing the crop intensity under rainfed agriculture:
   Millets, and pearl millet in particular, are amenable for incorporation into the cropping system for enhancing the cropping intensity, as these can adapt to a range of climatic variations including poor endowment conditions of water stress and high temperature.

   In India, an estimated extent of 12 million hectares of post-kharif paddy land is available, mostly in central India, with good residual moisture. This area can be used to grow pulses, oilseeds and millets, including pearl millet.
3. Improvement in total factor productivity (TFP):
   There should be due emphasis on attaining resource use efficiency to improve returns on investment, with a view to helping millet farmers gain higher income. ICAR-IIMR has estimated the Malmquist TFP mean of major millet-producing states of the country for the period 2004–2005 to 2013–2014, and reports the TFP growth rates to be positive. This indicates the scope that exists for increasing farmers' income. As per this study, the TFP for bajra worked out to a mean of 1.07, and varied from state to state: 1.05 (Gujarat), 1.12 (Haryana), 1.05 (Karnataka), 1.02 (Maharashtra), 1.04 (Rajasthan) and 1.13 (Uttar Pradesh).

4. Seed management:
   Seeds play a critical role in enhancing the yield levels, as a good seed can contribute to a positive change in yield by around 25%. The strategy should be to identify location-specific, short-duration and high-yielding varieties; building a robust seed production chain; and ensuring easy access and at affordable prices to the farmers. It is also important that identified varieties are not only high yielding but also resistant to drought and other biotic and abiotic stresses.
5. Soil management:
   A healthy soil can support good growth. Since bajra and other millets are grown under harsh conditions, moisture-holding capacity of the soil is important. From this perspective, soil organic carbon should be taken care of. Also, nutrient management should be evidence based, which implies use of soil health card issued to the farmers (as in India) after the soil sample is tested for various nutrients and physico-chemical properties.
6. High-yielding varieties and hybrids:
   Among the millets, the use of HYVs is more pronounced in case of sorghum and bajra. To enhance access of farmers to these, promoting a healthy seed village programme would help.
7. Generate authentic data on nutritional benefits of pearl millet and undertake bioavailability studies.

23.2.8 Research and Development (R&D) Interventions

1. Pearl millet, as also other millets, needs to benefit from science the same way as wheat, rice and maize have stood to gain from the breeding programme among other interventions. For a pole-vault in yield levels and developing resilience against biotic and abiotic stresses, the R&D relating to pearl millet must address a few important aspects as follows:
   (a) *Product-specific cultivars* may be bred, and pending this (since it involves a certain lag period), already bred and released cultivars may be screened based on end-product requirement. This process will pave the way for developing and popularising ready to carry (RTC) and ready to eat (RTE) food products. This is one way of harvesting low-hanging fruits.
   (b) *Standards and grades* must be developed to enable market integration of the marketable surpluses. Uniform standards and grades will help trading across the market space both physically and online.
   (c) *Fabrication of primary processing machinery* is important to improve the efficiency of dehulling of the pearl millets and other small millets, as their outer seed coat is indelible making it unfit for consumption. One of the many reasons for reduced consumption of the millets is the drudgery that is associated with their primary processing.
2. Breeding, being an important component of R&D, deserves a little more of detailed examination. Most of breeding programmes have failed to deliver hybrids on account of vast variation in microclimate (day and night temperature

and humidity), rainfall and soil. The variations need to be quantified. Another reason for failure is the narrow cultivar diversity in drought-prone ecology. In order to enhance total output of pearl millet and its utilisation, some focus is required on development of hybrids/varieties with better regenerative capacity on reversal of dry spell for harsh environment/drought-prone areas (for A1 zone in India). The priority areas to be addressed are:

(a) Development of hybrids/varieties resistant/tolerant to salt and high temperature.
(b) Shift in linear focus of breeding from productivity-centricity to the identification of end-product-specific traits.
(c) Mainstreaming of biofortification in pearl millet for iron and zinc.
(d) Enhancement of shelf life of pearl millet flour and overcoming rancidity to promote its products.
(e) Development of screening protocols and control measures against different diseases such as downy mildew, blast, rust, ergot and smut.

23.2.9 Post-Harvest Interventions

1. Generation of demand for millets:
   By creating awareness among the consumers about the nutritional advantages of millets including pearl millet, demand will come to grow and the farmers are then likely to realise higher real remunerative prices on their market sales. This will serve as an incentive for them to enhance capital and technology investments, paving way for better yields. A virtuous cycle would come to be created.
2. Promotion of value system model:
   This will encourage integration of disconnected value chains, and thereby the stakeholders who produce and market the value-added products from millets will realise greater efficiency to the advantage of all including the farmers. Various research institutions including ICAR-IIMR have developed and commercialised a number of value-added millet products like atta, multigrain atta, pasta, vermicelli and sweetmeats. It would help to build linkages of farmers–entrepreneurs besides catering to the infrastructural gaps that exist in the value chain.
3. A study on demand survey on pearl millet would be useful in planning for production.

23.2.10 Policy Interventions

1. These include price and procurement policy, promoting farmers' producer organisations and their federations, facilitating institutional credit and coverage under crop insurance scheme.
2. Of these, reasonably estimated minimum support price (MSP) and procurement operations seem to be very useful in imparting the pearl millet farmers an

immediate incentive. The Government of India in its new MSP Policy adopted in the year 2018 assures a minimum of 50% of the cost of production as profit. With this, the millets apart from pulses and oilseeds have come to benefit with a good elevation in the MSP notified from year to year. As regards procurement, the momentum is yet to pick up. Under the Government's new procurement policy called Pradhan Mantri Annadata Aay SanraksHan Abhiyan (PM-AASHA), there is a slow but growing attention to procurement of millets. There has been some progress in the states of Karnataka and Odisha that have been procuring finger millet and sorghum and using the same for distribution through public distribution system (PDS). Pearl millet is yet to be covered. Major cereals like wheat and paddy have benefitted from high MSPs and procurement by the Food Corporation of India. Millets including pearl millet can also benefit through a similar support.

23.3 Conclusion and Recommendation

The data and discussions around millets as a group, and pearl millet as a specific crop, show up several advantages with respect to nutrition for the consumers, livelihood and income opportunities for the farmers and resilience in the face of climate change adversities.

Given the latitude that improvements in agricultural technology and policies have brought about in terms of food grain status in India as also in various other countries, there exists scope to have a close look at the nutritional needs of the growing population. The food system necessitates adoption of an ecosystem approach from farm gate to consumption plate. This entails reconfiguration of the production matrix, so as to generate adequate quantum of various macro- and micronutrients. Further, the production environment having been compromised on account of deep mining of soil nutrients and water, a more sustainable system of production is warranted. This latter demand has acquired urgency in the light of adverse influence expected from climate change.

Given that various demands from the new agriculture, namely, nutrition, incomes and sustainability, seem apparently contradictory, there does exist enough scope for reconciling these by adopting a well-thought production basket and concomitant production system. Pearl millet, along with other millets, which has been sidelined over the last half a century dominated by a few major cereals, can be brought into focus. A comprehensive response based on R&D, policy framework, extension, consumer awareness etc. will help in better acceptance of pearl millet. In rainfed systems, particularly in harsher areas, crops like pearl millet constitute the best option and therefore deserve attention from all concerned.

References

Directorate of Economics & Statistics (n.d.) Ministry of Agriculture and Farmers Welfare, Govt. of India, New Delhi. https://aps.dac.gov.in/LUS/Index.htm

FAO (2019) The state of food and agriculture 2019. Moving forward on food loss and waste reduction. Rome. Licence: CC BY-NC-SA 3.0 IGO
FAO (2021) World food and agriculture—statistical yearbook 2021. Rome. https://doi.org/10.4060/cb4477en
Kumar RR, Rai GK, Goswami S, Singh SP, Satyavathi CT, Praveen S (2020) Millets the orphan crop for uncertain future. Scientific International Pvt. Ltd., New Delhi, p 194
Manley C, Mayer J (2012) Lipase. In: Mayer J, Donnelly TM (eds) Clinical veterinary advisor: birds and exotic pets. Elsevier, Amsterdam. https://doi.org/10.1016/B978-1-4160-3969-3.00427-3
Michael Raj PSJ, Shanmugam A (2013) A study on millets cultivation in Karur district of Tamil Nadu. Int J Manage Res Rev 3(1):2167–2179
NIN (2018) National nutrition monitoring bureau. NIN, Hyderabad. www.nin.res.in
Rani S, Singh R, Sehrawat R, Kaur BP, Upadhyay A (2018) Pearl millet processing: a review. Nutr Food Sci 48(1):30–44. https://doi.org/10.1108/NFS-04-2017-0070
Report of the Committee on Doubling Farmers' Income (DFI) (2018) Report of the Committee for Doubling Farmers' Income Volume VIII "Production enhancement through productivity gains". Government of India, Ministry of Agriculture. 227 pp
Saxena R, Vanga SK, Wang J, Orsat V, Raghavan V (2018) Millets for food security in the context of climate change: a review. Sustainability 10:2228
Stanly JM, Shanmugam (2013) A study on millet-based cultivation and consumption in India. Int J Market 2(4):49–58

SPRINGER NATURE

GPSR Compliance

The European Union's (EU) General Product Safety Regulation (GPSR) is a set of rules that requires consumer products to be safe and our obligations to ensure this.

If you have any concerns about our products, you can contact us on ProductSafety@springernature.com

In case Publisher is established outside the EU, the EU authorized representative is:

Springer Nature Customer Service Center GmbH
Europaplatz 3
69115 Heidelberg, Germany

The manufacturer's authorised representative in the EU is Springer
Nature Customer Service Centre GmbH, Europaplatz 3, 69115 Heidelberg,
Germany. If you have any concerns regarding our products, please
contact ProductSafety@springernature.com

Printed and bound by CPI Group (UK) Ltd, Croydon, CR0 4YY
01/10/2026
02225867-0019